# Nobel Laureates in Chemistry
## 1901–1992

# Nobel Laureates in Chemistry
# 1901–1992

Laylin K. James, EDITOR
*Lafayette College*

History of Modern Chemical Sciences
Jeffrey L. Sturchio, SERIES EDITOR
*Merck & Co., Inc.*

1993

American Chemical Society
and the Chemical Heritage Foundation

**Library of Congress Cataloging-in-Publication Data**

Nobel laureates in chemistry, 1901–1992 / Laylin James, editor

p.   cm.—(History of modern chemical sciences, ISSN 1069–2452)

Includes bibliographical references and index.

ISBN 0–8412–2459–5 (cloth). ISBN 0–8412–2690–3 (pbk.)

1. Chemists—Biography. 2. Nobel prizes.

I. James, Laylin. II. Series.

QD21.N63    1994
540′.92′2—dc20                                  93–17902
[B]
                                                                    CIP

The paper used in this publication meets the minimum requirements of American National Standard for Information Sciences—Permanence of Paper for Printed Library Materials, ANSI Z39.48–1984. ∞

PRINTED IN THE UNITED STATES OF AMERICA

# Series Foreword

Why study the history of chemistry? Chemical perspectives inform a diverse array of contemporary disciplines, from molecular biology and clinical pharmacology, to materials science and chemical engineering: Isn't keeping up with expanding research frontiers challenge enough? Indeed, in a world of email, fax, and on-line journals, it often seems we are drowning in a sea of information about chemistry in the present. Why, then, be concerned with chemistry in the past? The answer to that question offers intriguing perspectives on our understanding of the practices of the chemical sciences and technologies.

Chemists have always had an interest in the history of their science, and traditional accounts share certain common traits. As historian John Hedley Brooke has recently characterized the genre, these traits may be grouped under the headings of celebration, anticipation, autonomy, and continuity.[1] Historians of chemistry in earlier generations celebrated the achievements of their intellectual forebears, largely in relation to how well their theories and experiments anticipated later views. Traditional approaches also assumed a continuous progression of ideas from past to present, with developments following an internal logic largely divorced from broader intellectual, social, economic, and political currents. A splendid example of this mind-set comes from the work of the Swedish mineral chemist Torbern Bergman (1735–1784). In a 1779 essay entitled "Of the Origin of Chemistry," he divided its history into three distinct periods: "the mythologic, the obscure, and the certain."[2] This faith in the progress of chemical knowledge, culminating in the sophisticated present, is a leitmotif of much of the literature in the history of chemistry.

We find echoes of Bergman's periodization in nineteenth-century treatises like Thomas Thomson's *History of Chemistry* (London, 1830–1831) and Hermann Kopp's *Geschichte der Chemie* (Braunschweig, 1844–1847), as well as the work of more recent authors like J. R. Partington, whose *A History of Chemistry* (London and New York, 1961–1970) is still a standard reference. Similar sensibilities have motivated the generations of textbook authors who include the obligatory historical summaries of great discoveries, precursors, and heroic individuals in introductory general chemistry textbooks.[3] As interesting as these vignettes may be, chronicle is not history; the cumulative effect on students and scientists has often been a deadening of interest in historical inquiry.

But the history of chemistry offers other perspectives on the science, perspectives that help to place the quotidian activities of chemical scientists and engineers firmly in historical context. In recent decades, historians of science and technology have begun to elaborate on the myriad social, political, and economic dimensions of chemistry. The investigation of experimental practice, the study of scientific controversies, and the analysis of the reception and diffusion of new schools of thought illustrate the complex interplay of social and intellectual factors in interpreting theory change in chemistry. Attention to the relationship between chemical texts and their audiences are also illuminating the rich interactions of professional and public cultures in the development of chemistry. And there is increasing attention to invention and innovation in the chemical industries, the connections between academic and industrial research, and the reciprocal effects of chemical science and economic, business, and political structures. The result of a contextual approach to the history of chemistry is a fuller appreciation of the more complex interactions of science, technology, and society in the development of the "central science."[4]

A key goal of this new series on the "History of Modern Chemical Sciences," published jointly by the American Chemical Society and the Chemical Heritage Foundation, is to bring these new perspectives to a wider audience. Through an eclectic mix of original monographs, biographies, autobiographical memoirs, edited collections of essays and documentary sources, translations, classic reprints, and pictorial volumes, the new ACS–CHF series will document the individuals, ideas, institutions, and innovations that have created the modern chemical sciences. In so doing, we hope to foster among our readers a balanced view of the contributions that the chemical sciences and technologies have made to creating modern culture.

Edgar Fahs Smith (1854–1928), three-time President of the American Chemical Society and one of the pioneers of history of chemistry in America, expressed these sentiments in his last book, *Old Chemistries* (New York, 1927), in words that provide a fitting epigraph for our new series[5]:

> The criticism that chemistry is absolutely commercialized is frequently heard and, further, that it is the commercial value of the science alone which claims the thought of chemists. Such views are widely prevalent.

But other ideas exist, and chemistry teachers especially know that to them the discarded "old chemistries" bring many other messages, messages in history, in philosophy, in economics, in social relations, in art, in international relations, in literature, and in a wide and extensive culture.

It is those "many other messages" about chemistry's past and future prospects that the ACS–CHF History of Modern Chemical Sciences Series is designed to explore. We welcome your thoughts on topics and directions for the series and, more importantly, your participation as an engaged audience.

Jeffrey L. Sturchio
Merck & Co., Inc.
July 1993

# References

1. John Hedley Brooke, "Chemists in Their Contexts: Some Recent Trends in Historiography"; In *Atti del III° Convegno Nazionale di Storia e Fondamenti della Chimica*; Abbri, F.; Crispini, F., Eds.; Edizioni Brenner: Cosenza, Italy, 1991; pp 9–27, at 9–11.

2. I owe this reference to William B. Jensen, "Gibber, Jabber, or Just Geber?"; *Bulletin for the History of Chemistry* **1992**, *No. 12 (Fall)*, 49–50.

3. For a lively review of the Victorian tradition in history of chemistry, with useful commentary on other properties and uses of history of chemistry in the period, see Colin A. Russell, "'Rude and Disgraceful Beginnings': A View of History of Chemistry from the Nineteenth Century"; *British Journal for the History of Science* **1988**, *21*, 273–294. See also Rachel Laudan, "Histories of the Sciences and Their Uses: A Review to 1913"; *History of Science* **1993**, *31*, 1-34; esp. pp 8–9, 16–17.

4. For an overview of some relevant recent studies, see Brooke, "Chemists in Their Contexts", esp. pp 16–24; and *Recent Developments in the History of Chemistry*; Russell, Colin A., Ed.; Royal Society of Chemistry: London, 1985. For a stimulating and perceptive guide to these trends in the history of science more generally, see Steven Shapin, "History of Science and Its Sociological Reconstructions"; *History of Science* **1982**, *20*, 157–211; and "Discipline and Bounding: The History and Sociology of Science as Seen Through the Externalism-Internalism Debate"; *History of Science* **1992**, *30*, 333–369.

5. Edgar Fahs Smith *Old Chemistries*; McGraw-Hill Book Company: New York, 1927; p 89.

# Acknowledgments

The publishers owe special thanks to the American Chemical Society librarians Araceli M. Domingo and Maureen W. Matkovich for their help in conducting literature searches to complete each biography.

The publishers also thank the Nobel Foundation for granting them permission to use the photographs of most of the laureates in this book and also for providing the portraits.

# Contents

Preface . . . . . . . . . . . . . . . . . . . . . . . . . . . . . . . . . . . . . . . . . . . . xv
1901   Jacobus van't Hoff . . . . . . . . . . . . . . . . . . . . . . . . . . . 1
1902   Emil Fischer . . . . . . . . . . . . . . . . . . . . . . . . . . . . . . 8
1903   Svante Arrhenius . . . . . . . . . . . . . . . . . . . . . . . . . . . 15
1904   William Ramsay . . . . . . . . . . . . . . . . . . . . . . . . . . . 23
1905   Adolf von Baeyer . . . . . . . . . . . . . . . . . . . . . . . . . . 30
1906   Henri Moissan. . . . . . . . . . . . . . . . . . . . . . . . . . . . . 35
1907   Eduard Buchner . . . . . . . . . . . . . . . . . . . . . . . . . . . 42
1908   Ernest Rutherford. . . . . . . . . . . . . . . . . . . . . . . . . . 49
1909   Wilhelm Ostwald . . . . . . . . . . . . . . . . . . . . . . . . . . 61
1910   Otto Wallach . . . . . . . . . . . . . . . . . . . . . . . . . . . . . 69
1911   Marie Curie . . . . . . . . . . . . . . . . . . . . . . . . . . . . . . 75
1912   Victor Grignard . . . . . . . . . . . . . . . . . . . . . . . . . . . 83
1912   Paul Sabatier . . . . . . . . . . . . . . . . . . . . . . . . . . . . . 88
1913   Alfred Werner . . . . . . . . . . . . . . . . . . . . . . . . . . . . 93
1914   Theodore William Richards . . . . . . . . . . . . . . . . . . . 100
1915   Richard Martin Willstätter . . . . . . . . . . . . . . . . . . . 108
1918   Fritz Haber. . . . . . . . . . . . . . . . . . . . . . . . . . . . . . . 114
1920   Walther Hermann Nernst . . . . . . . . . . . . . . . . . . . . 125
1921   Frederick Soddy . . . . . . . . . . . . . . . . . . . . . . . . . . . 134
1922   Francis William Aston . . . . . . . . . . . . . . . . . . . . . . . 140
1923   Fritz Pregl . . . . . . . . . . . . . . . . . . . . . . . . . . . . . . . 146
1925   Richard Zsigmondy . . . . . . . . . . . . . . . . . . . . . . . . . 151
1926   The Svedberg . . . . . . . . . . . . . . . . . . . . . . . . . . . . . 158
1927   Heinrich Wieland . . . . . . . . . . . . . . . . . . . . . . . . . . 164
1928   Adolf Windaus . . . . . . . . . . . . . . . . . . . . . . . . . . . . 169
1929   Hans von Euler-Chelpin . . . . . . . . . . . . . . . . . . . . . 175
1929   Arthur Harden . . . . . . . . . . . . . . . . . . . . . . . . . . . . 181
1930   Hans Fischer . . . . . . . . . . . . . . . . . . . . . . . . . . . . . . 187
1931   Friedrich Bergius . . . . . . . . . . . . . . . . . . . . . . . . . . . 192
1931   Carl Bosch. . . . . . . . . . . . . . . . . . . . . . . . . . . . . . . . 198
1932   Irving Langmuir . . . . . . . . . . . . . . . . . . . . . . . . . . . 205

| 1934 | Harold Urey | 211 |
| 1935 | Frédéric Joliot | 217 |
| 1935 | Irène Joliot-Curie | 223 |
| 1936 | Peter Debye | 228 |
| 1937 | Walter Haworth | 236 |
| 1937 | Paul Karrer | 242 |
| 1938 | Richard Kuhn | 248 |
| 1939 | Adolf Butenandt | 253 |
| 1939 | Leopold Ružička | 259 |
| 1943 | George de Hevesy | 266 |
| 1944 | Otto Hahn | 272 |
| 1945 | Artturi Virtanen | 282 |
| 1946 | James Sumner | 288 |
| 1946 | John Northrop | 294 |
| 1946 | Wendell Stanley | 300 |
| 1947 | Robert Robinson | 306 |
| 1948 | Arne Tiselius | 315 |
| 1949 | William Francis Giauque | 321 |
| 1950 | Kurt Alder | 328 |
| 1950 | Otto Paul Hermann Diels | 332 |
| 1951 | Edwin McMillan | 338 |
| 1951 | Glenn Seaborg | 344 |
| 1952 | Archer John Porter Martin | 352 |
| 1952 | Richard Laurence Millington Synge | 356 |
| 1953 | Hermann Staudinger | 359 |
| 1954 | Linus Carl Pauling | 368 |
| 1955 | Vincent Du Vigneaud | 380 |
| 1956 | Cyril Hinshelwood | 386 |
| 1956 | Nikolay Nikolayevich Semënov | 392 |
| 1957 | Alexander Robertus Todd | 399 |
| 1958 | Frederick Sanger | 406 |
| 1959 | Jaroslav Heyrovský | 412 |
| 1960 | Willard Libby | 419 |

| 1961 | Melvin Calvin | 422 |
| 1962 | John Kendrew | 428 |
| 1962 | Max Perutz | 435 |
| 1963 | Giulio Natta | 442 |
| 1963 | Karl Ziegler | 449 |
| 1964 | Dorothy Crowfoot Hodgkin | 456 |
| 1965 | Robert Woodward | 462 |
| 1966 | Robert Mulliken | 471 |
| 1967 | Ronald W. Norrish | 479 |
| 1967 | George Porter | 487 |
| 1967 | Manfred Eigen | 494 |
| 1968 | Lars Onsager | 500 |
| 1969 | Derek Harold Richard Barton | 507 |
| 1969 | Odd Hassel | 514 |
| 1970 | Luis Leloir | 520 |
| 1971 | Gerhard Herzberg | 525 |
| 1972 | Christian Anfinsen | 532 |
| 1972 | Stanford Moore | 538 |
| 1972 | William Stein | 546 |
| 1973 | Ernst Otto Fischer | 551 |
| 1973 | Geoffrey Wilkinson | 557 |
| 1974 | Paul Flory | 564 |
| 1975 | John Cornforth | 571 |
| 1975 | Vladimir Prelog | 578 |
| 1976 | William N. Lipscomb, Jr. | 584 |
| 1977 | Ilya Prigogine | 590 |
| 1978 | Peter Mitchell | 597 |
| 1979 | Herbert Charles Brown | 604 |
| 1979 | Georg Wittig | 611 |
| 1980 | Paul Berg | 618 |
| 1980 | Walter Gilbert | 626 |
| 1980 | Frederick Sanger | 633 |
| 1981 | Kenichi Fukui | 639 |

1981    Roald Hoffmann . . . . . . . . . . . . . . . . . . . . . . . . . . . . . . . . 648
1982    Aaron Klug . . . . . . . . . . . . . . . . . . . . . . . . . . . . . . . . . . . 654
1983    Henry Taube . . . . . . . . . . . . . . . . . . . . . . . . . . . . . . . . . . 660
1984    Robert Bruce Merrifield . . . . . . . . . . . . . . . . . . . . . . . . . 667
1985    Herbert Aaron Hauptman . . . . . . . . . . . . . . . . . . . . . . . . . 674
1985    Jerome Karle . . . . . . . . . . . . . . . . . . . . . . . . . . . . . . . . . . 678
1986    Dudley R. Herschbach . . . . . . . . . . . . . . . . . . . . . . . . . . . 686
1986    Yuan Tseh Lee . . . . . . . . . . . . . . . . . . . . . . . . . . . . . . . . . 693
1986    John C. Polanyi . . . . . . . . . . . . . . . . . . . . . . . . . . . . . . . . 700
1987    Donald J. Cram . . . . . . . . . . . . . . . . . . . . . . . . . . . . . . . . 708
1987    Jean-Marie Lehn . . . . . . . . . . . . . . . . . . . . . . . . . . . . . . . 715
1987    Charles J. Pedersen . . . . . . . . . . . . . . . . . . . . . . . . . . . . . 722
1988    Hartmut Michel, Johann Deisenhofer, and . . . . . . . . . . . 729
        Robert Huber
1989    Sidney Altman . . . . . . . . . . . . . . . . . . . . . . . . . . . . . . . . . 737
1989    Thomas R. Cech . . . . . . . . . . . . . . . . . . . . . . . . . . . . . . . 745
1990    Elias J. Corey . . . . . . . . . . . . . . . . . . . . . . . . . . . . . . . . . 750
1991    Richard R. Ernst . . . . . . . . . . . . . . . . . . . . . . . . . . . . . . . 759
1992    Rudolph A. Marcus . . . . . . . . . . . . . . . . . . . . . . . . . . . . . 766
A Bibliographical Guide . . . . . . . . . . . . . . . . . . . . . . . . . . . . . . 775
Index . . . . . . . . . . . . . . . . . . . . . . . . . . . . . . . . . . . . . . . . . . . . . 783

*No Nobel Prizes in chemistry were awarded in 1916, 1917, 1919, 1924, 1933, 1940, 1941, and 1942.*

# *Preface*

Since the inauguration of the Nobel Prizes in 1901, they have gradually gained in significance until the present era, in which they have come to symbolize superlative achievement in science. Certainly, Nobel laureates are considered by the public to be at the pinnacle of excellence in their fields. This book examines the scientific achievements in chemistry for which the Nobel Prize was given, together with the human side of the individuals so honored, their family backgrounds, and their lives. Where feasible, colleagues, co-workers, or former students of chemistry laureates have been enlisted as contributors.

A few observations about the chemistry laureates as a group are of interest. World War II represents a point of transition from the earlier European dominance of the field. Before 1946 the prizes were awarded predominantly to European chemists, with Germany leading the group. Britain and France were strongly represented as well. From 1946 to 1992 the prizes have been distributed in a somewhat more balanced way, with United States nationals receiving the greatest share—almost as many as all of Europe together—and with other continents such as Asia, South America, and Africa beginning to be represented. Before 1946 physical chemistry and organic chemistry were the most common fields of activity of the laureates; since 1946 biochemistry (including molecular biology) has assumed increasing significance and is now nearly equal to organic and physical chemistry in the list of areas in research.

Again contrasting the two groups of laureates (before and after World War II), the average age of laureates at the time of award has increased. For the 1901–1945 group, the median age at which the Nobel Prize was received was 50. For the period 1946 to 1992, the median age was 56. There are, no doubt, many explanations for this trend besides increasing life expectancy. (Nobel Prizes are awarded only to living individuals.) Some chemists had to wait a long time for the merits of their work to be recognized. For example, Hermann Staudinger was 72 in 1953 when the prize was finally awarded for his originally highly controversial theories on macromolecular chemistry. Odd Hassel shared the 1969 prize with D. H. R. Barton. The award came to Hassel at age 72 for his conformational analysis of cyclohexane, begun 30 years earlier. Hassel's work was interrupted by World War II and, according to Barton, was first summarized in 1946 in an obscure journal not readily available outside Hassel's home country of Norway, greatly slowing its appreciation in the wider scientific community. George Wittig received the Nobel Prize in 1979 at age 82 for the reaction bearing his name. The honor came to him 25 years after publication of the "Wittig reaction" and 12 years after his retirement after nearly 50 years of research. Finally, Charles Pedersen, an industrial chemist and rare non-Ph.D., was 83 when the Nobel Prize was awarded in 1987 for his work on crown ethers.

Common elements are also readily discernible in the family backgrounds of the laureates. In the pre-1946 group, their fathers were most commonly merchants or businessmen (about 40%), educators, or other professionals, such as physicians, engineers, chemists, or civil servants (totaling about 40%). The balance of the group of laureates, about one in five, came from working-class families. After 1946 the proportion of laureates from business backgrounds remained about the same, with a doubling of the proportion of children of educators and still fewer children of working-class origins (about one in ten).

Naturally the factors that influenced future laureates to pursue chemistry are varied, but some interesting common traits appear. Considering the 1901–1992 group as a whole, 13 laureates recall that childhood home-laboratory experiments (many in pyrotechnics) led them to chemistry. Nearly half of the group had teachers or research mentors who were themselves future Nobel laureates. A few reported having been diverted from other interests into scientific study by parental influence. Some were strongly encouraged by siblings or other family members, such as uncles or aunts. Whether sex biases have been at work or not, it is worth noting that between 1901 and 1992 only three women were awarded the Nobel Prize in chemistry: Marie Curie in 1911, her daughter Irène Joliot-Curie in 1935, and Dorothy Hodgkin in 1964. In the nearly 30 years since a woman chemist was last honored by the Nobel Foundation, it is hard to imagine that no other deserving women chemists have been put forth as candidates.

In spite of the increased sophistication of experimental techniques and deeper understanding of chemical theory that developed since the Nobel Prize began, even casual comparison between the earliest laureates and those who have received the prize more recently shows many points of commonality.

J. H. van't Hoff, named the first chemistry laureate for his work in chemical kinetics and osmotic pressure, also developed principles of stereochemistry and electrolytic solutions. As a boy he experimented with explosives and did chemical demonstrations for his parents and friends. Emil Fischer, the father of carbohydrate chemistry and recipient of the 1902 Nobel Prize, was a prodigy in mathematics and physics in school but was persuaded by his father to pursue the study of chemisty because of its economic potential. Svante Arrhenius, honored for his theory of electrolytic dissociation in 1903, was an outstanding student in school, particularly in mathematics and physics, and had an uncle who was a botanist. William Ramsay, honored in 1904 for discovery and isolation of the noble gases, came from a scientific family (his grandfather founded the Glasgow Chemical Society). Ramsay received a classical education but had his interest in chemistry aroused when he read a section on gunpowder manufacture in a chemistry textbook. Adolf von Baeyer, recipient of the 1905 award for the synthesis of dyes, did chemistry experiments at home beginning at age 9, discovered a new double salt at age 12, and experimented on his grandfather's farm with plants and chemical fertilizers.

More recently the 1989 prize was shared by Thomas Cech and Sidney Altman for the discovery of catalytic RNA. Cech was encouraged as a boy by his father's broad scientific interests. Altman found inspiration in reading popular books on nuclear physics as a youth. Elias J. Corey, recipient of the Nobel Prize in 1990 for transforming organic synthesis into a more logical science, showed an early mathematics aptitude and was inspired to study chemistry by the enthusiasm of his undergraduate professors. The 1991 laureate, Richard Ernst, honored for his contributions to the development of high-resolution nuclear magnetic resonance spectroscopy, as a boy had his interest in chemistry sparked by home experiments with a case of chemicals belonging to an uncle. And Rudolph Marcus, the 1992 laureate, was told by his mother as she pushed his baby carriage around the campus that he would someday attend McGill University which he did indeed do. The moral of the story: catalyzing an interest in science among children (boys and girls) can have unanticipated and dramatic consequences. The collective biography of the Nobel elite in chemistry is eloquent testimony to the importance of early exposure to science through mentorship and "hands-on" experience. Who knows what exciting results our current efforts at precollege science education reform will yield? Perhaps future editions of *Nobel Laureates in Chemistry* will tell us.

### Acknowledgments

I thank Bernard S. Katz of Lafayette College for suggesting that this volume be prepared and the staff of the American Chemical Society

Books Department—particularly Margaret Brown, Cathy Buzzell-Martin, Robin Giroux, and Cheryl Shanks—for their indefatigable help in the complex task of seeing it through to publication. I also thank the contributors for their diligence and patience during its long gestation. Indeed, G. R. Barker, former Professor of Biological Chemistry at the University of Manchester and Honorary Archivist of the Biochemical Society, died in June 1988, less than a year after submitting his articles. This book is dedicated to his memory.

Laylin K. James
107 Morrison Avenue
Easton, PA 18042

# Jacobus van't Hoff

## 1852–1911

Jacobus Hendricus van't Hoff, the first Nobel laureate in chemistry, was born on August 30, 1852, in Rotterdam, The Netherlands. He received the Nobel Prize on December 10, 1901, in recognition of his discovery of the laws of chemical kinetics and the laws governing the osmotic pressure of solutions. He married Johanna Francisca Mees in 1878 and was the father of two daughters and two sons. He died of tuberculosis on March 1, 1911, in Berlin. More than any other person, van't Hoff created the formal structure of physical chemistry. He developed chemical stereochemistry, and he formalized chemical kinetics. He developed the thermodynamics of electrolyte solutions and was cofounder and coeditor of the *Zeitschrift für physikalische Chemie,* the first journal of physical chemistry.

The son of a physician, van't Hoff was the third of seven children. As a boy he was interested in experimental chemistry, especially explosives, and is reported to have charged spectators a fee to watch his chemical demonstrations. He was also very much involved with music and poetry: singing, playing the pianoforte, and writing poetry in Dutch and in English.

After completing secondary school in Rotterdam, van't Hoff studied practical chemistry at the Polytechnic School of Delft. He distinguished himself at Delft, completing the three-year course in two years and placing first in the examinations in 1872, but his interests did not focus on applied science. He immersed himself in the positivist phi-

I

losophy of August Comte and the poetry of Lord Byron. He had a vision of chemical theory based on physics; it was a vision in which · positivism, poetry, and mathematics would have significant roles. To prepare for his part in this romantic vision, he needed to learn more mathematics and chemistry. He also needed a doctorate and a job.

To learn mathematics he spent the following year at the University of Leiden. To learn more chemistry, he spent the winter of 1872–1873 in Bonn with August Kekulé and the spring of 1874 in Paris with Charles-Adolphe Wurtz. He passed the chemistry doctoral examination at the University of Utrecht in 1873 and in the fall of 1874 prepared a doctoral dissertation on cyanoacetic and malonic acids. He received the Ph.D. in December 1874, but jobs were hard to find. Finally, in 1876 he was appointed lecturer in physics at the Veterinary School in Utrecht.

Young Jacobus van't Hoff led chemists to picture molecules as objects with structure and three-dimensional shape. He published his revolutionary ideas about chemistry in three dimensions just after his 22nd birthday, before he had completed his Ph.D. thesis, in a 15-page pamphlet.[1] He proposed models of organic molecules in which atoms surrounding a carbon atom are situated at the apexes of a tetrahedron. These spatial models explained the puzzling occurrence of isomers (compounds with identical chemical composition but different properties). He constructed illustrative sets of tetrahedral cardboard models and sent them to chemists most likely to be interested in structural chemistry. Expanded versions of his pamphlet were published in French[2] and in German.[3] His proposals initially met vehement opposition, but his ideas eventually prevailed. Models based on the tetrahedron are now part of the intellectual toolbox of every chemist. The visualization of molecules as three-dimensional objects has become a fundamental objective of every college chemistry curriculum.

Bitter criticism came from Hermann Kolbe, who saw the tetrahedral models as examples of "an overgrowth of the weed of seemingly learned and ingenious but in reality trivial and stupefying natural philosophy".[4] Van't Hoff met the challenge directly in his inaugural lecture

---

[1] *Voorstel tot uitbreiding det tegenwoordig in de scheikunde gebruikte structuurformules in de ruimte . . .* ; J. Greven: Utrecht, September 1874.

[2] *La chimie dans l'espace;* P. M. Bazendijk: Rotterdam, 1875.

[3] *Die Lagerung der Atom im Raume;* Vieweg & Sohn: Braunschweig, 1877.

[4] Quoted by van't Hoff in "De verbeeldingskracht in de wetenschap"; inaugural lecture, University of Amsterdam, October 11, 1878; translated as *Imagination in Science*; Springer-Verlag: New York, 1967.

as professor at the University of Amsterdam. He defended "the role of the imagination in investigating the connection between cause and effect", defining imagination as "the ability to visualize any object with all its properties so that one recognizes it with the same great certainty as by simple observation". He was proposing an equal partnership of theoretical science with experimental science, a partnership central to his conception of the new discipline of physical chemistry.

Van't Hoff expressed his conception of physical chemistry in detail in his *Ansichten über die organische Chemie*.[5] The book was intended to unify chemistry under general mathematical and physical principles that would be applicable to technical, pharmaceutical, zoological, and botanical chemistry. He wanted to present a rational, analytic basis for chemistry, showing how the properties of each compound are the consequence of its structure. Although van't Hoff did not achieve a universal physical chemistry with this book, writing it was a significant step in developing his most important book, *Études de dynamique chimique*.[6]

Systematic investigations of the rates of chemical reactions were conducted in the 1860s by Cato Maximilian Guldberg and Peter Waage in Oslo, Marcellin Berthelot and Léon Péan de Saint-Gilles in Paris, and A. G. Vernon Harcourt and William Esson in Oxford. Berthelot and Péan de Saint-Gilles observed that certain reactions ceased while reactants were still present. Guldberg and Waage emphasized the role of concentrations. Vernon Harcourt and Esson showed how calculus can be used to describe the rates. But it was van't Hoff who (in *Études de dynamique chimique*) created the unification of mass action and kinetics, which remains today the conceptual scheme for discussing reactions in solution. He seized on the central importance of the concentrations of reactants: He clarified the role of mass in chemical reactions. He showed how the molecular details of a chemical transformation can be used to predict an observable rate of reaction by assuming that the rate is proportional to the concentrations of each of the reacting molecules. This is the theory of mass action that is still taught and used by chemists. He emphasized the reversibility of chemical reactions, treating equilibrium as a mobile state that is a consequence of

---

[5]Vieweg & Sohn: Braunschweig; Vol. 1, 1878; Vol. 2, 1881.

[6]Frederik Muller: Amsterdam, 1884; revised as *Studien zur chemischen Dynamik* by E. Cohen; Frederik Muller: Amsterdam; and Wilhelm Engelmann: Leipzig, 1896; English trans. by T. Ewan, *Studies in Chemical Dynamics;* Frederik Muller: Amsterdam, and Williams & Norgate: London, 1896.

the dynamic balance of opposing reactions. He invented the symbol that is still used to signify opposing reactions: $\rightleftharpoons$. He related the kinetic rate constants for the forward and reverse reactions to the equilibrium constant and discussed the temperature dependence of the rate constants and the equilibrium constant. *Études de dynamique chimique* reads today like a modern book; the essential structure of mass-action theory remains as van't Hoff presented it in 1884.

Chemical thermodynamics is the science of relationships among apparently diverse phenomena. Van't Hoff introduced thermodynamic methods into chemistry as a way to understand solutions. It was characteristic of van't Hoff that his fundamental contribution to solution thermodynamics came as a result of his attempts to understand a scientific problem that did not appear to be chemical. Hugo de Vries, a plant physiologist at the University of Amsterdam, asked for van't Hoff's help in explaining osmotic pressure and directed him to the work of Wilhelm F. P. Pfeffer. Experimentally, Pfeffer had shown that osmotic pressure in dilute sugar solutions can be represented by an equation identical in form to the ideal gas law $PV = RT$, where $P$ is the pressure, $V$ is the volume, $T$ is the absolute temperature, and $R$ is the universal gas constant. Van't Hoff showed that the analogy between ideal gases and dilute solutions is fundamental and that osmotic pressure measurements can be used to determine molecular weights of substances in solution. When the measurement of the osmotic pressure of various solutions revealed that the ideal equation often did not hold, van't Hoff introduced an ad hoc variable $i$, wrote the equation as $PV = iRT$, and considered $i$ to be a measure of the abnormality of the substance. Physicochemical research on solutions, both experimental and theoretical, then concentrated on the measurement and interpretation of $i$. Svante Arrhenius (Nobel Laureate in chemistry, 1903) later argued that large values of $i$ are due to the dissociation of the dissolved substance into ions.

The contributions of van't Hoff to solution chemistry can scarcely be overestimated. He introduced osmotic pressure into thermodynamics and showed how osmotic pressure is related to a wide range of measurable properties of solutions. He then generalized his results and showed, for example, how the same results apply to the lowering of the freezing point by dissolved substances. He paid special attention to the significance of very dilute solutions as idealized chemical systems. He investigated fundamental connections between chemical thermodynamics and electricity and showed how electrical measurements can give insight into chemical phenomena. He created a conceptual structure that enabled a younger generation of physical chemists to imagine new experiments and to report their observations in a common language.

In exploiting the analogy between ideal gases and ideal solutions, in focusing on the properties of dilute aqueous solutions, in introducing the variable $i$ to express deviations from ideality, and in showing how electrochemical measurements are related to other properties of solutions, van't Hoff provided the conceptual framework for research programs that occupied many physical chemists throughout the first two-thirds of the 20th century. The influential textbook on chemical thermodynamics by Gilbert Newton Lewis and Merle Randall, *Thermodynamics and the Free Energy of Chemical Substances*,[7] is in many respects a direct extension of van't Hoff's ideas, as are the later books of Herbert S. Harned and Benton B. Owen[8] and of R. A. Robinson and R. H. Stokes.[9]

Physicochemical research in the early 1880s was published in many different journals, making an overview difficult. In 1886 van't Hoff and Wilhelm Ostwald (Nobel Laureate in chemistry, 1909) independently began negotiations with publishers for a journal of physical chemistry. They soon decided to collaborate, formed an international editorial board, and founded the *Zeitschrift für physikalische Chemie, Stöichiometrie und Verwandtschaftslehre*. The first issue appeared in February 1887; van't Hoff used it to expound his analogy between solutions and gases.[10] The *Zeitschrift* was certainly the first journal devoted to the new discipline of physical chemistry. During its first years it was the primary journal for most of its contributors. Van't Hoff continued as coeditor until his death.

Van't Hoff said that he disliked teaching and administration. He happily accepted a position in 1895 as a professor in the University of Berlin, where his only responsibility was one lecture a week. There he was able to collaborate with his former pupil Wilhelm

[7]McGraw-Hill: New York, 1923.

[8]Harned, H. S.; Owen, B. B. *The Physical Chemistry of Electrolytic Solutions*; Reinhold: New York, 1941, 1950, 1958.

[9]Robinson, R. A.; Stokes, R. H. *Electrolyte Solutions: The Measurement and Interpretation of Conductance, Chemical Potential and Diffusion in Solutions of Simple Electrolytes*; Butterworths: London, 1955, 1959.

[10]van't Hoff, J. H. "Die Rolle des osmotischen Druckes in der Analogie zwischen Lösungen und Gasen"; *Z. Phys. Chem. Stoechiom. Verwandtschaftsl.* **1887**, *1*, 481–508; English trans., "The Function of Osmotic Pressure in the Analogy between Solutions and Gases"; *Philos. Mag. Series 5* **1888**, *26*, 81–105.

Meyerhoffer in the thermodynamic analysis of salt deposition in the Strassfurt potash beds. The death of Meyerhoffer in 1906 affected him deeply, and his health began to deteriorate soon afterward. His research program was curtailed in his latter years, but his insight into the application of physico–chemical methods remained keen. In his last research, he showed that enzymatic action can be described by ordinary reversible chemical reactions that conform to the principles of mass action.[11] His last two papers were among the first publications in which anyone applied mass-action principles to physiology. Investigations of enzymatic reactions using methods that are fundamentally derived from the contributions of van't Hoff created the 20th century science of biochemistry and continue to this day.

**GEORGE FLECK**
*Smith College*

*Bibliography*

Cohen, E. *Jacobus Henricus van't Hoff: Sein Leben und Wirken;* Akademische Verlagsgesellschaft: Leipzig, 1912.

Cohen, E. "Jacobus Heinrich van't Hoff, 1852–1911"; In Farber, E. *Great Chemists;* Interscience: New York, 1961; pp 947–958.

Holleman, A. F. "My Reminiscences of van't Hoff"; *J. Chem. Ed.* **1952,** *29,* 379–382.

Laidler, K. J. "Chemical Kinetics and the Origins of Physical Chemistry"; *Archive for the History of Exact Sciences* **1985,** *32,* 43–75.

Partington, J. R. *A History of Chemistry;* Macmillan: London, 1964; Vol. 4, pp 656–662.

Ramsay, O. B. *Stereochemistry;* Nobel Topics in Chemistry: A Series of Historical Monographs on the Fundamentals of Chemistry; Heyden: London, 1981.

Root-Bernstein, R. S. "The Ionists: Founding Physical Chemistry, 1872–1890"; Ph.D. Dissertation, Princeton University, 1980.

Servos, J. W. *Physical Chemistry from Ostwald to Pauling: The Making of a Science in America;* Princeton University: Princeton, 1990; esp. Chapter 1.

Snelders, H. A. M. "J. H. van't Hoff's Research School in Amsterdam (1872–1895)"; *Janus* **1984,** *71,* 1–30.

---

[11] van't Hoff, J. H. "Über synthetische Fermentwirkung"; *Sitzungsberichte der Koenigliche-Preussische Akademie der Wissenschaften* **1909,** 1065–1076; **1910,** 963–971.

Snelders, H. A. M. "van't Hoff, Jacobus Henricus"; In *Dictionary of Scientific Biography;* Gillispie, C. C., Ed.; Charles Scribner's Sons: New York, 1976; Vol. 13, pp 575–581.

van Klooster, H. S. "van't Hoff (1852–1911) in Retrospect"; *J. Chem. Ed.* **1952,** *29,* 367–379.

van't Hoff, J. H. *The Arrangement of Atoms in Space;* Longmans, Green: London, 1898.

van't Hoff, J. H. *Physical Chemistry in the Service of the Sciences;* University of Chicago Press: Chicago, 1903.

van't Hoff, J. H. *Vorlesungen über theoretische und physikalische Chemie;* F. Vieweg und Sohn: Braunschweig, c. 1898–1900.

*Van't Hoff–Le Bel Centennial;* Ramsay, O. B., Ed.; ACS Symposium Series 12; American Chemical Society: Washington, DC: 1975.

Walker, J. "van't Hoff Memorial Lecture"; In *Memorial Lectures Delivered before the Chemical Society, 1901–1913;* Gurney & Jackson: London, 1914. Original article appeared in *J. Chem. Soc.* **1913,** *103,* 1127–1143.

# Emil Fischer

## 1852–1919

Copyright Nobel Foundation

Emil Hermann Fischer was born in Euskirchen, a small town near Bonn, Germany, on October 9, 1852. He died in Berlin on July 15, 1919. In 1902 he received the Nobel Prize for work on the carbohydrates and the purines. He was the son of Lorenz Fischer, a successful businessman, and Julie Poensgen. After a short attempt at a business career he entered the university. Emil wanted to become a mathematician or a physicist, but his father considered these professions to be without economic potential and persuaded his son to study chemistry.

In 1871 Emil Fischer entered the University at Bonn, where he attended the lectures of August Kekulé and Rudolf Clausius. In the following year he transferred to Strasbourg, where he studied chemistry with Adolf Baeyer and earned his doctorate in 1874. He followed Baeyer to Munich in 1875, where he became a *Privatdozent* in 1878 and a junior professor in 1879. He became Professor and Director of the Chemistry Institute at Erlangen in 1882 and took a similar position at Würzburg in 1885. In 1892 he succeeded A. W. von Hofmann as Director of the Chemistry Institute of Berlin. In 1888 Emil Fischer married Agnes Gerlach; she died after only seven years of marriage. Together they had three sons: Two of the sons became medical doctors and died as soldiers in World War I. The third son, Hermann Fischer, became a distinguished biochemist and completed his career at the University of California, Berkeley, where he died on March 9, 1960.[1]

---

[1]Bergmann, 1930.

Fischer's first publication (1875) concerned derivatives of hydrazine, $N_2H_4$, even though the parent compound had not yet been made. The related diazonium compounds had been formulated by Kekulé. However, because these were unstable compounds and because the proposed formulations were similar to the stable azo compounds, the whole matter was controversial. Fischer solved the problem by confirming the work of Kekulé and establishing the structure of phenylhydrazine. In the course of this effort he synthesized many derivatives of hydrazine that would become important in making dyes. He also discovered a method for making derivatives of indole. (Fischer synthesized indole in 1886.)

In 1884 Fischer learned that phenylhydrazine was a valuable reagent for investigating aldehydes and ketones because the reaction products were readily identifiable crystalline solids. Later he learned that carbohydrates, which are polyhydroxy aldehydes and polyhydroxy ketones, reacted with phenylhydrazine to form osazones, beautifully crystalline compounds that later enabled him to elucidate the chemistry and structure of the carbohydrates.

Emil Fischer's doctoral thesis reported on the chemistry of dyestuffs and colors. In 1862 August von Hofmann had prepared an important dye called rosaniline. The structure of the dye had been studied, but it was not known with certainty. In 1878, Emil Fischer, along with his cousin Otto, showed that rosaniline and related dyes were triphenylmethane derivatives.

In 1881 Fischer began a study of uric acid and its derivatives and continued this study until 1914, when he made theophylline-D-glucoside phosphoric acid, the first synthetic nucleotide. Nucleotides are biologically important substances that are found in the nuclei of animal cells. In 1882 he suggested formulas for uric acid, caffeine, theobromine, xanthine, and guanine, but he was not yet certain that he was correct. He synthesized theophylline and caffeine in 1895 and uric acid in 1897. He proposed in 1897 that uric acid and related compounds were oxygen compounds of a hypothetical base, which he named purine. By 1900 he had synthesized about 130 derivatives, including hypoxanthine, xanthine, theobromine, adenine, and guanine. Finally, in 1898 he demonstrated the correct structures by making purine itself, the parent substance of these compounds. In 1914 he made the glucose derivatives, and from them he made nucleotides.

Fischer's laboratory methods became the basis for the industrial production of caffeine, theophylline, and theobromine by the German drug industry. In 1903 he made 5,5-diethylbarbituric acid, a hypnotic and sedative known under the trade names of Barbital, Veronal, or Dorminal. The phenylethyl derivative was made in 1912, and it became widely known as Luminal or phenobarbital.

When Emil Fischer started his carbohydrate research in 1884 at Erlangen, only four monosaccharides were known: the aldohexoses, glucose and galactose, and the ketohexoses, fructose and sorbose. Three disaccharides were known: sucrose, maltose, and lactose. From the work of A. Le Bel and J. H. van't Hoff he knew that the structure then attributed to glucose must have at least 16 stereoisomers. In what has become a classic work he synthesized most of the stereoisomers and demonstrated the correct configuration of glucose. His earlier discovery of phenylhydrazine proved indispensable for the study of the carbohydrates. Because of his earlier discovery of phenylhydrazine as a reagent, he was able to identify and characterize the components of various intractable and viscous syrups encountered in this branch of chemistry.

In 1894 Fischer studied the action of enzymes on some sugars. He noted that the enzymes were very specific in their action. Maltase, for example, hydrolyzed $\alpha$-methylglucoside but not $\beta$-methylglucoside, and emulsin hydrolyzed $\beta$-methylglucoside but not $\alpha$-methylglucoside. He reasoned that an enzyme was active only if it had a specific configuration that would "fit" the substrate. He used the analogy of a lock and key to explain this hypothesis, comparing an enzyme with a key that would fit only one lock.

In 1908 Fischer studied the tannins, substances that are gallic acid derivatives of the sugars. In 1912 he showed that these materials are gallic acid esters of carbohydrates such as glucose and maltose.

When Emil Fischer started his work on the proteins in 1899, he knew that 13 $\alpha$-amino acids had been obtained as hydrolysis products of the proteins. He then proceeded to synthesize many of them. Laboratory syntheses of these materials produced mirror-image pairs of the amino acids, so it was necessary to separate the L-form from the D-form. Natural proteins contain exclusively the L-form amino acids. He succeeded in this separation by converting the mixtures to salts that were subsequently separated by fractional crystallization. The purified forms were then obtained from the salts.

In 1901 he modified a method devised by Theodor Curtius to separate mixtures of amino acids that were obtained by decomposing proteins. The complex mixture was converted to esters and then fractionally distilled. Using this method he identified valine, proline, and hydroxyproline. By 1907 he had developed a method for combining amino acids to form segments of proteins known as peptides. He then made a peptide consisting of 18 amino acids, an amazing feat for that time. He recognized that the proteins were very complex and that even simple peptides could have numerous isomers. He showed that a characterization of the proteins required the identity, the number, and the sequence of the constituent amino acids. He also recognized a pitfall, called the Walden inversion, and then used it to advantage in synthesizing amino acids.[2]

Emil Fischer's early years were normal for a boy of his day, although it was evident that he was exceptionally bright. His family had practiced the Lutheran faith for at least 200 years, but his father, Lorenz, declared himself to be an atheist. Lorenz Fischer was a successful businessman and he wanted Emil, his only son, to succeed him in directing the family enterprises. After a short apprenticeship in business, it was apparent that Emil was in the wrong vocation. Lorenz Fischer declared that "The kid is too dumb to be a businessman; he should go to school." After a year at Bonn he transferred to Strasbourg and studied under Adolf Baeyer. Baeyer had a profound influence on Fischer, an influence that endured for the rest of Fischer's life. Like Baeyer, Emil Fischer relied on experimental data and had a distrust of theoretical approaches. He became a skilled analyst and learned to conduct most laboratory procedures with very simple equipment. Most of his research was so thorough and complete that it remains unchallenged today. Except for some work on specific heats and a brief theoretical encounter with the Walden inversion, Fischer's methodology was essentially pragmatic. Approximately 200 of his students earned doctorates and many of them achieved distinction in industry or in education. Fischer was also very influential in establishing the Kaiser Wilhelm Institutes, now known as the Max Planck Institutes. He himself founded the first such institute in chemistry. In addition to the Nobel Prize, he received the highest honors that his own country and many foreign countries could confer.[3]

Emil Fischer made significant contributions to the improvement of laboratory safety and the design of laboratory buildings. Early in his career he was poisoned by accidentally inhaling the vapors of diethylmercury. For many years he suffered from the poisonous effects of phenylhydrazine, not realizing the source of his physical distress.

[2]Farber, 1970.
[3]Herneck, 1970.

Driven by these experiences he forcefully promoted the erection of safer buildings, stressing improved ventilation. He designed exhaust hoods to remove toxic and inflammable vapors, the first devices of their kind to be put into general use.

Although he was generally engrossed in his own research, Fischer had an unerring sense of the importance of other branches of chemistry. He was instrumental in bringing eminent inorganic chemists to the Berlin Institute, such as Alfred Stock, and Franz Fischer, discoverer of the Fischer–Tropsch synthesis of hydrocarbons. Even though he did not understand radiochemistry, he provided working space and financial support at his Berlin laboratory for Otto Hahn, who would later become a Nobel laureate. He also sponsored Lise Meitner, who, like Hahn, became a distinguished nuclear chemist.

Although he had a somewhat austere and formal mien and was not blessed with a good speaking voice, Emil Fischer was an exceptional teacher. His 550-seat lecture room in Berlin was filled daily. He also conducted a colloquium during which he generated much discussion and listened to one and all. His lectures were meticulously organized and were enlivened by demonstrations staged by his assistant, Johannes Wetzel. He made allusions to history and sometimes quoted poetry during his presentations.[4]

During World War I Emil Fischer, like many of his colleagues, was pressed into the war effort. He was active in organizing German chemical resources and headed a commission for the production of chemicals and food supplies. Through these years Fischer suffered many disappointments and became disillusioned with the policies of his country. Two of his sons died in the German army as medical officers. His own work was reluctantly put aside because of the war effort. After the war he had great difficulty in restarting his research because of material and manpower shortages. He lost most of his considerable personal wealth because of postwar inflation and then learned that he had inoperable cancer of the bowel. He died on July 15, 1919. Richard Willstätter wrote that "he was the unmatched classicist, master of organic-chemical investigation with regard to analysis and synthesis, and as a personality a princely man."[5]

JOHN J. LUCIER
*University of Dayton*

---

[4]Fischer, 1960.
[5]Willstätter, 1948.

# Bibliography

Bergmann, M. "Fischer, Emil"; In *Das Buch der grossen Chemiker;* Bugge, G., Ed.; Verlag Chemie: Berlin, 1930; Vol. 2, pp 408–420.

Cain, J. C. *The Chemistry of the Diazo-Compounds;* Longmans, Green: New York, 1908.

Chamberlain, A. H. *The Conditions and Tendencies of Technical Education in Germany;* C. W. Bardeen: Syracuse, NY, 1908.

Clapperton, G. *Practical Paper-Making: A Manual for Paper-makers and Owners and Managers of Paper Mills, etc.,* 2nd ed., revised and enlarged; D. Van Nostrand: New York, 1907.

Darmstädter, L.; Oesper, R. E. trans. *J. Chem. Ed.* **1928,** *5,* 37–42.

du Pont de Nemours, E. I., Powder Co. *Useful Information for Practical Men;* compiled for E. I. Dupont de Nemours, 1908.

Farber, E., "Fischer, Emil Hermann"; In *Dictionary of Scientific Biography;* Gillispie, C. C., Ed.; Charles Scribner's Sons: New York, 1970; vol. 5, pp 1–5.

Feldman, G. D. In *Deutschland in der Weltpolitik des 19. und 20. Jahrhunderts: Fritz Fischer zum 65. Geburtstag;* Geiss, I.; Witt, P.-C., Eds.; Bertelsmann Universitätsverlag: Düsseldorf, 1973; Vol. 1, pp 341–362.

Fischer, E. *Aus meinem Leben;* Julius Springer: Berlin, 1922.

Fischer, E. *Neuere Erfolge und Probleme der Chemie;* Julius Springer: Berlin, 1911; p 30.

Fischer, E. *Untersuchungen aus verschiedenen Gebieten;* In *Gesammelte Werke;* Bergmann, M., Ed.; Julius Springer: Berlin, 1924.

Fischer, E. *Untersuchungen über Aminosäuren, Polypeptide und Proteine (1899–1906);* Julius Springer: Berlin, 1906.

Fischer, E. *Untersuchungen über Depside und Gerbstoffe (1908–1919);* Julius Springer: Berlin, 1919.

Fischer, E. *Untersuchungen über Kohlenhydrate und Fermente (1884–1908);* Julius Springer: Berlin, 1909.

Fischer, E. *Untersuchungen über Triphenylmethanfarbstoffe, Hydrazine und Indole;* In *Gesammelte Werke;* Bergmann, M., Ed., Julius Springer: Berlin, 1924.

Fischer, H. O. L. "Fifty years 'Synthetiker' in the Service of Chemistry"; *Ann. Rev. Biochem.* **1960,** *29,* 3–4.

Forster, M. O. "Emil Fischer Memorial Lecture"; *J. Chem Soc.* **1920,** *117.*

Fruton, J. S. *Contrasts in Scientific Style;* American Philosophical Society: Philadelphia, PA, 1990.

Gilman, A. F. *Quantitative Chemical Analysis;* Chemical Publishing: Easton, PA, 1908.

Goerens, P. *Introduction to Metallography;* Ibbotson, F., Trans.; Longmans, Green: New York, 1908.

Heess, J. K. *Practical Methods for the Iron and Steel Works Chemist;* Chemical Publishing: Easton, PA, 1908.

Herneck, F. "Emil Fischer als Mench und Forscher"; *Z. Chem.* **1970,** *10,* 41–48.

Hill, L. *Recent Advances in Physiology and Bio-Chemistry, by Leonard Hill and others;* Longmans, Green: New York, 1908.

Hoesch, K. *Emil Fischer: Sein Leben und sein Werk;* Verlag Chemie: Berlin, 1922.

Holley, C. D.; Ladd, E. F. *Analysis of Mixed Paints, Color Pigments, and Varnishes;* John Wiley & Sons: New York, 1908.

Johnson, J. A. *The Kaiser's Chemists: Science and Modernization in Imperial Germany;* University of North Carolina: Chapel Hill, 1990.

Kauffman, G. B. "Emil Fischer's Role in the Founding of the Kaiser Wilhelm Society"; *J. Chem. Ed.* **1989,** *66,* 394–400.

Kauffman, G. B.; Priebe, P. M. "The Emil Fischer-William Ramsay Friendship: The Tragedy of Scientists at War"; *J. Chem. Ed.* **1990,** *67,* 93–101.

Landauer, J. *Spectrum Analysis,* 2nd ed., rewritten; Bishop, J., Trans.; John Wiley & Sons: New York, 1908.

Loeb, J. *A New Proof of the Permeability of Cells for Salts or Ions;* preliminary communication, University of California Press: Berkeley, CA, 1908.

Maire, F. *Modern Pigments and Their Vehicles;* John Wiley & Sons: New York, 1908.

Moy, T. D. "Emil Fischer as 'Chemical Mediator': Science, Industry and Government in World War I"; *Ambix* **1989,** *36,* 109–120.

Poincaré, J. H. *The Value of Science;* Halsted, G. B., Authorized Trans.; Science Press: New York, 1908.

Scholz, H. *NTM* **1984,** *21,* 91–98.

Standage, H. C. *Decoration of Metal, Wood, Glass, etc.;* Standage, H. C., Ed.; John Wiley & Sons: New York, 1908.

Stansfield, A. *The Electric Furnace: Its Evolution, Theory, and Practice;* Hill Publishing, 1908.

Talbot, H. P. *An Introductory Course of Quantitative Chemical Analysis, with Explanatory Notes and Stoichiometrical Problems;* 5th ed. rewritten and revised; MacMillan: New York, 1908.

Whipple, G. C. *Typhoid Fever, Its Causation, Transmission, and Prevention;* John Wiley & Sons: New York, 1908.

Willstätter, R. *Aus meinem Leben;* Verlag Chemie: Weinheim, 1948; p 212.

Fischer's papers are in the Bancroft Library of the University of California, Berkeley.

# Svante Arrhenius

## 1859–1927

Svante August Arrhenius was born on February 19, 1859, near Uppsala, Sweden. He received the Nobel Prize on December 10, 1903, in recognition of his theory of electrolytic dissociation. In 1894 he married Sofia Rudbeck, and the couple had a son. He married Maria Johansson in 1905, and they had two daughters and a son. He died in Stockholm on October 2, 1927. He introduced the idea that many substances dissociate in water to produce positive and negative ions, an idea central to the 20th century conception of chemical solutions. He was instrumental in establishing the separate discipline of physical chemistry and later extended physico-chemical methods to cosmology, meteorology, and biochemistry.

His father, Svante Gustaf Arrhenius, held a position at the University of Uppsala. His uncle, Johan Arrhenius, was a botanist, writer, and long-time secretary of the Swedish Agricultural Academy in Stockholm. Svante August Arrhenius was an outstanding student in school. He learned reading and arithmetic at a very early age and in secondary school excelled particularly in mathematics and physics. His physics teacher was M. M. Floderus, author of the most popular contemporary Swedish secondary school physics textbook. His university degrees were all from the University of Uppsala.

Arrhenius had no choice of doctoral programs if he wished to remain in Sweden; the University of Uppsala was the only Swedish institution permitted to award the doctorate. There were only two possibilities at Uppsala for a research mentor: Per Theodor Cleve, a chemist who synthesized inorganic chemicals, and Tobias Thalén, a spectroscopist. Arrhenius found both professors rigid and uninspiring; he believed that independent research under the direction of either man would be impossible. So in September 1881 he made arrangements to do his doctoral research in absentia at the Physical Institute for the Swedish Academy of Sciences in Stockholm under the direction of Eric Edlund, a physicist whose research ranged over many aspects of electricity.

Arrhenius's Uppsala teacher, Per Cleve, had argued that the molecular weight of cane sugar was both unknown and unknowable; Arrhenius took up the implicit challenge to measure the molecular weight as part of his doctoral research and decided to use electrical methods. Arrhenius tried to interpret the electrical conductivity of sugar solutions, but without success. However, he did learn valuable techniques for measuring the conductivity of solutions.

Eric Edlund and Sven Otto Pettersson (professor of chemistry at the Stockholm Högskola) together directed Arrhenius's research program for his dissertation. Pettersson showed Arrhenius that the new periodic table of Mendeleyev revealed ways to interpret physical measurements of many kinds. Moreover, Pettersson had personal contacts with Peter Waage in Oslo and passed along Waage's ideas about mass-action theory to Arrhenius. Arrhenius used Edlund's apparatus to measure the conductivity of very dilute solutions of many acids, bases, and salts (electrolytes), in each case measuring the conductivity within series of concentrations. He interpreted the dependence of conductivity on concentration in terms of mass-action theory, assuming that an equilibrium exists between active molecules that conduct electricity and inactive molecules that do not.

The warm relationships among Arrhenius, Edlund, and Pettersson promoted productive experimentation and fruitful speculation. The academic degree was another matter. The public defense of the dissertation was held at Uppsala in May 1884. Cleve and Thalén had not been pleased that Arrhenius bypassed them and went to Stockholm to find suitable research advisors; they were unimpressed by the dissertation and by Arrhenius's oral defense. The doctoral degree was awarded with only a fourth-class pass for the dissertation and a third-class pass for the defense. As a consequence, Arrhenius did not qualify for a docentship at the university, and he had no other prospects for a job.

Arrhenius's dissertation was written in French.[1] Arrhenius sent copies of his dissertation to most of the people in the world who were likely to understand his work, including Jacobus van't Hoff (Nobel Laureate in chemistry, 1901), Wilhelm Ostwald (Nobel Laureate in chemistry, 1909), Rudolf Clausius, Julius Thomsen, and Lothar Meyer. He published his conclusions in German.[2]

The reception that was given to Arrhenius's ideas was mixed. Three young men—Ostwald (age 31), van't Hoff (32), and Oliver Joseph Lodge (33)—found his ideas original and stimulating. Cleve (44), Meyer (54), Thalén (57), Thomsen (58), and Clausius (62) dismissed his work as uninteresting. The response of van't Hoff was detailed, sometimes critical, and led to an extraordinary collaboration (discussed later). Lodge was instrumental in making the work known in England via publications in the *British Association for the Advancement of Science Report*.[3] Ostwald, professor of chemistry at Riga (Latvia) Polytechnicum, traveled to Uppsala in August 1884 to visit Arrhenius and to offer him a position at Riga. Because of his father's terminal illness, Arrhenius could not move to Riga, but he used this offer to obtain a position in November 1884 as docent in physical chemistry at the University of Uppsala. Edlund was influential in obtaining a substantial postdoctoral travel grant for Arrhenius from the Swedish Academy of Sciences.

In February 1886 Arrhenius did go to Riga, and he worked with Ostwald. Later in 1886 he went to the laboratory of Friedrich Kohlrausch in Würzburg, where he met Walther Nernst (Nobel Laureate in chemistry, 1920). The following year he traveled to the laboratories of Ludwig Boltzmann in the city of Graz, and in 1888 he visited van't Hoff in Amsterdam. In these postdoctoral travels, Arrhenius met the men with whom he was to collaborate in revolutionizing the chemical theory of solutions. These scientists were later dubbed the "Ionists", a term of ridicule when used by their opponents and a term of respect when used by their followers.

[1] "Recherches sur la conductibilité galvanique des electrolytes"; Ph.D. dissertation, University of Uppsala, 1884; summarized and reviewed in English by Oliver Lodge, *British Association for the Advancement of Science Report* **1886**, *56*, 357–384; translated into German in *Ostwalds Klassiker der exakten Wissenschaften*, No. 160; W. Engelmann: Leipzig, 1907.

[2] "Über die Gültigkeit der Clausius-Williamsonschen Hypothese"; *Ber. Dtsch. Chem. Ges.* **1884**, *15*, 49–52.

[3] *British Association for the Advancement of Science Report.* **1886**, *56*, 310–312, 315–318, 344–348, 357–384, 384–387, 387–388.

In 1887 Ostwald became professor of physical chemistry at the University of Leipzig, and Nernst became his assistant; Arrhenius spent much of his time during the years 1888–1890 in Leipzig, although he taught a course in physical chemistry at the University of Uppsala. In 1891 he obtained a faculty appointment at Stockholm Högskola and was named professor in 1895. In 1905 he became director of the Nobel Institute of Physical Chemistry in Stockholm, a position he held until the year of his death.

The most significant event in chemistry of the mid-1880s for Arrhenius was the publication of van't Hoff's *Études de dynamique chimique* (Frederik Muller: Amsterdam, 1884). Pettersson probably showed the book to Arrhenius, who wrote an enthusiastic review.[4] Arrhenius expressed excitement at the fields of research van't Hoff had opened, predicting that many eager chemists would soon flock to these fields; he was one of those eager chemists. Arrhenius sent a copy of the review to van't Hoff, who replied on August 4, 1885. Within two years these two men had developed the modern theory of ions in solution.[5]

By 1887 Arrhenius had become a proponent of the theory of electrolytic dissociation. He said that many substances—acids, bases, and salts—dissociate when they dissolve in water, forming electrically charged ions. He said that, for instance, sodium chloride in water exists entirely as sodium cations (with positive charges) and chloride anions (with negative charges). An acid such as acetic acid in solution is a mixture of acetic acid molecules, acetate anions, and hydrogen cations; the equilibrium between molecules and ions, according to Arrhenius, depends on the concentrations of the molecules and ions, and the dissociation is greater when the solution is very dilute. The idea of separated electrical charges in ordinary solutions was resisted and ridiculed by many older chemists and physicists, but the new generation of scientists readily accepted the notion. The concepts of electrolytic dissociation easily accommodated the discovery of electrons and the incorporation of electrons into chemical theory during the early years of the 20th century.

---

[4]*Nordisk Revy* March 31, 1885; quoted in full in E. Cohen, *Jacobus Henricus van't Hoff: Sein Leben und Wirken*; Akademische Verlagsgesellschaft: Leipzig, 1912; pp 212–215.

[5]An analysis of the complex collaboration between Arrhenius and van't Hoff and their relationships with other physical chemists was given by Robert Scott Root-Bernstein, "The Ionists: Founding Physical Chemistry, 1872–1890" Ph.D. dissertation, Princeton University, 1980.

The dissociation of electrolytes in water is now universally accepted; it is presented in introductory chemistry books as a central, undisputed fact of chemistry. Active research programs during the first half of the 20th century developed the electrolytic dissociation theory of Arrhenius and van't Hoff into a sophisticated field of physical chemistry.[6] The qualitative features remain in much the same form set forth by Arrhenius.

Arrhenius's appointment to the faculty of Stockholm Högskola gave him the opportunity to organize his ideas into coherent courses in modern chemistry. His lectures in the autumn of 1897 were published as *Lärobok i teoretisk elektrokemi* (Norstedt: Stockholm, 1900).[7] In the summer of 1904 he delivered a series of lectures at the University of California at Berkeley, presenting "a coherent account of the development of theories in general chemistry" and placing the new physical chemistry in historical context; he published the lectures as *Theories of Chemistry* (Longmans, Green: London, 1907). In 1911 he gave the Silliman lectures at Yale, and these were published as *Theories of Solutions* (Yale University Press: New Haven, 1912).

When studying the rate of inversion of cane sugar in acidic solution, Arrhenius was impressed by the dramatic exponential increase in the rate of reaction when the temperature is increased: a rate increase of about 12% for each degree of temperature change near room temperature. He explained this temperature dependence by assuming that an equilibrium exists between inactive and active sugar molecules: The greater the energy required to activate individual molecules, the greater the temperature dependence.[8] The qualitative features of this explanation are retained in current theories that relate kinetic rate constants to temperature by means of an energy of activation.

Arrhenius spent his professional life as a physical chemist, but he did not confine his interests to electrolytic solutions. He anticipated the speculations of the late 20th century about the greenhouse effect in his long 1896 paper "On the Influence of Carbonic Acid in the Air

---

[6]Harned, H. S.; Owen, B. B. *The Physical Chemistry of Electrolytic Solutions*; Reinhold: New York, 1941, 1950, 1958; Robinson, R. A.; Stokes, R. H. *Electrolyte Solutions: The Measurement and Interpretation of Conductance, Chemical Potential and Diffusion in Solutions of Simple Electrolytes*; Butterworths: London, 1955, 1959.

[7]This appeared in English two years later as *Text-Book of Electrochemistry*; McCrae, J., Trans.; Longmans, Green: London, 1902.

[8]"Über die Reaktionsgeschwindigkeit bei der Inversion von Rohzucker durch Säuren"; *Z. Phys.Chem. Stöichiom. Verwandtschaftsl.* **1889**, *4*, 226–248.

upon the Temperature of the Ground".[9] In his *Lehrbuch der kosmischen Physik* (Hirzel: Leipzig, 1903), the first textbook on cosmic physics, he developed a theory of radiation pressure in outer space and discussed such phenomena as the solar corona and the northern lights in terms of charged particles. He had previously published papers on ball lightning, thunderstorms, and the electrical state of the atmosphere. He speculated about how the second law of thermodynamics (which, Clausius said, predicts a heat death for the universe because entropy is always increasing) might be compatible with an infinite lifetime of the universe; Arrhenius explored the idea that the law of increasing entropy in our galaxy might be balanced by galaxies where the law was reversed and entropy always decreased.

Arrhenius played a significant role in introducing the methods of physical chemistry into biochemistry. He showed how to describe the action of toxins and antitoxins quantitatively as simple chemical equilibria. His book *Quantitative Laws in Biological Chemistry* (G. Bell and Sons: London, 1915), based on his Tyndall lectures at the Royal Institution in London in 1914, can still be read with profit by students of biochemistry.

In spite of opportunities to move to other European universities, Arrhenius remained in Stockholm. He was loyal to the Högskola and was rector there from 1896 to 1902, helping to stabilize and develop this struggling institution. An attractive offer from the Prussian Ministry of Education prompted King Oscar II and the Swedish Academy of Sciences to take action to keep Arrhenius in Sweden. They established the Nobel Institute for Physical Chemistry with Arrhenius as director in 1905 and in 1909 completed a new laboratory for him. The laboratory was built in Experimentalfältet, a park on the outskirts of Stockholm. An official residence was attached to the laboratory, and there Arrhenius could work and write. He had an assistant, and guests often came to work in the laboratory. He devoted much of his later years to popularizing science and to exploring the interdisciplinary fields of physical chemistry, cosmic physics, geophysics, and immunochemistry.

GEORGE FLECK
*Smith College*

---

[9]"On the Influence of Carbonic Acid in the Air upon the Temperature of the Ground". *Philos. Mag.* **1896**, Series 5, *41*, 237–276.

# Bibliography

Arrhenius, S. A. *Die Chemie und das moderne Leben*; Finkelstein, B., Trans.; Akademische Verlagsgesellschaft: Leipzig, 1922.

Arrhenius, S. A. "On the Dissociation of Substances in Aqueous Solution"; *The Foundations of the Theory of Dilute Solutions*; Alembic Club Reprint No. 19; The Alembic Club: Edinburgh, 1929.

Arrhenius, S. A. *Immunochemie*; Akademische Verlagsgesellschaft: Leipzig, 1907 (*Immunochemistry*; Macmillan: New York, 1907).

Arrhenius, S. A. *Lehrbuch der Elektrochemie*; Euler, H., Ed.; Quandt & Händel: Leipzig, 1901; J. A. Barth: Leipzig, 1915.

Arrhenius, S. A. *Lehrbuch der Kosmischen Physik*; S. Hirzel: Leipzig, 1903.

Arrhenius, S. A. *Quantitative Laws in Biological Chemistry*; Bell, G.: London, 1915.

Arrhenius, S. A. *Theories of Chemistry*; Price, T. S., Ed.; Longmans, Green: London, 1907.

Arrhenius, S. A. *Theories of Solutions*; Yale University Press: New Haven, 1912.

Crawford, E. "Arrhenius, the Atomic Hypothesis, and the 1908 Nobel Prizes in Physics and Chemistry"; *Isis* **1984**, *75*, 503–522.

Dolby, R. G. A. "Debates Over the Theory of Solution: A Study of Dissent in Physical Chemistry in the English-Speaking World in the late Nineteenth and Early Twentieth Centuries"; *Hist. Stud. Phys. Biol. Sci.* **1976**, *7*, 297–404.

Farber, E. *Nobel Prize Winners in Chemistry, 1901–1961*; Abelard Schuman: London, 1963.

Jaffe, B. "Arrhenius: Three Musketeers Fight for Ions"; In *Crucibles: The Great Chemists, Their Lives and Achievements*; Tudor: New York, 1930; pp. 219–241.

*Nobel Prize Winners*; Wasson, T., Ed.; H. W. Wilson: New York, 1987.

Ostwald, W. "Svante August Arrhenius"; *Z. Phys. Chem. Stöechiom. Verwandtschaftsl.* **1909**, *69*, 4–20; followed by a bibliography, 21–27.

Palmaer, W. "Svante Arrhenius"; In *Das Buch der grossen Chemiker*; G. Bugge, Ed.; Verlag Chemie: Berlin, 1929–1930; Vol. 2, pp 443–462. Translated and abridged by Ralph E. Oesper in *Great Chemists*; Farber, E., Ed.; Interscience: New York, 1961; pp 1093–1109.

Root-Bernstein, R. S. "The Ionists: Founding Physical Chemistry, 1872–1890"; Ph.D. Dissertation, Princeton University, 1980.

Rubin, L. P. "Styles in Scientific Explanation: Paul Ehrlich and Svante Arrhenius on Immunochemistry"; *J. Hist. Med. Allied Sci.* **1980**, *35*, 397–425.

Servos, J. W. *Physical Chemistry from Ostwald to Pauling: The Making of a Science in America*; Princeton University Press: Princeton, 1990. See esp. chapter 1.

Snelders, H. A. M. "Arrhenius, Svante August"; In *Dictionary of Scientific Biography*; Gillispie, C. C., Ed.; Charles Scribner's Sons: New York, 1970; Vol. 1, pp 296–302.

*Swedish Men of Science*; Lindroth, S., Ed.; Anderson, B., Trans.; Swedish Institute: Stockholm, 1952.

Walker, J. "Arrhenius Memorial Lecture"; *J. Chem. Soc.* **1928**, 1380–1401.

# William Ramsay

## 1852–1916

In 1904 Sir William Ramsay was awarded the Nobel Prize in chemistry for his experimental work that included the discovery and isolation of the noble gas family.

Ramsay was born in Glasgow, Scotland, on October 2, 1852. He was the only child of civil engineer and businessman William Ramsay and his wife Catherine Robertson Ramsay. Sir William came from a family with scientific leanings: His grandfather founded the Glasgow Chemical Society, and an uncle, Sir Andrew Ramsay, was a well-known geologist. In his youth Sir William was interested in music, languages, mathematics, science, and athletics. His interest in chemistry began while casually reading about the manufacture of gunpowder in a chemistry text. At the time, he was recovering from a football injury.

In spite of the scientific background of his family, Ramsay received a classical education. He was originally expected to study for the ministry. His secondary education was completed at the Glasgow Academy, and from there he entered the University of Glasgow in 1866, at the age of 14, to take an arts course. In 1868 he began his work in chemistry by working as a chemist apprentice to Robert Tatlock in the laboratory of the Glasgow City Analyst. In 1870 Ramsay left Scotland to study in Germany, first under Robert Bunsen at Heidelberg, and then under Rudolf Fittig at Tübingen. At the latter city, Ramsay worked in the area of organic chemistry, specifically, studying nitrotoluic acids and the ammonium compounds of platinum. He received a Ph.D. from Tübingen in 1873.

Ramsay returned to Glasgow in 1874 as an assistant to Georg Bischof at Anderson's College (later called the Royal Technical College). This position was followed by a tutorial assistantship at the University of Glasgow. He continued his work in organic chemistry, investigating pyridic acids and the alkaloids of quinine and cinchonine. This research enabled him to become one of the first scientists to offer a logical explanation for Brownian movement.[1]

In 1880 Ramsay was appointed Professor of Chemistry at University College, Bristol (now Bristol University). With his assistant Sydney Young, Ramsay began a series of experiments on the determination of the physical properties of liquids such as water, alcohol, ethers, and various hydrocarbons. Their primary purpose was to determine the relationship of these properties to atomic or molecular weights. As a result of their studies, more than 30 papers were published by Young and Ramsay on the critical states of liquids and vapors. To accommodate the nature of their research endeavors, Ramsay learned the art of glassblowing. It enabled him to make the new kinds of laboratory equipment needed. Ramsay continued his successful pursuits in the laboratory, even though he was appointed Principal of the College in 1881, a position he held until 1887.

In 1887 Ramsay was appointed to succeed Alexander Williamson as Professor of Chemistry at University College, London. After reorganizing the out-of-date laboratory, he and his students studied diketones, the metallic compounds of ethene, and the atomic weight of boron. Ramsay conducted research on surface tension and densities at different temperatures. He developed one of the first experimental methods for the determination of the molecular weights of substances in the liquid state.[2]

By 1890 the composition of air had been very thoroughly studied and was found to be primarily nitrogen and oxygen, with traces of carbon dioxide and water vapor, and with minute traces of compound gases such as ammonia. There was no anticipation of the discovery of any new elements in the atmosphere, even though Henry Cavendish suggested in 1785 that there might be another element in air. Cavendish noted that when an electric spark was passed through air causing nitrogen and oxygen to combine, there was always a small unreactive residue. This colorless, odorless, insoluble gas would

---

[1] Trenn, 1972, p 278.
[2] Moureu, 1961, p 1000.

not form compounds with other more reactive elements. Cavendish recorded that the residue amounted to about 1 part in every 120 by volume.

In 1892, Lord Rayleigh (John William Strutt), then Professor of Natural Philosophy at the Royal Institution in London, began measuring the relative densities of the principal simple gases (nitrogen, hydrogen, and oxygen). Using several different methods in the preparation of oxygen, he found no density variations beyond experimental error. However, Rayleigh found that nitrogen prepared from ammonia was lighter than nitrogen prepared from air. Noting that the variation was beyond experimental error, Rayleigh reported the results of his work in *Nature* and asked for comments. In 1894, at a meeting of the Royal Society, Rayleigh showed again that chemically prepared nitrogen was less dense than atmospheric nitrogen. He suggested that the chemically prepared nitrogen might be contaminated by the presence of a lighter gas.

Ramsay discussed the subject with Rayleigh and asked if he might work on the problem. Ramsay thought that atmospheric nitrogen might contain a heavier gas. Rayleigh encouraged Ramsay to work on this problem from a chemical point of view.[3] Ramsay passed atmospheric nitrogen over hot magnesium in large-scale experiments and collected a small volume of unreacted gas. Cavendish had no spectroscope to analyze his unreactive residue, but Ramsay did; the spectroscope had recently been devised by Bunsen and Kirchhoff. Some of the unreactive gas was placed in a Plucker tube, the gas was heated to a glow, and the spectrum was analyzed. In addition to the characteristic lines of nitrogen, there were distinct lines that did not correspond to any known gas.

In the meantime, Rayleigh prepared nitrogen from several compounds and found that in all instances nitrogen was less dense than atmospheric nitrogen. Rayleigh then repeated Cavendish's experiments and found that the residual gas, which was neither oxygen nor nitrogen, was present to the extent of 1 part in every 107 by volume. This evidence showed again what a remarkable experimentalist Cavendish was, and it showed that there was indeed something else present in the atmosphere.

In 1894 Ramsay and Rayleigh made a preliminary announcement to the British Association that they had discovered a new element in the atmosphere. The name chosen for the gaseous element was *argon*, from a Greek word meaning *lazy*. Early in 1895, Ramsay and

---

[3]Ihde, 1964, p 370

Rayleigh presented their completed results in a joint paper to the Royal Society. Many scientists greeted the announcement with skepticism, thinking that the new gas might be a form of nitrogen cyanide, $CN_2$, or perhaps a triatomic form of nitrogen, $N_3$. After a few months, Ramsay, William Crookes, and Karol Olszevski of Kraków had gathered enough data to show that the gas actually was a new element.

When the discovery of argon was announced, Ramsay suggested that it might be placed in the periodic table in a new column between chlorine and potassium. The atomic weight of argon was greater than potassium, but the chemical properties dictated that it did not belong between potassium and calcium. This new element, argon, was inert; its valence could therefore be considered to be zero. Zero valence would not fit in with Mendeleev's general scheme for the periodic table, because the elements just before and just after argon had a valence of one. So a new group would have to be made for argon. Ramsay also suggested that there might be other inert gases in the atmosphere.

In 1895 Ramsay received a letter from Sir Henry Miers of the British Museum noting that in 1888 an American geochemist, William Hillebrand, had obtained an inert gas when heating a uranium-containing mineral with sulfuric acid. Ramsay immediately secured a related mineral, treated it with sulfuric acid, and analyzed the spectrum of the gas. The spectrum of argon was not found, but among the few lines that did appear there was a strong yellow line similar to the D line of sodium. Samples of the gas were examined by William Crookes and also by Joseph Norman Lockyer, who announced that the yellow line coincided with the $D_3$ line of helium. This announcement of the discovery of a new element came less than one week after Ramsay received Miers's letter.

The $D_3$ line of helium was first observed by the French astronomer Pierre Janssen during the solar eclipse of 1868. In England the same spectrum was analyzed by Lockyer and Edward Frankland. The line could not be attributed to hydrogen or to water vapor or to any other known element. Frankland proposed in 1868 that this element be named *helium*, after the Greek word *helios*, for the *sun*.[4] The official announcement of the discovery of helium on Earth was made at the annual meeting of the Chemical Society in March of 1895.

Ramsay tried to find the other suspected elements in minerals, but without success. In 1898, assisted by Morris Travers, he tried fractionating large amounts of argon. In a preliminary experiment, they

---

[4]Ibid., p 373

examined the last milliliter of liquid air that remained after 1 L had boiled away. With spectral analysis, the spectrum of argon was seen, but also new yellow and green lines appeared. The density of this new gas was greater than pure argon.

Ramsay and Travers then spent several months preparing 15 L of liquid argon. The clear liquid was carefully allowed to undergo fractional evaporation. The first fraction contained a new light gas, which was given the Greek-derived name *neon*, meaning *new*. The remaining two fractions contained traces of two new heavy gases, which were also given names derived from Greek: *krypton* for *hidden* and *xenon* for *stranger*. A new column of elements was now filled except for the missing noble gas in the sixth period.

The final noble gas was discovered by Friedrich Dorn in 1900. Ramsay did play a role, however, in the characterization of all the noble gas elements. Although he did not discover radon, he and Robert Whytlaw-Gray determined its atomic weight, and that determination was quite a remarkable experiment. The volume of radon that was available was about 0.005 mm$^3$, which is about the size of $\frac{1}{10}$ of the head of a pin. The weighing was done with a microbalance that was modified to detect differences in mass of several millionths of a milligram.

After the discovery of helium in a uranium mineral, Ramsay became more interested in radioactivity and the products of radioactive disintegrations. In 1903, in collaboration with Frederick Soddy, he showed that helium was continually produced by radium chloride.

In 1904 Ramsay was awarded the Nobel Prize in chemistry for his discoveries of the noble gases. In the same year, Lord Rayleigh was awarded the Nobel Prize in physics, primarily for his nitrogen studies with Ramsay. Ramsay continued at University College, London, until his retirement in 1912. Four years later he died at age 63 at High Wycombe, Buckinghamshire, on July 23, 1916.

Ramsay was an excellent teacher, described as having an elegant and picturesque manner.[5] He did not hesitate to use the most advanced theories of chemistry in his classes, being the first in England to introduce the works of Raoult, Arrhenius, and van't Hoff. He had a talent for experimentation and a boldness and enthusiasm in carrying out his ideas. He has been characterized as having great energy and persevering ardor in his scientific pursuits. Ramsay held the belief that original research should preoccupy the early years of the chemistry student. This belief was rooted in his distrust for ex-

---

[5]Moureu, op. cit., p 1010.

aminations. Often they were the only criterion for judging a student and could lead to the elimination of a particular student from his or her chosen vocation. Ramsay felt that the instructor should remain close to the student in the classroom and in the laboratory to fully appreciate the student's worth.

### NORMAN W. HUNTER AND KIMBERLY ZEIGLER
*Western Kentucky University*

## *Bibliography*

Asimov, I. *The Search for the Elements*; Basic Books: New York, 1962.

*British Chemists*; Findlay, A.; Mills, W. H., Eds.; The Chemical Society: London, 1947.

Brode, T. *Elemente d. Physikalischen Chemie*; Hanover, 1908.

Chernick, C. L. *The Chemistry of the Noble Gases*; U.S. Atomic Energy Commission: Oak Ridge, TN, 1967.

Dannerth, F. *The Methods of Textile Chemistry. Being a syllabus of a textile course for use in textile laboratories*; John Wiley & Sons, 1908.

David. *Ratgeber für Aufänger im Photographieren*; 42–4 Aufl.; W. Knapp: Halle.

*The Dictionary of National Biography*; Davis, H. W. C.; Weaver, J. R. H., Eds.; Oxford University Press: London, 1927.

*Dictionary of Scientific Biography*; Gillespie C. C., Ed.; Charles Scribner's Sons: NewYork, 1975; Vol. 11.

Dierbach, R. *Der Betriebschemiker;* 2 Aufl..

Duhem, P. *Ziel und Struktur der physikalischen Theorien;* Adler, Fr.; Joh. Ambr. Barth: Leipzig, 1908.

Eder *Rezepte und Tabellen für Photographic und Reproductionstechnik*; 7 Aufl:, W. Knapp: Halle.

Fischer, J. *Die Lebensvorgänge in Pflanzen und Tieren. Versuch einer Lösung d. physiologischen Grundfragen*; R. Friedländer, 1908.

Furman, H; Pardoe, W. D. *A Manual of Practical Assaying*; John Wiley & Sons: New York.

Hausen, E. *Die photographische Industrie Deutschlands*; W. Knapp: Halle.

Hiebert, E. "Historical Remarks on the Discovery of Argons: The First Noble Gas"; in *Noble Gas Compounds*; Hyman, H. H., Ed.; University of Chicago Press: Chicago, 1963; pp 3–20.

Hinrichsen, F. W. *Vorlesungen über Chem. Atomistik*; B. G. Teubner: Leipzig und Berlin, 1908.

Ihde, A. J. *The Development of Modern Chemistry*; Harper & Row: New York, 1964; Reprinted Dover: New York, 1984.

Jaffe, B. *Crucibles: The Story of Chemistry*; Dover: New York, 1976.

Jellett, J. H. *Chem-optische Untersuchungen*; L. Frank, Übersetz.

Jurisch, K. W. *Stickstoffdüngung*; L. Hirzeli Leipzig, 1908.

Kauffman, G. B.; Priebe, P. M. *J. Chem. Ed.* **1990**, *67*, 93–101.

Klöppel, E. *Patentrecht und Gebrauchs-musterrecht*; Carlerzmanns: Berlin, 1908;

König, A. *Oxydation des Stickstoffs im gekühlten Hochspannungsbogen bei Minderdruck*; W. Knapp: Halle, 1908;

Kümmel, G. *Photochemie*; B. Teubner: Leipzig, 1908;

Lehman, O. *Flüssige Krystalle und die Theorien des Leben*; Vortr. gehalten in der Vers. deutscher Naturforscher V. Arzte zu Stuttgart; J. A. Barth: Leipzig, 1908.

Lüppo-Crammer *Koloides Silber und die Photohaloide von Carey Lea*; Deutsche Übersetzung mit Anmerkungen neu herausgegeben; Theodore Steinkopf.

Meyer, H. *Jahrbuch d. Chemie. Bericht über wichtigsten fortschritte der reinen w. agnew. Chemie*; XVII Jahrg.; *Fr. Vieweg, & Sohn:* Braunschweig, 1908;

Miethe, A. *Dreifarben Photographie nach der Natur*; 2 Aufl.; W. Knapp: Halle, 1987.

Moureu, C. in *Great Chemists;* Farber, E., Ed.; Interscience: New York, 1961.

Morgan, J. L. R. *The Elements of Physical Chemistry*; John Wiley & Sons: New York, 1987.

Ramsay, W. *The electron as an element*; Lecture I of three lectures delivered at the inauguration of the Rice Institute, by Sir William Ramsay; Rice Institute: Houston, 1915;

Ramsay, W. *Elements and electrons*; Harper & Brothers: London & New York, 1912; pp 855–856.

Ramsay, W. *Modern Chemistry*; J. M. Dent & Co.: London, 1900–1903;

Ramsay, W. *Modern chemistry, theoretical and systematic;* Macmillan: New York, 1907;

Tilden, W. A. *Sir William Ramsay, K. C. B., F. R. S.*; Macmillan: London, 1918;

Travers, M. W. *A Life of Sir William Ramsay*; E. Arnold: London, 1956.

Trenn, T. "Ramsay, William"; In *Dictionary of Scientific Biography*; Gillispie, C. C., Ed., Charles Scribner's Sons: New York, 1972; Vol. 11.

# Adolf von Baeyer

## 1835–1917

Johann Friedrich Wilhelm Adolf (von) Baeyer was born in Berlin on October 31, 1835. He died at Starnberg, Germany, on August 20, 1917. Baeyer was awarded the Nobel Prize for his work in synthetic organic chemistry, particularly for his synthesis of indigo and the triphenylmethane dyes.

Baeyer's father, Johann Jacob Baeyer, had the rank of captain in the Prussian general staff and later became a general. He was responsible for the geodetic survey in Prussia and is credited with founding the European skeleton map. Adolf Baeyer's mother, Eugenie, was the daughter of a criminal court director, whose house in Berlin was a center of literary life. However, literature and the arts were not very attractive to young Adolf. He was more interested in natural science and the laboratory. It was on his grandfather's farm in Müggelsheim that he started to experiment with plants and mineral fertilizers, after his father bought him a book on plant nutrition. In his family home in Berlin, Baeyer started to do chemical experiments at the age of nine. In his memoirs he writes that his work on indigo began on his 13th birthday. For two talers, which he got as a present on that occasion, he bought a quantity of the blue plant dye. At the age of 12 he synthesized his first chemical compound, a copper sodium carbonate, unknown up to that time and described four years later by Struve.

At the Friedrich Wilhelm Gymnasium in Berlin, Baeyer's science teacher chose him as his assistant for the chemistry lectures. In 1853 he finished secondary school and started to study physics and mathematics at the University of Berlin. In 1856, after a year's service in the Prussian army, Baeyer decided to make chemistry the sole object of his studies. Because the University of Berlin had no laboratory for chemical instruction, he went to Heidelberg to study at Bunsen's laboratory, the most famous in Europe at that time. There he met August Kekulé, in whose private laboratory he started to work after a dispute with Bunsen.

In 1858 Baeyer returned to the University of Berlin to get his doctoral degree, with a thesis describing his work on cacodylic (methylarsenic) compounds. Unfortunately, the examination committee did not really understand or value his work. After the disappointing oral examination, Baeyer went back to work with Kekulé, who in the meantime had taken a position at Ghent, Belgium.

At Ghent, Baeyer worked mainly on uric acid and was able to put the purine derivatives in order. These were an important group of compounds, the structures of which were later determined by one of his most famous students, Emil Fischer (Nobel Laureate in chemistry, 1902). In 1860 Baeyer returned to Berlin, where he acquired the right to give academic lectures by submitting a paper on uric acid. There still was no laboratory at the University of Berlin, so he took a position as lecturer in organic chemistry at the Berlin Institute of Technology. There he was badly paid but was provided with a large laboratory. He kept this position for 12 years, a time of hard work, mainly on aromatic and polycyclic systems. In 1863 he discovered barbituric acid, the parent compound of the barbiturates, supposedly named for his girlfriend at that time, whose first name was Barbara. About 1865 he started work on indole, which was the first step along the path that led him to the solution of the "indigo problem". In 1868 Baeyer married Adelheid Bendemann, the daughter of a friend of his father; their first two children, Eugenie and Hans, were born in 1869 and 1870.

Baeyer was allowed to take over Heinrich Rose's large lecture class at the University of Berlin after the latter's death in 1864, but the position was officially filled by August Wilhelm Hofmann, who arrived in Berlin the following year. Although Baeyer was a successful scientist, he found it very difficult to obtain a position as a professor. At Marburg, where the faculty chose him, the government took someone else; at Königsberg the situation for research was so bad that he declined the offer to go there. Finally Baeyer was granted the title of professor at the University of Berlin but without pay.

Another problem continually troubled him: There were no institutional arrangements for discussions on chemistry and chemical research outside the universities. If he wanted to know what was going on in another laboratory, he had to read its publications. He thought that Germany needed a forum to discuss the chemical problems that were being investigated, something like the Chemical Society of London. Thus, following the lead of Hofmann and in concert with others, Baeyer became a founder of the German Chemical Society in 1867.

In 1872 Baeyer's career began to improve at last: He was appointed professor in Strasbourg, a city that had just become a part of Germany after the Franco-Prussian War. At first Baeyer had to build a laboratory in the garden of a former pharmaceutical institute. In the fall of the same year he began to work in the new building, continuing with the problems he had begun in Berlin and entering some new fields of research. Among these were his research on indigo and his research on fluorescein, in which he collaborated with his student Emil Fischer.

In 1875, the year of the birth of his third child, Otto, Baeyer was asked to come to Munich as Justus Liebig's successor. Munich was a lively place, a center of music and art, as well as a city of renowned philosophers and scientists, such as Schelling, Wagner, Pettenkofer, and Fraunhofer. Baeyer accepted the offer even though after only two years at Strasbourg, he had to start over again, beginning with the construction of a new laboratory. The building was finished in 1877, when Baeyer moved in and began his work with students and assistants, many of whom had accompanied him from Strasbourg. After extensive further work with indigo, along with many other problems, Baeyer's group was the first to achieve a total synthesis of the dye, finally determining its structure in 1883.

Even though he had not been able to find a method to produce indigo on a large scale for industrial purposes, his work was considered important enough for King Ludwig II of Bavaria to raise Baeyer to the hereditary nobility in 1885, acquiring the honorific "von". During the next years he was showered with honorary degrees and honors, the foremost of which was the Nobel Prize in chemistry, which he received on December 20, 1905, to acknowledge his work on organic dyes and hydroaromatic compounds. At that time he had already reached the age of 70 but had not stopped working, although his research productivity was declining. In 1915, the second year of World War I, at the age of 80, he left the university. On August 20, 1917, Adolf von Baeyer died in his country house near Lake Starnberg surrounded by his family.

In evaluating Baeyer's scientific contributions one must take into account the state of chemical knowledge in his time. Modern analytical methods did not yet exist: Liebig had just improved a method to determine the relative amounts of different elements in an organic compound. There was no method of molecular mass analysis and no spectroscopy for determining the structure of a molecule. Structure itself was just beginning to become a problem for chemists.

The determination of the structure of indigo and its total synthesis in the laboratory are perhaps the foremost contributions of Baeyer. He also synthesized many other important dyes such as alizarin and Congo red. In addition, he studied terpenes and from their reactions with peroxides concluded that hydrogen peroxide had the structural formula HO·HO and that oxygen could have weakly basic properties in some compounds, which he termed "oxonium compounds". He originated methods for synthesizing compounds containing small rings of carbon atoms and produced a theoretical explanation of ring formation that is still known as "Baeyer strain theory".

Adolf von Baeyer was a devoted lecturer and laboratory instructor, so many students found their way into chemistry under his guidance. His famous students included not only Emil Fischer but also Victor Meyer, Otto Fischer, Eduard Buchner (Nobel Laureate in chemistry, 1907), Richard Willstätter (Nobel Laureate in chemistry, 1915), Carl Duisberg, Arnold Hollemann, and Paul Walden.

FRANK STEINMÜLLER
*Tyska Skolan, Stockholm, Sweden*

# Bibliography

Baeyer, Adolf von. *Gesammelte Werke*; Fr. Vieweg und Sohn: Braunschweig, 1905.

Bugge, Gunther. *Das Buch der grosse Chemiker II*; Verlag Chemie: Weinheim, 1974.

Farber, E. *Great Chemists*. Interscience: New York, 1961.

Farber, E. *Nobel Prize Winners in Chemistry, 1901–1961*; Abelard-Schuman: London, 1963.

Fruton, J. S. *Contrasts in Scientific Style: Research Groups in the Chemical and Biochemical Sciences*; American Philosophical Society Memoirs 191; American Philosophical Society: Philadephia, 1990.

Gienapp, R. A. "Baeyer, Adolf"; In *Dictionary of Scientific Biography*; Gillispie, C. C., Ed.; Charles Scribner's Sons: New York, 1970; Vol. 1, 389–391.

Huisgen, R. "Adolf von Baeyer's Scientific Achievements—A Legacy"; *Angew. Chem.* **1985**, *25*, 297–311.

*Nobel Prize Winners*; Wasson, T., Ed.; H. W. Wilson: New York, 1987.

Nobelstiftelsen. *Nobel Lectures: Chemistry*; Elsevier: New York, 1966; Vol. I, p 83–88.

Schmorl, Karl. *Adolf v. Baeyer*; Stuttgart, 1952.

Willstätter, Richard. *From My Life*; Benjamin: New York, 1965.

Willstätter, Richard. *Jahrb. der Bayer. Akad. Wiss.*; Munich, 1918, p 33.

# Henri Moissan

## 1852–1907

Ferdinand Frédéric Henri Moissan was born in Paris on September 28, 1852. He died in Paris at the age of only 54, on February 20, 1907, just two months after receiving the 1906 Nobel Prize in chemistry "for his investigation and isolation of the element fluorine and for... the electric furnace [the Moissan furnace] named after him".

Henri Moissan's father, François Ferdinand Moissan, was born in Toulouse on December 16, 1825, and his mother, Joséphine Théodorine Almédorine Mitel, was born at Checy, near Orléans, on May 18, 1826. The family moved from Paris to Meaux in 1864, and Henri entered college in October of that year. His parents had only modest means; his father was employed by a railway company, and his mother contributed her earnings as a seamstress. Consequently Henri could enroll only for an abbreviated secondary education program, which did not include Latin or Greek and did not lead to the baccalaureate examination, the requirement for university admission.

In July 1870, after finishing school, Moissan was apprenticed to M. Godaillier, a watchmaker, and, had it not been for the Franco-Prussian War, he might have stayed a watchmaker for life. But in the autumn of 1870 Meaux was threatened with a Prussian invasion, and the Moissan family returned to Paris. During the siege of Paris the 18-year-old boy was conscripted and saw action in the battle of the Plateau d'Avron, in December 1870, which helped relieve Prussian pressure on Paris. After the surrender of Paris and the signing of the armistice, Moissan decided to undertake the study of pharmacy and acquire the diploma of pharmacist, second-class—the only qualification in pharmacy open to students who had not passed the baccalaureate examination. From February 1871 to June 1874 he was an apprentice at the Baudry pharmacy in Paris, and then he enrolled in the École Supérieure de Pharmacie in Paris for a three-year course.

The orientation of Henri Moissan's scientific career and his eventual vocation as a researcher were profoundly influenced by Jules Plicque, a classmate and friend at Meaux.[1] From 1871 to 1872 Plicque was a student in Edmond Frémy's School of Experimental Chemistry at the Natural History Museum in Paris; afterward he became a research chemist in Professor P. P. Dehérain's laboratory at the Museum. Plicque's enthusiasm for research inspired Moissan, who was finding only a limited interest in his work at the Baudry pharmacy. Moissan joined his friend at the Museum, where he registered as a student in Frémy's School for the 1872–1873 academic year and then joined Dehérain's laboratory group, while still retaining his position at the pharmacy. Encouraged by Professor Dehérain, who noted his keen intelligence and aptitude for research, Moissan took the courses in classics that would open the way to a university. In 1874 he obtained the baccalaureate and also published his first piece of research, carried out with Dehérain, on the respiratory activity of various plants in darkness. In 1878 and 1879 he published two more papers on plant physiology and in 1879 submitted a thesis, entitled "On the amounts of oxygen absorbed and carbonic acid emitted during plant respiration", to obtain the diploma of pharmacist, first-class.

Beginning in November 1875 Moissan served his year of required military service as a hospital attendant at Lille. In a letter written to Jules Plicque on September 20, 1876, he said that while keeping current in plant physiology, he had also begun to study ferrous oxide. This new interest in inorganic chemistry and his gradual withdrawal from plant chemistry and physiology may be attributed to the fact that in his year at the School of Pharmacy he had been strongly influenced by Professor Riche's lectures on inorganic chemistry, as Moissan's own later writings would reveal. Initially he studied pyrophoric iron and oxides of the iron

---

[1] Viel, C., "Sur une correspondance de jeunesse inédite d'Henri Moissan avec ses amis meldois Jules et Théodore Plicque"; *Bulletin de la societé littéraire et historique de la Brie* **1982**, *38*, 97–123.

family. After he obtained the degree of Licencié ès Sciences (roughly equivalent to a B.S.) he wrote a thesis that he presented for the degree of Doctorat ès Sciences in 1880.[2] In the research for this thesis he discovered a new allotrope of magnetic iron oxide and also identified a similar type of polymorphism in the oxides of manganese and nickel. These findings led him to study chromium oxides and salts, and during this research, in 1879, he developed the first process for the preparation of pure chromium by the distillation of its amalgam in a stream of hydrogen.

Moissan became senior lecturer and head of practical studies at the School of Pharmacy in Paris and, after defending a thesis entitled "Cyanogen Series", was appointed Associate Professor there in November 1880. That same year he married Marie Léonie Lugan, the daughter of a Meaux pharmacist. The Moissans had only one child, Louis Ferdinand Henri, who was born on January 3, 1885. He became a pharmacist and a chemical engineer, but died on active service on August 10, 1914, in one of the first battles of World War I, while serving as a sub-lieutenant in the infantry reserves.

In 1884 Henri Moissan began working on the isolation of elemental fluorine, a major experimental challenge to inorganic chemists since the late 18th century, when Lavoisier had speculated that the "fluoric radical" found in such minerals as fluorspar (calcium fluoride) might be a new element. Because of the great reactivity of fluorine, its isolation proved to be a daunting task. Several generations of chemists, including such famous names as Humphry Davy and Michael Faraday, had failed in their attempts, but it was generally believed that the electrolysis of some electrolyte based on fluorides would be the way to prepare the element. Edmond Frémy, under whom Moissan had studied in Paris, had come close to isolating fluorine. He succeeded in one of the key steps in the eventually successful project, the preparation of anhydrous hydrogen fluoride. Frémy claimed credit for the first isolation of fluorine because, by the electrolysis of molten calcium or potassium fluorides, he observed the production of a gas that decomposed water and produced hydrogen fluoride. However, his high-temperature fused electrolytes were hard to handle, and he could not isolate or characterize fluorine under the conditions of his experiments, even if it was produced.

---

[2]Moissan, H. "Sur les oxydes métalliques de la famille du fer"; Thèse de Doctorat ès Sciences Physiques, Paris, 1880.

Moissan's first forays into the chemistry of fluorine compounds led to new fluorides of phosphorus and arsenic. Their solutions in anhydrous hydrogen fluoride proved to be nonconducting. Finally, on June 26, 1886, Moissan isolated fluorine by electrolyzing a solution of potassium fluoride in anhydrous hydrogen fluoride at $-50°C$. The electrolytic cell body and electrodes were made of platinum (in later work, copper was found to be satisfactory for the cell body), with insulating stoppers of calcium fluoride. The electric current was supplied by a battery of 50 Bunsen cells. For the first time, the most reactive of the elements, fluorine, had been prepared in quantity and characterized unambiguously. One of the most challenging problems of the inorganic chemistry of the period had been solved.

From 1886 to 1891 Moissan's research was devoted completely to the study of fluorine and its derivatives. He studied the reactions between fluorine and a wide range of elements and compounds, and he characterized many new materials, including some simple alkyl fluorides. In January 1887 he became Professor of Toxicology at the School of Pharmacy and in 1899 assumed the Chair of Inorganic Chemistry at the same school. In 1900 he was appointed to the Faculty of Sciences at the Sorbonne, in Paris, where he found better working conditions than at the School of Pharmacy. Moissan was made a member of the Academy of Medicine in 1888 and of the Academy of Sciences in 1891.

Moissan's association with the School of Pharmacy did lead to some related research. He carried out studies on carbon monoxide poisoning, the toxicity of opium smoke, the anaesthetic properties of methyl and ethyl fluorides, and the detection of pathogenic bacillae in mineral waters. This work was secondary to his research in inorganic chemistry, which remained the major interest of his career. After his successes with fluorine he turned to the preparation of pure elemental boron and to a study of its main compounds. He then moved to an extraordinarily difficult problem, the synthesis of artificial diamonds. He hoped that the inclusion of catalytic amounts of fluorine compounds would help transform amorphous carbon into crystalline diamond, but experiments along these lines were unsuccessful. Moissan then tried to duplicate in the laboratory the conditions under which diamonds are formed in nature, namely, slow crystallization at high temperatures and pressures, as J. Werth postulated in 1893. Moissan tried to achieve these conditions by heating a saturated solution of carbon in iron to $3000°C$ and then rapidly chilling the incandescent iron in water. After dissolving away the iron, he collected some small crystals with the density and hardness of diamond, which gave carbon dioxide on combustion. He announced these results on February 6, 1893. This controversial experiment has been repeated a number of times by other groups with varying success. Modern studies of the iron–carbon phase diagram suggest that the transformation of graphitic carbon to diamond requires a pressure of some 50,000 atmospheres, while in Moissan's experiments the pressure was only about 10,000 atmospheres. Regardless of the outcome of the diamond synthesis, Moissan's experiments were extremely fruitful in another way. They turned his attention to the chemical effects of very high temperatures.

Before Moissan's time, the highest recorded temperature reached in a laboratory was around 2000°C. The diamond synthesis required a temperature of around 3000°C. To reach this temperature Moissan designed a simple but effective furnace. An electric arc was confined between two blocks of either lime (calcium oxide) or natural limestone (calcium carbonate). This furnace was first reported in 1892, and later improvements made it possible to reach temperatures of around 5000°C. With this intense source of energy available Moissan opened the new and important chapter of high-temperature inorganic chemistry. In addition to the experiments on the synthesis of diamond, he prepared crystalline samples of metal oxides that had previously been thought to be infusible. He isolated pure samples of some highly refractory metals by reducing their oxides with carbon; he prepared many new metal borides, nitrides, silicides, and carbides. His preparation of calcium carbide opened the way to the development of acetylene chemistry. He also purified many metals and refractories by very high temperature distillation.[3]

Moissan's interests in metal compounds led him to study the metal hydrides. He prepared the novel calcium hydride, and the alkali metal hydrides, showing that they reacted with carbon dioxide to form alkali formates. In 1898 he developed a process for making pure calcium by the high-temperature reduction of calcium iodide with excess sodium and studied the properties of the pure metal and many of its compounds. Moissan's next and, as it turned out, final research area was his attempt to isolate the ammonium radical. He worked with the so-called ammonia metals (solutions of metals in liquid ammonia) and with the electrolysis of solutions of ammonium salts and mercury iodide in

[3]Moissan, H. Le Four électrique; G. Steinheil: Paris, 1897.

liquid ammonia. These experiments were unsuccessful. By decomposing sodium amalgam with a solution of ammonium iodide in liquid ammonia at $-80°C$, he did obtain a pure ammonium amalgam, from which he hoped to isolate the ammonium radical.

Unfortunately on February 6, 1907, he suddenly developed appendicitis. His condition worsened rapidly in the succeeding days, and he died at the age of 54, on February 20, 1907, only two months after receiving the Nobel Prize in chemistry. His prolific research had led to the publication of more than three hundred articles and notes; two famous books, *Fluorine and its Compounds* and *The Electric Furnace*; and the five-volume *Treatise on Inorganic Chemistry*, of which he was editor-in-chief and to which he contributed greatly as an author.

The scope of Moissan's work is far-reaching and has shaped many aspects of modern chemical industry. Elemental fluorine is still produced by variants of Moissan's original method, the electrolysis of solutions of potassium fluoride in anhydrous hydrogen fluoride, and is used, for example, in the production of uranium hexafluoride for the nuclear industries, the production of sulfur hexafluoride as a transformer dielectric, and the production of graphite fluorides for long-life batteries. His research on organic derivatives of fluorine led to the modern fluorocarbon industry, with products ranging from refrigerant gases and inert polymers to medicinal and phytopharmaceutical compounds. The electric arc furnace was used in the industrial production of calcium cyanamide and calcium carbide, from which many other important chemicals were derived. Even industrial diamond synthesis was finally achieved in 1953, and current production is around 15,000 kg per year.

In addition to being an outstanding chemist, Henri Moissan was also a highly cultured man with a lively sense of humor. He was a knowledgeable lover of painting and was eclectic in his tastes, provided the work was well executed. His collection of paintings by contemporary masters was left to the city of Meaux by his son Louis and now forms almost the entire collection of 19th-century painting in the town's art museum.[4] In his early twenties he was a member of a close circle of like-minded friends in which literature, art, and science were cultivated. Like many other young intellectuals he wrote poetry and dabbled in the theater. He wrote a comedy, *An Unexpected Marriage,*

[4]Ruiz, J.C.; Viel, C.; "Henri Moissan, amateur avisé et collectioneur de tableaux"; *Bulletin de la société littéraire et historique de la Brie* **1985**, *41*, 49–58.

which he submitted to the management of the Odéon Theater, but it was rejected by the selection committee. Realizing that he was not destined to be a popular author, he would later say of this episode, "I think I was right to study chemistry". Henri Moissan was one of the greatest scientists of his time; he stimulated a rebirth of inorganic chemistry and contributed greatly to the revitalization of its methods and theories.

CLAUDE VIEL
*Tours University, Tours, France*

HAROLD GOLDWHITE
*California State University, Los Angeles*

# Bibliography

Flahaut, J.; Viel, C. "The Life and Scientific Work of Henri Moissan"; In *Fluorine— The First Hundred Years;* Sharp, D.W.A.; Banks, R.E.; Tatlow, J.C., Eds.; Elsevier Sequoia: Lausanne, Switzerland, 1986; pp 27–43.

Harrow, B. *Eminent Chemists of Our Time;* Books for Libraries: Freeport, NY, 1968.

Lebeau, P. "Notice sur la vie et les travaux d'Henri Moissan"; *Bull. Soc. Chim. Fr.* **1908**, 1–38.

Moissan, H. *The Electric Furnace;* de Mouilpied, A. T., Tr.; Edward Arnold: London, 1904.

Moissan, H. *Le Fluor et ses composés;* G. Steinheil: Paris, 1900.

Moissan, H. *Le Fluor: Mémoires;* A. Colin: Paris, 1914.

Moissan, H. *Le Four électrique;* G. Steinheil: Paris, 1897.

Ramsay, W. "Moissan Memorial Lecture"; *Trans. Chem. Soc.* **1912**, *101*, 477–488.

Viel, C. "Notes généalogiques sur les familles Moissan et Lugan"; *Héraldique et Généalogie* **1986**, *18*, 403–406.

# Eduard Buchner

## 1860–1917

It is difficult to imagine that a series of experiments performed almost a century ago, which today might appear as astonishingly simple as carefully opening an egg, would shake the very foundations of chemistry, lead to a Nobel Prize, and open the door to the era of modern biochemistry. But this was exactly what was to occur as a consequence of Eduard Buchner's startling and quite accidental discovery that the age-old phenomenon of yeast alcoholic fermentation (the process by which living yeast cells convert fruit juice into wine and leaven dough for bread) can be made to occur artificially in a test tube with a cell-free juice extracted from the yeast.

Awarded the Nobel Prize in chemistry in 1907 "for his biochemical researches and his discovery of cell-free fermentation", Eduard was born on May 20, 1860, in Munich, Germany, the son of Ernst and Friederike (Martin) Buchner. He died during World War I at Focsani, Rumania, on August 13, 1917, as a result of wounds received while serving on frontline duty as a volunteer officer in the German army. His father was a physician and professor of forensic medicine at the University of Munich and editor of the *Ärztliches Intelligenzblatt* (which later became the *Münchener medizinische Wochenschrift*). His older brother, Hans, became a bacteriologist and professor of hygiene at University of Munich; Hans was to have a profound impact on Eduard's career as a chemist and on his pioneering research activities.

Buchner became interested in chemistry while attending the Munich *Technische Hochschule* (Polytechnic) and, while there, worked in the chemistry laboratory of Emil Erlenmeyer, senior. After a brief period at the Polytechnic, Buchner was forced to interrupt his studies because of financial problems and worked for four years in canning factories in Munich and Mombach. Through the intercession of his brother Hans, Eduard was able to resume his studies in chemistry at the University of Munich in 1884, working in the laboratory of the great synthetic organic chemist, Adolf von Baeyer (Nobel laureate in chemistry, 1905).

During this same period, Buchner also studied botany in the laboratory of Karl von Nägeli at the Institute for Plant Physiology. While studying under the special supervision of his brother Hans, then an assistant in Nägeli's laboratory, Buchner became interested in the phenomenon of alcoholic fermentation. This is the process, previously described by Pasteur as "la vie sans l'air" (life without air), that involves the conversion of sugar into alcohol and carbon dioxide by a yeast microorganism. In his very first publication, Buchner concluded that yeast cells were capable of causing fermentation either aerobically or anaerobically.[1] The discovery of the facilitative nature of the yeast contradicted Pasteur's notion that the absence of oxygen was a prerequisite for fermentation, and it was also of great importance to the fermentation industry as in, for example, the brewing of beer and other alcoholic beverages. Buchner's early scientific work on fermentation was a prelude to his pioneering studies on the chemical origin of the important biological phenomenon of yeast alcoholic fermentation.

Under Baeyer's mentorship at the Bavarian Academy of Sciences, Buchner's primary focus over the next several years was on his studies of preparative organic chemistry. He was awarded his doctorate in 1888 for his research on pyrazole and became one of Baeyer's teaching assistants in 1890. He also established close friendships with two of Baeyer's assistants, Theodor Curtius and Hans von Pechmann, both of whom would later become distinguished chemistry professors. Although Buchner's subsequent work as an organic chemist resulted in numerous publications between 1885 and 1905 in synthetic organic chemistry (including the preparation of cycloheptatriene and cycloheptanecarboxylic acid), he was to remain primarily interested in the problems of fermentation.

---

[1]"Über den Einfluss des Sauerstoffs auf Gärungen;" *Zeitschrift für physiologische Chemie* **1886**, *9*, 380.

Following Buchner's appointment as a *Privatdozent* in 1891, Baeyer provided Buchner with funds to establish his own small laboratory for fermentation studies. At the suggestion of his brother, Buchner began experiments on rupturing yeast cells to extract their fluid. George Nuttall had just discovered that blood serum has an antiseptic effect on bacteria, and the two brothers speculated that yeast fluid might have similar therapeutic value. Unfortunately, the Board of the Laboratory that supervised Buchner's work was quoted by him as saying that "nothing is coming from this". They considered his experiments to be unproductive and a waste of time and required that Buchner discontinue the very studies that, when continued later, eventually led to his Nobel Prize.

The Board's objection to Buchner's work grew out of an ongoing conflict that had been dividing the scientific community for over four decades. On one side, the mechanists sought a purely chemical explanation for yeast fermentation, based on Berzelius's thesis in 1835 that all chemical reactions within living organisms were initiated and regulated by biological catalysts, substances that speed up chemical reactions without being consumed in the process. Although the mechanists, who included among their members such distinguished chemists as Liebig and Wöhler, differed in their interpretation of the phenomenon of fermentation, they all considered it to be a purely chemical process.

The vitalists, on the other side of the conflict, contended that yeast fermentation was inextricably linked to some type of "life force", inseparable from the living organism and governed by a set of laws applicable only to living cells and not inanimate objects (such as biological chemicals). Berthelot, Mayer, Nägeli (Buchner's mentor in botany), and other great scientists including Pasteur had already attempted, by numerous methods, to separate and identify a fermentation-producing agent from yeast cells. Their inability to experimentally observe alcoholic fermentation in the absence of living yeast cells led to the vitalistic notion of an unalterable connection between the process of life and fermentation. This view was generally accepted by the scientific community at the start of Buchner's experiments in 1893, so much so that the term "enzyme" (named by Kühne) was intended to refer only to the extracellular water-soluble biological catalysts, such as pepsin from gastric juice, which had been known for some time to digest protein in the alimentary canal of animals. Substances thought to be responsible for the catalysis of biological reactions that are required to sustain life within living cells were referred to as "ferments", in association with the presumed essential life process of cellular yeast fermentation.

Following the rejection of Buchner's experiments by the Board at the University of Munich, he accepted an appointment to succeed his old friend Curtius as head of the Section for Analytical Chemistry at the University of Kiel in 1893. While there he was promoted to associate professor. In 1896 Buchner became professor of analytical pharmaceutical chemistry in the laboratory of his other long-time friend, Pechmann, who had become established as a chemistry professor at the University of Tübingen. While on vacation that same year in Munich, Buchner was able to continue his previously interrupted studies on fermentation, as his brother Hans was at that time a professor and member of the Board of Directors at the Hygienic Institute.

The problem Buchner faced in attempting to extract yeast cell fluid was particularly intriguing and might best be described in an imaginary way. Suppose you wish to extract the contents of imaginary chicken eggs, the microscopic size of yeast cells, from their eggshells without simultaneously destroying the natural chemical state of the yolk or egg white (presumably you would be allowed to physically mix the yolk and egg white in the process). How could you accomplish the separation? Buchner was keenly aware that any attempt to rupture the yeast cell membrane (i.e., crack the eggshell) might somehow destroy the natural chemical state of the yeast fluid (i.e., the yolk and egg white) in much the same fashion as frying an egg causes the yolk and egg white to denature. He also needed a method to separate the contents of his "crushed eggs" from his broken "eggshells".

At the suggestion of his brother's assistant, Martin Hahn, Buchner succeeded in rupturing the yeast cell membranes by grinding the cells in a mortar with a mixture of sand and diatomaceous earth. The pulverized material was then wrapped in canvas, and the fluid content was squeezed out using a hydraulic press. Thus, by purely mechanical means, he was able to obtain the yeast juice in as pure a chemical form as possible without the exposure of the cells to harsh chemicals or high temperatures. Then, in an effort to preserve the easily decomposed juice for future experiments, Buchner fortuitously added to it a concentrated solution of sucrose. This is the same procedure used in the kitchen to "preserve" fruits and juices against bacterial contamination. Quite astonished, he observed that this sugar–juice mixture began to froth, looking like the head on a freshly poured glass of beer.

In an extraordinary feat of scientific imagination, Buchner recognized this process as analogous to cellular fermentation by intact yeast cells. Subsequent experiments verified that carbon dioxide and alcohol were being produced from his mixture in just the same manner as alcoholic fermentation by living yeast. He had provided the first direct evidence that the same chemical processes that take place within living cells (in vivo) can also occur in a test tube (in vitro) using cell-free extracts of the organism. His first papers on this sensational discovery, "Alkoholische Gärung ohne Hefezellen [Alcoholic Fermentation without Yeast Cells]" were published in 1897. Fermentation was thus, as Berzelius had initially proposed, caused by the chemical catalytic activity of cellular enzymes and not, as the vitalists had insisted, caused by any "life force" phenomena.

In 1898 he accepted an appointment as full professor at the College of Agriculture in Berlin and also became director of the Institute for the Fermentation Industry. Between 1898 and 1902 he continued his experiments on the properties of "zymase", the term he used to describe the active fermentation–producing agent in the yeast juice. During this period, he undertook very careful and laborious experiments to defend his thesis on the chemical nature of zymase. In particular, a number of scientists, including the physiologist Max Rubner and the biochemist Hans von Euler-Chelpin (who would later share the 1929 Nobel Prize in chemistry with Arthur Harden for their discovery of the coenzymatic nature of Buchner's zymase), initially argued that the fermentation activity of the yeast juice might somehow be due to "living" remnants or yeast cells that remained unseparated from the filtered juice. In a series of remarkably convincing experiments, Buchner defended against this argument by demonstrating unequivocally that zymase alone was responsible for the sugar fermentation, because it retained its fermentation activity even under conditions known to kill living yeast cells.

The first comprehensive presentation of his seminal achievements was published in 1903 in his book Die Zymase-Gärung.[2] Following his Nobel Prize in 1907, Buchner accepted faculty positions at the University of Breslau in 1909 and then at the University of Würzburg in 1911. Until his death in 1917, he continued his investigations on the zymase and fermentation processes. In particular, he focused on attempts to

---

[2]Buchner, E.; Buchner, H.; Hahn, M.; R. Oldenbourg: Munich, 1903.

elucidate the chemical properties and specific nature of zymase and on the nature of the intermediate chemical reactions and products in the overall fermentation process.

A number of factors contributed to Buchner's success as a pioneering scientist. First, he was open-minded, nontraditional, and imaginative in his thinking, despite the initial rejection of his ideas by the scientific community of his day. Second, he had the extraordinary ability to develop new and uncomplicated experimental methodology and to incorporate the methodology in his research. Third, he had an especially keen sense of observation, a particularly important scientific attribute. Fourth, to confirm the validity of his conclusions, he carefully and meticulously excluded other possible interpretations by further experimentation. Finally, he benefited enormously from the inspiration, environment, and exchange of ideas provided by his family, friends, colleagues, and the institutions that supported his work.

Buchner's contributions to science were of immense value. Not only did he disprove the vitalistic doctrine of his era, but his revolutionary discovery set the very foundation for the emergence of biochemistry as a legitimate interdisciplinary science incorporating the fundamental fields of physics, chemistry, and biology. The modern biochemical study of enzymology is rooted in Buchner's initial yeast-crushing experiments with mortar and pestle.

It was to take this newly found discipline and the accomplishments of scores of talented biochemists nearly half a century to completely unravel the chemistry of alcoholic fermentation, now recognized as glycolysis (the Embden–Meyerhof pathway), the central pathway of intermediate metabolism. The glycolysis of sucrose to alcohol and carbon dioxide is now known to consist of not 1, but 14 linked chemical reactions, each of which is catalyzed by a specific enzyme in Buchner's "zymase". One of these enzymes, alcohol dehydrogenase, is used today to test blood alcohol levels of motorists suspected of driving while intoxicated.

Perhaps one of Buchner's greatest legacies to modern society is the development of today's fermentation industry. This industry provides us with a number of the foods and beverages we enjoy daily, with organic products for chemical and agricultural industries, and with antibiotics and other important biologicals for medicine and research. Most remarkably, by demonstrating at the close of the 19th century that the biochemical reactions of an organism involve enzymes that can function chemically outside the living cell, Buchner built the foundation for the astonishing scientific achievements of today's recombinant DNA and genetic-engineering technologies (see Paul Berg, Nobel Laureate in chemistry, 1980). Ironically, toward the close of the 20th century, we now see the utilization of intact living cells to chemically manufacture medicines and other biological products in a manner and of an abundance not previously possible by even the best extracellular chemical methods.

**CHRISTIAN G. REINHARDT**
*Rochester Institute of Technology*

# Bibliography

Buchner, E. "Cell-Free Fermentation"; In *Nobel Lectures Chemistry 1901–1921;* Elsevier: New York, 1966; pp 99–122.

Buchner, E.; Buchner, H.; Hahn, M. *Die Zymase-Gärung;* R. Oldenbourg: Munich, 1903.

Buchner, E.; Rapp, R. *Ber. Dtsch. Chem. Ges.* **1897**, *30*, 117–124; the first in a series of 10 papers on the topic published between 1897 and 1901.

Farber, E. *Nobel Prize Winners in Chemistry, 1901–1961;* Abelard-Schuman: London, 1963.

Harries, C. In *Ber. dtsch. Chem. Ges.* **1917**, *50*, 1843–1876; appendix contains a complete bibliography of Buchner's work.

Kohler, R. E. "The Background to Eduard Buchner's Discovery of Cell-Free Fermentation"; *J. Hist. Biol.* **1971**, *4*, 35–61.

Kohler, R. E. "The Enzyme Theory and the Origin of Biochemistry"; *Isis* **1973**, *64*, 181–196.

Kohler, R. E. "The Reception of Eduard Buchner's Discovery of Cell-Free Fermentation"; *J. Hist. Biol.* **1972**, *5*, 327–353.

Schriefers, H. "Buchner, Eduard"; In *Dictionary of Scientific Biography;* Gillispie, C. C., Ed.; Charles Scribner's Sons: New York, 1970; Vol. 2, pp 560–563.

# Ernest Rutherford

## 1871–1937

Born near Nelson, New Zealand, on August 30, 1871, and raised in an ethos of hard work and common sense, Ernest Rutherford made his way to J. J. Thomson's Cavendish Laboratory in 1895 upon accepting an 1851 Exhibition scholarship. This award gave Rutherford the opportunity to hone his research skills, and he obtained a first-rate physics professorship at age 27. Rutherford earned the Nobel Prize in chemistry in 1908 "for his investigations into the disintegration of the elements, and the chemistry of radioactive substances." His pioneering work on atomic structure was grounded in a Victorian style of mechanics that emphasized semiliteral mechanical models rather than abstract mathematical representation. His work at Cambridge, Montreal, and Manchester gave Rutherford the reputation of being the greatest experimental physicist of his time. Rutherford died in Cambridge on October 19, 1937, after an operation for a strangulated hernia. His ashes were interred at Westminster Abbey.

Rutherford's parents came to New Zealand from Great Britain in the mid-nineteenth century as children. His Scottish father, James Rutherford, often spent much of the working week pursuing such small business ventures as logging, milling, bridge construction, and flax farming and processing. His English mother, Martha Thompson, was a schoolmistress. Rutherford sent weekly letters after leaving New Zealand and maintained a close relationship with his mother until her death in 1935. Many features of Rutherford's childhood echoed throughout his adult life, including the singing, or humming, of church hymns with his 11 siblings. At age 15, he earned a record high score on a scholarship examination that enabled him to attend a small secondary school, Nelson College.

Financed by a series of scholarships, Rutherford attended Canterbury College, Christchurch, from 1889 to 1894. He distinguished himself in mathematics and physical science, earning his B.A. degree in 1893 and his M.A. in 1894. Although his professors at Canterbury recognized him as a successful student and a talented experimentalist, they did not consider him a genius. In 1893, he considered studying medicine in Edinburgh, but a Canterbury alumnus, John Stevenson, helped extinguish this desire by warning him that "some of the most selfish, uncouth, brutal fellows" attended Edinburgh.[1] From the tone of this letter and from Rutherford's relationship with a woman who campaigned against drinking—his future mother-in-law, Mrs. Arthur de Renzy Newton—it is evident that Rutherford socialized in conservative circles. Unsure of his future, Rutherford investigated several topics in magnetism. (Some have claimed that this work earned him a B.S. degree at Canterbury, but there is no record of this degree.) As part of this research, he produced high-frequency radio waves and constructed a receiver. His scholarship to the Cavendish Laboratory was awarded largely in recognition of these magnetism and radio studies.

Soon after arriving in Cambridge in 1895, Rutherford put aside further development of his radio-wave detector as J. J. Thomson directed his attention to the behavior of ions produced in gases by X-rays. This collaborative project with Thomson not only resulted in a theory of gaseous ionization, but also confirmed in Rutherford a desire to pursue physics beyond the practical orientation already familiar to him. Working with Thomson, he established a repertoire of favorite instruments that included electrometers, X-ray tubes, parallel-plate condensers, and devices for creating air currents. In fact, Thomson credited Rutherford with helping to make the study of X-rays and ionization a "metrical" rather than merely descriptive enterprise, because he "devised very ingenious methods for measuring various fundamental quantities."[2]

After radioactivity was discovered in 1896 by Henri Becquerel, Rutherford soon dominated its study by using the metrical approach

[1] Wilson, D. *Rutherford: Simple Genius*; MIT Press: Cambridge, 1983; p 48.
[2] Wilson, p 116.

he had pioneered in X-ray research. In 1897, Rutherford considered uranium radiation as merely an alternative means of ionizing gases, and thus a tool for the development of gaseous ionization theory. By 1898, however, he had shifted his attention to the nature of uranium radiation itself. Working with uranium radiation transmitted through foils, he distinguished two types, designating as "alpha" the part that was easily absorbed and heavily ionizing, and "beta" the part that was penetrating and weakly ionizing. In September 1898, Rutherford reported this research, still placing greater emphasis on his method of measuring the electrical discharge of ionization caused by radiation than on the "alpha" and "beta" components themselves. Although in 1898 he was not fully aware of the significance of his partial shift in focus from the study of gaseous ionization to the study of radiation itself, shortly before his death Rutherford identified this change in research agenda as the most important moment of his life. Rutherford first devoted himself to investigations of radioactivity after he left the Cavendish Laboratory for a professorship in Montreal.

Lured by promises of ample time and money for research at McGill University, Rutherford moved to Montreal in the fall of 1898. The MacDonald Physics Building at McGill was one of the best equipped research facilities in the world, a resource made possible by the lavish support of tobacco millionaire Sir William MacDonald. Because John Cox, senior professor of physics and director of the physical laboratories, preferred administration and teaching, he put Rutherford in charge of research. During his nine years at McGill, Rutherford demonstrated his skill as an organizer of scientific research by building up the infrastructure for physics research at the university. In addition to McGill students, Rutherford collaborated with many others, including such talented chemists as Frederick Soddy, Bertram Boltwood, and Otto Hahn.

His financial security and professional stability enabled Rutherford to return to New Zealand in the summer of 1900 to marry Mary Newton. His correspondence with his fiancée from 1895 to 1900 reveals how thoroughly his personal life revolved around the progress of his work and reputation as a physicist. In his first letter to Mary from Cambridge he noted, "My success here will probably depend entirely on the research work I do."[3] In other Cambridge letters, he mentioned his

---

[3]Rutherford to Mary Newton, October 3, 1895, as cited in Eve, A. S. *Rutherford;* Cambridge University Press: Cambridge, 1939; p 17.

struggle to gain the cooperation of the physics demonstrators at the Cavendish Laboratory. At Cambridge, Rutherford had to overcome prejudice against his colonial origins and antagonism toward his special status as the first postgraduate researcher from another university, but such obstacles were not present at McGill. Despite his salary of £500—twice what Pierre Curie was earning—Rutherford insisted on postponing their wedding for 18 months.[4]

They finally married after Rutherford completed two academic years at McGill. During this period, he prepared his first lectures, acted as a consultant in a court case concerning the damaging vibrations of the local tramway company's power house, and pursued his radioactivity studies. His most important research during this period was stimulated by a study of thorium radiation he had assigned to the electrical engineer R. B. Owens. Rutherford was intrigued by Owens' report of the odd sensitivity of the radioactivity of a piece of thorium oxide (thoria) to air drafts. After Owens left Montreal in the spring of 1899 to work at the Cavendish Laboratory, Rutherford found an explanation for the anomaly. "In addition to this ordinary radiation [from thorium], I have found that thorium compounds continuously emit radio-active particles of some kind, which retain their radio-active powers for several minutes", Rutherford announced.[5] Passing this gaseous "emanation" of particles, mixed with air, through an open tube connected to an electrometer, he showed that its radioactivity (assumed to be proportional to ionic current) decreased to half the original value in one minute. It then decreased by half again in the next minute, and so on. Rutherford's was the first determination of what came to be called a half-life.

In the final paragraph of this September 1899 paper on the air-draft anomaly, Rutherford announced a conclusion that implied what he later called "the newer alchemy". He discovered that the gaseous positive ions produced by the thorium oxide emanation have the "power of producing radio-activity" in all substances they touch. The radiation of this process was "of a more penetrating character than that given out by thorium or uranium". He concluded that "the emanation from thorium compounds thus has properties which the thorium itself does not possess."[6] This investigation, which was still in progress, pointed to the transmutation of matter.

[4]Wilson, p 130.

[5]Rutherford, E. "A Radio-Active Substance Emitted from Thorium Compounds"; *Philosophical Magazine* **1900**, series 5, *49*, 1-14, p 1.

[6]Rutherford, p 14.

Rutherford completed another important research project before the end of his second academic year at McGill by establishing the quantitative laboratory procedures needed to determine the magnitude of radioactive processes.[7] With the assistance of an undergraduate, R. K. McClung, he developed a method of measuring the energy expended by X-rays in creating a pair of ions in a gas. From this measurement, Rutherford extrapolated the energy given off by radioactive substances such as thorium and uranium on the assumption that their emanations also lose energy in making gaseous ions. He concluded that, because of its great intensity, radioactivity energy could not be an ordinary chemical process. Through this statement, he was one of the first to imply the existence of what he and Soddy called "atomic energy" in 1903.[8] At the beginning of a major conceptual transformation in chemistry and physics, Rutherford set sail for New Zealand in April 1900 to get married. His work came to a halt for nearly six months.

The next major step in Rutherford's studies in radioactivity required the help of a chemist, and Frederick Soddy was available at just the right time. He had arrived in Montreal in May 1900 after a failed attempt to secure a chemistry position in Toronto. In October 1901, six months after an extensive debate with Rutherford at the local Physical Society over "the existence of bodies smaller than an atom,"[9] he agreed to help Rutherford determine the chemical nature of thorium's emanation. Soddy subjected the gaseous emanation of thorium to potent chemical reagents such as platinum black, zinc, magnesium, and lead chromate, but to no avail. They concluded that it must be an inert argon gas, and thus a new element. "Rutherford, this is transmutation", Soddy reportedly proclaimed on the day of discovery.[10] However, in the paper they sent to the Chemical Society in December 1901, they refrained from mentioning transmutation.

Based on William Crookes's preparation of uranium X, the radioactive impurity believed to be found in all uranium ore, Soddy separated thorium X from thorium just before Christmas 1901. Upon his return in January, he was confronted with bewildering results. The "purified" thorium had regained its original level of activity and the

[7]Rutherford, E.; McClung, R. K. "Energy of Röntgen and Becquerel Rays, and the Energy Required to Produce an Ion in Gases"; *Philosophical Transactions*, **1901**, series A, *196*, 25-59.

[8]Rutherford, E.; Soddy, F. "Radioactive Change"; *Philosophical Magazine*, **1903**, series 6, *5*; 576-591, p 590.

[9]Wilson, p 148.

[10]Howorth, M. *The Life Story of Frederick Soddy*; New World: London, 1958, p 83.

extract of thorium X had become virtually impotent. Rutherford and Soddy, however, were comforted to learn that Becquerel found the same phenomenon in uranium and uranium X. In a paper finished in February 1903, the month of Soddy's return to England, Rutherford and Soddy proposed a general theory of radioactive change based on experimental findings such as the regeneration of thorium X by thorium. In other words, they showed the connection between the emission of alpha-rays from thorium and the production of thorium X (and the emission of alpha-rays from thorium X and the production of thorium emanation, and so on). Accordingly, they concluded "not merely that the expulsion of a charged particle accompanies the [radioactive] change, but that this expulsion actually *is* the change."[11] This transformation theory of radioactivity, which Rutherford and Soddy formulated in 1902 and refined in 1903, eclipsed the phosphorescence theory of Becquerel as well as the views of Crookes and the Curies.

After Rutherford and Soddy formulated the transformation theory of radioactivity, Rutherford and his research staff pursued radioactivity studies in a more orderly fashion, interrupted by Rutherford's extensive writing and speaking activities from 1904 to 1906. The transformation theory stated that each radioactive decay amounts to the transmutation of a parent into a daughter element (with the expulsion of one or more kinds of subatomic particles) and that each element transmutes a definite percentage of its atoms in a characteristic period. The daughter element may in turn disintegrate in a similar manner, producing successive unstable elements until a more stable element is reached. The half-life of each radioelement, varying from a few seconds to billions of years, was recognized as the most reliable constant upon which to establish elemental identity, a status that atomic weight had long held for ordinary stable elements. Accordingly, the primary research agenda in radioactivity studies was to order all known radioelements into decay series and to fill in the gaps by discovering new radioelements. In 1913, Soddy defined the term "isotope" in order to save the periodic law from the implications scientists were drawing from the 30 radioelements that had been identified. In short, Soddy avoided the conclusion of the existence of as many different elementary species as atomic masses by defining isotopes as atomic masses occupying the "same place" (Greek) in the periodic table.

---

[11]Rutherford and Soddy, p 580. Emphasis in the original.

Behind the decay series research that revolutionized chemistry stood a fundamental alteration in the meaning of a "radioactive substance" that has persisted to the present day. Rutherford and his colleagues at McGill had produced quantitative evidence that a radioactive substance is a substance made up not of particles emanating radiation, but of atoms that radiate simultaneously with spontaneous transformation into new chemical species.[12] Rutherford codified this advance in physics and chemistry in his first book, *Radioactivity,* published in 1904 and greatly expanded in 1905. During this period, he had numerous speaking engagements in which to promote his research agenda, including his Silliman lectures at Yale University (1905), later published as *Radioactive Transformations* (1906).

Beyond the central task of ordering radioelements into decay series, the study of alpha rays formed another important component of Rutherford's research program. After having demonstrated that alpha rays are not wavelike radiations, as previously thought, but corpuscular particles, in 1903 he deflected alpha particles in electric and magnetic fields, giving indirect evidence of their positive charge. In 1905, Rutherford first obtained direct evidence of the positive charge of the alpha particle, but it took him until 1908 to prove that it was a positively charged helium ion. Beyond the scintillation-counting method he and Hans Geiger used to establish the double charge of the alpha particle, Rutherford directed his glassblower, Otto Baumbach, to construct a device that offered a more dramatic confirmation of the identity of the alpha particle. Baumbach made a glass tube thin enough to permit the penetration of the majority of rapidly moving alpha particles. This tube was filled with radium emanation and placed in a large evacuated tube of thicker glass. The alpha particles emitted from the decaying radium emanation collected in the space between the inner and outer tubes and, when sparked, gave off the spectrum of helium. Rutherford enjoyed this research so much that he spoke of alpha particles as being "his" particles. Accordingly, he chose as the topic of his December 11, 1908, Nobel lecture, "The Chemical Nature of the Alpha-Particles from Radioactive Substances". After establishing their identity beyond reasonable doubt in 1908, Rutherford often used "his" particles as a tool for exploring the interior of other atoms, a general approach that persists in current atom-smashing research.

---

[12]Heilbron, J. L. "Physics at McGill in Rutherford's Time" In *Rutherford and Physics at the Turn of the Century*; Bunge, M.; Shea, R. Eds; Dawson: New York, 1979, p 53.

Among the applications of radioactivity studies that Rutherford proposed, the determination of the age of rocks has been the most important in disciplines outside of physics and chemistry. In 1905, based on the idea of radioactivity as a random set of discrete events, he offered the experimentally determined rates of radioactive decay of radium and uranium as indicators of geological time. The chemist Bertram Boltwood made similar proposals based in part on his correspondence with Rutherford after Soddy's departure from McGill. The scientific community eventually accepted the necessary assumptions of the radiometric clock because its outcome accorded well with a growing conviction that the earth's age was greater than physicists had previously allowed. Although Darwin's theory did not require any particular time scale, William Thomson's 20-million-year estimate of the age of the earth was generally perceived as a telling objection to prevailing views of evolution and geological time.

Rutherford left Montreal for Manchester in 1907, primarily because he desired more contact with his scientific peers. He also wanted to eliminate the time lag in publication caused by 3000 miles of ocean; he alluded to these problems in Montreal as early as 1902, when he wrote to his mother, "I have to keep going, as there are always people on my track. I have to publish my present work as rapidly as possible in order to keep in the race."[13] Although Rutherford supervised an average of only 20 research students annually (fewer than in his later years) and lived relatively isolated from the physics community, he reached his peak research productivity at Manchester—70 papers in nine years.

After he received the Nobel Prize, Rutherford's most notable achievement altered theories of atomic structure. Before leaving Montreal, and in his early years at Manchester, Rutherford studied the behavior of alpha particles in electric and magnetic fields and in their passage through thin metal foils. "See if you can get some effect of alpha particles directly reflected from a metal surface", he said one day in 1909 to his undergraduate assistant, Ernest Marsden, who was working under Hans Geiger.[14] Geiger and Marsden reported the results of Rutherford's hunch in a paper submitted in May 1909. They concluded that about 1 in 8000 of the incident alpha particles was re-

---

[13]Rutherford to Martha Thompson, January 5, 1902, as cited in Eve, p 80.

[14]Marsden, E. "Speeches at the Commemorative Session: Rutherford at Manchester"; In *Rutherford at Manchester*; Birks, J. B., Ed. Heywood: London; 1962, p 8.

flected. Rutherford's response to this back-scattering effect has become a classic: "It was quite the most incredible event that has ever happened to me in my life. It was almost as incredible as if you fired a 15-inch shell at a piece of tissue paper and it came back and hit you."[15] In view of J. J. Thomson's "plum pudding" model, which described an atom as negatively charged corpuscles embedded in a sphere of uniformly distributed positive electricity, physicists in 1909 had good reason to stagger at the discovery of the deflection of alpha particles through angles of 90 degrees and more. Rutherford used this anomaly as an occasion for an advance in physics. He outlined a new theory of atomic structure in April 1911, theorizing that "the atom consists of a central charge supposed concentrated at a point, and that the large single deflexions of the alpha and beta particles are mainly due to their passage through the strong central field."[16] The nuclear model of the atom was born. Although Niels Bohr initially adopted the nuclear atom in 1911 as a semiliteral mechanical model of the atom (as Rutherford had intended it), he reformulated it in 1913 on a quantum theoretical basis, launching a new generation of studies of the atom.

World War I drew Rutherford into research on sonic methods for detecting submarines, but he returned to his own research shortly before succeeding Thomson as Cavendish Professor of Physics in 1919. He soon demonstrated that nuclei could be artificially disintegrated by natural means. An alpha particle given off by naturally decaying radioactive material could knock a proton out of a nitrogen atom, leaving a different and lighter nucleus. After the war, Rutherford and James Chadwick showed that many light atoms could be disintegrated in this manner. Between 1919 and early 1927, Rutherford developed a sophisticated, though largely qualitative, satellite model of the nucleus that not only explained the results of alpha-particle scattering experiments, but also accounted both for the artificial disintegration of light nuclei and for the natural disintegration of heavy nuclei. Although Rutherford fashioned a quantified, and partially quantum-mechanical, version of his satellite model in 1927, his model was superseded the following year by the work of 24-year-old George Gamow. After visiting with Gamow in 1929, Rutherford reluctantly abandoned his visualizable

---

[15] Andrade, E. N. da C. *Rutherford and the Nature of the Atom*; Anchor Books: Garden City, N.Y.: 1964; p 111.

[16] Rutherford, E. "The Scattering of α and β Particles by Matter and the Structure of the Atom"; *Philosophical Magazine*, **1911**, series 6, *21*, 669-688, p 686.

model for Gamow's quantum mechanical theory, which did not address the detailed structure of the nucleus.[17] Upon the first disintegration of an atom by artificial means, accomplished in Rutherford's own laboratory by John Cockcroft and E. T. S. Walton, Rutherford more fully embraced the quantum mechanics that formed the basis of this work.

Rutherford's anecdotes, discoveries, and predictions, whether inspiring or misleading, have repeatedly captured the imagination of our century. In 1903, Rutherford and Soddy proposed a research strategy that a number of subsequent scientists exploited (some thereby adding their names to the list of Nobel Laureates in chemistry):

> If elements heavier than uranium exist it is probable that they will be radioactive. The extreme delicacy of radioactivity as a means of chemical analysis would enable such elements to be recognized even if present in infinitesimal quantity.[18]

At the September 11, 1933, meeting of the British Association for the Advancement of Science, Rutherford ventured a prophecy that proved disconcertingly wrong:

> The energy produced by the breaking down of the atom is a very poor kind of thing. Anyone who expects a source of power from the transformation of these atoms is talking moonshine.[19]

Ernest Rutherford's path of scientific discovery and his witticisms reflecting common sense displayed an unwavering confidence in metaphysical realism—a view increasingly under fire from nonrealist perspectives. Like most humorists, Rutherford intended something serious beneath his jokes, including his observation that no physical theory is worth much if it cannot be explained to a barmaid.

MICHAEL N. KEAS
*Oklahoma Baptist University*

[17]Stuewer, R. H. "Rutherford's satellite model of the nucleus"; *Historical Studies in the Physical and Biological Sciences* **1986**, *16*, 321-352, pp 351-352.

[18]Rutherford and Soddy, p 585.

[19]Rutherford had expressed this opinion a decade earlier in a less prominent setting: "The energy in the atom. Can man utilize it?" In *Popular Research Narratives*, vol. 2, *Fifty Brief Stories of Research, Invention or Discovery*; Williams & Wilkins: Baltimore, MD, 1926, pp 109–111. The article is dated July 1, 1924.

# Bibliography

Andrade, E. N. da C. *Rutherford and the Nature of the Atom;* Anchor Books: Garden City, N.Y., 1964.

*Rutherford and Boltwood, Letters on Radioactivity;* Badash, L., Ed.; Yale University Press: New Haven, 1969.

Badash, L. "Rutherford, Ernest"; In *Dictionary of Scientific Biography;* Gillispie, C. C., Ed.; Charles Scribner's Sons: New York, 1975, Vol. 12, pp 25–36.

Badash, L. "The Influence of New Zealand on Rutherford's Scientific Development" In *Scientific Colonialism: A Cross-Cultural Comparison;* Reingold, N.; Rothenberg, M., Eds.; Smithsonian Institution Press: Washington, DC, 1987; pp 379–389.

*Rutherford at Manchester;* Birks, J. B., Ed. Heywood: London, 1962.

Boltz, C. L. "Ernest Rutherford"; In *The Great Nobel Prizes;* Heron Books: London, 1970.

*Rutherford and Physics at the Turn of the Century;* Bunge, M.; Shea, W. R., Eds.; Dawson: London, 1979.

*The Collected Papers of Lord Rutherford of Nelson;* Chadwick, J., Ed. George Allen and Unwin: London, 1962–1965; 3 vols.

Evans, I. B. N. *Man of Power: the Life Story of Baron Rutherford of Nelson;* Stanley Paul: London, 1939.

Eve, A. S. *Rutherford;* Cambridge University Press: Cambridge, 1939.

Feather, N. *Lord Rutherford;* Priory Press: London, [1940] 1973.

Heilbron, J. L. "Rutherford-Bohr Atom"; *Am. J. Phys.* **1981,** *49,* 223–231.

Heilbron, J. L. "The Scattering of $\alpha$ and $\beta$ particles and Rutherford's Atom"; *Arch. Hist. Exact Sci.* **1968,** *4,* 247–307.

Howorth, M. *Pioneer Research on the Atom;* New World: London, 1958.

Lowood, H. *Ernest Rutherford: A Bibliography of his Non-technical Writings;* Berkeley Papers in the History of Science; 4, Office for History of Science and Technology, University of California: Berkeley, 1979.

Oliphant, M. *Rutherford: Recollections of the Cambridge Days;* Elsevier: New York, Amsterdam, 1972.

Rowland, J. *Ernest Rutherford, Atom Pioneer;* Werner Laurie: London, 1955.

"Rutherford by Those Who Knew Him: Being the Collection of the First Five Rutherford Lectures of the Physical Society"; Reprinted from *Proceedings of the Physical Society* 1943–1951.

Rutherford, E. *The Newer Alchemy;* Cambridge University Press: Cambridge, 1937.

Rutherford, E. *Radio-activity;* Cambridge University Press: Cambridge, 1904.

Rutherford, E. *Radioactive Substances and their Radiations;* Cambridge University Press: Cambridge, 1913.

Rutherford, E. *Radioactive Transformations;* Yale University Press: New Haven, CT, 1906.

Rutherford, E. *Rutherford Correspondence Catalog*; Badash, L. Ed., National Catalog of Sources for History of Physics, Report 3; American Institute of Physics: New York, 1974.

Rutherford, E.; Chadwick, J.; Ellis, C. D. *Radiations from Radioactive Substances;* Cambridge University Press: Cambridge, 1930.

Stuewer, R. H. "Rutherford's Satellite Model of the Nucleus"; *Hist. Stud. Phys. Biol. Sci.* **1986,** *16,* 321–352.

Trenn, J. "The Justification of Transmutation: Speculations on Ramsay and Experiments of Rutherford"; *Ambix* **1974,** *21,* 53–77.

Trenn, T. J. *The Self-Splitting Atom: The History of the Rutherford–Soddy Collaboration;* Taylor & Francis: London, 1977.

Wilson, D. *Rutherford: Simple Genius;* MIT Press: Cambridge, 1983.

# 1909

NOBEL LAUREATE

# *Wilhelm Ostwald*

## 1853–1932

Copyright Nobel Foundation

Friedrich Wilhelm Ostwald was born September 2, 1853, in Riga, Latvia. He married Helene von Freyher in 1880; they had five children. He received the Nobel Prize on December 10, 1909, in recognition of his work on catalysis, chemical equilibrium, and rates of chemical reactions. He died in Leipzig on April 4, 1932. Ostwald had a marked ability for recognizing important scientific ideas and for popularizing those ideas. Effective teacher, creative experimentalist, prolific writer, indefatigable editor, and consummate organizer, he was a central figure in the birth, development, and consolidation of physical chemistry.

Ostwald enjoyed making things by hand, and his parents permitted him to have a home workshop and laboratory. As a youth he developed a method (decalcomania) of transferring images to ceramic surfaces, repeating with only slight variations the same process over and over again to find the best conditions. He experimented with fireworks, often making many versions of the same firecracker to determine the best recipe. He recalls in his autobiography, *Lebenslinien, eine Selbstbiographie* (Berlin: Klasing, 1926–1927, 3 vols.), how excited he was at age 14 to discover for himself that a sharp optical image is produced at *two* places between object and lens. An avid reader, he recalls having devoured a three-volume novel in an afternoon.

He entered the university at Dorpat in 1872 as a chemistry student. His interests often strayed from the curriculum, and he spent much of his time collecting insects, painting, binding books, and equipping a private chemical laboratory where he made fireworks and experimented with photography. He played in a student chamber music ensemble, learning the viola parts for all 83 string quartets of Haydn. He developed a deep interest in music theory and lectured on music at Dorpat; he met his future wife at one of those lectures.

His first research publication reported work on the composition of solutions of bismuth chloride in water, a chemical project he chose himself. Bismuth chloride dissolves in water to produce hydrochloric acid and a precipitate of bismuth oxychloride. Ostwald found that the amount of solid bismuth oxychloride increases when the amount of water is increased and interpreted these results quantitatively in terms of the concentrations of the dissolved substances.[1] This research was in the tradition of Claude-Louis Berthollet (1748–1822), who had explored the relationship between the mass of chemical reactants and what he called "affinity". Cato Maximilian Guldberg and Peter Waage had recently developed Berthollet's ideas as the mass-action description of chemical equilibria.[2] Ostwald was one of the first chemists to understand, exploit, and popularize the implications of the work of Guldberg and Waage.

His dissertations for the master's and doctoral degrees at Dorpat described the extent of chemical reactions by comparing the competing affinities of reactants and products. Reading thermochemistry papers by Julius Thomsen, Ostwald was struck with the idea that reactions could be studied by measuring properties other than heat. Having the simple equipment needed for the precise measurement of the liquid volumes, he decided for his master's dissertation to study the volume changes that accompany reactions in solution. These innovative studies involved many repetitive experiments. He then extended his research program, measuring refractive index changes that accompany reactions. From extensive measurements of densities and refractive indices, he developed tables of affinities for a representative series of acids. These tables formed the substance of his doctoral dissertation, "Volumchemische und optisch-chemische Studien" (Kaiserlichen Universität Dorpat, 1878).[3]

[1] Ostwald, W. "Über die chemische Massenwirkung des Wassers"; Journal für praktische Chemie 1875, 2 (12), 264–270.

[2] Guldberg, C. M.; Waage, P. Études sur les affinité chimiques; Christiana: 1867; German trans. in Ostwald's Klassiker der exakten Wissenschaften; W. Engelmann: Leipzig, 1899; no. 104.

[3] Journal für praktische Chemie 1878, 2 (18), 328–371. Both dissertations appear, with a bibliography of his books and a biography by Herhard Harig and Irene Strube, in Ostwalds Klassiker der exakten Wissenschaften no. 250; Akademische Verlagsgesellschaft: Leipzig, 1966.

Ostwald stayed at Dorpat as a *Privatdozent* after receiving his doctorate and lectured on physical chemistry; at a local school he taught physics, chemistry, and mathematics. He continued his research on affinities, began writing a textbook of general chemistry, and used his considerable skills in glassblowing, woodworking, and metal fabricating to design and construct simple yet elegant apparatus for physicochemical measurements.

In 1881 he became the professor of chemistry at Riga Polytechnicum, where he stayed until 1887. At Riga he completed the *Lehrbuch der allgemeinen Chemie,*[4] his monumental, 1764-page first attempt to systematize the important facts of chemistry. This grand enterprise was to yield a series of textbooks by Ostwald in inorganic, analytical, and physical chemistry, and eventually his *Principien der Chemie,*[5] a textbook in general chemistry that dwelt on basic ideas rather than specific details. *Principien der Chemie* had considerable influence on American chemistry curricula, laying the foundation for a theoretically oriented college course that would precede inorganic, organic, and analytical chemistry. Such courses in general chemistry are now standard throughout the United States. Ostwald spent much of his professional life organizing chemistry "in the form of a rational scientific system, without bringing in the properties of individual substances", showing students "the great connections by which these separate facts are bound together in a unit . . . united into a great, simple whole".[6]

He published more than 500 research papers, 45 books, and 5,000 reviews during his career. Ostwald's extensive literary output transcended physical chemistry, including also books on the history and philosophy of science and on color theory. He founded or edited six journals.

Perhaps Ostwald's greatest contributions to physical chemistry stemmed from his many professional relationships. He identified talented and productive young scientists, encouraged them, and popularized their work. During his graduate research he discovered Guldberg and Waage and used their theories. In June 1884 Ostwald read the doctoral dissertation of Svante Arrhenius (Nobel Laureate in chemistry, 1903) on the electrical conductivity of solutions. Sensing a close connection between conductivity and affinity, Ostwald traveled to Uppsala

---

[4] W. Engelmann: Leipzig; Vol. I, *Stöchiometrie,* 1885; Vol. II, *Verwandtschaftslehre,* 1887.

[5] Leipzig, 1907; English trans. by Harry W. Morse; *The Fundamental Principles of Chemistry: An Introduction to All Text-Books of Chemistry;* Longmans, Green: New York, 1909.

[6] "Preface"; *The Fundamental Principles of Chemistry.*

in August 1884 to meet Arrhenius and begin a lifelong friendship and partnership. Ostwald invited Jacobus van't Hoff (Nobel Laureate in chemistry, 1901) to be cofounder in 1887 of the *Zeitschrift für physikalische Chemie, Stöichiometrie und Verwandtschaftslehre,* the first journal of physical chemistry. Ostwald, Arrhenius, and van't Hoff together developed the modern theories of the equilibria in solutions of acids, bases, and salts, exploiting the idea that many substances dissociate in solution to form ions. The idea of ions was highly controversial; the three men were dubbed the Ionists, a term of ridicule by their opponents and a term of respect by their followers. Ostwald translated Josiah Willard Gibbs's seminal paper on the mathematical structure of thermodynamics into German.[7] He popularized the scientific work of many others in his series of translations and reprints, *Klassiker der exakten Wissenschaften* (W. Engelmann: Leipzig, 1889–); 173 volumes had appeared by the time he received the Nobel Prize. The collaboration of Ostwald and Leipzig publisher Wilhelm Engelmann was fruitful. Engelmann became the world's foremost publisher in physical chemistry; his list included the *Zeitschrift für physikalische Chemie.*

In 1887 Ostwald became professor of physical chemistry at the University of Leipzig (the only chair in physical chemistry in Germany, established only six years earlier and previously held by a physicist). He immediately organized the teaching and research in physical chemistry, appointing Walther Nernst (Nobel Laureate in chemistry, 1920) as his assistant. He emphasized independent work and critical discussion of experimental data. The practical aspects of teaching a physical chemistry laboratory were important to Ostwald, and he collaborated with Robert Luther to write a manual on its pedagogy.[8] This book set the format both of such manuals and of the laboratory itself for the following century throughout the world. It was published almost simultaneously in English.[9] Leipzig became the international center for physical chemistry, attracting graduate students particularly from Britain and the United States. In 1897 a new laboratory was built, and Ostwald became director of the Physikalisch-chemischen Instituts der Universität Leipzig.

---

[7] *Thermodynamische Studien;* W. Engelmann: Leipzig, 1892.

[8] *Hand- und Hülfsbuch zur Ausführung physiko-chemischer Messungen;* W. Engelmann: Leipzig, 1893.

[9] *Manual of Physico-Chemical Measurements;* Walker, Sir James, Trans.; Macmillan: London, 1894.

Ostwald's influence on chemistry curricula extended beyond physical chemistry. In 1891 he developed a description of colored acid–base indicators based on mass-action principles, incorporating these ideas into a beginning college textbook.[10] Translated into Hungarian, Polish, French, Italian, and English within 10 years, this book revolutionized the undergraduate teaching of analytical chemistry. He also wrote a chemistry book for youngsters.[11]

Ostwald and his associates helped remove mystery from the idea of catalysts. Jöns Jacob Berzelius had introduced the word in 1836 for substances that bring about chemical changes, even when present in minute amounts, without themselves decomposing. Ostwald emphasized that a catalyst can accelerate a reaction but cannot change the direction of a reaction; otherwise chemists could use catalysts to achieve perpetual motion. With E. A. Bodenstein and Ernst Brauer, Ostwald used hot iron wire to catalyze the oxidation of ammonia to nitric acid. This commercially successful method of manufacturing nitric acid was instrumental in his nomination for the Nobel Prize.

His association with Arrhenius led to many investigations involving electrical interactions with solutions, including the oxidation or reduction of substances at electrodes and the conductance of electricity by solutions. Ostwald developed the dropping mercury electrode for studying electrochemical reactions. In 1894 he helped found the Elektrochemische Gesellschaft and the new society's *Zeitschrift für Elektrochemie*. He was president of the society for its first four years.

Ostwald spoke of energy and its transformations in his inaugural lecture at Leipzig.[12] From his early days at Dorpat he had been interested in thermodynamics, and he was intrigued by the emphasis that Gibbs placed on energy. Ostwald became convinced that molecules, atoms, and ions were only convenient fictions and that the universe was formed from energy in its various forms; the processes of chemical change were transformations of energy. He aroused strong emotions in an 1895 Lübeck lecture ("Die Überwindung des wissenschaftlichen Materialismus") when he attacked the 19th-century mechanical models of scientific reality, asserting that it is fruitless for chemists to reduce all phenomena to matter and to mechanical laws relating motion to

---

[10] *Die wissenschaftlichen Grundlagen der analytischen Chemie;* W. Engelmann: Leipzig, 1894.

[11] *Die Schule der Chemie;* Vieweg & Sohn: Braunschweig, 1903.

[12] "Die Energie und ihre Wandlungen"; W. Engelmann: Leipzig, 1888.

matter. Ostwald urged that the building block of chemical materialism, the atom, be rejected! His proposal to substitute energy for matter was too simplistic, but he had the vision to see that mechanistic descriptions were inadequate for modern chemistry. He applied energy principles to diverse fields, envisioning a *Naturphilosophie* that would be the unifying science of the sciences. In 1901 he founded yet another new journal: *Annalen der Naturphilosophie*.

He retired in 1906 to his house, "Villa Energie", in the village of Gross Bothen near Leipzig. There he painted, worked on his color theories, campaigned for universal peace, and worked in his laboratories. He had four laboratories in Villa Energie facing east, west, north, and south so that he could choose the conditions of sunlight best suited to his project and mood. Artists remember Ostwald for his theory of colors.[13] His own company, Verlag Unesma, published most of Ostwald's work on color. He encouraged efforts to develop a world language and spoke Ido, a simplified version of Esperanto, with visitors to Villa Energie.

Ostwald preached his energy imperative: "Dissipate no energy, but strive to use energy by converting it into more useful forms." The second law of thermodynamics asserts that no being can live without dissipating energy, but the energy imperative focuses attention on responsible ways of living in an energy-based world. Ostwald proposed a theory of happiness, mathematized as

$$G = k\,(A - W)\,(A + W)$$

where $G$ is *Glück* (happiness), $A$ is *Arbeit* (energy expended in doing useful work) and $W$ is *Widerstand* (energy dissipated in overcoming resistance).

GEORGE FLECK
*Smith College*

---

[13]See, for instance, *Die Farbenfibel;* Verlag Unesma: Leipzig, 1916; English trans. by Faber Birren, *The Color Primer;* Van Nostrand Reinhold: New York, 1969. A bibliography of his many books on color is given by Edmund P. Hillpern ("Some Personal Qualities of Wilhelm Ostwald Recalled by a Former Assistant"; *Chymia* **1949**, *2*, 57–64).

# Bibliography

Aus dem wissenschaftlichen Briefwechsel Wilhelm Ostwalds; Körber, H.-G., Ed.; Akademie-Verlag: Berlin, 1961–1969; 2 vols.

Bancroft, W. D. "Wilhelm Ostwald, The Great Protagonist"; J. Chem. Ed. **1933**, 10, 539–542, 609–613.

Clark, P. "Atomism versus Thermodynamics"; in Method and Appraisal in the Physical Sciences: The Critical Background to Modern Science, 1800–1905; Howson, C., Ed.; Cambridge University: Cambridge, 1976; pp 41–105.

Dolby, R. G. A. "Debates Over the Theory of Solution: A Study of Dissent in Physical Chemistry in the English-Speaking World in the late Nineteenth and Early Twentieth Centuries"; Hist. Stud. Phys. Sci. **1976**, 7, 297–404.

Donnan, F. G. "Ostwald Memorial Lecture"; J. Chem. Soc. (London) **1933**, 316–332.

Farber, E. "A Study in Scientific Genius — Wilhelm Ostwald's Hundredth Anniversary"; J. Chem. Ed. **1953**, 30, 600–604.

Farber, E. "Wilhelm Ostwald: 1853–1932"; In Great Chemists; Interscience: New York, 1961; pp 1019–1030.

Hiebert, E. N. "The Energetics Controversy and the New Thermodynamics"; In Perspectives in the History of Science and Technology; Roller, D. H. D., Ed.; University of Oklahoma: Norman, OK, 1971; pp 67–87.

Hiebert, E. N.; Körber, H.-G. "Ostwald, Friedrich Wilhelm"; In Dictionary of Scientific Biography; Gillispie, C. C., Ed.; Charles Scribner's Sons: New York, 1978; Vol. 15, pp 455–469.

Holt, N. R. "A Note on Wilhelm Ostwald's Energism"; Isis **1970**, 61, 386–389.

Jacobson, E. Basic Color: An Interpretation of the Ostwald Color System; P. Theobald: Chicago, 1948.

Leegwater, A. "The Development of Wilhelm Ostwald's Chemical Energetics"; Centaurus **1986**, 29, 314–337.

Lotz, G.; Dunsch, L.; Kring, U. Forschen und Nutzen: Wilhelm Ostwald zur Wissenschaftlichen Arbeit; Akademie-Verlag: Berlin, 1978.

Ostwald, G. Wilhelm Ostwald, mein Vater; Berliner Union: Stuttgart, 1953.

Ostwald, W. Abhandlungen und Vorträge allgemeinen Inhaltes; Veit: Leipzig, 1904.

Ostwald, W. Grosse Männer; Akademische Verlagsgesellschaft: Leipzig, 1909.

Ostwald, W. Lebenslinien: Eine Selbstbiographie; Klassing: Berlin, 1926–1927; 3 Vols.

Ostwald, W. Lehrbuch der allgemeinen Chemie; W. Engelmann: Leipzig, 1885–1887.

Ostwald, W. Les principes scientifiques de la chimie analytique; Hollard, A., Jr.; Nand: Paris, 1903.

Ostwald, W. Prinzipien der Chemie; Akademische Verlagsgesellschaft: Leipzig, 1907.

Ostwald, W. The Scientific Foundations of Analytical Chemistry Treated in an Elementary Manner; M'Gowan, G., Jr.; Macmillan: London, 1895.

Partington, J. R. "Chemical Affinity"; In *A History of Chemistry;* Macmillan: London, 1964; Vol. 4, pp 569–607.

Partington, J. R. "Wilhelm Ostwald"; *Nature* **1953,** *172,* 380.

Root-Bernstein, R. S. "The Ionists: Founding Physical Chemistry, 1872–1890"; Ph.D. Dissertation, Princeton University, 1980.

Servos, J. W. *Physical Chemistry from Ostwald to Pauling: The Making of a Science in America;* Princeton University Press: Princeton, NJ, 1990.

Slosson, E. E. "Wilhelm Ostwald"; In *Major Prophets of Today;* Little, Brown: Boston, 1914; pp 190–241.

Walden, P. *Wilhelm Ostwald;* Wilhelm Engelmann: Leipzig, 1904.

# Otto Wallach

## 1847–1931

Otto Wallach was born in Königsberg, Prussia, on March 27, 1847, and died in Göttingen, Germany, on February 26, 1931. His Nobel Prize in chemistry was awarded in 1910 to acknowledge his work in organic chemistry in the area of alicyclic compounds, cyclic compounds that possess chemical properties similar to compounds like methane and ethylene.

Wallach was born into a family of lawyers, and his father served as a state official in Prussia. This position necessitated that the family move to Stettin and later to Potsdam in 1855. Otto's elementary education was undertaken at the Potsdam Gymnasium. His early interests in chemistry, art, and the history of art were clear. Chemistry was practiced in his home, where he prepared a variety of products using very simple equipment. Despite a number of health problems, Wallach continued his interests at school and successfully completed all requirements.

The year 1867 found him as a beginning student at the University of Göttingen. The chemistry laboratory was under the directorship of Friedrich Wöhler. After beginning work there, he spent a semester in Berlin and returned to work again under Wöhler. The students, including Wallach, were expected to put in a 10-hour day. The serious students would continue their work by candlelight after the gas was turned off at five o'clock. His doctorate was awarded in 1869 after his research studies were completed concerning position isomerism of a series of toluene-related compounds.

His career brought him to spend a brief period of time with Hans Wichelhaus in Berlin, and in 1870 he began to assist the distinguished chemist August Kekulé in Bonn. Wallach, the great experimentalist, had an interesting relationship with Kekulé, the great theoretician of molecular architecture. Over the period of their collaboration, Kekulé spent less and less time in actual laboratory manipulation, while Wallach continually preferred to work in the laboratory. Their collaboration would continue for 19 years with the exception of one brief period when Wallach became the only chemist at an industrial site, the Aktien-Gesellschaft für Anilin-Fabrikation. His experience at this plant was not pleasant because the disagreeable fumes were clearly deleterious to his health, and he quickly returned to Kekulé's laboratory. Wallach remained at Bonn until 1889, when he became director of the Chemical Institute of Göttingen. His stay at Göttingen continued through his retirement in 1915 and included additional work done until a few years prior to his death.

Wallach became acquainted with his field of future recognition in a surprising manner. Kekulé had presented Wallach with an increasing number of responsibilities during his years at Bonn. Wallach assumed responsibility for instruction in pharmacy in 1879. It became necessary for him to learn a great deal of new chemistry to assume responsibility in this unfamiliar field. In these efforts to learn about pharmacy, he came across a number of sealed bottles containing various oils that had been distilled from some interesting plants. These oils had been isolated originally by Kekulé, who had intended to investigate them as part of a research program but had not yet had an opportunity. Wallach became curious about these oils and requested permission from Kekulé to undertake his own studies on them. Kekulé readily granted permission. Wallach's first paper in this field, published in 1884, was the beginning of a long series of successful researches in a field that until then had largely been an enigma.

An interest in plant oils goes back centuries. Plant oils contain volatile components, which impart a strong odor or taste. Many were noteworthy for their distinctive and recognizable odors; this fact was of great importance in human affairs. Oils with a pleasant fragrance were valued for their uses as incense, perfumes, or ceremonial oils. Their roles as commercially valuable spices and seasonings spread as caravans moved them around the world; nations competed to dominate their trade or control their sources. Technology provided the means to remove the essence from a plant and provide it in concentrated form. Techniques such as steam distillation and the removal of the desired components by extracting them in selected liquids were used and improved to provide the purest oil in as large an amount as possible at the most reasonable cost. The number of essential oils increased over the years, and commercial enterprises grew and expanded their operations to increase the availability of these valuable commodities.

Wöhler and Justus Liebig in 1832 brought about one of the first breakthroughs in the analysis of vegetable oils when they determined how the aroma of bitter almonds originated. After the nature of aromatic compounds was revealed by Kekulé in 1865, this bitter almond oil was chemically identified as a simple benzenelike compound known as benzaldehyde. Other benzene-related oils were later identified, and these oils are generally described as aromatic. Some simple oils not related to benzene were also identified and termed nonaromatic. A large group of nonaromatic oils—the turpentine oils, orange peel oil, and the oils from plants such as the pines and eucalyptus—contained structures that seemed to be incapable of being determined by the chemical approaches then in use. Oils in this group, which remained liquids when cooled below room temperature, were termed terpenes, whereas those that tended to become solids were termed camphors. This approach, in which the terpenes were sorted out according to physical properties, was the best approach at that time, but not everyone was pleased with the system.

The investigation of individual terpenes was initiated by Berthelot, who described the pinenes, a group of terpenes obtained from turpentine oil. Additional terpenes were characterized, and soon a great number of terpene compounds were known that contained 10 carbon atoms and 16 hydrogen atoms. The field became increasingly confused as the number and chemical variety of compounds continued to increase without any link that could tie them together. Wallach approached this field from several perspectives. He maintained that any distinct compound would have properties rendering it identifiable and that the interrelationships among different terpenes could be established by using precise identification techniques. Nor did it seem likely to him that there were as many different terpene compounds as the literature suggested. He resolved to determine whether oils from different plants might contain the same chemical component or similar chemical components.

Wallach began his work by attempting to form derivatives from the various distinct fractions of terpene mixtures. He employed a variety of simple reagents, conventional approaches, and several of his own newly developed techniques designed to carry out the procedures more rapidly. The first plateau in his work arrived when he described the existence of eight different terpenes and the characteristics of each one. These eight different terpenes were pinene, camphene, limonene, dipentene, sylvestrene, terpinolene, terpinene, and phellandrene. Simultaneously he studied the terpene interconversions, which were essential to his identification and chemical characterization of these terpenes. Wallach must be credited for the discovery of terpinolene and the first identification and characterization of terpinene. This list was revised by Wallach and other chemists over the next several decades as new compounds were added to the list. Clearer procedures also made it necessary to remove dipentene, a racemic mixture of (+)- and (−)-limonene, and necessary to reclassify several of the other materials as either mixtures or as a rearranged product in one case. Wallach next discovered the structures of other terpenes and continued to indicate the manner in which closely related terpenes could be converted into each other.

His interest would eventually take him to two important publications. The first was a chart published in 1891 that illustrated the interrelationships among 12 different terpenes related to pinene. These compounds were pinene, camphene, borneol, camphor, terpin, terpineol, cineole, pinol, dipentene, limonene, terpinene, and terpinolene. A summary of his work that included a number of charts illustrating the relationships between important terpenes and their families appeared in book form in 1909. This book, *Die Terpene und Campher,* pointed out the relationships between his work and the work of other natural-product chemists; it presented the most comprehensive picture of terpene chemistry ever done up to that point.

Another important contribution from Wallach was the establishment of the isoprene rule. The 10-carbon compound dipentene and other important terpenes had been shown to be built up from the basic five-carbon isoprene compound. It appeared that numerous terpenoid compounds were biosynthesized by using a pathway that derives the terpenoids in principle from the five-carbon isoprene fragment. The original isoprene rule, which indicated that all terpenoids are multiples of the isoprene fragment, appears not to be followed strictly by natural products, even though these compounds do seem to be put together using isoprenelike units. Natural products that apparently form from the isoprene unit include the monoterpenes (two units), sesquiterpenes (three units), diterpenes (four units), triterpenes (six units), tetraterpenes (eight units), and the polyterpenes (many units). In fact, isoprene is not the actual biological precursor of these compounds. The biological precursor of the various plant terpenoids and steroids is mevalonic acid, which is itself formed from the basic metabolic building block known as acetyl coenzyme A. Nevertheless, Wallach's observation on the isoprene buildup of structures was a very useful and integrating insight that has been highly regarded by natural-product chemists and biochemists. The rule can be utilized today in the analysis of structures such as vitamin A and squalene, which are found to be built from these isoprene units.

Wallach's work received much recognition from his peers. He was elected an honorary fellow of the Chemical Society of London in 1908 and honorary member of the Verein Deutscher Chemiker and the Belgian Chemical Society, as well as the recipient of honorary doctorates from the University of Manchester in 1909 and the University of Leipzig in 1912. Also in 1912 he was chosen to serve as the president of the Deutsche Chemische Gesellschaft. Besides the Nobel Prize, he received the Davy Medal from the Royal Society. He was described as personable and seemed to possess a quiet dignity. Wallach's life reflected a strong dedication to chemistry and an interest in a wide variety of activities.

In judging Wallach's contributions, it is misleading to focus on any one particular experiment. The value of his work is seen only in its impact as a whole. He converted the field of natural-product chemistry from a disorganized, confused collection of facts into a reasonably complete, organized, and integrated field that made sense and could provide a basis for progress. The practical applications of his work in Germany resulted in a prosperous industry in supplying these essential oils. Wallach should be remembered as a practical and dedicated experimentalist whose precision and thoroughness brought about progress in his field and helped to create new fields for future chemists.

ROBERT H. GOLDSMITH
*St. Mary's College of Maryland*

# Bibliography

Leicester, H. "Wallach, Otto"; In *Dictionary of Scientific Biography*; Gillispie, C. C., Ed; Charles Scribner's Sons: New York, 1976; Vol. 14, pp 141–142.

Nobel Foundation. *Nobel Lectures in Chemistry, 1901–1921*; Elsevier: New York, 1966; pp 175–196.

Patridge, W.; Schierz, E. *J. Chem. Ed.* **1947**, *24*, 106–108.

Ruzicka, L. *J. Chem. Soc.* **1932**, 1582–1597.

Ruzicka, L. "Otto Wallach, 1847–1931"; In Farber, E., Ed; *Great Chemists*; Interscience: New York, 1961; pp 833–851.

Wallach, O. "On the Action of Concentrated Sulfuric Acid on Azoxybenzene"; *Ber. Chem. Ges.* **1891**, *24*, 1525.

Wallach, O. "Papers on Terpene"; *J. Liebigs Ann. Chem.* **1887**, *238*, 78–89.

Wallach, O. "Zur Constitutionsbestimmung des Terpinols"; *Ber. Chem. Gesell.* **1895**, *28*, 1773–1777.

Wallach, O. *Die Terpene und Campher*; Veit: Leipzig, 1909.

# Marie Curie

## 1867–1934

Marie Curie (née Maria Skłodowska) was born in Warsaw, Poland, on November 7, 1867. She emigrated to Paris in 1891 and married Pierre Curie in 1895. Together they discovered polonium and radium in 1898. In 1903, they shared the Nobel Prize in physics with Henri Becquerel in recognition of the discovery of radioactivity—a word coined by Marie Curie. After Pierre Curie's accidental death in 1906, Marie Curie pursued the chemical investigation of radioactive substances and was awarded a second Nobel Prize in 1911, for chemistry. She was the first woman appointed as a professor at the Sorbonne and the first scientist to receive two Nobel Prizes. She died on July 4, 1934, six months after Irène and Frédéric Joliot-Curie had discovered artificial radioactivity.

Maria Skłodowska was the fifth child of a family of intellectuals. She had three sisters—Zofia, Bronia, Helena—and one brother, Josef. Her mother Bronisława was the headmistress of a private school for girls in Warsaw. Her father Władisław was assistant director of a gymnasium. The failure of the 1863 insurrection of Poles against the Russian domination over Poland was followed by harsh repressions and an increased tzarrist control on all aspects of everyday life, particularly on education. Władysław Słodowski was demoted from his teaching position in 1873 and the family had to leave the apartments that the job provided. Maria's mother died from tuberculosis in 1878. Because the family resources were meager, Maria had to earn money. When she was 18 years old, she left Warsaw to work as a governess. Back in

Warsaw in 1889, she attended a liberal clandestine university and supported a political positivist movement. Her sister, Bronia, and Bronia's husband, Casimir Dluski, lived in Paris, where both had graduated as doctors in medicine. They insisted that Maria join them in Paris. Maria came and rented a student room in the Latin Quarter, not far from the Sorbonne. She passed the exams in physics in July 1893 and those in mathematics in July 1894.

In April 1894 she met Pierre Curie; they married in 1895. Pierre Curie was then a 35-year-old physicist of great esteem. When he was 18, Pierre and his brother Jacques Curie, who was assistant to Charles Friedel at the Sorbonne mineralogy laboratory, had discovered the important phenomenon of piezoelectricity. In 1882 the Ecole Municipale de Physique et de Chimie Industrielle was created to train engineers for the chemical and electrical industries, and Pierre Curie was appointed director of laboratory work and placed in charge of laboratory equipment. He demonstrated his technical skill by designing and constructing a highly sensitive electrometer. In 1895, the year he married Marie, he published a dissertation on magnetism in which he formulated the well-known Curie's law, dealing with the variations of magnetic properties according to the temperature of substances.

Marie Curie's first publication, prepared at the request of the Society for the Encouragement of National Industry, also dealt with magnetic properties but specifically with those of tempered steel. In 1896 she spent the entire year studying for the physics *agrégation*, a competition that could allow her to get a teaching position in a secondary school. She placed first, and then had to choose between teaching and research. Excited by the mysterious, recently discovered Roentgen rays, she decided to prepare a doctorate. Although Henri Poincaré suggested that Roentgen rays were a phenomenon of "fluorescent bodies", Henri Becquerel, of the Muséum d'Histoire Naturelle, discovered in 1896 that uranium salts spontaneously emitted rays very similar to Roentgen rays.

Choosing as her starting point a remark of Lord Kelvin on the "electrification of air by uranium and its compounds", Marie first set up to determine the ionizing power of various substances like pitchblende, thorium oxide, and chalcolithe with Pierre Curie's highly sensitive electrometer. She noticed that thorium and pitchblende were more active than uranium itself. The rays that Becquerel named "uranic" appeared to be less specific. As early as April 1898, in a note published in the *Comptes-rendus* of the Academy of Sciences, Marie Curie boldly generalized the phenomenon described by Becquerel and created the word "radioactive" to characterize the property she observed in thorium. She also ventured the idea that there could be an unknown, more radioactive, element in pitchblende.

The next step was to identify the mysterious radioactive element. For this purpose, Marie had to concentrate the active substance by chemical separation, to measure the various products with the electrometer, and to compare them with a sample of uranium. Pierre interrupted his own research on crystals—just for a few weeks, he thought—to collaborate with her. After a few months of hard work with the assistance of Gustave Bémont, a chemist and laboratory chief at the Ecole Municipale de Physique et de Chimie, they were able to announce the discovery of a new element 400 times more active than uranium. They proposed naming it *polonium,* from "the homeland of one of us."[1] Another radioactive substance, 900 times more active than uranium, was discovered in November 1898 and named *radium.*[2]

Pierre and Marie Curie had the conviction that polonium and radium were both new elements. However, a lot of work was required to persuade the scientific community of their elementary nature. The Curies undertook four years of laborious research in order to isolate the new radioactive substances and determine their atomic weight. They had to treat tons of pitchblende supplied by the Austrian government in an attempt to extract a few milligrams of radium. They were assisted by André Debierne from the Ecole Municipale de Physique et de Chimie. The task was divided within the team; interstingly, the exciting part of the job, hunting for new radioactive substances, was assigned to Debierne, who discovered actinium in 1899. Pierre did the physical measurements of radiation and Marie took the chemical part of the research: Through repeated filtrations, crystallizations, and precipitations, she tried to isolate polonium and radium. In 1902, Marie Curie presented her doctoral thesis, published in 1903. She had isolated 0.1 dg of pure radium and determined its atomic weight, 225.[3]

The International Congress of Physics held in Paris on the occasion of the World Exhibition in 1900 offered the Curies an opportunity to present their work before a crowd of renowned scientists. Radium raised great interest in the scientific community. While the Curies

---

[1] *Comptes rendus* of the Academy of Sciences; 1898, 127, 175.

[2] *(Comptes rendus* of the Academy of Sciences; 1898, 127, 1215.

[3] Recherches sur les substances radioactives; Thèse: Paris, 1903; Trans. in Chem. News, 1903, *88*, 85, 97, 134, 159, 169, 175, 187, 199, 211, 223, 247, 259, 271.

dedicated themselves to long and painstaking experimental work, a number of physicists and chemists all over the world were trying to solve the enigma of the energy of radium, and racing to find a theoretical explanation for radioactivity. Marie Curie had suggested that the energy of radium came from the outside—for instance, from the sun. In 1900, Pierre ventured that radioactivity violated the law of conservation of energy. Both conjectures were proved incorrect when Rutherford provided his explanation. Trained at the Cavendish laboratory in Cambridge, where J. J. Thomson conducted a research program on the structure of the atom, young Rutherford was more theoretically inclined than the Curies. As early as 1899, he identified two different kinds of rays, labeled α and β rays. In 1902, after studying thorium emanation with Soddy at McGill University in Montreal, he was able to prove that radioactive substances like thorium produced another radioactive substance, independent and different from the parent-element. Transmutation, the ancient alchemists' dream, was achieved, the enthusiastic Soddy announced.

After resisting for some time, the Curies eventually adopted Rutherford and Soddy's atomic explanation of radioactivity, but they never shared Rutherford's view of the phenomenon. Rutherford immediately used radioactivity as a tool for investigating the structure of the atom, whereas the Curies remained more concerned with the purification and identification of radioactive elements. Rutherford developed radioactivity as a branch of nuclear physiscs. In contrast, the Curies' main interest lay in radiochemistry.

Though it proved very successful, Marie Curie's intense laboratory work in this period was exhausting for a young woman and mother. Her first daughter, Irène, was born in 1897 and the second, Eve, in 1904. To the laboratory and housework, Marie added lectures at the Ecole Normale Supérieure de Jeunes Filles, in Sèvres, in order to increase the family income. There she prepared future women teachers for the physics agrégation. She managed, however, to save a few happy vacation days for bicycle rides around Paris with her husband Pierre, while Eugène Curie, her father-in-law, took care of young Irène.

Toward the end of 1903, this busy but quiet life was disturbed by honors. On November 5, Pierre and Marie were awarded the Humphry Davy Medal from the Royal Society, and on November 12, a Nobel Prize shared with Henri Becquerel. Immediately the Curies became popular, and radium appeared to be a magic substance that could release fantastic power. Newspapers described Marie laboring for years in a poorly equipped laboratory. Thus was shaped the public image of

Marie Curie as a heroic, self-sacrificing scientist. (In fact, because the Curies were awarded various prizes and observations from the Academy of Sciences, they were not really working in extremely poor conditions.)

Because Pierre's and Marie's health was seriously affected by radiation, the acceptance ceremony in Stockholm was delayed until 1905. In his address Pierre Curie suggested that radium, in the hands of criminals, could be dangerous. But he added that he believed, as did Nobel, that new discoveries would bring greater benefit than damage to mankind.

Recognition came later in France than abroad. In 1903 Pierre was still in search of an academic position. The fame of the Nobel Prize hardly compensated for the fact that he had not been trained at the Ecole Normale Supérieure like most leading academic scientists of France. His chances in Paris were so uncertain that he thought for a while of accepting a lucrative position in Geneva. Pierre eventually got a chair at the Sorbonne in 1904 and was elected a member of the Academy of Sciences in 1906.

The radioactive material needed for the Curies' research was provided by a French manufacturer who opened a factory at Nogent-sur-Marne for the treatment of pitchblende residues. The Curies did not patent the separation process for radium. Did they fail to recognize the future applications of radium before publishing in a journal? Or was it, as they later claimed, that they chose not to patent because they shared the conviction that scientific research must be pure and unconcerned with financial profit?

In April 1906 Pierre Curie was accidently killed, crushed by a carriage on the Place Dauphine. After mourning for just a few weeks, Marie decided to go forward in the research initiated with Pierre. She resigned the widow's pension given by the state and succeeded Pierre in the chair at the Sorbonne (May 1906). Because she was the first woman to teach at the Sorbonne, a large crowd attended her first lecture, on November 5, 1906. But the crowd left rather disappointed because she delivered a professional lecture, titled "Les Théories modernes relatives à l'électricité et à la matière", in an impersonal voice without stylistic panache.

Marie Curie did her best to develop radioactivity as a new discipline in physics. With the help of five assistants, she studied the effects of radioactivity and developed the atomic explanation, introducing the terms *désintégration* and *transmutation*. She encouraged the teaching of radioactivity by writing a treatise (1910). She also emphasized the need for an official standard of measurement of radioactivity. A unit was named the curie (defined as the quantity of radiation emitted by 1 g of radium), and the standard was deposited with the Bureau of Weights and Measurements in 1910. She founded the Radium Institute, linked with both the Paris Faculté des Sciences and the Pasteur Institute for research on the medical applications of radium.

In 1911 she applied for membership in the Paris Academy of Science. However, Edouard Branly, a catholic physical scientist famous for his work in electricity (which proved decisive for the development of wireless telegraphs), was elected, and Marie felt so discouraged that she never applied again. A few months later, she was awarded a second Nobel Prize "for services to the advancement of chemistry by the discovery of the elements radium and polonium". The same year she was invited to the first Solvay Conference on Physics, held in Brussels. Fame and scandal came together, however. Even though she was recognized by the international scientific establishment as a leading figure of modern chemistry, she was the target of slanderous attacks in Parisian newspapers for her private life. In November 1911 the rumor of a scandal spread like wildfire at the Sorbonne. It was caused by the publicity of a love affair between Marie Curie and the French physicist, Paul Langevin, who was married and the father of four children. Several condemnatory articles were published, including letters from Marie Curie to Langevin. They became a pretext for a xenophobic, antifeminist, antiscience campaign. Marie Curie replied immediately. Langevin also tried to defend her reputation. He challenged a journalist to a duel and wrote a letter to the Nobel Prize Committee president, Svante Arrhenius. But Curie was so deeply affected by the public insults that she had a nervous breakdown, accompanied by serious somatic troubles. Toward the end of December 1911, exhausted, she had to spend a couple of weeks in a nursing home.

During the First World War, Marie Curie proved her attachment to France. Thanks to private gifts, she organized military mobile radiological units. She obtained a driver's license in July 1916 and helped wounded soldiers at the front. When the French Red Cross named Marie to head the Military Radiological Service, her eldest daughter Irène became her assistant; Marie was able to train the army doctors in the new methods of radiology.

After the armistice of 1918, the Radium Institute, built in 1914, was ready to function but lacked enough money to buy laboratory equipment. Marie spent most of her time after the war raising money for her beloved Institute. Marie Meloney, an American journalist, organized a press campaign in favor of Marie Curie in 1920. Marie became so popular among American women that they started a campaign to provide her with a gram of radium. In 1921, though she disliked public relations and mundane receptions, Marie Curie crossed the Atlantic with her daughters. She made a tour of America and received the precious radium at the White House. This present encouraged a French magazine to organize a gala in her honor at the Paris Opera, which began with an appeal for donations to the Radium Institute. The Curie Foundation was created in 1921 with important subsidies from the Lazard Brothers bank and Baron Henri de Rothschild. Marie Curie's public image was completely transformed. From then on she embodied the moral power of science.

In 1922 Curie became one of the 12 members of the International Committee for Intellectual Cooperation of the League of Nations, created in Geneva in August 1922. She did her best to persuade her friend Albert Einstein to change his mind after he resigned. Curie never again had further political involvements, except occasional visits to French ministers to request money from the state for the Radium Institute. She devoted most of her energy to directing the Institute. She had many collaborators, including her daughter Irène and Irène's husband Frédéric Joliot. During the early 1930s she was writing a second treatise on radioactivity and still working in her laboratory preparing actinium derivatives, but her health was declining quickly. She suffered from tuberculosis and she had serious lesions on her fingers caused by radiation; she needed several operations on her eyes. In 1932 she made a last trip to Poland to inaugurate the Maria Skłodowska-Curie Radium Institute in Warsaw. Marie's health was significantly weakened and she was nearly blind when Irène and Frédéric Joliot-Curie discovered artificial radioactivity in January 1934. Marie Curie died in a sanatorium in the French Alps on July 4, 1934.

BERNADETTE BENSAUDE VINCENT

## Bibliography

Cunningham, M. E. *Madame Curie (Sklodowska) and the story of radium*; The Saint Catherine Press: London, 1918.

Curie, E. *Madame Curie*; Gallimard: Paris, 1938; Sheean, V., Trans.; Heinemann: London, 1939.

Curie, M. *L'isotope et les éléments isotopes*; Paris, 1922–23.

Curie, M. *Pierre Curie*; Payot: Paris, 1924.

Curie, M. *La Radiologie et la guerre*; Librairie Félix Alcan: Paris, 1921.

Curie, M. *Traité de radioactivité*; Paris, 1910, 2 vols.

DeLeeuw, A. *Marie Curie, woman of genius*; Garrard: Champaign, IL.

Giroud, F. *Une Femme honorable*; Hachette: Paris, 1981; *Marie Curie, a life*; Davis, L., Trans.; Holmes & Meier: New York, 1986.

Ivimey, A. *Marie Curie: pioneer of the atomic age*; Arthur Barker: London, 1964.

Marbo, C. *Souvenirs et rencontres*; Grasset: Paris, 1968.

Marquand, R. *A Love Story* [video recording]; [Centron Educational Films.]

Pyecior H. M. "Marie Curie's Anti-Natural Path: Time only for Science and Family"; in Abir-Am, P. G.; Outram, D., Eds.; *Uneasy Careers and Intimate Lives. Women in Science, 1789–1919*; Rutgers University Press: New Brunswick, NJ, 1987; pp 191–215.

Reid, R. *Marie Curie*; American Library: New York, 1974.

Rubin, E. *The Curies and radium*; F. Watts: New York, 1961.

Skłodowska-Curie, M. *Oeuvres*; Academy of Science: Warsaw, 1954.

# Victor Grignard

## 1871–1935

F. A. Victor Grignard was born on May 6, 1871, at Cherbourg, where his father Théophile Henri Grignard was foreman and sailmaker at the marine arsenal. After studies at the Ecole Normale Spéciale in Cluny and at the Faculty of Sciences in Lyon, Victor Grignard completed a doctoral thesis in 1901. The thesis was unusually significant, laying out the applications of the method of chemical synthesis that has come to be called the Grignard reaction. In 1912 the Nobel Prize in chemistry was awarded jointly to Paul Sabatier and Grignard, recognizing Grignard for his discovery of the role of organomagnesium halides in organic synthesis. From 1909 until 1919 Grignard taught at the Sciences Faculty in Nancy, where he directed one of the most productive chemical research laboratories in France. In 1919 he returned to Lyon as professor of general chemistry, and in 1921 he also became director of the School of Industrial Chemistry of Lyon. He died at Lyon on December 13, 1935.

While a boy at Cherbourg, Grignard prepared for the baccalaureate at the Collège de Cherbourg from 1883 to 1887. For financial reasons he was unable to prepare for entrance examinations to the prestigious Ecole Polytechnique or Ecole Normale Supérieure. Instead, he successfully took the less-demanding qualifying examinations for the Ecole Normale Spéciale at Cluny, a training school for teachers of modern secondary education founded in 1866. When the school closed in 1891, Grignard and other students were enrolled as scholarship students at the Faculties at Lyon, some 75 kilometers from Cluny. After a year of military service from 1892 to 1893 and after obtaining his licence in mathematics in 1894, Grignard was persuaded by his Cluny comrade Louis Rousset to accept a post as *préparateur* in the general chemistry laboratory of the Sciences Faculty at Lyon. He worked first with Louis Bouveault and then with Philippe A. Barbier, both of whom were innovative and productive chemists and teachers.

In 1898 Grignard and Barbier published a paper on a stereochemical problem. Grignard next wrote papers on hydrocarbons with ethylene and acetylene linkages and on hydrocarbons with three adjacent double bonds. At the time, Barbier was working on the problem of converting natural methylheptenone into the lemon-smelling dimethylheptenol by substituting magnesium for zinc turnings and covering the ketone with ether solution. This procedure was a variation of the Saytzeff method, which uses methyl iodide, a ketone, and zinc, followed by hydrolysis, to produce an alcohol. Barbier reported the success of his synthesis but pursued the method no further because he found the results to be undependable and the yields unsatisfactory. He encouraged Grignard to take up this line of research for his doctoral thesis.

On the assumption that the reaction involves an organometallic intermediary compound, Grignard scoured chemical literature for information on organozinc compounds and organomagnesium compounds. He found Edward Frankland's observations in the 1850s that organozinc compounds prepared in anhydrous ether are stable and not flammable in air. This led Grignard to try the method of slowly adding to magnesium a mixture of isobutyl iodide and anhydrous ether. On cooling, the product was a limpid, colorless liquid that was not flammable in air; it furnished an excellent yield of phenylisobutyl carbinol after the addition of benzaldehyde. The reaction took place spontaneously in ether at room temperature under ordinary pressure and worked with aromatic halides as well. In his first paper on this discovery of the organomagnesium halides, he suggested the formula $R$MgI or $R$MgBr, where $R$ might be an alkyl or phenyl radical.

Despite encouragement by Henri Moissan to present his doctoral thesis before the Paris Sciences Faculty, Grignard took his degree at Lyon in the summer of 1901. His thesis listed 29 new compounds prepared by his method, including phenylisobutyl carbinol, dimethylphenyl carbinol, and dimethylbenzyl carbinol. He described the synthesis of carboxylic acids by the action of carbon dioxide on the ether solution of the organomagnesium halide; the preparation of secondary alcohols from aldehydes or formic esters; the synthesis of tertiary alcohols from ketones, carboxylic acid esters, and acid halides or anhydrides; and the preparation of unsaturated hydrocarbons from tertiary alcohols. By 1905 approximately 200 papers were available on organomagnesium synthesis, 80 of them in major French journals and 91 in reports of the German Chemical Society. By 1912 more than 700 papers had appeared; by 1926, 1800; and by 1950, 4000. The importance of the method lies in the fact that the "Grignard reagent" reacts with all functional groups except the ethylene and acetylene linkages. The large scale of the future industrial use of Grignard's process could hardly have been foreseen in 1900: One American firm recently was manufacturing 45 tons of the Grignard reagent each day.

The organomagnesium halide is a valuable tool theoretically, for example, for estimating the numbers and positions of replaceable hydrogen, since magnesium is reactive toward active hydrogen. For Grignard, the most pressing theoretical issue was the composition of the active intermediary compound in the reaction, a question that embroiled him in controversy with a number of chemists. One of them was E. E. Blaise, who from 1901 to 1907 studied reactions of organo-magnesium compounds with nitriles. Blaise argued that in the intermediary organomagnesium etherate, the halide is directly attached to oxygen. Grignard's more accurate formulation viewed $MgX$ as the entity attached to oxygen. Grignard's testing of these hypotheses resulted in synthesis with ethylene oxides, glycol oxides, and glycol derivatives, including his preparation of the rose-smelling 2-phenyl ethyl alcohol, the production of which became the first commercial application of his method in France.

Grignard's scientific reputation became firmly established in the early 1900s. He shared or received prizes from the French Academy of Sciences, including the Prix Cahours (1901 and 1902), the Berthelot Medal (1902), and the Prix Jecker (1906). At the celebration of the fiftieth anniversary of the French Chemical Society, in 1907, he was identified by Armand Gautier as the originator of one of the three most powerful methods of synthesis in modern organic chemistry: the methods of Friedel and Crafts, Sabatier and Senderens, and Grignard. Despite such recognition, it was not until 1909 that Grignard was named to fill temporarily, and then permanently, a position in organic chemistry at the University of Nancy. A factor that impeded his successful candidacy to prestigious posts was the fact that he had never prepared and passed the postbaccalaureate, predoctoral *agrégation* examination, which in France is viewed as a teaching credential for lycée and higher education.

Except for one year as lecturer in general chemistry at Besançon, where little research was possible, Grignard served until 1909 as laboratory instructor (1898) and lecturer (1906), and associate professor (1908) in general chemistry at Lyon. He and Barbier collaborated in publishing 11 papers on terpene chemistry from 1907 to 1914, despite some resentment on the part of Barbier at the credit that Grignard was receiving for synthetic methods using magnesium. At Nancy, Grignard became associated with another vigorous research school, both in the Sciences Faculty and in applied-science institutes for chemistry and for brewing. The Nancy Chemical Institute boasted approximately 150 students, active staff members, and excellent facilities, and its chemistry program was one of the most outstanding in France.

In 1912 Grignard shared the Nobel Prize with Paul Sabatier. Along with Henri Moissan in 1906 and Marie Curie in 1911, Grignard and Sabatier were thus the only French scientists to receive the Nobel Prize in chemistry until Frédéric and Irène Joliot-Curie in 1935. The award to Grignard and Sabatier also recognized the recent successes of provincial sciences laboratories in France. Elected a corresponding member of the Academy of Sciences in 1913, Grignard became one of the six non-resident-section members in 1926, replacing Pierre Duhem. By this time Grignard had returned to Lyon, twice refusing the offer of a chair at the Collège de France, an institution that did not require its professors to prepare students for the *agrégation* examination. Despite the entreaties of Georges Urbain, he refused to become a candidate for a position at the Sorbonne.

When war began in August 1914, Grignard initially was assigned a corporal's sentry duties at a railroad bridge in Normandy, becoming perhaps the only corporal wearing the red ribbon of the Legion of Honor. After assignments to Navy laboratories and his own laboratory at Nancy, he was sent to Paris in 1915 to work in the chemical materials division for the war effort. His war work before 1915 had to do with studies of the cracking of petroleum; in Paris, he worked on chemical gases, including the preparation and detection of phosgene, lachrymatory gases, and dichloroethyl sulfide (mustard gas). He served on the Tardieu committee in the United States in late 1917 and early 1918 to coordinate the research and manufacture of explosives and gases between America and France.

In 1919 Grignard returned to Lyon upon Barbier's retirement from the chair of general chemistry. He became director of a reorganized School of Industrial Chemistry in 1921 and served as dean for the

Faculty of Sciences from 1929 until his death in 1935. He directed state, university, and engineering doctoral research programs on topics that included, in addition to Grignard reactions, the condensation of aldehydes and ketones, cracking of hydrocarbons with an aluminum chloride catalyst, catalytic hydrogenation and dehydrogenation under reduced pressures, and the determination of the constitution of unsaturated compounds by quantitative ozonization. In 1930 he became the general editor for the multivolume *Traité de chimie organique*, which was edited after his death by René Locquin and Georges Dupont.

In 1910 Grignard married the recently widowed Augustine-Marie Boulant Paindestre, whom he had known from school days in Cherbourg. They had two sons, Robert Paindestre and Roger Grignard, born in 1911. Roger Grignard, a chemist, collaborated with his father's student Jean Colonge to publish in 1937 Grignard's lecture course in organic chemistry, *Précis de chimie organique*. Victor Grignard died at Lyon in 1935 after a six-week illness.

MARY JO NYE
*University of Oklahoma*

# *Bibliography*

Courtot, C. "Notice sur la vie et les travaux de Victor Grignard (1871-1935)"; *Bulletin Société Chimique de France* **1936**, *5*, 1433–1472.

Gilman, H. "Victor Grignard"; *J. Am. Chem. Soc.* **1937**, *59*, 17–19.

Grignard, R. *Centenaire de la naissance de Victor Grignard, 1871–1971*; Audin: Lyon, 1972.

Grignard V. "Sur les combinaisons organomagnésiennes mixtes et leur application à des synthéses d'acides, d'alcools et d'hydrocarbures"; *Annales de l'Université de Lyon* **1901**, *6*, 1–116.

Nye, M. J. "Lyon: Applied Chemistry, Victor Grignard, and the Experience of the First World War"; In *Science in the Provinces. Scientific Communities and Provincial Leadership in France, 1860–1930;* University of California Press: Berkeley, 1986; pp 154-194.

Rheinboldt, H. "Fifty Years of the Grignard Reaction"; *J. Chem. Ed.* **1950**, *27*, 476–488.

# *Paul Sabatier*

## 1854–1941

Paul Sabatier was born on November 5, 1854, in Carcassonne, one of France's oldest cities. After studies in Paris, Sabatier began teaching in 1882 at the Faculty of Sciences at Toulouse, less than 100 kilometers from his birthplace. From 1905 to 1929 he was dean of the Faculty of Sciences, where he established applied-science institutes in chemistry, electrotechnology, and agriculture. In 1912 Sabatier and Victor Grignard shared the Nobel Prize in chemistry; Sabatier received recognition for his contributions to the chemistry of catalysis and especially the hydrogenation of unsaturated organic compounds. The discovery by Sabatier and Jean Baptiste Senderens of the activity of finely divided metals, freshly reduced from their oxides, in saturating organic compounds with hydrogen opened the way to many modern industrial procedures. Sabatier died on August 14, 1941, at Toulouse.

Sabatier and his older brother Théodore both studied mathematics and scientific subjects under the tutorship of their uncle, who was a professor at the *lycée* of Carcassonne. Théodore Sabatier in turn became a physics professor at the lycée. After three years as a boarding student at the Collège Sainte-Marie in Toulouse, Paul Sabatier entered the Ecole Polytechnique in Paris in 1874, having passed entrance examinations both for the Ecole Polytechnique and for the Ecole Normale Supérieure. In 1877 he received a first place in the physics *agrégation* and taught one year at the *lycée* of Nîmes. He then was recommended to the post of *préparateur* in Marcellin Berthelot's laboratory of organic chemistry at the Collège de France.

In 1880 Sabatier completed a doctoral thesis on the thermochemistry of sulfides, and he continued to work in the general field of thermochemistry and physical chemistry until 1897. His researches included analyses of metallic and alkaline-earth sulfides and chlorides, the preparation of hydrogen disulfide by vacuum distillation, the isolation of selenides of boron and silicon, the definition of basic tetracupric salts, and preparations of the deep blue nitrosodisulfonic acid and the basic mixed argentocupric salts. As well, he studied the velocity of transformation of metaphosphoric acid and the distribution of a base between two acids, using the spectrophotometric change of the coloration of chromates and dichromates as an indicator of acidity. By the time he began the experiments that led to his Nobel Prize winning work in organic chemistry, Sabatier had published over ninety articles or essays in organic chemistry, for which he was awarded the 1897 Prix LaCaze by the French Academy of Sciences.

During 1880–1881 Sabatier taught at the Faculty of Sciences at Bordeaux and then successfully posed his candidacy for a position in his native Midi at Toulouse. After serving as a temporary replacement for the physics professor in 1882, he became *chargé de cours* in physics, then in chemistry, and in 1884, when he was thirty years old, professor of chemistry. From the beginning of his tenure at Toulouse, Sabatier exerted administrative and intellectual leadership. Innovations included the introduction into his courses of contemporary atomic theory and of Mendeleev's periodic table at a time when Berthelot and many other French chemists had accepted neither their pedagogical usefulness nor their theoretical accuracy.

Sabatier's attention turned from inorganic to organic chemistry as a result of researches in collaboration with Senderens on the fixing of unsaturated gas molecules, like nitrogen dioxide, on metals such as copper, cobalt, nickel, and iron. An aim of these studies was a better understanding of chemical affinity. These researches began in 1892, the year that Senderens, who taught chemistry and directed chemical laboratories at the Institut Catholique in Toulouse, completed his doctoral thesis in Sabatier's laboratory.

Turning from their initial experiments with nitrogen oxides, Sabatier and Senderens attempted to prepare compounds of ethylene or acetylene gas combined with finely divided metals. Learning that Henri Moissan and Charles Moureu would no longer pursue their failed attempts to prepare an acetylene-metal compound, Sabatier and Senderens set to work first with ethylene, heating an oxide of nickel to 300 °C in a current of hydrogen gas and directing a current of ethylene onto slivers of freshly reduced nickel. They recognized that, in addition to a deposit of black carbon, gases were generated that contained ethane. They also recognized that almost pure ethane could be obtained by directing equal volumes of ethylene and hydrogen on reduced nickel with temperatures of only 30–45 °C. Cold acetylene with nickel similarly produced ethane; cobalt, iron, copper, and platinum black gave analogous but less intense results. It was not only the easy hydrogenation of ethylene and acetylene that was striking, but the unexpected success of using metals outside the platinum family for catalysts.

In 1901 Sabatier and Senderens successfully prepared cyclohexane by using reduced nickel and hydrogen at 180 °C in a vertical U-tube cooled by melting ice. Colorless crystals of cyclohexane with the odor of roses appeared as the tube cooled down to 6.5 °C. The two continued to work together until 1905, when they jointly received the Prix Jecker from the Academy of Sciences. By then, they had effected the transformation of unsaturated ethylenic and acetylenic carbides into saturated carbides; the transformations of nitrate derivatives and of nitriles into amines; the change of aldehydes and ketones into the corresponding alcohols; the reduction of carbon monoxide and carbon dioxide into methane; and the change of phenol into cyclohexanol and of aniline into cyclohexane. In addition, they succeeded in producing the major types of natural petroleum by modifying conditions of the hydrogenation of acetylene.

In 1903 Sabatier began collaborating, as well, with Alphonse Mailhe, who took his doctorate in chemistry at Toulouse in 1902. Other students and co-workers included Marcel Murat, Léo Espil, Georges Gaudion, Bonasuke Kubota, and Antonio Fernandez. Sabatier and Senderens had discovered that they could dehydrogenate primary alcohols and obtain the corresponding aldehyde simply by varying conditions. Sabatier and Mailhe now found that some powdered metallic oxides, such as thoria, silica, and alumina, were catalysts for hydration and dehydration, not hydrogenation and dehydrogenation. For example, reduced copper acts as a catalyst for splitting alcohol vapors into hydrogen and aldehyde, but replacing copper with alumina results in a division of alcohol into water and ethylene. Sabatier and co-workers also learned to manipulate experimental conditions

so that they could carry out both oxidation and reduction simultaneously, removing hydrogen in one part of a molecule and fixing it in
another.

Although he filed eight patent applications, Sabatier generally paid
little attention to commercial applications of his work, which included
the transformations of nitrobenzene into aniline, acetone into isopropyl
alcohol, and carbon monoxide into methane, as well as the preparation
of cyclohexanes, especially cyclohexanol and paramethylcyclohexanol
from phenol and paracresol, respectively. He filed a patent for the conversion of liquid fatty acid (oleic acid) into solid acid (stearic acid). It
was not until after the First World War that the hydrogenation of oils
into solid fats became a large industry.

In explaining the mechanism of catalysis, Sabatier preferred the
theory of the formation of a temporary unstable intermediary between
catalyst and reactant, for example, a nickel hydride, which forms on
the surface of the catalyst. The intermediary's combination with the
second reactant regenerates the catalyst, for example, nickel. This interpretation was at odds with the physical theory of catalysis preferred
by some chemists, including Wilhelm Ostwald. However, Sabatier
demonstrated the advantages of his theory by successful predictions of
the outcome of reactions, for example, the decomposition of alcohols
into hydrogen and aldehyde on copper, and into water and ethylene
on alumina or thoria. Sabatier's theory differed, as well, from Irving
Langmuir's later theory of "chemisorption", whereby the intermediate
compound $M_xG_y$ may vary as $x$ depends on the mass of the catalyst
and $y$ on its surface, pressure, and temperature. Sabatier in the late
1920s still preferred a theory of distinct, individual intermediates, with
little emphasis on physical conditions. His book *La Catalyse en chimie
organique,* first published in French in 1913, quickly became a classic in
organic chemistry.

In 1907, shortly after receiving the Jecker Prize, Sabatier made the
unlikely decision to decline chemistry chairs in Paris. Offered the chair
previously held by Henri Moissan at the Sorbonne and the chair of
Berthelot at the Collège de France, Sabatier turned them both down,
preferring to stay in Toulouse where he directed vigorous research
programs and new laboratories. These included programs associated
with three applied-science institutes, which Sabatier founded soon after becoming dean of the Sciences Faculty in 1905: the Chemical Institute (1906), Electrotechnological Institute (1907), and Agricultural
Institute (1909).

The year following the joint award of the Nobel Prize in chemistry to Sabatier and Grignard, the Academy of Sciences inaugurated a new six-member section expressly for scientists not resident in Paris. Sabatier became the first new member. His contributions to chemistry as a provincial scientist have been considered especially influential and significant for the decentralization of scientific institutions in France. His many honors include the Davy Medal of the Royal Society in 1915, the Legion of Honor in 1907, and the Franklin Medal of the Franklin Institute of Philadelphia in 1933, as well as foreign memberships in many national scientific academies. Widowed in 1898, he was the father of four daughters.

MARY JO NYE
*University of Oklahoma*

# Bibliography

Paul Sabatier's most important publication is *La Catalyse en chimie organique* (C. Béranger: Paris, 1913; 2nd ed., 1920). E. Emmet Reid's translation *Catalysis in Organic Chemistry* (Van Nostrand: New York, 1923) has been amended and reprinted in *Catalysis Then and Now* (Franklin Pub. Co.: Englewood, NJ, 1965).

Babonneau, L. "Paul Sabatier"; *Génies occitans de la science;* Privat: Toulouse, 1947; pp 167–189.

Camichel, C.; Dupouy G., et al. *Centenaire Paul Sabatier. Prix Nobel. Membre de l'Institut. 1854–1954;* Privat: Toulouse, 1956.

Nye, M. J. "Sabatier, Paul" In *Dictionary of Scientific Biography;* Gillispie, C. C., Ed.; Scribner's Sons: New York, 1975; Vol. 12, pp. 46–47.

Nye, M. J. "Non-Conformity and Creativity: A Study of Paul Sabatier and the French Scientific Community"; *Isis,* **1977,** *68,* 375–391.

Nye, M. J. "Toulouse: Politics, Entrepreneurship, and Sabatier's Chemistry Program"; *Science in the Provinces: Scientific Communities and Provincial Leadership in France: 1860–1930.* University of California Press: Berkeley, CA, 1986; pp 117–153.

Partington, J. R. "Paul Sabatier"; *Nature* **1954,** *174,* 859–860.

Taylor, H. S. "Paul Sabatier"; *J. Am. Chem. Soc.* **1944,** *66* (October 9), 1615–1617.

Wojtkowiak, B. *Paul Sabatier, un chimiste indépendant (1854–1941): Naissance del'hydrogénation catalytique:* Jonas: Argueil, 1989.

# Alfred Werner

## 1866–1919

Alfred Werner was born in Mulhouse, France, on December 12, 1866. His family was of moderate means; Jean, his father, was a foreman in an ironworks and his mother was a housewife. By the age of 18 he was experimenting with chemicals, and at 20 he studied at the Technische Hochschule in Zurich. After completing his studies there, he went on to the Swiss Federal Institute of Technology (Eidgenössische Technische Hochschule) in Zurich to complete his doctorate. For his doctorate, which was completed in 1890 under the tutelage of Professor Arthur Hantzsch, he studied the structure and stereochemistry of organic nitrogen-containing compounds. He then went to Paris to work with Marcellin Berthelot for a year. In 1893, at the age of 27, he was appointed associate professor at the University of Zurich. The next year he married, and the year after that he became a Swiss citizen and was appointed a full professor at the University of Zurich. At first he lectured on organic chemistry and then in 1902 on inorganic chemistry. Throughout his university years he continued to study and confront the issues of stereochemistry with originality; he published over 150 papers. By 1913 the import of his work was so well recognized that he received the Nobel Prize in chemistry. Six years later, on November 15, 1919, he died.

The important aspects of Werner's work can most effectively be viewed by studying how he did the work; he was an intuitive and bold scientist who expanded and challenged many of the long-standing and widely accepted theories of his day. His doctoral thesis, "Über raumliche Anordnung der Atome in Stickstoffhaltigen Molekulen"(1890), extended to the tetrahedral-trivalent nitrogen the widely accepted ideas of Le Bel and van't Hoff on the tetrahedral-tetravalent carbon. Thus, the stereochemistries of both organic and inorganic molecules were now seen as closely related to each other.

Shortly after receiving his doctorate, he rejected Friedrich August von Kekulé's idea (1861) that the valence of an element was fixed; he replaced it in 1902 with his own concept of principal and auxiliary valence. In the original theory, Kekulé claimed that there was a fixed number of directed valencies: An atom like carbon was "tetratomic" or had four "affinity units", and these four units were free to unite with four single-affinity units on other atoms (e.g., hydrogen) or two double-affinity units (e.g., oxygen). In addition, carbon could unite with affinity units on other carbons. Werner was concerned that the Kekulé concept could not properly explain geometric isomers, the racemization of optically active substances, and the chemical reactions of cyclic compounds. Werner differentiated between "affinity" (force) and "valence" (capacity); furthermore, he saw atoms as having a variable force emanating in all directions.

The work for which he received the Nobel Prize was built on the ideas of C. W. Blomstrand and P. T. Cleve on metal–ammonia compounds. Blomstrand linked the ammonia molecules in long chains (as was suggested by J. J. Berzelius for organic molecules). Thus, Blomstrand suggested that $CoCl_2 \cdot 6NH_3$ would be arranged as

$$Co \underset{\diagdown \cdot NH_3 \cdot NH_3 \cdot NH_3 \cdot Cl}{\overset{\diagup \cdot NH_3 \cdot NH_3 \cdot NH_3 \cdot Cl}{}}$$

Werner's work was considerably challenged by the ideas of S. M. Jørgensen, who was able to show two kinds of $NH_3$ groups: one directly attached to the metal and one unattached. Thus, when an acid was reacted with a hexammine metal–ammonia molecule, up to two ammonia molecules could be replaced by the "acid residue". Jørgensen was therefore able to draw the molecule as

$$\underset{X \cdot H_3N \cdot \diagup}{\overset{X \cdot H_3N \cdot \diagdown}{}} Co \cdot NH_3 \cdot NH_3 \cdot NH_3 \cdot NH_3 \cdot X$$

Werner was able to show that all six ammonias could be replaced on a trivalent metal atom. He concluded that the ammonia molecules were not attached in a chain fashion, as Blomstrand and Jorgensen claimed. Rather the ammonia molecules were each directly attached to the metal such that they could be seen as

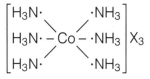

and that the ammonia molecules could be replaced as in

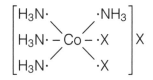

He also found that he could get analogous compounds with platinum, chromium, and other metals. To prove that the ammonia or the acid residue (anion) was not attached ionically, Werner, Arturo Miolati, and Charles Herty tested the recently proposed Arrhenius electrolyte theory and found that few or no ions formed in solution. The basis for the Werner theory was now in place.

Werner claimed that for inorganic chemistry the coordination number of six is very important. Compounds having complex radicals with the formula $MA_6$ predominate. The reaction of the compound platinum tetrachloride indicates this principle:

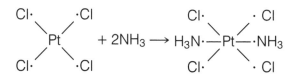

or

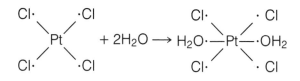

Werner was able to establish that a wide range of simple addition reactions occurs as illustrated in the preceding structures.

Furthermore, he was able to show that a wide range of "intercalation compounds" is formed when the acid residue is replaced by another group. The result is that the acid residue becomes ionic once again.

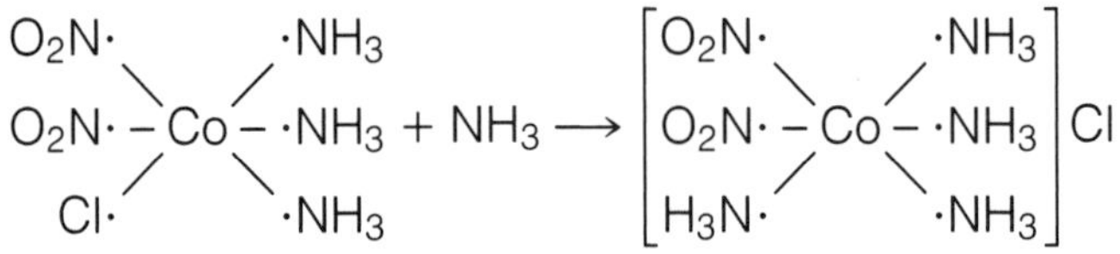

$$
\begin{array}{c}
O_2N\cdot \quad \cdot NH_3 \\
\diagdown \quad \diagup \\
O_2N\cdot - Co - \cdot NH_3 \\
\diagup \quad \diagdown \\
Cl\cdot \quad \cdot NH_3
\end{array}
+ NH_3 \longrightarrow
\left[
\begin{array}{c}
O_2N\cdot \quad \cdot NH_3 \\
\diagdown \quad \diagup \\
O_2N\cdot - Co - \cdot NH_3 \\
\diagup \quad \diagdown \\
H_3N\cdot \quad \cdot NH_3
\end{array}
\right] Cl
$$

The Werner concept of two spheres of chemical action (the inner sphere, where a direct connection exists between metal and addend, and the outer sphere, where addends are ionically attached) was a major breakthrough. The extension of the model from ammonia-containing inner-sphere molecules to water-containing inner-sphere molecules was also quite important. Now chemists could see hydration in a new context. Werner was able to show that chromium, chloride, and water could be seen in two isomeric combinations, the difference being based on the mode of water and chloride linkages with the metal. One formula was

$$[Cr(H_2O)_6] Cl_3$$

and the other formula

$$[CrCl_2(H_2O)_4] Cl + 2H_2O$$

Werner extended the concept of the metal–oxygen linkage in water to other oxygen-containing groups such as the hydroxyl group and the acetate group. This metal–oxygen model was then expanded to include two metal atoms or ions connected to a single oxygen, as in

$$
\left\{ Al(OH)_3Al_2 \right\}
\begin{array}{l}
SO_4 \\
SO_4K
\end{array}
$$

This concept of "multinuclear" compounds was a major breakthrough for geologists and others interested in minerals such as atacamite and alunite.

Werner increased his understanding of complexes with a study of aquo-ammonia metal salts and their reactions with acids to become hydroxo compounds. He saw these reactions as revealing new insights into hydrolysis and salt formation. To Werner ". . . simple bases are water addition compounds which in aqueous solution dissociate into hydroxyl ions." Werner discovered that he could control the degree of base character in the complex by altering the central metal and the attaching groups. Werner obtained compounds that ranged from neutral through strongly basic. He reasoned that this change in character was due to the shift in the ability of the hydroxyl ions present in the complex to attract the hydrogen ions in the water: The stronger the basic character, the greater was the hydrogen attraction. Using this reasoning he redefined bases as compounds that have varying degrees of ability to attract hydrogen ions in water. The shift in the mobility of the hydrogen ions resulted in increases in hydroxyl ion concentration in the water. This redefinition of acids and bases is probably one of the lesser known contributions of Alfred Werner to chemistry.

Among the most famous contributions was his approach to the stereochemistry of inorganic molecules. Through experimentation Werner reasoned that the arrangements of radicals around the metal must be symmetrical. He further reasoned that based on the number of isomers found in platinum and cobalt molecules that the octahedron was the most likely three-dimensional arrangement. Among the first compounds that Werner used to unravel this isomerism issue was $[Cl_2Co(NH_3)_4]$ X, in which the two chlorine atoms could reside in either the cis or the trans position. Confirmation was partly provided with the realization that the cis position would allow for one group to occupy two adjacent coordination points whereas the trans position would not allow this. Thus, $[(CO_3)Co(NH_3)_4]$ X must have cis arrangement for the double attachment of the carbonato group to the central metal. On replacement of the carbonate group with chlorine, the cis dichloro complex was always obtained.

Werner went on to find that cis complexes did not always have cis substitution and trans complexes did not always have trans substitutions. In some cases the complexes substituted vice versa. This finding led Werner to theorize that the entering group was controlled by the nature of the central metal atom and the nature of the entering group rather than by the position of the leaving group. A parallel was established by Werner between the organic Walden inversion and inorganic chemistry. Werner was thus able to explain why spatially identical molecules or mirror-image arrangements occurred.

The issue of optically active mirror-image isomers was dealt with when he resolved the complex

$$[Cl(NH_3)Co(en)_2]X_2$$

where "en" stands for the molecule ethylenediamine. There are two forms of this complex that are optically active mirror-images:

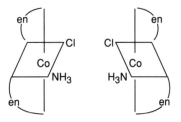

In this complex the cobalt atom is asymmetric, as is carbon in some organic compounds. The asymmetry is lost if the $NH_3$ is replaced in the complex by Cl, so that we have

$$[Cl_2Co(en)_2]X$$

In the case of the compounds like $[Co(en)_3]X_3$, we have a situation in which the structure is totally determined by the geometry rather than by the nature of the entering group. The en group will only span the cis positions, and there are only two geometries possible, each a mirror image of the other:

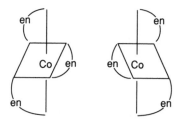

The implications of the work of Alfred Werner are many; not only did he change the thinking of the chemists of his day, but he altered the course of chemistry for many years to come. Geologists and mineralogists now see their disciplines with new insights. Chemists now see the boundary between organic and inorganic chemistry as blurred, and coordination chemistry itself is a field worthy of much study. Biochemists and molecular structure specialists have fresh approaches to their disciplines. We have come to understand crystal structure in a new light; the rules of nomenclature have been revised to accommodate the new and still-growing field. New understandings about molecular structure and bonding are constantly being offered in the literature. The work of Alfred Werner, done through traditional wet-chemistry techniques, has been validated by X-ray crystallographers in recent years.

PAUL S. COHEN
*Trenton State College*

# *Bibliography*

*The Chemistry of the Coordination Compounds;* Bailar, J. C., Ed.; Reinhold: New York, 1956.

*Classics in Coordination Chemistry: Part I;* Kauffman, G. B., Ed.; Dover: New York, 1968.

*Dictionary of Scientific Biography;* Gillespie, C. C., Ed.; Charles Scribner's Sons: New York, 1976; Vol. 14, pp 264–272.

*Inorganic Coordination Compounds;* Nobel Prize Topics in Chemistry; Kauffman, G. B., Ed.; Heyden: London, 1981.

Jorgenson, S. M. "Beiträge zur chemie der Kobaltammoniakverbindung"; *J. Prackt. Chem.* **1878,** *2(18),* 209.

Kauffman, G.B. "Alfred Werner"; *Journal of Chemical Education,* **1959,** *36,* 521–527; and **1966,** *43,* 155–165, 677–679.

Kauffman, G.B. *Alfred Werner, Founder of Coordination Chemistry;* Springer-Verlag: New York, 1966.

Kekulé, A. *Lehrbuch der organischen Chemie;* Enke: Erlangen, Germany, 1866.

Le Bel, J. A. "Sur les relations qui existent entre les formules atomiques des corps organique et le pouvoir de leurs dissolutions"; *Bull. Soc. Chim. Fr.* **1874,** *2(22),* 337–347.

*Nobel Prize Winners;* Wasson, T., Ed.; H. W. Wilson: New York, 1987.

Partington, J. R. *A History of Chemistry;* Macmillan: New York, 1964.

van't Hoff, J. H. "Sur les formules de structure dans l'espace"; *Arch. Neerl. Sci.* **1874,** *9,* 445–454.

Werner, A. In *Nobel Lectures: Chemistry 1901–1921;* Elsevier: New York, 1966.

Werner, A.; *Berichte* **1907,** *40,* 1433.

Werner, A.; Herty, C. H. "A Contribution to the Constitution of Inorganic Compounds"; *Z. Physik Chem.,* 1901, 38, 331.

Werner, A.; Miolati, A. "Beiträge zur Konstitution anorganischer Verbindungen, II. Abhandlung"; *Z. Physik Chem.* **1894,** *14,* 506–521.

*Werner Centennial;* Gould, R. F., Ed.; American Chemical Society: Washington, DC, 1967.

# Theodore William Richards

## 1868–1928

Theodore William Richards was born on January 31, 1868, in Germantown, Pennsylvania. He received the 1914 Nobel Prize for his precise determinations of atomic weights. In 1896 he married Miriam Stuart Thayer; they had three children. He died in Cambridge, Massachusetts, April 2, 1928.

A consummate experimentalist and a gifted teacher, Richards was influential in establishing academic chemical research in the United States and in applying physico-chemical principles to analytical and inorganic chemistry. His research initially seemed to produce convincing evidence for the universality of atomic weights, but later (by showing that atomic weights can be variable) it provided chemical evidence for the existence of isotopes. He was both praised and criticized for insisting on the highest precision in his atomic weight measurements; the methods he developed remain standards for excellence in experimental technique.

Richards was the fifth of six children of seascape painter William Trist Richards and author Anna Matlack. His Quaker mother distrusted the Germantown public schools and felt that the pace was set by the slowest child in the class. She taught her younger children at home, and Theodore received all his formal education until age 14 from his mother. It was not, however, an isolated education. During a summer vacation at Newport, Josiah Parsons Cooke, a professor of chemistry at Harvard, introduced the six-year-old Theodore to astronomy and showed him the rings of Saturn through a telescope. The relationships with Cooke and Harvard continued for many years. The family spent two years (1878–1880) in England, where Theodore developed a love for the English countryside and a fascination for experiments he could do with his new chemistry set. In 1881 John Marshall, a medical professor at the University of Pennsylvania and a friend of the family, invited Theodore into his laboratory to learn qualitative chemical analysis. At Marshall's suggestion he attended lectures in chemistry at the university.

In 1882, at the age of 14, Richards entered the sophomore class of nearby Haverford College. He later recalled that he had "little expectation of anything more than a mediocre success" in college. He "simply tried to do from day to day the tasks in college which presented themselves." This strategy of faithfully doing his homework produced an enviable academic record, and he graduated at the head of his class in 1885 with a specialty in chemistry.

He decided to spend a year at Harvard studying chemistry with Josiah Cooke. Harvard considered this a repeat of his college senior year; the registrar required him to pass the Harvard entrance examinations, which included a knowledge of Greek. He studied Greek with his mother for six weeks during the summer of 1885, passed the entrance exam, and entered Harvard as the youngest member of the senior class. In June 1886 he received his second baccalaureate degree, this time summa cum laude, with highest honors in chemistry. He stayed at Harvard as a graduate student with Cooke and received a Ph.D. in June 1888. Richards was then 20.

Josiah Cooke, the Erving Professor of Chemistry at Harvard from 1850 to 1894, had introduced laboratory instruction in chemistry into the Harvard College curriculum and had pioneered in the involvement of undergraduates in chemical research. Because Cooke was interested in the relationships between atomic weights and the properties of the elements, he proposed a periodic classification.[1] 15 years before Mendeleev presented his periodic table. This interest led to a series of research papers by Cooke (in 1877, 1878, 1879, and 1882) on the atomic weight of antimony, and it led to Richards's thesis research on the ratio of the atomic weights of hydrogen and oxygen.

William Prout had argued in 1815 that the atomic weights of all elements are exact integral multiples of the atomic weight of hydrogen

[1]Cooke, J. P. "The Numerical Relation between the Atomic Weights, with Some Thoughts on the Classification of the Chemical Elements"; *American Journal of Science and Arts,* Ser. 2 **1854,** 17, 387–407.

and that all elements are probably composed of hydrogen atoms. The ratio of the atomic weights of hydrogen and oxygen is about 1:16; Prout's hypothesis required that the ratio be *exactly* 1:16. Jean-Baptiste Dumas in 1842 reported a ratio of 1:15.96 that gave some credence to Prout's hypothesis. Cooke was skeptical and convinced Richards to determine the ratio by synthesizing water (and then weighing the water) from a weighed quantity of hydrogen. Very careful techniques are required for transferring, reacting, and weighing gases; such care became the hallmark of the research of T. W. Richards. Richards's ratio was 1:15.869; it gave evidence that Prout's hypothesis is not strictly valid for oxygen.

Richards received a postdoctoral fellowship for the 1888–1889 year, allowing him to visit the important analytical chemistry laboratories in Germany, France, Switzerland, and England. In the fall of 1889 he returned to the United States to teach analytical chemistry at Harvard. He had assisted in teaching quantitative analysis as a graduate student, and he taught that subject at Harvard College from 1889 to 1902.

Josiah Cooke died in 1894, and Richards was chosen to teach Cooke's physical chemistry course. To prepare for this assignment, Richards spent the spring and summer of 1895 in Germany studying with Wilhelm Ostwald (Nobel Laureate in chemistry, 1909) and Walther Nernst (Nobel Laureate in chemistry, 1920). The next year Richards introduced the course in physical chemistry that he taught to Harvard graduate students for the rest of his life, a course that focused on the underlying causes of chemical phenomena with little emphasis on mathematics. He also taught an undergraduate course in physical chemistry from a historical standpoint.

Richards quickly attracted able research students to Harvard; many of his students became leaders in the chemical profession. He was named by Visher as one of America's six most influential teachers of chemistry;[2] among the other five was Gilbert Newton Lewis, who obtained his Ph.D. with Richards in 1899. Richards's students greatly influenced both the graduate and undergraduate teaching of chemistry throughout North America; they joined the faculties of at least 44 colleges and universities in Canada and the United States. James Bryant Conant obtained his Ph.D. with Richards in 1916, married Richards's daughter Grace Thayer Richards, and became president of Harvard. Several significant textbooks in physical, analytical, and organic chemistry were

---

[2]Visher, S. S. *Scientists Starred 1903–1943 in "American Men of Science"*; Johns Hopkins Press: Baltimore, 1947; p 177.

written by Richards's students, including J. B. Conant, Farrington Daniels, Norris Folger Hall, G. N. Lewis, Joseph H. Mathews, William Buell Meldrum, Charles Phelps Smyth, and Hobart H. Willard.

His first scientific paper reported on research conducted as a Harvard undergraduate, under the direction of Josiah Cooke.[3] The reaction described in his paper, the precipitation of silver chloride when a solution of soluble silver nitrate is mixed with a solution of a soluble chloride salt, was the basis for many of his later determinations of atomic weights. Richards was extending an experiment begun as a course exercise: to make physical measurements on a chemical system and to look for consistencies transcending the particular system.

As Richards completed his doctoral research, Cooke posed another problem. A revised value of the atomic weight of copper had just been reported. The new atomic weight was significantly higher than earlier values; Richards decided to resolve the discrepancy by direct chemical methods. He was meticulous in purifying his reagents and in searching for unexpected sources of error. Exploring the possibility that the source of copper might affect its atomic weight, he found that copper from near Lake Superior and from Germany gave the same atomic weight value within the error of his experiments.

After his postdoctoral year in Germany, Richards began an exhaustive investigation of the atomic weight of copper. He used a variety of chemical methods, including the analysis of copper sulfate by the precipitation of barium sulfate. Calculations involving barium sulfate required a knowledge of the atomic weight of barium, and Richards questioned the accepted value. Since the atomic weights of 18 other elements depended on these same sorts of calculations, he redetermined the atomic weight of barium. It was the work of a lifetime to sort out these interrelated problems and then to design and perform the rigorous experiments required for the highest possible accuracy. Richards and the students in his laboratory determined the atomic weights of 25 elements, and the atomic weights of an additional 30 elements were determined in the laboratories of Gregory P. Baxter and Otto Hönigschmidt, two of Richards's students.

Richards critically assessed the methods used by previous investigators and concluded that the most precise analytical methods involved the precipitation of silver chloride and other "insoluble" silver salts. This was the reaction that he had studied as a Harvard senior. He

[3]Richards, T. W. "On the Constancy in the Heat Produced by the Reaction of Argentic Nitrate on Solutions of Metallic Chlorides"; *Proceedings of the American Academy of Arts and Sciences, (Boston)* **1886**, *22*, 162–164.

spent several years investigating ways to improve the accuracy of analyses based on the precipitation of silver salts. He noted that no salt is absolutely insoluble and used physicochemical principles to correct for the slight solubility of the silver salts. He invented the nephelometer, an optical instrument that measures the quantity of the microscopic particles of solid that do not settle from solution. He developed many procedures for ensuring the purity of the precipitates and for increasing the accuracy of critical weighing operations.

Richards remarked that he chose to determine atomic weights in part because he was familiar with the experimental techniques, "but also because atomic weights seemed to be one of the primal mysteries of the universe. They are values which no man by taking thought can change. They seem to be independent of place and time. They are silent witnesses of the very beginning of the universe, and the half-hidden, half-disclosed symmetry of the periodic system of the elements only enhances one's curiosity about them. Moreover, among the many properties possessed by an element, the atomic weight seems one of the most definite and precise. Hence in trying to satisfy a desire which had as its object the discovery of more knowledge concerning the fundamental nature of things, one naturally assigns to the atomic weights an important place."[4] The Smithsonian Institution affirmed the fundamental significance of atomic weights by devoting a volume to them in its series *The Constants of Nature*. Results achieved by Richards and his students dominated the 548-page 1910 edition of the book.[5]

The atomic weights of the elements are still central to the practice of chemistry, but their significance has changed. By 1910 several scientists had data suggesting that for some elements there were atoms of different masses; we now know that every element has such *isotopes*. The atomic weights that generations of chemists had determined with such care are now known to be averages dependent on the proportions of the various isotopes in the material. Richards provided chemical support for this new idea by showing that the atomic weight of lead depends on the source of the lead.[6] Richards summarized this transfor-

[4]Quoted by Harold Hartley, "Theodore William Richards Memorial Lecture"; *Journal of the Chemical Society, London* **1930**, 1937–1969.

[5]Clarke, F. W. "A Recalculation of the Atomic Weights"; in *The Constants of Nature*, part IV; The Smithsonian Institution: Washington, 1882, 1897, 1910.

[6]Richards, T. W.; Lembert, M. E. "The Atomic Weight of Lead of Radioactive Origin"; *J. Am. Chem. Soc.* **1914**, *36*, 1329–1344; Richards, T. W.; Wadsworth, C., III. "Further Studies of the Atomic Weight of Lead of Radioactive Origin"; *J. Am. Chem. Soc.* **1916**, *38*, 2613–2622; Kaufmann, George B. *J. Chem. Ed.* **1982**, *59*, 3–8, 119–123.

mation of the meaning of atomic weights in the first article of *Chemical Reviews*.[7]

Richard's interest in the ultimate nature of the material universe was not restricted to atomic weights. He challenged contemporary theoretical models in which atoms were considered to be hard billiard balls or structureless mass points. In more than 20 research papers he reported experiments on the equations of state of gases, liquids, and solids as he developed and promoted his alternative theory of compressible atoms.[8]

A grand theme of 19th-century chemistry was the search for the modern chemical elements and the quantitative determination of their properties; a periodic classification of the elements on the basis of atomic weights became the organizing principle. The work of Richards and his students in determining atomic weights with precision unsurpassed by chemical methods was the culmination of this monumental work. The corresponding 20th-century theme is based on a modern version of Prout's hypothesis in which all atoms are composed of neutrons and equal numbers of protons and electrons. The *atomic number* has replaced atomic weight as the organizing principle. Richards sought to determine numbers in the laboratory that would reveal primal mysteries of the universe; it turned out that atomic weights do not have such fundamental significance. The legacy of Richards is not those precise numbers, but rather his strong influence on his students, many of whom became leaders in shaping chemical education and research in the 20th century.

GEORGE FLECK
*Smith College*

## *Bibliography*

Baxter, G. P. "Theodore William Richards"; *Science* (Washington, DC) **1928**, *68*, 333–339.

Farber, E. *Nobel Prize Winners in Chemistry, 1901–1961;* Abelard-Schuman: London, 1963.

Forbes, G. S. "Investigations of Atomic Weights by Theodore William Richards"; *J. Chem. Ed.* **1932**, *9*, 452–458.

---

[7]Richards, T. W. "Atomic Weights and Isotopes"; *Chemical Reviews* **1924**, *1*, 1–40.

[8]Richards, T. W. "The Present Aspect of the Hypothesis of Compressible Atoms"; ACS Presidential address. *J. Am. Chem. Soc.* **1914**, *36*, 2417–2439. Contains bibliography of papers by Richards on compressible atoms.

Ihde, A. J. "Richards—Corrector of Atomic Weights, 1868–1928"; in *Great Chemists;* Farber, E., Ed.; Interscience: New York, 1961; pp 822–826.

Ihde, A. J. "Theodore Williams Richards and the Atomic Weight Problem"; *Science* (Washington, D. C.) **1969**, *16*, 647–651.

Kaufmann, George B. *J. Chem. Ed.* **1982**, *59*, 3–8, 119–123.

Kopperl, S. J. "Richards, Theodore William"; in *Dictionary of Scientific Biography;* Gillispie, C. C., Ed.; Charles Scribner's Sons: New York, 1975; Vol. 11, pp 416–418.

Kopperl, S. J. "The Scientific Work of Theodore William Richards"; Ph.D. Dissertation, University of Wisconsin, 1970.

*Nobel Prize Winners;* Wasson, T., Ed.; H. W. Wilson: New York, 1987.

Richards, T. W. "Die Bestimmung der Kompressibilität flüssiger und fester Substanzen"; In *Handbuch der Arbeitsmethoden in der anorganischen Chemie;* Stähler, A., Ed.; Leipzig, 1912; p 247.

Richards, T. W. "The Chemical Significance of Crystalline Form"; *J. Am. Chem. Soc.* **1913**, *35*, 617.

Richards, T. W. "The Fundamental Properties of the Elements, Faraday Lecture"; *J. Chem. Soc.* **1911**, *99*, 1201; *Science, N. S.* **1911**, *34*, 537; *Proc. Chem. Soc.* **1911**, *27*, 177; *Rev. Sci.* **1912**, *50*, 321; *Smithsonian Report* **1912**, p 199.

Richards, T. W. "Further Remarks on the Chemical Significance of Crystalline Form"; *J. Am. Chem. Soc.* **1914**, *36*, 1687.

Richards, T. W. "New Method of Determining Compressibility, with Application to Bromine, Iodine, Chloroform, Bromoform, Carbon Tetra Chloride, Phosphorus, Water and Glass"; *J. Am. Chem. Soc.* **1904**, *26*, 399; *Z. Phys. Chem.* **1904**, *49*, 1.

Richards, T. W. "Note on the Efficiency of Centrifugal Purification"; *J. Am. Chem. Soc.* **1905**, *27*, 104–111.

Richards, T. W. "The Possible Significance of Changing Atomic Volume"; *Proc. Am. Acad. Arts Sci.* **1901**, *37*, 1; *Z. Phys. Chem.* **1902**, *40*, 169; *Chem. News* **1902**, *96*, 81.

Richards, T. W. "The Relation of the Hypothesis of Compressible Atoms to Electrochemistry"; *Trans Internat. Electr. Congress* **1905**, *2*, 7.

Richards, T. W. "The Significance of Changing Atomic Volume. II. The Probable Source of the Heat of Chemical Combination and a New Atomic Hypothesis"; *Proc. Am. Acad. Arts Sci.* **1902**, *37*, 399; *Z. Phys. Chem.* **1902**, *40*, 597.

Richards, T. W. "The Significance of Changing Atomic Volume. III. The Relation of Changing Heat Capacity to Change of Free Energy, Heat of Reaction, Change of Volume and Chemical Affinity"; *Proc. Am. Acad. Arts Sci.* **1902**, *38*, 293; *Z. Phys. Chem.* **1902**, *42*, 129.

Richards, T. W. "The Significance of Changing Atomic Volume. IV. The Effects of Chemical and Cohesive Internal Pressure"; *Proc. Am. Acad. Arts Sci.* **1904**, *39*, 581; *Z. Phys. Chem. Arts Sci.* **1904**, *49, 15.*

Richards, T. W. "The Significance of the Quantity $b$ in the Equation of Van der Waals"; *J. Am. Chem. Soc.* **1914**, *36*, 617.

Richards, T. W. "The Theory of Compressible Atoms"; *Harvard Graduate's Magazine* **1913**, *21*, 595.

Richards, T. W.; Behr, G., Jr. "The Electromotive Force of Iron under Varying Conditions, and the Effect of Occluded Hydrogen"; *Pub. Carnegie Inst. Wash.* **1906**.

Richards, T. W.; Brink, F. N. "Densities of Lithium, Sodium, Potassium, Rubidium, and Caesium"; *J. Am. Chem. Soc.* **1907**, *29*, 117.

Richards, T. W.; Brink, F. N. "The Linear Compressibility of Copper and Iron, and the Cubic Compressibility of Mercury"; *Pub. Carnegie Inst. Wash.* **1907**, *76, 43*; *Z. Phys. Chem.* **1907**, *61*, 171.

Richards, T. W.; Forbes, G. S. "Energy Changes Involved in the Dilution of Zinc and Cadmium Amalgams"; *Pub. Carnegie Inst. Wash.* **1906**.

Richards, T. W; Jackson, F. G. "The Specific Heats of the Elements at Low Temperatures"; *Z. Phys. Chem.* **1909**, *70*, 414.

Richards, T. W; Jones, G. "The Compressibilities of the Chlorides, Bromides, and Iodides of Sodium, Potassium, Silver, and Thallium"; *J. Am. Chem. Soc.* **1909**, *31*, 158; *Z. Phys. Chem.* **1910**, *71*, 152.

Richards, T. W.; Köthner, P.; Tiede, E. "Further Investigation of the Atomic Weights of Nitrogen and Silver"; *J. Am. Chem. Soc.* **1909**, *31*, 6.

Richards, T. W.; Lacy, B. S. "Electrostenolysis and Faraday' s Law"; *J. Am. Chem. Soc.* **1905**, *27*, 232–233.

Richards, T. W.; Matthews, J. H. "The Relation Between Compressibility, Surface Tension and Other Properties of Material"; *J. Am. Chem. Soc.* **1908**, *30, 8*; *Z. Phys. Chem.* **1908**, *61*, 449.

Richards, T. W.; Mueller, E. "A Revision of the Atomic Weight of Potassium. The Analysis of Potassium Bromide"; *J. Am. Chem. Soc.* **1907**, *29*, 639–656.

Richards, T. W.; Speyers, C. L. *J. Am. Chem. Soc.* **1914**, *36*, 491.

Richards, T. W.; Staehler, A. "A Revision of the Atomic Weight of Potassium. The Analysis of Potassium Chloride"; *J. Am. Chem. Soc.* **1907**, *29*, 623–639.

Richards, T. W.; Stull, W. N. "The Compressibilities of the More Important Solid Elements and Their Periodic Relations"; *Pub. Carnegie Inst. Wash.* **1907**, *76*, 55; *Z. Phys. Chem.* **1907**, *61*, 183; *J. Am. Chem. Soc.* **1909**, *31*, 154.

Richards, T. W.; Stull, W. N.; Bonnet, F., Jr. "The Compressibility of Lithium, Sodium, Potassium, Rubidium, and Caesium"; *Pub. Carnegie Inst. Wash.* **1907**, *61*, 77.

Richards, T. W.; Stull, W. N.; Brink, F. N. "The Compressibility of Carbon, Silicon, Phosphorous, Sulfur, and Selenium"; *Pub. Carnegie Inst. Wash.* **1907**, *76*, 29; *Z. Phys. Chem.* **1907**, *61*, 100.

Richards, T. W.; Stull, W. N.; Matthews, J. H.; Speyers, C. L. "The Compressibilities of Certain Hydrocarbons, Alcohols, Esters, Amines, and Organic Halides"; *J. Am. Chem. Soc.* **1912**, *34*, 971.

Richards, T. W.; Wells, R. C. *A Revision of the Atomic Weights of Sodium and Chlorine*; Publication No. 28; Carnegie Institution: Washington, 1905.

Servos, J. W. *Physical Chemistry from Ostwald to Pauling: The Making of a Science in America*; Princeton University Press: Princeton, 1990.

Söderbaum, H. G. "The Nobel Prize in Chemistry for 1914"; In *Les Prix Nobel*; P. A. Norstedt & Söner: Stockholm, 1920; pp 40–44.

# Richard Martin Willstätter

## 1872–1942

Richard Martin Willstätter was born in Karlsruhe, Baden, Germany, in August 1872. He was the son of Max Willstätter and Sophie Ulmann, the daughter of a textile merchant. Richard's father worked overseas at his clothing business in New York during most of Richard's youth, so he and his brother were reared by his mother and her family.

Until the age of 10 he lived quite happily in Karlsruhe. He found school distasteful at first, but he enjoyed learning and was an avid collector of stamps, coins, and minerals. In 1883 his family moved to Nuremberg. He was not as happy there; he found school more demanding and had little opportunity for the outdoor activities he had previously enjoyed. Although he had experienced some anti-Semitism from the street children in Karlsruhe, in Nuremberg he found it in middle-class families and in his classmates. This prejudice was to shadow him much of his life. Although he did well in his math, geology, and history classes, he did very poorly in Latin, which was used in many classes. A family council thus decided that he should enter the more practical *Realgymnasium* and become a businessman.

One of his uncles ran a factory in Nuremberg and allowed the young Willstätter to dabble in chemistry; he was allowed to mix acids and metals to produce gases and to perform other similar experiments.

He had read too little to understand the results of his experiments, however, and soon abandoned the lab. He still had an interest in science, however, as later events showed.

In October 1890 Willstätter matriculated at the University of Munich and also enrolled in some classes at the Munich Institute of Technology. He took his first year of chemistry at the institute and was disappointed to find that many of the problems he found intriguing had already been solved. He had to repeat this first year of chemistry when he started full-time at the University of Munich the next year, because his professor refused to accept the work done at the institute. After passing exams in inorganic chemistry and analytical chemistry during his third year, he began organic chemistry under Adolf Baeyer. Baeyer, who found time for him even at the beginning, was to be a great influence on Willstätter's life.

Finding that Baeyer no longer took graduate students, Willstätter started his doctoral work under Alfred Einhorn, who was investigating the structure of cocaine and its derivatives. Although Baeyer had not taken him as a student, he became involved with directing Willstätter, and a close friendship developed between them during this time. Willstätter received his doctorate early in 1894 and, quite against his parents' wishes, decided on a career in academic research. After graduation he stayed with Einhorn for another six months and wanted to investigate the structure of cocaine further, a project that Einhorn forbade. Willstätter then did the project in a roundabout manner by looking instead at the structure of a related compound, tropine. When Einhorn eventually discovered the ruse, it angered him, and he remained aloof for years. In the fall of 1894 Willstätter began work independently as a research associate in the organic division, continuing to look at tropine derivatives. He retained a bench in the lab but had to register and pay for classes in order to do so.

In July 1896 Baeyer suggested that he qualify for a private lectureship (*Privatdozent*), the next rung on the academic career ladder. As a summa cum laude graduate of the institution, he was not required to take the usual exam for this, but he had to give a lecture. He managed to finish his talk, even though he had influenza at the time and overturned the easel-mounted blackboard during the lecture. After Willstätter qualified, Baeyer twice urged him to allow himself to be baptized a Christian, because it generally eased the path to success in those days. Willstätter refused, finding a conversion without true feeling repugnant.

In 1902 Willstätter was promoted to professor at Munich and took up the new responsibility for teaching the field of coal tar dyes. He had new privileges as well, such as the use of the laboratory glasswasher and a laboratory key so that he could do research whenever he wished. About this time he took up horseback riding and often spent his vacations in rural areas or resorts where he could indulge in his new hobby. On his 1903 Easter vacation at Wiesbaden he met and became friends with a Heidelberg economics professor and his family. He had dinner with them often and was particularly taken with the professor's daughter, Sophie Leser. They were married that August and had two children, Ludwig and Ida Margarete, before his wife's death from a ruptured appendix in May 1908.

In his last years as a professor at Munich, Willstätter began experiments on chlorophyll, a plant pigment involved in photosynthesis. He continued these studies after moving to a professorship at the Zurich Institute of Technology in the fall of 1905. Because he decided to isolate the chlorophyll directly from the plant rather than purchasing it from a chemical company, the initial experiments developed techniques for the separation of nitrogenous compounds from leaves. He was a pioneer in this respect, as most chemists at that time did not do their own separations of natural products. Even though Michael Tswett had introduced chromatographic separations in 1906, his work was published in Russian and was largely ignored by European scientists; Willstätter's use of chromatography made the technique more popular. In addition to developing the separation techniques that made possible the investigation of many natural products, Willstätter went on to show that the magnesium in chlorophyll was not an impurity but a component of the compound itself, and that there were only two types of chlorophyll rather than the dozens of types previously believed. He and his students developed the techniques to separate the blue-green chlorophyll *a* and the yellow-green chlorophyll *b* and also deduced much about their structures by examining the products of their reactions with acids and bases. These studies provided the conclusive proof of the chemical relationship between blood pigment and chlorophyll. They also found that chlorophyll was always accompanied by a yellow pigment called carotene and one of its derivatives. Interested in whether these carotenoids might also play a part in photosynthesis, they developed methods to isolate and purify them. These pigments were found not only in leaves, but in such diverse places as a tomato, an egg yolk, and the bovine corpus luteum.

Lured by a promise of complete personal and financial independence, in the summer of 1911 Willstätter moved to the Kaiser Wilhelm Institute in Dahlem, Germany (then a suburb, now a part of the city of Berlin). While there he finished his work on chlorophyll extractions and purifications and authored *Investigations on Chlorophyll*. He continued the preparation and isolation of flower and fruit pigments, again following the practice of isolating them directly from the plants. He began work on the degradation of cellulose to sugar using hydrochloric acid. (This technique was eventually applied in industry, and it provided food material in Germany during World War II.)

The German and Allied forces of World War I introduced gas warfare, and Willstätter developed the chemical basis of a gas mask to protect the soldiers from chlorine and phosgene. The design was so successful that he received a congratulatory letter from the Prussian Minister of War and was awarded an Iron Cross Second Class for his part in the war effort. However, the war hampered his research: for example, several of his assistants were drafted.

At the Kaiser Wilhelm Institute he did not receive the promised personal and financial freedoms, and he discovered that at such a nonuniversity institution there was less interaction with the country's other scientists. Although he maintained friendships with prominent chemists by giving lectures in Berlin and elsewhere, the isolation and disruptions of his work made staying at the institute less attractive.

Early in 1915 Baeyer retired, and Willstätter became his successor as director of the laboratory at the University of Munich. He accepted the position with the condition that a new auditorium was to be built and a professorship for physical chemistry established. The document of appointment was signed by King Ludwig III with the remark to the Bavarian Minister of Education that it was the last time that he would approve a Jew for such a position. Thus, 25 years after he came to the Munich laboratory as a student, Willstätter returned as its director. The happiness of returning was ruined when shortly afterward his son died from what was probably a diabetic coma.

In November 1915 a newspaper contacted Willstätter to report that the Royal Swedish Academy had awarded him the Nobel Prize in chemistry for his investigations of plant pigments, particularly chlorophyll. Because of the war, however, he did not attend the ceremonies until 1920.

While in Munich, Willstätter and his co-workers synthesized cocaine, thus opening the door for further anesthetic research. They began research on enzymes, looking at catalase and peroxidase, which react with hydrogen peroxide. Willstätter became convinced that enzymes were not composed of protein, and this view held sway for ten years until James Sumner and John Northrop showed that it was incorrect.

German society was in a state of upheaval after World War I. Returning soldiers streamed to the universities, creating overcrowded conditions. The Bavarian monarchy was overthrown in November 1918 and a republic was proclaimed. Munich was occupied by Communist rebels and then retaken by the leaders of the republic. Inflation was rampant, and by 1923 the German mark had devalued a millionfold. The Munich lab stayed open only through gifts from Theodor Haebler, a German-American humanitarian whose son had studied briefly with Willstätter.

It was popular at the time to blame many things on the Jews, including the overthrow of the Bavarian monarchy, the defeat of Germany in the war, and the horrendous inflation that followed. This attitude became more and more prevalent, even inside the university. In the summer of 1924 the faculty at the University of Munich opposed the appointment of a Jewish academic to replace one of the retiring professors. Despite pleas from the faculty and students, Willstätter resigned his position in protest of the obvious anti-Semitism. Instead of accepting any of the many positions he was offered elsewhere, he decided to retire.

For several years Willstätter remained involved in research through his assistant, Dr. Margarete Rohdewald, who carried out experiments during the day and then discussed them on the telephone with Willstätter in the evening. He continued his associations with many scientific societies and committees until Hitler came to power. Then, in his forced retirement from scientific activities in Germany, Willstätter found that he had time for hobbies. He began an art collection and he traveled, sometimes for pleasure and sometimes to scientific meetings outside Germany.

In 1938 the effects of anti-Semitism became more severe. After being required to turn in his passport, in November the Gestapo went to his home to arrest him. Luckily, he was in the garden at the time, and they did not think to look for him there. He decided to emigrate, and after a great deal of trouble from the authorities he was allowed to go to Switzerland in March 1939. All but a very small fraction of his belongings were confiscated, including the gold Nobel Medal.

He built a home in Muralto-Locarno and began a relatively isolated life. He continued to take an active interest in science, writing a paper on the enzyme systems of muscles. He traveled only within Switzerland, because he was restricted by Swiss rules controlling the movements of aliens. He began to experience heart problems in 1941 and on August 3, 1942, he died in his sleep of heart failure.

ZEXIA BARNES
*Morehead State University*

# Bibliography

Farber, E. "Richard Willstätter, 1872–1942"; In *Great Chemists;* Farber, E., Ed.; Interscience: New York, 1961; pp 1365–1374.

Farber, E. *Nobel Prize Winners in Chemistry, 1901–1961;* Abelard-Schuman: New York, 1963; pp 65–70.

Fruton, J. S. "Willstätter, Richard"; In *Dictionary of Scientific Biography;* Gillispie, C. C., Ed.; Charles Scribner's Sons: New York, 1970; Vol. 14, pp 411–412.

Hammarsten, O. In *Nobel Lectures, Chemistry 1901–1921;* Elsevier: Amsterdam, 1966; pp 297–300.

Nachmansohn, D. *German-Jewish Pioneers in Science*; Springer-Verlag: Berlin, New York, 1979.

*Nobel Prize Winners*; Wasson, T., Ed.; H. W. Wilson: New York, 1987.

*Obituary Notices of Fellows of the Royal Society*; The Royal Society of London: London, 1953; Vol. 8.

Westgren, A. "The Chemistry Prize"; In *Nobel: The Man & His Prizes*; Nobel Foundation and W. Odelberg, Eds.; Elsevier: New York, 1972; pp 347–438.

Willstätter, R. *Nobel Lectures, Chemistry 1901–1921;* Elsevier: Amsterdam, 1966; pp 301–312.

Willstätter, R. *From My Life;* Stoll, A., Ed.; Hornig, L. S., Trans.; W.A. Benjamin: New York, 1965.

# *Fritz Haber*

## 1868–1934

Following two years in which no Nobel Prize in chemistry was given, Fritz Haber was named to receive the prize for 1918. Because of the estrangements and travel problems deriving from World War I, the presentation of the prize to Haber and some other laureates did not take place until June 1920. That ceremony was unusual in that no member of the Swedish royal family participated, because of the recent death of Crown Princess Margaret. It was unusual as well, for a reason that will be discussed later, because the award to Haber generated considerable controversy, something rare in cases of the physical sciences. Fritz Haber was born on December 9, 1868, in Breslau (now Wrocław) in lower Silesia (now western Poland, but at the time, part of Prussia, the largest state in the German Reich). He died of a heart attack in Basel, Switzerland, on January 29, 1934, following a period of poor physical health coupled with intense emotional stress.

The citation for his Nobel prize was "for the synthesis of ammonia from its elements", but as this brief account of his career will reveal, his interests and accomplishments spanned a remarkably wide range of chemistry and physics. In his 1939 book, *The Social Function of Science*, J. D. Bernal declared Haber to have been "the greatest authority in the world on the relations between scientific research and industry." This may have been an overly generous summation, but it does direct attention to Haber's exceptional skill in spanning the sometimes unbridgeable gap in the physical sciences between theory and application.

He was not a narrow specialist: As a young man, Haber acquired a classical education, with emphasis on literature and philosophy, and throughout his life he sustained an interest in and love of the humanities. He was a brilliant conversationalist and a fun-loving companion, often given to expressing his sentiments in light verse or to participating in fun-filled skits for the amusement of his friends and colleagues. Yet for all that, Haber's life was beset with, and unquestionably saddened by, a number of disappointments and tragedies that cannot have failed to darken his private life.

The first such tragedy, which occurred at the time of his birth, was the death of his young mother, who was also his father's first cousin. His father, Siegfried, a well-to-do manufacturer's agent and something of a public figure in Breslau, eventually remarried and had three daughters; he retained toward his son an attitude that has been described as severe and sometimes harsh. Young Fritz was provided with excellent preparatory schooling and then pursued his university training in the "nomadic" way then customary in Germany. It included a semester at Berlin, with instruction in chemistry from A. W. von Hofmann; a year and a half at Heidelberg, with chemistry from R. W. Bunsen; and then, after a period of military service, a year and a half at the Charlottenberg Technical College in Berlin. At this last institution Haber first undertook experimental chemical research, under the direction of Karl Liebermann. This work was in organic chemistry and culminated in a thesis submitted in 1891 for which, after the usual examination, he was awarded a doctorate.

Thereafter followed a period of seemingly aimless activity, quite out of character with his otherwise energetic and sharply focused career. In the span of just over a year he successively tried three industrial jobs, mostly in companies belonging to his father's friends. He then spent a semester at the Federal Technical College in Zurich where, under the guidance of Georg Lunge and to his subsequent great advantage, he learned much about the technological application of scientific principles. Then followed a disastrous few months of working in his father's business, after which he accepted a position as junior assistant to Ludwig Knorr, an aging professor at the University of Jena. Following a year and a half of uninspiring work there, he sought and eventually obtained in 1894 an assistantship in the department of Chemical and Fuel Technology at the Fredericiana Technical College in Karlsruhe. This proved to be the turning point in his career; he remained at Karlsruhe until 1911, by which time he had become a very eminent chemist worldwide.

His professor at Karlsruhe, Hans Bunte, had suggested as a topic of research the breakdown of hydrocarbons at elevated temperatures. Haber applied himself to this with skill and energy and within two years had entirely reshaped the interpretation of the thermal decomposition of such compounds and given new insight concerning the strength of bonds linking atoms in hydrocarbons. His work would come to be of great value in understanding the "cracking" of hydrocarbons during the distillation and other heat treatment of petroleum. By 1896 Haber could publish a habilitation thesis summarizing this work: "Experimental Studies of the Decomposition and Combustion of Hydrocarbons". On the strength of this, he was granted status as a *Privatdozent.*

Soon after his arrival at Karlsruhe, Haber struck up a warm friendship with Hans Luggin, a former pupil of Svante Arrhenius and the only physical chemist among Haber's colleagues. Physical chemistry was still then a comparatively recent extension of the subject, whereby chemical events could be rationalized through unifying theoretical principles, especially transfers of energy. Luggin was able to assist Haber in interpreting his experimental results with the hydrocarbons and thereby aroused the latter's interest in mastering the applications of physical chemistry. This proved to be work with an intellectual component that appealed more to Haber than synthesizing and characterizing organic compounds, and so he redirected his subsequent research efforts. Having at last found his true *métier,* Haber applied himself to the mastery of physical chemistry with an intensity and commitment not previously seen in his work but that would become characteristic of most of the rest of his career.

His new interest led to investigations in electrochemistry, of which the first concerned the reduction of nitrobenzene, a common organic substance, at the cathode of an electrochemical cell. Recent work by others appeared to have shown that a variety of different products were obtained thus, according to the conditions used. Haber was able to show, not only in this case but in general, that the reducing power of a cathode depended on its electrical potential, and that, by varying the latter in this instance, he could demonstrate the reduction of nitrobenzene in stages. He also showed that secondary (nonelectrolytic) reactions among the primary products formed by electrolysis accounted for some of the substances identified following the passage of current. Thenceforth, the control of cathodic potential was established as a requirement for carrying out specific reduction processes electrochemically.

In recognition of the value of this work Haber was promoted in 1898 to a professorship *extraordinarius* (comparable to an associate professorship). In the same year he published an important book, *Outline of Technical Chemistry on a Theoretical Basis*. The significance of this book is suggested by its title, in that it sought to provide a theoretical basis for many of the essentially empirical practices in new industries for carrying out chemical processes by the expenditure of electrical energy. It must be kept in mind that the generation and transmission of electric power on a scale sufficient for industrial purposes was not carried out until late in the 1880s, and thus that in 1898 "technical electrochemistry" was still comparatively undeveloped.

Of the numerous electrochemical researches conducted by Haber and his students, three others can be briefly mentioned here. One was a study of the reversible oxidation/reduction of hydroquinone/quinone, the balance of which was dependent on electrode potential and on hydrogen-ion concentration in solution. Based on this investigation later came, at the hands of Einar Biilman, the quinhydrone electrode as an indicator of hydrogen-ion activity. Another, which arose from Haber's studies of phase-boundary potentials, was the development of the glass electrode, which served the same purpose. The widespread use of the latter was delayed for want of a convenient device for measuring null-current balances, but Haber and Klemensiewicz carried out pH titrations with this device and a quadrant electrometer as early as 1909. The third important electrochemical investigation, which arose as an unexpected outgrowth of some inconclusive work on fuel cells, led to the development of an electrode responsive to gaseous oxygen. Haber immediately applied this to the study of combustion reactions and to the estimation of the free energies for oxidation for various elements and compounds. Free-energy values, through the application of thermodynamic reasoning, made it possible to evaluate the degree of completeness of chemical reactions.

Linked to this development in his experimental researches came Haber's next book: *The Thermodynamics of Technical Gas Reactions* (1905). It was widely acclaimed, and an updated English translation appeared three years later. Haber provided a clear exposition of thermodynamic theory and its application and a critical compilation of established thermodynamic data whereby yields of most technically important gas reactions could be calculated for a wide range of temperature. The authors of a later classic treatise on thermodynamics, G. N. Lewis and M. Randall, described Haber's volume as a "model of accuracy and critical insight."[1] Shortly after the publication of Haber's book, Walther Nernst, a fellow countryman, proposed what he called his "heat theorem". This principle made possible the calculation of the absolute value of free-energy change (and hence the yield) of a chemical reaction from measured thermochemical data. (See the chapter on Nernst, Nobel Laureate in 1920.) It is a fact that in Haber's book much of the evidence upon which Nernst based his conclusions was also discussed, but Haber was not prepared to make the leap of faith inherent in Nernst's proposals; for the English version of his book Haber did include an account of the heat theorem.

The work on gas reactions led directly to Haber's most celebrated achievement: the synthesis of ammonia from the elements. At the turn of the century, threats of a worldwide shortage of nitrogen compounds were circulating. Their importance as agricultural fertilizers had been appreciated since Liebig's time, and an increasing sophistication in weaponry was fueling demand for chemical explosives, which required nitric acid for manufacture. Many chemists were attracted by the possibility of converting ("fixing") atmospheric nitrogen into usable compounds, and a few such were produced but not by very practical methods. In 1904 Haber was engaged as a consultant for one such scheme and undertook to measure the equilibrium constant, which, at a particular temperature, enabled a calculation of the degree of conversion for the following reaction:

$$N_2(g) + 3H_2(g) = 2NH_3(g)$$

The slowness of the reaction required work at high temperatures, and it required a catalyst, for which he chose iron. The yields were disappointingly low and did not encourage further investigations. Two years later, Nernst calculated the equilibrium constant for the same reaction by means of his heat theorem; he found a value appreciably different from Haber's, and subsequently confirmed experimentally the yields of ammonia predicted by his calculations. Haber responded with a new series of improved measurements, the results from which were

[1]Lewis, G. N.; Randall, M. *Thermodynamics and the Free Energy of Chemical Substances;* McGraw-Hill: New York, 1923.

still at variance with the conclusions reached by Nernst. Neither of the Prussian prima donnas was disposed to give way to the other on the question of who was right in this case. Nernst argued, quite properly, that as his experimental studies had been made with the gases under pressure (about 50 times atmospheric) the resulting greater degree of conversion to ammonia should increase the certainty of his estimates of the equilibrium constant compared to Haber's, measured at atmospheric pressure. For him, as well, the integrity of the heat theorem could be at risk if his results were disproved.

But Haber refused to be talked down, and following a stormy session at a meeting of the Bunsen Society in Hamburg in 1907 at which Nernst had challenged his results publicly, he returned to Karlsruhe determined to have the last word. With the help of his assistant Robert le Rossignol, and his mechanic Kirchenbauer, an apparatus was built to conduct experiments at 30 atmospheres of pressure. His new measurements validated his earlier results at atmospheric pressure. Nernst, for once, was wrong and ultimately had to acknowledge an error in calculation. Haber and his associates had actually gone much further than merely gaining an intellectual victory. They had discovered catalysts other than iron, and they had literally built a machine capable of making liquefied ammonia on a continuous (not batch) basis. Eagerly the Badische Anilin und Soda Fabrik (BASF) took on the transformation of this laboratory prototype into industrial-scale production.[2]

Other work conducted by Haber at about the same time dealt with the water-gas equilibrium and with processes occurring in flames. As a homely example of the sort of system he examined, one concerned what takes place in the flame of the ubiquitous Bunsen burner. Here the familiar blue cone was unexpectedly shown to be the cooler part of the flame and to be made up of the gases inherent to the water-gas equilibria, in the proportions appropriate to the prevailing temperature. In the outer zone, the burning of hydrogen and carbon monoxide in the surrounding atmospheric oxygen provided most of the heat of the flame. Haber's numerous successes in the study of technically important gaseous processes by thermodynamic reasoning led to his appointment as professor and director of a new Institute of Physical Chemistry and Electrochemistry at Karlsruhe in 1906.

The year 1911 marked the establishment of the Kaiser Wilhelm Society for the Advancement of Science (which would later become

---

[2]Haber, 1971.

the Max Planck Society), comprising a number of research institutes, mainly in Berlin. These were to involve the cooperation of scientists, industry, and government in the prosecution of original research, with the obvious intention of sustaining for Germany the scientific and technological leadership achieved throughout the previous half-century. One of the institutes, proposed with Haber in mind as director, was for Physical Chemistry and Electrochemistry. Such an appointment could not fail to attract Haber's interests and enthusiasm, and so in 1911 he left Karlsruhe to immerse himself in the planning and specifying of the facilities for the new and prestigious foundation, the official opening of which took place in 1912. Once installed there, Haber had to finish odds and ends in the ammonia work and to resume studies begun in 1908 with G. Just, studies concerned with the emission of free electrons during certain chemical reactions. This kind of process focused on events taking place at the level of individual atoms and involved assumptions and explanations comparable to those introduced only a few years previously by Einstein to account for the photoelectric effect. Such work was later to earn worldwide acclaim for Haber's institute and to create the new subdiscipline of chemical physics.

But first came the Great War. When hostilities in Europe broke out on August 1, 1914, the German plan called for a quick attack to encircle Paris and to defeat France in a matter of weeks. By mid-September the plan had been thwarted, and the antagonists were reduced to the near stalemate of trench warfare. It soon became apparent that Germany was ill-prepared for a long struggle; the country lacked stocks of essential war materials and, being virtually surrounded in Europe, had no means of securing these. The ever-patriotic Haber had volunteered his own and his institute's services on behalf of the national struggle and was soon put in charge of mobilizing Germany's scientific and industrial resources.

One urgent problem was the production of nitric acid with which to manufacture explosives. The customary raw material had been sodium nitrate, of which the only significant source was Chile. Haber advocated and worked for much augmented production of synthetic ammonia by his own and other available methods, and he facilitated huge increases in the manufacture of nitric acid by the catalytic oxidation of ammonia. Most authorities agree that, but for this rapid expansion of the means for producing these chemicals synthetically, Germany's pursuit of the war would have foundered much earlier owing to a lack of explosives and essential agricultural chemicals.

Another part of Germany's war effort with which Haber's name has been inextricably linked, and for many people tarnished, was his involvement in the introduction of poison gas on the battlefield. The rationale for doing so was to try to break the deadlock of trench warfare. At first, lachrymators and irritant powders were dispersed from bursting shells, but then Haber proposed the use of chlorine gas liberated from cylinders at the battle line. This was first done in the second battle of Ypres, in April 1915. Inevitably, retaliation followed a year later when the French introduced phosgene for the same purpose. The gas respirator became an indispensable component of every soldier's kit, and the range of chemical agents introduced by both adversaries grew and grew. In 1916 Captain Haber was appointed director of the new Chemical Warfare Service of the Ministry of War, and his institute in Berlin was wholly involved in developing new and "better" poison gases.

The war dragged on until Germany finally capitulated late in 1918. It left in its wake a legacy of economic disasters, political upheavals, and unprecedented social changes. It also left Haber diminished in health and spirit, a condition from which he never fully recovered for the remaining 15 years of his life, yet in that period he continued to work hard for his country and for science. The terms of the Treaty of Versailles (1919) required Germany to make unrealistically large reparative payments to the countries violated. One consequence was that the German mark suffered almost uncontrollable devaluation in 1923, and because of that Haber's efforts to rebuild and broaden the scope of his research institute were delayed and limited. In due course, sufficient funding was provided to afford it a few years of glory before its key men were dispersed.

Linked to the problem of war reparations was a somewhat quixotic quest to which Haber devoted a great deal of time and effort throughout a period of almost eight years, the object of which was to extract a great quantity of gold believed to be dispersed throughout the oceans. He was convinced that this wealth could be extracted by chemical means and could be made available to his government to discharge its international obligations. He based his plans on a number of supposedly reputable analyses, but when the methods of recovery fell dramatically below expectations he was obliged to develop and perfect his own method of microanalysis. By this he was able to show that the true gold content of the oceans was about one-thousandth of the supposed value. The project proved to be a costly and unproductive adventure from which Haber emerged deeply chastened.

Readers with no more than a passing interest in the history of physical science will be aware of the dramatic changes brought about early in the 20th century by the recognition of the validity of quantum phenomena, the internal structure of atoms and molecules, the interpretation of atomic and molecular spectra, and so on. Little wonder then that the Kaiser Wilhelm Institute, into which Haber as director had brought promising new men after the war, became one of the important centers for further investigation of these new ideas. Among the prominent men conducting research there (during the periods indicated) were Herbert Freundlich (1919–1933), whose theoretical and applied work did much to rid colloid and surface chemistry of empiricism; Rudolf Ladenburg (1921–1931), a physicist of wide-ranging interests but principally remembered for his work on the effects of electronic excitation on the properties of gases; and Michael Polanyi (1922–1933) who made significant contributions, including the application of quantum theory, to the interpretation of chemical reaction rates. Haber himself, after the gold fiasco, carried out with K. F. Bonhoeffer a number of important investigations dealing with chemiluminescence in flames and its interpretation in terms of band spectra of molecules and radicals involved in or formed by the combustion process.

As well as the experimental work in progress in the Institute, a feature that never fails to be mentioned by all who experienced it was the fortnightly Colloquium. Accounts of work in progress inhouse or lectures from invited speakers, many of whom were chosen from outside the Institute or outside physical chemistry, were presented formally. The level of the papers read was of such a high caliber that the meetings drew a wide audience from all the scientific establishments in the city. Haber had a rare genius for summing up or clarifying the discussion, and the events were lively and inspiring.

But all this came to an abrupt end in 1933 when Haber, Freundlich, and Polanyi tendered their resignations to protest against the zealotry of the National Socialist Party, which had come into power in January of that year. In April laws were enacted forbidding the employment of Jews in government establishments (except in some circumstances). Although a Jew, Haber actually qualified for exemption from this enactment, but he and his colleagues acted on a point of principle. It is now commonly agreed that this Nazi policy robbed German science of its indisputable world leadership prior to the 1939–1945 war.[3]

---

[3] Mendelssohn, 1973; Nachmansohn, 1979.

Haber left Germany in the summer of 1933. In London he arranged to meet Chaim Weizmann, the great chemist, Zionist, and future president of the State of Israel, with whom he discussed future plans. Weizmann already then had an interest in establishing a research institute at Rehovot in Palestine (now in Israel), and he suggested that a senior position for Haber could be created there. However, before a firm decision on that proposal was made, Haber spent the autumn of 1933 in the laboratory of Sir William Pope (who had worked with him in Karlsruhe). In January he set off for Switzerland, en route to Palestine and a decision about his future. Regrettably, he died of a severe heart attack in Basel.

As remarked previously, Haber's life was tinged with stress and sorrow. He married twice, yet each marriage ended in divorce because his almost obsessive commitment to his work appeared to rule out normal domestic relationships. He continued to see each wife on a friendly basis subsequently. His first wife (a chemist) bitterly opposed his promotion of poison gas, and finally in the spring of 1915 took her own life in protest. Not surprisingly, in view of the effort that he had invested, he took very hard his country's defeat in the war and all the ensuing difficulties. His receipt of the Nobel Prize in 1920 was the occasion for numerous protests and snubs. French laureates who were to have attended this postponed ceremony boycotted the event owing to Haber's presence there. In America a swarm of editorials and letters challenged the suitability of the award to Haber; the point of many of these protests was not the gas warfare, but the extended duration of the war made possible by the manufacture of nitric acid from synthetic ammonia. Many people (Rutherford, for example) continued to censure Haber for years. His failure to harvest gold from the oceans was a massive disappointment to him, but the ultimate sorrow must have been his decision to surrender the directorship of the research institute in Berlin, which literally had been created for him in 1911. Time and the horrors of a subsequent, even more deadly world war have largely dispelled such contemporaneous attitudes, and we are able today to see Haber as a man of rare scientific genius and integrity.

**W. A. E. MCBRYDE**
*University of Waterloo*

# Bibliography

Bernal, J. D. *The Social Function of Science;* Macmillan: New York, 1939.

Bonhoeffer, K. F. In *Great Chemists;* Farber, E., Ed.; Interscience: New York, 1961; pp 1301–1311.

Coates, J. E. "The Haber Memorial Lecture"; *J. Chem. Soc.* **1939,** 1642–1672.

Farber, E. *Nobel Prize Winners in Chemistry, 1901–1961;* Abelard-Schuman: London, 1963.

Goran, M. *The Story of Fritz Haber;* University of Oklahoma Press: Norman, OK, 1967.

Goran, M. "Fritz Haber"; In *Dictionary of Scientific Biography;* Gillispie, C. C., Ed.; Charles Scribner's Sons: New York, 1972; Vol. 5, pp 620–623.

Günther, P. *Fritz Haber;* R. Oldenbourg: Munich, Germany, 1969.

Haber, L. F. *The Chemical Industry 1900–1930;* Clarendon: Oxford, England, 1971; pp 76–107.

Haber, L. F. *The Poisonous Cloud. Chemical Warfare in the First World War;* Clarendon: Oxford, England, 1986.

Mendelssohn, K. *The World of Walther Nernst: The Rise and Fall of German Science;* Macmillan: London, 1973.

Nachmansohn, D. *German–Jewish Pioneers in Science 1900–1933;* Springer-Verlag: New York, 1979; pp 166–195.

Nobel Foundation. *Nobel Lectures 1901–1921;* Elsevier: Amsterdam, Netherlands, 1966; pp 321–344.

*Nobel Prize Winners;* Wasson, T., Ed.; H. W. Wilson: New York, 1987.

Stern, F. D. "Fritz Haber: The Scientist in Power and in Exile"; In *Dreams and Delusions: The Drama of German History;* Alfred A. Knopf: New York, 1987; pp 51–76.

# Walther Hermann Nernst

## 1864–1941

Copyright Nobel Foundation

In 1920 the Nobel Prize in chemistry was awarded to Walther Hermann Nernst. The citation for the Nobel Prize was "in recognition of his work in thermochemistry." On the basis of modern usage, the term *thermochemistry* too narrowly describes his interests and activities, which are more appropriately denoted by *chemical thermodynamics*. Actually, in his education and subsequent career he could justifiably claim as much affinity with physics as with chemistry, and certainly many of his pupils earned great distinction as physicists. Nernst, though a few years younger, fell in with a group of scientists—van't Hoff, Arrhenius, and Ostwald were the most evident leaders—who sought to embody chemistry with a theoretical basis, and by so doing created the new subdiscipline of physical chemistry. It was in this frontierland, in which chemists and physicists had found common ground at last, that Nernst was to apply his skill and imagination to formulate new ideas and in turn to attract new talent to share in the advance of physical science.

Nernst was born on June 25, 1864, in Briesen, a small town in what was then West Prussia and that now lies within Poland. He died on November 18, 1941, at his country estate "Zibelle", near Muskau in lower Silesia (also now part of Poland). His ashes were returned to Germany in 1949 and are buried in Göttingen. In the town where

Nernst was born, and subsequently at Graudenz (now, too, in Poland), his father sat as a judge. In Graudenz young Nernst attended the *Gymnasium*, where in time he became Head Boy. The education he received there was based on classics, literature (to which he was greatly attracted throughout his life), and natural science. It was apparently during this period that his interest in chemistry was kindled, for we learn that his family provided him with laboratory space in the cellar at home, where he carried out experiments. At the age of 19, Nernst embarked on a university education in physics, following the pattern typical among continental students of that day whereby periods of residence were spent at several universities. Thus Nernst spent semesters at Zurich (twice), Berlin, Graz, and a lengthier stay at Würzburg, where he prepared and presented his inaugural dissertation. He had the good fortune to be taught by men such as Wilhelm Edward Weber, Hermann von Helmholtz, Ludwig Boltzmann, and Friedrich Wilhelm Geary Kohlrausch. While he was working with Kohlrausch at Würzburg his fellow investigators included Svante Arrhenius and Emil Fischer; Arrhenius in turn was able to introduce Nernst to the visiting Wilhelm Ostwald, with whom the former had recently been working at Riga. Before he completed his doctorate in 1887, Nernst already knew and was known by many of the men who were adding new dimensions to the study of chemistry.

In that same year, 1887, Ostwald was appointed professor of physical chemistry at the University of Leipzig. Here for the next 18 years he was to create a center of great importance for the development of his subject, to which research students came from around the world. In that same year, with van't Hoff, he founded the *Zeitschrift für physikalische Chemie*, one of the earliest journals for a chemical specialty and a successful vehicle for promoting the ideas of this new branch of the subject. Also in 1887 Ostwald completed his massive *Lehrbuch der allgemeine Chemie,* begun two years earlier; in it he emphasized the importance of general principles, rather than the attributes of individual substances, by which chemistry could be interpreted and unified. Ostwald lost no time in recruiting Nernst and Arrhenius as assistants in his new department at Leipzig, where each remained for the next four years, apart from a semester spent by Nernst at Heidelberg in the summer of 1889.

During his Leipzig years, amid the almost daily stimulus provided by his associates, Nernst gained in stature and recognition in consequence of his development of new ideas in the growing domain of physical chemistry. These included the important relationship between the electromotive forces of chemical cells and the concentration of the electrolyte substance(s) in these cells. Every undergraduate student in chemistry nowadays is familiar with the so-called Nernst equation:

$$E = \text{constant} + (RT/nF) \ln C$$

governing the potential of an electrode in a chemical cell ($E$ = electrode potential; $R$ = gas constant; $F$ = Faraday constant; $n$ = number of charge units changed; $C$ = concentration of ion participating in the electrode process). Nernst arrived at such a relationship by linking two original ideas: first, that changes in concentration of the electrolyte during the cell processes were paralleled by changes in osmotic pressure; and, second, that the electrode substance itself possessed a characteristic "solution pressure" (analogous to vapor pressure, as a measure of its "escaping tendency" into a separate phase). In the form of the Nernst equation given above, the latter quantity is incorporated in the (temperature-dependent) constant preceding the logarithmic term. Although a general thermodynamic theory of concentration cells had been developed 10 years earlier by Helmholtz, Nernst's publication of the foregoing ideas in 1888–1889, backed by a comprehensive compilation of evidence, quickly brought him into considerable prominence.

Even before tackling the work just described, Nernst had studied factors governing the rate of the diffusion of electrolytes in solution, and in 1888 he derived an equation relating the diffusion coefficient, ionic charge, and mobility, applicable to solutions of single electrolyte substances. In the following year he described what he called the reciprocal effect in the solubility of salts, today known as the common-ion effect, and he demonstrated the validity of what is now referred to as the solubility-product principle. By this he was able to provide a unified explanation of a large number of already recorded experimental observations.

In 1872 Claude Louis Berthollet and Émile Clement Jungfleisch had investigated the way in which dissolved substances became distributed between two immiscible solvent phases in contact, when a steady state had been obtained. They concluded that for most substances the distribution was a fixed ratio, which was also the ratio of their solubilities in the single respective solvents. However, some departures from this relationship were known to exist; one such was the partition of benzoic acid between water and benzene. In 1890–1891,

Nernst investigated this system further and was able to show that the anomalous behavior in this case was due to side reactions affecting the molecular state of the benzoic acid in each phase. In water the acid underwent partial dissociation into ions, and in benzene it formed dimers. By making allowance for these competing processes, especially the dimerization, he was able to show that the concentrations of the benzoic acid monomer in the two phases were in accord with a suitable mass-law relationship.

Accomplishments such as the foregoing earned for young Nernst recognition as one of the up-and-coming scientists and created new opportunities for his career. In 1891 he accepted an appointment at the University of Göttingen as assistant professor of physics. He had also in the same year been offered a chair at the University of Giessen, the institution where chemistry had flourished so brilliantly a generation earlier under Justus Liebig.

Soon after his arrival in Göttingen, Nernst undertook the writing of a textbook entitled *Theoretische Chemie*, published in 1893. This book ran through 15 editions and for at least 35 years was reputed to be the most effective presentation of physical chemistry in print. As a point of interest concerning that book it should be mentioned that only five years prior to its publication, Ostwald, the usually acknowledged father of physical chemistry, had published his *Outlines of Physical Chemistry,* incorporating the new ideas developed by himself, van't Hoff, and Arrhenius in what today would be recognized as classical physical chemistry. There was, however, a marked contrast between the writings of Ostwald and of Nernst concerning this new area of chemical theory. Ostwald was at that time, and indeed until about 1912, quite skeptical about the reality of atoms and molecules, and so he developed a theoretical approach to chemistry that was devoid of the use of intellectual models. Nernst, on the other hand, acquired most of his scientific education in schools of physics, and in particular had been exposed to the ideas of Ludwig Boltzmann and of his pupil Albert von Ettingshausen, at Graz. Accordingly, from the outset Nernst sought to interpret the phenomena of physical chemistry convinced of the reality and usefulness of a particulate theory. The full title of Nernst's textbook included the phrase *vom Standpunkte der Avogadroschen Regel und der Thermodynamik* ("from the standpoint of Avogadro's law and of thermodynamics"). In acknowledging the importance of Avogadro's molecular interpretation of physical phenomena on an equal footing with that of energy, Nernst went further and was perhaps to have a greater impact on the direction in which physical chemistry grew, than had Ostwald.

In the same year in which his textbook appeared Nernst, and independently J. J. Thomson, drew attention to the importance of the dielectric constant of solvents as a factor governing the extent of dissociation of electrolytes in solution. This qualitatively useful idea was later refined and incorporated into the more exact theories of electrolyte solutions developed in the 1920s by Peter Debye and Erich Hückel, and extended by Lars Onsager.

In 1894 Ludwig Boltzmann left the chair of theoretical physics at the University of Munich in Bavaria, and Nernst was invited to succeed him. However, because the Ministry of Education in Prussia was desirous of retaining his services, Nernst found himself in a position to maneuver to his own advantage, with the result that he was made professor of physical chemistry at Göttingen, and director of the newly created Institute for Physical Chemistry and Electrochemistry at that university. There, for the next 11 years, he managed to develop a large and impressive research school to which students and scholars from many countries came to work.

The titles of papers published during that period and those papers that bore his name as author or coauthor reveal the extent and diversity of the work accomplished during his years at Göttingen. A good deal of experimental work was directed toward better understanding electrode processes and other electrochemical phenomena, including polarization, liquid junction potentials, overvoltage, operation of lead storage cells, conductivity through solid bodies at high temperatures, and so forth. About the turn of the century a gradual shift of interest becomes apparent. Nernst and his students were paying increasing attention to measurements and observations at high temperatures, especially in gaseous systems. Thus there was much work recorded on the determinations of molecular weights, leading to inferences about dissociation or association reactions or leading to the effect of temperature on chemical equilibria. In retrospect it is easy to recognize that Nernst's attention and interest were beginning to focus on what would turn out to be his most celebrated contribution to chemical theory: a method of evaluating chemical equilibrium constants solely on the basis of thermal data.

Before enlarging on the previous statement, however, I shall record that in 1905 Nernst was offered and accepted the professorship of physical chemistry at the University of Berlin. In this appointment he succeeded Hans Landolt, a respected but very conservative chemist, on whose work the new physical chemistry had made little impact. It is said that Max Planck, who held the chair of theoretical physics at

Berlin, had strongly supported Nernst's appointment in the hope of stimulating some action in this transdisciplinary field. As things were to work out, the names of these two men became permanently linked to the thermodynamic principle known as the third law.

Thermodynamics is a body of theory developed during the 19th century to relate transfers of heat and mechanical energy. By mid-century its principles had become summarized in two laws. The first established that heat was a form of energy and that in all processes energy is conserved. The second law identified certain limitations in the transfer of heat to work, it invested the absolute temperature scale with new theoretical significance, and it introduced a third factor, entropy, to be involved in physical transformation. Entropy can be conceived as an innate tendency toward molecular disorder or randomness. Processes will occur spontaneously even when neither work is done nor temperature altered, as when gas expands into a vacuum; the driving force in that case is the gain in entropy of the system.

These ideas did not have much effect on chemical thinking for some time. However, claims were made by Julius Thomsen in 1854, and more comprehensively by Marcellin Berthelot in 1873, that the heat evolved during chemical reactions is a measure of the chemical affinity between reacting substances. Such claims could not stand up to the test of further evidence, however, and it was left to physicists, notably Josiah Willard Gibbs between 1874 and 1878 and Helmholtz in 1882, to extend thermodynamic principles to chemical systems. To do so, they invented new thermodynamic quantities, usually called free energies, which can more usefully be applied to data from chemical investigations. The free energy of a system consists of the total energy $E$ (Helmholtz), or of the heat content $H$ (Gibbs), *diminished* by the entropy $S$, multiplied by the absolute temperature. For any process undergone by that system the last term, $T$ multiplied by $\Delta S$, represents *unavailable* energy equivalent to the system's entropy. The driving force in a chemical reaction, then, is the equalization of the free energy content of the reactants and products, and this is achieved at equilibrium. The equilibrium constant, $K$, can thus be defined in terms of free energies which, in turn, are the parameters that determine chemical affinity. Furthermore, van't Hoff had shown in 1884 that the temperature coefficient of an equilibrium constant was related to the heat gained in the reaction in accordance with an equation that he termed an isochore:

$$d \ln K / dT = \Delta H / RT^2$$

This defines the temperature coefficient of $\Delta G$ (measured at constant pressure) in terms of other thermodynamic quantities. Although this relationship permitted the comparison of an equilibrium constant at one temperature with that at another, the lack of any suitable integration constant made it impossible to fix the absolute value of an equilibrium constant at a particular temperature. The isochore itself had been derived from another relationship known as the Gibbs–Helmholtz equation:

$$(\partial \Delta G / \partial T)_p = -\Delta S = \frac{\Delta G - \Delta H}{T}$$

At about the turn of the century, evidence from various quarters began to appear showing that at very low temperatures this temperature coefficient, and a corresponding quantity for $\Delta H$, becomes small and tends to converge toward a common value of zero, which value, it might be reasonable to assume, they both attain at a temperature of absolute zero. Nernst saw in this evidence the germ of a new principle, which he called his *heat theorem*. Essentially he proposed that the values of $\Delta G$ and $\Delta H$ must become the same and that hence the value of $\Delta S$ must become zero, when the temperature is lowered to absolute zero. By reasoning that is given in detail elsewhere,[1] these assumptions led to an expression of the Gibbs–Helmholtz equation in integrated form in which $\Delta G$ appears as a function that can be evaluated from thermochemical data. The troublesome integration constant that had plagued earlier efforts to derive $\Delta G$ values from the Gibbs–Helmholtz equation turned out to be zero in the light of the assumed equality of $\Delta G$ and $\Delta H$ at absolute zero.

On the basis of what many people have regarded as meager experimental evidence Nernst announced his heat theorem at the end of 1905. He then spent the next decade marshaling his forces at the University of Berlin to secure better supporting evidence. This entailed a great body of measurements of thermochemical properties at very low temperatures. Nernst's ideas embodied in his heat theorem initially rested solely on empirical evidence, but they received what for him must have been very agreeable confirmation from the work by Einstein that applied quantum theory to the calculation of entropies for gases or of specific heats for solids. In 1907 Einstein was prompted by

---

[1]For example, Hiebert, 1978, p 437.

such considerations to predict that the heat capacities of solids should vanish at absolute zero. Subsequently in 1910, Planck offered a restatement of Nernst's heat theorem, to the effect that all entropies become zero at absolute zero. Planck himself was not comfortable with this statement and two years later proposed yet another, to the effect that the attainment of absolute zero was not possible. By this time the heat theorem, however stated, had become endowed with the title *third law of thermodynamics*.

Before concluding this account of Nernst's various investigations it is appropriate to mention a major contribution to photochemistry. It was known that in the reaction between hydrogen and chlorine, induced by light, the number of molecules of hydrogen chloride formed vastly exceeded the number of inducing photons. Nernst proposed what today is known as a chain mechanism whereby secondary reactions were capable of sustaining a chain of supply of fresh atoms without the requirement of photochemical equivalence for the amounts of products. Thus a photochemically treated chlorine atom could start a chain of secondary reactions: $Cl + H_2 = HCl + H$, $H + Cl_2 = HCl + Cl$, and so forth.

The foregoing is an abridged summary of Nernst's principal scientific achievements. He was, beyond doubt, an ingenious, creative scientist, with a flair for imaginative investigations and explanations. Many of the ideas credited to him in this account form part of the essential framework of modern physical chemistry. The award of a Nobel Prize was only one of many honors that came his way, and such recognition is clear evidence of the respect in which his work was held. It is less evident that this respect was matched by personal affection for the man himself. Former students and associates, in writing about Nernst,[2] are sympathetic and tend to emphasize his good qualities, personal as well as intellectual. However, any disinterested inquirer cannot fail to discover evidence of Nernst's arrogance and insensitivity toward the viewpoint of others. He was notoriously jealous concerning his priority, if not exclusivity, with regard to the discovery or enunciation of new scientific ideas, and he was repeatedly unwilling to give due credit to the work of others when it related to his own. For this reason the reader may wish to consult an alternative account of the heat theorem given by a prominent contributor to chemical thermodynamics.[3] Although no one would dispute the correctness of the

---

[2]Lindemann, 1945; Mendelssohn, 1973.
[3]Lewis, 1923, pp 438–454.

award of a Nobel Prize to Nernst, one is left to wonder if it might have come earlier in his career and been more applauded had he been a person of more gracious disposition.

W. A. E. MCBRYDE
*University of Waterloo*

# *Bibliography*

Cooper, W. H. "Walther Nernst and the Last Law"; *J. Chem. Ed.* **1987,** *64,* 3–8.

Findlay, A. *A Hundred Years of Chemistry;* Methuen: London, 1965; pp 71–112.

Hiebert, E. N. "Nernst, Hermann Walther"; In *Dictionary of Scientific Biography;* Gillispie, C. C., Ed.; Charles Scribner's Sons: New York, 1978; Vol. 15, pp 432–453.

Hiebert, E. N. "Walther Nernst and the Application of Physics to Chemistry"; In *Springs of Scientific Creativity: Essays on Founders of Modern Science*; Aris, R. T.; Davis, H. T.; Stuewer, R. H., Eds.; University of Minnesota Press: Minneapolis, 1983; pp 203–231.

Johnson, J. A. *The Kaiser's Chemists: Science and Modernization in Imperial Germany*; University of North Carolina Press: Chapel Hill, 1990.

Lewis, G. N.; Randall, M. *Thermodynamics and the Free Energy of Chemical Substances;* McGraw-Hill: New York, 1923.

Lindemann, F. A. (Viscount Cherwell); Simon, F. "Walther Nernst"; *Obituary Notices of Fellows of the Royal Society of London,* **1945,** *4,* 101–112.

Mendelssohn, K. *The World of Walther Nernst: The Rise and Fall of German Science;* Macmillan: London, 1973.

Nernst, W. *Experimental and Theoretical Applications of Thermodynamics to Chemistry*; Yale University Press: New Haven, CT, 1907.

Nernst, W. "Studies in Chemical Thermodynamics;" *Nobel Lectures: Chemistry, 1901–1921;* Elsevier: Amsterdam, the Netherlands, 1966; pp 349–364.

Nernst, W. *Theoretische Chemie vom Standpunkte der Avogadroschen Regel und der Thermodynamik*; F. Enke: Stuttgart, Germany, 1893 (*Theoretical Chemistry from the Standpoint of Avogadro's Rule and Thermodynamics*; Revised in accordance with the fourth German edition by R. A. Lehfeldt; Macmillan: London, New York, 1905, 1908).

# Frederick Soddy

## 1877–1956

Copyright Nobel Foundation

The term isotope may not mean anything to many people, but to Frederick Soddy it meant the Nobel Prize in chemistry. The Nobel citation of 1921 states that Soddy received this award for "his contributions to our knowledge of the chemistry of radioactive substances, and his investigations into the origin and nature of isotopes".[1] Soddy was born at Eastbourne, England, on September 2, 1877, the youngest of seven sons of a prosperous corn merchant. Soddy's mother died before he was two years old. The Soddy family was Calvinist, and Frederick was indoctrinated in the tenets of truthfulness, sobriety, industry, and responsibility by his dominant half-sister. His rather turbulent childhood resulted in an introverted and sometimes difficult character. Soddy wrote that "this environment possibly produced in me an exaggerated deference to others more happily brought up in childhood, and to my superiors in social accomplishments".[2]

From 1892–1894, Soddy attended Eastbourne College where R. E. Hughes encouraged him to study chemistry. He studied for a year at University College of Wales at Aberystwyth, and won an open science scholarship to Merton College, Oxford, in 1895. In 1898, Soddy received a first-class honors degree, with William Ramsay serving as external examiner. He remained at Oxford for two years doing inde-

---

[1]Frederick Soddy(1877–1956): *Early Pioneer in Radiochemistry*, p xi.
[2]Keller, 1983, p 98.

pendent chemical research in synthetic organic chemistry. Soddy later said that this was wasted time.

In 1900, he applied for a professorship in chemistry at the University of Toronto. Soddy was only twenty-three and had no teaching or administrative experience and very little research training. He was turned down. Soddy made a personal visit to Toronto, and then to Montreal, where he accepted a position as demonstrator (laboratory instructor) in chemistry at McGill University. There he met Ernest Rutherford, who was a professor of physics. Rutherford had been investigating the radioactive properties of uranium, thorium, and radium. He had noticed that thorium gave off a small amount of a very radioactive gas that could not be identified as any element then known. Rutherford named this thorium gas "emanation". He needed a chemist to help him determine the identity of this emanation. Soddy was an appropriate choice.

The first contribution resulting from their collaboration was the Disintegration Theory, proposed in 1902. They explained that radioactivity was really the result of the spontaneous disintegration of radioactive elements into new elements. By expelling a small amount of mass or electricity from its center, an atom of an element changes into an atom of a different element. Soddy used the term transmutation, an old alchemy term, in describing their theory.[3] They noted that only a portion of the atoms disintegrate at any one time, and that the portion is definite for each element and is characteristic of that element.

Soddy and Rutherford also proposed that there were two radioactive decay series, one beginning with uranium and ending with lead, and the other beginning with thorium and also ending with lead. They also predicted that the decay product of radium should be helium. They published their model of radioactive change in May 1903. Soddy and Rutherford were among the first to calculate the large amount of energy associated with radioactive change.[4] This suggested to Soddy that if the way could be found to rapidly release this energy by artificially accelerating radioactive change, it could be used for practical purposes.

In the spring of 1903, Soddy left McGill to work with William Ramsay at University College, London. One day, Soddy noticed a sign in a shop window saying that pure radium compounds were on sale. Soddy quickly bought 20 milligrams and rushed back to inform

---

[3]Nagel, 1982, p 739.
[4]Freedman, 1979, p 259.

Ramsay. After testing the radium for purity, Soddy and Ramsay saw spectroscopically that helium was present. To make sure that helium had not been attached to radium in some other way, they isolated the emanation, condensed it, and allowed it to vaporize again. They found no spectroscopic lines for helium in the gas. After four days, the spectroscopic lines for helium were found again, proving that the radium emanation was spontaneously producing helium. Soddy concluded that the helium gas being produced was coming from α-particles. Rutherford later proved that this conclusion was correct.

Soddy spent the next few months traveling and giving lectures. He developed a "picturesque and attractive lecture style, full of vivid and exciting images guaranteed to provoke the imagination".[5]

In 1904, Soddy was appointed lecturer in physical chemistry and radiochemistry at the University of Glasgow. Before leaving for Glasgow, he had proposed writing a book based on his lectures about radioactivity. This upset Rutherford and his friends because Rutherford was also planning to publish a book on radioactivity. They felt that Soddy's book might steal some of Rutherford's thunder, especially if it were written in a more popular style. Soddy deferred publication of his book until Rutherford's had been published. The books were actually quite different; Rutherford's book summarized the work that had been done in radioactivity over a five-year period, while Soddy's book, *Radioactivity* (1904), was an elementary treatise.

At Glasgow, Soddy spoke and wrote widely about practical implications of radioactivity research. He envisioned atomic energy as a basis for a new civilization and the solution to the depletion of coal reserves inevitable in an industrial society. One of Soddy's books, *The Interpretation of Radium* (1908), was the inspiration for the science fiction novel *The World Set Free,* by H. G. Wells. Soddy realized that it was important that a reliable source of radium for research be available.

In 1909, Soddy learned that the newly discovered radioelement mesothorium, with properties similar to radium, could be extracted from thorium ore. However, the method to produce mesothorium was being kept secret by its discoverer, Otto Hahn. By 1910, Soddy succeeded in finding the method and had it patented. The patent was approved in 1911, and the German firm Knoefler and Company had to pay Soddy for the patent.[6]

---

[5]Keller, 1983, p 102.
[6]Freedman, 1979, p 258.

Soddy's father-in-law, George Beilby, general manager of the Cassel Cyanide Company of Glasgow, involved Soddy in some of the scientific problems of the company. In return, Beilby's company assisted Soddy in his work on the radium substitute.

During his investigations with mesothorium, Soddy found that he was also getting radium chloride and the chloride of what was called thorium X from a sample of uranium. He concluded that mesothorium, radium, and thorium X were really the same element. In 1912, Soddy asked Alexander Fleck to work with him to determine the chemical properties of radioactive intermediates. They discovered many chemically identical but short-lived radioactive elements. Some were chemically inseparable and spectroscopically identical, but disintegrated in different ways.

In 1913, Soddy announced the General Displacement Law. When an α-particle is expelled, the element moves two places in the direction of lower charge or atomic number on the periodic table. The element moves forward one place on the periodic table with the emission of a β-particle. Soddy was the first to suggest the change caused by the emission of the α-particle. With respect to the change caused by the emission of the β-particle, Soddy (with Fleck), Alexander Russell, Kasimir Fajans, and George de Hevesy can claim a share of the credit.[7]

When an element loses an α-particle and then loses two β-particles, it returns to the same place on the periodic table as it originally occupied. It is the same element with all of the same chemical properties, except that it has a different mass. At a dinner party at the Beilby home, a discussion of Soddy's ideas on the displacement of elements led a friend, Dr. Margaret Todd, to suggest the word *isotope* (from the Greek, meaning same place) to describe chemically identical atoms of different weights.

By 1914, Soddy had shown that lead was the final stable element that resulted from radioactive substances. Lead found in the ores of uranium and thorium did not have the same atomic weight as lead found in ores that were not radioactive. The different samples of lead were the same chemically, but had different atomic masses.

Soddy was named to the Chair of Chemistry at Aberdeen University, Scotland, in 1914, but World War I interrupted his teaching and all but stopped his research. While part of his time was spent on war work, he continued some research that he had started in Glasgow,

---

[7]Aston, 1923, p 11.

which led to the independent discovery of protoactinium. He was very upset by the waste and destruction of war, and especially by the death of Henry G. J. Moseley in combat. The topics of his interest began to shift to social issues and economics.

In 1919, Soddy was called to the Dr. Lee's Chair of Chemistry at Oxford. Some sources suggest that the selection committee could not agree on Harold Hartley or Neville Sidgwick, both of whom were already at Oxford, so they chose Soddy, who had not even applied for the position.[8] For almost all of his tenure at Oxford, Soddy was not even a member of the Board of the Faculty because of his refusal to pay the fee for the required Oxford M.A. The facilities assigned to Soddy at Oxford were old, not well kept, and not equipped to conduct the research that was expected of him. He was also assigned a very heavy lecture schedule of instruction in both physical and inorganic chemistry. Soddy devoted his energies to the improvement of chemistry teaching and to the modernization of the laboratories.

In 1921, Soddy was awarded the Nobel Prize in chemistry for his contributions to knowledge of the chemistry of radioactive materials and for his studies into the origin and nature of isotopes. His other honors include Fellow of the Chemical Society (1899), Fellow of the Royal Society (1910), Cannizzaro Prize (1913), and the Albert Medal (1951).

Soddy did little work in chemistry after his appointment to Oxford, and turned more and more to writing on finance, economics, business, and sociology. In 1923, he developed a process for the extraction of thorium, which he patented in 1940. It seems that he was becoming more disillusioned with science and his place in it.

Soddy's wife died in 1936. He never fully recovered from this loss, and he resigned his position. After his retirement, he isolated himself from his colleagues. Some of the reasons for this withdrawal may be suggested in his very complex personality. Soddy was never satisfied with compromise. He was not very patient and even suspected ill-will at slow progress. He was such a good writer that it was easy for him to give his essays the sting he wanted. He was bitter over the way in which he and other radiochemists seemed to be overshadowed by the later work of physicists, especially at the Cavendish Laboratory. He was very outspoken about affairs concerning war, and spoke often in favor of greater control of atomic weapons. In 1955, he joined seventeen other Nobel Prize winners to warn about the dangers of nuclear weapons.

---

[8] *Frederick Soddy (1877–1956): Early Pioneer in Radio Chemistry*, p 159.

On September 21, 1956, at the age of 79, Frederick Soddy died at Brighton. As Fritz Paneth noted, "He was gifted in many ways, perhaps too many ways."

NORMAN W. HUNTER AND RACHEL ROACH
*Western Kentucky University*

# Bibliography

Aston, F. *Isotopes*; Edward Arnold: London, 1923.

Fleck, A. "Prof. Frederick Soddy, F. R. S."; *Nature* **1956**, *178*, 893.

*Frederick Soddy (1877–1956): Early Pioneer in Radiochemistry*; Kauffman, G. B., Ed.; D. Reidel Publishing: Dordrecht, the Netherlands; Boston, MA, 1986.

Freedman, M. "Frederick Soddy and the Practical Significance of Radioactive Matter"; *The British Journal for the History of Science* **1979,** *12(42)*, 256–260.

Howorth, M. *The Life Story of Frederick Soddy;* New World: London, 1958.

Keller, A. *The Infancy of Atomic Physics: Hercules in His Cradle;* Clarendon Press: Oxford, England, 1983.

Nagel, M. "Frederick Soddy: From Alchemy to Isotopes"; *J. Chem. Ed.* **1982**, *59,* pp 739–740.

Paneth, F. "A Tribute to Frederick Soddy"; *Nature* **1957,** *180,* 1085–1087.

Romer, A. *The Restless Atom;* Anchor Books: Garden City, NY, 1960.

Russell, A. "Soddy, Interpreter of Atomic Structure"; *Science* **1956,** *124,* 1069–1070.

Soddy, F. *The Chemistry of Radioelements;* Longmans, Green: London, New York, 1911–1914.

Soddy, F. *The Interpretation of the Atom*; Putnam: New York, 1932.

Soddy, F. *The Interpretation of Radium*, 2nd ed.; John Murray: London, 1909.

Soddy, F. *Matter and Energy;* Holt and Company: New York, 1912.

Soddy, F. *Radioactivity; An Elementary Treatise from the Standpoint of the Disintegration Theory;* D. Van Nostrand: New York; "The Electrician": London, 1904.

Soddy, F. *Radioactivity and Atomic Theory: Presenting Facsimile Reproductions of the Annual Progress Reports on Radioactivity 1904–1920 to the Chemical Society;* Trenn, T. J., Ed.; Taylor and Francis: London, 1975.

Soddy, F. *Science and Life;* J. Murray: London, 1920.

Soddy, F. *The Story of Atomic Energy;* Nova Atlantis: London, 1949.

Trenn, T. J. *The Self-Splitting Atom: The History of the Rutherford-Soddy Collaboration;* Taylor & Francis: London, 1977.

Wise, L. *Frederick Soddy;* Holborn: London, 1946.

# Francis William Aston

## 1877–1945

Francis William Aston was awarded the Nobel Prize in chemistry in 1922 for his contributions to analytical chemistry and the study of atomic structure. He is primarily associated with the design and use of the mass spectrograph.

Aston was born on September 1, 1877, at Harborne, near Birmingham, England, and died on November 20, 1945, at Cambridge. His father, William, was a merchant and farmer. His mother, Fanny Charlotte Hollis, was the daughter of a gunmaker. Aston's early education was at the Harborne Vicarage School and at Malvern College, where he graduated at the top of his class in science and mathematics.

In 1893 he went to Mason College (later to become Birmingham University), where he studied under chemists W. A. Tilden and P. F. Frankland and physicist J. H. Poynting. While there Aston became a skilled glassblower, an important aid to him later in his career. In 1898 Aston received a Forster scholarship to work with Frankland on the preparation and optical rotation properties of dipyromucyltartaric acid esters.

For financial reasons Aston left Mason College in 1900 to work at a brewery in Wolverhampton. (During the period 1898–1900 he had taken a course in fermentation chemistry.) He continued his scientific investigations, however, in a home laboratory, where he worked on gas discharge phenomena. Aston built several pieces of apparatus in his laboratory, including vacuum pumps of his own design and small discharge tubes from test tubes. He discovered a type of rectifying valve (vacuum tube) during his research.[1]

In 1903 Aston received a scholarship and returned to Birmingham University to work with Poynting; he continued his study of gaseous discharge at low gas pressure. He was particularly interested in the variation of the length of the dark space with current and pressure changes. This dark space occurs between the cathode and the negative glow and is named *Crookes's dark space,* after W. C. Crookes. In 1907 Aston detected a new dark space directly adjoining the cathode, which is now known as the "Aston dark space".

In 1909 Aston left Birmingham to accept the invitation of J. J. Thomson to work at the Cavendish Laboratory in Cambridge. He was the personal assistant to Thomson at first and then became a Clerk Maxwell Scholar. Thomson was investigating the positive rays from a special discharge tube that he had developed. This discharge tube consisted of an evacuated glass bulb with an anode at one end and a perforated cathode at the other end. As positive rays came through the perforations in the cathode, they were deflected both horizontally and vertically by perpendicularly arranged electric and magnetic fields. The particles formed parabolas of constant $e/m$ (charge/mass) ratio on a fluorescent screen. The exact position on a parabola was determined by the velocity of the particle. Whereas Wilhelm Wien in 1902 had obtained feathery parabolas in his $e/m$ studies of positive rays, Thomson obtained sharp parabolic lines. Thomson suggested that Wien's experiments gave feathery parabolas owing to collisions between the positive rays and gas molecules in the tube.[2] Thomson's tube was under very low pressure, thus decreasing the number of collisions and resulting in sharp lines. With hydrogen Thomson obtained two parabolas, at positions corresponding to $H^+$ and $H_2^+$. Oxygen gave three lines, corresponding to $O^+$, $O^{++}$, and $O_2^+$. This was clear evidence that atoms and molecules were of definite weights for a particular substance.

Aston improved Thomson's apparatus by using a spherical discharge tube, a more efficient vacuum pump, a finely engineered slit, a coil to detect vacuum leaks, and a camera to photograph the parabolas. When using neon gas in the discharge tube, two parabolas were obtained, corresponding to weights of 20 and 22 units. When William

---

[1] Thomson, 1946, p 290.
[2] Ihde, 1964, p 495.

Ramsay and Morris Travers, the discoverers of neon, had determined its weight in 1898, they obtained a value of 20.2 units. Crookes had suggested that the nonintegral atomic weight might be due to the gas being a mixture of two very similar but different atoms. Both of the gases would be expected to have a whole number atomic mass; the mixture would be an average of the two numbers.

Frederick Soddy had only recently introduced the idea of isotopes. Aston believed that different isotopes of neon were present in the tube, but there had never before been any evidence for isotopes among nonradioactive substances. Aston attempted to separate neon from "metaneon" by fractional distillation, fractional diffusion, and differential adsorption. During these investigations he invented a quartz microbalance that was sensitive to $10^{-9}$ g. Partial separation was announced in 1913, but since the results were just outside experimental error, Thomson was not yet convinced of the existence of two isotopes of neon.

Aston's research was then interrupted by World War I. During the war he worked at the Royal Aircraft Factory, at Farnborough, as a chemist. He studied the properties of lacquered canvas and its preservation by using pigments. Aston could not forget about the neon problem and took the opportunity to discuss scientific matters with others who worked at Farnborough. Aston and F. A. Lindemann, who was rather skeptical about the existence of natural isotopes, talked about physics and the new quantum theory and its relationship to isotopes. They published two joint papers on the subject shortly after the war. In 1914 Aston escaped injury in the crash of an experimental aircraft. It is interesting to speculate the course of chemistry if Aston had not survived.

After the war, Aston again tried to separate the isotopes of neon by diffusion. Two successive pieces of apparatus failed to achieve the desired results, but this did not discourage Aston. In 1919 he devised the apparatus with which his name is so closely associated: the mass spectrograph.

As in the previous instruments, a beam of positive particles was produced by passing a high-voltage discharge between two electrodes in a glass tube that contained gas particles at a very low pressure. The accelerated positive ions were allowed to pass through the perforations in the cathode and then between two electrically charged plates. The ions were deflected according to their $e/m$ ratio, with lighter ions being deflected more than heavier ones. Next the stream of ions was passed through a magnetic field that was arranged at right angles to the electric field. It deflected the stream in the opposite direction, but in the same plane. Ions of the same $e/m$ could then be focused at the same spot regardless of velocity. When focused on a photographic plate, the stream of ions would produce a series of lines that formed a mass spectrum. The position of the line depended on the mass of the ions, and the intensity of the line depended on the relative abundance of the ions. The resolving power of the mass spectrograph was 1 in 100, and it had an accuracy of 1 part in 1000.[3]

When neon was placed in the apparatus, two lines appeared corresponding to masses of 20 and 22, but no line appeared at a mass of 20.2, the weight of neon reported by Ramsay and Travers. The line corresponding to a mass of 20 was about 10 times darker than the line corresponding to a mass of 22, giving a value of 20.2 as a weighted average. This was the first conclusive proof that isotopes were not restricted to radioactive elements but could be found in stable elements of low atomic mass as well.

Aston studied many other elements and obtained evidence for three isotopes of sulfur (32, 33, 34), two isotopes of chlorine (35, 37), and three isotopes of silicon (28, 29, 30). With the original mass spectrograph, which can be seen at the Science Museum in London, Aston analyzed more than 50 elements and found that all except hydrogen had integral isotopic masses. Hydrogen's mass appeared to be slightly greater than one. Using this instrument he was able to discover 212 of the 287 stable isotopes. In December of 1919 he announced the "whole-number rule": atoms have masses that are integral on a scale that assigns a mass value of 16 to oxygen. This was similar to William Prout's hypothesis of 1816 that all elements are built up from the hydrogen atom and that their atomic masses are integrals of hydrogen.

Aston's research resulted in his receiving many awards. He was named a Fellow of Trinity College in 1920 and elected a Fellow of the Royal Society in 1921. He received the Nobel Prize and the Hughes Medal of the Royal Society within two days in 1922.

In 1927 Aston introduced a new model of the mass spectrograph that was capable of handling solids and had an accuracy of 1 in 10,000.[4] This instrument showed that the masses of the elements are very close but are not exact whole numbers. In general, the exact masses of the elements through fluorine are a little larger than the nearest whole number, whereas neon through osmium have exact masses a little smaller

[3]Thomson, 1961, p 87.
[4]Thomson, 1946, p 291.

than the nearest whole number. From iridium through the rest of the elements, the exact masses rise steadily above the nearest whole number.

Aston explained that the missing mass is really not missing but is the Einstein mass-energy equivalence of the energy that is binding the nucleus together. For example, the particles that make up one atom of helium are the same as the particles that make up four atoms of hydrogen, but in helium these particles are "packed" and have less mass. Aston introduced the term "packing fraction", which he calculated by dividing the difference between the nearest whole number and the exact mass number by the nearest whole number (using oxygen equal to 16 units as the standard) and then multiplying by 10,000. The packing fraction was a measure of the stability of the atom and the energy that would be required to break up or change the nucleus. By plotting packing fractions against mass numbers, Aston obtained a simple curve that gave valuable information on nuclear abundance and stability. This mass defect is now known to be the source of energy released during the fusion of hydrogen to form helium.

It should be noted here that this work suggested that nuclear reactions could become an important source of energy. Aston believed that this would eventually take place, and in his Nobel speech he warned of the dangers that were possible.

A third mass spectrograph was built by Aston and introduced in 1935. It was to have a resolving power of 1 in 2000 and an accuracy of 1 in 100,000.[5] This instrument was very difficult to adjust and never achieved the resolving power desired. World War II began before any important work was carried out. By that time other scientists were able to build instruments that surpassed Aston's.

The mass spectrograph has been developed into an instrument that is widely used today for scientific research as well as for industrial applications. It has been important in the study of nuclear physics and chemistry and has become an important tool in structural determinations and analysis of organic compounds.

Outside the laboratory, Aston's interests were mainly sports, travel, and music. A bachelor, Aston was an avid traveler, especially sea travel. He loved to ski and was an above-average golfer. He played the piano, violin, and cello. He was interested in astronomy, particularly solar eclipses, and was such an outstanding photographer that

[5]Thomson, 1946.

he became a valuable member of the expeditions that studied the solar eclipses viewed in Sumatra (1925), Canada (1932), and Japan (1936). He was a poor lecturer and had no gift for teaching.[6] Aston acquired an interest in finance, as did most of those who worked with J. J. Thomson, and showed great skill as an investor. Upon his death, he left a large estate to Trinity College and several scientific beneficiaries.

NORMAN W. HUNTER
*Western Kentucky University*

# *Bibliography*

Aston, F. W. "The Constitution of Atmospheric Neon"; *The London Edinburgh, and Dublin Philosophical Magazine* **1920,** *39,* 449–455.

Aston, F. W. *Isotopes;* E. Arnold: London, 1922.

Aston, F. W. "The Masses of Some Light Atoms Measured by Means of a New Mass Spectrometer"; *Nature* **1936,** *137,* 357–358.

Aston, F. W. "The Mass-Spectra of the Elements (Part II)"; *The London Edinburgh, and Dublin Philosophical Magazine* **1920,** *40,* pp 628–634.

Brock, W. H. "Aston, Francis William"; In *Dictionary of Scientific Biography;* Gillispie, C. C., Ed.; Charles Scribner's Sons: New York, 1970; Vol. 1, pp 320–322.

*The Dictionary of National Biography (1941–1950);* Davis, H. W. C.; Weaver, J. R. H., Eds.; Oxford University Press: London, 1959.

Farber, E. *Nobel Prize Winners in Chemistry 1901–1961;* Abelard-Schuman: London, 1963.

Ihde, A. J. *The Development of Modern Chemistry;* Harper & Row: New York, 1964; reprinted Dover: New York, 1984.

Keller, A. *The Infancy of Atomic Physics: Hercules in His Cradle;* Clarendon: Oxford, 1983.

Hevesy, G. C. D. "F. W. Aston"; *Obituary Notices of Fellows of the Royal Society* **1945;** *5,* 635–651.

Thomson, G. P. "Dr. Francis William Aston, F. R. S."; *Nature (London)* **1946,** 157, 290–292.

Thomson, G. P. "Francis Wilhelm Aston, 1877–1945"; In *Great Chemists;* Farber, E., Ed.; Interscience: New York, 1961; vol. 2, pp 1455–1462.

Thomson, G. P. *The Inspiration of Science;* Oxford University: London, 1961.

Thomson, G. P. *J. J. Thomson and the Cavendish Laboratory in His Day;* Doubleday: Garden City, NJ, 1965.

---

[6]Thomson, 1946.

# Fritz Pregl

## 1869–1930

Copyright Nobel Foundation

Fritz Pregl was born in Laibach, which is now Ljubljana (Slovenia), on September 3, 1869. When he was 18 years old, his father, a bank official, died. He and his mother then moved to Graz, Austria, where he studied medicine at the University of Graz, receiving his M.D. in 1894. He worked as an assistant lecturer of physiology and histology in the Physiological Institute at the University of Graz from 1893 to 1903 and became an associate professor in 1904. In that year, Pregl took a long leave to Germany, spending his time studying under Gustav V. Hufner in Tübingen, Wilhelm Ostwald in Leipzig, and Emil Fischer in Berlin. When he returned to Graz in 1905, Pregl worked at the Institute of Medical Chemistry. From the years 1910 to 1913 he was full professor of applied medical chemistry at Innsbruck University. He returned to stay at the University of Graz in 1913, where he remained until his death. He was appointed dean of the Institute of Medical Chemistry for the year 1916–1917 and vice-chancellor of the University of Graz for 1920–1921. Pregl devoted himself to the cause of science and education. He never married. After a short illness, he died at the age of 61 at Graz on December 13, 1930.

He was awarded the Lieben Prize for chemistry from the Imperial Academy of Science in Vienna in 1914 and an honorary doctorate in philosophy from the University of Göttingen in 1920. He was elected corresponding member by the Academy of Sciences in Vienna in 1921. He was awarded the Nobel Prize for chemistry in 1923 for his development of microanalytical methods for organic substances.

At first, Pregl had been interested in physiology. When he undertook physiological research, particularly on bile acids, the substances could only be obtained in extremely small quantities. He had to decide whether to process tons of raw material or to search for new methods that would enable him to obtain correct analytical results using much smaller quantities. Pregl chose the latter approach, devoting himself to developing methods of quantitative microanalysis for organic substances.

The determination of carbon and hydrogen is a fundamental problem of organic elemental analysis; it requires the greatest skill and the most exacting conditions. Previously, organic substances were burned under suitable conditions, and the carbon and hydrogen were converted into carbon dioxide and water, respectively. Carbon dioxide was absorbed by potassium hydroxide, and water was absorbed by calcium chloride. There were, however, difficulties with the microdetermination of carbon and hydrogen. The first problem was to absorb the combustion products free from foreign admixtures. Secondly, the increase in weight of the absorbing agent must be related only to the resulting carbon dioxide and water produced.

In developing his microanalysis method for the determination of carbon and hydrogen, Pregl improved and introduced new apparatuses and techniques to reduce the amount of sample and to obtain accurate results. Kuhlmann's assay balance was improved in accordance with Pregl's suggestions. Thus, it was possible to weigh with an accuracy of ± 0.001 mg over a range of 20 g. This type of balance, later known as the microchemical balance, has been used by most microchemists since Pregl, and it has contributed significantly to the rapid development of quantitative microanalysis.

Pregl developed a universally effective packing mixture for his combustion tube that would retain all components except carbon dioxide and water. It consisted of a mixture of copper oxide and lead chromate between two sections of silver and finally a section of lead peroxide on asbestos. By using this filling and by heating at 180 °C, the gaseous products from the sample other than carbon dioxide and water were prevented from entering the absorption apparatus.

Pregl used the mercury gasometer to avoid losses of carbon dioxide that occasionally happened because of incomplete combustion. The gases coming from the combustion tube were collected in the gasometer and then passed through the heated combustion tube a second time. The gasometer was also used to regulate the rate of the combustion process. Later Pregl employed a Mariotte bottle instead of the mercury gasometer to regulate the velocity of the gas stream and the pressure conditions in the apparatus.

The absorption tubes were also improved by fitting capillary taperings to prevent inward diffusion of water vapor. Soda lime was taken to replace potassium hydroxide as the absorbing agent for the collection of carbon dioxide. Pregl succeeded in the determination of carbon and hydrogen with accurate analytical results for 2–4 mg of the substance. The smallest quantity ever used was 1 mg, and the deviation still was within normally permitted error.

Pregl developed two methods for the microdetermination of nitrogen. The Dumas method for determination of nitrogen was converted to an accurate microanalysis method by increasing the length of the packing mixture in the combustion tube, controlling the carbon dioxide flow rate, and using an accurate azotometer. The Kjeldahl method was also improved for the microdetermination of nitrogen. Microdetermination methods for halogens, sulfur, phosphorus, and a number of metals in organic compounds were also developed by Pregl or by others under his guidance.

When Pregl determined halogens and sulfur with the Carius method, he recognized that the presence of barium chloride was essential for the decomposition of sulfur-containing organic samples with nitric acid. Then he discovered a completely new analytical process for the decomposition of organic substances: to burn them with the aid of red-hot platinum catalysts in a current of oxygen. The combustion products were collected in an alkaline sulfite solution for the determination of halogens. With sulfur, the combustion products were collected in perhydrol. Correct analytical results were obtained in all cases.

In addition to quantitative microdetermination of the elements in organic substances, microdeterminations of functional groups were studied. Pregl developed a method for the determination of carboxyl groups by acidimetry using phenolphthalein as the indicator, and he extended the procedure for the determination of methoxy and methylimide groups. Moreover, the determination of molecular weight by the boiling point method could be performed through microanalysis. He obtained reliable values for these adaptations with quantities as small as 7 mg.

Pregl's microanalysis methods were a great contribution to chemistry, biochemistry, and medical science. With them it is possible to carry out quantitative analysis on a few milligrams of substance, an amount about one-hundredth of that required previously with micro-analytical methods. The microanalysis methods resulted in a great savings in time, labor, and expense.

Pregl applied his microanalysis methods in his investigations on enzymes, sera, and bile acids, and in forensic analyses to identify alkaloids. Pregl's methods were invaluable and indispensable to the organic chemists and biochemists who adopted them. Microanalysis methods have been successfully applied to investigations in many research fields in which the original materials or products are often obtainable only in limited quantities. These methods greatly aided and accelerated the study and elucidation of the chemical character and structure of many biologically significant substances; among them are enzymes, hormones, and vitamins, all of which are important for vital processes.

Pregl was highly skilled in the design, construction, and modification of apparatuses for experimental research. He was also good at glassblowing, which allowed him to make or modify many ingenious items of glass apparatus to suit the needs of microanalysis. He willingly helped colleagues perform analyses of their samples, using his microanalysis methods, which he continually tested and improved.

Pregl dealt with his microanalysis methods with a rigorous scientific approach. He avoided publishing individual research papers in an early stage, until he had convinced himself that his methods were generally applicable and successful when carried out in other laboratories. Pregl described the elemental methods for organic substances that he had developed by the end of 1911 in E. Abderhalden's *Handbuch der biochemischen Arbeitsmethoden,* published in 1912. He gave a successful address on his methods at the meeting of the Verein deutscher Naturforscher (German Scientific Society) in Vienna in September 1913. His monograph entitled *Die quantitative organische Mikroanalyse* was published in Berlin in 1917. The second edition and the third, revised and enlarged, edition were published in 1923 and 1930, respectively. The seventh edition was published in Vienna in 1958. The monograph has been translated into various languages including English, French, and Russian.

Pregl was enthusiastic in his educational undertakings. His students found him an inspiring teacher who knew how to flavor his lectures with instructive experiments as well as with humor. In his laboratory, he always provided the personnel with instruction and allowed them the opportunity to demonstrate for visiting chemists.

Later, a special course on quantitative microanalysis was developed and given several times each year. Particularly after Pregl was awarded the Nobel Prize in chemistry, chemists from all over the world came to his laboratory to study Pregl's microanalysis under his guidance. As a result, quantitative microanalysis methods were quickly disseminated and widely applied in industrial, research, and university laboratories. After Pregl died, the Vienna Academy of Sciences established the Fritz Pregl Prize for outstanding work in microchemistry by Austrian microchemists.

MIMLIANG XU
*Tianjin Institute of Light Industry*

# Bibliography

*Fritz Pregl an Karl Berthold Hofmann: Briefe aus den Jahren 1904–1913*; Holasek, A.; Kernbauer, A., Eds.; Publikationen aus dem Archiv der Universität Graz, Bd. 25; Akademische Druck und Verlagsanstadt Graz, Austria, 1989.

Lieb, H. "Fritz Pregl"; *Mikrochemie* **1931,** *3,* 105–116.

Lieb, H. "Fritz Pregl (1869–1930)"; *Ber. Dtsch. Chem. Ges.* **1931,** *64A,* 113–118.

Lieb, H. "Fritz Pregl: Nobel Prize Winner for Chemistry in 1923"; *Mikrochemie* **1923,** *1,* 63–71.

Pregl, F. "Die quantitative Mikroanalyse organischer Substanzen"; In *Handbuch der biochemischen Arbeitsmethoden;* E. Abderhalden, Ed.; Urban & Schwarzenberg: Vienna, Austria, 1912; Vol. 5, pp 1307–1356.

Pregl, F. *Die quantitative organische Mikroanalyse,* 3rd ed.; Julius Springer: Berlin, Germany, 1930.

Pregl, F.; Fyleman, E. *Quantitative Organic Microanalysis,* 2nd English ed.; P. Blakiston's Son: Philadelphia, PA, 1930.

Roth, H. *Pregl-Roth quantitative organische Mikroanalyse,* 7th ed.; Julius Springer: Vienna, Austria, 1958.

Roth, H.; Daw, E. B. *Quantitative Organic Microanalysis of Fritz Pregl,* 3rd English ed.; P. Blakiston's Son: Philadelphia, PA, 1937.

# 1925

# *Richard Zsigmondy*

## 1865–1929

Richard Adolf Zsigmondy was born April 1, 1865, in Vienna, Austria. He received the 1925 Nobel Prize for his pivotal role in elucidating the nature of colloids. His experiments from 1900 to 1905 provided data to convince a skeptical scientific world of the reality of atoms and molecules. He married Laura Luise Müller in 1903; they had two daughters. He died September 24, 1929, in Göttingen, Germany, from arteriosclerosis.

The term *colloid* was coined by Thomas Graham (1805–1869) to describe such substances as clays, starch, gelatin, albumin, and gums.[1] Colloids often form gels; they seldom crystallize. Graham described parchment membranes, impermeable to colloids, through which *crystalloids* (such as sodium chloride, water, alcohol, and sugar) readily diffuse. Many quantitative methods for studying crystalloids and their solutions were developed by physical chemists in the closing decades of the 19th century. However, physical chemists found the intriguing colloids difficult to fit into their new theories. Richard Zsigmondy played a central role in bringing the study of colloids into the realm of physical chemistry.

Richard Zsigmondy was the fourth child of Irma von Szakmáry and Adolf Zsigmondy. Of Hungarian ancestry, Adolf Zsigmondy was senior physician at the Vienna General Hospital, a university lecturer on dental surgery, and an inventor of surgical instruments. Both mother

---

[1]"Liquid Diffusion to Analysis"; *Philos. Trans. R. Soc. London* **1861,** *151,* 183–224.

and father supervised home laboratory experiments in physics and chemistry for their children. Their elder sons, Emil and Otto, became physicians, and their youngest son, Karl Ernst, became a professor of mathematics at the Vienna Technische Hochschule. Several of their children, including Richard, were enthusiastic Alpine mountain climbers.

Richard attended the public secondary school near the Vienna hospital, passing the graduation examinations in 1883. He was a student at the Technische Hochshule in Vienna from 1883 to 1887. There he studied chemistry with Alexander Anton Emil Bauer (1836–1921), Johann Oser (1833–1912), and Rudolf Benedikt (1852–1896). He also attended an introductory course in analytical chemistry given by Ernst Ludwig (1842–1915) at the University of Vienna. Bauer had written a chemical technology textbook, and his research had ranged throughout organic and analytical chemistry; by 1883 his scholarly interests had moved to history of chemistry and alchemy. Oser's research was in inorganic chemistry, agricultural chemistry, and chemical technology. Zsigmondy was particularly attracted to Benedikt, who at 31 had just published a major book on the chemistry of coal-tar dyes and had an active research program. Benedikt was writing a book on chemical analysis of fats and waxes (published in 1886) and was on his way to becoming dean of the school of chemistry at the Technische Hochschule. Benedikt encouraged Richard Zsigmondy's lifelong interest in the colors of materials. With Benedikt, Zsigmondy extended a method for the chemical analysis of glycerin in fats and in dilute water solutions; this procedure was the subject of his first research publication.[2]

While a student at the Vienna Technische Hochschule, he also collaborated with the Prague chemist C. Haller in experimental investigations at a glass factory in Bohemia. Zsigmondy reported his systematic studies of the relationships between the luster and color of glasses and their chemical composition in *Dinglers Polytechnischen Journal*.[3]

Organic chemistry dominated the chemistry faculties of the German universities, and Zsigmondy took the next step toward an academic career by beginning graduate research in organic chemistry. From 1887 to 1889 he was an assistant in the organic chemistry laboratory of Wilhelm von Miller (1848–1899) at the Technische Hochschule in Munich. There he synthesized a chlorine derivative of indene, an

[2]*Chemiker Zeitung* **1885**, *9*, 975; *Repertorium der analytischen Chemie* **1885**, *5*, 266–269.
[3]**1887**, *266*, 364–370; **1889**, *271*, 36–44, 80–88; **1889**, *273*, 29–37.

aromatic hydrocarbon found in coal tar. His synthesis was routine, published in his 29-page dissertation and as an abstract.[4] The dissertation was the basis for the doctor of philosophy degree awarded to Zsigmondy by the University of Erlangen in December 1889.

He needed the doctorate for an academic career, but in the 1880s a doctorate alone did not qualify a person to become a university professor in Austria or Germany; evidence of further independent scholarly achievement was required. This continued research, called *Habilitation,* culminated with defense of a second dissertation and a formal inaugural lecture. The most significant portion of Zsigmondy's *Habilitation* was two years of postdoctoral research from 1890 to 1892 as assistant to August Adolph Eduard Eberhardt Kundt (1839–1894) in the Institute of Physics of the University of Berlin. Kundt was an expert on optical properties of colored materials, including fluids and thin metal layers. Zsigmondy and Kundt collaborated on an investigation of the luster colors produced when organic suspensions of colloidal gold are applied to porcelain before firing. During this period Zsigmondy developed filters (using ferrous salts) to absorb radiant heat from the intense light sources used in microscopy.

The years of Zsigmondy's formal education coincided with the development of thermodynamics, electrochemistry, and physical chemistry. The chief spokesman for these new areas of chemistry was Friedrich Wilhelm Ostwald (Nobel laureate in chemistry, 1909), who led a vigorous debate on the reality of atoms and molecules. He argued that chemists should abandon atomic and molecular hypotheses because there was no evidence for the existence of atoms. Nevertheless, most German organic chemists used molecular structural formulas in their daily work. In analytical chemistry and chemical technology, science was carried forth with little regard for the debate. Zsigmondy's education in chemistry and physics took place outside this controversy; unlike most physical chemists of his generation, he avoided taking a public philosophical position on atoms. Later, when his observations produced experimental evidence about the ultimate constitution of matter, he was free to interpret that evidence objectively.

In 1893, certified as habilitated, Zsigmondy was appointed *Privatdozent* at the Technische Hochschule in Graz. As *Privatdozent* he gave lectures on chemical technology. He received no salary from the Technische Hochschule, but instead collected fees from the students

---

[4]*Chem. Zentralbl.* **1890**, *61*, 312.

who attended his lectures. Zsigmondy had a wide range of expertise. Although nominally an organic chemist, he had skills in analytical chemistry, glass technology, microscopy, and spectroscopy. He was a physical chemist, but he was not a protégé of Ostwald or his associates. At the age of 28, Richard Zsigmondy was in a position at Graz to define the area of his own research. He chose to focus on colloidal gold and its colors.

Why was Zsigmondy so fascinated with gold? He must have watched his father working with dental gold amalgams. Perhaps Professor Bauer introduced him to the alchemical investigations of Andreas Cassius (d 1673), the reputed discoverer of the gold-containing pigment Cassius purple. He probably witnessed the use of Cassius purple in making ruby glass at the Bohemian glassworks. His interest in gold increased the attractiveness of postdoctoral research with Kundt.

At Graz he conducted a systematic study of colloidal gold. Using a Zeiss spectroscope, he determined which wavelengths of light are absorbed by gold particles when the colloidal system is a water solution and when it is a glass.[5] He exhaustively studied the preparation and properties of Cassius purple.[6] Zsigmondy devoted much of his 1905 book on colloids to discussions of gold systems as prototypes for understanding colloids in general.

Zsigmondy maintained his contacts with glassmakers. He developed a particularly fruitful relationship with the glass technology laboratory of Schott Glass Manufacturing Company in Jena. Otto Friedrich Schott, Ernst Abbe, Carl Zeiss, and Roderich Zeiss founded the laboratory in 1883. Jena specialty glasses developed in the laboratory became highly prized for scientific apparatus, including optical instruments designed and manufactured by the Carl Zeiss Company in Jena. In 1897 Zsigmondy left Graz to join the staff of the Schott laboratory for three years. There he systematically studied colored glasses, extending investigations he had begun while a student in Vienna. Many of these colored glasses were colloidal systems.

To investigate colloids, he needed new instrumentation. He sought technical collaborators, and together they invented or developed those new instruments. One collaborator was Henry Friedrich Wilhelm Siedentopf (1872–1940), who joined the Carl Zeiss Company in 1899.

---

[5] Zsigmondy, R. "Ueber Wässrige Lösungen metallischen goldes"; *Justus Liebig's Ann. Chem.* **1898,** *301,* 29–54.

[6] Zsigmondy, R. "Die chemische Natur des Cassius' Schen Goldpurpurs"; *Justus Liebig's Ann. Chem.* **1898,** *301,* 361–387.

Together Zsigmondy and Siedentopf invented the slit ultramicroscope, a microscope in which a colloidal system was viewed at right angles to the direction of its illumination. Just as small dust particles can be seen in a ray of sunlight viewed at right angles to the ray, so individual colloid particles can be observed using this technique of "dark-field illumination". The ultramicroscope permitted particles to be viewed that had only 1/100,000 the volume of the smallest particles that had hitherto been seen. By counting the number of colloidal gold particles in a small volume of a dilute solution of known concentration, one could calculate the average size and mass of individual particles. Zsigmondy and Siedentopf made preliminary observations on solutions of glue, gelatin, gum tragacanth, stannic acid, and a silver hydrosol in spring of 1900. They publicly presented chemical results from the ultramicroscope at a meeting of the Deutschen elektrochemischen Gesellschaft in Würzburg in 1902, and they published a lengthy paper on the ultramicroscope in (*Ann. Phys. (Leipzig)* **1903**, *Series 4, 10*, 1–39). Just two years later Zsigmondy published a book on the use of the ultramicroscope in gaining an understanding of colloids. Zeiss manufactured the commercial version of the instrument.

Direct observation of solute particles in a solution marked a turning point in chemistry. Chemistry as a quantitative molecular science dates from these experiments. Observers were readily convinced that a colloidal solution in the ultramicroscope was not a homogeneous, continuous liquid. The random and lively Brownian motion of individual particles could be followed, and the movements of gold particles were described in detail by Siedentopf and Zsigmondy in their 1903 *Annalen der Physik* paper. Colloidal gold, exploited by Zsigmondy as a model for colloidal systems, also provided a general model for all chemical solutions. The ultramicroscope provided the experimental data used by Zsigmondy, Jean Perrin (1870–1942), Maryan Smoluchowski (1872–1917), Paul Langevin (1872–1946), Albert Einstein (1879–1955), and The Svedberg (1884–1971) to convince skeptical physical chemists that molecules are real and that matter at the molecular level is discontinuous. Ostwald maintained that atoms are hypothetical as late as 1906, but because of the research on Brownian motion even he eventually changed his mind; a recantation appears in the preface to the fourth edition of his *Grundriss der allgemeinen Chemie*.[7]

---

[7] W. Engelmann: Leipzig, 1908; translated by W. W. Taylor as *Outlines of General Chemistry*, 3rd ed.; Macmillan: London, 1912.

The ultramicroscope also made possible for the first time the investigation of proteins and other biological macromolecules as molecular entities. Many physiologists became colloid chemists, and many physical chemists turned their attention to investigations of protoplasm and its constituents. Molecular biology as an experimental science dates from this time.

Zsigmondy left the laboratory staff in 1900, and until 1907 he had no formal institutional connection. He described himself as a private researcher in Jena. His professional relationships with the staff of the glass technology laboratory and the Zeiss optical company continued, but he was free to focus his research as he wished. He spent considerable time at his vacation home near Terlano, in the Tirol, writing numerous research papers and two books on colloid chemistry that were published in 1905 and 1906. His writing was clear and concise. He understood the implications of his work and effectively communicated with a broad scientific audience.

In 1907, Zsigmondy accepted an appointment as assistant professor of inorganic chemistry and director of the inorganic chemistry institute at the University of Göttingen. In Göttingen he worked with the R. Winkel Optical Works and the Vereinigte Göttinger Werkstätten (V. G. W.). In 1913 he described an improved ultramicroscope that he called an immersion ultramicroscope.[8] Winkel manufactured the instrument. He also developed apparatus for studying colloids, including specialized filtration equipment and a dialyzer. Some of this apparatus was manufactured at V. G. W., where Zsigmondy pursued an active research program in colloid chemistry.

In 1919 he was promoted to professor, a position he held for the rest of his life. He was elected to the academies of science of Göttingen, Vienna, Uppsala (Sweden), Zaragoza (Spain), Valencia (Spain), and Haarlem (Netherlands). He received honorary doctorates from the Technische Hochschule in Vienna and Graz and from the University of Königsberg.

GEORGE FLECK
*Smith College*

---

[8]Zsigmondy, R. "Über ein neues Ultramikroskop"; *Phys. Z.* **1913**, *14*, 975–979.

# Bibliography

Chamot, E. M. *Elementary Chemical Microscopy;* John Wiley: New York, 1916; Chapter 4.

Freundlich, H. "Richard Zsigmondy (1865–1929)"; *Ber. Dtsch. Chem. Ges.* **1930,** *A63,* 170–175.

Kerker, M. "Zsigmondy, Richard Adolf"; In *Dictionary of Scientific Biography;* Gillispie, C. C., Ed.; Charles Scribner's Sons: New York, 1976; Vol. 14, pp 632–634.

Lottermoser, A. "Zsigmondy, Richard"; In *Deutsches biographisches Jahrbuch;* Deutsche Verlags-Anstalt: Stuttgart, 1929; Vol. 11, pp 335–338.

Lottermoser, A. "Richard Zsigmondy zum Gedächtnis"; *Zeitschrift für angewandte Chemie* **1929,** *42,* 1069–1070.

Nye, M. J. *Molecular Reality: A Perspective on the Scientific Work of Jean Perrin;* American Elsevier: New York, 1972.

Thiessen, P. A. "Richard Zsigmondy—Nobelpreisträger für Chemie im Jahre 1925"; *Mikrochemie* **1927,** *6,* 1–7.

Wegscheider, R. *Almanach für das Jahr 1930;* Akademie der Wissenschaften: Vienna, 1930; Vol. 80, pp 262–268.

Weyl, W. A. Part IV of "The Colours Produced by Metal Atoms"; *Coloured Glasses;* Society of Glass Technology: Sheffield, 1951.

Zsigmondy, R. *Zur Erkenntnis der Kolloide: Über irreversible Hydrosols und Ultramikroskopie,* 2nd ed.; G. Fischer: Jena, 1905, 1919. English translation: *Colloids and the Ultramicroscope: A Manual of Colloid Chemistry and Ultramicroscopy;* Alexander, J., Trans.; John Wiley & Sons: New York, 1909.

Zsigmondy, R. *Über Kolloid-Chemie: mit besonderer Berücksichtigung der anorganischen Kolloide;* Stuttgart, 1906; J. A. Barth: Leipzig, 1907, 1925.

Zsigmondy, R. *Kolloidchemie: ein Lehrbuch;* Otto Spamer: Leipzig, 1912; 1925, Vol. 1; 1927, Vol. 2. Translated as "Kolloidchemie," Part I of *The Chemistry of Colloids;* Spear, E. B., Trans.; John Wiley & Sons: New York, 1917.

Zsigmondy, R.; Thiessen, P. A. *Das kolloide Gold;* Akademische Verlagsgesellschaft: Leipzig, 1925.

Zsigmondy, R. "The Immersion Ultramicroscope", Chapter 54; and "Membrane Filters and Their Uses", Chapter 60; In *Colloid Chemistry: Theoretical and Applied;* Alexander, J., Ed.; The Chemical Catalog Company: New York, 1926.

# The Svedberg

## 1884–1971

The (Theodor) Svedberg was born near Gavle, Sweden, on August 30, 1884, and died in Orebro, Sweden, on February 25, 1971. He was awarded the Nobel Prize in chemistry for 1926.

Svedberg's invention of the ultracentrifuge and its application to the study of proteins from the mid-1920s has secured his niche as a principal founder of molecular biology and biophysical chemistry. Other modern instrumental techniques, such as X-ray diffraction and moving boundary electrophoresis, were first applied to the study of protein structure only from the mid-1930s. Actually, Svedberg initiated the work on moving boundary electrophoresis for which his student Arne Tiselius won the 1948 Nobel Prize.

It surprises many to learn that the 1926 Nobel Prize in chemistry was awarded to Svedberg, then in mid-career, for his work in colloid chemistry, most particularly for studies of Brownian motion. Indeed his first results on the ultracentrifugation of proteins were only published in 1926. The first optical centrifuge—not yet an ultracentrifuge—was built in 1923 in collaboration with graduate student J. B. Nichols to study gold hydrosols, while Svedberg was a visitor at the University of Wisconsin.

Svedberg was the only child of Elias Svedberg, a civil engineer, and Augusta Alstermark. Even though he had already demonstrated great talents during his school years, his success at the University of Uppsala was amazing. He arrived in January 1904, passed the various courses and examinations for the *Fil. kand.* degree in record time, and plunged directly into research in the new colloid chemistry in September 1905.

Although he had earlier displayed strong interests in biology, Svedberg avoided a trend that was fashionable in colloid chemistry: diffuse speculations about colloids as models of living systems. Rather, he tackled the practical problem of preparing stable colloids reproducibly in order to permit quantitative studies of the relation between particle size and other physical properties. In December 1905 he published his first paper, in which he refined Georg Bredig's electric arc technique for the preparation of gold and platinum hydrosols by introducing alternating current discharges and performing the experiments in various organic solvents. He next assembled an ultramicroscope according to the design of Zsigmondy and Siedentopf.

His doctoral dissertation was submitted in 1907. There does not appear to be any indication of collaboration with a member of the faculty. Brownian motion is certainly the most dramatic aspect of colloidal behavior made apparent by the ultramicroscope, so it is no wonder that a study of Brownian motion comprised a major portion of his dissertation. He also studied the effect of particle size on optical absorption, the effect of both temperature and added electrolyte on the coagulation of hydrosols, and the electrical synthesis of sols.

His Brownian motion work displayed great imagination. Following an erroneous suggestion by Zsigmondy that the motion of larger particles was of an oscillatory nature rather than translation by a random walk, he devised an experiment in which particles were observed in the ultramicroscope while the sol as a whole flowed through the apparatus at a known linear velocity. Svedberg reasoned that the superposition of the oscillatory Brownian motion and the linear motion of the fluid would result in a sinusoidal curve. He reported what appeared to be waveforms, and from the velocity and wavelength he calculated a period of oscillation. From the amplitude and the period he then calculated a velocity that he identified with the gas kinetic velocity of the colloid, considered as a very large molecule.

Shortly thereafter he encountered Albert Einstein's 1905 paper on Brownian motion, and, identifying his measured amplitude with Einstein's displacement and his period with Einstein's characteristic time, he claimed that his data, obtained in liquids of various viscosities, verified Einstein's treatment. This was challenged immediately by Einstein, who pointed out that the motion was not oscillatory and that the relaxation time for the random motion made observation of a velocity quite impractical. Somewhat later he was attacked very sharply by Jean Perrin. Svedberg actively defended this work for nearly two decades, and it continued to be cited in the colloid literature. Indeed, it provided the main basis for the award to Svedberg of the Nobel Prize in chemistry in the same year that Perrin was honored for his Brownian motion experiments in physics.

Svedberg was appointed docent after completing his dissertation. He and his students expanded his initial projects, and Uppsala became known as a center for quantitative work on colloids. He provided further support for the kinetic molecular view through experimental tests of Marian Smoluchowski's theory of thermal fluctuations. This support was obtained by observing, for a microscopically small fixed volume, the variation of concentrations of particles in a gold sol by using the ultramicroscope and of polonium ions in a solution by counting scintillations. In other experiments, colloidal particle size calculated from diffusion data was compared with values obtained with the ultramicroscope.

All of Svedberg's work on Brownian motion, thermal fluctuations, and diffusion were subjects in which young Einstein was also intensely interested at that time. Svedberg had a talent for relating colloidal phenomena to that most fundamental, current scientific problem: the question of the reality of atoms and molecules. His work on radioactivity, in which he retained a lifelong interest and to which he was to dedicate the last phase of his career, provides another instance of his ability to tackle fundamental problems. In addition to exploring the effect of X-rays on the stability of gold colloids, he and D. Strömholm in 1909 investigated the isomorphic coprecipitation of radioactive compounds. They discovered that thorium X (radium) coprecipitated with lead and barium salts, indicating the existence of isotopes.

From about 1914 through 1922, Svedberg's published work was devoted mainly to two new subjects. There were studies of anisotropies produced by electric and magnetic fields in substances today known as liquid crystals, including measurements of electrical conductivity, reaction rates, and diffusion. Starting in 1920, there were several papers published in photographic journals devoted to the relation between the size and the sensitivity of the grains in a photographic emulsion.

Then came the Wisconsin interlude, which marked his career's watershed. In 1922 J. H. Mathews invited Svedberg to the University of Wisconsin to lecture and to do research during the 1923 spring semester and then to participate in a symposium that evolved into the National Colloid Symposium of the American Chemical Society. The Wisconsin colloid symposium inaugurated the oldest such specialized annual meeting within the American Chemical Society and stimulated the formation of the Division of Colloid Chemistry.

For Svedberg, the Wisconsin sojourn led to his development both of the ultracentrifuge and, through his student Tiselius, of the moving boundary electrophoresis apparatus, two instruments that have had a profound influence on the recent spectacular advances in macromolecular chemistry, biochemistry, and molecular biology. Svedberg had just developed a technique for following the variation of optical density with height in a sedimenting system in order to determine the actual distribution of colloidal particle size, rather than merely the average size. However, only relatively large particles settled in the gravitational field, and, in order to study the formation and growth of particles, it would be necessary to increase the rate of sedimentation by centrifugation.

This was done in Wisconsin, where J. B. Nichols, one of the six graduate students assigned to Svedberg, built what was to evolve into the ultracentrifuge. The centrifugal force attained by the Wisconsin ultracentrifuge was only 150 times gravity, but, after returning to Uppsala, Svedberg and Herman Rinde built a centrifuge that attained a force of about 7000 times gravity. With this instrument, they were able to determine the size of gold particles much smaller than those that would be visible in the ultramicroscope. In analogy with the ultramicroscope and ultrafiltration they proposed the name *ultracentrifuge*. In 1925–1926 Svedberg was able to obtain funds to build a still larger instrument, which attained 100,000 times gravity and made it possible to cover a still broader range of sizes. Thus, by using the sedimentation velocity technique, he was able to penetrate into the macromolecular domain of proteins. His first paper on a protein appeared in 1926, the year in which he was awarded the Nobel Prize. In this paper he confirmed earlier studies by G. Adair, using osmotic pressure, that the molecular weight of hemoglobin was 68,000 rather than 16,700 based on the iron content, making that molecule a tetramer consisting of four monomeric subunits.

By then Svedberg had left the domain of colloidal particles that comprised Wolfgang Ostwald's "world of neglected dimensions", and he was to devote his career mainly to proteins and other biological macromolecules, using the ultracentrifuge as his principal tool. A key finding of Svedberg's that in many cases soluble proteins had molecules with well-defined uniform size was first treated with skepticism, yet later confirmed. However, his proposal that the molecular weights of proteins were multiples of a basic unit did not have the generality that he had expected. Svedberg remained involved in the continued development of greater centrifugal fields and in the study of new macromolecular systems up to the outbreak of World War II, when he was charged with developing Sweden's production of synthetic rubber.

In 1949 he retired from his chair of physical chemistry and became head of the Gustaf Werner Institute of Nuclear Chemistry. His early interest in radiation chemistry had actually been revived in the late 1930s in connection with the effects of ultrasonics, ultraviolet light, and alpha particles on hemocyanines. After the war, having succeeded in getting a wealthy industrialist to build the Werner Institute, he plunged into a new phase of related research that dealt with problems that overlapped chemistry, biology, and high-energy physics. The last paper bearing his name, published in 1965, dealt with high-energy proton radiotherapy.

Svedberg's work belongs to three broad areas of science: colloid chemistry, physical biochemistry, and radiation chemistry. We can claim him as a colloid chemist not only because of his training and contributions to that discipline, but also because his work in the latter two areas flowed from his early work with colloids.

**MILTON KERKER**
*Clarkson University*

## *Bibliography*

Claesson, S.; Pedersen, K. O. "Svedberg, The (Theodor)": In *Dictionary of Scientific Biography;* Gillispie, C. C., Ed.; Charles Scribner's Sons: New York, 1975; Vol. 13, pp 158–164.

Claesson, S.; Pedersen, K. O. "The Svedberg 1884–1971"; *Biogr. Mem. Fellows R. Soc.* **1972,** *18,* 595–627.

Kerker, M. "The Svedberg and Molecular Reality"; *Isis* **1976,** *67,* 190–216.

Kerker, M. "The Svedberg and Molecular Reality, An Autobiographical Postscript"; *Isis* **1986,** 77, 278–282.

Nye, M. J. *Molecular Reality: A Perspective on the Scientific Work of Jean Perrin;* Elsevier: New York, 1972.

Svedberg, T. *Colloid Chemistry*, 2nd ed.; Chemical Catalog Co.: New York, 1928.

Svedberg, T. "The Formation of Colloids": In *Monographs on the Physics and Chemistry of Colloids*; Van Nostrand: New York, 1921.

Svedberg, T., et al. *The Ultracentrifuge*; Clarendon Press: Oxford, 1940.

Svedberg, T.; Fåhraeus, R. "A New Method for the Determination of the Molecular Weight of the Proteins"; *J. Am. Chem. Soc.* **1926**, *48*, 430–438.

Svedberg, T.; Jette, E. R. "The Cataphoresis of Proteins"; *J. Am. Chem. Soc.* **1923**, *45*, 954–957.

Svedberg, T.; Nichols, J. B. "Determination of Size and Distribution of Size of Particle by Centrifugal Methods"; *J. Am. Chem. Soc.* **1923**, *45*, 2910–2917.

Svedberg, T.; Nichols, J. B. "The Molecular Weight of Egg Albumin I. In Electrolyte-Free Condition"; *J. Am. Chem. Soc.* **1926**, *48*, 3081–3092.

Svedberg, T.; Stein, B. S. "Density and Hydration in Gelatin Sols"; *J. Am. Chem. Soc.* **1923**, *45*, 2613–2620.

*The Svedberg, 1884–1944*; Tiselius, A.; Pedersen, K. O., Eds.; Almqvist & Wiksell: Uppsala, Sweden, 1944.

# Heinrich Wieland

## 1877–1957

Heinrich Otto Wieland was born in Pforzheim, Germany, on June 4, 1877, and died in Starnberg, Germany, on August 5, 1957. He was awarded the Nobel Prize in chemistry for his work on the bile acids, an achievement characterized as one of the most difficult in organic chemistry.

In pursuing chemistry as a profession, Heinrich Wieland was following in his father's footsteps. Dr. Theodor Wieland was a pharmaceutical chemist. Both father and mother, Elise Blom, were from Württemberg. Heinrich Wieland's sons also continued the family tradition in chemistry: Wolfgang, another pharmaceutical chemist like his grandfather; Theodor, a professor of chemistry at the University of Frankfurt; and Otto, a professor of medicine at the University of Munich. Even the only daughter of the family, Eva, maintained the family predilection for chemistry by marrying a professor of biochemistry, Feodor Lynen, a Nobel Laureate in 1964. Wieland's education was broad. He was a student at the Universities of Munich, Berlin, and Stuttgart. Under Johannes Thiele at the Baeyer laboratory in Munich, he received his doctorate in 1901.

By 1913 he obtained a senior position at the University Chemical Laboratory in Munich. Only four years later, in 1917, he became a full professor at the Technical College in Munich. During the war years of 1917 and 1918, he worked at the Kaiser Wilhelm Institute in Berlin-Dahlem. In 1921 he went to Freiburg University, but in 1925 he went back to the University of Munich to accept the chair vacated by the eminent chemist Richard Willstätter. For 27 years he was the head of the Munich laboratory.

The awards given to him in his professional life were many and stand as a testament to his stature as a chemist. Besides the Nobel Prize, he received the Order of Merit and the Otto Hahn Prize. He was a member of most of the important scientific societies of the world and edited *Justus Liebigs Annalen der Chemie* for 20 years. In addition Wieland was noted throughout his career as a wonderful teacher and academician. Heinrich Wieland was a devoted family man. He married Josephine Bartmann in 1908, and they had the three sons and one daughter mentioned above.

Although the scope of Wieland's work in chemistry is hard to understand in the modern context of specialization, he won the Nobel Prize on the basis of his work on the bile acids, a problem that the presenter of the Nobel Prize, Professor H. G. Söderbaum, Secretary of the Royal Swedish Academy of Sciences, called "one of the most difficult which organic chemistry has had to tackle." Other phrases in the presentation speech also bring insight into this achievement: "striking success," "remarkable skill in experimentation," "years of hard, diligent and resourceful work." It was also pointed out that the problem of the bile acids had been attacked for 100 years by many eminent researchers, but Wieland overcame the complexity of structure and the difficulty of obtainment and purification to bring real elucidation to the field.

In his Nobel lecture, on December 12, 1928, which he titled "The Chemistry of the Bile Acids," Wieland summarized his Nobel Prize winning research. The chemical structure of the bile acids was evaluated essentially through the use of decomposition studies. Although large molecules, the bile acids contain only carbon, hydrogen, and oxygen. The bile acids are linked to many other fundamental and important biological compounds, such as the sterols, the vegetable cardiac poisons, toad skin poisons, and certain vitamins. Cholic acid (24 carbons, 40 hydrogens, 5 oxygens) is made up of four condensed carbon rings with three branches. The carboxyl group, which is responsible for the acidic characteristics of the substance, is on a side chain. The carboxyl group contains a carbon and two oxygens. The other three oxygens are found as alcohol groups in positions 3, 7, and 12.

Adolf Windaus and Wieland isolated one of the other bile acids at the same time. This particular bile acid (24 carbons, 40 hydrogens, 4 oxygens) has two alcoholic groups at positions 7 and 12. What this information about structure illustrated was the close relatonship of all the bile acids to one another.

Wieland clarified the ring structure by opening the first ring with nitric acid. Two isomeric acids resulted and by chemical means were converted to a cyclic ketone. The formation of the cyclic ketone from the two acid groups indicated that the ring contained six carbons. Further decomposition opened the other rings, except one. The product with one ring was solanellic acid, which was converted to biloidanic acid (23 carbons, 32 hydrogens, 12 oxygens). Biloidanic acid was an important decomposition product, as both cholic acid and deoxycholic acid decompositions led to it. Biloidanic acid unfortunately proved to be a dead end, however. Attempts to degrade the ring in biloidanic acid proved unsuccessful.

Interest then shifted to the side chain associated with the fourth ring. This side chain had been investigated by Adolf Windaus in his work on cholesterol. By using the ester of cholanic acid and the Grignard reaction, Wieland was able to split off one carbon at a time. The third and fourth carbon split off together, indicating that the third carbon contained a methyl group. By oxidation of the Grignard intermediate alcohol, the fourth ring opened, leaving the first three rings intact. Again, by reclosing the ring, the formation of an anhydride indicated that the ring contained five carbons.

Once a structure is deduced for a naturally occurring substance, the chemist's attention turns to laboratory synthesis. Wieland admitted, however, that in cholanic acid there were seven asymmetric carbon atoms, and ten in cholic acid. To an organic chemist in the late 1920's, this seemed insurmountable. Even Wieland did not see how the bile acids could be synthesized, but at the same time he was not a man to turn his back on a challenge, and he concluded his Nobel lecture with the thought that he had a "duty" to follow a line of research right through to the end.

Although Wieland never did synthesize the bile acids, he continued to work on and refine the data on their structures. In 1927 he believed that the first three rings in the structure surrounded a common carbon, making the molecules globular. In 1932 J. D. Bernal showed by X-ray studies that the sterol molecule is long and thin. Wieland and some English workers were soon able to come up with the structure that we support today.[1]

---

[1]Fieser, L. F. "Steroids"; in *Bio-organic Chemistry;* W. H. Freeman: San Francisco, 1968, pp 158–166.

In addition to his Nobel Prize winning research on the bile acids, Wieland was interested and involved in many areas of chemical experimentation. He published more than 400 books and articles in his lifetime. Wieland was very active in the research of organic nitrogen-containing compounds and was the first to produce stable organic nitrogen radicals. The importance of that breakthrough in radical chemistry can hardly be overemphasized. He also studied the reaction of nitrogen oxides with alkenes and aryl compounds.

Wieland also worked with furoxanes and the polymerization of fulminic acid. Many of his later efforts were centered on natural products. He contributed to the discovery of the structures of morphine and strychnine. His research on the alkaloids lobelia and curare are considered "masterpieces". The cyclopeptides phalloidine and amanitine were isolated because of Wieland's work on the "death cap" mushroom, and Wieland's group discovered an important class of pterin compounds while investigating the pigment of butterflies.[2]

Another very important area of research for Wieland was that of biological oxidations. He thought that the removal of hydrogens from organic compounds was the main oxidative process. Experimentally, he found that substances oxidized biologically were very easily dehydrogenated and in fact could proceed in the presence of metal catalysts such as palladium without any involvement of oxygen at all. Wieland used grape sugar at biological temperatures with oxygen-free palladium and effected a very rapid oxidation. Wieland went on to include reduction as another expression of one process of dehydrogenation.[3]

Some anecdotes about Wieland that give insight into his personality, sense of humor, and ideals can be found by reading *Was Nicht in den Annalen Steht*[4] Once Wieland was questioned about his experimental findings that showed the presence of phosphorus in the brain. He asserted that phosphorus did indeed occur in the brain, but he added that modern Germany was a phosphorus-poor country. Like his friend Adolf Windaus, Wieland was not totally in tune with the ideals of the Third Reich. A friend commented that some new government build-

[2]Wieland, H. *Nobel Lectures: Chemistry, 1922–1941;* Elsevier: Amsterdam, 1966; pp 103–104.

[3]*Source Book in Chemistry 1900–1950;* Leicester, H. M., Ed.; Harvard University: Cambridge, 1968; pp 335–340.

[4]Josef Hausen, Verlag Chemie: Weinheim, Germany, 1958.

ings were unfinished in the top floors as yet, and Wieland quipped that it was symbolic of the Nazi government.

One time an assistant was joking to some other students, and he announced loudly a list of famous chemists of the day including his mentor, Heinrich Wieland, but then added his name, Fritz von Werder, at the end of the list. Wieland had in the meantime come into the room quietly and unnoticed; he quickly added that Fritz could be right but advised him not to say it so loudly.

When Wieland died in 1957, the scientific community lost one of the truly great chemists of all time. Although he was recognized during his lifetime for his remarkable achievements, especially by the award of the Nobel Prize, his life stands now and for the future as an example to aspiring young scientists.

JOANNE M. FOLLWEILER
*Lafayette College*

## *Bibliography*

Farber, E. "Heinrich Wieland, 1877–1957"; In *Great Chemists*; Farber, E., Ed.; Interscience: New York, 1961; pp 1442–1451.

Hausen, J. *Was Nicht in den Annalen Steht;* Verlag Chemie: Weinheim, Germany, 1958.

Jones, D. P. "Wieland, Heinrich Otto"; In *Dictionary of Scientific Biography;* Gillispie, C. C., Ed.; Charles Scribner's Sons: New York, 1976; Vol. 14, pp 334–335.

Karrer, P. *Biographical Memoirs of Fellows of the Royal Society* **1958,** *4,* 341–352.

Lunde, G. "The 1927 and 1928 Nobel Chemistry Prize Winners, Wieland and Windaus"; *J. Chem. Educ.* **1930,** *7,* 1763–1771.

*Die Naturwissenschaften* **1942,** *30.* Issue commemorates Wieland's 65th birthday.

*Nobel Prize Winners;* Wasson, T., Ed.; H. W. Wilson: New York, 1987.

Wieland, H. "The Chemistry of the Bile Acids"; *Nobel Lectures: Chemistry 1922–1941;* Elsevier: Amsterdam, the Netherlands, 1966.

Wieland, H.; Dane, E. "Studies on the Constitution of the Bile Acids. XXXIX. Concerning 12-Hydroxycholanic Acid"; Wieland, H. "The Mechanism of Oxidation Processes"; *Source Book in Chemistry 1900–1950;* Leicester, Henry M., Ed.; Harvard University Press: Cambridge, Massachusetts, 1968; pp 288–294, 336–340.

Witkop, B. "Remembering Heinrich Wieland (1877–1957): Portrait of an Organic Chemist and Founder of Modern Biochemistry"; *Med. Res. Rev.* **1992,** *12 (3),* 195–274.

# *Adolf Windaus*

### 1876–1959

Copyright Nobel Foundation

In 1876 on Christmas Day, Adolf Windaus was born to Adolf Windaus, a textile manufacturer, and Margarete Elster in Berlin, Germany. The son, Adolf, was the recipient of the Nobel Prize in chemistry in 1928. The research that won him this highest honor was on sterols and their relation to other important biological chemicals such as the vitamins. He died on June 9, 1959.

Young Windaus attended an excellent grammar school in Berlin (*Französisches Gymnasium*). His interests at that time were settled on literature, but in 1895, at the age of 19, he undertook the study of medicine in Berlin and Freiburg.

While studying in Berlin, Windaus attended lectures by Emil Fischer, the Nobel laureate in chemistry for 1902. Fischer worked on many biological compounds such as proteins and sugars. An interest in chemistry prompted Windaus to study the subject under the well-known chemist Heinrich Kiliani. Fischer and Kiliani devised the famous synthetic method for making sugars from smaller sugars by the use of a cyanohydrin intermediate. Kiliani must have seen some promise in Windaus, because it was he who directed the young scientist to begin work in the sterol area, the work for which Windaus eventually received the Nobel Prize. Windaus continued to study medicine along with his chemical interests. True to the image of most successful scientists, Windaus exhibited a curiosity and energy rarely found. Besides medicinal and chemical studies, he also pursued an interest in zoology.

Windaus obtained his doctorate at the beginning of the new century. His dissertation was on cardiac poisons, particularly those available from the plant genus *Digitalis*. Most people today recognize the word *digitalis* and think of it as the name of a drug to control heart rate. After obtaining his degree, Windaus returned to Fischer in Berlin. At that time he became very good friends with Otto Diels, another Nobel Prize winner in the making. Diels shared the 1950 Nobel Prize in chemistry with Alder for their work in organic synthesis. The friendship between Windaus and Diels was to last until Diels's death in 1954.

Windaus also carried out important collaborative work with Franz Knoop, who is perhaps best known for his work on fatty acid degradation. Windaus and Knoop worked together on imidazole derivatives. From this work the hormone histamine was discovered. Since histamine was important as a drug, Windaus became an important chemist in the eyes of the German chemical industry.

In 1901 Windaus returned to work at Freiburg, and using Kiliani's suggestion he began work on cholesterol, the parent compounds, along with other related compounds. The structure of cholesterol was unknown at the time, and work in that area proved very fruitful. In 1919 Windaus converted cholesterol to cholanic acid. By that time, cholanic acid had been isolated by Heinrich Wieland, another friend and future Nobel Prize winner whose work interconnected with that of Windaus.

Alfred Hess invited Windaus to come to New York in 1925 to study vitamin D with him; another vitamin, $B_1$, was investigated by Windaus in cooperation with I. G. Farbenindustrie. Windaus proved that a thiazole ring and a pyrimidine ring were parts of the structure of the vitamin. An imidazole ring had been suggested previously by work done elsewhere. Chemotherapy became the major research interest of Windaus in the latter part of his career.

The research on sterols for which Windaus won the Nobel Prize was, according to the presentation speech by Professor H. G. Söderbaum, the Secretary of the Royal Swedish Academy of Science, intended to yield information on the essence of organic materials, their interconversions and transformations. The terms "diligent" and "resourceful" were applied by Professor Söderbaum. Windaus was also described in this speech as "overcoming even the greatest experimental difficulties" and using "lucidity in interpreting".

In his Nobel lecture on December 12, 1928, Adolf Windaus detailed his work on sterols. He defined sterols as large alcohols with no nitrogen content that are widely distributed in both the animal and

plant kingdoms, but not found in bacteria. The parent sterol, cholesterol, was first isolated from human gallstones. Cholesterol is found
throughout the human body and is an unsaturated compound containing 27 carbons, 46 hydrogens, and 1 oxygen. As found in nature, its
single alcohol group is, in part, esterified to various fatty acids. Although it is present in particularly large concentrations in the brain,
cholesterol can exhibit wide differences in the quantities found in human blood serum. High levels of cholesterol accumulate in and damage
the liver, and Windaus believed data gathered at the time that supported
a link between hardening of the arteries and a diet containing medium-
to-high levels of cholesterol.

Another important sterol that Windaus worked with extensively
was ergosterol, found in yeast and other fungi. Ergosterol has a formula of 27 carbons, 42 hydrogens, and 1 oxygen and contains three
double bonds (unsaturated centers) compared to the single double bond
in cholesterol.

The work of Windaus and his good friend Wieland became related
through a dihydrosterol: coprosterol. It occurs naturally, but does not
form from catalyzed hydrogenation of cholesterol. In the laboratory,
Windaus converted cholesterol to coprosterol by an initial treatment
with hydrochloric acid, to form allocholesterol, and a subsequent hydrogenation yielded coprosterol. Then, using coprosterol as the starting material, Windaus converted it to cholanic acid by utilizing an
oxidizing agent. The common bile acids derive from cholanic acid.
Wieland then went from cholanic acid to the steroids.

Windaus proved a connection between the steroids and vitamin
D. The ultraviolet irradiation of ergosterol between the wavelengths
of 248 nm and 313 nm gave vitamin D. Windaus also explored the conditions of irradiation. He found that even at very low temperatures the
reaction occurred, and, if irradiation continued too long, the vitamin
D was destroyed. Windaus tried to convert ergosterol into vitamin D
by chemical means alone, but was unsuccessful.

By continued experimental refinement, Windaus proved quantitatively that ergosterol and the active vitamin D have the same empirical formula. To differentiate between polymerization and isomerization, Windaus did molecular weight studies that showed the sterol
and vitamin to have the same molecular weight within experimental
error.

Windaus then investigated the possibility of an alcohol and double
bond in ergosterol that might isomerize into a ketone group in the

vitamin. This could certainly happen during the irradiation procedure. Careful laboratory work showed that the hydroxyl group did not disappear during the irradiation, so the alcohol could not be converted to a ketone. Not satisfied with just one method, Windaus prepared the ketone compound synthetically and found that it was inactive as a vitamin.

Another possibility was that there was isomerization of the double bond to a different position. Using quantitative methods, Windaus showed that the number of double bonds does not change in ergosterol and vitamin D, proving that the isomerization is a function of a change in position of a double bond or a steric rearrangement. Today irradiated ergosterol is added to milk to enhance its nutritional value by supplementary vitamin D.

What areas of research other than those responsible for his Nobel Prize did Windaus pursue? *Chemical Abstracts,* in its first 10-year index, lists some of the following subjects as the topics of publications by Windaus alone: grape sugar decomposition, thermal decomposition of sterols, imineazole ring synthesis, and fission. With other authors Windaus is listed as reporting on amines, cardiac poisons, imineazole derivatives, oxidation reactions, and hydroxide interactions with glucose and other sugars.

As an educator, Windaus rose steadily through the professorial ranks. In 1906 he was an assistant professor and by 1913 had obtained the title of Professor of Applied Medicinal Chemistry at Göttingen, and he retired in 1944 with the title of Director of the Laboratory for General Chemistry (Wöhler Institute). Two prominent students of Windaus were Adolf Butenandt (Nobel Prize in chemistry, 1939) and Hans Brockmann.

Besides the Nobel Prize in 1928, Windaus was awarded the following: Pasteur Medal (1938); Goethe Medal (1941); Grand Order of Merit (1951); Grand Order of Merit with Star (1956); and Order for Merit, Peace Class (1952).

Adolf Windaus married Elizabeth Resau in 1915, and they had two sons, Gunter and Gustav, and a daughter Margarete.

Windaus was a brilliant chemist, and some of the personal characteristics he exhibited in order to do the kind of research he did were mentioned before in reference to his Nobel Prize presentation speech. What other personality traits were known about the man? Did he have a sense of humor? Were his ideals and beliefs of a lofty nature? Some of these questions can be answered by the book *Was Nicht in den Annalen*

*Steht.*[1] It contains several anecdotes about Windaus that give some insight into his character. For instance, Windaus was well-known to oppose Nazi ideology. The university library's copy of *Mein Kampf* by Adolf Hitler had handwritten in it: "by order of the government". Many people believed that the handwriting was that of Windaus.

Another incident tells of Windaus's returning from an out-of-town meeting and addressing his students. He emphasized to them that he would never oppose the Third Reich with force of arms. He made this statement several times. Literally, if reported to the government, there could be no fault found with his statement, but it was obvious to his students that he was reaffirming his pacifism and showing that he would resist the Third Reich in every way except a show of force.

Windaus was a careful teacher with a sense of humor; he had a special relationship with his students. In an anecdote about the labeling of chemicals in his lab, we find that he insisted on strict terminology and did not like to see the term "substance" applied unless the chemical was pure, with a constant melting point, for instance. Once he found a student with a chemical that wasn't purified enough, yet it was labeled "substance". Windaus responded sadly, while shaking his head, that the chemical was not a substance but, at best, only a product or a material.

Windaus was obviously a great experimental chemist. His work formed a building block of modern biochemistry. His strength was not so much in theoretical predictions, but in practical knowledge, based on experimental facts.

JOANNE M. FOLLWEILER
*Lafayette College*

# *Bibliography*

Bailey, H. *The Vitamin Pioneers;* Rodale Books: Emmaus, PA, 1968.

Butenandt, A. "Adolf Windaus, December 25, 1876 to June 9, 1959"; *Bayr. Akad. Wiss: Jahrb.* **1960,** 157–64.

---

[1] Hausen, Josef; Verlag Chemie: Weinheim, Germany, 1958; pp 99–100.

Fieser, L. "Steroids"; In *Bio-organic Chemistry;* W. H. Freeman: San Francisco, CA, 1968; pp 158–166.

Hausen, J. *Was Nicht in den Annalen Steht;* Verlag Chemie: Weinheim, Germany, 1985.

Ihde, A. J. *The Development of Modern Chemistry;* Harper & Row: New York, 1964.

Leicester, H. M. *Development of Biochemical Concepts from Ancient to Modern Times;* Harvard University Press: Cambridge, MA, 1974.

Leicester, H. M. "Windaus, Adolf Otto Reinhold"; *Dictionary of Scientific Biography;* Gillispie, C. C., Ed.; Charles Scribner's Sons: New York, 1976; Vol. 14, pp 443–446.

*Nobel Prize Winners;* Wasson., T., Ed.; H. W. Wilson: New York, 1987.

*Source Book in Chemistry 1900–1950;* Leicester, H. M., Ed.; Harvard University: Cambridge, MA, 1968.

Windaus, A. "Constitution of Sterols and Their Connection with Other Substances Occurring in Nature"; In *Nobel Lectures: Chemistry 1922–1941;* Elsevier: Amsterdam, the Netherlands, 1966.

# Hans von Euler-Chelpin

## 1873–1964

Hans Karl August Simon von Euler-Chelpin was born in Augsburg, Germany, on February 15, 1873, and died in Stockholm, Sweden, on November 6, 1964. Although born in Germany and always a German patriot, Hans von Euler-Chelpin played a critical role in the development of chemistry in Sweden. His efforts to develop and institutionalize biochemistry from within chemistry proper helped make Stockholm an internationally significant center in this field.

Born to Rigas von Euler-Chelpin, a captain in the Royal Bavarian Regiment and of the same familial lineage as the 18th-century Swiss mathematician Leonhard Euler, and to Gabrielle Furtner, Hans was exposed to and became part of upper-class German culture. As a young man, he moved comfortably between the worlds of the aristocrats and the bohemians. After receiving primary education in Munich, Würzburg, and Ulm, he studied art between 1891 and 1893 at the Munich Academy of Art. His keen interest in the theory of colors brought him into contact with science, which he decided to pursue professionally.

Euler-Chelpin left Bavaria and enrolled in 1893 at the University of Berlin, the crown jewel of the Reich's academic world. There he followed the lectures of Emil Warburg and Max Planck in physics and of Emil Fischer in chemistry. He then turned to investigators who attempted to employ physics to explore chemical problems. First he spent two years in Göttingen working with Walther Nernst, and then in the summer of 1897 he moved to Stockholm to become the assistant to another of the founding Ionists, Svante Arrhenius. After qualifying as a *Privatdocent* in physical chemistry at the fledgling Stockholm Högskola (later the University of Stockholm), he spent the summers of 1899 and 1900 with J. H. van't Hoff in Berlin.

Although Euler-Chelpin was working with Arrhenius and had close ties to other leaders of the new specialty of physical chemistry, he gradually turned his interest to organic chemistry. The turn might well have been professionally expedient: Organic, not physical, chemistry was generally the specialization needed to obtain a professorship. He visited the laboratories of a number of leading German organic chemists; that of Eduard Buchner in Berlin proved especially eventful for his future research interests. Euler-Chelpin's early research had primarily focused on the inorganic catalytic reactions. He began working on organic substances in collaboration with his wife Astrid Cleve and finally he took up problems derived from Buchner's work on fermentation. In 1906 he became professor of general and organic chemistry at the Stockholm Högskola.

Euler-Chelpin was married twice. His first marriage (1902–1912) to Astrid Cleve, daughter of Uppsala University professor of chemistry P. T. Cleve, resulted in five children. One son from this marriage, Ulf Svante von Euler, became a noted physiologist and also received a Nobel Prize. In 1913 Euler-Chelpin married Elisabeth, Baronness Uggla, with whom he had four children. Although he became a Swedish citizen in 1902, his devotion to Germany endured throughout his life. During World War I he reported for duty in the German army. In 1915 he entered the air force and eventually became the commander of a bomber squadron. He endeavored to meet his annual teaching obligations in Stockholm by compressing his lectures into one semester. In the 1930s he accepted leadership positions in various German-Swedish friendship societies and actively worked to promote and defend German interests. He considered accepting a high-level administrative position in the German research establishment, but the offer became ensnarled in political-bureaucratic intrigues; eventually he worked in a diplomatic capacity to assist Germany during World War II. His affection and loyalty was to Germany as a nation and a cultural entity; although he was not an adherent to the National Socialist party, he seems to have been unable successfully to separate party from nation. He aided some Jewish refugees and accepted a select handful into his laboratory; in science, ability and status mattered the most to him.

Euler-Chelpin shared the 1929 Nobel Prize in chemistry with Arthur Harden "for their investigations on the fermentation of sugar and fermentative enzymes". The work derived from Buchner's discovery in 1896 that cell-free yeast juice could still ferment sugar; the enzyme responsible, which he precipitated from the juice, was called "zymase". Harden, working with William John Young, separated the juice into two components, neither of which alone was able to induce fermentation but when mixed together fermented sugar normally. Harden asserted that fermentation required the presence of an enzyme consisting of relatively large molecules, the actual zymase, and a complementary substance of relatively small molecules. Few researchers pursued Harden's finding; many expressed skepticism regarding the existence of the coenzyme. Euler-Chelpin nevertheless began in 1923 what turned into a decade of study of the low-molecular-weight component, cozymase.

To clarify the chemistry of fermentation, Euler-Chelpin recognized the need to identify the substrates at each phase of the process. Working with Ragnar Nilsson, he devised the means of using various toxic inhibitors to arrest the fermentation at specific stages, thus allowing an investigation of each step. He began with Harden and Young's observation that phosphoric acid plays a role in fermentation by giving rise to sugar phosphates. First showing that cozymase played a role leading to the formation of the hexose diphosphoric acid ester that Harden had identified, Euler-Chelpin and Nilsson claimed that cozymase was involved in the transformation of aldehydes into alcohol and acid. Its significance for carbohydrate and aldehyde reactions had to be the same in fermentation and in living processes. Together with Karl Myrbäck, Euler-Chelpin managed to do that which Harden could not: obtain a sufficiently concentrated sample of cozymase to be able to determine its chemical nature. Through repeated purifications they obtained a sample many hundred times more active than that of the original yeast juice. Consequently its molecular weight was determined to be 490; its nature seemed to be that of a nucleotide since it lost its activity when under the influence of nucleosidase, an enzyme that splits up such substances. Further, cozymase's base was assumed to be adenine and its carbohydrate largely a pentose. Finally, Euler-Chelpin established that a coenzyme had to be present to activate the enzymes responsible for cellular biological processes of oxidation/reduction. This result indicated why cozymase occurs frequently in vegetable and animal tissue.

Nominations for Euler-Chelpin to receive a Nobel Prize came exclusively from a small coterie of German and Nordic colleagues. Richard Willstätter and Planck had nominated him on several earlier occasions. The single nomination to split a prize between Euler-Chelpin and Harden came from Carl Neuberg, who himself had been nominated on several occasions and whom Euler-Chelpin later championed, unsuccessfully, for a prize.

Euler-Chelpin continued research during the next decade at an almost feverish pace. His enthusiasm frequently turned to impatience with his co-workers and assistants who could not produce results on demand. Among the many projects that he and his collaborators worked on was the demonstration of the relationship between vitamin A and carotenoids. Much of this work was done with Paul Karrer in Zurich; together they made important advances in determining the structure of ascorbic acid. When Otto Warburg showed nicotinamide to be a cofactor in erythrocytes, Euler-Chelpin picked up this lead and identified nicotinamide in cozymase. One of a stream of German assistants, Fritz Schlenk, then showed the chemical structure of cozymase to be a diphosphopyridine nucleotide. Further outgrowth of the research entailed elucidating the role of nicotinamide and thiamine (vitamin $B_1$) in metabolically active compounds. In 1935 Euler-Chelpin began collaborating with George de Hevesy on a technique for labeling the nucleic acid present in tumors so that their behavior could be traced.

Euler-Chelpin began his professional career at an institution lacking in laboratory and material resources; by the time he retired he had created two major research institutes that commanded international attention and financing. His career reveals a series of strategies for overcoming the difficulties of working in a relatively impoverished scientific community. When he came to the Stockholm Högskola, he came to a privately funded institution almost devoid of laboratory facilities. Although the situation gradually improved, he still considered leaving in 1921 for either Berlin or Vienna, but he reconsidered when Germany's and Austria's postwar economic inflation destroyed the possibility of establishing a research laboratory and a vital research school. During the 1920s he turned to the Rockefeller Foundation and the Knut and Alice Wallenberg Foundation for money to build an institute for biochemistry in Stockholm. Pointing to the practical benefits of such an institute, he convinced the Stockholm city council to donate land for the building. His Nobel Prize brought him greater prestige and greater access to gifts and grants. Still, even after the institute opened in 1930 the question of funding the ever-growing number of projects and assistants constantly haunted Euler-Chelpin. To obtain funds he worked with health officials on the vitamin enrichment of foods and with pharmaceutical cooperatives on the production of artificial mineral waters and vitamin pills. Perhaps the need to seek funding for the laboratory and to obtain the means for an additional institute for organic chemical research prompted Euler-Chelpin to transform laboratory results, preliminary and completed, into a never-ending torrent of publications carrying his own name and those of co-workers. In 1936 he managed to establish an institute for organic chemistry, relying once again chiefly on assistance from the Rockefeller and Wallenberg Foundations.

Euler-Chelpin was elected to the Swedish Academy of Sciences in 1914 and to the Nobel Committee for Chemistry in 1929. His elections, and his receipt of the Nobel Prize, were not universally applauded by his chemist colleagues. At stake was the extent to which biochemistry should be considered part of chemistry proper. Indeed, on the committee Euler-Chelpin frequently championed biochemists. He especially supported those who worked in areas close to his own or those with whom he collaborated, such as Paul Karrer. Although his efforts, not always successful, prompted calls for rewarding achievements in general chemical principles, such as the work of Peter Debye and Niels Bjerrum, Euler-Chelpin did manage to integrate biochemical research yet further into the disciplinary structure of chemistry.

Euler-Chelpin received many honors in addition to the Nobel Prize. He was awarded the Grand Cross for Federal Service of the German Federal Republic as well as honorary degrees from universities in Stockholm, Zurich, Athens, Kiel, Bern, Turin, and New Brunswick. He was a member or foreign member of several national and professional academies.

ROBERT MARC FRIEDMAN
*University of California, San Diego*

# Bibliography

Euler-Chelpin, H. von. *Biokatalysatoren;* F. Enke: Stuttgart, Germany, 1930.

Euler-Chelpin, H. von. *Chemie der Enzyme;* J. F. Bergmann: Munich, 1920–1934; 2 vols. published in 4 parts.

Euler-Chelpin, H. von. *Chemotherapie und Prophylaxe des Krebses*; Thieme: Stuttgart, Germany, 1962.

Euler-Chelpin, H. von. *Grundlagen und Ergebnisse der Pflanzenchemie*; 3 vols.; F. Vieweg und Sohn: Braunschweig, Germany, 1908–1909.

Euler-Chelpin, H. von; *Homogene Katalyse;* F. Enke: Stuttgart, Germany, 1931.

Euler-Chelpin, H. von; Hasselquist, H. *Redictone, ihre chemischen Eigenschaften und biochemischen Wirkungen;* F. Enke: Stuttgart, 1950.

Euler-Chelpin, H. von; Lindner, P. *Chemie der Hefe, und der alkoholischen Gärung;* Akademische Verlagsgesellschaft: Leipzig, 1915.

Euler-Chelpin, H. von; Skarzynski, B. *Biochemie der Tumoren;* Enke: Stuttgart, Germany, 1942.

Ihde, A. J. "Euler-Chelpin, Hans von"; *Dictionary of Scientific Biography;* Gillispie, C. C., Ed.; Charles Scribner's Sons: New York, 1971; Vol. 4, pp 485–486.

Lynen, F. "Hans von Euler-Chelpin"; *Bayerische Akademie der Wissenschaften, Jahrbuch* (1965), pp 206–212.

Schlenk, F. "The Dawn of Nicotinamide Coenzyme Research"; *Trends in Biochemical Sciences* **1984,** *9,* 383–386.

# *Arthur Harden*

## 1865–1940

Copyright Nobel Foundation

Arthur Harden was born in Manchester, England, on October 12, 1865, and died on June 17, 1940, at Bourne End, Buckinghamshire. He shared the Nobel Prize in chemistry for 1929 with Hans von Euler-Chelpin. The second son and the third of nine children of Albert Tyas Harden, a Manchester merchant, and his wife Eliza (née Macalister) of Paisley, Scotland, Harden received his early education at Victoria Park School in Manchester and then at Tettenhall College in Staffordshire. At the early age of 16 he entered Owens College (the forerunner of the University of Manchester), where he qualified with first-class honors in chemistry in 1885. He immediately began his postgraduate work there. Following a paper published with J. B. Cohen, the college awarded him in 1886 the Dalton scholarship, which allowed him to study under Otto Fischer at Erlangen; his work there on synthetic organic chemistry resulted in the award of a doctoral degree in 1888.

Harden returned to Manchester as assistant lecturer in the chemistry department and remained there for nine years; during this time he published only two research papers and "had not manifested an overpowering desire for scientific research".[1] He concentrated on teaching and literary work, which included collaboration in writing and revising textbooks. After two unsuccessful attempts to move from Man-

---

[1]Hopkins, F. G.; Martin; C. J. "Arthur Harden"; *Obituary Notices of Fellows of the Royal Society, 1942*, **4,** 3–14.

chester, Harden was eventually appointed chemist at the Jenner Institute of Preventive Medicine in London in 1897. The choice was made on the grounds of his wide knowledge of chemistry and his proven ability as a teacher. He was responsible for the chemistry course for medical doctors wishing to become medical officers of health. This course was eventually taken over by the Medical Schools of the University of London, which offered a Diploma in Public Health, and thus Harden became free to carry out full-time research at the Lister Institute (as the Jenner Institute was renamed in 1903); he remained there for the rest of his professional life. When the head of the biochemistry department left in 1905, it was combined with the chemistry department, then headed by Harden, who was put in charge of the new expanded Biochemical Department. In recognition of his distinguished work on bacterial chemistry and alcohol fermentation, to be discussed later, Harden was appointed the first Professor of Biochemistry at the University of London in 1912; it was only the second time that a chair of biochemistry was created in the United Kingdom; the first was the Johnston Chair of Biochemistry founded at the University of Liverpool in 1902.

Harden began his work on alcohol fermentation in 1898 at the suggestion of a senior colleague, Alan Macfadyen. The starting point was the recent dramatic discovery by Eduard Büchner that cell-free juice obtained from yeast would ferment sugar. Harden's two germinal observations that led to his election to the Fellowship of the Royal Society in 1909 and to the Nobel Prize for chemistry were published in 1904–1905.

Great progress has often been made in biochemistry by the development and use of new, sensitive, and sophisticated methodology, and this was the case with Harden's work on fermentation. Today it seems a very simple development, but the abandonment of the standard, tedious, and time-consuming gravimetric method for determining $CO_2$ and the development of a quick and simple volumetric method allowed him to assay samples at 10-minute intervals and thus determine the reaction rates on which his two memorable observations were based.

The first discovery was that of a cofactor, which he called "coferment", without which the enzyme (complex) zymase would not carry out fermentation at a significant rate. The addition of boiled, autolyzed yeast juice to a mixture of zymase and sugar greatly increased the fermentation rate. The factor responsible was thermostable, dialyzable, precipitated by 75% alcohol, and destroyed by incineration. Apart from discovering phosphate in the ash, Harden did not push this aspect of the work further, and it remained for Euler-Chelpin to demonstrate the nature of the components (adenine, nicotinic acid, ribose) of the cofactor, which he named cozymase and which we now know as nicotinamide-adenine dinicelotide phosphate. However, the new concept that a small, heat-stable cofactor was necessary for the activity of an enzyme was to have an enormous influence on the future study of metabolic biochemistry: Very few known reaction sequences can function without at least one step requiring a cofactor in addition to the specific substrate for the enzyme concerned.

The discovery of phosphate in the ash of co-ferment led to experiments perhaps even more important than the discovery of co-ferment itself. The effect of adding potassium phosphate to yeast juice led to a startling increase, by at least sevenfold, in the rate of liberation of $CO_2$ within 15 minutes of the start of the experiment; thereafter it rapidly fell to the original rate measured before the addition of phosphate. Moreover, he found that the inorganic phosphate had been "fixed", and then with Young showed that it had been fixed into a hexose diphosphate, which Young eventually identified as fructose-1, 6-diphosphate.

Harden was intrigued by the difference in behavior between intact yeast cells and the cell-free extracts obtained therefrom: In the former, fermentation proceeded at a uniform and faster rate than in the latter. This indicated that bound phosphate had to be continuously liberated so that a stimulatory amount of inorganic phosphate was always available. Support for this view, which incidentally had in it the germ of the idea of cyclic metabolic processes to be demonstrated so elegantly some years later by Hans Krebs, was obtained when Harden with Young found in yeast a phosphatase that would hydrolyze their hexose diphosphate. Furthermore, the addition of phosphatase to a preparation of zymase stimulated the fermentation rate much as if more inorganic phosphate had been added. However, the rate could not be maintained, as expected, but it did last longer than when phosphate was added. The rate was, however, maintained if arsenate was added together with the phosphatase, an unexpected result noted by Harden in his Nobel lecture as an "undeserved reward for thinking chemically about a biochemical problem, a dangerous thing to do". The experiment was, in fact, carried out to see if arsenate, which has very similar chemical properties to phosphate, would mimic phosphate in biological systems. Such was not the case, and Harden ascribed the action of arsenate on his cell-free system as being due to the inhibition of the destruction of phosphatase by the system itself. This discovery of the involvement of phosphate was the experimental breakthrough that allowed the details of the steps in glycolysis to be fully elucidated by Meyerhof, among others, and it led finally to the demonstration that phosphate esters were essential in almost all metabolic processes involving energy coupling and eventually to the demonstration that adenosine triphosphate was the universal molecule for energy transfer in living cells.

Apart from the two wellspring observations from which so much flowed, Harden's later work does not appear to have had the penetration of his 1907 work. Perhaps the reason was that in 1914 he abandoned glycolysis research at a critical stage in order to carry out investigations relevant to the war effort: He worked on the accessory food factors, the structures of which were then unknown. The lack of knowledge of those factors caused beriberi and scurvy, diseases that were observed among British troops in Asia and Africa. Another reason perhaps was that once the breakthrough had been made there were many extremely able biochemists around ready to exploit it. Indeed it was the ensuing developments that impressed on the Nobel Prize committee the great significance of Harden's early discoveries, which became fully appreciated during the 22 years before the award of the Nobel Prize. Harden had no doubt that the award was of much greater significance to biochemistry than to him personally: "If I may for a moment yield to the Scandinavian atmosphere of saga and fairy tale, Biochemistry was for long the Cinderella of the Sciences, lorded over by her elder—though I will not say ugly—sisters, Chemistry and Physiology. But now the secret visit to the ball has been paid, the fur slipper has been found and brought home [no doubt to the Nobel Institute] by the Prince, and Biochemistry, raised to a position of proud independence, knocks boldly at the door of the Palace of Life itself".[2] According to his obituary, written by colleagues who knew him well, Harden's outstanding qualities as an investigator were clarity of mind, precision of observation, and a capacity to analyze dispassionately the results of an experiment and define their significance. He mistrusted the use of his imagination beyond a few paces in advance of the facts. "Had he exercised less restraint, he might have gone further; as it was he had little to withdraw".[3] One appreciates the viewpoint, but it is a rather unenthusiastic comment on a Nobel Prize winner.

---

[2]Harden, 1929, p 1–11.
[3]Hopkins and Martin, op. cit.

Harden never lost his early interest in scientific literature, and from 1913 until 1937 he was editor of the *Biochemical Journal*. When he retired it was calculated that he had read 18 million words in proof form during his tenure as editor.[4]

Apart from the Nobel Prize and the Fellowship of the Royal Society, many other awards came Harden's way, including the Davy Medal of the Royal Society in 1935, a knighthood in 1936, and Honorary Membership of the Biochemical Society in 1938. In addition, he received honorary degrees from the Universities of Manchester, Liverpool, and Athens and was elected a member of the Kaiserlich Leopold Deutsche Academie der Naturforschung of Halle. His memory is perpetuated by the Biochemical Society in their annual Harden Conferences, founded with the help of money he bequeathed to the Society.

In 1900 Harden married Georgina Sydney, the elder daughter of Cyprian Wynard Bridge of Christchurch, New Zealand. She died in 1928; they had no children.[5]

According to those who knew him.[6] Harden was very reserved, generally cheerful but never exuberant, of an equable temperament, and had a dry humor. "He was preeminently a fair man."

T. W. GOODWIN
*The University of Liverpool*

# Bibliography

Goodwin, T. W. *A History of the Biochemical Society;* Biochemical Society: London, 1987.

Harden, A. "The Function of Phosphate in Alcoholic Fermentation"; *Les Prix Nobel, 1929;* Norstedt & Sons: Stockholm, Sweden, 1930; pp 1–11.

Harden, A.; Young, W. J. "The Alcoholic Ferment of the Yeast Juice [Part I]"; *Proc. Roy. Soc.* **1906,** 77B, 405–420.

---

[4]Goodwin, 1987.

[5]Martin, C. J. *Dictionary of National Biography, 1931–1940;* Wickham, L. G., Ed.; Legg: Oxford, 1949; p 395.

[6]Hopkins and Martin, op. cit.

Harden, A.; Young, W. J. "The Alcoholic Ferment of Yeast Juice [Part II]; The Co-Ferment of Yeast"; *Proc. Roy. Soc.* **1906,** *78B,* 309–313.

Harden, A.; Young, W. J. "The Alcoholic Ferment of Yeast Juice [Part III]. The Function of Phosphates in the Fermentation of Glucose by Yeast Juice"; *Proc. Roy. Soc.* **1907,** *80B,* 299–311.

Harden, A.; Young, W. J. "The Alcoholic Ferment of Yeast-Juice [Part V]. The Function of Phosphates in Alcoholic Fermentation"; *Proc. Roy. Soc.* **1909,** *82B,* 321–330.

Harden, A.; Young, W. J. "The Alcoholic Ferment of Yeast-Juice. [Part VI]. The Influence of Arsenates and Arsenites on the Fermentation of Sugars by Yeast-Juice"; *Proc. Roy. Soc.* **1910–1911,** *83B,* 451–475.

Hopkins, F. G.; Martin, C. J. "Arthur Harden"; *Obit. Notices of Fellows of the Royal Society* **1942,** *4,* 3–14.

Ihde, A. J. *The Development of Modern Chemistry*; Harper & Row: New York, 1964.

Ihde, A. J. "Harden, Arthur"; In *Dictionary of Scientific Biography;* Gillispie, C. C., Ed.; Scribner: New York, 1972; Vol. 6, pp 110–112.

Kohler, R. E. "The Background to Arthur Harden's Discovery of Cozymase"; *Bull. Hist. Med.* **1974,** *48,* 22–40.

Leicester, H. M. *Developments of Biochemical Concepts from Ancient to Modern Times*; Harvard University Press: Cambridge, MA, **1974.**

Martin, C. J. "Sir Arthur Harden (1865–1940)"; *Dictionary of National Biography, 1931–1940*; Wickham, L. G., Ed.; Legg: Oxford, 1949; pp 395–397.

Smedley-Maclean, I. "A. Harden"; *Biochem. J.* **1941,** *35,* 1071–1081.

# Hans Fischer

## 1881–1945

Hans Fischer received the 1930 Nobel Prize in chemistry for his research on the structures of hemin and chlorophyll. Born to Eugene and Anna Herdegen Fischer on July 27, 1881, he died on March 31, 1945.

Fischer was exposed to chemistry at an early age, as his father was a chemist at Meister, Lucius, and Brüning in Höchst-am-Main, Germany, when Hans was born. The family moved a few years afterward when Eugene Fischer became the director of the laboratory of the Kalle Dye Works in Biebrich.

Fischer went to primary school in Stuttgart and then to a classical *humanistische Gymnasium* in Wiesbaden, matriculating in 1899. He then read chemistry and medicine at the University of Lausanne and at Marburg, where he received his doctorate in chemistry in 1904 under the direction of Theodor Zincke. Fischer continued to study medicine, obtaining his license to practice in 1906 and qualifying for his M.D. in 1908.

He began his working years studying peptides and sugars under the direction of Emil Fischer at the First Berlin Chemical Institute. He was invited to work at the Second Medical Clinic in Munich in 1910 by Friedrich von Müller, who was interested in the similarities of blood and bile pigments. Here Fischer began to investigate the composition of the bile pigment bilirubin, and so began an investigation that he was to continue for the rest of his life.

He returned to Munich in 1911 and there qualified as a lecturer on internal medicine in 1912. In 1913 he became a lecturer in physiology at the Physiological Institute in Munich, where he had no teaching duties but worked with the physiologist Otto Frank. In 1916 Adolf Windaus left his position at the University of Innsbruck, and Fischer succeeded him as a Professor of Medicinal Chemistry before moving to a similar position at the University of Vienna in 1918. The years during World War I and the following period of reconstruction were difficult for researchers in Germany because finances, working conditions, and the availability of students were poor; Fischer accomplished little of consequence during this time.

The beauty of the countryside surrounding the University of Innsbruck appealed to Fischer, and he enjoyed skiing and mountain climbing. Unfortunately, Fischer's father tragically died during a climb in the Alps when, before the eyes of his son, the elder Fischer fell irretrievably into a crevice.[1]

Working conditions had improved considerably by 1921, when Fischer succeeded Heinrich Wieland as the head of the Institute of Organic Chemistry at the Technische Hochschule in Munich. At the time of his appointment doubts were voiced about the wisdom of appointing a professor interested in medicine to a position at a technical university, but the doubts were soon quelled as Fischer organized a large and productive laboratory.[2]

In 1921, at the Munich Technische Hochschule, Fischer began research on hemin, the pigment derivative from blood hemoglobin. Work by Nencki, Küster, Piloty, and Willstätter had demonstrated that bilirubin and hemin were chemically similar, and in 1912 Küster had shown that the hemin molecule had the formula $C_{34}H_{32}O_4N_4FeCl$. The molecule contained more than 70 atoms, and there were a multitude of ways in which they could be assembled to form molecules, so Fischer set out to determine which way they were arranged in hemin.

It was found that removing the iron from hemin resulted in the formation of one of the porphyrins, a class of molecules now known to contain many naturally occurring pigments and enzymes. The degradation of hemin resulted in the formation of pyrroles, which are pentagonal closed-ring systems with four carbon atoms and one nitrogen atom. Fischer began a systematic study of porphyrins and their constituent pyrroles and found that the principal differences among the

[1]Wieland, H. "Hans Fischer, 1881–1945"; *Great Chemists*; Farber, E., Ed.; Oesper, R. E., Tr.; Interscience: New York, 1961; pp 1527–1533.
[2]ibid.

pyrroles were governed by which atoms were attached to the carbon corners of their rings: Some pyrroles had $-CH_3$ groups; others had $-CH_2CH_3$, $-CH_2CH_2COOH$, or other groups.

The synthesis of pyrroles was a very underdeveloped field, and Fischer and his co-workers developed methods to assemble the pyrroles that had been identified as constituents of various porphyrin molecules. The methods developed were tedious and time-consuming, and they yielded only small amounts of material, so Fischer later employed "cooks" to do these preparations, reserving his doctoral students' energies for the less-routine aspects of the research. Microanalytical techniques were also developed, and a busy staff of microanalysts eventually did over sixty thousand analyses in his lab. Using reactions discovered by Fischer, four of the pyrroles were combined to form the first synthesized porphyrin in 1926. The synthesis of hemin, for which Fischer won the 1930 Nobel Prize, eluded him until 1929.

After the structure of hemin was determined, Fischer turned his attention to chlorophyll. Fischer's chlorophyll research built on that of Richard Willstätter, who received the Nobel Prize for developing purification methods and doing preliminary structural research on chlorophyll. Fischer looked at the pyrroles that resulted from the decomposition of chlorophyll, determined their composition, and developed techniques to synthesize them. He partially succeeded in combining these pyrroles to synthesize chlorophyll, but had not completed the work before his death.

After starting the chlorophyll research, Fischer also returned to the bile pigment bilirubin, the first porphyrin he had studied. Its synthesis proved to be almost as difficult as that of chlorophyll because of the difficulty of correctly joining the pyrrole units in the molecule, but he succeeded in 1944, thus finishing a project started by his old supervisor, Friedrich von Müller. In addition to determining the bilirubin structure, he showed that the other bile pigments were linear tetrapyrroles that were formed in nature by oxidative fission of the porphyrin ring.

Fischer worked with other porphyrins and in fact synthesized more than 130 of them. He also studied the relationships among various naturally occurring porphyrins and found how some of them could be interconverted. He was the author of over 300 papers, including 129 on chlorophyll alone. His work cleared the way for further porphyrin research, much of which continues today.

Fischer received the Nobel Prize of 1930 for his work with porphyrins, including the 1929 synthesis of hemin. The title of Privy

Councillor was conferred on him by Germany in 1925, and he received the Liebig Medal in 1929, an honorary doctorate from Harvard in 1936, and the Davy Medal in 1937.

By reputation Fischer was retiring and unassuming.[3] He did not lead a pretentious life and for many years carried a knapsack as luggage, even on overseas lecture trips. His travel habits became more socially acceptable after his 1935 marriage to Wiltrud Haufe, who was 30 years his junior.

Like many scientists at the time, he did not discuss his work until it was published. Although he was very sensitive to criticism of his scientific endeavors and sometimes left a forbidding first impression, another side of his personality was evident to those who knew him well: He was very concerned for his students and they in turn were devoted to him. Even though the people in his large laboratory worked industriously, there were lighter moments as well, often directed by Fischer himself. A new doctoral candidate was likely to return to his desk after his oral exam to find a cartoon reminding him that he was mortal, and the stock of large flasks was used at Christmastime to synthesize "Christmas pyrrole" by a process using alcohol.[4]

These lighter moments and the devotion of his students were overshadowed by events during World War II, particularly the almost total destruction of his institute by bombing. On Easter Sunday, March 31, 1945, a depressed Hans Fischer took his own life.

ZEXIA BARNES
*Morehead State University*

## Bibliography

Fieser, L. F.; Fieser, M. *Reagents for Organic Synthesis;* John Wiley: New York, 1967.

Findlay, A. *A Hundred Years of Chemistry;* Gerald Duckworth: London, 1965; p 184.

Fischer, H. "On Haemin and the Relationships between Haemin and Chlorophyll"; *Nobel Lectures, Chemistry, 1922–1941*; Elsevier: Amsterdam, 1966; pp 165–183.

---

[3]Wieland, H. "Hans Fischer, 1881–1945"; *Great Chemists*; Farber, E., Ed.; Osper, R. E., Tr.; Interscience: New York, 1961; pp 1527–1533.

[4]MacDonald, S. F. "Prof. Hans Fischer"; *Nature* **1947,** *160* (October 11), pp 494–495.

Fischer, H.; Orth, H. *Die Chemie des Pyrrols*; Akademische Verlagsgesellschaft: Leipzig, 1934–1940; 4 vols.

Leicester, H. M. "Hans Fischer"; In *Dictionary of Scientific Biography;* Gillispie, C. C., Ed.; Charles Scribner's Sons: New York, 1980; Vol. 15, pp 157–158.

Macdonald, S. F. "Prof. Hans Fischer"; *Nature (London)* **1946,** *160 (October 11),* 494–495.

Roberts, J. D. *Basic Principles of Organic Chemistry*; W. A. Benjamin: New York, 1965.

Wieland, H. "Hans Fischer, 1881–1945"; *Great Chemists;* Farber, E., Ed.; Oesper, R. E., Tr.; Interscience: New York, 1961; pp 1527–1533.

# *Friedrich Bergius*

## 1884–1949

Copyright Nobel Foundation

Friedrich Bergius shared the 1931 Nobel Prize in chemistry with Carl Bosch for their work on chemical high-pressure methods. He was born to Heinrich and Marie Hasse Bergius near Breslau, Germany, on October 11, 1884. After a previous marriage, he married Ottilie Krazert. He had two sons and a daughter. After World War II, Bergius left Germany and died in Buenos Aires, Argentina, in March 1949.

Bergius was exposed to chemistry and academics at an early age, as his father owned a chemical factory in Goldschmeiden, and his maternal grandfather was a professor of economics. During high school Bergius took a personal interest in chemistry. He spent a considerable amount of time at the laboratory and works of his father's chemical factory and carried out experiments under the supervision of his father. Before he entered the university, his father sent him to the laboratory of a foundry in the Ruhr industrial district, where he spent six months learning some aspects of practical metallurgical chemistry. These experiences gave Bergius an early exposure to industrial chemistry,[1] the field that most influenced his later choice of research projects.

In 1903 he studied chemistry further under Ladenburg, Abegg, and Herz at Breslau University. After one year of military service, he

---

[1]Kerstein, 1970.

entered Leipzig University in 1905 and received his doctorate under Hantzsch in 1907. His thesis work dealt with the use of absolute sulfuric acid as a solvent and was actually finished in the laboratory of Abegg in Breslau. It was the atmosphere of inquiry in Hantzsch's and Abegg's laboratories that led Bergius to choose a career in academic research. To further this end, he spent two years at the Nernst Institute in Berlin, where he assisted in the laboratories of Nernst, Haber, and Bodenstein. In Haber's laboratory he attempted syntheses with imperfect apparatuses and had little success,[2] but it was during that time that Bergius developed an interest in high-pressure reactions.[3]

In 1909 he became a university lecturer at the Technische Hochschule Hannover and was then able to set up a private laboratory where he began to systematically investigate the effect of high pressures and high temperatures on chemical reactions. He taught for a short time in Hannover, lecturing about metallurgy, gas reactions with technical applications, and equilibrium theory, but he soon quit lecturing in order to devote more time to research. He was particularly interested in finding practical ways to synthesize existing substances that had potential for industrial use.[4]

The first problem that Bergius considered was the dissociation equilibrium of calcium peroxide. In retrospect, the results of the experiment were not as important as the equipment that was developed to carry it out. The reaction container, with its numerous pipes and valves that remained leakproof while at elevated temperatures and pressures for several months, was used in subsequent investigations of high-pressure chemical reactions. After the use of a stirring motor in the proximity of high-pressure oxygen resulted in an explosion, an agitation method involving the rotation of the complete pressure vessel was developed.[5]

He obtained moderate success in his next investigation, an attempt to produce hydrogen gas by reacting carbon and water at lower temperature but higher pressure than had been used before. These conditions helped to prevent the formation of carbon monoxide along with the hydrogen gas, and they produced purer hydrogen than was previously obtainable. They also produced hydrogen at a lower cost than other

---

[2]Bergius, 1966.
[3]Farber, 1963.
[4]ibid.
[5]Bergius, 1966, op. cit.

methods. Much of the work on hydrogen gas production was done for Chemische Fabrik AG (formerly, Moritz Milch and Company) in Posen.[6]

In the course of the carbon-water reaction experiments, Bergius noticed that when peat was used as a source of carbon the reaction residue had an elemental analysis resembling that of coal. Bergius then considered what the composition of coal signified, and he concluded that it might be possible for coal to combine chemically with hydrogen.[7] Thus it was this study, along with a later one investigating the combination of oil and hydrogen, that led to his important coal-liquefaction experiments.

In the summer of 1913 it was apparent that the dependence on gasoline and oil as fuels was increasing along with the use of automobiles. Even at that time the limitations of petroleum supply and distribution were evident, so it seemed important that an alternative source of gasoline and oil be developed. To do this, Bergius again turned his attention to peat and coal.

The only previous noncombustive transformation of coal was a process known as destructive distillation, in which coal was heated in a sealed vessel, resulting in the release of the hydrogen and a large part of the carbon in the coal as methane. Small quantities of cooking gas and tar were thus obtained, but the process was useless in terms of the production of the fuel and oil needed for automobiles.

It was known that if the distillation was carried out slowly at low temperatures, then the yield of oil was improved, but it was still poor. Bergius set about to improve on this low-temperature distillation using the knowledge he had gained in his previous high-pressure reaction studies; he attempted the reaction at low temperatures in the presence of high-pressure hydrogen gas. The high pressure had the effect of forcing more hydrogen to combine chemically with the coal so that a higher yield of oil resulted. Even the first test resulted in 80% of the coal being converted to useful substances, and Bergius and his co-workers, Hugo Specht and John Billwiller, immediately applied for and received a patent. Later trials also demonstrated that the addition of small quantities of metal oxides improved the process, but Bergius had neither the expertise nor funds available to probe this phenomenon in more detail.[8]

---

[6]Stranges, 1984.

[7]Bergius, 1966, op. cit.

[8]Stranges, 1984, op. cit.

In 1914 it became apparent that more funds and equipment were needed to bring the process to an industrial scale, so Bergius moved his laboratory to the Goldschmidt Corporation, which was owned by his friend and supporter, Karl Goldschmidt. Shortly thereafter he became the head of the corporate research laboratories. After World War I it was difficult to find funding for research in Germany, and more money was required than one company could provide, so Bergius obtained funds from a variety of German, British, and American corporations, particularly from those in the coal industry.

Bringing the process to an industrial scale required further development of the technology of high-pressure reactions. To avoid localized overheating, the coal was pulverized and suspended in oil, which served to transfer heat uniformly. The use of this coal–oil slurry had a second advantage in that it could be pumped into the reaction apparatus, thereby enabling a continuous operation, as reactants were pumped into one end and products were removed at the other end. The continuous processing, the use of the slurry, and a method to heat and agitate large quantities of the slurry were important innovations in the field of high-pressure reaction chemistry. In addition new instruments were used to measure gases flowing at high pressure, and new valves were developed to withstand the extreme operating conditions.

Several thousand systematic small-scale tests were done on over two hundred types of coal. Operating conditions were varied to optimize the yield of oil from these coals, and preliminary experiments were done with catalysts. An apparatus was developed to allow a more precise control of the temperature of the reaction mixture.[9] In 1925, when the coal liquefaction process had developed to the point of being industrially practical, Bergius sold the patent rights.

I.G. Farbenindustrie and Imperial Chemical Industries in Billingham took up the work on a truly industrial scale in 1927. The Bergius process was used by Germany during World War II to make the relatively scarce gasoline from their abundant coal reserves. Only under special economic conditions was the Bergius process profitable, but its development, along with the methods developed by Carl Bosch, paved the way for the now-widespread use of high-pressure techniques in the chemical industry. These techniques are now used, for instance, in the production of urea, which is used as fertilizer and also to produce resins. After selling the coal-liquefaction patent, Bergius continued

---

[9]Ibid.

work on the conversion of cellulose to sugar through hydrolysis reactions, a study he actually started in 1915. This work was to occupy him for another 15 years. His work culminated when an industrial plant began utilizing his methods. The process provided a sizable amount of food material to Germany during World War II.[10]

It was while working on the cellulose conversion that he received the Nobel Prize for the previous work. Bergius was awarded the Liebig Medal, a Ph.D. from the University of Heidelberg, and an honorary doctorate from the University of Hanover. He was a member of the board of directors of many associations and companies with an interest in coal.

After World War II, Bergius could not find suitable work in Germany, and he moved to Austria for a short time. He then began a company in Madrid, but in 1947 moved to Argentina, where he was a scientific adviser to the Argentine government.[11] He remained in Argentina until his death on March 30, 1949.

ZEXIA BARNES
*Morehead State University*

## Bibliography

Beck, H. *Friedrich Bergius, ein Erfinderschicksal*; R. Oldenbourg: Munich, 1982.

Bergius, F. *Die Anwendung hoher Drucke bei chemische Vorgangen und eine Nachbildung des Entstehungsprozesses der Steinkohle*; Wilhelm Knapp: Halle, 1913.

Bergius, F. "Chemical Reactions under High Pressure"; *Nobel Lectures, Chemistry 1922–1941;* Elsevier: Amsterdam, 1966; pp 244–276.

Bergius, F. "An Historical Account of Hydrogenation"; *Proceedings of the World Petroleum Congress* **1934**, 282–289.

Bergius, F. "The Transformation of Coal into Oil by Means of Hydrogenation"; *Proceedings of the International Conference on Bituminous Coal 1926*; Pittsburgh, 1927; pp 102–131.

Farber, E. *Nobel Prize Winners in Chemistry, 1901–1961;* Abelard-Schuman: New York, 1963; pp 123–131.

---

[10]Farber, 1963, op. cit.
[11]Kerstein, 1970, op. cit.

Findlay, A. *A Hundred Years of Chemistry;* Gerald Duckworth: London, 1965; p 292.

Kerstein, G. "Bergius, Friedrich"; In *Dictionary of Scientific Biography;* Gillispie, C. C., Ed.; Charles Scribner's Sons: New York, 1970; Vol. 2, pp 3–4.

Schmidt-Pauli, E. von. *Friedrich Bergius: Ein deutscher Erfinder kämpft gegen die englische Blockade;* E. S. Mittler und Sohn: Berlin, 1943.

Stranges, A. N. "Friedrich Bergius and the Rise of the German Synthetic Fuel Industry"; *Isis* **1984,** *75,* 643–667.

Westgren, A. "The Chemistry Prize"; In *Nobel, the Man and His Prizes;* The Nobel Foundation; Odelberg, W., Eds.; Elsevier: New York, 1972; pp 347–348.

# 1931

NOBEL LAUREATE

# Carl Bosch

## 1874–1940

Carl Bosch was born in Cologne, Germany, on August 27, 1874, and died in Heidelberg on April 26, 1940. He received the Nobel Prize in 1931 for the invention and development of chemical high-pressure methods. His father, Carl Friedrich, owned a wholesale firm with a workshop for gas and water installations. His uncle, Robert Bosch, was the founder of one of the leading companies in the German electrical industry.

Carl was the eldest of six children, and owing to his father's educational principles he grew up in an open-minded atmosphere. In contrast to what was common at that time, Carl's father would never punish the children; instead he would discuss interesting problems and explain his point of view. Carl's father was a great lover of nature and was able to explain to his children a great number of phenomena they saw in the garden or out in the fields. At an early age Carl started to collect minerals and animals, which became a passion he never lost. His father also introduced him to his craftsmanship. He bought Carl tools and the material needed to build the things he liked, such as bird cages, a terrarium, and even a boat. When Carl was 16, his interest gradually shifted to chemistry. Already a skillful craftsman, he was able to construct his own equipment, sometimes together with a foreman from his father's workshop.

In 1893 Bosch finished *Gymnasium,* and although he was mainly interested in chemistry, he started an apprenticeship in a metallurgical plant and after one year became a journeyman. With this practical background Bosch started to study metallurgy and mechanical engineering at the Technische Hochschule in Berlin. In pursuit of his special interest he went to chemistry lectures, worked in a chemistry laboratory, and practiced spectral analysis.

In 1896 Bosch finally decided to switch from engineering, a field which was mainly based on practical applications, to pure science. He went to the University of Leipzig, at that time one of the leading German centers of chemistry. As in Berlin, Bosch not only worked in one subject but also went to lectures on physics, mineralogy, and the natural history of animals. In chemistry he worked under Johannes Wislicenus, and in 1898 received his doctoral degree with a thesis in organic chemistry. After that he became a laboratory assistant.

Bosch actually wanted to start a university career, but his father persuaded him to look for a position in the chemical industry. He applied for a position at the Badische Anilin und Soda Fabrik (BASF) in Ludwigshafen, one of the leading chemical companies, and was accepted. He was assigned to the main laboratory and at first worked on azo dyes and phthalic acid anhydride, a compound needed for the synthesis of artificial indigo.

In 1900, after one year at Ludwigshafen, Bosch was confronted with a new problem. He was asked to redo some of Wilhelm Ostwald's experiments on the catalytic synthesis of ammonia from its elements. Ostwald was one of the most prominent physical chemists at that time, and he wanted to sell his ideas to BASF. Bosch could not verify Ostwald's results. Certainly it was not easy for a young laboratory chemist like Bosch to speak up against such an authority as Ostwald, but his excellent experiments and his clear way to prove where and how Ostwald had obtained incorrect results convinced Ostwald and the BASF directors of Bosch's ability to solve difficult problems. This was the first time Bosch was confronted with the nitrogen problem, the problem of how to convert molecular nitrogen into a nitrogen compound that could be used as fertilizer. Later he would become famous for the solution of this problem. His excellent work in disproving Ostwald's findings had won Bosch the favor of his superiors and of his own laboratory group at BASF. Therefore, they never hesitated to support him when he was working on the nitrogen conversion besides his work at the phthalic acid plant.

At the end of the 19th century the natural sources of nitrogen compounds, including Chilean saltpeter (sodium nitrate), were diminishing, and the chemical industry was desperately searching for a process to replace saltpeter as a base of fertilizer. Several methods of fixing nitrogen had been developed, and Bosch tried all of them. He even tried a large-scale method, to produce 90 tons of ammonia, but finally he discarded all the methods as being too expensive. Nevertheless, by testing those methods he had accumulated a huge amount of knowledge and experience as far as the solving of technical and chemical problems was concerned.

In 1909 the successful partnership between Bosch and Fritz Haber began. Haber had worked on the direct formation of ammonia, but contrary to Ostwald he used high pressure and found promising amounts of ammonia. The directors of BASF agreed to finance Haber's research. When Haber was able to find a good catalyst, osmium, and got a yield of 6–10% ammonia at a pressure of 200 atm and 600 °C, Bosch was convinced that the system could be adjusted to technical dimensions. At that time it was absolutely unthinkable to work on a technical level at 200 atm and at a temperature at which steel just started to glow, but Bosch had enough courage and confidence in the engineering capacity of modern technology to trust the impossible.

He started by choosing Alwin Mittasch as head of the chemical research and Franz Lappe as chief engineer. The problem was to find a suitable catalyst. Osmium, the catalyst Haber had used, was too scarce, so the chemists started to study practically every element's catalytic activity. After half a year of research with more than 20,000 experiments, they found a catalyst that gave results as good as osmium: iron contaminated with alumina (aluminum oxide). Engineers at BASF had to construct a reaction tube able to withstand 200 atm and 600 °C. The worst problem they found was that after a few hours of contact with hydrogen steel became very brittle, lost its ductility, and caused the reaction tube to explode under the high pressure. Bosch's solution was to separate the two functions of the tube, a tight seal for the gas and a great strength for pressure resistance: The inner lining of the reaction tube was soft iron, gas-tight except for the diffusing hydrogen. The outer part functioned to withstand the high pressure. The outer tube continued to be made of steel, but now it was punctured with thousands of holes so that the diffusing hydrogen could escape without reacting with the steel and thus destroy it. On March 5, 1911, the first specimen of the newly constructed tube worked, and the most important step in solving the nitrogen problem was taken. The commercial production of ammonia, however, did not start until

September 1913, when 20 metric tons of ammonia could be produced per day.

Bosch did not relax when the big plant worked at last. He knew that agriculture needed different types of fertilizer and that ammonium sulfate was not the best of them. Therefore he immediately started to work on the synthesis of urea and the oxidation of ammonia to nitric acid. At that time he did not know how important his research on nitric acid would be for his country. At the beginning of the century, nitric acid, produced until then from saltpeter, was an important ingredient for gunpowder manufacture. Before World War I nearly all saltpeter was imported, but when Germany's international connections were cut by the British fleet at the beginning of the war, the country needed new sources of nitric acid for the production of gunpowder. When Bosch heard about this problem he promised the German military that BASF would find a way to start the large-scale production of nitric acid. Based on his experiments on the oxidation of ammonia he worked out a procedure for the industrial production of nitric acid using atmospheric oxygen to oxidize $NH_3$. In May 1915 BASF was able to produce 150 metric tons of nitric acid per day.

When Bosch started to work on the production of ammonia he was an excellent chemist. However, the way he solved the arising problems and the way he coordinated all the steps needed to make the plant work showed that he was not only a very good scientist but also an outstanding organizer. The BASF board of directors appreciated that and always supported him and his often-unusual ways of handling things. His knowledge and capability soon made him one of the leading figures at BASF, and in 1916 he became one of its directors. Bosch's reputation for being an excellent organizer grew when he built a second ammonia plant, the famous "Leuna Werke", in 1916. He had to deal with all parts of German industry and the military to get the material he needed despite the war's needs. Nevertheless, it took him only 11 months from the beginning of construction to the first running ammonia reactor.

In 1916 the *Interessengemeinschaft der deutschen Teerfarbenfabriken* (Interest Group of the German Coal Tar Dye Industry) was founded. Bosch, already a member of many advisory boards and committees and one of the founders of the employers' association of the German chemical industry, took part in most of the meetings and thus from the beginning knew what was going on in the largest congregation of power within German industry.

His influence kept growing, and he was chosen to become one of the consultants for the German diplomats who had to conclude the treaties of Spa and Versailles after the end of World War I. The Versailles treaty forced the Germans to hand over the whole chemical industry with all its patents to the victors. Besides that, all war factories were to be destroyed, among them, of course, the ammonia production, important in the manufacture of explosives. Yet Bosch was able to persuade the French industrialists and generals that cooperation among the industries could be more valuable than confiscated patents without practical know-how. In addition they finally agreed to Bosch's suggestion of the Germans' helping to build an ammonia plant in Toulouse and sending all the machinery and even trained people who would help to run it. In exchange the German ammonia plant would not be destroyed. Despite the German public's resentment of this agreement, these transactions were the beginning of Bosch's career as a captain of industry. In those days he became acquainted with the leaders of foreign industries; he already knew the leaders of German industry, was a member of many of their boards and committees, and also knew quite a few politicians.

In the meantime Bosch had been promoted to head of the board of directors of BASF, and from that position he was able to increase his influence in the center of the coordinating committee of the German chemical industry. In 1925 he reached his next goal, the fusion of the eight big dyestuff companies of Germany to form IG Farben, whose head Carl Bosch became. It was by far the biggest joint-stock company in Germany and one of the leading chemical companies worldwide. Contracts between IG Farben and foreign chemical companies consolidated the international chemical market.

Besides all this political and economic work, Bosch worked hard on the hydrogenation of coal. IG Farben bought the patents of Friedrich Bergius, whose knowledge of high-pressure reactions enabled that company to start the commercial hydrogenation of tar oil and coal in less than a year and to produce about 100,000 metric tons of gasoline per year. Unfortunately, the world depression was very harmful to international trade, which diminished considerably and was no longer thought of as a guarantee of peace.

In 1931, at the peak of the depression, when Carl Bosch was overburdened with economic problems, the Swedish Academy decided to honor him with the Nobel Prize for chemistry. It was the first time that industrial achievement rather than pure science was honored.

Together with Bergius he received the prize for "their contribution to the invention and development of chemical high-pressure methods". Thus, in the middle of the economic depression, Bosch's mind was turned back to the beginning of his career, to his love of scientific research and its technical application. The Swedish Academy especially honored the start of high-pressure chemistry, because that was the revolutionary beginning of a new age in chemical production methods. After the production of synthetic ammonia, Bosch had started the high pressure synthesis of methanol from carbon monoxide and water, soon to be followed by the high-pressure production of urea and gasoline. A consistent use of the theory of thermodynamic equilibrium, together with the use of an adequate catalyst and of high pressure to move the equilibrium in the right direction, and the application of all that on a large scale, constituted Bosch's prominent achievement. Most of the processes developed by Bosch and his co-workers are still used today without having been substantially changed. Some of his methods might even be of increased importance these days: for example, the production of gasoline from coal and tar.

When the Nazis came to power in Germany, Bosch's relation to them was very ambivalent and difficult to judge: On the one hand IG Farben donated large sums to the Nazi party; on the other hand Bosch resented their interference with free world trade and the freedom of scientific research, and he voiced that resentment. Soon after he had met Hitler in person, it became clear that they both deeply resented each other. Nevertheless, Bosch could not give up his position in German industry; IG Farben was his personal achievement, and, even when the Nazis took away most of his freedom of action, he stayed with IG Farben and tried to guide it through its difficulties. Very often, however, he could not find a solution to a particular problem and therefore suffered considerably from all the compromises he had to make with the regime. He was often depressed and acquired various diseases of the intestinal organs. During the years from 1935 to 1939 his health gradually deteriorated. After the beginning of World War II, Bosch's depressions became worse until he died in Heidelberg on April 26, 1940, full of dark visions about the future of his country and the future of his ideas.

During his lifetime Bosch was most famous as a captain of industry. To chemists, the work for which he received the Nobel Prize, namely, the techniques of high-pressure catalytic processes, is most important, but to most people Bosch is better known for another of his achievements. His name cannot be separated from Haber's, because their "Haber–Bosch process" solved the nitrogen problem, the artificial generation of nitrogen compounds to be used as fertilizer, which became more and more important to the rapidly increasing human population.

FRANK STEINMÜLLER
*Tyska Skolan,*
*Stockholm, Sweden*

# Bibliography

Borkin, J. *The Crime and Punishment of IG Farben*; Free Press: New York, 1978.

Goran, M. *The Story of Fritz Haber*; Univ. of Oklahoma Press: Norman, OK, 1967.

Haber, L. F. *The Chemical Industry, 1900–1930*; Clarendon: Oxford, 1971.

Hayes, P. "Carl Bosch and Carl Krauch: Chemistry and the Political Economy of Germany, 1925–1975;" *Journal of Economic History* **1987**, *47,* 353–363.

Hayes, P. *Industry & Ideology: IG Farben in the Nazi Era*; Cambridge University Press: Cambridge, 1987.

Holdermann, C. In *Banne der Chemie. Carl Bosch, Leben und Werke*; Econ: Düsseldorf, Germany, 1953.

Hughes, T. P. "Technological Momentum in History: Hydrogenation in Germany, 1898–1933"; *Past and Present* **1969**, *44,* 106–132.

Kerstein, G. "Bosch, Carl"; *Dictionary of Scientific Biography*; Gillispie, C. C., Ed.; Charles Scribner's Sons: New York, 1970; Vol. 2, pp 323–324.

Mittasch, A. *Geschichte der Ammoniak-synthese*; Verlag Chemie: Weinheim, Germany, 1951.

# Irving Langmuir

## 1881–1957

Copyright Nobel Foundation

Irving Langmuir, who received the Nobel Prize in chemistry in 1932, was only the second American to be so honored. He was the first American employed by an industrial laboratory to receive the prize. It was awarded "for his outstanding discoveries and inventions within the field of surface chemistry". Langmuir was born on January 31, 1881, in Brooklyn, New York. His parents were Charles Langmuir, a prosperous insurance man, and Sadie Comings Langmuir; Irving was the third of their four sons. He died suddenly on August 16, 1957, at Woods Hole, Massachusetts, where he was visiting his nephew's family. Langmuir received his undergraduate degree in metallurgical engineering from Columbia University in 1903. He chose that course because of its strong combination of mathematics, physics, and chemistry. He then went to Germany for graduate study and completed his Ph.D. dissertation at Göttingen in 1906; his advisor, Walther Nernst (Nobel Prize in chemistry, 1920), has been called "the father of modern physical chemistry". Langmuir's thesis research concerned the dissociation of gases on hot filaments.[1]

After receiving his Ph.D., Langmuir took a teaching position in the chemistry department at Stevens Institute of Technology, a fledgling institution in Hoboken, New Jersey. He remained there three years, apparently with little opportunity for research (he published only

---

[1]Langmuir, I. "The Dissociation of Water and Carbon Dioxide at High Temperatures"; *J. Am. Chem. Soc.* **1906**, *28*, 1357–1379.

one paper, which contained no experimental work, during the Stevens period). In 1909 Dr. Willis R. Whitney, Director of the General Electric Research Laboratory in Schenectady, New York, offered Langmuir a summer job, and this began the association that continued for the rest of Langmuir's life.

Langmuir's first research at the General Electric laboratory was motivated by two phenomena that were observed in the early tungsten-filament high-vacuum electric lamps. The first of these, the gradual blackening of the glass envelope as the lamp was used, reduced the efficiency of light production. The second was the fact that the vacuum inside the lamp actually improved with time. In seeking to understand these effects, Langmuir undertook a series of experiments with various gases in contact with hot tungsten filaments. Within a few years, he had proved that blackening was due simply to the evaporation of tungsten, which led to his greatest practical invention, the gas-filled electric lamp.[2] At the same time, the study of the heat transfer in various gases showed that molecular hydrogen ($H_2$) was dissociated into atomic hydrogen at the hot filament;[3] this discovery in turn led eventually to Langmuir's invention of atomic hydrogen welding.

Clearly, the first several years at the General Electric laboratory produced practical results of great value; it has been stated that the gas-filled lamp, incorporating Langmuir's filament redesign to reduce heat loss, meant savings of millions of dollars in electricity costs for users of incandescent lamps. However, the same studies led to the theory of gas adsorption and heterogeneous catalysis that established Langmuir as the pioneer of modern surface chemistry.[4] In the paper just cited he rejected the assumption made by most earlier workers that catalysis occurred in thick layers of adsorbed gases, postulating instead that a reaction in a single molecular layer on the metal surface was all-important. He also assumed that the reacting molecules were held in a regular array on the surface by chemical forces, and on that basis he derived a mathematical expression for the surface concentration as a

---

[2]Langmuir, I., U.S. Patent 1,180,159, April 18, 1916; and Langmuir, I.; Orange, J. A. "Tungsten Lamps of High Efficiency: I. Blackening of Tungsten Lamps and Methods of Preventing It and II. Nitrogen-Filled Lamps"; *Proceedings of the American Institute of Electrical Engineers* **1913**, *32*, 1894; *Transactions of the American Institute of Electrical Engineers* **1913**, *32*, 1915.

[3]Langmuir, I. "Thermal Conduction and Convection in Gases at Extremely High Temperatures"; *Transactions of the American Electrochemical Society* **1911**, *20*, 225–237.

[4]Langmuir, I. "Chemical Reactions at Low Pressures"; *J. Am. Chem. Soc.* **1915**, *37*, 1139–1167.

function of the pressure or concentration in the gas phase: the famous "Langmuir adsorption isotherm".

With the publication of his two monumental papers on solids and liquids in 1916 and 1917, the groundwork of his contribution and the basis of his Nobel Prize was established.[5] In the first of these, he re-iterated and expanded on the theory of the adsorption, catalysis, and evaporation and condensation of atoms and molecules. Turning his attention to liquid surfaces in the second paper, Langmuir described the surface film balance that permits the measurement of the properties of molecular layers on water and developed his concepts of molecular orientation at surfaces (arrived at independently by William D. Harkins at the University of Chicago). He also showed that the shapes and sizes of molecules could often be determined with simple tools. To this day his set of experiments remains one of the simplest and most elegant demonstrations of the reality and properties of molecules.

Scientifically important and practically valuable as this Nobel Prize winning work was (and it is worth noting that the prize citation specified "his outstanding discoveries and inventions"), it constituted only a small part of Irving Langmuir's scientific and practical contributions. During the 1912–1917 period, he also made substantial contributions to understanding heat transfer, both in gases and in mechanical equipment, as well as the phenomenon of the thermionic emission of electrons from hot filaments. He determined the melting point of tungsten and the vapor pressures of tungsten, platinum, and molybdenum at high temperatures. His work began to be acknowledged through major scientific awards, beginning with the Nichols Medal of the American Chemical Society in 1915 and the Hughes Medal of the Royal Society of London in 1918.

During the First World War, Langmuir was involved in programs on the sonic detection of submarines, which contributed substantially to ending the German U-boat menace, and near the end of the war he began work on nitrogen fixation, which was of concern for the supply of explosives. With the coming of peace in November 1918 the wartime work was put aside. During the war, however, Langmuir had devoted considerable thought to questions of atomic structure and valence, and during the period 1919–1921 he published a dozen papers on the subject. Building on G. N. Lewis's ideas of the electron pair

---

[5]Langmuir, I. "The Constitution and Fundamental Properties of Solids and Liquids: I. Solids"; *J. Am. Chem. Soc.* **1916,** *38,* 2221–2295; and "II. Liquids"; *J. Am. Chem. Soc.* **1917,** *39,* 1848–1906.

bond and the octet theory, Langmuir developed a detailed theory that accounted for much of the chemical information known at that time, and until the development of quantum mechanics a few years later the Lewis–Langmuir theory provided the best interpretation of chemical bonding available.

During the 1920s, in addition to continued work on adsorption, heat transfer, and the invention of atomic hydrogen welding, Langmuir made major contributions to the understanding of electrical discharges in gases (he adopted the term "plasma"). He discovered the effect of thorium on the electronic emission of tungsten, a discovery that led to a substantial improvement in the efficiency of radio-broadcasting tubes. A wide variety of other kinds of thermionic vacuum tubes were also improved by his work. (In addition to his substantive technical contributions, Langmuir participated in the naming of vacuum tubes after the fashion that has been called "Schenectady Greek", owing to the -*tron* endings.)

Shortly after his receipt of the Nobel Prize in 1932, Langmuir's interest returned to the monomolecular films on liquid surfaces that he had first studied in 1916 and 1917. For most of the 1930s, Langmuir with his colleagues Katharine Blodgett and Vincent Schaefer studied the properties of a wide range of organic compounds in monolayers on water. From fatty acids, which had been the subject of the original studies two decades earlier, they progressed to sterols, proteins, and plant pigments. In 1936 Dr. Blodgett discovered how to transfer successive monomolecular layers from the water surface to a solid substrate to form the built-up layer structures that are called "Langmuir–Blodgett films".

During the Second World War, Langmuir was again called on for work related to national defense. Among the projects to which he contributed were improved gas mask filters and techniques to decrease the tendency for ice formation on aircraft. Probably the greatest success was his development of a greatly improved smoke generator, which was extensively used by allied armies for smoke screens, including smoke for the protection of ground installations from air attack.

Langmuir's work on droplet nucleation, work that led to the smoke generator during the war, was the inception of his last great scientific enthusiasm, which continued until the end of his life. Intrigued by the possibilities of inducing nucleation in natural systems, Langmuir at the age of 65 set out to deliberately modify weather. His experiments with cloud seeding (using dry ice and silver iodide as nucleating agents) attracted tremendous public, as well as scientific, attention. Because of the plethora of natural variables that were beyond the experimenter's control, a proof of Langmuir's success at weather modification did not come during his lifetime. Nevertheless, the subject remains an area of active research many years after his pioneering effort.

By the late 1920s, Langmuir had become a prominent scientific statesman. He served as president of the American Chemical Society in 1929 and as president of the American Association for the Advancement of Science in 1942. Beginning when he was 40 years old, he was awarded a total of 15 honorary degrees by universities in the United States, Canada, Germany, and England. In addition to the Nobel Prize, he received a large number of medals and awards for his work. Yet among the awards that he prized most were two not related to science: the Silver Beaver and Silver Buffalo, awarded by the Boy Scouts of America for his contributions to that organization and his efforts on behalf of boys in Schenectady.

When the American Chemical Society established a journal for the publication of research in surface and colloid science and decided to name it "Langmuir", Dr. Karol Mysels wrote the following in a prefatory article:

> Not many people have advanced as many novel scientific ideas that did *not* survive the test of time as did Irving Langmuir. But there are few, if any, who have contributed as many new and original ideas that became obvious truths, and often basic truths, for his contemporaries and for succeeding generations.

Albert Rosenfeld ended his definitive biography describing Langmuir as "the man who was perhaps as great a scientist as has ever been born in America."

GEORGE L. GAINES, JR.
*Rensselaer Polytechnic Institute*

# Bibliography

Bacon, E. K. "Irving Langmuir, 1881–1957"; In *American Chemists and Chemical Engineers*; Miles, W. D., Ed.; American Chemical Society: Washington, D.C., 1976.

Blodgett, K. B. "Irving Langmuir"; *J. Chem. Educ.* **1933**, *10*, 396–399.

Gray, G. W. "A Summer Vacation"; *The Atlantic Monthly* **1933**, *152*, 732–743.

Hall, R. N. "Pathological Science" (Lecture by Irving Langmuir, 1953); *Speculations in Science and Technology* **1985**, *8*, 77–94; *Physics Today*, October 1989.

Kastens, M. L. "Weather to Order"; *Chem. Eng. News* **1951**, *29*, 1090.

Kohler, R. E. "Irving Langmuir and the Octect Theory of Valence"; *Historical Studies in the Physical Sciences* **1974**, *4*, 39–87.

*The Collected Works of Irving Langmuir*, 12 vol 5.; Suits, C. G., Ed.; Pergamon: New York, 1962.

Langmuir, A. C. "My Brother Irving"; *Industrial and Engineering Chemistry, News Edition* **1932**, *10*, 305–306.

Langmuir, I. *Phenomena, Atoms and Molecules*; Philosophical Library: New York, 1950.

Reich, L. "Irving Langmuir and the Pursuit of Science and Technology in the Corporate Environment"; *Technology and Culture* **1983**, *24*, 199–221.

Rosenfeld, A. *The Quintessence of Irving Langmuir*, Pergamon: Oxford, England, 1966.

Suits, C. G.; Martin, M. J. "Irving Langmuir"; *Biographical Memoirs of the National Academy of Sciences* **1974**, *45*, 215–247.

Taylor, H. S. "Irving Langmuir"; *Biographical Memoirs of Fellows of the Royal Society* **1958**, *4*, 167–179.

"Weather or Not"; *Time*, Aug. 28, **1950**, *56*, 52–56.

Westervelt, V. V. *Incredible Man of Science*; Messner: New York, 1968.

Whitney, W. R. "Langmuir's Work"; *Science* **1928**, *20*, 329.

Wise, G. "Ionists in Industry: Physical Chemistry at General Electric, 1900–1915"; *Isis* **1983**, *74*, 7–21.

Wise, G. "A New Role for Professional Scientists in Industry: Industrial Research at General Electric, 1900–1916"; *Technology and Culture* **1980**, *21*, 408–429.

# Harold Urey

## 1893–1981

Harold Clayton Urey, recipient of the Nobel Prize for chemistry in 1934, was born on April 29, 1893, in Walkerton, Indiana. He was one of the three children of Cora Rebecca (Reinoehl) and Samuel Clayton Urey. His father, a schoolteacher and lay minister, died when Harold was six. His mother later remarried and Urey had two stepsisters. Urey married Frieda Daum, a bacteriologist, in 1926; they had four children. He died at the age of 87 on January 5, 1981, in La Jolla, California.

After graduation from high school in 1911, Urey taught in country schools in Indiana and in Montana, to which his family had moved. In 1914 he entered the University of Montana at Missoula, graduating three years later with a B.S. in zoology; he minored in chemistry. Although he intended to be a biologist (his first research problem dealt with protozoa), the needs of the war effort led him to work as a chemist. After his experience in industry, he chose an academic career, and in 1919 he returned to the University of Montana as an instructor in chemistry. Two years later he entered graduate school at the University of California at Berkeley. There he studied under G. N. Lewis, doing research on the calculation of thermodynamic properties from molecular spectra and on the distribution of electrons among the orbits of excited hydrogen atoms.

After obtaining his doctorate in 1923, Urey studied for a year with Niels Bohr and H. A. Kramers at the Institute for Theoretical Physics in Copenhagen on an American–Scandinavian Fellowship. For five years after his return to the United States in 1924, he taught chemistry at Johns Hopkins University and published papers on chemical kinetics, quantum mechanics, and molecular spectra. It was during this time that he and Ruark wrote a book describing the recent advances in quantum mechanics.[1]

At Columbia University, to which he was called as an associate professor in 1929, Urey continued work in these fields (the Urey–Bradley force field is still used in the analysis of the infrared spectra of tetrahedral molecules), but also became interested in isotopes. He became the first editor (1933–1940) of the *Journal of Chemical Physics* and helped develop it into a major scientific journal.

Early in July 1931, Urey read a letter in *Physical Review*[2] that pointed out that a heavy hydrogen isotope could explain the anomaly, later shown to arise from errors of measurement, that the atomic weight of hydrogen was the same on both the chemical and physical atomic weight scales. Within a day or so of reading this, Urey had planned a series of experiments to determine whether a heavy hydrogen isotope really existed. He saw that the difference in mass between the two kinds of hydrogen atoms should lead to detectable changes in the wavelengths of lines in the spectrum of atomic hydrogen. Believing that the amount of the heavy hydrogen isotope, which Urey later named *deuterium*, was too small to be detected spectroscopically without prior concentration, Urey decided on the basis of theoretical calculations of the vapor pressures of hydrogen and deuterium that the latter would be concentrated in the residual liquid as liquid hydrogen evaporated. Because liquid hydrogen in the requisite amounts was not available in New York City, he asked Ferdinand Brickwedde, whom he had known as a graduate student in physics at Johns Hopkins and who was working at the National Bureau of Standards in Washington, D. C., to reduce five or six liters of liquid hydrogen to one or two milliliters by evaporation at 20 °K (−424 °F).

While waiting for that to be done, Urey and his assistant, George Murphy, with whom he had designed and built a grating spectrograph at Columbia, were surprised to find spectroscopic evidence of deuterium in samples of commercial hydrogen when testing their spectroscopic procedures. Afraid that these results were spurious, arising from imperfections in their apparatus, Urey did not immediately announce

---

[1]Ruark, A. E.; Urey, H. C. *Atoms, Molecules, and Quanta*; McGraw-Hill: New York, 1930.

[2]Birge, R. T.; Menzel, D. H. "The Relative Abundance of the Oxygen Isotopes and the Basis of the Atomic Weight System"; *Phys. Rev.* **1931**, *37*, 1669–1671.

the discovery of deuterium. When the first sample of "enriched" hydrogen arrived from Brickwedde, it showed no increase in deuterium content. This failure proved, on later analysis, to be due to the procedure used for producing the hydrogen prior to its liquefaction. A second sample, prepared by evaporation at 14 °K (−434 °F), did (on Thanksgiving Day, 1931) show an enrichment in deuterium. Convinced now that their results were valid, Urey announced the discovery in a 10-minute paper at a meeting of the American Physical Society in New Orleans at the end of December. This was followed by a letter and then a detailed paper.[3] From conception to announcement, the investigation had taken only six months.

The existence of isotopes had been accepted since 1913, but it was believed that the isotopes of a given element could not be differentiated or separated by any chemical process. Urey showed that existing techniques were able to separate isotopes in favorable cases. In 1932 he and Edward Washburn of the National Bureau of Standards showed that deuterium could be concentrated in the form of deuterium oxide ($D_2O$, or heavy water) by the electrolysis of water.[4] The availability of deuterium and, especially, heavy water made possible new varieties of scientific investigation. The deuterium nucleus proved to be an effective, high-energy projectile bringing about different nuclear transformations than the lighter hydrogen nucleus (proton). Since molecules containing deuterium behave chemically like their hydrogen counterparts but can be differentiated from them, it became possible to follow the changes of particular molecules through complicated chemical processes. This was extraordinarily fruitful in studying the mechanisms of chemical reactions and of biochemical processes. Although the distribution of radioactive lead in plants had been studied, the availability of deuterium afforded the first opportunity of studying the fate of a normal biochemical constituent in vivo. The first use of deuterium in this way (1935) resulted from a suggestion by a former graduate student of Urey's whom Urey had urged to take up biochemistry.[5]

---

[3]Urey, H. C.; Brickwedde, F. G.; Murphy, G. M. "A Hydrogen Isotope of Mass 2"; *Phys. Rev.* **1932**, *39*, 164–165; "Hydrogen Isotope of Mass 2 and Its Concentration"; *Phys. Rev.* **1932**, *40*, 1–15.

[4]Urey, H. C.; Washburn, E. W. "Concentration of the $H^2$ Isotope of Hydrogen by the Fractional Electrolysis of Water"; *Proc. Natl. Acad. Sci. U.S.A.* **1932**, *18*, 496–498.

[5]Rittenberg, D. "The Influence of Nuclear Mass and Biological Systems"; In *Isotopic and Cosmic Chemistry;* Craig, H.; Miller, S. L.; Wasserburg, G. J., Eds.; North Holland: Amsterdam, 1964; pp 60–70; Kohler, R. E. "Rudolf Schuenheimer, Isotopic Tracers, and Biochemistry in the 1930s"; *Historical Studies in the Physical Sciences* **1977**, *8*, 257–298, 267–277, 290–296.

In recognition of his discovery of heavy hydrogen, the Royal Swedish Academy of Sciences awarded Urey the Nobel Prize for chemistry in 1934. Earlier the same year, he received the American Chemical Society's Willard Gibbs Medal and was promoted to full professor by Columbia.

Urey's expertise in isotope separation put him among the leaders in the American scientific community involved with the development of atomic energy. Even before the government had developed interest in the subject, he had been working on the separation of $^{235}U$, the fissionable uranium isotope, from natural uranium. As a member of the S-1 (Atomic) Committee of the Office of Scientific Research and Development, he was appointed to direct the work on isotope separation in 1940. Initially working primarily on separation by centrifugation, he became deeply involved with the gaseous diffusion process for concentrating $^{235}U$. By the end of 1943 Urey had become convinced that the gaseous diffusion process could not succeed in time to benefit the war effort. Indeed, most of the $^{235}U$ that went into the Hiroshima bomb was produced by an electromagnetic separation procedure. He saw the decision, made against his advice, to continue the work along this line as an indication that the government was looking to the production of nuclear weapons beyond the immediate needs of the war. He then began to work for the control of atomic energy, though he continued to provide nominal direction of the nuclear research at Columbia until 1945.[6]

At the end of the war, Urey left Columbia to go to the University of Chicago as Distinguished Service Professor of Chemistry at the Institute for Nuclear Studies, becoming the Martin A. Ryerson Distinguished Service Professor of Chemistry in 1952. He left Chicago in 1958 to become Professor of Chemistry-at-large at the University of California—San Diego until he retired in 1970.

At Chicago, he continued his work with isotopes[7] and predicted that the temperature effect on the equilibrium of exchange of oxygen isotopes between carbonate ion and water could be used as a way to measure ocean temperatures in the distant past. By 1951, he and

[6]Urey's role in the development of the atomic bomb is described in Groueff, S. *Manhattan Project;* Little, Brown: Boston, 1967; pp 179–186 and 261–272; Hewlett, R.; Anderson, O. E. *The New World, 1939–1946;* Pennsylvania State University Press: University Park, PA, 1962.

[7]Urey, H. C. "The Thermodynamic Properties of Isotopic Substances"; *J. Chem. Soc.* **1947**, 562–581.

his co-workers had verified this experimentally.[8] It is now possible to determine the temperature of an ancient ocean by measuring the ratio of oxygen-18 to oxygen-16 in the calcium carbonate from shells of creatures living in the ocean at that time.

Urey had measured the oxygen-16 and oxygen-18 contents of stony meteorites, and his interest was rekindled at Chicago by Harrison Brown's work on meteorites. He saw that the chemical and isotopic composition of meteorites and other planetary bodies limited the characterization of the possible processes by which they had been formed. He concluded that the moon had never been molten and that its composition provided a better approximation than did the Earth's of the composition of the dust cloud from which both had been formed. Because of the importance this gave to the lunar composition, Urey urged that manned lunar exploration should take priority in the U.S. space program. Although he strongly argued for his hypothesis, he abandoned it when it was contradicted by the evidence in the lunar samples brought back by the Apollo 11 mission. His book on the planets[9] was the first systematic application of chemical principles to the problem of the origin of the solar system. His approach remains a paradigm for those in the field.

Urey's work was characterized by a strongly quantitative approach, an insistence on considering the entire problem rather than a piece of it, and a willingness to follow his conclusions into areas beyond his initial specialization.

RUSSELL F. TRIMBLE
*Southern Illinois University*

[8]Urey, H. C.; Epstein, S.; Buchsbaum, R.; Lowenstam, H. A. "Carbonate Water Isotopic Temperature Scale"; *Bull. Geol. Soc. Am.* **1951**, *62*, 417–426; Urey, H. C.; Lowenstam, H. A.; Epstein, S.; McKinney, C. R. "Measurements of Paleotemperatures and Temperatures of the Upper Cretaceous of England, Denmark, and the Southeastern United States"; *Bull. Geol. Soc. Am.* **1951**, *62*, 399–416.

[9]Urey, H. C. *The Planets: Their Origin and Development;* Yale University Press: New Haven, CT, 1952.

# Bibliography

Brickwedde, F. G. "Harold Urey and the Discovery of Deuterium"; *Phys. Today* **1982**, *35*, 34–39.

Brush, S. G. "Nickel for Your Thoughts: Urey and the Origin of the Moon"; *Science (Washington, DC)* **1982**, *217*, 891–898.

Cohen, K. P.; Runcorn, S. K.; Suess, H. E.; Thode, H. G. "Harold Clayton Urey"; *Biographical Memoirs of Fellows of the Royal Society;* The Royal Society: London, 1983; Vol. 29, pp 623–659.

Libby, L. M. *The Uranium People*; Crane, Russak: New York, 1979.

Murphy, G. M. "The Discovery of Deuterium"; in *Isotopic and Cosmic Chemistry;* Craig, H.; Miller, S. L.; Wasserburg, G. J., Eds.; North Holland: Amsterdam, The Netherlands, 1964; pp 1–7.

*The National Cyclopedia of American Biography*; James T. White: New York, 1938; Vol. E, p 275.

Ruark, A. E.; Urey, H. C. *Atoms, Molecules, and Quanta*; McGraw-Hill: New York, 1930.

"Some Thermodynamic Properties of Hydrogen and Deuterium"; *Nobel Lectures: Chemistry, 1922–1941;* Elsevier: New York, 1966; pp 333–356.

Thomas, S. *Men of Space*; Chilton: Radnor, PA, 1963; Vol. 6.

Urey, H. C. "Significance of the Hydrogen Isotopes"; *Ind. Eng. Chem.* **1934**, *26*, 803–806.

Urey, H. C. "The Separations and Properties of the Isotopes of Hydrogen"; *Science* **1933**, *78*, 566–571.

Urey, H. C.; Brickwedde, F. G.; Murphy, G. M. "A Hydrogen Isotope of Mass 2"; *Physical Review* **1932**, *39*, 164–165.

Urey, H. C.; Brickwedde, F. G.; Murphy, G. M. "The Abundance of the Elements"; *Physical Review* **1952**, *88*, 248–252.

# 1935

# *Frédéric Joliot*

## 1900–1958

Frédéric Joliot, born in Paris on March 19, 1900, began his career in scientific research at the Radium Institute of the University of Paris in 1925 under the guidance of Marie Curie. In 1926 he married her elder daughter, Irène Curie. They had two children, Hélène, born in 1927, and Pierre, born in 1932, who in turn entered scientific careers. From 1931 to 1934 Frédéric Joliot collaborated with Irène Curie in a series of experiments on radioactivity; their collaboration culminated in the discovery of artificial radioactivity. In November 1935 they were awarded the Nobel Prize in chemistry "for their synthesis of new radioactive elements". Frédéric Joliot was appointed professor at the Collège de France and set up his own laboratory there.

Then began the second part of Joliot's career, devoted to research on nuclear reactions and energetically conducted amid the tension and competition of the Second World War. In 1943 Joliot was elected a member of the Paris Academy of Sciences. After the war, he was appointed to be the first High Commissioner of the French Atomic Energy Commission, founded in October 1945. In April 1950 Frédéric Joliot was suddenly removed from this position because of his political activities. He died on August 14, 1958.

Joliot's brilliant career exemplifies within the French physics community a general trend away from traditional individualistic research toward large state laboratories. Joliot was trained in an engineering school, the Ecole de Physique et de Chimie Industrielle, where the emphasis was on laboratory work. Thus Joliot developed great ability as an experimenter, especially in the fine electrometric methods required for radioactive measurements. The Ecole Municipale de Physique et de Chimie Industrielles, though founded in 1882 to provide the chemical industry with qualified engineers, praised and encouraged pure scientific research. It was in a laboratory of the Ecole that radium was discovered by Pierre and Marie Curie in 1898. When Frédéric Joliot was admitted in 1920, Paul Langevin was director of studies. Langevin, a former student of Pierre Curie, was renowned both as the leading theoretician who introduced relativity theory into France and as the skilled technician who built the first sonar during World War I. Langevin had a decisive influence on Joliot's career. He transmitted his humanistic, pacifist, and socialist ideals to Joliot and oriented him toward political action.

At the end of Joliot's military service, Langevin diverted him from industrial engineering to scientific research and introduced him to Marie Curie at the Radium Institute. Joliot got a grant from the Caisse Nationale de Sciences thanks to Jean Perrin, who was a close friend of both Marie Curie and Langevin. At the Radium Institute, Joliot learned how to prepare strong sources of alpha rays from polonium. His doctoral research on the electrolysis of polonium solutions trained him to prepare and use very thin metallic sheets. Joliot's training as an engineer proved very useful for improving the scientific instruments required for experiments on interactions between rays and particles. In 1931 Joliot designed his own Wilson chamber. Designed for visualizing trajectories of ionized particles through a gas saturated with water vapor, this apparatus allowed the accurate determination of the energy of particles at all pressures by measuring the tracks on photographs. Joliot also designed an ionization chamber connected to a highly sensitive electrometer. According to Joliot, all equipment and all experiments were intended to increase the chance of observing unexpected phenomena, "to open as many windows as possible on the unforeseen".

Joliot's strategy proved successful in the discoveries of positive electrons and artificial radioactivity (see Irène Joliot-Curie, Nobel laureate, 1935). But Joliot and his wife Irène Curie were not always ready to grasp the theoretical significance of the experimental data they had described. In 1931, while studying the properties of the Bothe–Becker penetrating radiation, then regarded as a form of gamma ray, through the ionization chamber with a window covered by a very thin sheet of aluminum, Irène and Frédéric observed a remarkable phenomenon:

When paraffin wax, a substance rich in hydrogen, was exposed to penetrating radiation, protons were emitted with a high velocity. They concluded that, unlike the gamma rays, the penetrating radiation ejected nuclei of light atoms.[1] James Chadwick who reproduced their experiments in the Cavendish Laboratory, identified a new particle in the penetrating radiation. The new particle had a mass nearly equal to the mass of the proton, but the particle could not be made visible in the Wilson chamber because it was electrically uncharged. Thus, the Joliots paved the way but missed the discovery of neutrons. When Joliot reported on their research a few years later, he admitted that Chadwick's discovery of neutrons proved the significance of the "spiritual and moral capital accumulated in the Cavendish Laboratory". The Joliots did give a precise determination of the mass of the neutron at rest (higher than the mass of a hydrogen atom), and that determination contributed to a better understanding of the stability of the hydrogen atom.

While still working intensively in his laboratory with Irène, Frédéric felt deeply concerned about political events. In 1934, as fascism became threatening in France, he joined the Comité de Vigilance des Intellectuels Antifascistes and in 1936 supported the Popular Front government. He shared Perrin's and Langevin's convictions that science and socialism could bring an age of peace and prosperity. Joliot was not inclined toward patenting the applications of artificial radioactivity, but he did develop the medical and biological applications, especially radioactive tracers.

In 1936 the close collaboration between Frédéric and Irène Joliot-Curie ended. Frédéric was appointed at the Collège de France and could turn to the study of nuclear reactions with the modern, heavy, and expensive equipment characteristic of the rise of "big science". He was offered all the facilities for equipping three laboratories. At the Collège de France, Frédéric had a cyclotron that could produce as many α-rays as 100 kilograms of radium. For his second laboratory in Arcueil, he had inherited the 1,200,000-volt Van de Graaff generator displayed at the Palais de la Découverte during the International Exhibition held in Paris in 1937. For his third laboratory of "atomic synthesis" at Ivry-sur-Seine, near Paris, he ordered a cyclotron and a 2-million-volt generator. He chose two young collaborators from the young foreign scientists recently arrived in Paris: Lev Kowarski, from Russia, and Hans Halban, from Austria.

---

[1] *Compts-rendus de l'Academie des Sciences* **1932**; Vol. 194, pp 273, 708, 876, 1229.

As soon as they read the article by Hahn and Strassmann on the fission of uranium nuclei, Joliot and his collaborators furiously set to work to master this phenomenon. In March 1939 they proved that the fission of uranium implied an emission of neutrons, and they tried to calculate their number. Like other nuclear physicists in the United States, Joliot's team, assisted by Francis Perrin, tried to get a continuous process of successive fissions linked in divergent chains by neutrons. In spite of Szilard's demands for secrecy in nuclear research, they decided to publish their results in April 1939, and in May they obtained a patent. An agreement was signed between the French National Center for Scientific Research (C.N.R.S.) and a Belgian company, the Union Minière du Haut Katanga, to provide Joliot's laboratory with 5 tons of uranium.

Joliot and his colleagues realized that the nuclear chain reaction would require a "moderator" to slow down the neutrons emitted during fission. Heavy water looked like a convenient moderator, but the only important stock available in 1940 belonged to the Norsk Hydro Company, located in Norway. Then began an incredible adventure, later described in a movie titled *La Bataille de l'eau lourde*. Joliot convinced Raoul Dautry, the minister in charge of military equipment, to order six tons of heavy water from the Norsk Hydro Company. A secret expedition of French ski troops went to Norway and flew the precious barrels back in spite of the efforts of German spies to pirate the stock. Nonetheless, the German army arrived in Paris a few days after the arrival of the heavy water. Under Joliot's supervision, the barrels were transferred outside the occupied territories to Riom, then Bordeaux, from where they were sent with Halban and Kowarski to Great Britain.

Joliot chose to remain in France and returned to Paris to oversee his laboratory. Fortunately, the German officer who occupied his laboratory, Wolfgang Gentner, was a former assistant of the Joliots at the Radium Institute. Joliot joined the resistance, and in 1942 he became a member of the clandestine Communist party.

Joliot's postwar activities revealed him as a dynamic state manager daily confronting scientific, technical, political, and financial issues. As a director of the C.N.R.S. since August 1944, he was in charge of reorganizing French science. Before the end of the war in 1945, he had shaped a national science policy focused on fundamental research and had doubled the number of people appointed by the C.N.R.S. He heard about the tremendous developments of nuclear science in North America from his former colleagues Hans Halban, Jules Guéron, and Bertrand Goldschmidt, who joined the French-Canadian atomic energy team in Montreal during World War II. Joliot tried to convince De Gaulle, the president of the provisional government in 1945, that France should enter the race for the mastery of nuclear power. After the two atomic bombs were dropped on Hiroshima and Nagasaki, De Gaulle set up the Atomic Energy Commission (C.E.A.), which was officially created on October 18, 1945. C.E.A. was given great autonomy and substantial funding. As a High Commissioner responsible for scientific and technical activities, Joliot supervised the making of the first French atomic pile, named "Zoé", and the building of a second, larger center of atomic research located in Saclay. But he publicly declared that he would never work on bombs. Although he was recognized as the leader of the French nuclear project, Joliot was known as a communist who was continually criticizing the French government and openly supporting the Soviet Union. As tensions between West and East increased in 1950, Joliot's provocative attitude sounded like treason. In 1950 Joliot was ousted from the Atomic Energy Commission. He thus encountered the same destiny as other pioneers of atomic energy, J. Robert Oppenheimer in the United States and Andrey Sakharov in the U.S.S.R., also dismissed for political reasons. Frédéric Joliot and Irène Curie were sometimes the victims of humiliations and purges after 1950. Far from renouncing his political activities, Joliot threw himself more decisively into pacifist movements. Together with Albert Einstein and Bertrand Russell, he organized the International Conference of Scientists for Peace. He also was elected president of the World Organization for Peace. He died in 1958, two years after his wife's death. In spite of political disagreements, the French government decided that Joliot deserved a state funeral.

**BERNADETTE BENSAUDE-VINCENT**
*Université Paris-X*

# Bibliography

Biquard, P. *Frédéric Joliot-Curie;* Seghers: Paris, 1961.

Biquard, P. *Frédéric Joliot-Curie: The Man and His Theories*; Strachan, G., Tr.; Souvenir: London, 1965.

De Broglie, L. *La Vie et l'oeuvre de Frédéric Joliot;* Editions Albin Michel: Paris, 1959.

Del Regato, J. A. "Jean-Frédéric Joliot-Curie"; *J. de Biophysique et Médecine Nucléaire* **1983**, *7*, 61–74.

Goldschmidt, B. *Pionniers de l'atome;* Seuil: Paris, 1989.

Goldsmith, M. *Frédéric Joliot-Curie: A Biography;* Lawrence and Wishart: London, 1976.

Goldsmith, M. *Three Scientists Face Social Responsibility: Joseph Needham, J. D. Bernal, F. Joliot-Curie;* Centre for the Study of Science, Technology, and Development: New Delhi, India, 1976.

Joliot, F. *La Paix, le désarmement et la coopération internationale;* Défense de la Paix: Paris, 1959.

Joliot, F.; Joliot-Curie, I. *Oeuvres scientifiques complètes;* Presses universitaires de France: Paris, 1961.

Perrin, F. "Joliot, Frédéric"; *Dictionary of Scientific Biography;* Gillispie, C. C., Ed.; Charles Scribner's Sons: New York, 1973; Vol. 7, pp 151–157.

Pestre, D. *Physique et Physiciens en France, 1918–1940;* Archives Contemporaines: Paris, 1984.

Weart, S. R. *Scientists in Power;* Harvard University Press: Cambridge, 1979.

# *Irène Joliot-Curie*

## 1897–1956

Irène Curie, born in September 1897, was the elder daughter of two Nobel Prize winners, Pierre and Marie Curie. Irène Curie assisted her mother at the Radium Institute and collaborated with her husband Frédéric Joliot. They were awarded the Nobel Prize in 1935 for the discovery of artificial radioactivity. She died in Paris, on March 17, 1956.

Irène Curie spent her childhood in a highly stimulating scientific environment. As a baby, she may have suffered from loneliness because of her parents' involvement in laboratory work, but as a child she was able to benefit from learned conversations at home. Her parents shared close friendships with Jean Perrin and Paul Langevin, two leading physicists working in atomic physics, which was rather unusual at the time. Charles Seignobos, a powerful historian and a professor at the Sorbonne, and Emil Borel, a famous mathematician and Director of the Ecole normale supérieure, were also close friends of Joliot and Joliot-Curie. They used to spend their summer vacations together at l'Arcouest, a little port in Brittany. Instead of attending the state primary school as most French children do, the Perrin, Langevin, and Curie children were taught by their parents. Marie Curie taught physics, Paul Langevin taught mathematics, and Jean Perrin taught chemistry. When she was about 12 years old, Irène attended a private school, the Collège Sévigné, from which she graduated in July 1914. While she was a chemistry student at the Sorbonne, she became more and more involved in the war effort. She assisted her mother on the

Northern Front with the X-ray mobile units and served as an army nurse until the end of the war.

In 1921, Irène Curie entered the Radium Institute organized and managed by Marie Curie. Following the research initiated by her parents, she investigated the fluctuations of α-rays emitted by polonium and γ-rays emitted by radium. She presumably began to collaborate with her husband Frédéric Joliot in 1929. Using a large stock of radium belonging to the Radium Institute, they performed dangerous and painstaking experiments to prepare strong sources of α-rays. Unfortunately, the theoretical explanation of their observations was given by James Chadwick, who identified the neutron in 1932 (see Frédéric Joliot, Nobel laureate, 1935). Far from being dispirited, they continued their experimental studies of the mysterious interactions between radiation and particles. Using a Wilson chamber designed by Frédéric Joliot, they proved that Bothe and Becker radiation contained neutrons and γ-rays capable not only of ejecting electrons from matter but also creating light particles with nearly the same mass as the electrons but running back toward the emitting source in a magnetic field. They concluded that they had factual evidence of the positive electrons predicted by Dirac and just discovered in cosmic rays. In October 1933, Irène and Frédéric Joliot-Curie attended the seventh Solvay Conference in Physics, chaired by Paul Langevin. In their presentation to the world's leading atomic physicists, they ventured a bold explanation of their experimental data. The emission of a positive electron, Irène Joliot-Curie declared, was the result of an induced transmutation accompanied by the creation of a neutrino. Their paper was immediately criticized by a German physicist working on the same subject with Otto Hahn in Berlin. Lise Meitner raised serious objections about the Joliot-Curies' experimental results.

Back in Paris, Irène and Frédéric Joliot-Curie resumed their experiments in order to provide incontrovertible proof of their original interpretation. On January 11, 1934, as they were bombarding a thin aluminum sheet placed over the window of the Wilson chamber with α-particles, they were surprised to observe that the emission of positive electrons did not stop when the neutrons stopped with the removal of the source. The emission of positive electrons lasted a few minutes and decreased in a way that suggested a radioactive phenomenon. Realizing that it was an important observation, Irène and Frédéric repeated their experiment with a Geiger counter and confirmed that radioactive atoms with a 3.5-minute half-life were produced. Irène and Frédéric Joliot-

Curie inferred that there should be phosphorus atoms generated by the transmutation of the aluminum atoms associated with the emission of neutrons. Because ordinary phosphorus atoms are not radioactive, they concluded that they had observed an unknown radioactive isotope of phosphorus that was quickly transmuted by the emission of positive electrons and neutrinos into a stable isotope of silicon.

"The first atomic nucleus created by man", Frédéric Joliot said proudly to Marie Curie, Langevin, and Perrin, whom they had immediately called to show their prodigious experiment. Irène and Frédéric rushed to send a note to the Paris Academy of Science.[1] Soon they were able to prove that the radioisotope they had created had the chemical properties of ordinary phosphorus.

The discovery of artificial radioactivity began a new era not only in physics but in chemistry, biology, geology, and medicine. Irène and Frédéric Joliot-Curie were immediately selected for the Nobel Prize for chemistry—the third Nobel Prize for the Curie family. In their Nobel address, Irène Curie summarized the process of the discovery while Frédéric Joliot described the chemical identification of the artificial isotope.

With the honor of the Nobel Prize also came academic positions that relieved the Joliot-Curie family of economic troubles. Frédéric Joliot organized his own laboratory at the Collège de France. Irène Curie followed the same career as her mother. She stayed at the Radium Institute and was appointed to a chair at the Sorbonne in 1937. She pursued her research with new collaborators such as Savitch from Yugoslavia and Tsien-San-Tiang from China.

In 1938, while bombarding uranium with neutrons, she noticed the production of a radioelement with a half-life of 3.4 hours. She identified chemical properties very similar to those of actinium and lanthanum. But once again, as in 1932 with the neutron, the theoretical interpretation of her observation was provided by others. Otto Hahn of Berlin, reproducing Irène Curie's experiments, established that a radioisotope of barium was also produced, and suggested that the phenomenon observed was, in fact, a partition of the uranium nucleus into two nuclei with nearly the same mass. Irène Curie was very close to discovering nuclear fission.

While investigating nuclear phenomena, Irène Curie also had great political responsibilities. Thanks to Jean Perrin's efforts to get funds

---

[1] *Comptes-rendus de l'Académie des Sciences*, **1934**, *158* (Jan. 15), 254.

from the state for pure scientific research, the position of Secretary of Research had been created in 1936 under Leon Blum's Popular Front government, and Irène Curie was named to the position. After four months of endless meetings in the ministry, she resigned and left her post to Jean Perrin. This gesture did not mean that she was not interested in politics, however. Between 1936 and 1939, she publicly supported anti-fascist movements and worked with pacifist women's associations. After the defeat of the French army in 1940, she decided to stay in Paris with her husband, who had entered the Resistance movement. In 1944, the Resistance clandestine organization, fearing reprisals from the Nazis if Frédéric Joliot were arrested, sent Irène with her children, Hélène and Pierre, to Switzerland.

In 1946, Irène Curie was nominated director of the Radium Institute and supervised the building of new laboratories at Orsay. She was also appointed as "chef de la section chimie" in the French Atomic Energy Commission. After her husband's dismissal from the Atomic Energy Commission in 1950, she suffered from the political tensions of the Cold War. For instance, the U.S. Immigration Service denied her entrance in 1952, and the American Chemical Society rejected her application for membership in 1953. Her health weakened. She had recovered from tuberculosis a few years earlier, but leukemia, caused by the radiation to which she had been exposed since the age of 18, claimed her life in 1956, at the age of 58.

BERNADETTE BENSAUDE-VINCENT
*Université Paris-X*

## *Bibliography*

Cotton, E. *Les Curies;* Seghers: Paris, 1963.

*Dictionary of Scientific Biography;* Gillispie, C. C., Ed.; Charles Scribner's Sons: New York, 1973; Vol. 7, pp 152–159.

Goldsmith, M. *Frédéric Joliot-Curie: A Biography;* Laurence & Wishart: London, 1976.

Joliot, F. "Sur l'excitation des rayons gamma nucléaires du bore par les particules alpha. Energie quantique du rayonnement gamma du polonium"; *Comptes Rendus* **1931,** *193.*

Joliot, F. "Preuve experimentale de la rupture explosive des noyaux d'uranium et de thorium sous l'action des neutrons"; *Comptes Rendus* **1939,** *208.*

Joliot, F.; Joliot-Curie, I. "Un nouveau type de radioactivité"; *Comptes-rendus de l'Académie des Sciences* **1934,** *198,* 254.

Joliot, F.; von Halban, H.; Kowarski, L. "Liberation of Neutrons in the Nuclear Explosion of Uranium"; *Nature* (London) **1939,** *143.*

Joliot, F.; von Halban, H.; Kowarski, L. "Sur la possibilite de produire dans un milieu uranifere des reactions nucleaires en chaine illimite, 30 Octobre 1939"; *Comptes Rendus* **1949,** *299.*

Joliot, F.; Kowarski, L.; Perrin, F. "Mise en evidence d'une reaction nucleaire en chaineau sein d'une masse uranifere"; *J. Phys.* Orsay, Fr., **1939,** *10.*

Joliot-Curie, I. "Autoradiographie par neutrons. dosage separe de l'uranium et du thorium"; *Comptes Rendus* **1951,** *232.*

Joliot-Curie, I. "Extraction et purification du depot actif a l'evolution lente du radium"; *J. Phys. Radium* (France) **1929,** *10.*

Joliot-Curie, I. "Sur la vitesse d'émission des rayons alpha du polonium"; *Comptes Rendus* **1922,** *175.*

Joliot-Curie, I.; von Halban, H.; Preiswerk, P. "Sur la creation artificielle d'eléments appartenant a une famille radioactive inconnue, lors de l'irradiation du thorium par les neutrons"; *J. Phys. Radium* (France) **1925,** *6.*

Joliot-Curie, I.; Yamada, N. "Sur les particules de long parcours emisea par le polonium"; *Journal de Physique et le Radium* **1925,** *6.*

*Cinquantenaire de la radioactivité artificielle 1934–1984;* Laberrigue, J.; Ershaïdat, N., Eds. Université Pierre et Marie Curie: Paris, 1984.

Loriot, N. *Irène Joliot-Curie;* Presses de la Renaissance: Paris, 1991.

McKown, R. *She Lived for Science: Irène Joliot-Curie;* Messner: New York, 1961.

Joliot, F.; Joliot-Curie, I. *Oeuvres scientifiques complètes;* Presses universitaires de France: Paris, 1961.

Reid, R. *Marie Curie;* American Library: New York, 1974.

# Peter Debye

## 1884–1966

Copyright Nobel Foundation

Petrus (Pie) Josephus Wilhelmus Debije was born on March 24, 1884, in Maastricht, Limburg, The Netherlands. He received the 1936 Nobel Prize in chemistry for his contributions to knowledge about molecular structure, including investigations of the dipole moments of molecules and about the electron diffraction and X-ray diffraction of gases. He was a dominant figure in physical chemistry and chemical physics during the first half of the 20th century. In 1913 he married Mathilde Alberer; they had two children. He died on November 2, 1966, in Ithaca, New York.

He was the elder child of Joannes Wilhelmus Debije and Maria Anna Barbara Ruemkens. A sister was born in 1888. Joannes was a metalworker; Anna was the cashier of the theater that was a center of social and cultural life in Maastricht. Three grandparents and his parents were born and died in Maastricht. Until the age of six he spoke only the Maastricht version of the Limburg dialect of Dutch; throughout his life he spoke this dialect with his Maastricht friends and used it in correspondence with them. He shared with many citizens of Maastricht a love for the city and a detachment from Netherlands nationalism. Although he received many awards, the recognition he most appreciated was that shown when the citizens of Maastricht placed his bust in front of their town hall in 1939. His personal and professional papers are in the town archives.

Pie attended elementary school in Maastricht. In 1896 he entered the Hoogere Burger School in Maastricht, where he studied French,

German, English, and standard Dutch. Greek and Latin were university entrance requirements in The Netherlands; not planning to continue his education, he studied neither language. After placing among the top students in Limburg in the graduation examinations, he changed plans. Unprepared for a Netherlands university, he enrolled in 1901 at the Technische Hochschule in Aachen, about 20 miles away in Germany. In Aachen he called himself Peter Debye.

A year before, Arnold Johannes Wilhelm Sommerfeld joined the faculty of the Aachen Technische Hochschule as professor of technical mechanics. Sommerfeld, soon to be one of the world's leading theoretical physicists, was a student of the distinguished mathematicians Felix Klein and David Hilbert. He had already developed mathematical theories for electromagnetic waves along wires and for the diffraction of X-rays by slits. Finding little interest among his engineering colleagues at Aachen in theoretical physics, Sommerfeld talked to his students instead. Several times a week he invited two bright students, Peter Debye and Walter Rogowski, to dinner and then talked to them about mathematical physics for two or three hours. Debye became Sommerfeld's assistant in 1904, continuing in that capacity after receiving his degree in electrical engineering in 1905. In the summer of 1906 Sommerfeld accepted a professorship in theoretical physics at the University of Munich. Debye went to Munich as Sommerfeld's first assistant.

At Aachen, Debye began his life's work: investigating the interactions of radiation and matter. He wrote an elegant mathematical description of electrical eddy currents in rectangular conductors.[1] He also analyzed the diffraction of light by cylindrical and spherical objects. He continued these studies at Munich, where his doctoral dissertation (presented on July 1, 1908) combined a mathematically sophisticated analysis of the electromagnetic field in a cylinder with contributions to the mathematical theory of rainbows. He stayed at Munich as an assistant and then as a *Privatdozent*.

Debye avidly read works of Ludwig Boltzmann, Paul Drude, James Clerk Maxwell, Max Planck, Lord Rayleigh, and Bernhard Riemann on vibrations, thermodynamics, and the new quantum theory. Two months after Planck proposed an empirical radiation law, Debye deduced that law by considering modes of vibration in a radiation cube.

---

[1] *Zeitschrift für Mathematik und Physik* **1907,** *54,* 418-437.

In 1911 Albert Einstein resigned his professorship of theoretical physics at the University of Zurich; Debye accepted a one-year replacement appointment. Einstein had just published a paper showing his own 1907 quantum theory of solids to be inadequate. Debye saw a solution: He calculated vibrational modes for an elastic sphere by extending mathematical methods he had used at Aachen and Munich to treat waves in cylinders and spheres.

In the next year Debye returned to The Netherlands as professor of theoretical physics at the University of Utrecht and hoped to begin experimental research in physics to complement his theoretical work. Without forsaking mathematics, he was becoming a physicist. But he found the prospects for experimentation at Utrecht bleak.

Debye had become acquainted in Munich with Mathilde Alberer, daughter of the proprietor of his boarding house. She became a Netherlands citizen, and they were married on April 10, 1913. Two weeks later he was at the Wolfskehl Conference on the kinetic theory of matter and electricity, a conference that was pivotal for Debye's career.

David Hilbert, organizer of the Wolfskehl conference, invited his student Sommerfeld to give one of the six lectures, and Sommerfeld in turn persuaded Hilbert to invite Debye. Debye was by far the youngest lecturer. The other four, each a distinguished mathematical physicist, were Max Planck, Walther Nernst, Maryan Smoluchowski, and Hendrik Antoon Lorentz. Impressed by Debye's lecture on the relationship between the quantum hypothesis and the equation of state of a system, Hilbert convinced the University of Göttingen to establish a new professorship of theoretical and experimental physics for Debye. With two generations of mentors helping, Debye was in the right place at the right time.

Peter and Mathilde Debye moved to Göttingen. Peter joined a stellar faculty in mathematics and physics that included Max Born, Constantin Carathéodory, Richard Courant, David Hilbert, Felix Klein, Edmund Landau, Ludwig Prandtl, Eduard Riecke, Carl David Runge, Herman Simon, Gustav Tamman, Otto Toeplitz, Woldemar Voigt, and Emil Wiechert. Facilities for Debye's experimental program were available in new laboratories.

The fortnightly journal *Physikalische Zeitschrift* had been founded at Göttingen in 1899 by Eduard Riecke and Herman Simon, who served as joint editors. Debye became coeditor, replacing Riecke, for Volume 15 (1914). In 1919 Debye became the sole editor. As editor, Debye began a three-decade association with S. Hirzel of Leipzig, publisher of the journal.

Sommerfeld's assistant Walter Friedrich, by discovering in 1912 that crystals diffract X-rays, initiated a flurry of research activity focused on determining how individual atoms in crystals are arranged. Debye, extending methods he had used studying heat capacities, published an analysis[2] of the temperature dependence of diffraction data expected from atomic motion in crystals. Convinced that all crystals have irregularities and that even "random" liquids and gases have some regularity, Debye predicted that X-rays would be diffracted by gases, liquids, and noncrystalline solids.

In the first test, his assistant Paul Scherrer used powdered lithium fluoride, composed of randomly oriented microcrystals. The results were spectacular: The diffraction pattern of sharp spots revealed the symmetry of the arrangement of individual atoms in lithium fluoride. The Debye–Scherrer powder diffraction method proved to be general for most crystals. Further research revealed that detailed information about the geometry of molecules can be obtained by the X-ray diffraction of liquids and gases.

Debye was also investigating the electrical asymmetry of polar molecules in solution. His student Luise Lange was the first to measure dipole moments of molecules in solution by determining the concentration dependence of molar polarization.

Debye's work at Göttingen continued to impress Hilbert, who nominated him for the Nobel Prize in physics in 1916 and 1917. Debye was at Göttingen during World War I. The war did not unduly complicate Debye's life. Subject of Netherlands Queen Wilhelmina, he was exempt from German military duty. Some of his students were also exempt: For example, Luise Lange was a woman and Paul Scherrer was Swiss. Economic conditions in postwar Germany made an offer of a senior appointment in Switzerland attractive.

In the spring of 1920 Debye became professor of experimental physics and director of the physics laboratories at Eidgenössische Technische Hochschule in Zurich. He remained the senior editor of *Physikalische Zeitschrift* and appointed Max Born and Erich A. A. J. Hückel to the editorial staff in Göttingen. Scherrer moved with Debye, and Erich Hückel followed shortly.

Henri Samuel Sack became a Debye assistant in 1925 and received the doctor of science degree in 1927. Debye and Sack were close as-

---

[2] *Verhandlungen der deutschen Gesellschaft für Physik* **1913**, Series 2, *15*, 678, 738, 857; *Annalen der Physik* **1914**, *43*, 49–95.

sociates for the following 40 years. Sack recalled[3]: "He was not only endowed with a most powerful and penetrating intellect and an unmatched ability for presenting his ideas in a most lucid way, but he also knew the art of living a full life. He greatly enjoyed his scientific endeavors, he had a deep love for his family and home life, and he had an eye for the beauties of nature and a taste for the pleasure of the out-of-doors as manifested by his hobbies such as fishing, collecting cacti, and gardening, mostly in the company of Mrs. Debye." The Debyes had two children: a son, Peter Rupprecht, born in 1916 in Göttingen and a daughter, Mathilde Maria, born in 1921 in Zurich.

Debye's interests began to include chemistry, stimulated by colloquia at which both chemists and physicists participated. A lecture by E. Bauer on colligative properties of solutions prompted Debye to revisit work by physicist Samuel Roslington Milner.[4] After two years of collaboration, Debye and Hückel published the mathematical consequences of an interionic attraction model.[5] The results were of enormous significance for chemists, although most readers were not prepared for either the physics or the mathematics. Debye wrote a summary for chemists,[6] as did Arthur A. Noyes.[7] For a half century this model served as the basis for extensive physicochemical experimentation on dilute aqueous solutions. It inspired the research of many, including Raymond Fuoss, John Kirkwood, Herbert S. Harned, Lars Onsager, Benton B. Owen, R. A. Robinson, and R. H. Stokes. Also at Zurich Debye developed the new concept of magnetic cooling.

In 1927 Debye moved to Leipzig, accepting an appointment as professor of experimental physics at the University of Leipzig and director of the Institute of Physics. Sack accompanied him. With the interionic attraction theory of solutions and the study of molecular structure by diffraction and dielectric methods, Debye was moving into chemistry, using the powerful tools of mathematics and physics. At the same time, many chemists were becoming comfortable with the methods and vocabulary of mathematical physics. Debye at Leipzig was instrumental

---

[3] "In Memory of Professor Peter Debye"; two pages bound with the *J. Am. Chem. Soc.* **1968**, *90*, (12).

[4] *The London, Edinburgh, and Dublin Philosophical Magazine and Journal of Science*, Series 6 **1912**, *23*, 551–578; **1913**, *25*, 742–751.

[5] *Physikalische Zeitschrift* **1923**, *24*, 185–206, 305–325.

[6] *Chemisch Weekblad* **1923**, *20*, 562–568.

[7] *J. Am. Chem. Soc.* **1924**, *46*, 1080–1097, 1098–1116.

in defining the nature of physical chemistry for the middle years of the 20th century.

Never a solitary researcher, Debye knew the value of bringing together scientists with a similar interest for discussion. At Leipzig he encouraged this process. For example, Debye and John Warren Williams of the University of Wisconsin-Madison executed a faculty exchange. Debye lectured at the University of Wisconsin early in 1927 on methods of determining the shapes of electrically asymmetric molecules such as the inorganic substances water and ammonia and organic substances such as substituted benzenes. Debye described Lange's methods for determining dipole moments of molecules in solution. Williams had conducted similar experiments at Wisconsin. Williams then spent several months with Debye in Leipzig. Both Debye and Williams published results in the research journals. In addition, the Wisconsin lectures were published for a wider audience.[8] Williams introduced Lange's method into the undergraduate curriculum.[9]

Debye convinced the Saxony Ministry of Education to support a conference in 1928 that was a model for five other small conferences in Leipzig on specialized subjects in chemical physics. Conference proceedings were edited by Debye and published by S. Hirzel; three were translated into English by Winifred M. Deans and published in London by Blackie & Son.

Debye was an effective lecturer at conferences, in courses, in talks to general audiences, and in demonstrations for children. His introductory physics lectures at Leipzig often had an attendance of 400 students.

In 1934 when unrest in Leipzig began to interfere with his work, he moved, becoming professor of theoretical physics at the University of Berlin and director of the new physics institute of the Kaiser Wilhelm Gesellschaft in Dahlen.

Peter Debye received the Nobel Prize in chemistry in 1936, having been nominated for the physics prize in 15 of the years from 1916 to 1936 and for the chemistry prize in every year from 1927 to 1936. He was elected to 22 academies of science and received 12 medals and 18 honorary degrees.

Cornell University (Ithaca, New York) invited Debye to give the 1939–1940 George Fisher Baker chemistry lectures. The invitation

---

[8]Debye, P. *Polar Molecules;* The Chemical Catalog: New York, 1929; translated by R. Sänger as *Polare Molekeln;* S. Hirzel: Leipzig, Germany, 1929.

[9]Daniels, F.; Mathews, J. H; Williams, J. W. *Experimental Physical Chemistry*, 1st ed.; McGraw-Hill: New York, 1929; Experiment 69.

coincided with increasing Nazi interference with the institute. His visit to Ithaca turned out to be a permanent stay. His wife followed him, and he was appointed to the Cornell faculty for the spring semester 1940 as professor of chemistry. Sack soon joined him at Cornell.

Debye was editor-in-charge of the *Physikalische Zeitschrift* for its September 1940 issue. He was replaced by Ludwig Dewilogua for the October 1 issue. Debye continued as nominal senior editor, but it is unlikely that he had any active role after he left Germany. The final issue of the journal was published on March 15, 1945.

Debye served as chairman of the chemistry department at Cornell from 1940 until 1950 and was named Todd Professor in 1948. He became a United States citizen in 1946. He became professor emeritus in 1952 but continued active research at Cornell. His bibliography includes 34 publications from 1940 through 1952 and 64 more after 1952. His research at Cornell on light scattering and other aspects of macromolecular chemistry moved him into both biochemistry and industrial polymer chemistry.

GEORGE FLECK
*Smith College*

# *Bibliography*

Baker, W. O. "Peter Joseph Wilhelm Debye, 1884–1966" in *Proceedings of the Robert A. Welch Foundation Conferences on Chemical Research XX. American Chemistry — Bicentennial*; Milligan, W. O., Ed.; Welch Foundation: Houston, 1977; pp 154–199.

*The Collected Papers of Peter J. W. Debye;* Fankuchen, I.; Fuoss, R. M.; Mark, H.; Smyth, C. P.; Sack, H. S., Eds.; Interscience: New York, 1954. A selection of Debye's papers presented to him on his 70th birthday; papers originally written in other languages have been translated into English. It includes a biography by Fuoss, pp xi–xiv, comments by Mark, Smyth, and Fuoss, and a portrait.

*Current Biography*; Moritz, C., Ed.; H. W. Wilson: New York, 1963; pp 102–104.

Davies, M. "Peter J. Debye (1884–1966)"; In *Biographical Memoirs of Fellows of The Royal Society;* The Royal Society: London, 1970; Vol. 16, pp 174–232. The article includes a portrait by W. Mantz of Maastricht and a bibliography. An expansion of Davies, M. "Peter J. W. Debye (1884–1966)"; *J. Chem. Ed.* **1968,** *45,* 467–473.

Davies, M. "Peter J. W. Debye (1884–1966): A Centenary Appreciation"; *J. Phys. Chem.* **1984,** *88,* 6461–6463.

Debye, P. *Polar Molecules;* The Chemical Catalog: New York, 1929. Lectures presented at the University of Wisconsin in early 1927; translated by R. Sänger as *Polare Molekeln;* S. Hirzel: Leipzig, 1929.

"Debye named Gibbs Medal Winner"; *Chem. Eng. News* **1949,** *27,* cover and p 1210.

*Dipolmoment und chemische Struktur;* Debye, P., Ed.; S. Hirzel: Leipzig, 1929. Proceedings of the 1929 Saxony conference; translated by Winifred M. Deans as *The Dipole Moment and Chemical Structure;* Blackie & Son: London, 1931.

Eicke, H.-F. "Peter J. W. Debye's Beiträge zur makromolekularen Wissenschaft—ein Beispiel zukunftsweisender Forschung"; *Chimia* **1984,** *38,* 347–353.

*Elektroneninterferenzen;* Debye, P., Ed.; S. Hirzel: Leipzig, 1930. Proceedings of the 1930 Saxony conference; translated by Winifred M. Deans as *The Interference of Electrons;* Blackie & Son: London, 1931.

*Kernphysik;* Debye, P., Ed.; S. Hirzel: Leipzig, 1935.

*Magnetismus;* Debye, P., Ed.; S. Hirzel: Leipzig, 1933. Proceedings of the 1933 Saxony conference.

*Molekülstruktur;* Debye, P., Ed.; S. Hirzel: Leipzig, 1931. Proceedings of the 1931 Saxony conference; translated by Winifred M. Deans as *The Structure of Molecules;* Blackie & Son: London, 1932.

*Probleme der modernen Physik: Arnold Sommerfeld zum 60. Geburtstage;* Debye, P., Ed.; S. Hirzel: Leipzig, 1928.

Sack, H. S. "Peter J. W. Debye: A personal appreciation"; 2 pp, bound at front of the Peter J. W. Debye 80th Anniversary issue of *Journal of the American Chemical Society* **1964,** *86*(17).

Smyth, C. P. "Debye, Peter Joseph William"; In *Dictionary of Scientific Biography;* Gillispie, C. C., Ed.; Charles Scribner's Sons: New York, 1971; Vol. 3, pp 617–621.

Williams, J. W. "Peter Joseph Wilhelm Debye, March 24, 1884–November 2, 1966"; In *Biographical Memoirs;* National Academy of Sciences: Washington, 1975; Vol. 46, pp 22–68. Includes portrait and bibliography.

# Walter Haworth

## 1883–1950

Walter Norman Haworth was born in Chorley, England, on March 19, 1883 and died in Birmingham, England, on the same day in 1950. He received the Nobel Prize in chemistry in 1937. His award was granted for his research into carbohydrates and vitamin C. The award for 1937 was shared with Paul Karrer, who had worked on vitamins A and B$_2$ and other natural products. Dr. Haworth was born to Thomas and Hannah Haworth in the northwest area of England. His early years were spent at a local school and later at the local linoleum factory, which his father managed. His interest in chemistry probably began at that factory, where he became familiar with simple dyes. His education continued under a private tutor, and his abilities enabled him to enter the University of Manchester, where he was a student of W. H. Perkin, Jr. Quiet, serious, and steady study habits earned him first-class honors in chemistry in 1906. After an additional three years of research experience with Perkin, a scholarly award enabled him to spend one year at Göttingen, where he received his Ph.D. under Otto Wallach, an expert in terpene and natural products chemistry; the second year of his award allowed him to return to Manchester and receive his Doctor of Science degree.

His professional career in academe began in 1911 at the Imperial College of Science and Technology in South Kensington, London, where he continued his research interest in terpenes. The turning point in his research came with his appointment as a lecturer at the University of St. Andrews in 1912. He developed an interest in carbohydrate chemistry while at St. Andrews. After a period of research work for the military during World War I, Haworth became a professor of chemistry at the University of Durham in 1920. He became involved in research, general administrative work, planning for new laboratories, and building up the department. His last move occurred in 1926 when he became the Mason Professor of Chemistry at the University of Birmingham. He remained there for 23 years, with the exceptions of a short medical leave of absence and his essential research for World War II, which included his chairmanship of the Chemical Panel on atomic energy research. His retirement in 1948 from Birmingham did not diminish his professional efforts. He continued work as an advisor and an active member of many boards and committees. He was a representative of the Royal Society at the South Pacific Science Congress in early 1949. This meeting provided him the chance to visit New Zealand, and he extended his visit to Australia to become acquainted with a number of laboratories and chemists there. His last year showed continued activity, and even within a few days of his death he chaired the meeting of the Committee on Carbohydrate Chemistry of the Chemical Society. He died quickly and without distress of heart failure at home shortly after he returned from a visit to one of his sons.

Haworth's impressive research in carbohydrate chemistry produced over 300 published communications. His book on the composition of sugars, which clearly and methodically explained the world of carbohydrates, was published in 1929 and became a classic in the field. His honors included service as the President of the Chemical Society from 1944 to 1946 and Vice-President of the Royal Society from 1947 to 1948. He was the recipient of several distinguished awards including the Longstaff Medal from the Chemical Society in 1933, and from the Royal Society he received the Davy Medal in 1934 and the Royal Medal in 1942. Honorary degrees were awarded to him by several universities, including Zurich, Oslo, and his own Manchester.

He displayed personal characteristics that were greatly admired. His gentleness, consideration, and kindness endeared him to his colleagues, his associates, and his students. On first appearance he seemed somewhat formal and difficult to know, but this impression was superseded by the reality of a friendly and compassionate person. One particularly distinguished characteristic was his ability to inspire his coworkers, to see the big picture and help them to reach their goals. His was a farsighted approach that made sure his laboratory had the tools necessary to solve real problems. These qualities also made him an effective and dedicated teacher. He was a very precise lecturer, spending much time in preparing a well-organized lecture and delivering it in an interesting and comprehensible manner. An illustration of the fine interaction between himself and his associates can be shown by noting that at the time of his retirement, when his students and colleagues wished to honor him with a gift, his only request was for an album with the photographs of all those workers who had been with him during his tenure at Birmingham.

Haworth's interests extended beyond science into the areas of art, architecture, and antiquities. He had spent much time in his early days visiting museums and acquiring knowledge of these areas. An illustration of the application of his interests is revealed in his design of buildings and gardens. His home at Barnt Green was planned and carefully built to his specifications in 1933, and his attractively planned gardens were a source of great pride and joy to him, his wife, and their two sons. His wife, whom he married in 1922, was the former Violet Chilton, daughter of Sir James Dobbie. Haworth assisted in the creation of a new building on the Edgebaston site of Birmingham University to house the teaching and research activities of the chemistry department.

Haworth's research efforts had begun in the area of terpene chemistry but had shifted into carbohydrate chemistry. It was for his efforts in determining the structural and chemical properties of carbohydrates and for his determination of the structure of vitamin C that he received the Nobel Prize in 1937. His initial research involved the establishment of a new method of preparing a group of compounds related to sugars known as methyl ethers. This method became important since these compounds had great value in making structural determinations. Haworth would make a basic contribution in determining the structures of the most simple and basic carbohydrates, the monosaccharides. Emil Fisher, the pioneer in carbohydrate chemistry, had originated various straight chain structures of five or six atoms. Several cyclic formulas for these simple sugars were developed without an experimental basis and derived from several inappropriate assumptions. Haworth and his associates, using data such as the ability of various simple carbohydrates to rotate the plane of polarized light, discovered that the commonly used formulas were incorrect. Haworth introduced in 1925 a new cyclic model for glucose that indicated a ring containing five carbon atoms and one oxygen atom and to which an additional carbon was attached. Structures for mannose, galactose, fructose, and other monosaccharides were also developed.

The establishment of the basic ring structure of the monosaccharides influenced his studies on the structure of the disaccharides, those sugars containing two simple sugar units, or monosaccharides. A major concern was the manner in which the two simple sugar units could be joined together to form the disaccharide. For example, two glucose molecules could form maltose, and they might also form cellobiose, a disaccharide with very different properties from maltose. Haworth eventually determined the structure of a number of important disaccharides, including maltose, cellobiose, lactose, and other sugars. The bond holding the two sugar rings together consisted of an oxygen atom that was simultaneously attached to two carbons, one from each ring. The bond was formed by the combination of the two monosaccharide rings, with loss of water. In the investigation of the disaccharides formed from glucose, it was determined that glucose had two distinctive forms: the alpha form, containing a hydroxyl group -O-H above the plane of the ring, and the beta form, which had this same grouping below the plane. When two alpha forms combined, a direct bond was formed with the rings remaining in the same orientation to form maltose. If the beta forms of glucose combined, this direct bond was formed with the inversion of one of the rings, so that cellobiose was formed. This latter bond, unlike maltose, could not be broken down by human enzymes.

Haworth extended his studies to include the polysaccharides, those sugars containing many units of a simple sugar. This work embraced the important polysaccharides, including starch, cellulose, inulin, glycogen, and xylan (wood gum). Starch and glycogen were found to be built up from glucose using the pattern found in maltose, whereas cellulose used glucose in a cellobiose pattern. Xylan was determined to be similar to cellulose, except that it employed the monosaccharide known as xylose. The length and shape of the various polysaccharide chains were also studied. Haworth imagined polysaccharides to be chains of a fixed length that came together to form a larger unit. Haworth's work in carbohydrate chemistry signified the establishment of clear patterns for the various types of carbohydrates besides the determination of many specific structures.

The second part of his Nobel Prize winning work (from 1932 to 1933) involved vitamin C. Albert Szent-Györgyi had isolated a sample of the vitamin and had given some of it to Haworth. These two scientists suggested that the vitamin be named *ascorbic acid*. Haworth prepared various compounds from the vitamin and also broke it down into simpler identifiable compounds. This work enabled him to identify the specific chemical structure and also establish that it had a flat structure. The structural identification received wide acclaim from the scientific community.

Haworth became one of the first chemists to prepare ascorbic acid. The synthesis involved the reaction of xylosone with hydrogen cyanide and the treatment of the intermediate compound formed with water. His initial synthesis in 1933 was followed by the development of simpler and more effective methods of preparation. This was the first synthesis of any vitamin. The theoretical significance of this work was the establishment of the reality that a vitamin could be prepared by artificial chemical manipulation. The practical significance lay in the fact that vitamin C could be directly prepared and would become available to the public in large quantities at a reasonable cost. One no longer had to depend upon the fresh supply of lemons and limes to eliminate scurvy, the dread disease of vitamin C deficiency, which had plagued humanity for centuries.

Professor Haworth's work had helped to organize and clarify the field of carbohydrate chemistry. The structures of many individual carbohydrates and the relationships between the different types of carbohydrates had been clearly spelled out. The identification and synthesis of vitamin C had made it possible to understand the material and make it accessible. For these distinguished accomplishments, Walter Haworth became the first British organic chemist to receive the Nobel Prize in chemistry.

ROBERT H. GOLDSMITH
*St. Mary's College of Maryland*

## Bibliography

Farber, E. *Nobel Prize Winners in Chemistry, 1901-1961,* rev. ed.; Abelard-Schuman: New York, 1963.

Frejka, J. *Chem. Listy* **1938**, *32*, 175–181.

Haworth, W. N. *The Constitution of Sugars;* Longmans, Green: New York, 1929.

Hirst, E. L. "Walter Norman Haworth"; *J. Chem. Soc.* **1951**, 2790–2806.

Kopperl, S. J. "Haworth, Walter Norman"; In *Dictionary of Scientific Biography;* Gillispie, C. C., Ed.; Charles Scribner's Sons: New York, 1972; Vol. 6, pp 184–186.

*Nobel Prize Winners*; Wasson, T., Ed.; H. W. Wilson Co: New York, 1987; pp 422–423.

"Walter Norman Haworth"; *Nobel Lectures: Chemistry, 1922–1941*; Elsevier: New York, 1966; pp 407–432.

# Paul Karrer

## 1889–1971

Copyright Nobel Foundation

Paul Karrer was awarded the Nobel Prize in chemistry in 1937, along with Walter Haworth, for research in the carotenoids, flavins, and vitamins A and $B_2$. He was born on April 21, 1889, in Moscow, but his parents were Swiss. His father, Paul Karrer, and mother, Julia Lerch, returned to their homeland in 1892, and Karrer was educated in rural schools at Wildegg and at Lenzburg, Aarau. In 1908, he went to the University of Zurich to study under Alfred Werner (Nobel Prize in chemistry, 1913), the "father" of coordination chemistry. He received his Ph.D. in 1911 with a thesis on nitrosopentamine cobalt salts. After a year as a research assistant in Zurich, during which time he became interested in arsenic compounds, Karrer went to Frankfurt-am-Main to work with Paul Ehrlich (Nobel Prize in physiology or medicine, 1908) in the chemistry section of the Georg Speyer-Hauses. For five and a half years Karrer studied organic compounds of arsenic and other metals, such as antimony and bismuth, with the discoverer of Salvarsan, the "magic bullet" against syphilis.[1]

In 1914, Karrer married Helen Froelich, daughter of the director of the Royal Psychiatric Clinic. They had three sons, one of whom died in childhood. Karrer served as an artillery officer in the Swiss army during World War I and, after the death of Ehrlich in 1915, assumed the direction of the chemistry laboratory in Frankfurt, beginning new

---

[1] Wettstein, A. "Paul Karrer"; *Helv. Chim. Acta* **1972**, *55*, 313–328.

research on alkaloids and hydrocarbons. As conditions there became unfriendly for non-Germans, Karrer returned to Zurich as reader in chemistry at the university in 1918. The following year he replaced Werner as a professor of chemistry and director of the Chemical Institute, remaining in these positions until his retirement 40 years later. Karrer was instrumental in bringing to the university physical chemists trained in the latest methods of analysis. A demanding teacher, he had a great concern for his students, and his laboratory attracted researchers from around the world. He wrote that his happiest years were those he spent working with his students.

Karrer published over 1000 papers during his lifetime, and his *Lehrbuch der Organischen Chemie* became a classic in the field. It went through 13 editions and was translated into English, Italian, Spanish, French, Polish, and Japanese. His book on the carotenoids, written with Ernst Jucker, was the definitive work on this group of compounds named after carotene, a pigment in carrots. Karrer received 20 honorary degrees in Europe and America, including M.D., Ph.D., and Pharm.D. degrees, and was a member of many societies, including the Royal Society, the U.S. National Academy of Sciences, the Académie des Sciences, the Royal Academies of Belgium and Sweden, and the Indian Academy of Science. In 1955, he presided over the 14th International Congress on Pure and Applied Chemistry and from 1924–1926 was president of the Swiss Chemical Society. He served on the editorial board of *Helvetica Chimica Acta* for over 50 years. In addition to the Nobel Prize, he was awarded the Marcel Benoist Prize (1922) for the study of polymeric carbohydrates and the Cannizzaro Prize (1935) of the Italian Chemical Society. He died in Zurich on June 18, 1971.

Karrer's research in chemistry was enhanced by an ability to identify the most fruitful synthetic path, the most correct relationship, or the best analysis.[2] He was able to isolate, analyze, and synthesize hundreds of natural products, using early twentieth-century methods. Karrer was responsible for many improvements in these methods. He was a pioneer in the use of chromatography and spectroscopy, and he discovered many new ways to synthesize organic compounds.

The research program in Karrer's laboratory at Zurich was varied and extensive. During the early years, he investigated aromatic nitriles, aldehydes, and ketones. While studying sugars and amino acids,

---

[2]*A Biographical Dictionary of Scientists;* Williams, T. I., Ed.; John Wiley and Sons: New York, 1987, pp 596–597.

he found a new way to synthesize glucosides. His investigation of alkaloids led to the discovery of additional local anesthetics, and he also
worked on toxins and phosphatides, particularly lecithin, isolating its α
and β forms. Karrer became more and more interested in natural polymers, proteins such as albumin, and polysaccharides. His study of the
surface properties of cellulose helped improve the dyeing process of cotton and some related artificial fibers. He expanded the method of using
enzymes to split macromolecules in order to elucidate their structure.

In 1926, Karrer began his study of plant pigments, the anthocyanins, flavins, and carotenoids, using degradation experiments to determine where sugar molecules were attached. While investigating the
saffron pigment crocetin, he found that the carotenoids contain many
conjugated double bonds, and his study of the tomato pigment lycopene
showed that these molecules are composed of eight isoprene units, arranged symmetrically so that both ends of the molecule have the same
structure. By 1930, Karrer had determined the structure of lycopin and
carotene, two important members of the group. The determination of
these structures was accomplished through degradation and through the
synthesis of the perhydro derivatives. Although carotene had been isolated a hundred years before, it was not until 1931 that Hans von Euler-
Chelpin (Nobel Prize in chemistry, 1929) with Karrer discovered that it
has the same biological activity as vitamin A. Adding two molecules of
water to carotene can produce two molecules of vitamin A. Karrer was,
therefore, able to elucidate the structure of a vitamin before it had been
isolated. Vitamin A was the first vitamin to have its chemical structure
understood, although many scientists in Karrer's time believed that the
effects of these growth factors were the result of a special colloidal state
of matter, not of a specific compound.[3] The work of Karrer and his contemporaries showed that the study of these compounds was within the
province of chemistry. It was in Karrer's laboratory that George Wald
(Nobel Prize in physiology or medicine, 1967) showed that vitamin A
plays an important part in the chemistry of sight.

During his 40 years of research on carotenoids, Karrer isolated
and determined the structure of many compounds, such as zeaxanthin
from corn; mutatoxanthin, antheraxanthin, auroxanthin, and chrysan-
themaxanthin from flowers; and xanthophyll, the yellow pigment in
autumn leaves. He introduced a method of converting one carotenoid

[3]Karrer, P. "Carotenoids, Flavins and Vitamin A and B$_2$"; *Nobel Lectures: Chemistry, 1922–1941;* Elsevier: Amsterdam, 1966, pp 443–448.

to another, and developed a general method of synthesis. The total synthesis of the carotenoids was accomplished in 1950. Carotene was made by synthesizing the $\beta$ ion from the $C_{16}$ fragment and condensing two such fragments with the $C_8$ midsection of the molecule. Karrer also studied the stereoisomerism of these compounds, noting the small number of possible isomers that occur naturally.

In 1931, Karrer succeeded in synthesizing squalene $C_{30}H_{50}$, another important hydrocarbon with conjugated double bonds, which is present in large quantities in fish liver. Squalene is used in the synthesis of certain terpenes and the biosynthesis of cholesterol, gall acids, sex hormones, and adrenal hormones.

Karrer continued to study vitamins other than vitamin A. He confirmed the structure Albert Szent-Györgyi proposed for vitamin C (ascorbic acid) and prepared the dimethyl derivative. In 1934, he accomplished the total synthesis of vitamin $B_2$, riboflavin, at about the same time as Richard Kuhn (Nobel Prize in chemistry, 1938). The research of these two men often overlapped, and they were rivals in the race to understand the carotenoids and vitamins. Karrer's method led to the industrial production of riboflavin. Karrer showed that irradiation of vitamin $B_2$ in acid or neutral solution removed the side chain, yielding the colorless compound lumichrome. He succeeded in developing two methods of synthesis of riboflavin and prepared synthetic derivatives that exhibited riboflavin activity.

In 1937, Karrer began work on vitamin E, the tocopherols. He performed the first synthesis of a tocopherol in 1938 through the condensation of trimethylhydroquinone with phytyl bromide. Contributions to the structure, determination, and synthesis of $\alpha$-tocopherol and $\beta$-tocopherol, as well as other derivatives, were also made in the Zurich laboratory. Karrer studied the complicated stereoisomerism of these compounds, differentiating the four different cis–trans isomers of racemase and tocotrienol.

Karrer and his group played an important part in research on the antihemorrhagic vitamin K. In 1939, the light-sensitive compound phylloquinone (vitamin $K_1$) was isolated from alfalfa meal independently by Karrer, Henrik Dam (Nobel Prize in physiology or medicine, 1943) in Copenhagen, and Edward Doisy (Nobel Prize in physiology or medicine, 1943) in St. Louis, Missouri. Karrer reported the synthesis of 2–demethylphylloquinone and a derivative of the related compound, menadione ($K_3$), and developed a color test using sodium ethoxide with Dam.

While studying riboflavin, Karrer became interested in a study of nicotinic acid amide. This compound had been recognized by Warburg (Nobel Prize in physiology or medicine, 1931) as a constituent of a coenzyme, a biological cofactor necessary for chemical reactions in nature, and by Hans von Euler-Chelpin as part of the cohydrase in which adenine, a pentose (a sugar with five carbons), and phosphoric acid are also present. In 1936, Karrer prepared the iodine salt of nicotinic acid amide, which on reaction with $Na_2S_2O_4$ had produced a substance with the same spectra as $NADPH^+$, the reduced ion. With Warburg, Karrer was able to show that the hydrogen-transferring role of the coenzyme and the cohydrase was due to the nicotinic acid amide derivatives, which had a pyridine nitrogen with quaternary character. The placement of the sugar in the molecule was determined and the mechanism of hydrogen transfer was delineated. Understanding the role this compound plays in metabolism was an important step in understanding biochemical reactions. In 1942, the pentose in the molecule was identified as D-ribose. Later, Karrer synthesized other coenzymes, thiamine pyrophosphate and pyridoxal-5-phosphate ester.

At the presentation of the Nobel Prize, Karrer was described as a scientist characterized by his ability to visualize great and important problems as well as their smaller parts and by the unique way in which he approached problems and pursued new ideas by using his own methods.[4] Throughout his long career, Karrer continued to be at the forefront of the study of natural products.

JANE A. MILLER
*University of Missouri—St. Louis*

## Bibliography

*A Biographical Dictionary of Scientists.* Williams, T. I., Ed.; John Wiley & Sons: New York, 1987; pp 596–597.

Farber, E. *Nobel Prize Winners in Chemistry, 1901–1961;* Abelard-Schuman: London, Rev. ed.; 1963; pp 152–164.

Farber, E. *The Evolution of Chemistry,* 2nd ed. Ronald Press: New York, 1969.

---

[4]Palmer, W. "Presentation Speech"; *Nobel Lectures: Chemistry, 1922–1941;* Elsevier: Amsterdam, 1966, pp 407–413.

Ihde, A. *The Development of Modern Chemistry;* Harper & Row: New York, 1964; reprinted by Dover: New York, 1984.

Karrer, P. *Einfuhrung in die Chemie der Polymeren Kohlenhydrate: Ein Grundriss der Chemie der Stärke, des Glykogens, der Zellulose und anderer Polysaccharide*; Akademische Verlagsgesellschaft: Leipzig, 1925.

Karrer, P. *Organic Chemistry;* 4th English ed.; Elsevier: New York, 1950.

Karrer, P.; Jucker, E. *Carotinoide;* Birkhauser: Basel, 1948 (*Carotenoids;* Braude, E. A., Trans. and Rev.; Elsevier: New York, 1950).

Leicester, H. M. "Karrer, Paul"; In *Dictionary of Scientific Biography;* Gillispie, C. C., Ed.; Charles Scribner's Sons: New York, 1978; Vol. 15.

*Nobel Lectures: Chemistry, 1922–1941;* Elsevier: Amsterdam, 1966, pp 407–450. Includes Karrer's Nobel lecture "Carotenoids, Flavins, and Vitamins A and $B_2$", delivered December 11, 1937.

Sebrell, W. H., Jr.; Harris, R. S. *The Vitamins,* 2nd. ed.; Academic Press: Orlando, FL, 1968, Vols. 1–5.

Wettstein, A. "Paul Karrer"; *Helv. Chim. Acta.* **1972,** *55,* 313–328.

# 1938

# Richard Kuhn

## 1900–1967

Richard Kuhn was awarded the Nobel Prize in chemistry in 1938. He was born on December 3, 1900, in Döbling, a suburb of Vienna, Austria, the son of Richard Clemens Kuhn and Angelika Rodler Kuhn. He died of cancer on July 31, 1967. His father was an engineer who planned water projects such as canals and harbor facilities, and his mother was a teacher. Kuhn received his basic elementary education at home from his mother until he was nine, then attended the Döbling *Gymnasium* for eight years. He was a member of a class of geniuses that, of its 27 students, boasted three university professors, two chiefs of large medical clinics, two famous actors, and two Nobel laureates, Kuhn and Wolfgang Pauli (Nobel Prize in physics, 1945), who were friendly rivals at the *Gymnasium*.[1]

Upon graduation, Kuhn was drafted into the Army signal corps for a short time, obtaining his release soon after the Armistice in November 1918. He studied at the Institute for Medicinal Chemistry at the University of Vienna for four semesters and then moved to Munich to work in the laboratory of Richard Willstätter (Nobel Prize in chemistry, 1915), receiving his Ph.D. in 1922 with a thesis on the specificity of enzymes in carbohydrate metabolism. He introduced physical chemistry methods into Willstätter's laboratory. In 1925 he was invited by the University of Munich to lecture on the basis of his habilitation thesis describing a study of the mechanism of action of amylases. In 1926,

---

[1] Grassmann, W. "Richard Kuhn"; *Bayer. Akad. Wiss. Jahrb.* **1969**, 231–253.

he moved to Zurich as professor of general and analytical chemistry at the Eidgenössische Technische Hochschule. In Zurich, he married Daisy Hartmann, one of his pharmacy students. The couple had two sons and four daughters.

In 1929, Kuhn was appointed head of the department of chemistry of the newly established Kaiser Wilhelm Institute (now the Max Planck Institute) for Medical Research at Heidelberg and professor in the faculty of medicine at the University. He remained at Heidelberg for the remainder of his life, becoming chief administrator of the Kaiser Wilhelm Institute in 1937 and professor of biochemistry at the University in 1950. Although he spent one year as a visiting research professor of physiological chemistry at the University of Pennsylvania, he refused offers of positions in the United States and in Berlin, Munich, and Vienna.

The Nobel Prize in chemistry was awarded to Kuhn in 1938 for "his work on carotenoids and vitamins". However, Hitler was angered by the presentation of the 1935 Nobel Peace Prize to Carl von Ossietzky, who was imprisoned in a German concentration camp, and forbade Germans to accept Nobel prizes. Kuhn received his gold medal and diploma after the war in 1949.

From 1946 to 1948, Kuhn helped to transform the Kaiser Wilhelm Society for Scientific Research into the Max Planck Society for the Advancement of Science. A charter member of the Society's Senate, he served as vice-president under Otto Hahn (Nobel Prize in chemistry, 1944) and Adolf Butenandt (Nobel Prize in chemistry, 1939). He was president of the German Chemical Society and the Society of German Chemists, and a member of several academies and societies. He received honorary degrees from the Technische Hochschule of Munich (1960), the University of Vienna (1960), and the University of St. Maria, Brazil (1961). He served as an editor of *Liebig's Annalen der Chimie* and as a member of the Board of the Badische Anilin- und Soda-Fabrik. He was awarded the Schule medal of the Stockholm Chemical Society in 1950, and the Exner medal in 1952. Kuhn was the author of a textbook on enzymes and wrote over 700 papers with 150 students and collaborators.

Kuhn was known as a scientist who worked long hours, had a prodigious memory, and insisted that all around him be familiar with all chemical literature. He was an excellent and meticulous lecturer who offered profound insights into future developments.[2] He was a fun-loving person and an excellent marksman and tennis player.

---

[2]Baer, H. H. "Richard Kuhn"; *Advances in Carbohydrate Chemistry and Biochemistry* **1969,** *24,* 1–12.

Throughout his research career, Kuhn worked on natural products, an interest stimulated during his work at Willstätter's laboratory. He was part of a group of young chemists who enthusiastically applied new physicochemical principles and methods to organic chemistry and was instrumental in developing new methods of synthesis and analysis. Kuhn is said to be responsible for the reintroduction of chromatography into the research laboratory,[3] improving enzyme adsorption through the use of more efficient elution carrier materials. With Brockman, he developed a micro method for isolating carotenoids, and with Moeller, he developed a process for microhydrogenation. He also improved methods of measuring kinetic and thermal properties of sugar derivatives.[4] In spectroscopy, he developed the use of the position of the longest wavelength maximum to determine the structures of carotenoids.

Kuhn is described as having a "remarkable understanding of stereoisomerism".[5] Early in his research, he studied the inhibited rotation among diphenyls, especially orthosubstituted derivatives, determining the activation energies of rotation, and put forward quantitative concepts of the spatial needs of groups on chemical molecules, which he designated "atropisomerism".

During his tenure in Zurich, Kuhn began his studies of conjugated hydrocarbons. He not only synthesized over 300 compounds belonging to this group, but studied their physical properties, determining relationships between properties and structure. For instance, he showed that in diphenyl polyenes with conjugated bonds, $C_6H_6(CH = CH)$ $C_6H_6$ when $n$ is greater than two, the compounds are colored. Continuing Willstätter's interest in pigments, he began a study of the carotenoids (naturally occurring Kuhn-type polymers). Kuhn, simultaneously with Paul Karrer (Nobel Prize in chemistry, 1937) and Otto Rosenheim in London found the dextrorotatory and optically inactive isomers of carotene in 1931. In 1933, he discovered the third or $v$ isomer. His group, as well as others in the field, succeeded in isolating carotenoids from many natural sources such as saffron, rose hips, crocus, palm oil, lobster shells, and the human placenta. In the course of his investigations, he determined the structure of bixin (an important

---

[3] Ihde, A. *The Development of Modern Chemistry;* Harper & Row: New York, 1964; reprinted by Dover: New York, 1984, p 573.

[4] Burk, D. "Kuhn, Richard"; *Dictionary of Scientific Biography;* Gillispie, C. C., Ed.; Charles Scribner's Sons: New York, 1973; Vol. 7, pp 517–518.

[5] Ibid, p 517.

natural dye), physalien, helenian, flavoxanthin, vialoxanthin, crypto-xanthin, and rubixanthin. He was also instrumental in determining the composition of rhodoxanthin in japonica and astaxanthian from lobster eggs. During his study of carotenoids, Kuhn and co-workers also found that crocin played a part in reproduction of certain unicellu-lar algae, so that the ratio of the cis to trans form determined whether the gametes would be male or female.

Kuhn's work in the structure of vitamins was primarily in the clarification of those in the B complex. He isolated approximately one gram of riboflavin (lactoflavin), the yellow vitamin $B_2$, from 5300 liters of skim milk. Otto Warburg (Nobel Prize in physiology or medicine, 1931) and Christian had discovered the "yellow enzyme" in 1932, which was found to be degraded into lumiflavin, itself a degradation product of riboflavin. Kuhn proposed a structural formula for lumi-flavin that aided in the determination of a formula for riboflavin, and he later synthesized both compounds. Kuhn's proof of the constitution of riboflavin-5-phosphate was the primary step in the first complete elucidation of the action of a prosthetic group, a tightly bound coen-zyme necessary for activation of enzymes. Combination of this group with the carrier protein gave the reversible yellow oxidation enzyme. Kuhn clarified the double role of this enzyme as vitamin and coenzyme and studied the pharmacological effects of riboflavin. In 1943, he dis-covered the first riboflavin antagonist, dichloro-D-riboflavin, which inhibits growth of certain staphylococci and streptobacteria.

Kuhn's research was also important in the discovery of vitamin $B_6$, isolating this antidermatitis compound (also named adermin and pyri-doxin), and establishing its composition and structure. Properties of the crude vitamin were reported in 1936, and it was isolated by five inde-pendent researchers in 1938. Kuhn's method was partial purification of a protein complex in freshly prepared yeast extract. Low-molecular-weight impurities were dialyzed away at low temperatures, and the remaining vitamin–protein complex was split by heating. It was crys-tallized as the hydrochloride. Kuhn solved the structure in 1939 with G. Wendt and Otto Westphal through identification of the oxidation product. The synthesis of the vitamin was accomplished in the same year by degradation of the methoxydicarboxylic acid derivative.

Research on the B-complex continued with identification of p-aminobenzoic acid and pantothenic acid. Recognizing that sulfanil-amide was a derivative of p-aminobenzoic acid, he attempted to mod-ify vitamin $B_2$ into an antibiotic. He succeeded in synthesizing many inhibiting compounds, or antivitamins.

After World War II, Kuhn initiated a study of the factors that resist infection. In 1947, in a South African wild tomato, he discovered an alkaloid that protects against the potato beetle larva. Knowing that human milk provides immunity for infants, he isolated and determined the constitution of various nitrogenous oligosaccharides from milk. He related resistance to the absence of a receptor, showing, for instance, that lactaminyl oligosaccharides in bovine milk were split by the influenza virus and a receptor-destroying enzyme of cholera vibro. Human milk did not contain lactaminyl oligosaccharide and therefore provided resistance to influenza. Kuhn investigated brain gangliosides during the 1960s, determining that these nitrogen-containing saccharides were analogous to those in milk. Kuhn's research on these compounds continued until his death.

JANE A. MILLER
*University of Missouri—St. Louis*

# Bibliography

Baer, H. H. "Richard Kuhn"; *Adv. Carbohydr. Chem. Biochem.* **1969**, *24*, 1–12.

Burk, D. "Kuhn, Richard"; In *Dictionary of Scientific Biography;* Gillispie, C. C., Ed.; Charles Scribner's Sons: New York, 1973; Vol. 7, pp 517–18.

Farber, E. *Nobel Prize Winners in Chemistry, 1901–1961;* Abelard-Schuman: London, 1963; pp 165–167.

Grassmann, W. "Richard Kuhn"; *Bayer. Akad. Wiss. Jahrb.* 1969, 231–253.

Karrer, P.; Jucker, E. *Carotenoids;* Elsevier: New York, 1950.

Kuhn, R. *Biochemie;* Kuhn, R.; Fischer, H., et al., Eds; Dieterich: Wiesbaden, 1947–1953.

Kuhn, R. *Richtlinien der diätetischen Therapie;* Enke: Stuttgart, Germany, 1947.

*Nobel Lectures: Chemistry, 1922–41;* Elsevier: Amsterdam, the Netherlands, 1966; pp 454–457.

Sebrell, W. H.; Harris, R. S. *The Vitamins;* Academic Press: Orlando, FL, 1968; Vols. 1–5.

Selchow, C. "Richard Kuhn"; *Arch. Gesch. Naturwissensch.* **1984**, *10*, 473–497.

# Adolf Butenandt

## 1903–

Adolf Friedrich Johann Butenandt was awarded the Nobel Prize in chemistry in 1939 "for his work on sex hormones"; he shared the prize with Leopold Ružička, who received it "for his work on polymethylenes and higher terpenes". Butenandt was forced by the German government to decline the prize, and it was not until after the war, in 1949, that he received the gold medal and the diploma. The son of a businessman, he was born in Bremerhaven-Lehe, Germany, on March 24, 1903, and went to school in Bremerhaven; thereafter he studied chemistry, biology, and physics at the Universities of Marburg and Göttingen. He received his doctorate in 1927 with a dissertation on the chemical constitution of rotenone under the direction of Adolf Windaus (a Nobel Prize in chemistry, 1928) at the University of Göttingen. In 1929 Butenandt isolated estrone, the hormone that determines sexual development in females, in pure crystalline form. This work led to the isolation of related hormones, and within a few years he had isolated androsterone (1931), a male sexual hormone, and progesterone (1934), a hormone important for the biochemical processes involved in pregnancy.

Adolf Windaus and Walter Schoeller, the latter from the Schering–Kahlbaum pharmaceutical company, had suggested that Butenandt work on the isolation of sex hormones. Butenandt began this research while an assistant in Windaus's laboratory at the University of Göttingen (1927–1930). The research at Göttingen had turned out to be fertile ground for Butenandt, because as a student he had been torn between the study of chemistry and biology. He had turned to chemistry for economic reasons. However, he experienced Göttingen at the height of its scientific flowering and was able to combine his dual-disciplinary interests; there he became interested in the chemical structure of biological substances. He had already worked on the hormone thyroxin as a first thesis topic but turned to the rotenone study when the constitution of thryoxin was elucidated in 1926.

By 1928 only a few hormones had been represented in their pure form: adrenaline, thyroxin, and insulin; little was known about the nature of sexual hormones, but there was much interest in the topic, and that led to instances of simultaneous discovery. Both Butenandt and Edward Doisy isolated estrone simultaneously but independently in 1929. Doisy announced his discovery in August 1929 at a lecture in Boston, and Butenandt had to wait until after summer vacation at his institute to publish his results in the October issue of Die Naturwissenschaften.[1] It was known that certain extracts from ovaries or placentas caused the characteristic "in-heat" phenomena in castrated female rats, but that was about all that was known. During the first two decades of the 20th century the stage was set for subsequent research through the development of a biological test in the United States (1923) by Edgar Allen and Edward Doisy that measured estrogenical material from castrated mice and rats. Another biological discovery that aided sex hormone research was Berlin gynecologists Selmar Aschheim and Bernhard Zondek's identification that estrus-producing activity appeared in the urine of pregnant women. This shifted chemical work from examinations of the ovaries to urine.

Butenandt took a big step forward in the history of biochemistry when he isolated estrone from the urine of pregnant women. It is one of the substances secreted by ovarian cells in small quantities and determines, in part, sexual development in women. He named it "progynon" in his first publication, and then "folliculine", which indicated its source, the follicle, in subsequent early articles.[2] The confusion in nomenclature reflected the novelty of the work and occurred in all the groups that isolated the same hormone independently. By 1932,

---

[1] " 'Progynon', ein kristallisiertes weibliches Sexualhormon"; Die Naturwissenschaften, 1929, 17, 879.

[2] See " ' Progynon', ein kristallisiertes weibliches Sexualhormon" and the lengthy treatise "Untersuchungen über das weibliche Sexualhormon (Follikel-oder Brunsthormon)"; Abhandlungen der Gesellschaft der Wissenschaften zu Göttingen; Weidmannsche Buchhandlung: Berlin, 1931; III. Folge, Heft 2.

after the discovery of the steroid nature of several sex hormones, he could determine its chemical structure, that its empirical formula was $C_{18}H_{22}O_2$, and that it was an oxyketone. Within two years after the isolation of estrone, Butenandt, together with Kurt Tscherning, isolated androsterone, a male sex hormone, from male urine. Earlier attempts had been made by other researchers to find a male hormone in the testes, but because it was difficult to obtain adequate amounts of active testis extract, other substances were looked for. Butenandt obtained 50 mg of a crystalline substance from about 4000 gallons of male urine. It had only been possible to work on this substance after a quantitative biological test, the capon comb test, was developed. After chemical purification of the substance, it was found that the male sex hormone behaved in many ways like estrone. Its chemical structure, determined ultimately in 1934, only differed from estrone by the addition of 1 methyl group and 5 hydrogen atoms: $C_{19}H_{30}O_2$. Another striking result of this research was the discovery that the substance, as noticed later with all three sex hormones, was closely related to steroids. This result led to the classification of sex hormones as steroids. It was then found by Ružička that cholesterol, a common animal sterol, could be transformed into androsterone. The steroid nature of these sex hormones made it easier to understand their chemical structure. It was fortunate that Butenandt had studied under Windaus, who had specialized in the chemistry of sterols.

After completing his *Habilitation* in 1931 on the follicular sex hormone,[3] Butenandt became head of the organic and biochemical department of the chemistry laboratory at the University of Göttingen. In 1933, at the relatively young age of 30, he received an appointment as professor of organic chemistry and director of the organic chemical institute at the Technische Hochschule in Danzig. After turning down an offer to become a professor of biological chemistry at Harvard University in 1935, Butenandt received a joint appointment as director of the prestigious Kaiser Wilhelm Institute for Biochemistry in Berlin-Dahlem in 1936 and honorary Professor at the University of Berlin, where he remained until the end of the war.

During his years in Danzig (1933–1936), Butenandt continued his research on the second ovarian hormone, progesterone. In 1934 Butenandt and Ulrich Westphal isolated progesterone from the corpus

---

[3] "Untersuchungen über das weibliche Sexualhormon", 1931.

luteum. They used extracts from sow ovaries—sow corpora lutea—prepared by the Schering–Kahlbaum pharmaceutical company. The corpus luteum hormone had already been obtained in crystalline form in 1931 and 1932 by various groups, but it was not isolated in its chemically pure form until 1934. Butenandt and Westphal also demonstrated progesterone's close relationship to pregnanediol, which had already been found independently by Guy Marrian and Butenandt in the urine of pregnant women in 1931. By the fall of 1934 Butenandt had converted pregnanediol into progesterone. By 1939 he had synthesized the pregnancy hormone from cholesterol.

It was also during his time in Danzig that Butenandt was awarded several grants from the Rockefeller Foundation. This recognition was a great deviation from the Foundation's policy during the 1930s because it had begun to withdraw from its German projects after the National Socialists' rise to power. In addition to a grant for equipment for his laboratory in Danzig, Butenandt was invited with a fellowship for a three-month stay in the United States in 1935 to visit laboratories specializing in hormone research.

In Berlin-Dahlem Butenandt continued his work on sex hormones. He also took up other areas of research: the new field of virus research, the synthesis of hormones, cancer research and the relationship between estrogenic hormones and tumors, the study of insecticides, and research on eye pigmentation in insects. In Dahlem there was much opportunity for cooperative research, and Butenandt was instrumental in creating a research group for virus research with members of the biology and biochemistry institutes. This work had been stimulated by Wendell Stanley's isolation of the tobacco mosaic virus in 1935, but Gerhard Schramm, the chief member of the group, contributed most of the original work. Schramm codiscovered the reversible decomposition of the tobacco mosaic virus in nucleic acids and units of protein. Butenandt also worked closely with Alfred Kühn, the zoologist, on genetic researches with the meal moth. Butenandt's major research program at the Kaiser Wilhelm Institute, however, took his already begun work on the isolation and purification of sex hormones one step further: He began work on hormone synthesis.

In 1943–1944, shortly before the war ended, Butenandt moved his institute to Tübingen, in Southwest Germany, where it was farther removed from the bomb raids in Berlin. After the war, the Kaiser Wilhelm Gesellschaft was dissolved and the Max Planck Gesellschaft, its successor organization, was founded. Butenandt remained in Tübingen as director of the Max Planck Institute for Biochemistry and was professor of physiological chemistry at the University of Tübingen from 1945 to 1956. In 1952 he was appointed professor of physiological chemistry at the University of Munich and became director of the physiological-chemical institute there from 1956 until 1960, when he became president of the Max Planck Gesellschaft, a position he held until 1972, a period of great expansion and institute building. Butenandt contributed much to the rebuilding of German science in the postwar period and became an influential science organizer.

Butenandt's most outstanding contribution to science was his work on sex hormones. Within five years (1929–1934) he had isolated two female sex hormones, estrone and progesterone, and one male sex hormone, androsterone. He found that they all belonged to the same class of substances, the steroids, and that their chemical structure was similar, although they all had different physiological functions. Butenandt was unique in that his work on sex hormones focused on their chemical constitution. He had an active mind, always searching for new knowledge about life's processes and functions. This led him to other areas of research. He also did very important work on the biochemistry of heredity, virus research, the relationship between cancer and hormones, and the active subtances of insects.

In his research on the biochemistry of heredity, undertaken together with Alfred Kühn, it was found that genes operated through ferments. This result was established by analyzing the genetic chain in the eye pigment of insects. In their work on the active substances of insects, Peter Karlson and Butenandt crystallized the first insect hormone, ecdysone, a chrysalis hormone, and established its structure as a steroid hormone. In addition, Butenandt and Erich Hecker found the first crystallized pheromone in bombykol, the sexual substance of the silk spinner.

Butenandt's isolation and elucidation of the structure of sex hormones led to the synthesis of other steroids; it also had far-reaching importance for medicine. It led to the production of cortisone on a large scale. Preparations were made that could be used therapeutically, such as estradiol for disturbances in the menstrual cycle and progesterone to prevent miscarriages. Estrogen is the basis for oral contraceptives, which have dramatically altered methods available for birth control in this century.

KRISTIE MACRAKIS
*Harvard University*

# Bibliography

Butenandt, Adolf. "Untersuchungen über das weibliche Sexualhormon (Follikel-oder Brunsthormon)"; *Abhandlungen der Gesellschaft der Wissenschaften zu Göttingen;* III. Folge, Heft 2, 1931.

Butenandt, Adolf. "Ergebnisse und Probleme in der biochemischen Erforschung der Keimdrüsenhormone"; *Die Naturwissenschaften* **1936,** *24,* 209–224.

Butenandt, Adolf. *Das Werk eines Lebens.* Max-Planck-Gesellschaft, Munich, Germany; Vandenhoeck & Ruprecht: Göttingen, Germany, 1981. All of Butenandt's papers are collected here, along with an autobiographical sketch and an introduction by Peter Karlson.

Hoppe, Brigitte. "Adolf Windaus, Heinrich Wieland, Richard Kuhn, Leopold Ružička, Alexander Todd und Adolf Butenandt: die Erforschung der Vitamine und der Hormone"; In *Die Grossen der Weltgeschichte;* Fassmann, K., Ed.; Kindler: Zurich, Switzerland, 1978; pp 337–381.

Karlson, P. *Adolf Butenandt: Biochemiker, Hormonforscher, Wissenschaftspolitiker;* Wissenschaftliche Verlagsgesellschaft: Stuttgart, Germany, 1990.

Maisel, A. Q. *The Hormone Quest;* Random House: New York, 1965.

*Nobel Lectures: Chemistry, 1922–1941;* Elsevier: New York, 1966.

# 1939

# *Leopold Ružička*

## 1887–1967

Copyright Nobel Foundation

Leopold Ružička was born in Vukovar, Croatia, on September 13, 1887, and died at Mammern, Switzerland, on September 26, 1976. He received the Nobel Prize in chemistry in 1939. Ružička attended public school and a classical *Gymnasium* in Osijek, where he had moved at age four with his mother after the death of his father. At first he had wanted to become a priest but then became interested in math and science. He decided to study chemistry in Germany on account of political unrest and the poor quality of education in his home country.

From 1906 to 1910 he studied chemistry at the Technical University of Karlsruhe, where he received his diploma and a doctorate in just four years. Among his teachers were Christian Bunte, Carl Engler, Fritz Haber and, most influential, Hermann Staudinger, who was only six years older. After he finished his dissertation under Staudinger on phenyl methyl ketene, he stayed on as Staudinger's assistant and followed him to Zurich, when Staudinger was called to the Swiss Federal Institute of Technology (ETH) in 1912. Before leaving for Zurich, Ružička married Anna Hausmann. In 1917 he became a Swiss citizen.

Staudinger was an enthusiastic teacher but a stern boss. When Ružička told him in 1916 that he wanted to pursue his own interests, he lost his assistantship. His opportunities were greatly reduced, and he had mixed feelings about his teacher.[1] However, in 1918 he was

---

[1]Prelog and Jeger, 1980.

awarded a position as lecturer at the ETH and in 1920 at the University of Zurich, but received no remuneration. Therefore, he accepted in 1921 an invitation to collaborate with a Geneva perfume factory, M. Naef et Cie. (later Firmenich), and embarked on an ambitious research program involving the synthesis of the sesquiterpene perfumes farnesol and nerolidol and clarification of the structures of the four important odorous ketones: civetone, muscone, irone, and jasmone. Odorous ketones later proved to be a lucky choice but still brought him no support from Staudinger.

In 1923, Ružička was appointed as a titular professor, but when Staudinger left for Freiburg, Ružička was passed over for the succession in favor of Richard Kuhn. He therefore decided to move to the laboratories of Naef in Geneva but stayed only a year, after which he accepted the chair of organic chemistry at the University of Utrecht because he had missed the academic life. He stayed in Holland from 1927 to 1929. When Kuhn was called to the Kaiser Wilhelm Institute for Medical Chemistry in Heidelberg, Ružička was offered the chair of inorganic and organic chemistry at the ETH; he returned to Zurich as Kuhn's successor.

His position presented a great challenge. He reorganized and reequipped the laboratories and revised the teaching program. He left inorganic chemistry to the professor of analytical chemistry (W. D. Treadwell). He also started a symbiosis with the Swiss chemical industry, a cooperation that enabled him to build up a first-rate institute with a team of prominent chemists, including Tadeus Reichstein.

The 1930s became his most successful period, and in 1939 for his work on polymethylenes and higher terpenes he was awarded the Nobel Prize, which he shared with Adolf Butenandt. Because of the war he had to postpone his trip to Stockholm and his Nobel lecture until 1945.[2]

During the war he supported refugees from the Nazi regime by enabling them to study in Switzerland and work in his laboratories. After the war he was involved in aid to Slavic countries even after they turned to communism. That brought him some criticism. His many outside activities diminished his interest in chemistry, and he delegated much authority to his co-workers. The sizable royalties from his patents in Switzerland and in the United States, where the royalties had accumulated during the war, enabled him to build a notable collection of 17th-century Dutch paintings which he presented to the Zurich

---

[2]Ružička, 1947.

Museum of Paintings (Kunsthaus). In 1950 he obtained a divorce from his first wife, and in 1951 he married Gertrud Acklin (née Frei).

In 1950 his interest in chemistry was rekindled by the new electronic theories, which demanded a reinterpretation of his terpene chemistry, and by the new physical methods—molecular spectroscopy, X-ray analysis, and chromatography—which required a reorganization of internal structure of the Institute. He retired in 1957 and was succeeded by his long-time associate, Vladimir Prelog.

From an early age Ružička was interested in the chemistry of natural products. This interest was reinforced in his postdoctoral work with Staudinger, on the isolation and constitutional elucidation of pyrethrins, the insecticide constituents of *Asteraceae cinerarifolium*. In Zurich he achieved the total synthesis of the monoterpenes fenchon and linalool,[3] but he considered his early work (1911–1920) training for his subsequent research programs.

His main work, which brought him worldwide fame and the Nobel Prize, was started in 1921. It involved three areas of research: macrocyclic compounds (many-membered rings), higher terpenes, and steroids, including male sex hormones. Ružička's work on large rings presented a novel approach, because Adolph Baeyer had predicted in his strain theory that many-membered rings were much too unstable to exist. All previous attempts to prepare alicyclic compounds with 9 or more carbon atoms had been unsuccessful. Ružička, who was interested in the ring ketones civeton and muscone on account of their musk odor, started a new chapter in alicyclic chemistry. He proved that civetone contained a ring of 17 and muscone 15 carbon atoms. By a series of degradation reactions he established the constitution of these compounds, which he then confirmed by synthesis. He explained the absence of strain by the fact that large rings are basically two parallel chains of $CH_2$ groups closed at each end and therefore analogous to open-chain compounds. One problem with the synthesis was the low yield, which was later improved by K. Ziegler.

In the work on terpenes Ružička used the isoprene rule as a fruitful working hypothesis and the method of dehydrogenation as an efficient tool in the structural elucidation of a great number of compounds. The isoprene rule had been proposed in 1887 by O. Wallach but not taken seriously until Ružička recognized its general significance. Ružička defined monoterpenes as compounds whose carbon skeleton is composed of two isoprene groups, each consisting of five carbons; sesquiter-

---

[3]Ružička, 1917; Ružička and Fornasir, 1919.

penes, of three isoprenes; diterpenes, of four isoprenes; and so on. He and his team investigated sesquiterpenes and synthesized farnesol and nerolidol. The method of dehydrogenation using sulfur and later selenium became a rapid method for determining the carbon skeleton in the alicyclic terpenes, because it allowed the conversion of a suitable substrate into "aromatic compounds which, being more rigidly constructed, should in turn give more easily recognizable degradation products and should also be easier to synthesize. Once the main carbon skeleton had been established . . . it would not be difficult to determine the position of the double bonds."[4] Central to his studies of diterpenes was the structure of abietic acid and the synthesis of a large number of aromatic dehydrogenation products.

The constitution of triterpenes (1929–1955) proved to be more difficult than mono-, sesqui-, and diterpenes, because until 1948 he had had no access to infrared spectroscopy, but used only microanalysis and ultraviolet spectroscopy. He relied again on the method of dehydrogenation, using selenium dioxide and a new reagent, N-bromosuccinimide. He gained valuable information from the intercorrelation among the reduction products as well as from pyrolytic cleavage and the identification of the easily recognizable fragments.

The Zurich group was instrumental in the structure elucidation of lanosterol and recognized the relationship between lanosterol and cholesterol. At this point Ružička realized that the structure of lanosterol contradicted the isoprene rule.[5] Ružička's research on steroids centered on three topics: structure elucidation by dehydrogenation methods, determination of configuration, and work on the male sex hormones. Based on X-ray studies in 1932, J. D. Bernal had shown that the cholesterol molecule was only 4.5–5 Å wide. Using space-filling models and stereochemistry Ružička proposed in 1933 a chair form for cholestanol (dihydrocholestanol), allowing for a flat molecule whose overall dimensions agreed with the X-ray analysis. The cholesterol problem was difficult and time-consuming, and all aspects of its structure and absolute configuration were not settled until around 1955.

The work on male sex hormones brought Ružička into competition with Butenandt, who had isolated androsterone from urine in 1932 and had boldy proposed its formula. Ružička was immediately tempted to test it (in 1934) by partial synthesis from cholestanol,

---

[4]Ružička et al., 1952.
[5]Ružička and Meyer, 1921.

which involved the oxidative removal of the side chain. He also established the physiological importance of the molecule's stereochemistry. He then set out to synthesize another male hormone, testosterone, which was accomplished by A. Wettstein in Ciba's laboratory in 1935 while Ružička was traveling in the United States. Upon his return he started a hectic program in the field of steroids to speed up his own research.

Between 1934 and 1939 he published some 70 papers and received many patents, but after 1940 chemistry took a back seat to his other interests and activities.

In the next decade, 1940–1950, he was sufficiently challenged by the "new chemistry" and by biogenetic interpretations of his isoprene rule that he again developed a renewed interest in his research now directed mainly toward the problematics of sterol and terpene biosynthesis, especially after Konrad Bloch, George Popjak, and John Cornforth had demonstrated the biosynthesis of squalene and cholesterol from acetate units.

His "biogenetic isoprene rule" became the crowning glory of his life's work (1953). It stated that among all possible structures of a particular terpene the favored one can be derived mechanistically by the cyclization of an aliphatic precursor such as geraniol, farnesol, geranyl geraniol, and squalene. The modified rule was now applicable also to compounds whose molecules could not be divided into discrete isoprene units. In 1955 the biogenetic isoprene rule was extended by a stereochemical amendment.[6]

Ružička had shown interest in biogenetic origin as early as 1926 when he viewed the formation of musk compounds in the civet cat as a product of fatty acid metabolism. In his later years he developed an intense interest in biochemistry and the origin of life, which he viewed as merely a complicated sequence of chemical reactions in accordance with Ernst Haeckel's materialistic philosophy. He strongly urged the establishment of a chair for biochemistry at the ETH, which he saw realized just before his retirement.

Ružička's team received much stimulation from his cosmopolitan worldview. During the war the ETH group had been isolated, but his close connection with American chemists, especially Roger Adams and R. B. Woodward right after the war, contributed to the rapid expansion of his laboratory.

---

[6]Ružička et al., 1955.

He was very skillful in building up a successful relationship with industry without compromising his scientific integrity. This fruitful symbiosis and his growing reputation allowed him to attract both young talent and experienced staff, among them Tadeus Reichstein, Vladimir Prelog, Moses Goldberg, Placidus Plattner, and Oskar Jeger, to name just a few. He also secured support from the Rockefeller Foundation. At one point Morris Kharasch offered him a chair at the University of Chicago, but Ružička decided to stay in Zurich.

As a teacher he was somewhat unconventional. He lectured in an unorthodox way, scribbling the blackboards full of reactions and for mulas at a very fast pace at eight o'clock in the morning, six days a week. He considered copying formulas important, asserting that they entered the subconscious of the student! He was enthusiastic but low-keyed and sometimes cynical. His students admired him for his great success and feared him for his demanding standards and severe oral exams. He had little patience with mediocrity and had almost no contact with students other than doctoral candidates, from whom he expected the highest performance.

In addition to the Nobel Prize and the Marcel Benoist Prize (Switzerland) he won numerous awards, medals, and honorary degrees. From 1945 to 1967 he published almost 600 papers and was active in the Swiss Chemical Society. For forty years he was a member of the editorial board of *Helvetica Chimica Acta* and one of the joint chief editors of the periodical series *Ergebnisse der Vitamin und Hormonforschung*. Just before the war he gave up this position in protest over the firing of two Jewish executives by the publisher. He was also one of the editors of the new Swiss journal *Experientia*. Leopold Ružička leaves a legacy not only as one of the most successful pioneers in classical organic chemistry but also as a man who was not afraid to stand up for human rights.

TONJA KOEPPEL

## *Bibliography*

Eschenmoser, A. "Leopold Ružička—From the Isoprene Rule to the Question of Life's Origin"; *Chimia* **1990**, *44*, 1–21.

Farber, E. *Nobel Prize Winners in Chemistry, 1901-1961*; Abelard-Schuman: London, 1963.

Kenner, G. W. "Leopold Ružička"; *Nature* **1977**, *266*.

*McGraw-Hill Modern Men of Science*; Greene, J. E., Ed.; McGraw-Hill: New York, 1968; Vol. 2.

Ohloff, G. "Leopold Ružička: Scientist as Inventor"; *Chimia* **1987**, *41*.

Prelog, V.; Jeger, O. "Leopold Ružička 1887–1976"; *Helv. Chim. Acta* **1983**, *66*, 1307–1342.

Prelog, V.; Jeger, O. "Leopold Ružička"; *Biographical Memoirs of Fellows of the Royal Society* **1980**, *26*, 413.

Ružička, L. "Bedeutung der theoretischen organischen Chemie für die Chemie der Terpenverbindungen"; In *Perspectives in Organic Chemistry;* Todd, A., Ed.; Interscience: New York, 1956, p 256.

Ružička, L. "In the Borderland Between Bioorganic Chemistry and Biochemistry"; *Ann. Rev. Biochem.* **1973**, *42*, 1.

Ružička, L. "Fundamentals of Odour Chemistry: A Summary, in Molecular Structure and Organoleptic Quality"; *Soc. Chem. Ind.* (London) **1957**, 116.

Ružička, L. "History of the Isoprene Rule"; Faraday Lecture. *Proc. Chem. Soc.* **1959**, 341.

Ružička, L. "Nobelpreise und Chemie des Lebens"; *Naturwiss. Rundsch.* **1971**, *24*, 50–56.

Ružička, L. "Rolle der Riechstoffe in meinem chemischen Lebenswerk"; *Helv. Chim. Acta* **1971**, *54*, 1753.

Ružička, L. "Die Totalsynthese des Fenchons"; *Ber. Dtsch. Chem. Ges.* **1917**, *50*, 1362.

Ružička, L. "Vielgliedrige Ringe, Höhere Terpenverbindungen und Männliche Sexualhormone"; in *Les Prix Nobel en 1945;* P. A. Norstedt & Söner: Stockholm, Sweden, 1947; pp 177–201.

Ružička, L.; Eschenmoser, A.; Jeger, O.; Arigoni, D. "Eine stereochemische Interpretation der biogenetischen Isoprenregel bei den Triterpenen"; *Helv. Chim. Acta* **1955**, *38*, 1890.

Ružička, L.; Fornasir, V. "Uber die Totalsynthese des Linalool"; *Helv. Chim. Acta* **1919**, *2*, 182.

Ružička L.; Meyer, J. "Überführung des Cadinens in einen Naphthalinkohlenwasserstoff"; *Helv. Chim. Acta* **1921**, *4*, 505.

Ružička, L.; Voser, W.; Mijovic, M. V.; Heusser, H.; Jeger, O. "Über die Konstitution des Lanostadienols [lanosterin] und seine Zugehörigkeit zu den Steroiden"; *Helv. Chim. Acta* **1952**, *35*, 2065.

# George de Hevesy

## 1885–1966

Copyright Nobel Foundation

George Charles de Hevesy is remembered in two distinct areas, as codiscoverer of hafnium in 1923 and for his extensive work with radiochemical isotopes. It was for the latter that he received the 1943 Nobel Prize in chemistry.

Born in Budapest on August 1, 1885, he received his Ph.D. degree at the University of Freiburg in 1908, at which time he accepted a position at the Eidgenössiche Technische Hochschule of Zurich. In 1911 he traveled to Manchester to work with Ernest Rutherford. Following service in the Austro-Hungarian army during World War I, he collaborated in Copenhagen with D. Coster on the discovery of element 72, hafnium. For a time, beginning in 1926, he was professor of physical chemistry at Freiburg. He returned to the Bohr Institute, Copenhagen, in 1934, where he remained until, during World War II, he moved to Stockholm. Following many years of active research in Stockholm, he died in Freiburg on July 5, 1966.

George de Hevesy traveled widely during his career, spending time in the United States and South America. He was the recipient of numerous honorary degrees. Among his awards, in addition to the Nobel Prize, were the Cannizzaro Prize (Academy of Sciences, Rome), Copley and Faraday Medals (Royal Society, London), Baily Medal (Royal College of Physicians, London), Silvanus Thompson Medal (British Institute of Radiology), Niels Bohr Medal, Rosenberger Medal (University of Chicago), and the Atoms-for-Peace Award. He was a member of several learned societies and an honorary member of numerous others. In 1924 he married Pia Riis, a union that produced five children.

The early years of the 20th century saw a tremendous growth in science. George de Hevesy was a part of that growth. He writes, in an autobiographical reflection, of his experiences including conversations with Einstein while at Zurich. At Manchester under Rutherford's guidance, he met Niels Bohr, Harold Moseley, Hans Geiger, and others. He knew Harold Urey, E. O. Lawrence, both Madame Curie and her daughter Iréne, and other pioneers in the rapidly expanding fields of nuclear chemistry and radiochemistry. Hevesy himself was a significant contributor to that expansion. In his later investigations, dealing as they did with biological and medical applications of radiochemical tracers, he also became associated with prominent researchers in these areas.

Inspired by his early interest in molten salts and electrical conductivity, Hevesy traveled to Manchester, England, in 1911 to study under Rutherford. While there, he was directed, among several projects, to study the separation of radium-D, one of the natural disintegration products of the uranium series. Because radium-D was always associated with large amounts of lead, Hevesy undertook the task of isolating the radioactive component from its matrix. An understanding of isotopes was at that time unknown. How frustrating it must have been, for no matter what chemical means were employed, radium-D could not be differentiated from ordinary lead.

We know now, of course, radium-D is properly labeled as a radioactive isotope of lead, lead-210, and behaves chemically like its nonradioactive counterpart. Realizing finally that chemical separation was impossible, Hevesy devised a series of experiments in which the course of a reaction involving lead could be followed by monitoring a minute amount of radioactive lead (radium-D) added to the reaction mixture. Thus, the concept of radioactive tracers was born. In collaboration with others, he applied this concept to the solubility determination of sparingly soluble lead salts and to the electrochemistry of lead and bismuth, including verification of Nernst's equation relating electrode potential to concentration. Later experiments along these lines related to the study of rates and the extent of diffusion of solids in solids. For example, a piece of lead foil containing radioactive lead was placed in contact with an identical piece of nonradioactive foil. After a period of time, the radioactivity of both pieces was measured. Arrhenius's theory of electrolytic dissociation of ions in solution was supported by Hevesy's work on the interchange of radioactive metal ions for nonradioactive metal ions when salts containing both were dissolved in solution and then recrystallized. Hevesy and co-workers were the first to apply radioactive tracers to biological samples. Using radium-D (lead-210) and thorium-B (lead-213), the uptake of labeled compounds by bean seedlings was investigated in 1923. Later studies, using radioactive lead and bismuth, focused on the uptake of these ions by animal organisms. These early studies were limited to some extent by the availability of only naturally occurring radioactive isotopes. The extensive studies of Hevesy using artificially produced radionuclides would come later, after the development of proper techniques by which they could be produced.

Also in the early 1920s Hevesy addressed himself to the physical separation of various isotopic mixtures. He was able to separate a light from a heavy mercury fraction using vacuum distillation. Similar studies dealt with isotopes of chlorine in hydrochloric acid.

The adoption of radioisotope tracer studies by analytical chemistry did not go unnoticed by Hevesy, who pioneered the technique of isotopic dilution. Using this procedure, a known activity of the element under study is added to a sample prior to commencing the analytical sequence of steps. This addition contributes nothing to component weight but does allow one to monitor the extent to which the added activity is recovered in the final steps of the sequence. If, for example, only 80% of the added activity appears in the measured component, then one assumes only 80% of the component was recovered and, working backward, calculates the weight associated with 100% activity. This procedure was used by Hevesy and his co-workers to ascertain, for example, the lead content of various rock and mineral samples. In later years, it was applied again in his extensive studies on biological samples. Later, too, in conjunction with studies on the rare earth elements, he utilized neutron bombardment of gadolinium oxide to ascertain the concentration of europium present in the sample. This was accomplished by monitoring the emissions of artificially induced radioactive europium present in the sample. The reader will recognize this approach as the genesis of neutron activation analysis, a presently accepted and widely used analytical tool.

In addition to his radioactivity studies, Hevesy in 1922 collaborated with D. Coster in a search to identify and characterize element atomic number 72. According to Bohr's atomic theory, this element should have properties similar to those of titanium and zirconium and should properly be placed with them in the periodic table. In Paris a claim to the discovery of element 72 had been made previously in 1911 by Georges Urbain, who identified a supposedly new component from the fractional crystallization of ytterbium salts. He named this material celtium, a new rare earth element. Hevesy and Coster examined zirconium minerals and were able to identify from them hitherto unreported X-ray lines, which they properly assigned to a new element, atomic number 72. They named the element *hafnium* after *Hafnia,* the Latin name for Copenhagen, where the work was done.

Hevesy soon turned his full attention to radiochemical tracer techniques and especially to the application of these techniques to biological and medical concerns. Although he was nominated earlier, World War II interrupted the awarding of any Nobel Prize. The Foundation designated him as the 1943 recipient in chemistry although the prize was actually awarded in 1944 and without the usual ceremony associated with the event. As stated by the Nobel Foundation, the prize to George de Hevesy was "for his work on the use of isotopes as tracers in the study of chemical processes."

The early biological studies of lead uptake in bean seedlings have been cited. Similar investigations were made with animal tissues in hope of differentiating between normal and cancerous cells. Other studies involved bismuth isotopes and potassium isotopes, both naturally occurring.

It was not until artificially produced radioactive isotopes became available that Hevesy and his group were able to significantly extend the scope of their studies. Soon after the discovery of deuterium in 1931, Hevesy and his co-workers utilized dilute heavy water (HOD) to study the uptake of water by fish, various other animals, and humans. By monitoring urine levels, they were able to ascertain the average lifetime of a water molecule in the human body and to estimate the total water content, about $10^{27}$ molecules. The rate of exchange between extracellular water and water that had penetrated cell walls was also studied. With the knowledge of artificially induced radioactivity, Hevesy directed the preparation of phosphorus-32 by neutron bombardment of sulfur, in the form of carbon disulfide. Using radioactive phosphorus in various forms, studies beginning in 1935 were undertaken pertaining to phosphorus metabolism in various plant, animal, and human tissues. The uptake of phosphorus in various forms by various organs within the body and the exchange of phosphorus in tissue and bone were studied. The role of adenosine triphosphate (ATP), the renewal of other phosphorus-containing compounds, the estimation of total body blood volume, and other studies pertaining to phosphorus were pursued by Hevesy and his co-workers. The phosphorus content of eggs and milk was studied, along with the role of phosphorus in the development of chick embryos.

With the availability of still other radioactive elements suitable for incorporation into compounds possessing biological activity, Hevesy was able to expand his studies to include the effects of halogens, nitrogen, and metals. These studies continued after he received the Nobel Prize, and he remained active until the time of his death. An example of these studies is the incorporation of a precursor into deoxyribonucleic acid (DNA) using carbon-14-labeled adenine as a means of studying the rate of cell formation. Iron metabolism, using iron-59, was extensively studied by Hevesy and co-workers. They monitored iron absorption and excretion, the uptake and release of iron from blood erythrocytes, variations of iron activity between normal and diseased cells, iron abnormalities in cancerous tissue, and other factors in which iron plays an important role.

George de Hevesy authored or coauthored nearly 400 publications, including articles, review papers, and books. (His journal publications through 1961 are available in a two-volume set of collected works.) He traveled extensively during his career, giving lectures and receiving awards throughout the world. As reflected in his many publications, his interests were broad, embracing many fields: The recognition of his isotope studies as given in the Nobel citation was but one facet of his interests.

GORDON A. PARKER
*University of Toledo, Ohio*

## Bibliography

Cockcroft, J. D. "George de Hevesy (1885–1966)"; In *Biographical Memoirs of Fellows of the Royal Society* **1967;** *13,* 125–133.

*Current Biography Yearbook, 1959*; H. W. Wilson: New York, 1959 (April).

Hevesy, G. de. *Adventures in Radioisotope Research. The Collected Papers of George Hevesy;* Pergamon: New York, 1962.

Hevesy, G. de. *Chemical Analysis by X-Rays and Its Applications;* McGraw-Hill: New York, 1932.

Hevesy, G. de. *Das Element Hafnium;* Springer: Berlin, 1927.

Hevesy, G. de. *Radioactive Indicators: Their Application in Biochemistry, Animal Physiology and Pathology;* Interscience: New York, 1948.

Hevesy, G. de; Alexander, E. *Praktikum der chemischen Analyse mit Röntgenstrahelen;* Akad. Verl.-Ges: Leipzig, Germany, 1933.

Hevesy, G. de; Paneth, F. *Lehrbuch der Radioaktivität;* Oxford University Press: London, 1926, 1938.

Howard, A. V. *Chambers Dictionary of Scientists;* E. P. Dutton: New York, 1951.

Ingle, D. J. *A Dozen Doctors; Autobiographic Sketches;* University of Chicago Press: Chicago, 1963.

Levi, H. *George de Hevesy: Life and Work;* A. Hilger: Bristol, Boston, 1985.

Levi, H. *Nuclear Phys.* **1967,** *A98.*

*Nobel Lectures Chemistry 1942–1962;* Elsevier: Amsterdam, The Netherlands, 1964.

Marx, G. *George de Hevesy 1885–1966 Festschrift;* Akadémiai Kiadó: Budapest, Hungary, 1988.

Szabadváry, F. "Hevesy, György"; *Dictionary of Scientific Biography;* Gillispie, C. C., Ed.; Charles Scribner's Sons: New York, 1972; Vol. 6, pp 365–367.

# Otto Hahn

## 1879–1968

Copyright Nobel Foundation

The Nobel Prize in chemistry for 1944 was awarded to Otto Hahn "for his discovery of the fission of the heavy nuclei". It was in recognition of distinguished work in his chosen field of science; it was also, as this account will stress, for work of enormous interest and consequence. Hahn was born on March 8, 1879 in Frankfurt-am-Main, in what had only recently become a united Germany following the Franco-Prussian war. He died on July 28, 1968, in Göttingen, in what had by then become the Federal Republic of Germany. His lifetime spanned a period of great economic and industrial expansion of Germany into what was undoubtedly the greatest power on the continent of Europe, a zenith that was shattered by two world wars and a period of political fanaticism that might more appropriately have belonged to the Middle Ages. The second of these wars led to the political and ideological division of Germany and to a diminution in the economic and political importance of Europe for many years to come.

The youngest of four boys in his family, Otto was the only one to acquire a university education. He was able to persuade his father, who had wanted him to study architecture, to permit him instead to study chemistry, a subject in which he had become interested during his high school years. By his own and other accounts,[1] Hahn was not a particularly conscientious student during his undergraduate years

---

[1]Hahn, 1966; Spence, 1970.

at Marburg and Munich (where he spent his second year). He was evidently prepared to devote more time and energy to laboratory work than to various of his lectures in science. Because he neglected much of his instruction in mathematics and physics, his want of skill in these disciplines proved to be a handicap in his later career. On the other hand, he was evidently quite happy to devote time to lectures in art and philosophy, even at the expense of his principal subjects. Distractions notwithstanding, Hahn devoted his last year of university education to research on the bromination of isoeugenol, the results of which were embodied in his Ph.D. thesis, submitted in the summer of 1901. Thereafter, following a year of military service, Hahn was appointed lecture assistant to his thesis supervisor, Professor Theodor Zincke. In September 1904, motivated by the prospect of a job in the chemical industry requiring facility in English, Hahn went to London to spend a few months in the laboratory of Professor Sir William Ramsay. He actually remained almost a year, during which he became versed in the handling of radioactive materials and actually discovered a new radioactive element to which he gave the name radiothorium.

This discovery was fortuitous. One of Hahn's first assignments had been to separate what was presumed to be radium from a relatively large quantity of barium salt. To do so he followed the customary method of fractional crystallization from a solution prepared from the starting material; but he added the novel procedure of monitoring the radioactivity of the gaseous emanation given by the successive fractions. From these measurements he was able to deduce the presence of a substance with unexpectedly high radioactivity in the mother liquor remaining after the removal of radium. This fraction should have contained thorium, but its radioactivity was much too great to be due to that element alone, and accordingly he postulated the presence of the new element. It must be remembered that at the time of this work the recognition of radioactive elements was based solely on rates of decay and intensities of radiation measured with fairly crude electroscopes. It was greatly to Hahn's credit that with little or no prior experience he was able to interpret his experimental observations so effectively.

Resolved to augment his skill and knowledge of radioactivity, he successfully applied to spend a year in Montreal in Ernest Rutherford's laboratory at McGill University. With him he brought various radiothorium and actinium preparations from London. His first hurdle, successfully surmounted, was to convince a skeptical Rutherford and his friend Bertram Boltwood in New Haven, Connecticut, of the validity of his discovery of radiothorium[2] . From his London material he managed to identify three additional species, and also he planted a seed for subsequent discoveries by determining the half-life of his radiothorium. More to the point, however, he was able to learn in Montreal the thinking processes and state-of-the-art methods of research in radioactivity, thereby becoming equipped with the skills of mind and hand that in time led him to a preeminent position in this field.

Hahn returned to Germany in the summer of 1906 to take up an appointment in the Chemical Institute at the University of Berlin, an appointment that had been offered to him a year previously by the director, Professor Emil Fischer. There, in a spartan basement room, he set up his laboratory, for which equipment such as electroscopes had to be constructed to his own design. He also set about acquiring additional radioactive materials, in particular, thorium preparations from the German firm, Knöfler. These he subjected to traditional wet-chemical separations, checking each fraction for radioactivity. Such investigations soon revealed that Hahn's new substance, radiothorium, could not be separated from thorium itself, but that the radioactivity of the various preparations of these depended significantly on their age. To account for these observations, Hahn was obliged to postulate the existence of yet another radioactive element, to which he gave the name mesothorium. This was deduced to have been formed from thorium, itself nonradioactive, and to be the parent of radiothorium. By taking into account the decay rates and specific activities of both thorium and radiothorium, he was able to interpret his observations on the variability of *net* activity according to the age of the specimens examined. On the basis of this interpretation, he could reconcile differences of opinion that had arisen between himself and Boltwood during the previous year in Montreal. A year later, Hahn reinterpreted his results in terms of *two* intermediates between thorium and radiothorium; he designated these mesothorium-1 and mesothorium-2, the latter being a rapid β-emitter and the parent of radiothorium. Hahn subsequently assisted the Knöfler company to produce mesothorium commercially as a radioactive source for medical purposes.

During this work with thorium and its radiochemical derivatives, Hahn encountered traces of radium emanation, the amounts of which increased with time. He inferred from this that the thorium must con-

---

[2]Trenn, 1983.

tain small amounts of a progenitor of radium. This was something the existence of which had already been suggested by Frederick Soddy, Bertram Boltwood, and others. However, even as Hahn was at work following up his observations, Boltwood in America announced his discovery of the element sought, to which he gave the name *ionium.*

In an account of this length it is impossible to provide particulars of all of Hahn's experimental investigations and his interpretations of these. In any case, their significance can only be judged within the context of the entire evolution of radiochemistry as an experimental and intellectual subdiscipline. Throughout the first dozen years of the present century, fragmentary evidence concerning three sequences of radiochemical transformations were coming to light. New "elements" were being discovered in "mother" or "daughter" relationships to each other or to some already known elements. The criteria for the identification of these were largely the kind and intensity of the radiation emitted by each new species. Before long it became apparent that various of these new so-called elements were chemically indistinguishable and inseparable, and that there were far too many of these to be accommodated within the periodic table. Out of this predicament evolved the concept of isotopes, credit for which in the first instance belongs to Kasimir Fajans, with subsequent reinforcement from Soddy.[3] Hahn, though himself aware of the prevailing confusion, acknowledged his own lack of vision to hit upon the ultimately obvious explanation of such evidence as the apparent chemical identity of radium and mesothorium-1, which, on the basis of different radiative characteristics, he regarded as distinct elements. It is probably not unfair to remark, at this point, that Hahn's strength lay more in his experimental skill than in his theoretical insight.

In the spring of 1907, Hahn had been appointed to the teaching staff in the Chemical Institute as a *Privatdozent,* though he evidently found the atmosphere more congenial and stimulating in the Institute of Physics. In the latter he made a number of good friends, and through these he was afforded the opportunity to meet a colleague to share in his research work. Her name was Lise Meitner. She had come from Vienna with a Ph.D. in physics and some postdoctoral research experience in radioactivity. She had come to study theoretical physics in Berlin with Max Planck and to do some experimental work. Because of her previous experience involving radioactivity she elected to do

---

[3]Romer, 1970; Trenn, 1977.

her laboratory work with Hahn, even though he was in the Chemical Institute. Thus began a collaboration that lasted for 30 years and a friendship that endured for more than 60 years until terminated by Hahn's death.

Their first joint project involved a study of the absorption of β-rays by metallic aluminum. Hahn by this time had accumulated quite a number of different sources of β-emission suitable for this investigation. By analogy with the behavior of α-rays, it had been expected that β-radiation from a single radioelement would be homogeneous in energy. Operating within this assumption, they did find that a number of elements, previously considered to be nonradiative, were actually weak β-emitters. However, when deflected by passage through a magnetic field, the β-rays were found not to be uniform and, instead, produced line spectra. In summary, β-emission proved to be much more complicated than had been originally supposed and was later intensively studied by Lise Meitner and her students. In other work on the active deposit formed from actinium, which Hahn identified as actinium-X, he interpreted his own and other related observations by introducing the idea of radioactive recoil. This principle, when applied to thin-layer specimens of a number of other α-emitting radioelements, led to the recognition of several new products, such as actinium-C″, thorium-C″, and radium-C″.

In 1911 the German emperor proposed the establishment of a new body for scientific research, to be known as the Kaiser-Wilhelm Gesellschaft and located in Dahlem, a suburb of Berlin. This was to consist of a number of institutes, including an Institute for Chemistry, with Emil Fischer as director, and an Institute for Physical Chemistry, with Fritz Haber as director. These two units were opened in October 1912. In the former, Hahn was given charge of a small, independent Department of Radioactivity, to the staff of which Lise Meitner was appointed.

As a departure from the work that he and Meitner had done up to this point, Hahn chose to take advantage of the uncontaminated rooms of the new Institute to examine the comparatively feeble radioactivity of the alkali metals potassium and rubidium. With a new assistant, Martin Rothenbach, Hahn succeeded in estimating the radioactive characteristics of strontium; attempts to secure comparable data for potassium were not particularly successful. A quarter of a century later Hahn and his co-workers returned to work with radioactive rubidium and were successful in applying their earlier information to the estimation of the age of certain minerals in which strontium-87, the daughter of radioactive rubidium, occurred. This was pioneering work in radiochemical dating; by the 1950s, as commercial mass spectrometers and much-improved counting equipment became available, such practices became widespread.

Soon after this, World War I broke out and, as a reservist, Hahn was called up for active duty. Within a few months he was transferred to serve with the gas-warfare corps, of which Fritz Haber served as scientific leader, and so for the duration of the war he was "involved in research, development, testing, manufacturing and using new weapons".[4] The nature of this war service was apparently elastic enough to permit Hahn at times to participate in some work in Berlin at his Department of Radioactivity. Of this the most significant was the discovery by Meitner and himself of what they called *protoactinium* (later *protactinium,* the progenitor of actinium). A shorter-lived isotope of this element 91 had already been discovered in 1913 by Fajans and Gohring, and by them named *brevium.* That name, however, seemed inappropriate for the comparatively long-lived isotope that Hahn and Meitner had found, and their name for the element (*protactinium*) was eventually adopted.

Following this discovery, Hahn made another discovery, related to the sequential $\beta$-decay of uranium-$X_1$ (thorium-234) to uranium-$X_2$ (protactinium-234) to uranium-II (uranium-234). Careful measurements revealed the presence of yet another source of $\beta$-activity, ultimately identified as an isotope of protactinium, but provisionally labeled uranium-Z. After much painstaking investigation, Hahn was compelled to conclude that uranium-$X_2$ and uranium-Z were in fact *both* formed in a constant ratio from uranium-$X_1$, and that *both* decayed to uranium-II. This was the first recognized instance of nuclear isomerism, now recognized as a not-infrequent occurrence in disintegration series.

As time went by, radiochemistry and its interpretation became increasingly easy to rationalize on the basis of new discoveries. Of these, the work of Rutherford in 1919, whereby nitrogen atoms were converted to oxygen atoms by the action of $\alpha$-particles from radium, established the possibility of nuclear transformations. In his discovery of the neutron in 1932, James Chadwick not only established the characteristics of the missing nuclear particle but opened the door to its

---

[4]Badash, 1972; see also Spence, 1970.

potential use as an atomic projectile. And then in 1934 Frédéric and Irène Joliot-Curie achieved the first instance of artificially produced radioactivity: They found that, under bombardment by α-particles, boron or aluminum produced some atoms of nitrogen or phosphorus, respectively, and that the latter were spontaneous positron emitters. (The positron itself was then something of a novelty, having been first identified by Carl Anderson only in 1932.) Then, although he was not the first to achieve a nuclear transformation by means of neutrons, Enrico Fermi pointed out in 1934 that neutrons, being electrically neutral, could penetrate to the nuclei of other atoms. Thereupon he proceeded to show that a wide range of elements in the periodic table were converted into radioactive isotopes by neutron bombardment. The new isotopes formed were β-emitters, which, at some characteristic rate, changed into the element of next higher atomic number. This brief account of significant events in nuclear physics serves as an introduction to the work for which Hahn was awarded the Nobel Prize.

Among the elements that Fermi and co-workers exposed to neutrons was uranium, and from it they obtained a β-emitting material that Fermi proposed might include isotopes of an element of atomic number greater than uranium, at that time the last element in the periodic table. Considerable controversy arose from this proposal and, as an authority in the radiochemistry of the heavier metals, Hahn undertook to investigate the matter further. In this he was assisted by Fritz Strassmann, as well as by Lise Meitner, but before the work had been completed the latter had found it necessary to leave Germany owing to the racial laws being imposed by the National Socialists. Soon afterward she was able to relocate in Sweden. Limitations of space preclude a detailed account of the work done by Hahn and his associates, but of such accounts there is no dearth.[5]

In brief, the Berlin group exposed uranium to a suitable source of neutrons and sought to identify the product or products formed. They first disposed of an early misconception that protactinium had been formed. Then, for some time, they were convinced that they had produced a series of transuranic elements with atomic numbers up to 96. Further work, however, caused them to suppose that they were dealing with radium isotopes. This conclusion was drawn from evidence that their new radioactive material was inevitably found with barium salts used as carriers in their separations. But all attempts to separate the

---

[5]See Hahn, 1956, 1958, 1966; Spence, 1970; Trenn, 1981.

putative radium salts from the barium failed, even though in parallel experiments natural radium was consistently separated from the barium carrier salts. After four years and twenty publications describing their work, Hahn was finally driven to a conclusion, which he initially published with great hesitancy in January 1939, that they had produced radioactive barium in consequence of "bursting" uranium atoms.

At regular and frequent intervals he kept Lise Meitner (now in Sweden) informed of the details of this work and the conclusions drawn from it. Thus it came about that she received his proposition about the cleavage of uranium prior to Christmas 1938. A few days later her nephew, O. R. Frisch, who was then working at the Niels Bohr Institute in Copenhagen, arrived to share the Christmas vacation with her. Together they discussed Hahn's suggestions and found them reasonable. In fact, they even sketched out an hypothesis why such "fission" (as it soon was called) might occur among nuclei of large atomic number, owing to the greater repulsive forces of the surface charge. Their ideas were immediately conveyed to Hahn and subsequently published in February 1939. As physicists, more familiar with nuclear properties than Hahn, they pointed out the capability of this nuclear fission to generate enormous quantities of energy. Fortified by this endorsement and refinement of his ideas, Hahn prepared a more complete and more confident report of his work, published in February 1939, in which he was able to add evidence that krypton, along with barium, had been formed by the fission of uranium. He mentioned the possibility of extra neutrons being formed in the process, but it was left to others to confirm the reality of that idea.

Like a brush fire, news of such extraordinary scientific importance could not long be contained. Frisch, on his return to Copenhagen in early January 1939, discussed the new information with Niels Bohr. The latter was about to leave for the United States to meet with many of the theoretical physicists there and felt compelled to announce the Hahn–Strassmann discovery and the Meitner–Frisch interpretation at a meeting in Washington, D.C. on January 26, 1939. American physicists were quick to repeat and confirm Hahn's work. Qualified scientists were not long in calculating that the fission of one gram of the appropriate isotope of the element uranium was capable of releasing about a hundred million kilojoules of energy. The fission process was initiated by the capture of a stray neutron, but when it occurred it generated two or three neutrons, so that the process was capable of self-propagation by a chain reaction. The combined masses of the products of fission amounted to less than that of the starting material. The production of such an amount of energy could be accounted for by the disappearance of an equivalent amount of matter in accordance with Einstein's relativity relationship. Thus the age of nuclear energy, with all its awesome consequences, was born.

Here was a new way to produce enormous amounts of energy, disclosed at a time when the major European nations were on the verge of war. The military significance of such a discovery was not lost upon the antagonists, and it was not long before much of the ensuing work on nuclear reactions became "classified". In this area of science, much of the customary publication and exchange of information was suspended "for the duration". Hahn and his group at the Kaiser Wilhelm Institute busied themselves with unraveling the complex chemistry associated with the fission processes and with isolating a large number of fission products derived from uranium. It was work carried out under difficulties, both personal and practical, and much of it was published in the open literature.

Shortly before the German surrender in 1945, Hahn and some other German nuclear scientists were "captured" by an Allied military intelligence party, taken to Britain, and detained there until early in 1946. It was a benign internment in a country home near Cambridge, and he was able to share visits by many friends and former colleagues. The news of the atomic bombs used against Japan that summer came as a great shock, and caused Hahn a period of great depression. Later, in November 1945, word came of the award of the Nobel Prize for 1944. He was not permitted to go to Stockholm to receive the prize then, however, so the presentation did not take place until December 1946. In fact, Hahn and the other detainees did return to Germany in January of 1946. Soon after, he took up the presidency of the Kaiser Wilhelm Institute, which was relocated in Göttingen and renamed the Max Planck Institute. In this work, taken up in his 67th year, he proved himself to be a very competent and effective administrator. He held this office with distinction until replaced at his own request at the age of 81. His many services to science led to a multitude of honors; his personal qualities earned him immense respect from his peers.

W. A. E. McBRYDE
*University of Waterloo*

# Bibliography

Badash, L. "Hahn, Otto"; In *Dictionary of Scientific Biography;* Gillispie, C. C., Ed.; Charles Scribner's Sons: New York, 1972; Vol. 6, pp 14–17.

Graetzer, H. G.; Anderson, D. L. *The Discovery of Nuclear Fission: A Documentary History*; Van Nostrand Reinhold: New York, 1971.

Hahn, O. *Applied Radiochemistry*; Cornell University Press: Ithaca, N.Y., 1936.

Hahn, O. "Personal Reminiscences of a Radiochemist"; *J. Chem. Soc.* **1956,** 3997–4003.

Hahn, O. *Mein Leben*; Bruckmann: München, 1968.

Hahn, O. *Otto Hahn: A Scientific Autobiography;* Charles Scribner's Sons: New York, 1966.

Hahn, O. "The Discovery of Fission"; *Scientific American* **1958,** 198, 76–84.

Hahn, O. *Vom Radiothor zer Uranspaltung* [Otto Hahn: A Scientific Autobiography]; Ley, W., Ed. and Trans.; MacGibbon & Kee: London, 1967.

Heisenberg, W. "Otto Hahn, Discoverer of Nuclear Fission, Dies"; *Phys. Today* **1968,** 21.

Irving, D. J. C. *The Virus House*; Kimber: London, 1967.

Romer, A. *Radiochemistry and the Discovery of Isotopes;* Classics of Science, Vol. 6; Dover: New York, 1970.

Smyth, H. D. *Atomic Energy for Military Purposes;* Princeton University Press: Princeton, New Jersey, 1945.

Spence, R. "Otto Hahn (1879–1968)"; *Biographical Memoirs of Fellows of the Royal Society* **1970,** *16,* 279–313.

*Trends in Atomic Physics*; Frisch, O. R. [et al.], Eds.; Interscience: New York, 1959.

Trenn, T. J. "Why Hahn's Radiothorium Surprised Rutherford in Montreal"; *Otto Hahn and the Rise of Nuclear Physics;* Shea, W. R., Ed.; D. Reidel: Dordrecht, The Netherlands, 1983; pp 201-212.

Trenn, T. J. *The Self-Splitting Atom;* Taylor and Francis: London, 1977.

Trenn, T. J. *Transmutation, Natural and Radioactive*; Heyden & Sons: London, 1981.

# *Artturi Virtanen*

## 1895–1973

Artturi Ilmari Virtanen won the Nobel Prize for chemistry in 1945 for his investigations of the effect of storage conditions on the preservation of agricultural fodder. He was the son of Kaarlo Virtanen and Serafiina Isotalo and was born in Helsinki, Finland, on January 15, 1895, and died on November 11, 1973.

After he completed a traditional lower-level education in the Classical Lyceum at Viipuri, Finland, Virtanen entered the University of Helsinki, where he obtained his master's degree in 1916. He served as first assistant in the Central Industrial Laboratory in Helsinki for one year before returning to school. His doctoral work was done at the University of Helsinki under the direction of O. Aschan and contributed to the elucidation of the structure of abietic acid, a major component of pine rosin.[1] In 1919 he completed this work and the following year married Lilja Moisio.

From 1919 to 1921 Virtanen was a chemist in a Finnish Cooperative Dairy Association laboratory that was concerned with the control of the manufacture of butter and cheese. This was a position that allowed him to travel to various laboratories to investigate new techniques: In Zurich he studied physicochemical techniques under G. Wiegner; in Stockholm he studied bacteriology under Chr. Barthel and enzymology under H. von Euler-Chelpin. In 1921 Virtanen became

---

[1]Farber, 1963.

director of the dairy association laboratory, a position at which he remained until he became director of the Biochemical Research Institute at Helsinki in 1931. He was also appointed Professor of Biochemistry at the Finland Institute of Technology at Helsinki in 1931 and a professor at the University of Helsinki in 1939.

Doubtless the time spent in the laboratories so directly associated with the dairy industry influenced his choice of research. He used his biochemical training to investigate some very practical problems faced by the dairy industry. During the course of many years of meticulous work, he developed better methods for the storage of butter and green fodder. He also investigated the assimilation of nitrogen by the bacteria found in legume root nodules.

In 1923 Virtanen began solving one of the dairy industry's problems. Without a protein supplement, dairy cattle need fodder with a high protein-to-carbohydrate ratio in order to produce high yields of milk. Legumes, plants such as peas, soybeans, and clover, have a high protein content and are suitable for fodder. These plants live in a symbiotic relationship with the bacteria in their root nodules; through a process known as nitrogen fixation, the bacteria convert the nitrogen in the air to chemical compounds that the plants can use. These compounds are secreted at the root nodule and are thus readily accessible to the plant. Virtanen felt that a better understanding of nitrogen assimilation might bring about better utilization of legumes.[2]

In the beginning, Virtanen investigated the relationship between the amount of nitrogenous compounds secreted from the root nodules and the growing conditions and maturity of a plant. He opted for a sterile cultivation technique, in which the only bacteria present were those purposefully introduced.[3] He found that the greatest quantity of nitrogenous compounds was given off during the initial stages of nodule growth and when the light was intense. The connection between these conditions and nitrogen fixation proved to be a red pigment that was found in the nodules. After confirming a disputed claim by Kubo that this red pigment was a compound known as leghemoglobin, Virtanen's investigations showed that under low levels of light and as the plant matured, the red pigment converted to a brown form known as methemoglobin, and nitrogen fixation slowed. When this brown pigment converted to a green form, the formation of nitrogenous com-

---

[2]Westgren, 1972.
[3]Virtanen, 1966.

pounds by the bacteria ceased.[4] Further evidence of the necessity of leghemoglobin for nitrogen fixation came with the discovery that on some plants these bacteria formed nodules with none of the red pigment and, under these conditions, they could not fix nitrogen.[5]

Legume bacteria were found to function best when the the soil pH was close to neutral, 6.5 to 7.0. This means that acidic soil is unsuitable for optimum production of legume crops unless it is treated with lime to decrease the acidity. The greater rates of nitrogen fixation found when the soil was well aerated mean that the soil should be loosely packed for optimum legume growth.[6]

Virtanen failed to achieve his original aim to determine more exactly how nitrogen fixation occurred, but his observations and improvements in research methodology became the foundation of further work. He was more successful in a different attempt at making the use of legumes more attractive, and it was this success that led to the receipt of the Nobel Prize in 1945. Difficulties in preserving protein-rich legume crops during the winter were a decisive factor limiting their use. Preserving legumes in the form of hay was inadequate because much of the nutrient value was lost during the drying and storage process, and the once-per-summer harvest of hay left the farmer at the mercy of the weather at harvest time. The alternative was the storage of fodder as a fermented mixture known as silage, but this resulted in a loss of 25–50% of the protein content. Virtanen thus reasoned that a silage preservation method that led to better retention of nutrients would lead to better utilization of legumes.[7]

When he began the development of an improved storage method, Virtanen knew some helpful things from biochemical studies. The loss of protein, vitamins, and carbohydrates during storage was due to the respiration of plant cells and the sugar fermentation brought about by the action of microorganisms. The fermentation also decreased the usefulness of the fodder because cattle found the butyric acid that was formed distasteful.

In the beginning, Virtanen and his co-workers concentrated on small-scale biochemical studies of protein decomposition and bacterial fermentation. They were the first to follow a sugar fermentation chemically from initial reactants to final products. In looking at the

---

[4]Farber, 1963.

[5]Virtanen, 1966.

[6]Ibid.

[7]Westgren, 1972.

fermentations caused by several types of bacteria, they found that the first stages in the decomposition of sugar were very similar in many cases. This similarity suggested that one set of conditions might inhibit many types of fermentation.

It had been known that acidic conditions inhibited the respiration and fermentation processes, but no systematic studies had been done on what conditions were most effective for preserving silage, and few tests had been made to determine the quality of the silage thus preserved. Virtanen and his co-workers began systematic, thorough studies to find conditions that would sufficiently retard these processes and also retain the nutrients in silage. They discovered that the pH of the silage must be less than 4 to prevent the formation of the butyric acid. [8] Analyses of the plants before and after storage showed that under these conditions the protein and vitamin losses were minimal. [9]

After failing to obtain conditions that were acidic enough through the natural production of acids during fermentation, they turned to the addition of mineral acids. They investigated the effects that the type of crop, soil acidity, and various physical treatments had on how much acid was needed to bring the pH below 4. Results of large-scale tests under actual farm conditions gave the same results, as well as confirming that large amounts of fodder and the acids could be mixed uniformly. The cattle fed large quantities of this silage over long time periods showed no ill effects, and analyses of their bones, blood, and tissues gave normal results. The cattle were disinclined to eat the silage when the pH was below 3, and when hunger drove them to do so, they showed symptoms of acidosis. Thus, the optimum acidity was determined to be at a pH between 3 and 4.

In the 1930s Virtanen became involved in extended field tests designed to optimize the utilization of legume crops. Analyses of the vitamin content of the plants were carried out at various stages of plant growth, and it was found that the plants lost vitamins as they approached the flowering stage. It was thus apparent that the silage should be harvested before this stage, and doing so led to a higher vitamin content in the milk produced by cattle fed this fodder. Optimum protein and fodder yields were in fact realized when the crop was harvested three times during the summer, a significant improvement over harvesting the less nutritious hay once per summer.[10]

---

[8]Farber, 1963.

[9]Leicester, 1970.

[10]Virtanen, 1966.

Virtanen developed the new fodder-storage method with the long Finnish winters in mind. It provided a constant supply of nutritious silage and made the farmers more independent of the weather. Animals fed this silage frequently had improved fertility and resistance to disease, and their wintertime milk had a higher vitamin content. In addition, the use of legume crops increased soil fertility by providing otherwise depleted nitrogen compounds.

A third problem that Virtanen investigated was the storage of butter. Until that time, stored butter inevitably developed a bad taste. Virtanen found that this was due to the oxidation of lecithin, a compound that occurs naturally in the butter. He also found that if a basic salt was used to increase the pH above 6 this oxidation did not occur. Butter storage was thus vastly improved.

Virtanen was a member of the Bavarian, Flemish, Norwegian, Swedish, and Danish Academies of Science, and he served as president of the State Academy of Science and Art in Finland. He held honorary degrees from the Universities of Lund, Paris, Giessen, and Helsinki, the Royal Technical College at Stockholm, and the Finland Institute of Technology. He was an honorary member of scholarly societies in numerous countries and received the Friesland Prize from The Netherlands in 1967, the Atwater Prize in 1968, the Siegfried Thannhauser Medal in 1969, the Gold Medal from Germany in 1971, the Gold Medal and Prize from Spain in 1972, and the Uovo d'oro from Italy in 1973.

ZEXIA BARNES
*Morehead State University*

# Bibliography

Dalton, L. "Virtanen, Arturri Ilmari"; In *The Who's Who of Nobel Prize Winners;* Schlessinger, B. S.; Schlessinger, J. H., Eds.; Oryx Press: Phoenix, AZ, 1986; pp 16–17.

Farber, E. *Nobel Prize Winners in Chemistry, 1901-1961;* Abelard-Schuman: New York, 1963; pp 185–191.

Leicester, H. M. "Virtanen, Artturi Ilmari"; In *Dictionary of Scientific Biography;* Gillispie, C. C., Ed.; Charles Scribner's Sons: New York, 1970; pp 45–46.

*Nobel Prize Winners;* Wasson, T., Ed.; H. W. Wilson: New York, 1987.

*The Nobel Prize Winners: Chemistry;* Magill, T. N., Ed.; Salem Press: Pasadena, CA, 1990; Vol. 2.

Virtanen, A. "The Biological Fixation of Nitrogen and the Preservation of Fodder in Agriculture, and Their Importance to Human Nutrition"; *Nobel Lectures, Chemistry 1901-1921;* Elsevier: Amsterdam, the Netherlands, 1966; pp 74–103.

Westgren, A. "The Chemistry Prize"; In *Nobel, The Man & His Prizes,* 3rd ed.; Nobel Foundation and Odelberg, W., Eds.; Elsevier: New York, 1972; pp 347–438.

# James Sumner

## 1887–1955

James Batcheller Sumner was born at Canton, Massachusetts, on November 19, 1887, into a prosperous family of cotton textile manufacturers. From 1900 to 1906 he attended Roxbury Latin School. In 1904 Sumner's left arm was amputated following an accidental shooting by his hunting companion. His left-handedness compounded the seriousness of the injury. He learned to use his right hand well, and, stimulated by the accident to excel in sports, he became an expert skier and tennis player. In 1906 he entered Harvard (B.A., 1910). After a year in the family business, he accepted a teaching position in chemistry and physiology at Mt. Allison College in New Brunswick, Canada, followed by one term as a research assistant at Worcester Polytechnic. In 1912, Sumner enrolled at Harvard Medical School as a graduate student in biochemistry. His research director was Otto Folin, renowned for the development of accurate biochemical tests for blood sugar and urine. Folin, certain that a person with Sumner's disability would not be adept at laboratory research, tried to dissuade him from becoming a scientist. Sumner, however, persisted and succeeded in his studies (M.A., 1913; Ph.D., 1914). While on a European vacation he received an appointment as an assistant professor of biochemistry at Cornell University Medical College in Ithaca, New York. He remained at Cornell until his retirement in 1955. Sumner married three times and had seven children. In 1946 he won the Nobel Prize for the discovery that enzymes could be crystallized. Following his retirement he had planned to go to Brazil to organize a research program on enzymes at the University of Minas Gerais. In the midst of this planning, and apparently in good health, a sudden illness related to cancer struck him. He died in a Buffalo, New York, hospital on August 12,1955.

Sumner was a pioneering investigator in the field of enzyme chemistry. At Cornell he gained a reputation as an excellent teacher. His exceptionally heavy teaching load did not diminish until after the award of the Nobel Prize, at which time he became director of a new Laboratory of Enzyme Chemistry (1947–1955). Sumner enjoyed teaching, but it so limited his research time that he felt that whatever research he did had to be something important. Folin's field, that of bioanalytical methods, he found unsatisfying. Daringly he decided to take a long shot: He would try to isolate and purify an enzyme, something that chemists had been trying to do for several decades. Sumner's interest in enzymes stemmed from a 1912 Harvard lecture by Lawrence Henderson, who declared that no enzyme would ever be isolated unless some new method was devised. Sumner began his quest in 1917; nine years later he succeeded in isolating for the first time a pure enzyme.

In 1917 the word *enzyme* was about 50 years old, although enzyme-catalyzed fermentations and other reactions had been long familiar. The nature of the ferment, however, was a mystery. Ferments of enzymes catalyzed specific transformations within the cells of organisms. They were studied in the form of concentrated extracts from organic sources. Some preparations gave protein tests, but the evidence was inconclusive so long as enzymes were not available in pure, homogeneous form.

Sumner selected urease as his enzyme for study, since it was one with which he was familiar, having used it as a quantitative reagent to estimate urea during his doctoral research. Urease was important in the nitrogen cycle, converting urea into ammonia and carbon dioxide. Sumner decided to try to purify urease by fractionation methods, assaying the enzyme activity in different fractions. Any fraction with a higher activity than in the original source was an indication that the enzyme had been enriched. By repeating the fractionation again and again he hoped to obtain the pure enzyme.

The richest source of urease was the jack bean. The task of isolation was immensely difficult and tedious owing to the low concentration of the enzyme, its sensitivity to reagents and conditions, and Sumner's lack of time, suitable apparatus, and funds. He purchased commercial jack beans, but they proved to be an unreliable source; he had to persuade individuals to grow beans with high urease content.

He so lacked apparatus that he used a coffee mill for the grinding. He would then attempt to extract the enzyme from the jack bean meal with different solvents. By the time of his sabbatical leave (1921–1922), little had been accomplished beyond the isolation of three globular proteins from the jack bean. On leave in Brussels, scientists there discouraged him from continuing the project.

On his return to Cornell, Sumner learned that Folin was using dilute alcohol to extract urease from organic sources. It was a better solvent, dissolving most of the enzyme and leaving behind much of the globular proteins. Sumner adopted its use, and, by keeping the extracts at low temperature in order to retain urease activity, he obtained enzyme concentrations 100 times greater than before. By the mid-1920s he had been able to separate the proteins, carbohydrates, lipids, and pigments from the jack bean. He worked his way to a residue from which he could not extract any additional matter. The isolation of the enzyme from the residue came in 1926 when he switched to a dilute acetone solvent. He put the filtered extract in an ice chest to refrigerate overnight. He then centrifuged it, and, instead of getting a precipitate as with alcohol, he observed the formation of tiny octahedral crystals. His preparation had high urease activity, some 200 times that of the original extract. Further purification gave him a sample with maximum activity. Furthermore, the crystals gave positive tests for protein.

His paper on crystalline urease appeared in 1926 with the claim that he had isolated urease as a crystalline protein.[1] The nine-year investigation was later reported by him in a lucid *Journal of Chemical Education* article in 1937. There he revealed the difficulties and tediousness, the use of large numbers of reagents and methods, the discouragement and temporary abandonment of the study. Yet he returned to it and finally with the acetone solvent obtained a filtrate with very high urease activity; from the filtrate he got crystals with specific enzyme activity.

The reception to his investigation, however, was not positive. Chemists and biochemists were either skeptical, hostile, or simply ignored him. The leading enzyme chemist was Richard Willstätter, in Germany. He had been trying to prepare pure enzymes for many years, and he believed he had compelling evidence that enzymes were not proteins. He had obtained highly enriched yeast enzyme preparations that were active solutions yet gave negative tests for proteins, as well as for

---

[1] "The Isolation and Crystallization of the Enzyme Urease"; *J. Biol. Chem.* **1926**, *69*, 435–441.

carbohydrates and lipids. To Willstätter, enzymes were small organic substances of unknown nature and carried through adsorption by proteins and by other colloidal carriers. Sumner's crystalline urease was therefore an adsorption compound of the unknown enzyme and a protein carrier. Willstätter's view of enzymes was popular among chemists because of his reputation, the importance of colloid chemistry, and the fact that Willstätter had provided new standards for assaying enzyme activity. What he did not realize was that his solutions were too dilute to register the presence of protein by standard chemical tests. Although he had achieved considerable purification of several enzymes, he had no idea how sensitive an enzyme test was relative to a chemical test. Thus, his active enzyme solutions gave no protein test, and he concluded that the enzymes were not proteins.

Sumner responded to the skepticism with 10 more papers over the next five years, providing additional data. By 1936 he had published 20 papers on urease. But it was only in 1937 that additional research by several investigators revealed that a protein had the structural units responsible for enzyme catalysis. The long 11-year controversy left Sumner embittered by what he thought was unjustified criticism. He had become an expert protein chemist and felt that the attacks on him by German chemists were due to personal animosity. Willstätter gave two lectures at Cornell in 1927. He told Sumner to digest the protein away from the urease with trypsin, a protein-splitting enzyme; it should remove the protein carrier and leave an active enzyme solution behind. Sumner replied that he had already done that, and his crystalline preparation was resistant to trypsin. Willstätter accused him of using an inactive trypsin preparation, and the two men parted in disagreement. In 1928 another visiting German chemist, Hans Pringsheim of the University of Berlin, told Sumner to go to Germany and learn from the great chemists there. Sumner felt that these criticisms arose from the fact that so many Europeans had failed to isolate the enzyme. The Europeans therefore scoffed at the possibility that an unkown American could have done it.

In 1929 Sumner took another sabbatical leave, this time to Sweden, determined to show others how to get urease crystals from jack beans. He brought 40 pounds of beans with him. He repeated his successful preparation there; in 1931 some Europeans used jack beans he left behind and also succeeded. Unfortunately, others tried and failed. In 1933 some German chemists claimed to have separated the enzyme activity from the protein, but Sumner demonstrated that their work was flawed. The controversy continued. The turning point came with work at the Rockefeller Institute by John Howard Northrop. Between 1930 and 1935 Northrop and his associate Moses Kunitz crystallized the protein-splitting enzymes pepsin, trypsin, and chymotrypsin. Northrop proved the protein nature of the crystals and established the catalytic specificity with regard to the peptide bonds in the proteins. By 1937 European scientists, such as Otto Warburg, had contributed more proof. Sumner himself made a compelling study of the enzyme catalase.[2] These researches showed that enzyme specificity was associated with the integrity of individual proteins.

Since Sumner's evidence had been insufficient to overcome opposition, the work of Northrop and others in the 1930s served to vindicate him. In 1946 Sumner and Northrop shared the Nobel Prize, along with Wendell Stanley, for their investigations of protein molecules.

Sumner continued his researches into the 1950s. He published over 100 papers despite the heavy teaching load and lack of support. He crystallized several additional enzymes and developed improved methods for protein isolation. He personally did most of the laboratory experiments reported in his papers. He was the author of several books in biochemistry, most importantly, the *Chemistry and Methods of Enzymes*, with G. Fred Somers (Academic Press: Orlando, FL, 1943), a popular and standard introduction to the subject, and, as coeditor with Kurt Myrbäck, *The Enzymes: Chemistry and Mechanism of Action* (Academic Press: Orlando, FL, 1950–1952), a four-volume survey with contributions by many eminent biochemists.

Sumner's isolation of urease and the researches that followed over the next 10 years put enzyme chemistry on a new level of significance. Indeed, his work is probably the single most significant American accomplishment in natural product chemistry between the two world wars. Enzymes were of unknown nature, and his discovery and advocacy of their protein nature, despite the lack of recognition of his work for many years, led to the study of the chemical structure of pure enzymes. Enzymes became central to biochemistry to the extent that much of modern biochemistry is an outgrowth of the work of Sumner and others from the 1930s.

**ALBERT B. COSTA**
*Duquesne University*

---

[2]"Crystalline Catalase"; *J. Biol. Chem.* **1937**, *121*, 417–424.

# Bibliography

Cori, C. "James B. Sumner and the Chemical Nature of Enzymes"; *Trends in Biochemical Science* **1981**, *6*, 194–196.

Farber, E. *Nobel Prize Winners in Chemistry*, 1901–1961, Rev. ed.; Abelard-Schuman: New York, 1963.

Maynard, L. C. "James Batcheller Sumner"; *Biographical Memoirs of the National Academy of Sciences* **1958**, *31*, 376–396.

*Nobel Prize Winners;* Wasson, T., Ed.; H. W. Wilson: New York, 1987.

*The Nobel Prize Winners: Chemistry;* Magill, F. N., Ed.; Salem Press: Pasadena, CA 1990; Vol. 2.

Sumner, J. B. "The Chemical Nature of Enzymes"; In *Nobel Lectures: Chemistry, 1942–1962*; Elsevier: Amsterdam, the Netherlands, 1964; pp 114–121.

Sumner, J. B. "The Story of Urease"; *Journal of Chemical Education* **1937**, *14*, 255–259.

Whorton, J. C. "James Batcheller Sumner"; In *Dictionary of American Biography*; Garraty, J. A., Ed.; Charles Scribner's Sons: New York, 1977; Supplement 5, pp 669–671.

*Who's Who of Nobel Prize Winners;* Schlessinger, B. S.; Schlessinger, J., Eds.; Oryx Press: Phoenix, AZ, 1986.

# John Northrop

## 1891–1987

John Howard Northrop, born in Yonkers, New York, on July 5, 1891, belonged to an eight-generation family of "Connecticut Yankees" that included several distinguished scholars. His father, John Isaiah Northrop, had a Columbia University doctorate and was a member of the zoology department. He was killed in a laboratory accident 10 days before the birth of his son. His mother, Alice Belle Rich Northrop, was a biologist who taught at Hunter College. She was responsible for the introduction of nature studies into the the public school curriculum of New York. Northrop attended public schools and then Columbia University (B.S., 1912; M.A. 1913; Ph.D., 1915), majoring in chemistry and studying biology with Thomas Hunt Morgan. He received his Ph.D. for work done with the enzyme chemist John M. Nelson. He held a captain's commission in the Chemical Warfare Service during World War I, devising a fermentation process for acetone production from potatoes and carrying the process into the first stage of manufacture. In 1917 he married Louise Walker; they had two children. From the war years until 1961 he was associated with the Rockefeller Institute for Medical Research, first in New York City and then in its Princeton, New Jersey, branch. He loved open country and hated the city. When the Princeton unit closed in 1947, he refused to return to New York but was allowed to remain a member of the Institute. He eventually moved to the University of California at Berkeley as professor of bacteriology and biophysics. He was a trustee of Woods Hole Marine Biological

Laboratory. In 1946 he won the Nobel Prize for his investigations of enzymes. He retired in 1970 and chose to live far away from urban noise in the remote area of Wickenburg, Arizona, where he died on May 27, 1987, at the age of ninety-five.

In 1916 Northrop joined the Rockefeller Institute staff under Jacques Loeb, a brilliant experimentalist in physiology. Loeb recognized Northrop's abilities as an ingenious experimenter, and together they explored a variety of subjects until Loeb's death in 1924. Among the topics were the kinetics of osmosis, the response of organisms to light, the diffusion of substances through wet and dry membranes, and fruit fly studies on the duration of life and the effect of temperature on its prolongation. Northrop began to study enzymes in 1919, pursuing the kinetics of enzyme-catalyzed reactions and the conditions affecting the action of the digestive enzymes pepsin and trypsin.

What first brought his work to the attention of scientists as being of major importance was his development in 1929 of the diffusion cell, a means to obtain diffusion coefficients of soluble substances and thereby an approximate molecular weight. This was followed by his development of solubility methods to study proteins in 1931, which provided a means to find whether one had a homogeneous preparation and which were important in proving that enzymes were proteins.

Following James Sumner's success in preparing crystalline urease in 1926, Northrop's interest in enzymes shifted to their isolation and crystallization. Sumner was the first to crystallize an enzyme, but Northrop's research was more extensive and more searchingly critical. Sumner's claim that urease was a protein had met with widespread skepticism. Between 1930 and 1935 Northrop crystallized pepsin, and his co-workers, especially Moses Kunitz, crystallized trypsin, chymotrypsin, ribonuclease, deoxyribonuclease, hexokinase, and pyrophosphatase; Mortimer Anson crystallized carboxypeptidase. The group established the physicochemical properties of these enzymes and correlated them with their catalytic action. Sumner's evidence of the protein nature of enzymes was debatable until these studies convinced scientists by their thoroughness that Richard Willstätter's view of enzymes as low molecular weight substances carried by proteins, then the most widely held theory of enzymes, was wrong.

Northrop's group crystallized three major proteolytic enzymes and with rigorous tests demonstrated that they were pure proteins. Preparation in crystalline form was not in itself an adequate criterion for purity. Northrop's test for purity involved careful solubility studies in well-defined media using the phase rule. A pure protein in such media yielded a constant solubility regardless of the amount of the crystalline phase in equilibrium with the solution. He showed that the proteins were pure by methods that went well beyond those of Sumner. The solubility experiments provided close correlation among enzymes, activity, and protein concentration. In further experiments Northrop was never able to separate an enzyme from the protein; the enzymic activity went with the protein and the identity of the enzyme and the crystalline protein had been established.

Northrop, Kunitz, and Roger Herriot also demonstrated the presence of precursors such as pepsinogen and trypsinogen; these converted into the respective enzymes in an autocatalytic reaction, a reaction in which the product catalyzed the formation of more of itself. In a 1932 lecture Northrop gave a clear account of the history of enzymes, the controversy provoked by Sumner's work, and his own evidence that the activity of enzymes depended on the arrangement of amino acids in the protein. He discussed the autocatalytic aspect, that enzymes had the power to form themselves from inert protein precursors, and his tests for homogeneity using solubility measurements.[1]

In 1935 Wendell Stanley at the Rockefeller Institute used Northrop's methods to crystallize the tobacco mosaic virus and establish its protein nature. Stanley's and Northrop's laboratories were but 100 yards apart and each shared with the other in their research. Stanley's feat spurred Northrop to return to the study of bacteriophages, which in the 1920s he had investigated briefly in terms of their kinetics and their infectious cycle. Northrop was convinced that like enzymes they would turn out to be proteins, and he regarded them as bacterial enzymes. His views on enzymes as self-producing systems governed his work on bacteriophages. In 1938 he reported the isolation of crystalline staphylococcus phages by fractional precipitation and showed them to be nucleoproteinaceous. This was the first isolation of a bacterial virus, and the fact that it contained nucleic acid was an important confirmation of the presence of nucleic acid in a virus, first revealed by Frederick Bawden and Norman Pirie in 1937 for the tobacco mosaic virus.

Northrop, however, missed the biological significance of the life cycle of bacteriophages, and his findings were later eclipsed by the

---

[1] "The Story of the Isolation of Crystalline Pepsin and Trypsin"; *Scientific Monthly* **1932**, *35*, 333–340.

work of molecular biologists. He had thought that his enzyme replication theory was applicable to phage reproduction: Enzymes possessed autocatalytic activity, catalyzing their own synthesis. Since phages increased dramatically in the host cell but had no metabolism of their own and no independent autocatalytic function, then in the cells there must be materials present that, with the addition of phages, are transformed into more phage molecules.

This theory left the problem of finding the phage precursors. Intense study failed to establish their existence, and Northrop's theory had to be abandoned in the 1950s in light of nucleic acid research whereby the characteristic of self-replication in viruses was not analogous to enzyme action but a consequence of the structure of nucleic acid. The molecular mechanisms that governed enzyme action did not govern the reproduction of viruses.

In 1946 Northrop won the Nobel Prize, sharing one-half the award with his colleague Stanley "for their preparation of enzyme and virus proteins in a pure form." The other half of the prize went to Sumner. During World War II Northrop isolated crystalline diphtheria antitoxin. In the postwar years at the University of California he isolated and crystallized pneumococcal antibody (1949). Continuing his bacteriophage work, he suggested that the nucleic acid was the essential autocatalytic part of the virus molecule, and in 1951 concluded that the protein part was necessary only to allow the entrance of the virus into the host cell. In the 1950s he studied the kinetics of phage action, the mechanism by which viruses arise in apparently healthy cells, the relation of phage growth to nucleic acid synthesis, and, in his last years of research, the origin of bacterial viruses.

His final statement as a scientist came in 1961 in the *Annual Review of Biochemistry,* where his philosophic attitude toward science and the nature of life received its clearest expression. He felt that the methods of biology never led to clear conclusions about the nature of living things and claimed that the book of nature was written in the language of physics and chemistry; these sciences would completely explain the properties of the living world. Both enzymes and viruses had a purely chemical nature, and he predicted that the mystery of the gene would be resolved by chemical means.

In 1961, the 70-year-old Northrop retired from the Rockefeller Institute. For his birthday the *Journal of General Physiology,* founded by Loeb in 1919, published a special issue in his honor, titled "Enzymes, Viruses, and Other Proteins". His association with the journal dated from an article in its first volume in 1919 to the year 1987, his name appearing in every volume as either author or editor. Northrop also wrote the influential *Crystalline Enzymes,* together with Kunitz and Herriot, describing their work at the Rockefeller Institute (Columbia University Press, New York; 1939).

Northrop did more than anyone to establish the protein nature of enzymes. By isolating several enzymes and devising rigorous tests for their purity, he greatly expanded the field of enzyme chemistry, opening the way to Stanley's study of viruses and to an understanding of the mode of action of enzymes. His studies demonstrated the usefulness of the chemical approach in the study of biological processes.

ALBERT B. COSTA
*Duquesne University*

# *Bibliography*

*A Biographical Encyclopedia of Scientists;* Facts on File: New York, 1981.

Bishop, G. H. et al. *The Excitement and Fascination of Science: A Collection of Autobiographical and Philosophical Essays;* Annual Reviews: Palo Alto, CA, 1965–1989; 3 vols.

*Current Biography;* Rothe, A., Ed.; H. W. Wilson: New York, 1947.

Edsall, J. T. "John Howard Northrop (1891-1987)"; *Nature (London)* **1987,** *329,* 396.

Farber, E. *Nobel Prize Winners in Chemistry, 1901–1961.* Abelard-Schuman: New York, 1963; rev. ed.

Herriot, R. M. "A Biographical Sketch of John Howard Northrop"; *Journal of General Physiology* **1962,** *45,* 1–16.

Herriot, R. M. "John Howard Northrop"; In *McGraw-Hill Encyclopedia of World Biography;* New York, McGraw-Hill: 1973; Vol. 8, pp 156–157.

Herriot, R. M. "John Howard Northrop"; *Journal of General Physiology* **1981,** *77,* 597–599.

Herriot, R. M. "John H. Northrop: The Nature of Enzymes and Bacteriophage" *Trends in Biochemical Sciences* **1983,** *8,* 297–298.

*The National Cyclopedia of American Biography;* J. T. White: New York, 1946; Vol. G, 1943–1946.

Northrop, J. H. "Biochemists, Biologists, and William of Occam"; *Annual Review of Chemistry* **1961,** *30,* 1–10.

Northrop, J. H. "Chemical Nature and Mode of Formation of Pepsin, Trypsin, and Bacteriophage"; *Science* (Washington, D.C.) **1937,** *86,* 479–483.

Northrop, J. H. "The Preparation of Pure Enzymes and Virus Proteins"; In *Nobel Lectures: Chemistry, 1942–1962*; Elsevier: Amsterdam, the Netherlands, 1964; pp 124–134.

Robbins, F. C. "John Howard Northrop"; *Proceedings of the American Philosophical Society* **1991,** *135,* 313–320.

# Wendell Stanley

## 1904–1971

Born on August 16, 1904, in Ridgeville, Indiana, to James and Claire Pessinger Stanley, publishers of the local newspaper, Wendell Meredith Stanley attended public school and served as delivery boy, reporter, and printer in the family enterprise. This experience may have contributed to his ability in later life to popularize science to the public. He attended Earlham College in Richmond, Indiana (B.S., 1926), with a chemistry and mathematics major. He excelled at football, being named to the all-Indiana team in his senior year. He had planned to become a football coach until his professor invited him on a trip to the University of Illinois to register another student in the graduate program. Stanley had hoped to meet the Illinois football coach, but instead he met Roger Adams, chairman of the chemistry department, and was so deeply impressed with Adams and the laboratory facilities that he entered the Illinois chemistry program (M.S., 1927; Ph.D., 1929). He married a fellow student of Adams, Marian Staples Jay, in 1929 with Adams as best man, and the three appear as coauthors on one paper. The Stanleys had four children, their only son becoming a well-known biochemist. Stanley spent a year in Munich working with Heinrich Wieland on sterols and then accepted an offer from the Rockefeller Institute for Medical Research in New York City, where he worked from 1931 to 1948. In 1946 he won the Nobel Prize for his isolation of a virus.

In 1948 on a trip to California, his plane was grounded by foul weather in Wyoming, and he discovered that a fellow passenger was

Robert Gordon Sproul, president of the University of California. Sproul was so impressed with Stanley that he offered him both a professorship and the opportunity to create and direct a virus laboratory (professor of biochemistry and director of the Virus Laboratory, 1948–1969). During the 1950s he devoted much of his time to the American Cancer Society in lecturing, fund-raising, organizing conferences, and campaigning for the support of cancer research. He also served as advisor to many governmental and international health organizations. Stanley died suddenly on June 15, 1971, following a heart attack in Salamanca, Spain, where he was attending a conference on viruses.

Stanley's earliest publications (1926–1929) were done under the direction of Adams at Illinois, including the structural determination of chaulmoogric oil acids and the synthesis of analogues, thus introducing him to chemistry in relation to medicine, since the constituents of chaulmoogric oil were the most promising chemical agents in the treatment of leprosy. Stanley synthesized several long-chain acids, made structural determinations, and tested them for bactericidal action.

This training made him a valuable member of the Rockefeller Institute, which was interested in a chemical approach to living systems. His first project there was to design a chemical model of the action of the marine plant *Valonia,* which has an unusual semipermeable membrane, enabling it to admit more potassium salts than sodium ones from seawater. Intensely examining reference material in the Institute library, he found information regarding a nonaqueous material in which potassium salts had greater solubility than sodium ones and constructed a working model of a living cell that selectively transported ions across a membrane, reproducing the action of the plant.

In 1932 Stanley requested a transfer to the Princeton branch because his wife was pregnant, and they wanted to raise their child in a rural setting. At first refused, he got the transfer only after the botanist Louis Kunkel, head of the new plant pathology department in Princeton, sought a chemist to work on plant viruses.

Kunkel was working on the tobacco mosaic virus (TMV) and needed someone chemically to purify the active material in the juice of diseased tobacco plants. At the time Stanley had never heard of viruses. The first to observe this pathogenic agent that passed through bacteria-retaining filters was the Russian Dimitri Ivanovskiy (1892). More important was the work in the 1890s of Martinus Beijerinck, who claimed that the agent was a different type of pathogen than any known, a *contagium vivum fluidum* (a nonbacterial, noncellular, living infectious fluid). The name *virus* replaced the various names subsequently given to filterable infectious agents. In addition to the discovery of several animal and plant viruses, filters had been developed by the 1930s that were graded in pore size, enabling one to determine which size held back the virus, and the filterability requirement disappeared. The term now meant a submicroscopic pathogenic agent that could multiply only within living cells. What a virus actually was—a small organism, a living fluid, a chemical molecule—remained a mystery.

Stanley began his study by assuming that viruses were protein molecules, an assumption based on the work in the 1920s of Carl Vinson and A. W. Petrie, whose concentrated preparation of TMV gave protein tests. He decided to investigate TMV with John Northrop's newly devised methods for preparing crystalline enzymes. He grew the tobacco plants in the Institute's greenhouses, infected them with the virus, and tried to obtain active juice extracts. Two and one-half years of hard work and 110 chemical reagents later, he isolated the virus on a filter and, after further purification, finally obtained needlelike crystals that were 1000 times more infectious than the original juice: a mere tablespoon of crystals from one ton of leaves. The crystals in solution, rubbed on leaves, produced the disease. His work appeared in *Science* in 1935 with an estimate of TMV's molecular weight and the conclusion that it was an autocatalytic protein requiring the presence of living cells for its multiplication.[1]

The virus had been isolated in crystalline form, contrary to the belief of many biologists that viruses were submicroscopic organisms. It was a giant chemical molecule. The publication created both a sensation and a controversy. The sensational aspect (it made the front page of the *New York Times*) was the notion that this was a missing link between the living and the nonliving. Kept in a test tube, the virus was not alive, merely protein crystals. Placed on a leaf of tobacco, however, the crystals came to life, invading the cells, reproducing millions of times. The *New York Times* article appeared on the same day as the *Science* one, and a follow-up article the next day examined the implications of TMV. Stanley had become a celebrity. Newspapers and magazines requested interviews; he was portrayed as a new Louis Pasteur.

---

[1]"Isolation of Crystalline Protein Possessing the Properties of Tobacco-Mosaic Virus"; *Science* **1935**, *81*, 644–645.

Shortly, more information about TMV became available. At the University of Uppsala, Theodor Svedberg and Arne Tiselius, inventors of the ultracentrifuge and electrophoresis apparatuses, took Stanley's crystals in 1936 and determined the molecular weight at 17 million, several times higher than any known order of magnitude, and determined that TMV was a rod-shaped molecule. Stanley used the Swedish contributions to laboratory instrumentation to extend his TMV work, since they provided the means to separate and isolate viruses from natural sources. He isolated 13 different strains of TMV, as well as other plant viruses. Stanley used another new instrument, the electron microscope, in the 1940s; he was the first to publish an electron microscopic portrait of the rod-shaped TMV. His researches included viscosity, sedimentation, and diffusion measurements, as well as structural differences and amino acid composition.

His work was repeated quickly. By 1936 British scientists had shown that Stanley's research was seriously flawed. His samples contained about 50% loose water. The TMV was not a pure protein. Nor was his preparation truly crystalline. Frederick Bawden and Norman Pirie by 1937 found the presence of nucleic acid in the virus; TMV was about 94% protein and 6% nucleic acid (RNA). J. D. Bernal and Isidor Fankuchen, using X-ray diffraction methods, found TMV to be a package of virus rods packed together side-by-side and end-to-end in layers. They were not three-dimensional crystals, but elongated molecules in a liquid crystalline state called "paracrystalline".

In the 1950s Stanley admitted "I have had to swallow a lot." He confirmed the Bawden and Pirie work, recognizing that he had missed the RNA component. Compounding the oversight, after recognizing the RNA, he continued for years to claim that nucleic acid was not part of the infectious particle and that it was possible to remove it and get a protein possessing viral activity, a position strongly upheld at the Rockefeller Institute by Stanley, Northrop, and others, who had developed a protein theory of the nature of life, enzymes, genes, and viruses. Stanley continued to refer to TMV as a protein, occasionally clarifying himself with the words "complex protein". Bound to a protein theory of viral activity, analogous to Northrop's enzyme autocatalytic theory, he minimized the nucleic acid as having no important function.

During World War II Stanley led a government project to develop a concentrated, multistrain, killed virus influenza vaccine. His success was such that inoculations in 1943 during a Type A epidemic reduced the incidence by 70%. By 1948 some 8 million soldiers had been vaccinated. In 1946 he won the Nobel Prize with Northrop and Sumner. His Nobel lecture contained the long-delayed admission of the importance of nucleic acid in viruses.

Settling in Berkeley in 1948, he gathered an outstanding staff of virus researchers at the Virus Laboratory. This institution, which he guided until 1969, did more for viral studies than any other, including the proof by Heinz Fraenkel-Conrat that RNA was the all-important component of TMV, forming its core and causing the infectivity of the virus. Stanley also served as spokesman of science to the public, notably in his national public television series on viruses, from which came *Viruses and the Nature of Life*, co-written with Evans Valens (Dutton: New York, 1961).

The importance of Stanley's research cannot be overestimated. He opened up a new field of inquiry, for critics and supporters alike. Some leading molecular biologists of the next generation asserted that his 1935 publication marked the effective beginning of molecular biology and inspired several of them to enter the field of biology from the molecular point of view. Owing to the self-correcting nature of science, his errors were corrected, his results reinterpreted in the 1950s, the old theories discarded and forgotten, and Stanley was reintegrated into the account of the development of virology.

ALBERT B. COSTA
*Duquesne University*

# Bibliography

Edsall, J. T. "Wendell Meredith Stanley (1904-1971)"; In *American Philosophical Society Yearbook 1971;* American Philosophical Society: Philadelphia, PA, 1972; pp 184–190.

Fraenkel-Conrat, H. "Wendell Meredith Stanley"; In *McGraw-Hill Encyclopedia of World Biography;* McGraw-Hill: New York, 1973; Vol. 10, pp 180–181.

Kay, L. E. "W. M. Stanley's Crystallization of the Tobacco Mosaic Virus, 1930-1940"; *Isis* **1986**, *77*, 450–472.

Pirie, N. W. "The Viruses"; In *Scientific Thought, 1900–1960;* Harré, R., Ed.; Clarendon Press: Oxford, England, 1969; pp 227–237.

Shope, R. E. "In Honor of Wendell M. Stanley"; *Perspectives in Virology* **1967**, *5*, xv–xxi.

Stanley, W. M. "The Reproduction of Virus Proteins"; *The American Naturalist* **1938**, 72, 110–123.

Stanley, W. M. "The Isolation and Properties of Crystalline Tobacco Mosaic Virus"; In *Nobel Lectures: Chemistry, 1942–1962*; Elsevier: Amsterdam, The Netherlands; Vol 3, pp 137–157.

Sullivan, N. *Pioneer Germ Fighters*; Atheneum: New York, 1962.

Vigneaud, V. du. "Scientific Contributions of the Medalist"; *Chemical and Engineering News* **1946**, 24, 752–755.

Williams, G. "Stanley: First Man to Crystallize Viruses"; In *Virus Hunters*; Alfred A. Knopf: New York, 1959; pp 88–107.

# Robert Robinson

### 1885–1975

Copyright Nobel Foundation

William Bradbury Robinson was a highly successful manufacturer of surgical dressings. In his factory in Chesterfield, England, were found many of his own inventions. The senior Robinson was, as described by his son Robert, "a tireless inventor who went every day to his mechanics shop where he had a small, skilled staff." One of his less successful ventures was the construction of a bleaching works. In the mind of this very practical man it seemed a good idea that his son Robert, born on September 13, 1885, should study chemistry when he went to a university. For the most practical of reasons Robert Robinson embarked on a career that would lead to the Nobel Prize in 1947 for his work on the synthesis of natural products, especially the alkaloids.

The Robinson family were prominent members of the Congregational Church in Chesterfield, yet Robert Robinson's crucial school days were spent at the Fulneck School, run by the Moravian Church in Britain. The quality and level of instruction was very high, and Robinson showed a high aptitude for mathematics and physics. "I wanted therefore to be a mathematician, but my father's wish was clearly expressed, and I decided to accept the inevitable and become a chemist."

In 1902 Robinson passed the matriculation exam of the Joint Board of the Universities of Manchester, Liverpool, and Leeds. He chose to enter the University of Manchester, then the dominant center in both teaching and research in chemistry in all of Britain.

Although at first a somewhat reluctant student of chemistry, it was during his second year at Manchester, when he attended the lectures on organic chemistry given by William Henry Perkin, Jr., that Robinson's lifelong love affair with organic chemistry began. "I think this is the right place to record that attendance at Perkin's lecture course decided my future career; I was fascinated by the beauty of the organic chemical system. Indeed, I am disposed to agree with Sir Frederick Gowland Hopkins, who once declared that the system of organic chemistry is one of the greatest achievements of the human mind."[1] Perkin was to be one of the two major influences on the direction of Robinson's career.

It was from Perkin that Robinson developed his interest in the synthesis of natural products, which was to lead to his Nobel Prize. Perkin and Robinson were to enter into a close collaboration that lasted for well over two decades as they perfected a complementary relationship. As Alexander Todd and John Cornforth have written, "Perkin's enthusiasm, vast practical experience, experimental skill and profound intuition blended perfectly with Robinson's comprehensive knowledge of the chemical literature, great theoretical interests, deep and incisive insight into reaction mechanism and into structure, and outstanding ingenuity and originality, to create a combination which has never been surpassed."[2] Robinson obtained his bachelor's in 1905 with first-class honors and was given a place in Perkin's private laboratory. This initial collaboration produced Robinson's doctoral thesis in 1909, concerned with the chemistry of brazilin and the related hematoxylin, the coloring matter of brazilwood. Robinson would pursue this subject for the next 60 years!

When Robinson was appointed to a junior post in the department in 1909, Arthur Lapworth, the second major influence on Robinson's career, accepted a senior position. The association with Lapworth led to Robinson's interest in the theoretical aspects of organic chemistry. Robinson was predisposed to this interest starting with his early aptitude in physics and mathematics. Lapworth was one of the pioneers in using physicochemical techniques as a tool to elucidate reaction mechanisms, as well as in the application of electronic theory to the mechanisms of organic reactions. In the period between 1909 and 1912

---

[1] Robinson, 1976.
[2] Todd, 1976.

Robinson and Lapworth became almost inseparable. The length and depth of their discussions led to Robinson's ultimate development of a general theory of organic reaction mechanism. In 1912 Robinson married Gertrude Maude Walsh, a research student of Chaim Weizmann at Manchester, and she was to become his lifelong collaborator. In this same year he was offered his first professorship, at the University of Sydney. By this time in Robinson's career, he had already published 50 papers dealing with synthetic methodology, as well as studies of a variety of routes leading to the synthesis of various natural products. Robinson's stay in Australia was to be a brief one, as he returned in 1916 to England when the newly created Heath Harrison Chair of Organic Chemistry at the University of Liverpool was offered to him.

At Liverpool Robinson had the chance to renew his collaboration with Perkin, working on various aspects of alkaloid chemistry. This included his synthesis of tropinone, as well as participating in research directly related to the war effort. His war work was related to rejuvenating the dyestuffs industry, which had entered a period of serious decline after the 1880s. The superiority of the organization of the German industry in terms of research and development as well as manufacturing efficiency made it the paramount producer of dyes. What had started as an experiment involving several academic institutions acting as satellites of British Dyes Ltd. led Robinson to accept a full-time position in 1919 as Director of Research at British Dyestuffs in Huddersfield. Owing to a feeling of frustration and to constant bickering in the management, Robinson soon left when the Chair of Chemistry in the University of St. Andrews became available in 1920. In the same year Robinson was elected a Fellow of the Royal Society.

Although his stay in St. Andrews was to be brief, Robinson was able to get back to work on the study of alkaloids and natural coloring matters. He also seriously started to formulate his ideas concerning an electronic theory of organic reaction mechanism based on ideas of Lapworth, G. N. Lewis, and Irving Langmuir.

In 1922 Arthur Lapworth became head of the department at Manchester, and Robinson was offered the Chair of Organic Chemistry, which he accepted. During the next six years at Manchester, an electronic theory of organic chemistry was to be fully developed, and a myriad of studies consolidating previous investigations on various alkaloids were to appear as well. This second period at Manchester would set the stage for Robinson's final moves: to University College, London, in 1928 and finally, with Perkin's death in 1929, to the Waynflete Professorship of Chemistry at Oxford in 1930, a post that he held until his retirement in 1955. By the time of his removal to Oxford, Robinson had published over 200 papers, and the Dyson–Perrins Laboratory was to become a Mecca for organic chemists who came from all over the world to work with this master of synthesis. Most of the early period at Oxford until the outbreak of war in 1939 was concerned with the completion of work begun earlier on anthoxanins and anthocyanins (plant pigments) and the initiation of exploratory work on the synthesis of steroids. His professional stature had grown to the point in 1939 that he was both knighted and elected President of the Chemical Society.

His major scientific contribution to the war was the collaboration with Howard Florey and Ernst Chain on penicillin chemistry. Following the end of hostilities, research began again in earnest at Oxford on the alkaloids strychnine and brucine and on the steroids. In 1947 Robinson was elected President of the Royal Society and awarded the Nobel Prize for chemistry. In the words of Professor A. Fredga, a member of the Nobel committee for chemistry, "Among organic chemists, you are today acknowledged as a leader and a teacher, second to none. In recognition of your services to science, the Royal Academy has decided to bestow upon you the Nobel Prize for chemistry for your investigations of plant products of biological importance and especially for your outstanding work on the structure and biogenesis of complicated alkaloids."[3]

Increasingly Robinson became more involved in activities outside Oxford, being one of the most notable spokespersons for science in postwar Britain. In 1949, he received the Order of Merit, the most prestigious decoration a civilian can receive in the United Kingdom. His tenure at Oxford was extended beyond the normal retirement age of 65, and he continued in his post until 1955. He retired from Oxford, but not from chemistry, as he became a director of Shell Chemical Co., Ltd. A small laboratory was placed at his disposal to continue his work on alkaloids and on other matters of interest. In his retirement he founded the journal *Tetrahedron,* which became one of the leading international journals of organic chemistry. In addition he traveled extensively around the Shell Research establishments as a consultant and continued his association with Shell until his death in 1975.

Robinson's major nonscientific interests were in mountaineering, which he pursued with great vigor, as well as chess. He served as President of the British Chess Federation from 1950 to 1953. Between 1905 and 1974 Robinson contributed over 700 papers to the scientific literature and was the mentor of or collaborator with many future

[3]Robinson, 1964.

Nobel Prize recipients. These included A. R. Todd (1957), D. M. Hodgkin (1964), E. Chain (1945), and H. W. Florey (1945). A. J. Birch has perhaps best stated the impact of Robinson's scientific work: "Seldom in a scientific discipline does the work of one man epitomize that of several generations. Sir Robert Robinson's long life spanned a vital period of organic chemistry when it was rapidly changing from a mostly empirical science into one soundly based on theory". A brief review of some of Robinson's most notable work follows.

His studies of the chemistry of brazelein and its oxidation product brazeilein, which is the red-colored component of brazilwood, were the first and last of Robinson's contributions to the chemical literature.[4] Perkin had begun the study of these compounds along with the monohydroxy derivative hematoxylin, the coloring matter of logwood, in 1903. Robinson's initial contribution was to prove the structure of brazilin by the synthesis of brazilinic acid, an oxidation product derived from the brazilin trimethyl ether without loss of carbon atoms. As Todd and Cornforth have stated in discussing Robinson's continuous interest in this compound, "Brazilin seems to have offered an ideal subject for academic research, being a fairly simple $C_{16}$ compound whose structure and reactions could be interpreted in the light of chemical understanding of the time. And yet enough was known about it to indicate that its varied reactions and transformations were likely to pose intriguing problems." Features of the synthesis of brazelein and derivatives were incorporated into his work on anthocyanins, steroids, and the alkaloids.

The alkaloids are widely distributed plant products with the common feature of being nitrogen bases. They usually have very marked physiological properties that can be of medicinal value (e.g., quinine, cocaine, atropine, and morphine) or make them act as potent poisons (e.g., strychnine and brucine). Robinson's interest in alkaloids was a result of earlier investigations in this area by his mentor Perkin.

One of Robinson's most notable early successes in this field was the synthesis of tropinone, an optically inactive cycloheptanone containing a bridge linking carbon atoms 3 and 6.[5] It is structurally related to the alkaloids atropine and cocaine. Robinson, utilizing his ability to analyze the structure of complex molecules, saw tropinone as being made up of several simpler components, in essence antici-

---

[4]*Proc. Chem. Soc.* **1906**, *22*, 160-161; *Tetrahedron* **1974**, *30*, 1295-1300.

[5]*J. Chem. Soc.* **1917**, *111*, 762-768.

pating the contemporary disconnection approach of synthetic chemists. He reacted methylamine, acetone, and succindialdehyde and obtained tropinone as its dipiperonylidene derivative. This approach has been used to synthesize other tropane alkaloids, such as pseudopelletierine, valeroidine, scopolamine, and teloidine. The tropinone paper was followed by Robinson's first speculations about the biogenesis of alkaloids.[6] For the first time it was shown that natural products could be made easily from simple precursors, using processes for which simple laboratory examples were known.

Although Robinson studied a great variety of alkaloids, his work on brucine and strychnine deserves special attention. They best illustrate his powers of analysis in terms of using bits and pieces of chemical information from his prodigious memory to propose a structure for a complex molecule. (His fondness for doing crossword puzzles further shows this ability, which resulted in the solution of the structures of many natural products.) The determination of the structures of the alkaloids strychnine and brucine represents the ultimate achievement of classical structural organic chemistry, considering that these compounds have seven rings with seven asymstereogenic centers and contain only 24 skeletal atoms.

Capitalizing on earlier work done on the degradation of strychnine and brucine by Hans Leuchs, Robinson was able, by a series of masterful experiments, to probe the structure of strychnine and brucine. His establishment of its various components such as a tetrahydroquinoline ring, a d-ketoamide, and a reduced carbozale ring all entered into his final solution. The structure was published in 1946,[7] although the work had been essentially completed in 1939. Robinson's structure was verified by the total synthesis published by R. B. Woodward in 1954. It is believed that the time spent by all the research groups in the investigation of strychnine, if totaled, would have added up to 500 years. Although Robinson contributed very little in terms of the experimental work, his ability to postulate a structure based upon minimal evidence was dramatically demonstrated in the strychnine work.

Robinson always believed that his contributions to the theory of organic chemistry constituted his most significant work. Yet his theoretical work spanned only the period from 1916–1932. This interest in the mechanism of organic reactions was probably due to Robinson's

---

[6] *J. Chem. Soc.* **1917**, *111*, 876–899.
[7] *J. Chem. Soc.* **1946**, 908–910.

association with Arthur Lapworth, which had begun in 1909. Robinson early on realized the need to have a systematic approach to organic synthesis rather than the intuitive approach as used by his mentor Perkin. Robinson's major interest was in the reactions of conjugated unsaturated systems, such as aromatic compounds. Initially Robinson used the concept of partial valencies (1917), using ideas introduced by Johannes Thiele in 1899. He was able, however, to make a transition to an electronic system based on the ideas that had been developed by G. N. Lewis and Irving Langmuir. With W. O. Kermack in 1922, Robinson published a landmark paper entitled "An Exploration of the Property of Induced Polarity of Atoms and an Interpretation of the Theory of Partial Valencies on an Electronic Basis".[8] This was one of the first successful applications of Lewis-Langmuir theory to organic chemistry, and the many insights into mechanism discussed in the paper became part of the general theory that developed between 1925 and 1927. In the 1922 paper in connection with the chemistry of trienes we see the first use of the "curly arrow" to show electron movement and the first discussion of what came to be known as the electromeric effect. A significant aspect of the 1922 paper was an electronic interpretation of aromaticity and a rationalization of the Kekulé formula based on the mobility of octets. It was Robinson who in 1925 introduced the term "aromatic sextet" to denote the six-electron system that accounted for the unique properties of benzene.[9] It must be said, however, that Robinson never showed any real interest in quantum chemistry, and this may have contributed to his withdrawal from any further theoretical work in the 1930s.

A general theory of reaction mechanism, especially as applied to aromatic systems, emerged between 1925 and 1927. Much of this was in response to a continuing argument with C. K. Ingold concerning the basis of how to use electronic interpretation in reaction mechanism.

Unfortunately, Robinson never presented his general theory of organic reaction mechanism in such a form that the vast majority of organic chemists could avail themselves of his insights. Two rather extended versions appeared in 1932, but in journals that were not readily available to most chemists. Over time, owing to the vast amount of work done by C. K. Ingold using Robinson's ideas and because of the numerous reviews Ingold wrote, most chemists considered the work

[8] *J. Chem. Soc.* **1922**, *121*, 427–440.
[9] *J. Chem. Soc.* **1925**, *127*, 1604–1623.

done by Robinson a minor contribution in the area. This disturbed Robinson greatly and caused a lifelong enmity between himself and Ingold.

Steroids are a class of compounds that serve several biological functions in animals. They are characterized by a basic skeleton, consisting of three six-membered rings and one five-membered ring. Work on the synthesis of these compounds began at Oxford in 1932 and continued for the next two decades, resulting in 53 papers. The initial work concentrated on the simpler aromatic steroids such as equilenin, as well as the estrones, in which the number of structural possibilities is limited as compared to the true steroids.

Robinson adopted two basic strategies to synthesize the ring system of the steroids. These were either to assemble all four rings together at an early stage and then make the appropriate modifications or to build tricyclic structures with the appropriate features that could lead to closure to form the fourth ring. The methodology of the modification of existing tricyclic systems led to the development of the highly versatile annulation reaction for which Robinson is best remembered by many chemists.[10] Sir John Cornforth has best summarized the significance of Robinson's work in steroids as follows: "Unquestionably Robinson was the pioneer of steroid synthesis and progress was made possible over the ground he broke."[11]

MARTIN D. SALTZMAN
*Providence College*

# Bibliography

Campbell, W. A. *Contemporary British Chemists;* Taylor and Francis: London, 1971.

*A Centenary Tribute to Sir Robert Robinson, 1886–1975;* Special issue of *Natural Product Reports* **1987,** *4* (February).

*Perspectives in Organic Chemistry;* Todd, Sir A. R., Ed.; International Publishers: New York, 1956.

Robinson, R. *Memoirs of a Minor Prophet;* Elsevier: New York, 1976.

Robinson, R. *The Structural Relations of Natural Product. Being the First Weizmann Memorial Lecture;* Claredon: Oxford, England, 1955.

---

[10] *J. Chem. Soc.* **1937,** 53–56.
[11] Todd, 1976.

Robinson, R. *Two Lectures on an Outline of an Electrochemical (electronic) Theory of the Course of Organic Reactions;* Institute of Chemistry Publications: London, 1932.

Robinson, R. *Nobel Lectures: Chemistry, 1942–1962;* Elsevier: New York, 1964; pp 161–187.

Saltzman, M. D. "Sir Robert Robinson: A Centennial Tribute"; *Chemistry in Britain* **1986**, *22*, 543–548.

Todd, A. R.; Cornforth, J. W. *Biographical Memoirs of Fellows of the Royal Society* **1976**, *22*, 415–527.

Williams, T. *Robert Robinson, Chemist Extraordinary;* Clarendon: Oxford, England, 1990.

# 1948

# *Arne Tiselius*

## 1902–1971

Arne Wilhelm Kaurin Tiselius, who was born in Stockholm, Sweden, on August 10, 1902, and died there on October 29, 1971, devoted his career to research and to improving the resources available to science in his native Sweden. His father, Hans Abraham J:son Tiselius, studied mathematics at Uppsala University and worked for an insurance company. His mother, Rosa Kaurin, was the daughter of a Norwegian clergyman. When Arne's father died in 1906, the family moved to Gothenburg, on the west coast of Sweden, where many members of the Tiselius family and friends resided.

Tiselius's interest in science was kindled by an inspiring biology and chemistry teacher at the *Gymnasium* in Gothenburg. Under the guidance of his teacher, Tiselius nurtured a growing preoccupation with chemistry. Like many other students drawn to chemistry, Tiselius chose to study at Uppsala University: The magnet was The Svedberg. Tiselius enrolled in the fall of 1921 and received a master's degree in chemistry, physics, and mathematics in 1924. He stayed at Uppsala University as Svedberg's research assistant in physical chemistry. In November 1930 he defended a doctoral thesis on electrophoresis, a problem given to him by Svedberg. Subsequently he was appointed *docent* in chemistry.

After several years devoted to chemical research on the zeolite mineral group, Tiselius returned to the problem of electrophoresis and developed a highly sophisticated apparatus that proved crucial for the further development of biochemistry and molecular biology. In 1938 he received a new, special chair in biochemistry at Uppsala University and held the position until retirement 30 years later. In 1944 he became active in government efforts to improve conditions for Swedish science; he was appointed chair of the new Swedish Natural Science Research Council in 1946. In 1947 he was elected one of the five members of the Nobel Committee for Chemistry and vice-president of the Nobel Foundation; in 1948 he received a Nobel Prize in chemistry "for his work on electrophoresis and adsorption analysis and especially for his discovery of the complex nature of the proteins occurring in blood serum."

Tiselius married Ingrid Margareta (Greta) Dalén in 1930. They had a son and a daughter. He was modest, quiet, and warm-hearted; many persons valued his witty but gentle sense of humor. Like his mentor, Svedberg, he was fond of botany and bird-watching. His lifelong good health began to fail during his last years. Not able to follow doctors' advice to slow down his many and varied activities in international science organizations, the Nobel Foundation, and government advisory councils, Tiselius suffered a heart attack at a meeting and died the next day, October 29, 1971.

When Tiselius became Svedberg's research assistant in July 1925, a new era in physical biochemistry was about to begin, one in which Uppsala would become an internationally leading center. When Svedberg returned to Uppsala in 1923, after spending eight months at the University of Wisconsin in efforts to develop colloid chemistry, he was committed to finding techniques for using physical and chemical principles to study biological colloids. He had constructed a low-speed ultracentrifuge; he now hoped to develop a high-speed model that might allow investigation of the size and shape of protein molecules. Svedberg also considered analyzing proteins using electrophoretic techniques, that is, methods to measure the movement of electrically charged molecules in solution when subjected to an electric field. Tiselius's first work entailed assisting Svedberg on the problem of measuring the mobility of proteins; they published a preliminary study in 1926. In that same year Svedberg received the Nobel Prize in chemistry, the main consequence of which was his newfound ability to attract much-needed financial resources to develop his ultracentrifuge and his proposed institute for physical chemistry at the university. These developments had profound significance for Tiselius's future career. First, Svedberg became so involved with designing, building, and running his ultracentrifuge models that he allowed his younger assistant to take over the electrophoresis problem. Second, Svedberg's increasingly close relation with the Rockefeller Foundation, especially its focused program on applying physical science to biological problems, eventually brought Tiselius into the network of Rockefeller-supported researchers and laboratories.

When Tiselius began his electrophoretic studies he recognized, as had others before him, that large molecules generally exhibited a charge when in solution and therefore, in principle, when placed in an electric field they migrated at different rates depending on their size, shape, and charge. The goal was to use this phenomenon to separate the differing molecular constituents of a solution. Like his mentor, Tiselius read biochemistry and turned to studying biologically and physiologically important molecules. The need for precise methods was underlined when he recognized that solutions that separated as homogeneous in an ultracentrifuge proved inhomogeneous when subjected to electrophoresis. Unfortunately, the task of creating precision techniques to allow accurate measurements proved extremely difficult. He found ways of controlling the temperature and electrical currents in an effort to control convection currents that distorted the migration. Tiselius began considering the problem of the definition and purification of the substances under investigation and convinced himself of the centrality of separation for the future development of physical biochemistry. He reasoned that additional physical and chemical methods, such as chromatographic and adsorption methods, were needed to supplement the ultracentrifuge and the nascent electrophoretic techniques. Still, he pushed ahead with electrophoresis. In November 1930 he defended his doctoral thesis, which for many years remained the definitive text on electrophoresis.

When retiring, Tiselius reminisced about his thesis and confessed openly about his dejection over the lack of success achieved in his doctoral work. "I remember very vividly that I felt disappointed. The method was an improvement, no doubt, but it led me just to the point where I could see indications of very interesting results without being able to prove anything definite. I can still remember this as an almost physical suffering when looking at some of the electrophoresis photographs, especially of serum proteins. I decided to take up an entirely different problem, but a scar was left in my mind which some years later would prove to be significant."[1] Although he received a *docent*

---

[1] "Both Sides of the Counter"; *Annual Review of Biochemistry* **1968**, *37*, 1–24.

title, Tiselius recognized that he needed to think about a permanent academic position. He also recognized that positions in biochemistry or what might be called biophysics did not exist. Even Svedberg's personal professorship was defined as physical chemistry. To compete for any upcoming position, he needed to expand his research competence to include more traditional specialties. His subsequent research project on the crystals of zeolite minerals was not sufficient for him to obtain the chair in inorganic chemistry at Uppsala, vacated by Daniel Strömholm in 1936, but it did prove significant in other unexpected ways.

Tiselius was attracted to the phenomenon by which the water of crystallization of some zeolites can be exchanged for other substances such as ethyl alcohol, bromine, and mercury; the crystal structure would remain intact even as the crystal was evacuated to remove its water. On rehydration, the dry crystals changed their optical properties; Tiselius devised a quantitative optical method to measure the diffusion of water vapor and other gases into the crystals. His work prompted him to work at Princeton University under Hugh S. Taylor; his sojourn during the academic year 1934–1935, the first of several visits, was financed by a fellowship from the Rockefeller Foundation.

While in Princeton, Tiselius came in contact with researchers at the Rockefeller Institute laboratories in Princeton and New York City. Rockefeller biological and medical researchers were already in contact with Svedberg, and the latter's largely Rockefeller-funded ultracentrifuge was also a resource for the American scientists. As Svedberg's protégé, Tiselius found entry into these circles greatly facilitated; he soon became a friend to M. L. Anson, J. H. Northrop, and W. M. Stanley as well as an acquaintance of M. Heidelberger, K. Landsteiner, and L. Michaelis, among others. Many of these researchers not only knew Tiselius's work on electrophoresis, but they recognized that the further perfection of this technique would be of great value for their own investigations. His own belief that efficient separation methods would be a key to biochemistry's development was reinforced by discussions with these scientists. Even before returning home to Sweden, he began to reconsider the design of electrophoretic apparatus, now benefiting from the insights derived from listening to the specific problems in protein research faced by his American colleagues.

On returning to Uppsala, Tiselius set about building a new electrophoresis apparatus. With the assistance of a grant from the Rockefeller Foundation in 1936, he built a new instrument, which amplified the resolving power and permitted reliable visual recording. His first published results from this research were important. Electrophoretically separated horse blood serum showed four separate bands. In addition to albumin, three other distinct serum components could be identified; Tiselius called them *alpha*, *beta*, and *gamma globulins*. Ironically, Tiselius's manuscript was rejected by a biochemical journal for being too physical. When it subsequently was published in the *Transactions of the Faraday Society* (1937), the article received an almost immediate and extremely favorable reception. Biochemical and biomedical researchers, especially from America, flooded Tiselius with requests for reprints and even requests for the apparatus. Many had been informed of Tiselius's innovation in advance as a result of Svedberg's comments while in the United States to reporters for the *New York Times* (October 14, 1937) concerning Tiselius's new apparatus for "separating proteins electrically". Subsequent close contact with Rockefeller Institute researchers enabled Tiselius to perfect the instrument and to extend the analyses of serum to include antibodies (immunoglobulin). Tiselius began to assist Rockefeller Institute researchers in building their own instrument, following his design. D. A. MacInnes and L. G. Longworth used an improved Tiselius apparatus to find differences in the electrophoretic patterns between normal and pathological human blood sera and plasmas.

Tiselius's innovative work brought just rewards. In 1937 Svedberg convinced Major Herbert Jacobsson and Karin (Broström) Jacobsson, the latter from a Gothenburg shipping family, to make a donation for a professorship at Uppsala University "for research and teaching in those fields of chemistry and physics which are of importance for the processes of life". Tiselius's need for a permanent position was clearly in Svedberg's thoughts. When Tiselius was named the first professor, he was able to turn to the Rockefeller and the Swedish Wallenberg Foundations for support. In 1946 a full department of biochemistry was created; Tiselius and his growing staff moved in 1952 from the rooms borrowed at Svedberg's Institute for Physical Chemistry into a new building. By this time, too, Tiselius had received a Nobel Prize in chemistry. His triumph transcended personal achievement. His work contributed to the instrumental foundation for developing molecular biology and physical biochemistry; the acknowledged significance of his endeavors for life science helped establish in Sweden a dynamic biochemical discipline. Uppsala, along with the Karolinska Institute in Stockholm and the Hans von Euler-Chelpin Institute for Biochemistry at the University of Stockholm, brought Sweden into a growing international multidisciplinary revolution in the life sciences.

Tiselius did not rest once he devised his electrophoretic apparatus. He recognized that, in spite of its abilities, the electrophoretic methods were not sufficiently specific for separating many biological substances; he sought a solution in adsorption processes. In the early 1940s he turned to chromatography, which relied on the tendency of dissimilar molecules to adhere differentially to the surface of certain substances. Over a period of several years, Tiselius and his assistants elucidated the fundamental processes involved in adsorptive analysis and paved the way for establishing protein chromatography as a precision method.

Tiselius was celebrated with a plethora of awards and honors in addition to the Nobel Prize. He received honorary degrees from 11 universities, including Paris, Oxford, Cambridge, and the University of California at Berkeley; he was made honorary member of 37 scientific academies and named a commander of the French Legion of Honor. When he became the president of the Nobel Foundation in 1960 he initiated the Nobel Symposia, bringing together leading scientists to discuss significant scientific problems, including the ethical and social implications of scientific advance.

ROBERT MARC FRIEDMAN
*University of California, San Diego*

# Bibliography

Kay, L. E. "Laboratory Technology and Biological Knowledge: The Tiselius Electrophoresis Apparatus, 1930–1945"; *History and Philosophy of the Life Sciences* **1988,** *10,* 51–72.

Pedersen, K. O. "Tiselius, Arne Wilhelm Kaurin"; In *Dictionary of Scientific Biography;* Gillispie, C. C., Ed.; Charles Scribner's Sons: New York, 1976; Vol. 13, pp 418–422.

Kekwick, R. A.; Pedersen, K. O. "Arne Tiselius 1902-1971", *Biographical Memoirs of Fellows of the Royal Society* **1974,** *20,* 401–428. This biography includes a bibliography of Tiselius's 161 publications.

Bishop, G., et al. *The Excitement and Fascination of Science: A Collection of Autobiographical and Philosophical Essays;* Annual Reviews: Palo Alto, CA, 1978; Vol. 2.

# *William Francis Giauque*

## 1895–1982

The 1949 Nobel Prize in chemistry was awarded to William Francis Giauque, professor of chemistry at the University of California at Berkeley, for his work in attaining and accurately measuring temperatures within one degree of absolute zero and for making accurate measurements of thermodynamic properties of materials at those low temperatures. Giauque was born in Niagara Falls, Canada, on May 12, 1895, and died at Berkeley on March 29, 1982. Although he was born in Canada, Giauque was a U.S. citizen.

Giauque attended public grammar schools in the United States and secondary school at the Niagara Falls Collegiate Institute in Canada. After two years of employment with the Hooker Electro-Chemical Company in Niagara Falls, New York, Giauque enrolled at the University of California to pursue a career in engineering, but, discovering that he enjoyed the challenges and rewards of fundamental research, he concentrated on pure science.

Giauque received his B.S. degree summa cum laude in 1920 and his Ph.D. degree in chemistry with a minor in physics in 1922. Appointed instructor of chemistry in 1922, he attained the rank of professor 12 years later and spent most of his professional life at Berkeley.

Milton Silverman, writing in the *Saturday Evening Post* in 1949, just after Giauque won the Nobel Prize, noted that he was not the stereotypical Californian.[1] In particular, he would not be likely to allow himself to experience the thrill of space travel should his work contribute to making such travel possible. Giauque was said to dislike speed in any form; he did not drive a car. He did not drink alcohol, smoke, or play cards, and rarely participated in campus affairs. In his own words, he was "a chronic nonattender of meetings". As was typical of experimentalists working with liquid helium, a highly volatile and fairly expensive material, once Giauque began an experimental run, he continued the experiment without interruption so as not to lose an opportunity to collect important data or waste a precious resource. These experimental runs frequently lasted all night or through the better part of a weekend.

Before the introduction of the magnetic cooling method for attaining temperatures within one degree of absolute zero, temperatures below room temperature were generally produced by liquefying materials that are gases at room temperature. For example, nitrogen liquefies at about 77 degrees above absolute zero, or 77 K. (Absolute zero, or 0 K, corresponds to $-273.16$ on the Celsius scale.) Once a way was found to liquefy nitrogen, lower temperatures could be produced by pumping on a liquid nitrogen bath, causing it to evaporate more rapidly and its temperature to fall. Such a lowering of temperature made it possible to cool another gas with a lower boiling point to such a temperature that it could be liquefied; pumping on that liquid allowed workers to attain still lower temperatures.

By 1908, the substance with the lowest boiling point, helium, had been liquefied in Leiden by the Dutch physicist H. Kamerlingh Onnes. The boiling point of ordinary helium at atmospheric pressure is 4.2 K; by pumping on a helium bath into which energy flow from the outside is minimized, one can reach a temperature of about 1 K. Many interesting experiments have been conducted in this temperature range; the random molecular motion of materials is much smaller than at room temperature, and some very interesting and unexpected results were discovered, including the discovery of superconductivity by Kamerlingh Onnes in 1911.

Giauque and others were interested in extending the temperature range even lower in order to take accurate measurements of the absolute entropy of substances. On a microscopic level, entropy can be thought

---

[1] *Saturday Evening Post*, Dec. 10, 1949, p 38.

of as the degree of disorder of a substance. From a macroscopic point of view, the change in the entropy of a substance can be determined by adding a small amount of heat to the sample and dividing the amount of heat added (in joules) by the absolute temperature of the sample. By definition, if one supplies heat to the sample, the entropy (disorder) increases; if one removes heat, the entropy decreases. Such measurements enable one to measure *changes* in the entropy of a sample but do not permit a determination of the *absolute* entropy of a sample—it is as if we could say that the temperature outdoors dropped twenty degrees today, but did not know whether the current temperature was above or below freezing.

Absolute values of entropy allow chemists to determine whether or not a proposed chemical reaction will occur. If one wants to know whether two substances will combine to form a third at room temperature (293 K), for example, one needs to know the entropy of the given amount of each of the two starting substances at 293 K and the entropy of the appropriate amount of the hypothesized product at the same temperature. If the entropy of the product is greater than that of the starting materials, then the reaction will spontaneously occur; if not, it will not. It is not enough to know that substance A has 5 more units of entropy at room temperature than it did at the freezing point, that B has 4 more, and that C has 6 more. One needs to know absolute values of entropies in order to make comparisons. (In fact, the appropriate quantity to look at is not the entropy but rather a quantity derived from the entropy known as the "free energy". It is still true that one must know the absolute entropies of the materials involved in order to know whether or not the reaction will occur.)

The key to obtaining absolute values of entropy lies in the so-called Third Law of Thermodynamics, which states that as the absolute temperature of any crystalline material falls toward zero, so does its entropy. To obtain the entropy of a substance at room temperature, one needs to measure the change in its entropy as its temperature is lowered toward absolute zero. The laws of thermodynamics do not permit one to reach absolute zero, but measurements indicate that entropy changes become smaller and smaller as one approaches absolute zero. At a temperature only one degree above absolute zero, it seemed reasonable (based upon extrapolated experimental results) to assume that virtually all of the entropy of a substance had disappeared.

At a seminar in 1924 at which Giauque presented calculations showing how magnetic fields affect the thermodynamic properties of various substances, some magnetic susceptibility measurements made at Leiden on a particular material were called to his attention. Using thermodynamic equations to calculate the change of entropy when a magnetic field is applied, he was greatly surprised to find that the application of a magnetic field removes a large amount of entropy from this substance, at a temperature so low that it had been thought that there was practically no entropy left to remove.

As Giauque stated in his Nobel lecture: "When a sufficiently powerful magnetic field is applied the magnets line up and the entropy is removed. The removal of entropy is accompanied by the evolution of heat. . . . [I]n principle any process involving an entropy change may be used to produce either cooling or heating. Accordingly it occurred to me that adiabatic demagnetization could be made the basis of a method for producing temperatures lower than those obtainable with liquid helium. Professor P. Debye also arrived at similar conclusions."[2]

To understand how the removal of entropy by magnetizing a substance can be used to cool that substance, it is helpful to consider a more familiar process, that of applying a pressure to a gas at room temperature. If one places a sample of gas in a cylinder closed at one end with a movable piston at the other end, and then exerts a force on the piston to compress the gas, heat will flow from the gas to the walls of the cylinder and ultimately to the surroundings. As indicated above, a flow of heat from a substance means, by definition, that the entropy of the substance has decreased. One can then thermally isolate the cylinder from its surroundings and permit the gas to expand back to its original volume. In order to return to its original state, heat needs to flow into the gas to return its entropy to its original value, but if the gas is prevented from exchanging heat with its surroundings (a so-called adiabatic process), the entropy of the gas cannot return to its original value and its temperature falls. This process is one of the principles used in liquefying gases.

The analogous adiabatic demagnetization process consists of lowering the temperature of a dilute magnetic salt to approximately 1 K and then applying a magnetic field while allowing the resulting heat to flow to the surrounding liquid helium bath, thus reducing the entropy of the salt. Once the salt has come into thermal equilibrium with the 1 K bath, the salt is thermally isolated from the bath to prevent further heat

---

[2]*Nobel Lectures: Chemistry, 1942–62;* Elsevier Publishing: New York, 1964, p 244.

exchange. The magnetic field is then removed and the salt, prevented from regaining the entropy it lost during the magnetization process, experiences an adiabatic drop in temperature to as low as 0.003 K.

While one cannot cause a measurable change in temperature by this process at ordinary temperatures, one can perform an analogous experiment with a thick rubber band. The long molecules within the rubber can be caused to align by stretching the rubber band in much the same way the magnetic dipoles in the magnetic salt are aligned by the magnetic field. If one lightly licks one's lower lip, then stretches the rubber band and waits for several seconds for it to come to equilibrium with its surroundings, and finally allows the rubber band to return rapidly to its unstretched state while in contact with the moistened lip, one can detect a very slight cooling. Similarly, if the rubber band is in contact with the lip during the stretching process, one can readily experience the transfer of heat to the surroundings, which is analogous to what happens when the magnetic field is applied in the case of the magnetic salt at low temperatures.

In any case, Giauque's idea of using the magnetic properties of a material at low temperatures to produce cooling was first mentioned at a meeting of the American Chemical Society on April 9, 1926. It took nearly seven years to build the equipment required to carry out the experiment. Giauque's first successful attempt was (characteristically) reported to have occurred between 3 a.m. and 9 a.m. on March 19, 1933.

As Giauque observed in his Nobel lecture: "The commonest question asked in the early days of this work was: 'How do you know it gets cold?' This was a fair question. Obviously no one had ever made thermometers which were calibrated at temperatures that had never been produced. Since even helium gas has negligible pressure at the low temperatures obtained, a gas thermometer is useless. Temperature can only be measured by some property of a substance which varies with temperature."[3]

The physical property Giauque used to determine the temperatures he and his collaborators had reached was a magnetic property—the magnetic susceptibility—of a sample of the very salt he was using to reach these low temperatures. Although the magnetic susceptibility of these materials is not a simple function of temperature, Giauque and

---

[3] *Nobel Lectures: Chemistry, 1942–62;* Elsevier Publishing: New York, 1964, p 244.

his collaborators were able to take advantage of basic thermodynamic relationships, which are independent of the nature of the material employed, and, by a series of very careful and difficult measurements, to determine accurately the temperatures attained.

In addition to his pioneering work in producing and measuring very low temperatures by the adiabatic demagnetization process, Giauque also is credited with discovering the fact that oxygen in the earth's atmosphere contains, in addition to the ordinary oxygen-16 isotope, small amounts of oxygen-17 and oxygen-18. As a consequence of this discovery, it became clear that chemists and physicists, both using oxygen as their reference material for mass determinations, were using slightly different scales. Chemists were using samples of naturally occurring oxygen, which included the heavier isotopes, whereas physicists, using mass spectroscopy to determine masses, were using a particular isotope, oxygen-16. In order to eliminate that slight discrepancy between the two scales, the reference substance has subsequently been changed to a particular isotope of carbon, carbon-12.

In addition to receiving the Nobel Prize in chemistry, Giauque received numerous other prestigious awards, including the Chandler Medal and an honorary Sc.D. from Columbia, an honorary LL.D. from the University of California, and the Elliott Cresson Medal from the Franklin Institute. In 1951, he was awarded the Willard Gibbs Medal of the American Chemical Society's Chicago Section, and in 1956 he received the Gilbert Newton Lewis Medal from the California Section of the ACS.

WILLIAM A. JEFFERS, JR.
*Lafayette College*

# Bibliography

Casimir, H. B. G. *Magnetism and Very Low Temperatures;* Dover: Mineola, NY, 1940.

Giauque, W. F. *Industrial & Engineering Chemistry* **1936**, *28.*

Giauque, W. F. "A Thermodynamic Treatment of Certain Magnetic Effect. A Proposed Method of Producing Temperatures Considerably below 1° Absolute"; *J. Am. Chem. Soc.* **1927**, *49,* 1864–1870.

Giauque, W. F. *Low Temperature, Chemical, and Magneto Thermodynamics; The Scientific Papers of William F. Giauque;* Dover: New York, 1969.

Giauque, W. F.; McDougall, D. P. "The Production of Temperatures below One Degree Absolute by Adiabatic Demagnetization of Gadolinium Sulfate"; *J. Am. Chem. Soc.* **1935**, *57*, 1175–1185.

Giauque, W. F.; McDougall, D. P. "Experiments Establishing the Thermodynamic Temperature Scale below 1° K. The Magnetic and Thermodynamic Properties of Gadolinium Phosphomolybdate as a Function of Field and Temperature"; *J. Am. Chem. Soc.* **1938**, *60*, 376–388.

Giauque, W. F.; McDougall, D. P. *Phys. Rev.* **1933**, *43*, 768.

Giauque, W. F.; McDougall, D. P. *Phys. Rev.* **1933**, *44*, 235.

Giauque, W. F.; Stout, J. W.; Clark, C. W. "Amorphous Carbon Resistance Thermometer. Heaters for Magnetic and Calorimetric Investigations at Temperatures below 1° K"; *J. Am. Chem. Soc.* **1938**, *60*, 1053–1060.

McDougall, D. P.; Giauque, W. F. "The Production of Temperatures below 1° A. The Heat Capacities of Water, Gadolinium Nitrobenzene Sulfonate Heptahydrate and Gadolinium Anthraquinone Sulfonate"; *J. Am. Chem. Soc.* **1936**, *58*, 1032–1037.

"Nobel Prize to U. S. Chemist; Japanese Physicist a Winner," *New York Times*, November 4, 1949, p. 1.

Silverman, M. *Saturday Evening Post*, Dec. 10, 1949, p 38.

Giauque's papers are in the Bancroft Library at the University of California, Berkeley.

# 1950

NOBEL LAUREATE

# *Kurt Alder*

## 1902–1958

Copyright Nobel Foundation

Kurt Alder was born on July 10, 1902, in Königshütte, Germany, now Chorzow, Poland. He attended the University of Berlin and the University of Kiel, where he studied under Otto Diels. In 1950 he and Diels were jointly awarded the Nobel Prize in chemistry for their diene synthesis. Kurt Alder died in Cologne, Germany, on June 20, 1958.

Alder's father was a schoolteacher in the Kathowitz (now Katowice) area of Upper Silesia. He attended the German schools in Königshütte, which was a heavily industrialized area. At the end of World War I, when this area became part of Poland, his family moved to Berlin. He completed his early schooling at the *Oberrealschule* in Berlin. In 1922 he entered the University of Berlin and later transferred to the University of Kiel, where he received his Ph.D. in chemistry in 1926. His dissertation, "On the Causes of the Azoester Reaction", was carried out under the direction of Professor Otto Diels. This thesis research involved a study of the reaction of the azoformic ester, $C_2H_5O-CO-N=N-CO_2-C_2H_5$, commonly called azocarboxylic ester, with such conjugated aromatic systems as styrene.[1]

He remained at the University of Kiel. In 1930 he was appointed a lecturer in organic chemistry and in 1934 extraordinary professor of Chemistry. During this time Alder studied the addition reaction

---

[1]Sidgwick, N. V. *The Organic Chemistry of Nitrogen*; revised and rewritten by L. W. J. Taylor and Wilson Baker; Oxford University: Oxford, England, 1937; pp 433-434.

328

of phenyl azide, $C_6H_5N_3$, with bicyclic unsaturated hydrocarbons. Phenyl azide did not react with stilbene or cinnamic acid esters, but it was found to react very rapidly in the cold with bicyclic systems containing a carbon-to-carbon double bond and a single methylene group in the bridge. The product of this addition to the carbon-to-carbon double bond, which crystallized in a few minutes, can be used as a general test for this type of bicyclic structure.

During this time Alder also published jointly with Otto Diels the reaction of azocarboxylic ester with styrene. In 1928 they published their work on the condensation of butadiene (the diene) and acrolein (the dienophile), which led to their jointly receiving the 1950 Nobel Prize in chemistry for their diene synthesis.

In 1936 Alder left academic life to become a research director at the Bayer Werke in Leverkusen, a branch of I. G. Farbenindustrie. At this time Alder's research interests seemed to concentrate on the stereochemical selectivity of the diene reaction. In 1937 he published, along with G. Stein, the Alder–Stein rules to predict the pronounced stereochemical selectivity of the diene reaction. Based on these rules a research worker could predict the stereochemical configuration or arrangement of the molecules produced by the diene reaction in three dimensions in space.[2] These rules are explained on the basis of product epimerization (conversion of the product from cis to trans isomer) or on reversibility of the Diels–Alder reaction.

Alder also investigated the reversibility of the diene reaction, the dissociation of the reaction products into the components from which they were prepared. Alder found that generally products resulting from five-carbon cyclic dienes were unstable, whereas those resulting from six-carbon dienes were considerably more stable, thus giving a means of differentiating between five- and six-carbon cyclic dienes.

During his time at the Leverkusen Werke he mainly studied the polymerization reaction which involved the preparation and synthesis of synthetic rubber or Bunatype synthetic rubber. The name Buna is derived from the original German name Perbunan. Buna-N is a polymer derived from Butadiene and acrylonitrile, and Buna-S is a polymer of butadiene and styrene.[3] The development of the diene synthesis gave a great deal of insight into the mechanism of polymerization

---

[2]*Organic Reactions*; Adams, R. et al., Eds.; John Wiley & Sons: New York, 1948; Vol. 4, pp 1–174.

[3]Wakeman, R. L. *The Chemistry of Commercial Plastics*; Reinhold: New York, 1947, p 552.

processes. Certain polymerization processes were known prior to the work of Diels and Alder such as the dimerization of butadiene with 4-vinyl-1-cyclohexene and the polymerization of isoprene to dipentene.[4] In light of the diene-synthesis these reactions were investigated further and the polymerization of isoprene and butadiene found to be a diene-type reaction.[5] In 1940 Alder accepted the Chair for Experimental Chemistry and Chemical Technology at Cologne University. During these difficult war years he carried on his research and was able to transform the diene synthesis from an additive process to a process of substitution. It was known that certain nitrogen-containing heterocycles react with maleic anhydride, but they fail to give the normal Diels–Alder adducts. In these compounds substitution products are produced at the active alpha position of the heterocycle. This type of substitution reaction was studied by Alder in 1943 when he reported the substitution reaction of furan and substituted furan with crotonaldehyde, methyl vinyl ketone, and vinyl phenyl ketone in the presence of a sulfur dioxide catalyst.[6]

Alder remained at the University of Cologne after World War II. From 1949 to 1950 he served as dean of the Faculty of Philosophy. He was at Cologne until his death on June 20, 1958.

In 1938 Alder received the Emil Fischer Memorial Medal from the Association of German Chemists and also became a member of the Deutsche Akademie der Naturforscher Leopoldina in Halle. In 1950 the Medical Faculty of the University of Cologne conferred on him the honorary degree of M.D. In 1954 he received an honorary doctorate from the University of Salamanca, Spain.

In 1955 Alder along with 17 other Nobel Laureates signed a declaration calling on the nations of the world to renounce war.

No biography of Kurt Alder or Otto Diels would be complete without a listing of some of the practical applications of the Diels–Alder diene synthesis. Among the many products that involve the use of this synthesis in one step of their preparation are the insecticides dieldrin, aldrin, and chlordane. Many natural products such as cantharidin involve this reaction as one of the steps in their synthesis.

[4] *Encyclopedia of Chemical Technology*, 2nd ed.; Kirk, R. E.; Othmer, D. F.; Eds.; Interscience: New York, 1965, Vol 7, p 677.

[5] *Organic Reactions*; Adams, R., et al., Eds.; John Wiley & Sons: New York, 1948, Vol 4, pp 7–8.

[6] *Organic Reactions*; Adams, R., et al., Eds.; John Wiley & Sons: New York, 1948; p 38, Vol. 4.

Robert B. Woodward (Nobel Prize in chemistry, 1965) used this synthesis as one of the many steps in his total synthesis of reserpine. In 1950 M. Gates used a Diels–Alder reaction as one of the steps in the total synthesis of morphine. It was also used in the synthesis of the unusual bicyclic compound given the trivial name "barrelene" and as one of the steps in the synthesis of cortisone.

**LEE R. WALTERS**
*Lafayette College*

# Bibliography

"Diels-Alder Reaction Proves Potent Chemical Ally"; *Chem. Eng. News* **1950**; p 4266.

*Encyclopedia of Chemical Technology;* Kirk, R. E.; Othmer, D. F., Eds.; Interscience: New York, 1965; Vol. 7.

Ihde, A. J. "Alder, Kurt"; in *Dictionary of Scientific Biography*; Gillispie, C. C., Ed.; Scribner's Sons: New York, 1971; Vol. 1, pp 105–106.

*McGraw-Hill Modern Scientists and Engineers*; Greene, J. E., Ed.; McGraw-Hill: New York, 1980; Vol. 1.

*Nobel Lectures: Chemistry, 1942–1962;* Elsevier: New York, 1964.

*Organic Reactions;* Adams, R., et al., Eds.; John Wiley & Sons: New York, 1948; Vol. 4, pp 1–174.

Sidgwick, N. V.; revised and rewritten by Taylor, L. W. J., Baker, W., *The Organic Chemistry of Nitrogen;* Oxford University: Oxford, England, 1937.

Wakeman, R. L. *The Chemistry of Commercial Plastics;* Reinhold: New York, 1947.

# 1950
NOBEL LAUREATE

# Otto Paul Hermann Diels

## 1876–1954

Otto Paul Hermann Diels was born on January 23, 1876, in Hamburg, Germany. He attended the University of Berlin and received his Ph.D. in chemistry in 1899. In 1916, he became professor and director of the Institute of Chemistry of the University of Kiel, and remained there until he retired in 1948. In 1950, he received the Nobel Prize jointly with Kurt Alder for the discovery and development of diene synthesis known as the Diels–Adler reaction.

Shortly after his birth, his family moved to Berlin where his father Hermann Diels, an eminent scholar, was a professor of classical philology at the University of Berlin. His mother, Bertha Dübell, was the daughter of a district judge. He had two brothers: Paul, who became a professor of Slavic philology at Breslau, and Ludwig, who became a professor of botany at the University of Berlin.

From 1882 to 1895, Diels attended the Joachimsthalsches *Gymnasium* in Berlin. He attended the University of Berlin from 1895 to 1899, where he studied chemistry under Emil Fischer (Nobel Prize in chemistry, 1902). In 1899, he received his Ph.D. magna cum laude and was appointed an assistant at the Institute of Chemistry at the University of Berlin. In 1904, he became a lecturer, and in 1913 he was appointed head of the department. The first public recognition of his work came

*(Copyright Nobel Foundation)*

in 1904. He participated in the Louisiana Purchase Exposition in St. Louis, Missouri, and received a gold medal for his exhibit.

In 1907, his text *Einführung in die organische Chemie* was published. A model of clarity and precision, it became one of the most popular textbooks in its field. By 1962, it had gone through 19 editions.

Diels' earliest research was on the borderline between organic and inorganic chemistry. In 1906 and 1907, he reported the preparation of carbon suboxide ($C_3O_2$) by the dehydration of diethylmalonic acid or malonic acid using heat and phosphorus pentoxide as the dehydrating agent.[1]

Diels married Paula Geyer in 1909. They had three sons and two daughters. In 1914, he was appointed associate professor at the Chemical Institute of the Royal Friedrich Wilhelm University, which is now Humboldt University. In 1916, he accepted the position of professor and director of the Institute of Chemistry at the Christian Albrecht University, Kiel, where he remained until his retirement in 1948.

Diels' research interests during this time were now directed to the field of organic chemistry. He became interested in the physiologically active cholesterol ($C_{27}H_{46}O$), which had been discovered in 1789 by the French chemist Michel Eugene Chevreul, who was noted for his research on fats, oils and the process of saponification. Very little was know about the structure of cholesterol except that it contained only one carbon-to-carbon double bond, an alcoholic –OH group, and was a complex hydro-aromatic system containing four fused rings. From his work on the oxidation of cholesterol, Diels was able to demonstrate that the –OH was a secondary alcohol and that it was bonded to one of the fused ring systems. He showed that oxidation of cholesterol with sodium hypobromite solution yielded a compound now commonly known as Diels acid, a high-melting, sparingly soluble dicarboxylic acid. After these discoveries, he states, he "withdrew for the time being from cholesterol research because...I had to familiarize myself with many other fields of organic chemistry."[2]

By 1925 the true structure of cholesterol had not yet been determined. Diels and his co-workers tried to "get through to the aromatic basic skeleton and determine its character by dehydration

---

[1] *Organic Reactions*; Adams, R. et al., Eds.; John Wiley & Sons: New York, 1946; Vol. III, p 113.

[2] Fieser, L. F.; Fieser, M. *Reagents for Organic Synthesis;* John Wiley & Sons: New York, 1967; p 489.

(dehydrogenation) of cholesterol."[3] Initial attempts using the extremely reactive sulfur or platinum black as the dehydrogenating agent were not promising. Using palladium on charcoal led to a mixture of products, one of which was chrysene, an aromatic compound containing four fused six-member rings.

Because the action of sulfur on cholesterol was too vigorous, in 1927 Diels used selenium, the element directly below sulfur in the periodic chart, as the dehydrogenating agent. Selenium did not combine with organic molecules, and the hydrogen was removed as hydrogen selenide gas ($H_2Se$). From cholesterol he obtained a complex mixture of products, one of which was a hydrocarbon. To this structure he assigned the formula $C_{18}H_{16}$, but published nothing further concerning it for five years.

The hydrocarbon that became known as Diels hydrocarbon was also reported by several other workers in the field of steroid research during this time; however, the exact structure of $C_{18}H_{16}$ was still open to question. Finally, in 1934 Diels and Hermann Klare put forth the correct structure as 3'-methyl-1,2-cyclopentaphenanthrene. This work was carried out to determine the structure of cholesterol, but during the nine-year period required for the study, the structure of cholesterol had been determined in 1932 by other methods. Diels's selenium dehydrogenation has been a valuable tool in the structure determination of estrogens and steroids.

From 1899 to 1923, Diels was also active in the field of organic nitrogen compounds.[4] He demonstrated that cyanuric acid ($C_3H_3N_3O_3$) is a heterocyclic ring formed by the polymerization of three cyanic acid (HOCN) molecules. It has a 1,3,5-triazine ring, which consists of three carbon and three nitrogen atoms arranged alternately.

In 1925, Diels reported the reaction of azoformic ester ($C_2H_5-O-CO-C-N=N-CO_2C_2H_5$), commonly called azodicarboxylic ester, with compounds that contain a system of conjugated double bonds. He found that the addition of azodicarboxylic esters takes place in the 1,4-position of the conjugated system as with cyclopentadiene and with butadiene. This work probably led to the famous Diels–Alder reaction described later.

---

[3]Fieser, L. F.; Fieser, M. *Reagents for Organic Synthesis;* John Wiley & Sons: New York, 1967; pp 990–992.

[4]Sidgwick, N. V., revised by Taylor, L. W. J.; Baker, W. *The Organic Chemistry of Nitrogen;* Oxford University Press: Oxford, 1937.

In 1927, Diels and his student Kurt Alder published a paper on the reaction of azodicarboxylic ester with styrene, a compound in which the conjugated system is part of an aromatic system. He was also engaged in the use of azodicarboxylic ester in the oxidation of unsymmetrically disubstituted hydrazines to produce tetrazenes. Tetrazenes consist of an unsaturated chain of four nitrogen atoms with a double bond in the middle of the nitrogen chain.

In 1928, Otto Diels and Kurt Alder first published the work that was to be known as the Diels–Alder reaction, for which they were awarded the 1950 Nobel Prize in chemistry. The reaction investigated by Diels and Alder was not new—examples had been known for several years. Early work had been done by Theodor Zincke in 1893 and 1897 on the dimerization of tetrachlorocyclopentadienone.[5] In 1906, Albrecht described the addition product of $p$-benzoquinone with one or two molecules of cyclopentadiene. Albrecht assigned erroneous formulas for these addition products, but they were later shown to be typical products of the diene synthesis by Diels and Alder.[6]

Euler and Josephson reported the addition products formed by isoprene and 1,4-benzoquinone in 1920. This research laid the groundwork for Diels and Alder. They demonstrated the great versatility of this synthetic organic reaction, which has been said to rank in importance with the Grignard reaction (for which Victor Grignard shared the Nobel Prize in chemistry in 1912). The Diels–Alder reaction is known as diene addition because it involves the addition to a molecule containing two conjugated carbon-to-carbon double bonds, which is called the diene. A conjugated system must contain at least four carbon atoms, two of which are bonded by a double bond, and in the center of the four-carbon-atom chain is a carbon-to-carbon single bond, $C=C-C=C$. The second molecule contains two carbon atoms bonded by one carbon-to-carbon double bond. It is called the dienophile, or plulodiene, which means diene-loving partner. The reaction involves addition of the dienophile across the 1,4-position of the conjugated diene to form the product, which in the case of an ethylenic dienophile is a cyclohexene. A cyclohexene is a compound containing six carbons in a ring with a single carbon-to-carbon double

[5]Ingold, C. K. *Structure and Mechanism in Organic Chemistry;* Cornell University Press: Ithaca, NY, 1953.

[6]*Organic Reactions;* Adams, R. et al., Eds.; John Wiley & Sons: New York, 1948; Vol. IV, pp 1–174.

bond between two of the ring carbons. The ethylenic dienophile is often conjugated with an activating group such as a carbonyl, carboxyl, cyan, or nitro group. The dienophile may contain an acetylenic bond between two carbons. This means that it contains two carbon atoms bonded by a carbon-to-carbon triple bond, $C \equiv C$. In this case, the product would be a cyclohexa-1,4-diene, a compound having a ring of six carbon atoms containing two carbon-to-carbon double bonds. The reaction gives a convenient and stereospecific method of obtaining the ubiquitous six-member ring.[7]

Diels and Alder first reported the reaction of butadiene with maleic anhydride in benzene solution to give a quantitative yield of *cis*-1,2,3,6-tetrahydrophthalic anhydride in 1928. Since that time, "The development of the Diels–Alder reaction has been of inestimable value not only in synthesis but also for the light it has cast upon the mechanism of polymerization."[8]

From 1928 to 1940, Diels, Alder, and their coworkers published several papers in German scientific journals. Most of Diels' work was concerned with the use of maleic anhydride as the dienophile, but he did not completely limit himself to this study. He also examined such dienophiles as crotonaldehyde, cinnamaldehyde, acetylenedicarboxylic acid, and azodicarboxylic acid, which he had studied earlier in his research. The extensive list of dienes investigated by Diels includes such compounds as butadiene, cyclopentadiene, anthracene, substituted furans, and α-pyrone; however, the common heterocycle thiophene does not react at all, and pyrrole reacts to give an abnormal product.

Professor Diels remained at Kiel during the difficult years of World War II. Near the end of the war, two of his sons were killed in action. Air raids completely destroyed the Chemical Institute, its library, and his home. Under these severe conditions, he applied for retirement in early 1945, but agreed to remain as director of the Chemical Institute. At age 70 he started to rebuild the institute, where he remained until his complete retirement in October 1948. In 1946, the Medical Faculty of Christian Albrecht University awarded him an honorary Doctor of Medicine degree.

In 1950, Otto Diels and Kurt Alder were awarded the Nobel Prize in chemistry for their diene synthesis, one of the few times the Nobel

---

[7]Lowry, T. H.; Richardson, K. S. *Mechanism and Theory in Organic Chemistry;* 3rd ed; Harper and Row: New York, 1987; p 919.

[8]*Organic Reactions;* Adams, R. et al., Eds.; John Wiley & Sons: New York, 1948; Vol. IV; p 7.

Prize has been awarded in the field of preparative organic chemistry. The other awards for this type of work went to Victor Grignard and Paul Sabatier in 1912. Illness prevented Diels from delivering his Nobel Lecture in Stockholm in May 1951. In 1931, he was awarded the Adolf von Baeyer Memorial Medal by the Society of German Chemists, and in 1952 he was awarded the Grosskreuz des Verdienstordens der Bundesrepublik Deutschland. Diels was a member of the Academies of Science of Göttingen and Halle and the Bavarian Academy of Science. He died on March 7, 1954.

LEE R. WALTERS
*Lafayette College*

# Bibliography

Farber, E. *Nobel Prize Winners in Chemistry, 1901–1961.* Abelard-Schuman, New York, 1963.

Fieser, L. F.; Fieser, M. *Reagents for Organic Synthesis;* John Wiley & Sons: New York, 1967.

*German Nobel Prizewinners;* Hermann, A., Ed.; Heinz Moos Verlagsgesellschaft: Munich, Germany, 1968.

Ingold, C. K. *Structure and Mechanism in Organic Chemistry.* Cornell University: Ithaca, NY, 1953.

Lowry, T. H.; Richardson, K. S. *Mechanism and Theory in Organic Chemistry;* 3rd ed.; Harper and Row: New York, 1987.

Olsen, S. "Otto Diels"; *Chemische Berichte* **1962,** *95(2),* v–xlvi.

*Organic Reactions;* Adams, R. et al., Eds.; John Wiley & Sons: New York, 1946, Vol. 3.

Schmauderer, E. "Diels, Otto"; *Dictionary of Scientific Biography;* Gillispie, C. C., Ed.; Charles Scribner's Sons: New York, 1972; Vol. 4, pp 90–92.

Sidgwick, N. V. *The Organic Chemistry of Nitrogen;* revised and rewritten by L. W. J. Taylor and W. Baker; Oxford University Press: Oxford, England, 1937.

# 1951

NOBEL LAUREATE

# Edwin McMillan

## 1907–1991

Edwin Mattison McMillan, a physicist, was awarded the 1951 Nobel Prize in chemistry jointly with Glenn T. Seaborg "for their discoveries in the chemistry of the transuranium elements." McMillan was born on September 18, 1907, at Redondo Beach, California, and died on September 7, 1991, at his home in El Cerrito, California. His parents were Edwin Harbaugh McMillan and Anna Marie McMillan, née Mattison. Edwin Harbaugh McMillan was a physician practicing for most of his long career in Pasadena, California, a fortunate circumstance for his son, as it provided contact with the California Institute of Technology. This contact in his early youth nurtured a wide and lasting interest in all things natural: rocks and minerals, electrical phenomena, chemistry, and botany.

McMillan entered the California Institute of Technology in 1924 as a physics student, but he also studied more chemistry than was usual among physicists. He came into contact with and was greatly influenced by Linus Pauling, who was a National Research Council fellow at the time. His first scientific publication, with Pauling, "An X-ray Study of the Alloys of Lead and Thallium" was published in the *Journal of the American Chemical Society*.[1] He received a bachelor of science degree from Caltech in 1928 and a master's degree in 1929

[1] **1927**, *49*, 666-669.

for his first-year graduate studies. He continued his graduate studies at Princeton University, where he did research on molecular beams under the general direction of Professor E. U. Condon. He received his Ph.D. in 1932 with a thesis titled "Deflection of a Beam of HCl Molecules in a Non-Homogeneous Electric Field."

On completion of his Ph.D., McMillan was awarded a two-year National Research Council fellowship. He elected to do his research in the physics department of the University of California at Berkeley. There he worked in spectroscopy and published several papers on hyperfine structure. At that time (1932–1934) the field of nuclear physics was being transformed by a succession of startling discoveries—the neutron, positron, induced radioactivity—and the technology of accelerators was entering a stage of rapid development. Lawrence's Radiation Laboratory, then an adjunct to the physics department with close ties to the chemistry department, was a leading laboratory in nuclear physics and was the scene of dynamic activity. The 27-inch cyclotron was producing a beam of unprecedented energy: 5 MeV deuterons. The burgeoning field of nuclear physics and the excitement at the Radiation Laboratory were a great attraction to McMillan, and he joined the small staff in 1934 as a research associate.

He distinguished himself immediately as an innovative and careful experimenter with an exceptional command of the theoretical background of nuclear physics. In an early investigation of gamma rays from nuclear disintegrations, he accomplished the first unambiguous demonstration of the anomalous absorption of gamma rays by the creation of electron pairs.

McMillan took a keen interest in the operation and improvement of the cyclotron, then an almost entirely empirical endeavor. He helped to rationalize the understanding of cyclotrons and was responsible for substantial improvements in ion sources, magnetic-field shaping, beam extraction, and power and control systems.

News of the discovery of fission by Otto Hahn and Fritz Strassmann in early 1939 created great excitement and stimulated a flurry of experimental activity in the Radiation Laboratory, as it did in many other places, to confirm and to elucidate the new process. McMillan undertook a simple experiment: measuring the range of fission fragments by their penetration of a stack of foils in contact with a thin layer of a uranium compound exposed to neutrons from a target bombarded by a beam of deuterons from the cyclotron. The results of the range measurement were routine. Investigation of the residue in the uranium layer proved to be another story. He observed a 23-minute half-life activity that came from a known uranium isotope, $^{239}$U. He also observed a previously unknown 2.3-day beta activity, which he immediately suspected to be a beta decay product of the uranium isotope and therefore an isotope of element 93. Emilio Segrè attempted to establish the identification of the new active body by chemical tests, but "because of the ease with which element 93 exhibits some of the chemical properties of the rare earths, they were interpreted as meaning that the 2.3-day body was a rare earth isotope".[2] Bothered by the inconsistency of this assignment with his observation that the active body did not recoil from the target, McMillan turned to a chemical investigation of the problem himself and enlisted the collaboration of Philip Abelson. They soon found that the key to understanding the chemistry was the state of oxidation of the material. They identified the 2.3-day activity as coming from $^{239}$93. The similarity of the chemistry of element 93 and uranium, which made the identification so difficult, led to the suggestion that there was a second "rare-earth" group of elements starting with uranium. It also led McMillan to propose the name "neptunium" for element 93 and to reserve the name "plutonium" for the not yet identified element 94. McMillan next sought to make this identification as a product of deuteron bombardment of uranium. He produced an alpha-active substance that he was able to say on the basis of chemical tests was not protactinium (91), uranium (92), or neptunium (93), but he could not make the final identification of element 94 because he left the Radiation Laboratory for the first of several urgent wartime positions. Subsequent work at Berkeley on the transuranium elements and the final identification of element 94 was carried out by a group including J. W. Kennedy, G. T. Seaborg, E. Segrè, and A. C. Wahl.

McMillan's wartime service began November 1940 with work on airborne microwave radar at the newly established Radiation Laboratory at the Massachusetts Institute of Technology. In 1941 he went to the U.S. Navy Radio and Sound Laboratory at San Diego, where he invented and developed an underwater echo repeater. In November 1942 he joined J. Robert Oppenheimer in searching for a site for a nuclear weapons laboratory and helping to organize that laboratory at Los Alamos, New Mexico. He had major responsibilities for the

---

[2]Seaborg, G. T.; Segrè, E. "The Trans-Uranium Elements"; *Nature* **1947,** *159,* 863.

development of the gun assembly method used in the uranium bomb and for test methods for the implosion assembly used in the plutonium bomb.

In the early summer of 1945, as activity at Los Alamos reached a climax and the end of the war was in sight, McMillan's thoughts reverted to the problem of accelerators. Up to the time just before the war, when peacetime research gave way to the mobilization of scientists and their equipment for war duty, a succession of accelerators and, in particular, cyclotrons of increasing energy were built and used in a wide variety of programs. The highest energy accelerator was the 60-inch Crocker Cyclotron at Berkeley with a beam of 16 MeV deuterons. Although the huge magnet for the next projected cyclotron, which was to produce beams of 100 MeV deuterons, was built, it was clear that the road to higher energies would be a very difficult one and, unless there was a new idea, would soon reach an absolute block. The reason was that the relativistic mass increase of the particles as they gained energy would destroy the cyclotron resonance condition, that is, that the rotational frequency of the particles in the magnetic field must be constant. The overriding problem of accelerator technology was how to circumvent the energy limit.

McMillan conceived a solution that was startling in its simplicity and that would prove to be far-reaching in practice. Under certain conditions, ions rotating in the magnetic field of a cyclotron are in stable, zero energy gain orbits. If then the magnetic field and/or the frequency of the electric field are slowly changed, the energy of the stable orbit may be increased, and the ions gain energy and oscillate about the higher energy stable orbit. This is called the principle of phase stability. There is no in-principle limit to this process. The door was opened to the design of accelerators for electrons, protons, or heavier ions of any energy, limited only by cost and by engineering factors, factors that yield to ingenuity and invention. The large accelerators of today, which are miles in circumference, have energies measured in TeV, and are used at the forefront of particle research, are based on three fundamentals: the principle of phase stability, the earlier invention of the cyclotron, and the more recent invention of strong focusing.

The principle of phase stability had been anticipated by V. Veksler in Russia. Because of a complete breakdown in communication during the war, the work of the two scientists was entirely independent. In 1963 McMillan and Veksler were chosen to share the Atoms for Peace

Award for their elucidation of the principle of phase stability and for their leadership in high energy physics.

In parallel with his remarkably productive research career, McMillan taught as a faculty member in the physics department at Berkeley, where he gained a reputation for clarity and simplicity in his teaching and for an astonishingly wide range of knowledge. He served as Associate Director of the Radiation Laboratory and, with the untimely death of Lawrence in 1958, he was appointed Director of the Ernest O. Lawrence Radiation Laboratory, a difficult position, which he held with grace and distinction until his retirement in 1973. In retirement he continued active research at the Lawrence Berkeley Laboratory and at the European Organization for Nuclear Research, and he wrote several important papers on aspects of the history of accelerators. He was awarded the National Medal of Science in 1990 in recognition of his long and distinguished career in science, education, and public service.

In 1984 McMillan suffered the first of a series of disabling strokes. He died at his home in El Cerrito, California on September 7, 1991.

EDWARD J. LOFGREN
*University of California, Berkeley*

## Bibliography

*Current Biography;* H. W. Wilson: New York, February 1952.

Farber, E. *Nobel Prize Winners in Chemistry, 1901–1961;* Abelard-Schuman: London, 1963.

McMillan, E. "Early History of Particle Accelerators"; In *Nuclear Physics in Retrospect;* Stuewer, R. H., Ed; University of Minnesota Press: Minneapolis, MN, 1979; pp 113–155.

McMillan, E. "History of the Cyclotron, Part 2"; *Physics Today* **1959,** *12,* 18–34.

McMillan, E. "Radioactive Recoils from Uranium Activated with Neutrons"; *Phys. Rev.* **1939,** *55,* 510.

McMillan, E. "Some Gamma Rays Accompanying Artificial Nuclear Disintegrations"; *Phys. Rev.* **1934,** *46,* 868–873.

McMillan, E. "The Synchrotron—A Proposed High Energy Particle Accelerator"; *Phys. Rev.* **1945,** *68,* 143–144.

McMillan, E. "The Transuranium Elements: Early History"; In *Nobel Lectures: Chemistry, 1942–1962;* Elsevier: New York, 1964; pp 314–322.

McMillan, E; Abelson, P. H. "Radioactive Element 93"; *Phys. Rev.* **1940,** *57,* 1185–1186.

*National Cyclopedia of American Biography*; J. T. White: New York, 1952; Vol. H.

*Nobel Prize Winners*; Wasson, T., Ed.; H. W. Wilson: New York, 1987.

*The Nobel Prize Winners: Chemistry*; Magill, F. N., Ed.; Salem Press: Pasadena, CA, 1990; Vol. 2.

Seaborg, G. T. *Man-Made Transuranium Elements*; Prentice-Hall: Englewood Cliffs, NJ, 1963.

Seaborg, G. T.; McMillan, E. M.; Kennedy, J. W.; Wahl, A. C. "Radioactive Element 94 from Deuterons on Uranium"; *Phys. Rev.* **1946,** *69,* 366–367.

# Glenn Seaborg

### 1912–

Glenn Theodore Seaborg, an American nuclear chemist and corecipient with Edwin M. McMillan of the 1951 Nobel Prize in chemistry, is known principally for his work on the synthetic transuranium elements. He was born on April 19, 1912, in Ishpeming, Michigan, to Herman Theodore Seaborg, a machinist, and Selma Erickson Seaborg. He received his bachelor's degree in 1934 from the University of California, Los Angeles, and his doctorate, in chemistry, from the University of California at Berkeley in 1937. He is the recipient of over 50 honorary degrees. Seaborg was married on June 6, 1942, to Helen Griggs. He has six children: Peter, Lynne, David, Stephen, John Eric, and Dianne. Throughout his distinguished career Seaborg has been presented with numerous awards. Some of the more prestigious ones have been the following: ACS Award in Pure Chemistry, 1947; Society of Chemical Industry Perkin Medal, 1957; Enrico Fermi Award, United States Atomic Energy Commission, 1959; Priestley Memorial Award, 1960; Franklin Medal, 1963; Gold Medal Award, American Institute of Chemists, 1973; John R. Kuebler Award, Alpha Chi Sigma, 1978; Priestley Medal, American Chemical Society, 1979; Henry DeWolf Smyth Award, American Nuclear Society, 1982; Actinide Award, 1984; Great Swedish Heritage Award, 1984; Vannevar Bush Award, National Science Board, 1988; and the National Medal of Science, 1991.

Influenced by an inspiring teacher in high school, Seaborg decided to major in chemistry. After graduation he continued his studies, obtaining his Ph.D.in 1937; he then served as a research assistant to Gilbert Newton Lewis, an American chemist remembered for his work on the chemical bond and thermodynamics. Seaborg joined the faculty of the University of California at Berkeley in 1939, and began rising through the ranks as research associate, instructor, and assistant professor. He worked on the Manhattan Project at the University of Chicago from 1942 to 1946. He was promoted to professor in 1945 and remained an active member of the professorial staff of the University of California until he was elevated to the chancellorship at Berkeley in 1958. He served as chancellor until 1961 when President John F. Kennedy appointed him chairman of the U.S. Atomic Energy Commission. Seaborg was reappointed by both Presidents Lyndon Johnson and Richard Nixon, serving in that position until 1971. On leaving Washington, he returned to the University of California at Berkeley to resume his position as professor of chemistry. Since 1972, he has also served as associate director of the Lawrence Berkeley Laboratory.

Seaborg's work on the synthetic transuranium elements was based on the 1934 research of Enrico Fermi, an Italian physicist, and coworkers, who had shown that nuclear transmutations could be brought about by irradiating uranium with neutrons. Fermi and his team tried to form elements beyond uranium, which would have atomic numbers greater than 92. Otto Hahn and Fritz Strassmann, working in Germany, showed that what Fermi believed to be transuranium elements were actually fission products. In 1940, working at the University of California at Berkeley, Edwin M. McMillan and Philip H. Abelson performed the first synthesis of the element with atomic number 93. While experimenting with fission, they discovered a beta particle activity with a half-life of about two days. After they were able to isolate it chemically and conclusively identify it as a transuranium element, they announced on June 8, 1940, that it was element 93. McMillan and Abelson proved that element 93 could be produced in very small quantities by a uranium reaction with neutrons that did not involve fission. Since uranium (element 92) had been named for the planet Uranus by Martin Heinrich Klaproth, McMillan and Abelson decided to follow his analogy. Thus they named element 93, lying beyond uranium, *neptunium* for Neptune, the planet beyond Uranus. Neptunium became the first of the transuranium elements.

Following McMillan and Abelson's discovery of the first transuranium element, Seaborg and his collaborators began the difficult search for element 94. In December 1940 Seaborg, along with his associates Joseph W. Kennedy and Arthur C. Wahl, bombarded uranium oxide with 16 MeV deuterons. They were successful in isolating chemically the unstable element 93 fraction of the resulting products, which decayed, according to their theory, by emitting a beta particle to yield element 94, with a mass number of 238. After careful study, element 94 was positively identified and named *plutonium,* after Pluto, the planet beyond Neptune.

Further research with plutonium led to the production of isotope $^{239}$Pu, which was found to be a potential source of nuclear energy. Only a very small specimen of $^{239}$Pu (0.0005 mg) was produced for the first experiments conducted. However, this small sample was sufficient to reveal that $^{239}$Pu was susceptible to fission by bombardment with slow neutrons and therefore that its production in substantial quantities was a matter of extreme importance to national defense. Seaborg, along with other members of the research team, sent secret communications concerning the discovery to the Uranium Committee in Washington, D.C., in January, March, and again in May 1941. The committee decided to withhold these reports from the general public. In Seaborg's book, *The Transuranium Elements* (Yale University Press: New Haven, CT, 1958), he states that "the announcement to the world of the existence of plutonium was in the form of the nuclear bomb dropped over Nagasaki."

In August 1942 Seaborg, along with a number of his colleagues, moved to the Metallurgical Laboratory of the University of Chicago (now the Argonne National Laboratory) to continue research on $^{239}$Pu and, in particular, to find a way to produce it in usable amounts for the production of an atomic bomb. During World War II, Seaborg remained at the University of Chicago as a section chief on the Manhattan Project. The first production of plutonium was undertaken in the newly devised uranium reactors. Seaborg's main responsibility was the isolation of the plutonium from the reaction products. Seaborg developed the process for separating plutonium. This process was later magnified 10 billion times in order to produce enough plutonium for the first atomic bomb. The microchemical aspects of this work were carried out chiefly by Burris B. Cunningham and Louis B. Werner under Seaborg's direct supervision. They were able to produce the first pure chemical compound of plutonium, in the form of $^{239}$Pu, free from carrier material and all other foreign matter. Their work provided the first sight of a synthetic element and was the first isolation of a weighable amount of an artificially produced isotope of any element.

With Cunningham and Werner concentrating on completing a major part of the plutonium work, Seaborg once again turned his attention to the production of further transuranium elements. Seaborg felt that the next element should be similar to plutonium, in that it should be possible to oxidize it to oxidation state VI and to use this in chemical isolation procedures. Through trial and error, Seaborg continued to work along these lines, only to end in failure. Finally, in the summer of 1944, Seaborg revised his thinking: he recognized that the next elements could be oxidized above state III only with extreme difficulty, and this new approach led him to the quick identification of an isotope ($^{242}$Cm) of element 96. Element 95 was identified shortly thereafter.

Seaborg recognized that oxidation state III was very stable and was the dominant feature of the chemistry of these new elements, which made them part of an actinide transition series. This meant that the chemical properties of these elements corresponded to the filling of the 5f electron shell and that they could be expected to resemble each other chemically, much like the rare earth or lanthanide series elements. They should also bear a strong similarity to the rare earth elements. Following this hypothesis, Seaborg noticed that starting with actinium (element 89) a second set of rare earth elements could be considered to exist. The two sets are commonly distinguished by using the term *lanthanide* for the older, which begins with lanthanum (element 57), and the term *actinides* for the newer. The actinide transition series hypothesis, along with its strong resemblance to the rare earth elements, proved to be so true that Seaborg and his team had great difficulties in separating elements 95 and 96 from each other and from the rare earth elements. Because of this, element 95 was named *americium* after the Americas, by analogy with the naming of its rare earth homologue *europium* after Europe. Using a similar analogy with the naming of its rare earth homologue gadolinium, element 96 was given the name curium, in honor of Pierre and Marie Curie.

After the discoveries of neptunium, plutonium, americium, and curium, Seaborg and his team went on to identify and name further elements. In 1949 berkelium (element 97) and californium (element 98) were identified and named after Berkeley, California. As Seaborg studied the chemistry and physics of the already-known transuranium elements, he was able to continue the discovery of even more actinides.

Glenn Seaborg and President John Kennedy on a tour of the Atomic Energy Commission and Air Force installations, Dec. 7 and 8, 1962. From *Kennedy, Kruschev, and the Test Ban* by Glenn T. Seaborg, University of California Press, 1981. (Credit: Office of Naval Aide to the President, R.L. Knudsen)

In 1952 einsteinium (element 99) and in 1953 fermium (element 100) were identified. Both were named in honor of Albert Einstein and Enrico Fermi, shortly after their deaths. Following those discoveries, elements 101 and 102 were found. In 1955 mendelevium was named in honor of Mendeleev, and in 1957 nobelium, in honor of Alfred Nobel. Seaborg was involved in the identification of nine (94 through 102) of the 13 transuranium elements known by 1970.

In addition to his extensive original research on the transuranium elements, Seaborg has published widely on the chemistry and history of the field (*see* Bibliography). His scientific surveys are clear and his historical accounts carry an immediacy and authority that comes from his position as one of the chief architects of actinide chemistry. A similar authenticity characterizes Seaborg's writings on science and public policy, which began to appear during his tenure as head of the Atomic Energy Commission, culminating in a trilogy on his experiences with arms control and nuclear regulatory affairs under Presidents Kennedy, Johnson, and Nixon. Seaborg's interests in the relations of chemical research, science education, and the politics of science are reflected as well in his presidencies of the American Association for the Advancement of Science (1972) and the American Chemical Society (1976), and in his leadership and advisory roles over the years with organizations ranging from the President's Science Advisory Committee and the Robert A. Welch Foundation to Science Service and the Lawrence Hall of Science.

GEORGE F. WOLLASTON
*Clarion University of Pennsylvania*

# *Bibliography*

*Actinides in Perspective*; Edelstein, N. M., Ed.; Pergamon Press: London, 1982.

*Facts on File*. Facts on File: New York, 1961; Vol. xxi, p 25.

Farber, E. *Nobel Prize Winners in Chemistry: 1901–1961;* Abelard-Schuman: Vol. xxi, New York, 1963, 219–231.

Henehan, J. F. "Glenn T. Seaborg, the Man from Ishpeming"; *Chemistry* **1978**, *51, (October)*, 26–28.

*McGraw-Hill Modern Men of Science;* Greene, J. E., Ed.; McGraw-Hill: New York, 1966; Vol. 1, p 423.

*McGraw-Hill Modern Scientists and Engineers;* McGraw-Hill: New York, 1980; Vol. 3, p 89.

*Nuclear Chemistry*; Benchmark Papers in Physical Chemistry and Chemical Physics, Vol. 5; Seaborg, G. T.; Loveland, W., Eds.; Hutchinson Ross Publishing Co.: Stroudsburg, PA, 1982.

*Production and Separation of U-233: Survey*; National Nuclear Energy Series, Manhattan Project Technical Section, Division IV, Vol. 17A; Seaborg, G. T.; Katzin, L. I., Eds.; McGraw-Hill: New York, 1951.

Seaborg, G. T.; "Chemistry—Key to Our Progress"; *Chem. Eng. News,* **1976,** *54,* 31–36.

Seaborg, G. T.; *Early History of Heavy Isotope Research at Berkeley, August 1940 to April 1942*; Lawrence Berkeley Laboratory Report PUB–97; Lawrence Berkeley Laboratory, University of California: Berkeley, CA, 1976.

Seaborg, G. T.; "Early Radiochemical Investigations of Plutonium"; In *Trends in Atomic Physics: Essays Dedicated to Lise Meitner, Otto Hahn, and Max von Laue on the Occasion of Their 80th Birthday*; Frisch, O. R. et al., Eds.; Interscience: New York; Braunschweig: Vieweg, 1959; pp 104–114.

Seaborg, G. T.; "The 40th Anniversary of the Discovery of Americium and Curium"; In *Americium and Curium Chemistry and Technology*; Edelstein, N. M.; Navratil, J. D.; Schulz, W. W., Eds.; D. Reidel: Dordrecht, Holland, 1985; pp 3–17.

Seaborg, G. T.: "Forty Years of Plutonium Chemistry: The Beginnings"; In *Plutonium Chemistry*; Carnall, W. T.; Choppin, G. R., Eds.; ACS Symposium Series 216; American Chemical Society: Washington, DC, 1983; pp 1–22.

Seaborg, G. T. "From Mendeleev to Mendelevium—and Beyond"; In *The Transuranium Elements—The Mendeleev Centennial*; Milligan, W. O., Ed.; Proc. Robert A. Welch Foundation Conferences on Chemical Research, 13;

Seaborg, G. T. *History of Met Lab Section C–I*; Lawrence Berkeley Laboratory Report PUB–112, 4 vols.; Lawrence Berkeley Laboratory, University of California: Berkeley, CA, 1977–1980.

Seaborg, G. T. "History of the Synthetic Actinide Elements"; *Actinides Rev.*, 1967, 1, 3–38.

Seaborg, G. T. *Man-Made Transuranium Elements*; Prentice-Hall: Englewood Cliffs, NJ, 1963.

Seaborg, G. T. "Nuclear Fission and Transuranium Elements—50 Years Ago"; *J. Chem. Ed.*, 1989, 66, 379–381.

Seaborg, G. T. *Nuclear Milestones*; W. H. Freeman: New York, 1972.

Seaborg, G. T. "Nuclear Synthesis and Identification of New Elements"; *J. Chem. Ed.*, 1985, 62, 392–395.

Seaborg, G. T. "The Periodic Table: Tortuous Path to Man-Made Elements"; *Chem. Eng. News*, 1979, 57, 46–52. Priestley Medal address.

Seaborg, G. T. "The Periodic Table of Today"; In *Structure and Dynamics of Chemistry*; Alhberg, P.; Sundeloef, L. O., Eds.; Almqvist & Wiksell: Stockholm, Sweden, 1978; pp 187–196.

Seaborg, G. T. "Reminiscences about the Joliots and Artificial Radioactivity"; In *Le Radioactivité Artificielle a 50 Ans, 1934–1984;* Editions de Physique: Paris, France, 1984; pp 79–90.

Seaborg, G. T. "Reminiscences on the Discovery of Berkelium and Californium"; In *Proceedings of a Symposium in Commemoration of the 25th Anniversary of Elements 97 & 98*; Lawrence Berkeley Laboratory Report LBL–4366; Lawrence Berkeley Laboratory: Berkeley, CA, 1976; pp 2–6.

Seaborg, G. T. "Some Recollections of Early Nuclear Age Chemistry"; *J. Chem. Ed.*, 1968, 45, 278–289.

Seaborg, G. T. *The Transuranium Elements*; Yale University Press: New Haven, CT, 1958.

Seaborg, G. T. "The Transuranium Elements"; *J. Chem. Ed.*1985, 62, 463–467.

Seaborg, G. T. "Transuranium Elements: A Half Century"; In *Transuranium Elements: A Half Century*; Morss, L. R.; Fuger, J., Eds.; American Chemical Society: Washington, DC, 1992; pp 10–49.

Seaborg, G. T. "The 25th Anniversary of the Discovery of Americium and Curium (Elements 95 and 96)—25 Years Ago"; In *The Transuranium Elements—The Mendeleev Centennial*; Milligan, W. O., Ed.; Proc. Robert A. Welch Foundation Conferences on Chemical Research, 13; Robert A. Welch Foundation: Houston, TX, 1970; pp 223–241.

Seaborg, G. T.; Corliss, W. R. *Man and Atom*; E. P. Dutton: New York, 1971.

Seaborg, G. T.; Katz, J. J. *The Chemistry of the Actinide Elements*; John Wiley and Sons: New York, 1958.

Seaborg, G. T.; Katz, J. J. "Remembering the Early Days of the Met Lab"; In *Fifty Years with Transuranium Elements*; Proc. Robert A. Welch Foundation Conferences on Chemical Research, 34; Robert A. Welch Foundation: Houston, TX, 1990; pp 224–251.

Seaborg, G. T.; Loeb, B. S.; *The Atomic Energy Commission under Nixon: Adjusting to Troubled Times*; St. Martin's Press: New York, 1993.

Seaborg, G. T.; Loeb, B. S.; *Kennedy, Krushchev & the Test Ban*; University of California Press: Berkeley, CA, 1981.

Seaborg, G. T.; Loeb, B. S.; *Stemming the Tide: Arms Control in the Johnson Years*; Free Press: New York, 1987.

Seaborg, G. T.; Loveland, W. D.; *The Elements Beyond Uranium;* John Wiley and Sons: New York, 1990.

Seaborg, G. T.; Wilkes, D. M. *Education and the Atom*; McGraw-Hill: New York, 1964.

*Transuranium Elements: Products of Modern Alchemy*; Benchmark Papers in Physical Chemistry and Chemical Physics, Vol. 1; Seaborg, G. T., Ed.; Dowden, Hutchinson, & Ross: East Stroudsburg, PA, 1978.

*The Transuranium Elements: Research Papers*; National Nuclear Energy Series, Manhattan Project Technical Section, Division IV, Plutonium Project Record, Vol. 14B; Seaborg, G. T.; Katz, J. J.; Manning, W. M., Eds.; McGraw-Hill: New York, 1949.

*Who's Who in Frontier Science and Technology;* A. Hast, Ed.; Marquis Who's Who: Chicago, IL, 1984; 1st ed., p 652.

*The Who's Who of Nobel Prize Winners;* Schlessinger, B.; Schlessinger, J. H., Eds.; Oryx Press: Phoenix, AZ, 1986; p 20.

# 1952

NOBEL LAUREATE

# Archer John Porter Martin

## 1910–

The use of chromatography can be traced back to a Russian botanist, M. Tswett, who separated plant pigments by this method in 1906. The rapid progress of this technique in the 1940s and early 1950s can be attributed, partly, to the pioneering work done by British biochemists Archer John Porter Martin and Richard Laurence Millington Synge. Martin and Synge extended the technique of countercurrent extraction to what is known as liquid–liquid partition chromatography in 1941. Their achievement in developing a new form of chromatography employing two immiscible liquids as stationary and mobile phases was recognized with the awarding of the 1952 Nobel Prize in chemistry. In addition, the credit for proposing a new form of gas chromatography, gas–liquid partition chromatography, which is one of the most widely used techniques to separate volatile compounds today, also goes to Martin and Synge. Gas–liquid partition chromatography was later developed by Martin with A. T. James in 1951.

Martin was born on March 1, 1910, in London. He is the only son of the four children born to Lilian Kate Brown, a nurse, and William Archer Porter Martin, a physician. He attended Bedford School in Bedford from 1921 to 1927. As a schoolboy, Martin was fascinated by the fractional distillation method of separation, and he built a frac-

352

tional distillation apparatus in the basement of his home using soldered coffee tins, with their bottoms removed, as the column and coke as the packing material. In 1927 he entered Peterhouse in Cambridge and studied chemistry, physics, mathematics, and mineralogy for the first two years. He took up biochemistry beginning in his third year, after meeting J. B. S. Haldane, who was the first head of the Lister Institute and who inspired an interest in the biological sciences in young Martin. He received a bachelor's degree in biochemistry in 1932.

As a Ph.D. student, Martin worked for a year in the physical chemistry department at Cambridge under C. P. Snow and T. Boden. In 1933 he switched to the Dunn Nutritional Laboratory of Cambridge University and developed an interest in the study of vitamin E. He received a master's degree in 1935 and a doctorate in 1936 from Cambridge. In 1938 he joined the Wool Industries Research Association Laboratory at Leeds as a biochemist. In 1946 he accepted the position of Head of the Biochemical Research Division of the Boots Pure Drug Company in Nottingham. From 1948 to 1959 he worked for the (British) Medical Research Council at Mill Hill. In 1959 he founded the Abbotsbury Laboratories in Borehamwood. He worked for the Wellcome Foundation as a consultant from 1970 to 1973. From 1973 to 1978, he worked as a professor of chemistry for the Medical Research Council of the University of Sussex. In 1974 he came to the United States and served as Robert A. Welch Professor of Chemistry at the University of Houston, Texas, until 1979 (part-time, 1974–1978). He served as a visiting professor at the Ecole Polytechnique Federale de Lausanne, Switzerland, from 1980 to 1983.

As a researcher at the Dunn Nutritional Laboratory of Cambridge, Martin gained valuable experience in the technique of separating and isolating closely related, biologically active compounds using countercurrent fractional extraction. As a research student, he built a liquid–liquid countercurrent extraction apparatus that was approximately equivalent to 200 separatory funnels. The liquid–liquid countercurrent extraction technique utilizes the partitioning of a solute between two immiscible solvents. As suggested by Sir Charles Martin, A. J. P. Martin's research adviser, he joined hands with Richard Synge, another gifted graduate student at Cambridge, to develop a separation technique for acetylamino acids. When their attempt to separate amino acids using countercurrent extraction failed to give satisfactory results, they decided to hold one liquid phase stationary on an inert support and move the other immiscible liquid. Thus a new type of chromatography germinated at the Wool Industries Research Association Laboratories. They used silica gel to hold the stationary phase (water) with chloroform as the mobile phase. They used methyl orange, which forms bright red bands over an orange background showing the separation of different acetylamino acids, as the indicator. Perhaps Martin and Synge made an even greater contribution to chromatography by suggesting that "the mobile phase need not be a liquid but may be a vapor".[1] Nobody picked up this suggestion until 1951.

In the early 1950s, Martin himself started working on this idea along with A. T. James at the Medical Research Council and developed what is known as gas–liquid partition chromatography.[2] Gas–liquid partition chromatography uses a gas, usually nitrogen or helium, as the mobile phase and a liquid as the stationary phase. The liquid stationary phase is coated on an inert solid stationary support (packed column gas–liquid chromatography) or on the inner walls of a capillary column (capillary column gas–liquid chromatography). In modern gas–liquid chromatography (GLC) columns, the stationary phase is chemically bonded to the inert stationary support or to the inner walls of the capillary column. The chemically bonded stationary phases have much higher physical stability than the stationary phases held by physical means such as adsorption. The separation of the components of a sample on a GLC column is based on the difference in their partition ratios between the stationary and mobile phases. Gas–liquid chromatography has now become an indispensable analytical instrument in most chemical laboratories.

In addition, Martin has been instrumental in developing the technique of paper chromatography, which is a simple but effective method for separating the components of a sample. In this method, a small volume solution of a sample is placed near the edge (approximately an inch from the edge) of a rectangular paper, in which the paper serves as the stationary phase. Then the edge near which the sample is placed is immersed in a suitable liquid mobile phase (the sample spot should not touch the liquid phase). The mobile phase is drawn up by the capillary forces. As the mobile phase advances, the sample components resolve depending on the difference in their partition ratios between the stationary and mobile phases.

---

[1] *Biochem. J.* **1941,** *35,* 1358–1368.

[2] *Biochem. J.* **1952,** *50,* 679–690.

Dr. Martin was elected a member of the Royal Society in 1950 and was awarded the Berzelius Medal of the Swedish Medical Society in 1951.

Acknowledgment: The author thanks Dr. Nora Wooster, A. J. P. Martin's sister, for her contribution.

PRABHAKARA H. SHETTY
*Ferris State University*

# *Bibliography*

"Interview with Archer J. P. Martin"; *J. Chem. Ed.* **1977**, *54*, 80–83.

James, A. T.; Martin, A. J. P. "Gas-liquid Partition Chromatography: the Separation and Micro-estimation of Volatile Fatty Acids from Formic Acid to Dodecanoic Acid"; *Biochem. J.* **1952**, *50*, 679–690.

Martin, A. J. P. "The Development of Partition Chromatography"; *Nobel Lectures: Chemistry, 1942–1962*; Elsevier: Amsterdam, the Netherlands, 1964; pp 359–371.

Martin, A. J. P.; Synge, R. L. M. "A New Form of Chromatography Employing Two Liquid Phases: 1. A Theory of Chromatography, 2. Application to the Micro-Determination of the Higher Monoamino Acids in Proteins"; *Biochem. J.* **1941**, *35*, 1358–1368.

Stahl, G. A. *A History of Analytical Chemistry*; Laitinen, H.A.; Ewing, G.W., Eds.; American Chemical Society: Washington, D.C., 1977; pp 296–321.

# Richard Laurence Millington Synge

## 1914–

Copyright Nobel Foundation

Richard Laurence Millington Synge is one of the leading biochemists of the twentieth century. He shared the 1952 Nobel Prize in chemistry with A. J. P. Martin for developing the technique of liquid–liquid partition chromatography in 1941 while working at the Wool Industries Research Association laboratory in Leeds, England. Synge later utilized this technique to separate the 20 amino acids, the building blocks of proteins and other biologically important compounds, and to elucidate the structure of an antibiotic known as gramicidin S.

Synge was born on October 28, 1914, in Liverpool, England. He is the eldest and the only son of three children born to Laurence Millington Synge, a stockbroker, and Katharine Charlotte Swan. He attended Winchester College in Hampshire, England, from 1928 to 1933, where he mainly studied classics until 1931 and natural sciences thereafter. In 1931 Synge won an Exhibition in classics from Trinity College of Cambridge University. He received a baccalaureate degree in 1936 and a Ph.D. in 1941 in biochemistry, both from Cambridge University. His graduate research at the Cambridge Biochemical Laboratory from 1936 to 1939 was supervised by N. W. Pirie. The University Biochemical Laboratory was headed by Sir Frederick G. Hopkins, another noted contemporary biochemist. Hopkins had shared the 1929 Nobel Prize

in physiology or medicine for his pioneering work on vitamins, carbohydrate metabolism, and muscular activity. Synge's graduate research
was in the area of separating acetylamino acids. About the same time,
A. J. P. Martin, at the Dunn Nutritional Laboratory of Cambridge
University, had built an extraction apparatus for the purification of
Vitamin E. Even though Martin's extraction apparatus was found unsuitable for the separation of amino acids, their common interest in the
area of separations led to one of the landmark collaborative research
achievements, the invention of liquid—liquid partition chromatography.[1] Additionally, part of the credit for the invention of gas-liquid
partition chromatography, the most widely used technique for the separation of volatile compounds today, goes to Synge. The idea of using
a vapor or gas as the mobile phase instead of a liquid in partition chromatography occurred during the course of his collaborative work with
A. J. P. Martin.

In 1941 Synge joined the Wool Industries Research Association
laboratory as researcher and continued working on the project of developing partition chromatography with Martin, who was also working
at the same laboratory. From 1943 to 1948, he worked for the Lister Institute of Preventive Medicine, in London, as a researcher. While there
Synge worked on the chemistry of antibiotic peptides. He collaborated with Martin in developing paper partition chromatography using
a two-dimensional development technique. In a two-dimensional development technique, the sample on a plane surface (stationary phase:
paper or a thin layer of an absorbent) is developed in two directions,
perpendicular to one another, using two different mobile phases. Two-
dimensional development of the stationary phase yields better and
sharper separation of the components in a sample. In addition Synge
had the opportunity to work with another eminent scientist, Professor
A. Tiselius (Nobel Prize in chemistry, 1948, for his pioneering work on
adsorption and electrophoresis) at the University of Uppsala, Sweden,
for eight months (1946–1947). From 1948 to 1967 he served the Rowett
Research Institute, in Aberdeen, Scotland, as a researcher and as an
administrator. For about one year (1958–1959) Synge worked with E.
P. White at the Ruakura Animal Research Station, at Hamilton, New
Zealand, on isolation of the toxic fungal component sporidesmin. In
1967 he joined the staff of the Food Research Institute, in Norwich,

---

[1] *Biochem. J.* **1941,** *35,* 1358–1368.

England, as a researcher. From 1968 to 1984, Synge served as Honorary Professor at the School of Biological Sciences at the University of East Anglia, England.

Dr. Synge is a Fellow of the Royal Society and the Royal Society of Chemistry. He is an honorary member of the American Society of Biological Chemists.

**PRABHAKARA H. SHETTY**
*Ferris State University*

# Bibliography

*A History of Analytical Chemistry;* Laitinen, H. A., Ewing, G. W., Eds.; American Chemical Society: Washington, D. C., 1977; pp 296–321.

Martin, A. J. P.; Synge; R. L. M. "A New Form of Chromatogram Employing Two Liquid Phases: 1. A Theory of Chromatography. 2. Application to the Micro-Determination of the Higher Monoamino-Acids in Proteins"; *Biochem. J.* **1941,** *35,* 1358–1368.

*Nobel Prize Winners;* Wasson, T., Ed.; H. W. Wilson: New York, 1987.

Synge, R. L. M. "Applications of Partition Chromatography"; *Nobel Lectures: Chemistry, 1942–1962;* Elsevier: Amsterdam, the Netherlands, 1964; pp 347–387.

*The Who's Who of Nobel Prize Winners;* Schlessinger, B. S., Schlessinger, J. H., Eds.; Oryx Press: Phoenix, AZ, 1986.

# Hermann Staudinger

## 1881–1965

Hermann Staudinger was born in Worms, Germany, on March 23, 1881, and died at Freiburg in Breisgau, Germany, on September 8, 1965. He was awarded the Nobel Prize in 1953 in recognition of his pioneering work, begun some 35 years earlier, in macromolecular chemistry.

Staudinger's private life has not been much reported, at least in the popular literature. He preferred to write his memoirs in the form of a "work remembrance", which appeared in the United States in 1961 when he was 70 years of age. The memoirs scrupulously avoided personal detail in favor of recounting the way that his work developed.[1]

His parents were Franz Staudinger, a professor of philosophy, and Auguste (Wenck) Staudinger. Young Staudinger graduated from the *Gymnasium* at Worms at age 18; at the time his interests lay in botany and microscopy. However, under the influence of his father, he studied chemistry at the University of Halle in order to better understand botany. When his father moved to a teaching post in Darmstadt in 1899, Staudinger transferred his studies there. He returned to Halle after a short period in the Baeyer laboratory at Munich, and in 1903 he completed a dissertation under D. Vorlander on the addition products of malonic esters. After a semester as private assistant to O. Dobner at

---

[1]Staudinger, H. *From Organic Chemistry to Macromolecules, A Scientific Autobiography;* Wiley: New York, 1961.

Halle, Staudinger became assistant to Johannes Thiele at the University of Strasbourg in the autumn of 1903. It was in Strasbourg that Staudinger began his important work on the highly reactive ketenes and, perhaps more importantly, formed the habits of diligence and punctuality that characterized his scientific work for the rest of his life. In 1907 he began a five-year period as associate professor under the noted chemical technologist Carl Engler, at the Technical University of Karlsruhe. Engler was consultant to Badische Anilin und Soda Fabrik (BASF), a company very interested in synthetic rubber. Hence, as one of his research projects Staudinger studied isoprene, the basic unit of rubber, and developed new processes for its synthesis that led to several patents. The association with Engler had a considerable influence on Staudinger beyond merely the conduct of chemical research. Years later when he was directing his own institute, if faced with an administrative problem, he would ask himself, "How would Engler have decided in this particular case?"[2]

In 1912 he made the important move to the Federal Institute of Technology at Zurich, where he succeeded Richard Willstätter as head of the general chemistry division. As a part of the new position, Staudinger was obliged to give lectures in organic chemistry, supervise study hours, and administer numerous examinations. Fritz Haber had warned him that his shining time of undisturbed work was over.

Yet the move to Zurich was made easier because Staudinger was able to take with him some outstanding co-workers including R. Endle, Emil Ott, and Leopold Ružička (Nobel Prize in chemistry, 1939). During the 12 years at Zurich he continued his research on ketenes and, with Ružička, studied a natural insecticide, the pyrethrines. During World War I he synthesized a pepper substitute with H. Schneider and isolated compounds that gave the odor to coffee with Tadeus Reichstein (Nobel Prize in physiology or medicine, 1950). Most importantly, he continued the work on the polymerization of isoprene that he had begun back at Karlsruhe. Around 1920 he started his investigation of macromolecular compounds, especially polyoxymethylene, natural rubber, and polystyrene.

In 1926 he was called to Freiburg as Director of Chemical Laboratories, succeeding Heinrich Wieland (Nobel Prize in chemistry, 1927), who had moved to Munich. Staudinger was to remain at Freiburg for the rest of his career, retiring in 1951 at age 70. The move from the ex-

---

[2]Yarsley, V. E., "Hermann Staudinger—His Life and Work"; *Chem. & Ind.* **1967**, 252.

cellent and well-equipped laboratories in Zurich involved a number of difficulties because the facilities at Freiburg were badly in need of modernization. The promised refitting and enlargement of the laboratories to house research in the new and expanding field of macromolecular chemistry did not occur until 1933 and 1937. During World War II, research at Freiburg became increasingly difficult to accomplish. It was finally brought to a halt by Allied bombing and the nearly complete destruction of the chemical laboratories on November 27, 1944.

Many important manuscripts were lost, but important papers stored in Staudinger's home were preserved. Finally, in 1947 the department was able to restore teaching and research activities, but the aging Staudinger was not able to pursue his long-held interests in the structure of biological macromolecules. He became Professor Emeritus in the spring of 1951, and in his later years devoted himself to writing and to the editorship of the journal he founded, *Die makromolekulare Chemie.*

Staudinger's scientific contributions were wide-ranging, beginning with extensive studies of the ketenes, which have been summarized in a book.[3] In 1910 at Karlsruhe he examined the diazo compounds with the goal of obtaining various methylene derivatives as decomposition products. Beginning in 1919 with co-worker J. Meyer he attempted to prepare pentavalent phosphorus compounds analogous to those of nitrogen studied earlier. The program was interrupted around 1924 so that Staudinger could devote all of his attention to high molecular weight compounds, which were assuming increasing importance. Later, the work on organic phosphorus compounds was taken up by others, notably Georg Wittig's group (Nobel Prize in chemistry, 1979).

Early in his career Staudinger studied oxalyl chloride with the goal of preparing the simplest ketene, the dimer of carbon monoxide, $O=C=C=O$. Pure oxalyl chloride was unknown until Staudinger prepared it by the reaction of phosphorous pentachloride with oxalic acid. The work was patented and represented one of the rare cases in which Staudinger became involved in industrial production. Oxalyl chloride and its derivatives were often very explosive, a fact that led Staudinger to examine the subject of chemical explosions in detail. He was interested in the pyrolytic decomposition of terpenes, such as limonene and dipentane, which yielded isoprene.

---

[3]Staudinger, H. *Die Ketene;* Verlag Enke: Stuttgart, 1912.

It was in Zurich in the 1920s that Staudinger began to concentrate on the main work of his life, the study of what were then called "high molecular compounds". After he moved to Freiburg in 1926, Staudinger concentrated exclusively on such materials. This was an act of great courage because the organic chemistry establishment of the time viewed the field with open contempt.[4] Even Staudinger wrote that attempts to work with such poorly defined, uncrystallizable substances were often called "grease chemistry". His distinguished colleague, Wieland, advised him, "Dear colleague: Drop the idea of large molecules; organic molecules with a molecular weight higher than 5000 do not exist. Purify your products, for example, your rubber, then it will crystallize and prove to be a low molecular compound".[5] Albert Frey-Wyssling is quoted as reporting that during one of Staudinger's lectures to the Zurich Chemical Society in 1925, he was attacked vigorously for his ideas about long-chain molecules by a number of great scientists until Staudinger repeated in defiance Martin Luther's famous pronouncement, "Here I stand and can do no other".[6]

Staudinger had to combat the commonly held opinions of a number of leading scientists that natural materials such as cellulose, starch, protein, and rubber were colloidal aggregates of small molecules. For instance, Reginald Herzog held that cellulose and the protein, silk fibroin, must consist of molecules smaller than the crystallographic unit cell. Paul Karrer (Nobel Prize in chemistry, 1937), Staudinger's former colleague at Zurich, was convinced that cellulose and starch had to be aggregates of small molecules. K. Hess held that the polymerization and depolymerization of cellulose were physical processes of association and dissociation. Max Bergmann believed that high molecular weight natural products consisted of the "three-dimensional ordering of simple units which are unstable when isolated". He also believed that proteins were aggregates that had to be dissociated by special enzymes before they could be hydrolyzed by other enzymes.[7] Rudolf Pummerer in the period from 1924 to 1928 published determinations of molecular weights of rubber indicating that it was composed of low

---

[4]Morawetz, H. *Polymers: The Origins and Growth of a Science*; Wiley: New York, 1985, p 86.

[5]Staudinger, H. *From Organic Chemistry to Macromolecules, A Scientific Autobiography*; Wiley: New York, 1961; p 79.

[6]Olby, R. *The Path to the Double Helix*; University of Washington Press: Seattle, WA, 1974; p 7.

[7]Morawetz, H. *Polymers*, p 89.

molecular weight molecules, in agreement with the ideas of the leading German expert on rubber, Carl Harries.

Staudinger's early work on isoprene and his interest in the structure of natural rubber may have fueled his later speculations that high molecular weight compounds consisted of long-chain molecules held together by ordinary covalent bonds. He first expressed this idea in a lecture on rubber to the Swiss Chemical Society in 1917 and amplified it in 1920. He asserted that unsaturated molecules combine in a manner similar to that of the formation of four- or six-membered rings but that for some reason, possibly steric, ring formation fails to occur. Instead, large numbers, possibly hundreds, of molecules assemble until equilibrium is established.

Working with J. Fritschi, Staudinger reduced natural rubber with hydrogen in an autoclave. Contrary to the expectations of proponents of the colloidal aggregate theory, the product was similar in properties to the original and could not be distilled in vacuum. Thus Staudinger concluded that natural rubber consisted of long-chain molecules and in the same paper originated the term "macromolecules" to apply to covalently bonded molecules the size of colloidal particles.[8]

With clear insight, Staudinger chose to focus his studies on model compounds that were easier to prepare as pure, well-defined materials. For example, he studied polyoxymethylene (or paraformaldehyde, as it was then known) as a model for cellulose. Formaldehyde polymers with a wide range of chain lengths were made, allowing Staudinger to examine the effects of degree of polymerization. Polystyrene was chosen as a model for the much too easily oxidized *Hevea* rubber (natural rubber). Some choices of systems—such as acrylic polymers as models for proteins—seem inappropriate in the light of modern knowledge.

Studies of polyoxymethylene begun at Zurich were continued after the move to Freiburg. The work of six of Staudinger's students was summarized in an extensive publication in 1929. It is considered to be the only instance in which the chemical and physical properties of a homologous series extending from short-chain molecules to high polymers were reported. Among the important conclusions were the following: First, the polymer consisted of chain molecules joined by covalent bonds and was not an association colloid. Second, at the ends of the chains were characteristic functional groups and not the "free valences" Staudinger had proposed earlier. Third, the analysis of end-

---

[8]Staudinger, H.; Fritschi, J. *Helv. Chim. Acta* **1922**, *5*, 785–809.

group concentrations made possible calculation of the average chain length of insoluble high polymers. And fourth, crystallographic studies showed that long-chain molecules could crystallize and that chain length bore no relation to the size of the unit cell.[9]

Staudinger found that polystyrene prepared under varying conditions had different physical properties and that such preparations could be fractionated into different molecular weight groupings. This was again inconsistent with the idea that the material was an association colloid. Staudinger continued to study various high molecular weight materials, and gradually opposition to his ideas weakened.

He used all available means of molecular weight determination including the electron microscope, the ultramicroscope, end-group determination, cryoscopic methods, osmotic pressure, and solution viscosity determination. The Svedberg (Nobel Prize in chemistry, 1926) and J. Burton Nichols developed the ultracentrifuge, permitting determination of the molecular weight of the protein hemoglobin from its equilibrium distribution in a centrifugal field. Refused funds to purchase his own badly needed ultracentrifuge, Staudinger turned instead to the simpler but less accepted determination of solution viscosity as his routine method of molecular weight determination. Molecular weights from viscosity measurements are usually higher than those based on cryoscopic or osmotic procedures because of influences of chain length.

Staudinger's well-known viscosity formula relating the solution viscosity to molecular weight of a macromolecular solute is given in nearly every textbook on macromolecules or polymer chemistry. The original expression, based on work with Werner Heuer from 1929 to 1932, was written simply as the specific viscosity (solution viscosity minus the solvent viscosity, all divided by the solvent viscosity) over the concentration of solution equaling a constant times the molecular weight of the solute. Staudinger considered the macromolecular solute or polymer molecules to be rigid rods, a considerable oversimplification. Nevertheless, Staudinger's equation advanced the study of macromolecules in that it represented a simple way of characterizing them as to chain length, even though the values calculated might be in error by significant amounts. Later the formula was revised to try to account for intermolecular interactions and for differences among types of molecules.

---

[9]Morawetz, H. *Polymers;* p 92–93.

When Staudinger originally offered his equation during a lecture at Frankfurt in 1929, there was for the first time no opposition, and in 1930 a meeting of the Colloid Society revealed the increasing acceptance of the idea of long-chain molecules. Wolfgang Ostwald, a leader in colloid theory, spoke of "chemical colloids or molecular colloids as particles held together by principal valences". Proponents of a colloidal aggregate or micellar theory of polymer solutions were not totally convinced, however, and a long, acrimonious controversy continued with Kurt H. Meyer, who insisted that long-chain polymers aggregated in dilute solutions. By 1935 Staudinger's ideas were largely accepted, allowing his efforts to be concentrated on extending the field of macromolecular chemistry rather than engaging in further polemical defenses of it. In 1939 Staudinger assumed editorship of the long-established *Journal für praktische Chemie,* published by Barth in Leipzig, and included macromolecular chemistry in its scope by 1940, eventually changing its title to *Journal für Makromolekulare Chemie.* After the war, when publication from Leipzig became impossible, he maintained continuity by starting a new journal, *Makromolekulare Chemie,* published by Wepf in Basel.

A tall, robust, yet genial and soft-spoken man, Staudinger was professionally a researcher, according to Herman Mark, a researcher, teacher, and apostle. He carried out extensive work in low molecular weight organic chemistry and was a true pioneer in macromolecular chemistry. He nurtured a generation of macromolecular chemists as teacher and colleague. He was also, Mark says, an apostle of a new creed, a role he neither chose nor sought. Staudinger often remarked that he would rather pursue his own work quietly in his laboratory or, indeed, tend his garden flowers, than forever have to defend his theories or take time to counter the erroneous work of others. His time in the laboratory was continually interrupted by the need to give lectures and to study the publications of critics, including Meyer and Mark, who supported the concept of long-chain molecules but whose ideas differed in detail from Staudinger's.

One of Staudinger's most important collaborators and supporters was his wife, Magda Woit Staudinger, a Latvian plant physiologist whom he married in 1928. She coauthored a number of publications, conducted morphological studies of macromolecules, and introduced phase-contrast microscopy and electron microscopy into macromolecular studies at Freiburg, beginning in 1937. The couple had one son, Hansjürgen, who became a medical doctor and professor of physiological chemistry.

When the Nobel Prize in chemistry came to Staudinger in 1953, he was two years past retirement and 72 years of age. An American news magazine headed its brief article reporting the award, "Better Late than Never". Many of his contemporaries had been honored decades earlier. Nonetheless, Staudinger was very appreciative of the honor. In his acceptance speech he expressed satisfaction that the field of macromolecular chemistry was being recognized, and he paid generous tribute to his many co-workers.

Although according to his wife Staudinger was never greatly interested in industrial processes and commercial applications of his work, the principles of macromolecular chemistry he developed and his fundamental studies of polystyrene, the polyesters, polyamides, vinyl polymers, amino plastics, and many others laid the foundation for the giant plastics industry of today.

He spent his last years in seclusion at his little house in Freiburg and died on September 9, 1965, at the age of 84.

LAYLIN K. JAMES
*Lafayette College*

# Bibliography

Farber, E. *Nobel Prize Winners in Chemistry, 1901–1961;* Abelard-Schuman: London, 1963.

Furukawa, Y. "Staudinger, Carothers, and the Emergence of Macromolecular Chemistry"; Ph.D. Thesis, University of Oklahoma, 1983.

Kern, W. "Zur Entwicklung der makromolekularen Chemie: H. Staudinger zum 70 Geburtstag"; *Angew. Chem.* **1951,** *63,* 229–231.

Livio, R. *Hermann Staudinger: Decouvreur de l'univers macromoleculaire;* Éditions da Pont royal: Paris, 1961.

Morawetz, H. *Polymers: The Origins and Growth of a Science;* Wiley: New York, 1985.

Morris, P. J. *Polymer Pioneers: A Popular History of the Science and Technology of Large Molecules;* Center for History of Chemistry: Philadelphia, PA, 1986.

*Nobel Prize Winners;* Wasson, T., Ed.; H. W. Wilson: New York, 1987.

*The Nobel Prize Winners: Chemistry;* Magill, F. N., Ed.; Salem: Pasadena, CA, 1990; Vol. 2.

Olby, R. "Staudinger, Hermann"; In *Dictionary of Scientific Biography;* Gillispie, C. C., Ed.; Charles Scribner's Sons: New York, 1976; Vol. 13, pp 1–4.

Olby, R. "The Macromolecular Concept and the Origins of Molecular Biology"; *J. Chem. Ed.* **1970**, *47*, 169–172.

Priesner, C. *H. Staudinger, H. Mark, and K. H. Meyer: Thesen zum Grösse und Struktur der Makromoleküle;* Verlag Chemie: Weinheim, 1980.

Quarles, W. "Hermann Staudinger: Thirty Years of Macromolecules"; *J. Chem. Ed.* **1951**, *28*, 120–122.

Staudinger, H. *Arbeitserinnerungen;* A. Hüthig: Heidelberg, 1961.

Staudinger, H. *Die hochmolekularen organischen Verbindungen*; Springer Verlag: Berlin, Germany, 1932; new edition, Edwards Brothers: Ann Arbor, MI, 1960.

Staudinger, H. *From Organic Chemistry to Macromolecules, A Scientific Autobiography;* Wiley: New York, 1961.

Staudinger, H. "Macromolecular Chemistry"; *Nobel Lectures: Chemistry, 1942–1962;* Elsevier: New York, 1964; pp 393–421.

*Who's Who of Nobel Prize Winners;* Schlessinger, B. S.; Schlessinger, J., Eds.; Oryx Press: Phoenix, AZ, 1986.

Yarsley, V. E. "Hermann Staudinger"; In *Pioneers of Polymers*; Plastics and Rubber Institute: London, 1981.

Yarsley, V. E. "Hermann Staudinger—His Life and Work"; *Chemistry & Industry* **1967,** 252–271.

# Linus Carl Pauling

## 1901–

Copyright Nobel Foundation

Linus Carl Pauling received the 1954 Nobel Prize in chemistry "for his research into the nature of the chemical bond and its application to the elucidation of the structure of complex substances." He was born in Portland, Oregon, on February 28, 1901, the first of three children and the only son of Herman W. Pauling, a pharmacist, and Lucy Isabelle (Darling) Pauling, a daughter of a pharmacist. He spent his early years in Condon, an arid Western town in the interior of Oregon where his father owned a drugstore and where young Linus encountered cowboys, one of whom showed him the proper way to sharpen a pencil with a knife, and Indians, one of whom showed him how to dig for edible roots. These two things impressed him deeply: that there was a correct technique for doing things and that there were people who had useful knowledge of nature. Condon, however, proved to be too restrictive on the ambitions of the energetic Herman Pauling, and in 1909 he moved his family to Portland. Not long after his new drugstore began to prosper, he died suddenly of a perforated gastric ulcer. He was only 33 years old.

Herman's death caused grave economic distress for his family, and "Belle" Pauling began to experience problems with her health. Linus became a shy adolescent who spent much of his time reading. In the light of his later career, the most important event in his youth occurred when Lloyd Jeffress, a friend, showed him how sulfuric acid could turn white sugar into a steaming mass of black carbon. So enthralled was Pauling by this reaction that he decided then and there to become a chemist. He began by studying one of his father's old chemistry books, and he set up a primitive laboratory in the basement of his home. At Washington High School in Portland, he took all the courses in science and mathematics available to him, and at Oregon Agricultural College (now called Oregon State University) he majored in chemical engineering. He was forced to leave college for a year to provide financial assistance for his mother and sisters, but after several months' service as a paving-plant inspector, he was able to return to Corvallis as an instructor in quantitative analysis. This hiatus from his heavy load of college courses gave him the opportunity to study the papers of G. N. Lewis and Irving Langmuir on the chemical bond.

After graduating summa cum laude from Oregon Agricultural College in 1922, he began his graduate work at the California Institute of Technology (CIT, the term preferred by Pauling), where his career was guided by Arthur A. Noyes, the head of the chemistry division; by Roscoe G. Dickinson, an X-ray crystallographer; and by Richard C. Tolman, a theoretical physical chemist. Dickinson, through the logical method he developed for his diffraction studies, made Pauling an enthusiast of X-ray crystallography. Pauling's first paper resulted from a study of the mineral molybdenite, which turned out to have a trigonal prismatic coordination of sulfur atoms around the molybdenum atom, an arrangement that pleasantly surprised him.[1] After a year at CIT, Pauling returned to Oregon to marry Ava Helen Miller, who had earlier been his student in Corvallis. She returned with him to CIT, where he took a heavy load of courses in physics, chemistry, and mathematics and where he began regularly publishing papers on crystal structures. He also grew interested in theoretical physics and began to carry out research in this field under the inspiration of Tolman, whose course in statistical mechanics so impressed Pauling that he continued to audit it even after he had passed it with high grades. In 1925, using the old quantum theory, Pauling and Tolman published a paper in which they concluded that the entropy of all perfect crystals, no matter how big or small their unit cells, should be zero at absolute zero.[2]

Following a successful defense of a thesis derived from his crystal-structure papers, Pauling was awarded his doctorate summa cum laude

[1]Dickinson, R. G.; Pauling, L. "The Crystal Structure of Molybdenite"; *J. Am. Chem. Soc.* **1923**, *45*, 1466–1471.

[2]Pauling, L.; Tolman, R. C.; "The Entropy of Supercooled Liquids at the Absolute Zero"; *J. Am. Chem. Soc.* **1925**, *48*, 2148–2156.

in 1925. After a brief period as a National Research Fellow at CIT, he spent a year and a half on a Guggenheim Fellowship in Europe to investigate the implications of the new quantum mechanics for the structure of molecules and the nature of the chemical bond. During his studies, which he carried out mainly at Arnold Sommerfeld's Institute for Theoretical Physics in Munich, he was able to apply this new approach to some important chemical problems, for example, the theoretical prediction of the properties of ionic crystals. As a result of this research he wrote a paper, which Sommerfeld himself submitted, that became one of his most cited publications.[3]

After his quantum chemical work in Europe, Pauling returned to CIT in 1927 to begin a long and fruitful career that first focused on the X-ray determination of crystal structures. In the late 1920s and early 1930s, he and his collaborators figured out the structures of many silicates (for example, sodalite and the micas) and several sulfide minerals (for example, sulvanite and enargite). Along with Lawrence Bragg, he made the silicates one of the best understood branches of structural chemistry. He also used his extensive knowledge of bond distances and bond angles to derive what came to be called his "coordination theory" of crystal structures. Before "Pauling's Rules", crystallographers had tried to establish the atomic arrangement of a crystal's unit cell by eliminating all but one of the possible configurations. Pauling decided to discard immediately all chemically unreasonable arrangements as well as those not in agreement with known interatomic distances. His rules (for example, the electrostatic valence principle which regulated local neutralization of charge in the crystal) helped crystallographers to select the most likely atomic arrangements to be tested experimentally.

In 1930, as a consequence of a meeting with Herman Mark in Germany, Pauling became interested in electron diffraction. This technique depends on the ability of the positively charged nuclei of gas molecules to scatter electrons, and it provided Pauling with an excellent way to collect essentially new information about such substances as benzene and cyclohexane. The many structures determined at CIT by X-ray and electron diffraction aided Pauling's theoretical work. For example, he found a way of assigning to elements certain numbers that represented the particular atom's power of attracting electrons in a

---

[3]Pauling, L. "The Theoretical Prediction of the Physical Properties of Many-Electron Atoms and Ions. Mole Refraction, Diamagnetic Susceptibility, and Extension in Space"; *Proc. Royal Soc. (London)* **1927,** *A114,* 181–211.

covalent bond (the so-called electronegativity). He used his electroneg-ativity scale, which chemists eagerly adopted, to determine the amount of ionic and covalent character in various chemical bonds.

Resonance was a central idea in his theoretical considerations dur-ing the 1930s. He used the interchange (or resonance) energy of two electrons in his treatment of bond hybridization, which was a pivotal idea in one of his most famous papers: "The Nature of the Chemical Bond. Application of Results Obtained from the Quantum Mechan-ics and from a Theory of Paramagnetic Susceptibility to the Structure of Molecules".[4] In resonance, the true state of a chemical system is neither of the component quantum states but some intermediate one, caused by an interaction that lowers the energy. This idea of resonance grew out of his studies in quantum mechanics. Another outcome was a very successful textbook written with one of his graduate students (who became a professor at Harvard), the aim of which was to intro-duce chemists to quantum mechanical reasoning.[5] Pauling's grasp of quantum mechanical principles was also a major factor in his develop-ment of valence-bond theory, in which he proposed that a molecule could be described by an intermediate structure that was a resonance combination (or hybrid) of other structures. His classic book, based on his George Fisher Baker Lectures at Cornell University, provided a unified summary of his own experimental and theoretical studies as well as those of structural chemists from around the world.[6]

The arrival of the geneticist Thomas Hunt Morgan at CIT in the late 1920s contributed to Pauling's growing interest in biologi-cal molecules, and by the mid-1930s he was performing successful magnetic studies on hemoglobin, which appealed to him because of its striking color and ability to combine reversibly with oxygen. He and his collaborators found that a magnet attracted hemoglobin from venous blood, whereas it repelled hemoglobin from arterial blood. Since hemoglobin is a protein molecule, interest in this molecule led logically to a more general interest in proteins, and with Alfred Mirsky he published a paper on protein structure in which they explained how a protein molecule is coiled into a specific configuration (stabilized by

[4] *J. Am. Chem. Soc.* **1931**, *53*, 1367–1400.

[5] Pauling, L.; Wilson, E. B., Jr. *Introduction to Quantum Mechanics, with Applications to Chemistry*; McGraw-Hill: New York, 1935.

[6] Pauling, L. *The Nature of the Chemical Bond and the Structure of Molecules and Crystals: An Introduction to Modern Structural Chemistry*; Cornell University Press: Ithaca, NY, 1939.

hydrogen bonds and weak intermolecular forces) and how it is denatured (when these bonds are broken and the molecule assumes a more random configuration).[7]

On one of his trips to the Rockefeller Institute for Medical Research in New York City to visit Mirsky, Pauling met Karl Landsteiner, the discoverer of blood types, who became his guide into the field of immunochemistry. Pauling was fascinated with antibody–antigen reactions, and he developed an influential theory of antibody structure.[8] In this, his first paper on antibodies, he offered an explanation of how the antibody protein molecule could obtain a unique three-dimensional configuration that would provide it with specificity for a particular antigen. In general, he proposed that this specificity was due to a unique folding of the antibody's polypeptide chain. In particular, he assumed that the ends of the polypeptide chain played a pivotal role in the process, because when these ends encountered an antigen molecule (which serves as a template), they assumed configurations complementary to the surface regions of the antigen. Thus, the specificity of interaction of antigen and antibody molecules is rooted in their structural complementarity.

During the Second World War, Pauling's work shifted toward the practical problems generated by the conflict. For example, he discovered an artificial substitute for blood serum that made more plasma available to wounded soldiers, and he invented an oxygen detector, a device based on oxygen's special magnetic properties, that found wide use in submarines and airplanes. He also spent time studying explosives, rocket propellants, and inks for secret writing. J. Robert Oppenheimer asked him to head the chemistry section of the atomic-bomb project at Los Alamos, but both the burden of the war work he had already assumed and the lingering effects of the glomerular nephritis that had earlier nearly ended his life induced him to refuse the offer. For his outstanding services during the war, he was later awarded the Presidential Medal for Merit.

Near the end of the war, owing to an encounter with Dr. William B. Castle, a co-worker on the Bush Report (whose central concern was to plan the evolution of American science in the postwar period), Pauling became interested in sickle-cell anemia. When he learned that the

---

[7]Mirsky, A. E.; Pauling, L. "On the Structure of Native, Denatured, and Coagulated Proteins"; *Proc. Nat. Acad. Sci.* (USA) **1936,** *22,* 439–447.

[8]Pauling, L. "A Theory of the Structure and Process of Formation of Antibodies"; *J. Am. Chem. Soc.* **1940,** *62,* 2643–2657.

red blood cells of patients with this hereditary disease became sickle-shaped only in venous blood, with its low oxygen content, he immediately speculated that the sickling was caused by a genetic mutation in the globin portion of the cell's hemoglobin. Three years later, he and his collaborators were able to prove that a defect in the protein portion of hemoglobin was indeed responsible.[9]

During these same postwar years Pauling developed an important theory of metals. He viewed the metallic bond as closely related to the covalent bond. Indeed, he thought that each atom in a metal could be considered as forming resonating covalent bonds with neighboring atoms. He even formulated an equation to express the change in covalent radius of an atom with a change in bond number, and he used his equation to derive interatomic distances for the elementary metals (chemists extensively applied his set of single-bond metallic radii).[10]

While an Eastman Professor at Oxford University in 1948, Pauling returned to the three-dimensional structure of proteins, a problem that had occupied him in the late 1930s. By folding a paper on which he had drawn a chain of linked amino acids, he discovered a cylindrical coil-like configuration (later called the alpha helix), in which the amino-acid groups were connected by hydrogen bonds. The key to his solution of the three-dimensional protein structure was his use of the coplanarity of all atoms in the bonding group (called the peptide bond). This put serious constraints on possible structures and led Pauling to a helical structure. The most significant element of Pauling's structure was its nonintegral number of amino acids per turn of the helix. Following an initial announcement of the discovery in 1950, Pauling and his collaborators published a series of eight articles on the alpha helix, the pleated sheet, and the structure of hair, muscle, collagen, hemoglobin, and other proteins. The most cited of these papers was the first one.[11]

During this period of intense scientific activity, Pauling was also involved in efforts to educate the public about the implications of nuclear weapons. When he became president of the American Chemical Society (ACS) in 1949, he used the position to speak out for values he deeply held. For example, in his presidential address, he asked the

---

[9]Pauling, L.; Itano, H. A.; Singer, S. J.; Wells, I. C.; "Sickle Cell Anemia, a Molecular Disease"; *Science* **1949**, *110*, 543–548.

[10]Pauling, L. "Atomic Radii and Interatomic Distances in Metals"; *J. Am. Chem. Soc.* **1947**, *69*, 542–553.

[11]Pauling, L.; Corey, R. B.; Branson, H. R. "The Structure of Proteins: Two Hydrogen-Bonded Helical Configurations of the Polypeptide Chain"; *Proc. Nat. Acad. Sci.* (USA) **1951**, *37*, April, 205–211.

nation's industrial corporations to create a well-endowed foundation for the support of basic scientific research. He also made it clear that he was not sympathetic with many of the aims of the American Medical Association, for he felt that the United States should have been moving toward a system of socialized medicine. His message caused much discussion among ACS members, some even suggesting that he should be dismissed from his office, but Pauling served out his term.

In the early 1950s, he became interested in deoxyribonucleic acid (DNA), and in February 1953, he and Corey published a structure for this molecule that contained three strands, twisted around each other in ropelike fashion, with the phosphoric acid groups triangularly arranged in the center and with the various bases pointed outward. Shortly after this structure's publication, James D. Watson and Francis H. Crick published their double helix, which turned out to be both correct and a powerful stimulus to future research. Watson and Crick profited from good X-ray photographs of DNA crystals taken by Rosalind Franklin and from free movement to attend various scientific meetings and to discuss their ideas with various scientists. Pauling, who was critical of his government's position on nuclear weapons, was unable to obtain a passport on several occasions, and this hampered his ability to gather information about the research going on in England (he surely would have profited from discussing with Rosalind Franklin her distinction between the wet and dry forms of DNA). Nevertheless, Pauling was very pleased with the new structure, and according to Watson, he and Crick would never have made their discovery without Pauling's prior work and his stimulation as a competitor (for his part, Pauling has said that he did not feel that he was in a race for the structure).

Pauling became well-known to the general public during the middle and late 1950s through his campaign to stop the testing of nuclear weapons in the atmosphere. He did not see these efforts as unrelated to his scientific work because he used scientific arguments to inform people of the dangers that these tests posed for human health and life. In January 1958 he and his wife presented to the United Nations an appeal for the end of testing that had been signed by more than 9000 scientists from 44 countries. Not all scientists agreed with the Paulings' appeal. For example, the physicist Edward Teller, an advocate of testing, felt that a little radiation might even be beneficial, or at most, as dangerous as being a few ounces overweight. Pauling, on the other hand, believed that the radioactive debris from these tests was causing defective children to be born and creating cancers in many people. Pauling and Teller debated the subject of nuclear fallout on public television in February 1958, but neither was able to modify the position of the other. So much time was being consumed by Pauling's crusade against nuclear testing that he resigned in 1958 from his administrative posts at CIT. In 1960, before a Congressional subcommittee, he had to defend his actions dedicated to bringing about a test ban. By refusing to reveal the names of those who had helped him collect the signatures for his United Nations appeal, he risked going to jail. He offered to explain all aspects of the petition with the exception of the names of those who gathered signatures because he felt that such information might be used against them. The committee eventually backed down, and Pauling was never charged with contempt of Congress. In the years after these hearings, legal scholars concluded that the senators had gone far beyond their congressional investigative power and that, in taking his stand, Pauling had made an admirable contribution to a free society. His work on behalf of world peace was also recognized when he was given the Nobel Peace Prize for 1962 on October 10, 1963, the date that the Nuclear Test Ban Treaty went into effect.

During the late 1950s and early 1960s, molecular medicine, the field he had helped found, consumed much of the time he was able to dedicate to his scientific researches. Through his work on antibodies and sickle-cell anemia, he had become convinced that it was molecular size and shape that were the important factors in biological reactions. In 1961 he put forward a new theory of anesthesia in which he explained how certain molecules cause anesthesia by aiding water molecules in the brain to build crystals, similar to ice, which interfere with the electrical activity connected with consciousness.[12] In 1962 he and Emile Zuckerkandl helped to initiate a new discipline, chemical paleogenetics, the goal of which is to discover how certain molecules, such as variant forms of hemoglobin, may have evolved from a common ancestral form.

Because of the negative reaction of CIT officials to his peace work and his Nobel Peace Prize and because of their removal of laboratory space for his work in molecular medicine, Pauling left CIT in 1963. During the mid-1960s he was a staff member at the Center for the Study of Democratic Institutions in Santa Barbara, where his work on behalf of peace was encouraged and where he was able to continue his theoretical scientific research. However, after a few years, he began

---

[12]Pauling, L. "A Molecular Theory of General Anesthesia"; *Science* **1961**, *134*, 15–21.

to miss his involvement in experimental research, and in 1967 he became a professor of chemistry at the University of California in San Diego, where he remained for two years, before becoming a professor at Stanford University, his last academic position.

In the light of later developments, perhaps the most significant occurrence during this final period of his academic career was a letter he received in 1966 from Irwin Stone, an industrial chemist who had a deep interest in vitamin C. Stone believed that the minimum daily requirement set by nutritionists and government agencies was far too low for many people, and, since he wanted Pauling to live a long and healthy life, he suggested that he take large amounts (megadoses) of this vitamin. Pauling and his wife followed Stone's advice, and they noticed an improvement in their overall health and a remarkably sharp decrease in the number and severity of their colds. This experience sent Pauling scurrying through the scientific literature on the protective effects of vitamin C against the common cold. From the published evidence, he concluded that most nutritionists were wrong and that vitamin C, provided that it is taken in megadoses, has a beneficial effect in helping the body fight off colds.

In 1968, developing still further his ideas on the importance of certain molecules in living things, Pauling wrote an influential paper on orthomolecular psychiatry.[13] His basic idea in this paper was that a healthy brain has an optimum molecular environment consisting of the proper amounts of a variety of chemical substances and that when a person's genes or diet cause improper concentrations, mental illness follows. His research on orthomolecular medicine led him to write a book on vitamin C and the common cold that became a best-seller, causing in turn a huge increase in vitamin C sales throughout the United States.[14] The book embroiled Pauling in a controversy, since many nutritionists and medical doctors disagreed with his claim that megadoses of this vitamin could lower the numbers and alleviate the symptoms of the common cold.

In the early 1970s Pauling became interested in using vitamin C for the treatment of cancer, largely through his collaboration with the Scottish physician Ewan Cameron. To pursue this and his other interests in molecular medicine, Pauling founded in 1973, after he had

---

[13]Pauling, L. "Orthomolecular Psychiatry"; *Science* **1968**, *160*, 265–271.

[14]Pauling, L. *Vitamin C and the Common Cold*; W. H. Freeman: San Francisco, CA, 1970.

## LINUS PAULING

become an emeritus professor at Stanford, the institute that now bears his name: the Linus Pauling Institute of Science and Medicine. This institute, both at its original location in Menlo Park and its present location in Palo Alto, has been the scene of much of his work in the 1970s and 1980s. For example, in the 1980s he obtained financial support to have his ideas about vitamin C and cancer tested in his own laboratory and in the laboratories of the Mayo Clinic. Workers at the Linus Pauling Institute obtained positive results in animal studies, but researchers at the Mayo Clinic obtained negative results in two studies with human patients. Pauling attacked both Mayo Clinic studies, the first because researchers had used heavy doses of highly toxic anticancer drugs along with the vitamin C and the second because of the short time that researchers had given cancer patients vitamin C. He therefore continues to maintain that cancer patients should be given megadoses of vitamin C. More recently, he has been studying vitamin C in relation to cardiovascular disease.

Throughout his career Pauling has enjoyed working on the frontiers of scientific knowledge. From his teenage years, when he posted lists of substances and their properties above his basement laboratory bench, to his most recent research on quasi-crystals, he has sought the simplest and most intellectually satisfying models to explain relevant observations. Experience taught him that the most insightful models are often elusive and accessible only to scientists with the most persistent and penetrating creative imaginations. He has said that without imagination a scientist will discover nothing, since the essence of discovery requires that a scientist look at what all people see but discern something nobody has ever seen. In all stages of his life, but particularly in his mature years, he has also believed in the importance of the moral imagination: that human beings have a profound duty to order their own lives in ways that minimize pain and maximize true happiness for all. Pauling's wife, who died in 1981, once said that what she admired most in her husband was that he worked with such passion for truth regardless of the awards and honors, and by truth she meant both scientific and moral truth. In this evaluation, Ava Helen, who lived with and loved Linus Pauling for well over half a century, captured the essence of a great scientist and a great human being.

ROBERT J. PARADOWSKI
*Rochester Institute of Technology*

## Bibliography

Crick, F. "The Impact of Linus Pauling on Molecular Biology: A Reminiscence"; In *The Chemical Bond: Structure and Dynamics*; Zewail, A., Ed.; Academic Press: San Diego, CA; 1992; pp 87–98.

*Crystallography in North America*; McLachlan, D., Jr.; Glusker, J. P., Eds.; American Crystallographic Association: New York, 1983.

Divine, R. A. *Blowing on the Wind: The Nuclear Test Ban Debate, 1954–1960*; Oxford University Press: New York, 1978.

Edelstein, S. J. *The Sickled Cell: From Myth to Molecules*; Harvard University Press: Cambridge, MA, 1986.

Goodstein, J. "Atoms, Molecules, and Linus Pauling"; *Social Research* **1984**, *51*, 691–708.

Hamner, W. B. A. B. Honors Thesis, Harvard University, 1983.

Judson, H. F. *The Eighth Day of Creation: Makers of the Modern Revolution in Biology*; Simon & Schuster: New York, 1979.

Kay, L. E. "Molecular Biology and Pauling's Immunochemistry; A Neglected Dimension"; *History and Philosophy of the Life Sciences* **1989**, *11*, 211–219.

*Linus Pauling: A Century of Science and Life* (videorecording); White, C. S., Executive Producer; Cabisco: Burlington, NC, 1987.

"Linus Pauling and the Future" (sound recording); *Man and Molecules: The Unique American Chemical Society Radio Series*; American Chemical Society: Washington, DC, 1979.

"Linus Pauling, Crusading Scientist" (videorecording); Richter, R., Producer; Corinth Films: New York, 1977.

*Linus Pauling on Science Education* (sound recording); *Dimensions in Science*; Vol. 1385; American Chemical Society: Washington, DC, 1987.

Miyazaki, F. *Linus Pauling: A Man of Intellectual Action*; Cosmos Japan International: Tokyo, 1991.

*Nova* (sound recording); Public Broadcasting System: broadcast Sept. 14, 1981.

Olby, R. *The Path to the Double Helix*; University of Washington Press: Seattle, WA, 1974.

Paradowski, R. J. "Linus Pauling"; In *The Nobel Prize Winners: Chemistry*; Magill, F. N., Ed.; Salem Press: Pasadena, CA, 1990; Vol. 2, p 619.

Pauling, L. "Fifty Years of Progress in Structural Chemistry and Molecular Biology"; *Daedalus* **1970,** *99, Fall,* 988–1014.

Pauling, L. "Linus Pauling on Science and Peace"; Nobel Peace Prize lecture; Center for the Study of Democratic Institutions: Santa Barbara, CA, 1964.

Pauling, L. "How I Became Interested in the Chemical Bond: A Reminiscence"; In *The Chemical Bond: Structure and Dynamics*; Zewail, A. H., Ed.; Academic Press: San Diego, CA, 1992; pp 99–109.

Pauling, L. "X-ray Crystallography and the Nature of the Chemical Bond"; In *The Chemical Bond: Structure and Dynamics;* Zewail, A., Ed.; Academic Press: San Diego, CA, 1992; pp 3–16.

*The Roots of Molecular Medicine: A Tribute to Linus Pauling*; Huemer, R. P., Ed.; Freeman: New York, 1986.

Serafini, A. *Linus Pauling: A Man and His Science*; Simon & Schuster: New York, 1989.

Servos, J. W. *Physical Chemistry from Ostwald to Pauling: The Making of a Science in America*; Princeton University Press: Princeton, NJ, 1990; pp 275–298.

Silverstein, A. M. *A History of Immunology*; Academic Press: San Diego, CA., 1989.

*Structural Chemistry and Molecular Biology: A Volume Dedicated to Linus Pauling by His Students, Colleagues and Friends*; Rich, A.; Davidson, N., Eds.; W. H. Freeman: San Francisco, CA, 1968.

White, F. M. *Linus Pauling, Scientist and Crusader*; Walker: New York, 1980.

# Vincent Du Vigneaud

## 1901 –1978

Vincent Du Vigneaud was born on May 18, 1901, in Chicago, Illinois, and died on November 11, 1978, in a hospital in White Plains, New York. He received the Nobel Prize in chemistry in 1955 for his work on biologically important compounds containing sulfur and particularly for the preparation of the polypeptide hormone oxytocin, which was the first synthesis of any such compound.

Vincent Du Vigneaud spent his early years in Chicago and attended Carl Schurz High School. He and his friends had a general interest in and curiosity about science. One of his friends had put together a tiny laboratory in his basement. The friends carried out simple experiments using chemicals supplied by the local druggist. Some of these experiments involved the use of sulfur; they gave Du Vigneaud his first exposure to this element, which was to become the major focus of his research. He spent a brief period of time working in a government farm program that was designed to supply farms with enough young workers to maintain production during the war. With the encouragement and some financial support from his sister Beatrice, he entered the University of Illinois to study chemical engineering. His interests quickly shifted to chemistry when he became curious about the laboratory work being carried out in the organic chemistry class at that time. He was fortunate enough to be exposed to the activities of Carl Marvel, the inspiring lectures of Howard B. Lewis, and the excellent teaching of William C. Rose. These persons solidified his interests in the relationship of chemical structures to biological activity. He received his B.S. in 1923 and his M.S. in 1924 from Illinois.

His years at the university were years of experience in many areas in addition to chemistry. In order to raise money for his education he held a number of jobs, including the position of headwaiter and the position of second lieutenant instructor in the U.S. Cavalry Reserves, which was his first teaching experience. Of special significance was his acquaintance with a young English major, Zella Zon Ford. He persuaded her to take courses in mathematics and science. They were married in 1924, and she worked as a science teacher during their first years of marriage. They had two children, a daughter, Marilyn, and a son, Vincent, who became a physician in New York State.

After briefly working in the analytical department of the DuPont Company, near Wilmington, Delaware, he spent about a year working in clinical chemistry at Philadelphia General Hospital. Du Vigneaud returned to academic life when he went to the University of Rochester to work with John R. Murlin on the chemical nature of insulin. In 1927 he received his Ph.D. from the University of Rochester. The work on insulin was continued at Johns Hopkins University under the direction of John Jacob Abel. The work was to become his first significant contribution to sulfur chemistry. Abel had isolated insulin in crystalline form and had expressed the view that it was a protein. This view contrasted with the dominant view that these hormones were not proteins. Du Vigneaud in collaboration with Abel demonstrated that several amino acids resulted from the breakdown of insulin. One of these was the sulfur-containing amino acid cystine, and the work showed that the sulfur content of insulin could be directly related to the sulfur contained in cystine. Sulfur was shown to be present as part of a bond in which the two sulfur atoms were directly bound to each other. Du Vigneaud, in agreement with Abel, believed that insulin was a protein, and his experimental work argued for the protein nature of insulin. His work had even described how sulfur was present in the complex structure as cystine. His efforts also demonstrated the need for additional studies on the sulfur-containing amino acids.

At the completion of his work at Hopkins, he received a fellowship that enabled him to study in Europe under a number of important chemists, especially Max Bergmann in Dresden. He returned shortly thereafter to the University of Illinois as an assistant in physiological chemistry, working under Rose, and spent three years there. He received a promotion to assistant professor in 1930 during this period. In 1932 he assumed the position of professor of biochemistry and department chairman at the George Washington University School of Medicine in Washington, D.C. After a six-year tenure there he became a professor of chemistry and department chair at Cornell University Medical College in New York City in 1938. This medical center would be his home for the rest of his academic career. After he reached normal retirement age and had to vacate his position in 1967, he assumed a research professorship at the Ithaca campus of Cornell University until a serious stroke forced him to halt activities in 1974. His wife took care of him until her death in December 1977. His final year was spent in a hospital in White Plains, New York, where he died at the end of 1978.

Du Vigneaud's research interests included the study of sulfur-containing amino acids and closely related compounds. Proteins, an essential part of the diet, are made up of various amino acids. Some of these amino acids are termed essential amino acids because they are necessary for carrying out important activities in the body. Cystine was the first amino acid to be discovered. Another sulfur-containing amino acid, methionine, was later shown to be essential, and it was noted that the cystine could be replaced by methionine in the diet. It was clear that cystine was derived from methionine in the body, but the relationship between these two amino acids and other biological materials would be investigated over the next few decades by Du Vigneaud and other biochemists. One of these materials, cysteine, had been shown to be a protein constituent and capable of being converted into cystine. Du Vigneaud used a new technique to convert cystine to cysteine. Another closely related material, homocystine, was prepared from methionine, and it appeared to be involved in the body's preparation of cystine. His experimentation using carefully prepared diets showed that under certain circumstances homocystine could replace cystine. When it was proposed that homocystine combined with another compound to yield an intermediate material that could break down to give cystine, he prepared his hypothetical intermediate in his laboratory and termed it cystathionine. The nonessential amino acid serine was shown to be involved in the compound that combined with homocystine. Because a sulfur atom was transferred, Du Vigneaud termed the process "transulfuration".

Choline is another dietary component that has been of interest to nutritionists over the years. A series of experiments conducted by Du Vigneaud and Rose collaboratively helped to establish the role of choline in the diet. Du Vigneaud had noted that diets containing homocystine in certain cases produced poor growth and fatty livers in rats, but with the addition of choline the growth resumed and the unhealthy signs disappeared. He reasoned that the choline donated methyl groups so that cystine could be produced from homocystine. Such methyl transfers were called *transmethylation,* and compounds such as choline that supplied methyl groups were termed "methyl donors" by Du Vigneaud. This area of transmethylation would be an active research area for many years to come.

Du Vigneaud was also involved in the establishment of biotin as a vitamin. In the 1920s and 1930s several investigators had noted that rats that received raw egg whites in large amounts without other proteins in their diet developed severe neurological problems and skin disorders. Certain foods seemed to contain a material that prevented these disorders. This substance was called vitamin H. The distinguished biochemist Paul György sought Du Vigneaud's help in identifying this substance. In 1936 a material had been isolated from egg yolks by Kögl and Tonnis that was a derivative of biotin. Noting some similarities between vitamin H and biotin, Du Vigneaud, György, and several other investigators carried out a series of experiments that proved that vitamin H and biotin were the same. Biotin was added to the list of essential B vitamins. Based on this work, others would prepare biotin synthetically. The factor in egg white that produced a biotin deficiency was determined to be a protein, avidin, that could link with biotin quite strongly and render it unavailable for use.[1]

Du Vigneaud's most recognized work was his detailed study on oxytocin, a posterior pituitary gland hormone. He began this study by attempting to isolate the hormone in a potent form. New techniques for isolating biologically active materials had been introduced in the early 1940s, and using one of these techniques Du Vigneaud was able to isolate a reasonably pure sample of the material. The next problem was the identification of the amino acids present in oxytocin. The presence of eight amino acids was established: aspartic acid, cystine, glutamic acid, glycine, isoleucine, leucine, proline, and tyrosine. It was also demonstrated that this octapeptide was a cyclic compound. After a long series of experiments, a structure for oxytocin was proposed in 1953. The structure included a sulfur-to-sulfur bond, a disulfide bond, that could be broken in order to open up the ring. The highlight of this work was the synthesis of oxytocin in the laboratory. It was the first synthesis of a natural protein hormone. Continued experimentation by

---

[1]Guthrie, H. A. *Introductory Nutrition;* C. V. Mosby: St. Louis, MO, 1975, pp 291–293.

Du Vigneaud involved the substitution of one particular amino acid for another amino acid in various biologically active hormones. The important conclusion was that a specific type of biological activity, such as the contraction of the uterus, was closely related to the particular amino acids present and the specific sequence of the amino acids.

Besides the Nobel Prize, Du Vigneaud was presented with honorary university degrees from Yale, Illinois, and Rochester and membership in the National Academy of Sciences. He was a popular lecturer, a member of various boards and scientific commissions, and a winner of the Mead Johnson Award and Mendel Award of the American Institute of Nutrition. His research papers individually and collaboratively numbered about 480. A large number of graduate students, postdoctoral fellows, and visiting professors became associated with him and his research efforts.

Du Vigneaud recognized the significance of his work and the evolution of this line of research, from the determination of the structure of insulin to the elaboration of the basics of structure-activity relationships in various hormones. These observations are recorded in his book *A Trail of Research in Sulfur Chemistry and Metabolism and Related Fields* (Cornell University Press: Ithaca, NY, 1952). The importance of sulfur in biochemistry (as a component of enzymes and important hormones and for its ability as a disulfide link to connect protein strands) and nutrition (as a component of several essential amino acids and vitamins) was clearly established, and the role of sulfur as an important constituent of various nutrients was described. Du Vigneaud was a pioneer in illustrating the relationship of chemical structure to biological activities, and his life's work illustrated the interdisciplinary nature of many scientific problems.

**ROBERT H. GOLDSMITH**
*St. Mary's College of Maryland*

## Bibliography

*Americana Annual;* Smith, J. J., Ed.; Americana: New York, 1956.

*American Men and Women of Science;* 12th ed.; Jacques-Cattell Press: R. R. Bowker Co., New York, 1972; Vol. 2, p 1578.

Bing, F. "Vincent du Vigneaud"; *Journal of Nutrition* **1982,** *112,* 1465–1473.

Current Biography Yearbook; Candee, M. D., Ed.; H. W. Wilson: New York, January 1956.

Current Biography Yearbook; Moritz, C., Ed.; H. W. Wilson: New York, 1979.

Du Vigneaud, V.; Melville, D. B.; György, P.; Rose, C. S. "Identity of Vitamin H with Biotin"; Science 1940, 92, 62–63.

Du Vigneaud, V. "The Role Which Insulin Has Played in Our Concept of Protein Hormones and a Consideration of Certain Phases of the Chemistry of Insulin"; Cold Spring Harbor Symposia on Quantitative Biology 1938, 6, p 275.

Du Vigneaud, V. "Biotin"; In Biological Action of the Vitamins, A Symposium; Evans, E. V., Ed.; University of Chicago Press: Chicago, IL, 1942; pp 144–168.

Du Vigneaud, V. "The Sulfur of Insulin"; J. Biol. Chem. 1927, 75, 393–405.

Du Vigneaud, V.; Chandler, J. P.; Cohn, M.; Brown, G. B.; "The Transfer of Methyl Groups from Methionine to Choline and Creatine"; J. Biol. Chem. 1940, 134, 787–788.

Du Vigneaud, V. A Trail of Research in Sulfur Chemistry and Metabolism and Related Fields; Cornell University Press: Ithaca, New York, 1952.

Farber, E. Nobel Prize Winners in Chemistry, 1901–1961; Abelard-Schuman: London, 1963.

Fruton, J. S. "Vincent du Vigneaud"; In American Philosophical Yearbook–1979; In American Philosophical Society: Philadelphia, PA, 1980; pp 54–58.

Hofmann, K. "A Tribute to Vincent du Vigneaud"; In Peptides: Structure and Biological Function; Gross, E., and Meierhofer, J., Eds.; Pierce Chemical: Rockford, IL, 1979; pp 5–24.

The New York Times. Dec. 12, 1978.

Plane, R. "Interview with Vincent du Vigneaud"; Journal of Chemical Education 1976, 53, 8–12.

Popenor, E. A.; Lawler, H. C.; Du Vigneaud, V. "Partial Purification and Amino Acid Content of Vasopressin from Hog Posterior Pituitary Glands"; J. Am. Chem. Soc. 1952, 74, 3713.

# Cyril Hinshelwood

## 1897–1967

Copyright Nobel Foundation

Sir Cyril Norman Hinshelwood was born in London, England, on June 19, 1897. He was educated at the Westminster City School. For three years during World War I, he worked at the Queensferry Royal Ordnance Factory, Department of Explosives Supply. He began his studies at Oxford University in 1919 with the Brackenbury Scholarship at Balliol College and received the Master of Arts and Doctor of Science degrees from Oxford University. He served as a fellow and tutor at Trinity College from 1921 to 1937, when he was appointed to succeed Frederick Soddy as the Dr. Lee's Professor of Chemistry at Oxford University. He served as a member of several advisory councils on scientific matters to the British government. He was elected Fellow of the Royal Society in 1929 and served as president from 1955 to 1960. He also served as president of the Chemical Society from 1946 to 1948 and as President of the Faraday Society from 1961 to 1962. In addition to the Nobel Prize in chemistry in 1956, Hinshelwood was awarded the Lavoisier Medal of the Société Chimique de France in 1935, the Davy Medal of the Royal Society in 1943, and the Guldberg Medal of Oslo University in 1952. He received numerous honorary degrees from universities in the United Kingdom (Cambridge, London, Dublin, etc.) and abroad. He held honorary memberships in the major scientific societies of the world. He was knighted in 1948 and appointed to the Order of Merit in 1960. He was fluent in many languages, including Chinese, and his main hobbies were painting, reading foreign literature, and collecting Chinese pottery. He remained unmarried throughout life; he died in London on October 9, 1967.

Hinshelwood is noted for his extensive contributions to the theoretical and experimental development of chemical kinetics. He elucidated the complex reaction system that contributes to the mechanism of explosive mixtures of hydrogen and oxygen. It was this work that earned him, jointly with Nikolay N. Semënov, the Nobel Prize in 1956.

His earliest work was undoubtedly influenced by his experience with explosives at Queensferry during the First World War. During his first few years at Oxford, he attempted to interpret the decomposition of solid mixtures containing oxidants such as potassium permanganate and ammonium dichromate. He was fascinated by the influence of catalysts, temperature, and the physical state of the system on the rate of reaction of these systems. By 1923 he had turned his attention to the study of reactions occurring in the gas phase. These studies occupied him for the greater part of his professional career.

One of his most important theoretical contributions to the development of chemical kinetics grew out of his work with H. W. Thompson on the decomposition of propionaldehyde.[1] The rate was found to fall off at low pressures, whereas at higher pressures the rate of decomposition was higher than could be accounted for on the basis of Lindemann's theory of collisional activation.[2] To the current Lindemann theory, Hinshelwood added the assumption that the internal energy of polyatomic molecules contributes to the activation energy. These reactions were termed "quasi-unimolecular". To account for the anomalous slopes observed in the reaction-rate-versus-pressure curves for the thermal decomposition of nitrous oxide, he extended the theory to embrace spontaneous and collisionally induced transitions to different internal states of the molecule. These ideas contributed significantly to the development of the theory of unimolecular reaction kinetics.[3]

During the 1920s kineticists turned their attention to the study of homogeneous unimolecular reactions, with an eye to developing a theory for what seemed to be the simplest of kinetic laws. Most of the reactions investigated during this period seemed to show simple

[1] *Proc. Roy. Soc.* **1927,** *114A,* 84–97.

[2] *Trans. Faraday Soc.* **1922,** *17,* 598.

[3] "The Theory of Unimolecular Reactions"; *Proc. Roy. Soc.* **1926,** *113A,* 230–233.

first-order kinetics. Nearly all were later proven not to be unimolecular at all but instead were shown to exhibit complex chain mechanisms. Two important discoveries resulted from this flurry of investigations of "unimolecular" reactions. First, molecular collisions were shown to play an important role in communicating activation energy to the molecules undergoing transformation. Second, activation energy was established as the *dominant*, but not the *sole* factor determining chemical reactivity.

In 1913 Max Bodenstein first suggested that a chemical reaction, especially one initiated by light, may proceed by a chain process.[4] Walther Nernst proposed a specific chain process for the $H_2$–$Cl_2$ photochemical reaction.[5] Johann Christiansen, Karl Herzfeld, and Michael Polanyi proposed a chain mechanism for the $H_2$–$Br_2$ reaction.[6] These reactions all depend on (1) the size of the reaction vessel, (2) the presence of inert gases, and (3) the presence of sensitizers and inhibitors.

Bodenstein had shown that reaction occurs at the surface under certain conditions.[7] Between the temperatures at which Bodenstein studied the $H_2$–$O_2$ reaction and the inflammation temperature, homogeneous processes seemed to be at work. Semënov was able to explain the lower pressure explosion limit for the $P_5$–$O_2$ reaction in terms of a chain-branching mechanism.[8] Hinshelwood showed that the $H_2$–$O_2$ reaction also exhibited a lower explosion limit.[9] Chain branching was shown to be controlled by deactivation of the chain carriers at the surface. The upper explosion limit of the $H_2$–$O_2$ reaction was accounted for by Hinshelwood in terms of three-body collisions in the gas phase. It was the study of the $H_2$–$O_2$ reaction that brought Hinshelwood's work in touch with that of Nikolay Semënov.

Francis Rice and Karl Herzfeld proposed theoretical reaction schemes for a series of hydrocarbon decomposition reactions based on free-radical mechanisms. They were able to explain reactions that showed simple first- and second-order kinetics in terms of chain mechanisms by assuming reasonable activation energies, by choosing

[4] Bodenstein, M., *Z. Physik. Chem.* **1913**, *85*, 329.

[5] Nernst, W., *Z. Elektrochem.* **1918**, *24*, 335.

[6] Christiansen, J. A. *Kgl. Danske Videnskab. Selskab. Mat.-Fys. Medd.* **1919**, *1*, 14; Herzfeld, K. F. *Z. Elektrochem.* **1920**, *26*, 50; *Ann. Physik.* **1919**, *59*, 635; Polanyi, M. *Z. Elektrochem.* **1920**, *26*, 50.

[7] Bodenstein, M. *Z. Physik. Chem.* **1899**, *29*, 665.

[8] Semënov, N. N. *Z. Physik* **1927**, *46*, 109; *ibid.* **1928**, *48*, 571.

[9] Hinshelwood, C. N.; Thompson, H. W. *Proc. Roy. Soc.* **1929**, *122A*, 610; Hinshelwood, C. N.; Moelwyn-Hughes, E. A. *Proc. Roy. Soc.* **1932**, *138A*, 311.

reasonable sets of elementary processes, and by varying the ways in which the free-radical chains were terminated.[10] Evidence for the existence of some of these schemes came from using small concentrations of NO, an odd electron molecule assumed to be capable of combining rather easily with free-radical species.[11] Hinshelwood and his colleagues used this technique extensively and tried other inhibitors such as propylene, ethylene, and isobutene. From these studies, it became clear that nearly all thermal decompositions occur by free radical mechanisms. This development considerably complicated the study of unimolecular reactions and unimolecular reaction rate theory. Most of the substances on which the theory was originally based, and which exhibited first-order kinetic behavior, were shown to decompose by chain mechanisms, rather than by simple unimolecular processes.

The thermal reaction of butane, however, was not completely suppressed by the addition of inhibitors, even in high concentrations. This suggested that a parallel, nonradical, unimolecular process was also occurring for the following reasons: (1) several different inhibitors gave the same limiting value; (2) added inert gases accelerated the reaction by the same amount as they did in the uninhibited reaction; and (3) the kinetics of the "residual" reaction showed peculiarities common to other unimolecular reactions. On the other hand, deuterium exchange in the $C_2D_6$–$C_2H_6$ system seemed to proceed to about the same extent in the uninhibited as in the inhibited reaction. Semënov and his colleagues accounted for this behavior in terms of the formation of free surface valences, which could act in a way similar to that in which radicals acted in the volume phase.

Beginning in the 1930s, Hinshelwood began to investigate heterogeneous and homogeneous catalytic reactions in the liquid phase. He also undertook systematic kinetic studies of substituted aromatic molecules in nonaqueous media.[12] These studies contributed to the development of Hammett's theories of the energy–entropy relations among rate constants for reactions of related series of substituted aromatic molecules.

---

[10]Rice, F. O.; Herzfeld, K. F. *J. Am. Chem. Soc.* **1934**, *56*, 284.

[11]Staveley, L. A. K.; Hinshelwood, C. N., *Nature* **1936**, *137*, 29–30; Staveley, L. A. K.; Hinshelwood, C. N. *Proc. Roy. Soc.* **1936**, *154A*, 335.

[12]Hinshelwood, C. N. "Mode of Action of Solvents on Chemical Reaction Velocity", *Trans. Faraday Soc.* **1936**, *32*, 970–972; Pickles, N. V. F.; Hinshelwood, C. N. "The Influence of Solvents on Reaction Velocity. The Interaction of Pyridine and Methyl Iodide and the Benzoylation of *m*-Nitro-Aniline"; *J. Chem. Soc.* **1936**, 1353–1357.

Shortly before World War II, he became interested in the kinetics of bacterial cells. For most of his work, he selected the nonpathogenic organism *Aerobacter aerogenes*. He pursued two lines of inquiry: (1) how do the bacteria adapt to the nutrients in a new medium in which they are placed? (2) how do antibacterial agents inhibit the growth of the bacteria? These studies led to the development of a "network theory" of interdependent enzyme balance mechanisms in the bacterial cell. He put forth this theory to supplement currently accepted theories of mutation and selection. These results were gathered together and published jointly with A. C. R. Dean in *Growth, Function, and Regulation in Bacterial Cells* (Oxford University Press: Oxford, England, 1966).

ERNEST G. SPITTLER , S.J.
*John Carroll University*

## *Bibliography*

Bowen, E. J. "Sir Cyril Hinshelwood, 1897–1967"; *Chem. in Britain* **1967**, *3.*

Farber, E. *Nobel Prize Winners in Chemistry, 1901–1961*; Abelard-Schuman: London, 1963.

Hartley, H. "Sir Cyril Norman Hinshelwood"; In *The Dictionary of National Biography, 1961–1970*; Williams, E. T.; Nicholls, C. S., Eds.; Oxford University Press: New York, 1981.

Hinshelwood, C. N. *Chemical Kinetics of the Bacterial Cell*; Oxford University Press: Oxford, England, 1947.

Hinshelwood, C. N. "Homogeneous Reactions"; *Chem. Revs.* **1926**, *3*, 227–256

Hinshelwood, C. N. *The Kinetics of Chemical Change*; Oxford University Press: Oxford, England, 1926.

Hinshelwood, C. N., "The More Recent Work on the Reaction Between Hydrogen and Oxygen"; *Proc. Roy. Soc.* **1946**, *188 A*, 1–9.

Hinshelwood, C. N. *The Structure of Physical Chemistry*; Oxford University Press: Oxford, England, 1951.

Hinshelwood, C. N. *Thermodynamics for Students of Chemistry*; Oxford University Press: Oxford, England, 1926.

Hinshelwood, C. N.; Dean, A. C. R. *Growth, Function, and Regulation in Bacterial Cells*; Oxford University Press: Oxford, England, 1966.

Hinshelwood, C. N.; Williamson, A. T. *The Reactions Between Hydrogen and Oxygen*; Oxford University Press: Oxford, England, 1947.

*The Nobel Prize Winners: Chemistry.* Magill, F. N., Ed.; Salem Press: Pasadena, CA, 1990; Vol. 2.

Oxbury, H. *Great Britons: Twentieth-century Lives;* Oxford University Press: Oxford, England, 1985.

Thompson, H. W. "Cyril Norman Hinshelwood"; *Biographical Memoirs of Fellows of the Royal Society* **1973,** *19,* 375–431.

# 1956

NOBEL LAUREATE

# Nikolay Nikolayevich Semënov

## 1896–1986

Copyright Nobel Foundation

Nikolay Nikolayevich Semënov was born in the Russian city of Saratov on April 3, 1896. He received his Ph.D. from Petrograd (St. Petersburg) University in 1917. In 1920 he took charge of the electron phenomena laboratory of the Leningrad Physico-Technical Institute. He was also appointed professor at the Polytechnical Institute in 1928. He became a corresponding member of the U.S.S.R. Academy of Sciences in 1929 and Academician in 1932. In 1931 he became director of the Institute of Chemical Physics of the U.S.S.R. Academy of Sciences. The Academy was moved in 1943 to Moscow, where he also held a professorship at the Moscow State University beginning in 1944. He was awarded five Orders of Lenin and the Order of the Red Banner of Labor. He was a member of the Chemical Society of London, a foreign member of the Royal Society, a member of the American Academy of Sciences, and of many other scientific academies. He held honorary doctorates from the Universities of Oxford and Brussels, among numerous others. He was married twice, first to Natalya Nikolayevna Burtseva, with whom he had one son and one daughter, and later to Lidiya Grigorievna Scherbakova. He was the first Soviet citizen to receive a Nobel Prize in chemistry. He wrote three important books concerned with his work: *Chemical Kinetics and Chain Reactions, Some Problems of Chemical*

*Kinetics and Reactivity,* and *Chain Reactions.* He died on September 25, 1986, at the age of 90.

Semënov's work revolves around two areas of interest: the application of chemical kinetics to problems of combustion and explosion processes and the mechanism of chemical reactions involving branching and nonbranching chains. It was for significant contributions in these two areas that he was awarded the Nobel Prize in chemistry jointly with Cyril Norman Hinshelwood in 1956.

Chemical transformation and the regularities found therein constitute the main problem of chemistry and chemical technology. The first efforts to deal successfully with this problem came from organic chemists in the mid-nineteenth century in their efforts to relate structure and reactivity. These relationships took the form of qualitative reactivity rules. They have since received a more theoretical basis in terms of electronic and quantum-mechanical considerations.

In the last quarter of the nineteenth century, physical chemistry became an established field of study. The introduction of thermodynamic and statistical methods into the investigation of chemical processes laid the groundwork for chemical kinetics. J. H. van't Hoff and Svante Arrhenius formulated mathematical laws governing the kinetics of systems free of perturbation effects. In this theory, chemical reactions ought to occur as simple mono- or bimolecular processes, exhibiting first- or second-order kinetic laws. Van't Hoff, especially, was well aware of the complexities of chemical reactions, but the emphasis for some time thereafter was on reactions that followed seemingly simple kinetic laws.

Some investigators did study complex reactions such as conjugated oxidation, homogeneous catalysis, and autocatalysis in electrolytic solutions (Ostwald, Menshutkin, Haber, and Goldschmidt, to name a few). They drew attention to the participation of unstable intermediates in chemical reactions.

In the early decades of the twentieth century, Max Bodenstein investigated many gas-phase reactions. Of particular significance were his 1907 study of the $H_2$–$Br_2$ system and his 1913 study of the photochemically induced reaction of the $H_2$–$Cl_2$ system.[1] The latter exhibited very high quantum yields that seemed to violate Einstein's photochemical

---

[1]Bodenstein, M. *Z. Physik. Chem.* **1913,** *85,* 329.

law. In 1919 Nernst explained the reaction on the basis of a chain mechanism involving the initial photochemical dissociation of $Cl_2$.[2]

$$Cl_2 \rightarrow 2Cl\cdot$$

$$Cl\cdot + H_2 \rightarrow HCl + H\cdot$$

$$H\cdot + Cl_2 \rightarrow HCl + Cl\cdot$$

Thus the chain theory of chemical reactions had its beginning. In 1923 Christiansen and Kramers applied the chain theory to thermal reactions exhibiting first-order behavior. The notion of energy chains was introduced but dropped for the time being for lack of further confirmation. In 1924 Christiansen connected the inhibiting action of additives with termination of the chain process by the inhibitor. In 1927 Bäckstrom proved the accuracy of Christiansen's ideas and established the chain mechanism of a variety of reactions in one of the most important series of studies in the history of chemical kinetics.

In the years 1926–1928, Hinshelwood and Semënov both began to take up the investigation of chain reactions, which ultimately led to their Nobel Prize-winning work.[3] Semënov and his colleagues uncovered the low pressure explosion limit of the $P_5$–$O_2$ and $H_2$–$O_2$ reaction systems, whereas Hinshelwood and his colleagues discovered the upper explosion limit of the $H_2$–$O_2$ system. Both of these phenomena were successfully interpreted in terms of the branching chain theory of reaction. This theory assumes that the reaction of a single radical results in the formation of more than one other radical, and thus an avalanche rapidly occurs, even under isothermal conditions. The lower explosion limit is explained in terms of the rapid destruction of radicals at the walls of the containing vessel, thus suppressing the avalanche. The upper limit is explained in terms of the destruction of the radicals in the volume phase by means of three-body collisions (termolecular processes).

---

[2]Nernst, W. Z. Elektrochem. 1918, 24, 335.

[3]For example, Hinshelwood, C. N.; Thompson, H. W. Proc. Roy. Soc. 1928, A118, 170; Gibson, C. H.; Hinshelwood, C. N. Proc. Roy. Soc. 1928, A119, 591; Thompson, H. W.; Hinshelwood, C. N. Proc. Roy. Soc. 1929, A122, 610.

Semënov's work resulted in the publication of his book *Chemical Kinetics and Chain Reactions* in 1934 (translated into English and published at Oxford in 1935). The work linked theory to experiment by successfully explaining long-known phenomena and by leading to quantitative laws and to the discovery of new phenomena. It was eventually possible to show that chain reactions, branched and non-branched, were the rule rather than the exception in gas-phase reactions, as well as in many liquid-phase reactions. Chain reactions were eventually shown to be important in polymerization processes[4] and in cracking processes as well.[5] It also became clear that the formulas of Arrhenius and van't Hoff should be applied only to elementary processes.

It was eventually shown that three basic types of chain mechanisms exist in chemical reaction systems. Most gas-phase chemical reactions proceed by means of unbranched chains. In an unbranched chain, only one free radical is formed for each one used up in the reaction. But this process can be repeated thousands of times for a single initiating free radical, which may be formed by thermal or photochemical dissociation of one of the components of the system. If more than one radical is formed for each one used up, chain branching can occur, followed by rapid ignition or explosion. A third type of branching mechanism was found to apply to the oxidation of hydrocarbons. In this case, the main chain reaction is unbranched, but one of the less-stable reaction products yields more than one radical. A good example is the formation of alkyl hydroperoxides.

$$R \cdot + O_2 \rightarrow RO_2 \cdot$$

$$RO_2 \cdot + RH \rightarrow ROOH + R \cdot$$

$$ROOH \rightarrow RO \cdot + OH \cdot$$

If such a molecular intermediate is formed, which reacts slowly to form more than one free radical, degenerate branching is said to occur. Degenerate branching was initially established in hydrocarbon oxidation

---

[4]Rice, F. O.; Rice, K. K. *The Aliphatic Free Radicals;* Johns Hopkins Press: Baltimore, MD, 1935.

[5]Frost, A. V. *Uspekhi Khim.* **1939**, *8,* 956; and Dintzes, A. I. *Proc. Acad. Sci. USSR* **1934**, *3,* 510.

reactions. The theory of degenerate explosions was developed in the years 1931–1934 and published in Semënov's book, *Chemical Kinetics and Chain Reactions.*

When branched chain reactions occur, two possibilities exist: The rate of branching exceeds the rate of termination, or the rate of termination exceeds the rate of branching. In the first case, an avalanche occurs, and ignition or explosion follows. In the second case, an avalanche is impossible, and reactions cannot even take place if the rate of free radical formation is low, as it usually is. Semënov and his colleagues showed that the transition from a completely inert to a violent reaction could be induced by (1) increasing the pressure of certain components, for example $O_2$; (2) changing the dimensions of the vessel; or (3) inserting tiny metal wires into, or withdrawing them from, the reaction vessel. The transition occurs if one of these parameters is varied within extremely narrow, often unmeasurable, limits.

The existing theory of combustion and explosion was thus developed and expanded. Early theory assumed that ignition took place when the "flashpoint", supposed to be a characteristic of each substance, was somehow attained. This theory, however, had to be modified to account for the fact that the flashpoint, or temperature of autoignition, was really a function of thermal and kinetic parameters such as reaction order, activation energy, density and composition of the mixture, heat of reaction, and rate of heat dissipation at the walls of the reaction vessel.[6]

With this development, it became possible to calculate the flashpoint as a function of pressure and to calculate the explosion delay time as a function of pressure and temperature. Thus, the flashpoint is not a constant characteristic of the particular combustible substance, as had been previously supposed. It was also possible to calculate the normal rate of flame propagation as a function of the same kinetic and thermal parameters. A further extension of the theory allowed one to interpret the existence of concentration limits for flame propagation as a function of the rate of heat loss at the walls and of the flame temperature. It was also shown that the flame propagation rate is proportional to the square root of the quadratic mean of the pulsation rate. Thus, the

---

[6]For example, Zel'dovich, Ya., B.; Frank-Kamenetskiy, D. A. *J. Phys. Chem. USSR* **1938**, *12* (1); Zel'dovich, Ya. B.; Semënov, N. N. *J. Phys. Chem USSR* **1940**, *10* (12); von Elbe, G. *Combustion, Flames and Explosions of Gases;* Cambridge University Press: Cambridge, England, 1938.

effect of turbulence on the rate of flame propagation was accounted for theoretically.[7]

These theoretical developments enabled Semënov to account for several phenomena inexplicable before then: (1) that low-energy mixtures (i.e., mixtures at low pressure or diluted by inert gases) do not detonate at all, (2) that there are concentration limits for detonation, and (3) that high-energy mixtures do not detonate in narrow tubes. Semënov and his colleagues verified various aspects of this theory during the years 1928 to 1940 in the $CO-O_2$ gas-phase system and in the combustion of liquid nitroglycol.

The explanation of the effect of inhibitors such as nitric oxide, propylene, and other unsaturated hydrocarbons on a variety of gas-phase reactions was particularly fraught with difficulty. Hinshelwood was of the opinion that inhibition took place in the gas phase. Semënov and Voevodskiy hypothesized that the walls could initiate as well as break radical chains.[8] They further theorized that free valences could be produced by appropriate preparation of the surface or by the process of chemisorption of free radicals. The chemisorption would lead to the rupture of a surface bond and the formation of a free valence on the surface. These free valences could in turn act as radical traps at the surface in the same way that inhibitors might act in the gas phase. The result of the process would be the exchange of a free radical in the gas phase for a free valence on the surface. These free valences would be destroyed only by interaction with one another. Free valences might also be created through the release of energy in chemical reactions occurring on the surface, for example:

$$V_2 + Cl\cdot \rightarrow VCl + V\cdot$$

$$V\cdot + Cl_2 \rightarrow VCl + Cl\cdot$$

$$2V\cdot \rightarrow V_2$$

This theory of surface free valences may also play an important role in the mechanism of heterogeneous catalysis in redox processes. The

[7]Damkohler, G. Z. Elektrochem. 1940, 46, 601; Shohelkin, K. I. J. Tech. Phys. USSR 1943, 13, 520; Shohelkin, K. I. Proc. Acad. Sci. USSR 1939, 23, 636; Zel'dovich, Ya. B. J. Tech. Phys. USSR 1947, 17, 3.

[8]Semënov, N. N. Some Problems of Chemical Kinetics and Reactivity; Acad. Sci. USSR: London, 1959.

greatest number of redox catalysts are semiconductors and metals. There seems to be a significant correlation between electrical conductivity and catalytic properties of semiconductors.

Thus, the outcome of this work is the certainty that a large number of chemical reactions in the gas, liquid, and solid phases, as well as heterogeneous catalytic reactions on solid surfaces, are aided by particularly labile forms, free radicals, which are produced during the course of the process and have considerably greater reactivity than even the least stable molecules with saturated valence bonds.

This work has also led to the development of theoretical and experimental methods of investigating free radicals and to successful attempts to relate their properties to their structure and to the structure of the molecules from which they are formed. It has opened a new path for understanding the connection between the reactivity and structure of particles entering into a chemical reaction. It has emphasized the importance of studying the kinetics of elementary processes in which a single bond is broken and a single new bond is formed. It has created the possibility of rationally regulating the rate and direction of chemical change. This, in turn, is bound to have profound consequences for the perfecting of established industrial processes and for the development of new processes, for example, in the fields of polymerization and direct oxidation as well as hydrocarbon cracking.

ERNEST G. SPITTLER, S.J.
*John Carroll University*

## Bibliography

Semënov, N. N. *Chemical Kinetics and Chain Reactions;* Leningrad, 1934; English translation, Clarendon Press: Oxford, England, 1935.

Semënov, N. N., *Chain Reactions;* Soviet Academy of Sciences: Moscow, USSR, 1986.

Semënov, N. N., *Chemical Physics: Physical Principles of Chemical Kinetics;* Soviet Academy of Sciences: Moscow, USSR, 1978.

Semënov, N. N.; Voevodskiy, V. V. *Heterogeneous Catalysis in the Chemical Industry;* Soviet Academy of Sciences: Moscow, USSR, 1955.

Semënov, N. N., *Some Problems in Chemical Kinetics and Reactivity;* Acad. Sci. USSR, Moscow, 1954; English translation, 1958.

# Alexander Robertus Todd

## 1907–

Lord Todd was born on October 2, 1907, in Newlands Crescent in Cathcart, a southern suburban area of Glasgow, Scotland. He was awarded the Nobel Prize in chemistry in 1957 for his work on the synthesis of nucleotides and nucleotide coenzymes. His father was a clerk in the head office of the Glasgow Subway Railway Company and later became cashier and secretary. In 1922 Lord Todd's father became managing director of the Drapery and Furnishing Co-operative Society, Limited. Although an enthusiastic supporter of the cooperative movement, he was strongly opposed to its political affiliation with the Labour Party. Lord Todd's parents, having had only an elementary-school education, did succeed in moving up to the lower-middle class. They held passionate beliefs in the benefits of education and were determined that their children would receive a good education. Todd had a brother, Robert, and a sister, Jean.

Todd's education began at Holmlea Public School in Cathcart in 1912, where he began in kindergarten, only to be promoted within a few days to a higher class because of his physical size rather than any mental precocity, according to his reminiscences in his book *A Time to Remember*. In 1918 he was admitted to Allan Glen's School in Glasgow. After passing the Scottish Education Department's Qualifying

Examination, he passed into the senior school in 1919. Later in his life he derived great amusement from the controversy in England over the "elevenplus examination", which supposedly had such an adverse effect on English children. Allan Glen's School was a high school of science, where no Greek was to be taught, and mathematics, physics, and chemistry were emphasized. This did not prevent Todd from mastering a number of foreign languages later on. He did attend Allan Glen's School with the intention of taking up medicine, but that changed when a physician botched up the repair of his left elbow, which he injured while climbing a tree.

His interest in chemistry probably began with a home chemistry set when he was eight and developed rapidly at senior high school, where his practical chemistry was done in a building across from Baird and Tatlock, Limited, a laboratory furnisher, from which he was able to purchase chemicals and supplies of a wide variety (thus depleting his lunch money). By his own autobiographical comments, he was the worst performer in school at freehand drawing, which led the art master to comment on his initials, A.R.T., and the sense of humor of his parents.

He entered the University of Glasgow without a scholarship because he was ineligible for the bursaries available and because his father refused to allow him to accept a scholarship that would have required his father to declare penury. At the end of his first year he was awarded the Joseph Black Medal and the Roger Muirhead Prize in chemistry, which provided funds for the rest of his course.

His courses in inorganic chemistry (Lecturer, G. G. Henderson) and organic chemistry (Lecturer, T. S. Patterson) were good and quite interesting. The practical courses in qualitative and quantitative analysis were another story. His most enjoyable course was a research project on the action of phosphorus pentachloride on ethyl tartrate and its diacetyl derivative. His objective was to determine if the nature of the leading group had any influence on the course of the Walden inversion. The results led to his first publication, which appeared in the *Journal of the Chemical Society*. He did have a second publication, with Patterson, on optical rotatory dispersion of mannitol and its derivatives. When he graduated in 1928 at age 20, he did so with first-class honors in chemistry and was awarded a Carnegie Research Scholarship of £100 per year to continue his work with T. S. Patterson.

Holding to the Berzelius definition of organic chemistry as the "chemistry of substances in living organisms" rather than Gmelin's "chemistry of carbon compounds", he found his work with Patterson uninteresting. In order to further his studies of organic chemistry and acquire a real command of a foreign language, he decided to study with Walther Borsche at the University of Frankfurt. Borsche had been a pupil of Otto Wallach and an associate of Adolf Windaus, who worked on natural products. Todd began working with Borsche in October 1929: He found the laboratories quite an improvement over those at Glasgow. Organic microanalysis was done as a routine service; catalytic hydrogenation with colloidal platinum and palladium at room temperature and atmospheric pressure was normal practice; standard interchangeable ground-glass joints were common; and other equipment like Jena sintered-glass filters was also widely used. Todd did some work on the structure of apocholic acid, a degradation product of cholic acid, and a proposed structure was published in the *Zeitschrift für physiologische Chemie*.[1] The structure proved to be erroneous, and he never worked on the bile acids again.

Todd had some very interesting and amusing experiences in the beginning at Frankfurt because his German was not fluent enough to make diplomatic comments at the appropriate time. For example, when Professor J. von Braun invited him to attend his group's seminar, Todd replied in his blunt German that he was busy on Saturdays and 8 A.M. was too early for him anyway. It is not surprising that Todd's relationship with von Braun, director of the Chemical Institute at Frankfurt, was rather cool thereafter. Todd completed his doctorate in 1931 after an oral examination in which the evaluator, a professor of physics, occupied most of the time reading the daily newspaper. Having gained a mastery of German while in Frankfurt, Todd was able to make close friendships with Germans, which gave him deep insight into German culture. He was an honorary member of *Freie Landsmannschaft Franco-Saxonia,* a fairly tame dueling club. He did attend one event in Sachsenhausen, at which a few facial cuts were inflicted followed by the consumption of vast quantities of beer, despite an early hour. He found the economic–political situation approaching chaos. He concluded that economic trouble coupled with a weak and vacillating government opens the door for totalitarianism, a regime to be ruled by an appropriate demagogue: Enter Adolf Hitler.

Todd was interested in working in the natural product field and made an abortive attempt to work with N. D. Zelinsky in Moscow.

---

[1] **1931,** *198,* 173.

Although he held a German doctorate, he obtained a senior studentship to work on a British Doctorate of Philosophy at Oxford under Robert Robinson. The research involved the synthesis of anthocyanins (red and blue coloring matter of flowers). He received his second doctorate in chemistry, a rare accomplishment, from Oxford in 1933. Todd's athletic ability in lawn tennis was a decisive factor in his selection to be a member of the Oriel Senior Common Room, which made a great difference in his social life at Oxford.

Todd's experience in Robinson's laboratory was very stimulating and productive. At one of their "tea and crossword puzzle" sessions, Todd made a real breakthrough in the synthesis of all the major diglucosidic anthocyanins. His attempts involved the condensation of acetobromoglucose with unprotected phloroglucinaldehyde, but only intractable syrups and gums resulted. One day, while he was concentrating the reaction gum in methanol and also having tea and working on a crossword puzzle, the flask overheated, slipped from his fingers, and ended in the water bath. He removed it from the water, set it aside with the dirty glassware, and was astounded to see crystals of 2-β-tetraacetyl-D-glucopyranosyl phloroglucinaldehyde the next morning. Todd's sojourn at the Dyson Perrin laboratory at Oxford was an extremely valuable experience; apart from establishing a permanent relationship with Robinson, he befriended many chemists. By 1932 Todd had completed the synthesis of hirsutin, pelargonin, malvin, and cyanin and thus had effectively rounded out the anthocyanins.

In 1934 Todd left Oxford and joined George Barger at the University of Edinburgh to work on the structure of vitamin $B_1$. B. C. P. Jansen had sent Barger 5 mg of $B_1$ that he had isolated from rice (which contains a few milligrams per ton). With the aid of Hoffmann-La Roche and the Rockefeller Foundation plus Dr. Franz Bergel (as well as Anni Jacob from Frankfurt, Juan Madinaveitia from Madrid, Karimullah from Lahore, Keller from Basle, and Fraenkel-Conrat from Germany), an industrially important synthetic route was discovered, but the first complete synthesis was accomplished by others. The synthesis allowed Hoffmann–La Roche to take a major commercial share of the world $B_1$ market.

Todd's stay in Edinburgh had another bright aspect as he met his future wife, Alison Dale, who was doing postdoctoral work with A. J. Clark in pharmacology. They married in 1937 after he had moved to the Lister Institute. At the young age of 28 it was decided not to make Todd a reader at the Institute; this was corrected a few months later.

Many years afterward, the correspondence relating to Todd's appointment fell into his hands. He found out that Robinson had commented that Todd was good for the Lister but that it was doubtful the Lister was good enough for Todd. These comments almost certainly had not been very well received in the Lister Institute. Barger, Robinson, and Dale (Sir Henry Dale, Nobel Laureate in physiology or medicine, 1936, and future father-in-law of Todd) were influential in Todd's move to the Lister Institute.

In his Pedler Lecture, on March 7, 1946, in London, Todd stated that "this approach must obviously begin with the simplest units and work up to the complex, and it has been and remains a major interest in my laboratory." An indication of his goal in the synthesis approach to nucleotides and specifically to the first successful synthesis of adenosine triphosphate (ATP) was his statement, again in the Pedler Lecture of 1946, that "although any synthetic approach to the nucleotide problem must ultimately lead, if successful, to the nucleic acids, my colleagues and I entered this field with more limited objectives."

The synthesis of the di-, tri-, and polynucleotides would require three steps according to Todd: first, the synthesis of the appropriate nucleoside; second, its special phosphorylation; and finally, the specific linking of nucleosides by phosphate or polyphosphate residues. At the time when Todd and his colleagues began their effort, no biologically significant purine or pyrimidine nucleoside had been synthesized. G. E. Hilbert and C. E. Rist had synthesized 3-D-ribopyranosidal uracil[2] and E. Fisher had made adenine glucuronate.[3] There was also a synthesis for adenine made available by W. Traube.[4] The Traube synthesis involved a formyl derivative of triaminopyrimidine. Because the succeeding reaction with the formyl triaminopyrimidine to form the ribofuranoside would lead to decomposition, Todd and colleagues—J. Baddiley, B. Lythgoe, and D. McNeil—devised a thio analogue that worked very well.[5] It might be of interest here to point out Todd's practice of placing his colleagues' names first on their publications. The first synthesis involved the synthesis of 9-alkylpurines as model compounds. Their next was a synthetic method for purines[6] that used a condensation of

[2] *J. Biol. Chem.* **1937,** *117,* 371.
[3] *Ber.* **1914,** *47,* 210.
[4] *Annalen* **1904,** *331,* 64.
[5] *J. Chem. Soc.* **1943,** 383.
[6] *J. Chem. Soc.* **1943,** 386.

formamidine and benzeneazomalonitrile, which was followed by reduction with Raney nickel, and then a treatment with dithioformate
to produce the adenine ring. In the actual synthesis of adenosine, di-
aminopyrimidine was condensed with ribose in the presence of an acid
catalyst in ethanol to produce a ribofuranoside (*l*). The ribofuranoside
(*l*) was treated with dichlorobenzenediazonium chloride and reduced to
introduce the third amino group. The addition of potassium dithioformate and heating produce adenosine. Periodate was used to establish
the ring size of the ribose moiety in all the synthetic nucleosides.

Todd and his colleagues made a very significant contribution to
the total structure elucidation of vitamin $B_{12}$ in 1956. They were able to
complete the unambiguous synthesis of the hexacarboxylate derivative
of vitamin $B_{12}$.[7] The vitamin $B_{12}$ hexacarboxylate was then used by
D. Hodgkin and her coworkers for the X-ray analysis of the total
three-dimensional structure of vitamin $B_{12}$.[8]

Todd went to the Lister Institute at age 28 in 1936 and set up a
laboratory at the University of Manchester in 1938, where he began
his studies of nucleotides. In 1944 he went to Cambridge to hold the
chair in organic chemistry. He was knighted in 1954 and received a
Life Peerage, becoming Baron Todd of Trumpington in 1962.

Lord Todd's membership on the Advisory Council on Scientific
Policy (ACSP) stimulated his interest in the interaction of science and
government usually referred to as "science policy". This council grew
out of the Scientific Advisory Committee in the War Cabinet of Great
Britain during the years 1940–1945. Todd's father-in-law, Sir Henry
Dale, then President of the Royal Society, was a member of that committee. So from 1947 onward Todd became active in the affairs of
science and technology and the government. It was membership on
the ACSP and his experiences with Solly Zuckerman and Henry Ti-
zard that taught him how to run a committee effectively. From 1952
to 1964 he served as chairman of the ACSP.

Todd served as managing trustee of the Nuffield Foundation, a
major charitable foundation in England. He served successively as managing trustee, deputy chairman, and chairman until 1979 and continues
to serve as chairman of the Ordinary Trustees.

Todd had some interesting experiences as a visiting professor in
California, at the Universities of Chicago and Sydney, and at the Mas-

---

[7] *Nature* **1954**, *174*, 1168.

[8] *Nature* **1956**, *178*, 64.

sachusetts Institute of Technology. After the completion of his work on nucleotide synthesis, he received many invitations to lecture abroad. In 1956 two Russian students, N. K. Kochetkov and E. A. Mistryukov, came and worked in Todd's laboratory. They returned to Russia with Kochetkov becoming an Academician and the director of the Zelinsky Institute of Organic Chemistry in Moscow. Earlier efforts to arrange visits of Russians with known friends of Russia—P. M. S. Blackett and J. D. Bernal—had failed. Todd was fairly fluent in Russian after some tutoring by Natasha Squire, who taught Russian at Cambridge University.

Todd's laboratory was always an international one, with visiting students (G. Khorana is one outstanding example) from the United States, Canada, Australia, and many other countries.

PAUL MELIUS
*Auburn University*

# Bibliography

Baddiley, J.; Lythgoe, B.; McNeil, D.; Todd, A. R. "Synthesis of Purine Nucleotides Part I. Model Experiments on the Synthesis of 9-Alkylpurines"; *J. Chem. Soc.* **1943**, 383–385.

Baddiley, J.; Lythgoe, B.; Todd, A. R. "Experiments on Synthesis of Purine Nucleotides Part II. New and Convenient Synthesis of Adenine"; *J. Chem. Soc.* **1943** 386–389.

Todd, A. R. "Organic Chemistry 1851-1951"; *Advancement of Science* **1952**, *8*, 393–396.

Todd A. R. "Synthesis in the Study of Nucleotides"; *J. Chem. Soc.* **1946, 647**–653.

Todd, A. R. *A Time to Remember: The Autobiography of a Chemist;* Cambridge University: Cambridge, England, 1983; New York, 1984.

Todd A. R. "Vitamins of the B Group"; *J. Chem. Soc.* **1941,** 427–432.

# Frederick Sanger

## 1918–

Frederick Sanger was born on August 13, 1918, in Gloucestershire, England, and after education at Bryanston School he studied at St. John's College, Cambridge, obtaining his B.A. degree in natural sciences in 1939. Next he followed an advanced course in biochemistry, in which he obtained a first-class award, and then he began research for the Ph.D. degree in the department of biochemistry under the supervision of A. Neuberger. His work leading to the award of the degree in 1943 was concerned with lysine metabolism and proved to be a sound basis for his subsequent work on insulin, for which he was awarded the Nobel Prize in chemistry in 1958. From 1944 to 1951 he held a Beit Memorial Fellowship for Medical Research, and from 1951 until his retirement in 1983 he was a member of the staff of the Medical Research Council. In 1962 the Council opened the new Laboratory for Molecular Biology in Cambridge, and Sanger moved there to continue his studies on nucleic acids, which he had begun following the award of the Nobel Prize for his work on insulin. These studies were brought to successful conclusions, and in 1980 he was again awarded a Nobel Prize in chemistry (together with Paul Berg and Walter Gilbert). He is one of the very few scientists who have been awarded two Nobel Prizes.

Sanger began his work on insulin at an exciting time in the study of protein chemistry. The work had been made possible partly by the preparation of proteins in pure form such that they could be expected to have a unique structure. The protein insulin had become readily available in crystalline form, and A. C. Chibnall, who had recently taken up the chair of biochemistry at Cambridge, suggested to Sanger that he should study the free amino groups in insulin.

Chibnall's work had shown that the number of free α-amino groups in insulin could be equated with the number of polypeptide chains in the protein molecule, and it had been shown that one such terminal group was phenylalanine.[1] Sanger first developed an improved method for characterizing amino-terminal residues using the reagent 1,2,4-fluorodinitrobenzene, which interacts with free amino groups under mild alkaline conditions to give a relatively stable derivative.[2] Using this method, he confirmed the presence of a free α-amino residue attached to phenylalanine and also showed the presence of some attached to glycine. Thus it appeared that insulin had two peptide chains of which phenylalanine and glycine were the respective N-terminal residues. Work in other laboratories had shown considerable variation in values for the molecular weight of insulin, and it was only during the course of Sanger's work that a firm value was obtained corresponding to the presence of two polypeptide chains.

Any attempt to determine the sequence of amino acid residues in the insulin molecule as a whole clearly required the separation and purification of the two polypeptide strands. Because insulin is rich in cystine residues, union of the strands through disulfide bridges seemed the most likely. The fission of the disulfide residues by reduction proved inappropriate because of the reversibility of the process and the formation of polymeric products. Sanger introduced a new technique whereby the disulfide residues were oxidized by performic acid, producing a cysteic acid residue at each side of the bridge.[3] This technique allowed the production of pure phenylalanine and glycine chains in the oxidized form and gave material suitable for meaningful experiments on the sequence of amino acid residues.

The interaction of a polypeptide with dinitrofluorobenzene followed by hydrolysis with acid gives only the nature of the amino-(N)-terminal residue. In order to determine sequences of residues beyond this, peptides were required resulting from partial hydrolysis. A mild

[1]Jensen, H.; Evans, E. A. "Studies on Crystalline Insulin XVIII: The Nature of the Free Amino Groups in Insulin on the Isolation of Phenylalanine and Proline from Crystalline Insulin"; *J. Biol. Chem.* **1935,** *108,* 1.

[2]Sanger, F. "The Free Amino Groups of Insulin"; *Biochem J.* **1945,** *39,* 507.

[3]Sanger, F. "Fractionation of Oxidized Insulin"; *Biochem. J.* **1949,** *44,* 126.

acid hydrolysis of the dinitrophenyl (DNP) peptide resulted in the isolation of products containing two, three, and a few more amino acid residues, thus giving the initial sequence, but a lack of suitable separation methods prevented this approach from giving more than very limited information. The technique of paper chromatography had been introduced in 1944, and it proved eminently adaptable to Sanger's problem. (The yellow color of DNP derivatives was an added advantage.) The hydrolysis of the polypeptides from insulin by enzymes yielded peptides of varying length. By using acid and enzymes to cause fission at different locations, the determination of the sequences of the resulting small polypeptides was carried out, and, by matching one sequence against another, longer and longer lengths of total polypeptide could be sequenced. This systematic approach led to the full sequences of both the phenylalanine and glycine chains.

The determination of the positions of the cystine $-S-S-$ linkages in the insulin molecule proved to be a difficult task. First, one $-S-S-$ linkage occurs between two cysteine residues in the glycine chain, and conditions for the fission of peptide linkages between these two residues proved difficult to find. The other two $-S-S-$ linkages join pairs of cysteine residues in the two chains, thereby linking them together. An unforeseen difficulty that had to be overcome was that the interchange of cystine-bridging links takes place during the acid hydrolysis of unoxidized insulin. However, by using $-SH$ inhibitors this difficulty was overcome, and the precise location of the $-S-S-$ linkages, including the two that link the two chains, completed the elucidation of the structure of the protein.

These researches differed from many equally noteworthy discoveries in which a single observation, or reaction by hypothesis, led to outstanding results. In the case of insulin, by virtue of the fact that it led to the first elucidation of the structure of a protein, many previously unexpected experimental problems arose. The outstanding success of the achievement depended on a clear enunciation of problems as they arose and on persistence in finding solutions. It showed for the first time that the determination of the chemical structure of a protein is an attainable objective. Because of the peculiar nature of the insulin molecule, it did not lay down methods of general applicability. It did, however, encourage others to tackle other proteins and to solve the problems intrinsic to the elucidation of their structures. Sanger's work was carried out using bovine insulin. Subsequent work showed small differences among insulins from different species that might result in immunological reactions in diabetic patients under treatment with the bovine hormone. This was an incentive very much later to find a way of producing a human insulin using the methods of genetic engineering. It must be pointed out at this stage that the achievement of that objective depended, in part, on Sanger's later researches for which he received his second Nobel Prize in 1980.

The beginning of Sanger's work on the sequencing of nucleic acids coincided approximately with his move to the newly established Laboratory of Molecular Biology. The problems associated with sequencing nucleic acids are different from those met with proteins, because only four monomer units are involved, necessitating longer units to be sequenced so that overlapping segments can be matched. This, in turn, increases the difficulty of separating closely similar oligonucleotides, and Sanger saw that radioactive labeling with $^{32}$P might hold the answer to this problem. To achieve the high degree of resolution required, high levels of labeling were needed together with the improved separation techniques, including two-dimensional ionophoresis.[4] Nucleases cutting RNA at specific positions were already available, and ribonuclease $T_1$ cutting after guanine nucleotides was used extensively. These methods led to the sequencing of the largest RNA up to that time, namely, 5S RNA.[5] However, before the general approach could be extended to determining the genetic code in m-RNAs, this had been achieved by quite different methods not involving determinations of sequences.

The application to DNA of the experimental methods of radioactive labeling and the ionophoresis of oligonucleotides presented greater difficulties because of the larger molecular weights and, hence, greater complexity of mixtures to be resolved. Experiments designed to overcome these difficulties led, almost accidentally, to what was to be perhaps the biggest step forward in the approach to sequencing nucleic acids.

Not only were efficient methods of separation required but also specific degradative enzymes analogous to ribonuclease $T_1$, used in the work on RNA. Moreover, the naturally occurring available single-stranded polydeoxyribonucleotides of unique structure were too large for the development of the necessary techniques. During the course of attempts to prepare suitable ogliomers synthetically, using DNA polymerase with the single-stranded ØX174 as template, high spe-

---

[4]Sanger, F.; Brownlee, G. G.; Barrell, B. G. "Two-Dimensional Procedure for Radioactive Nucleotides"; *J. Mol. Biol.* **1965,** *13,* 373–398.

[5]Brownlee, G. G.; Sanger, F.; Barrell, B. G. "The Sequence of 55 Ribosomal Ribonucleic Acid"; *J. Mol. Biol.* **1968,** *34,* 379–412.

cific radioactivity in the product was sought by using low concentrations of the highly labeled monomers. This resulted in the cessation of polymerization when all the labeled monomer was used up. Thus oligomers were produced, terminating in the nucleotide *preceding* the one that had been labeled and had been exhausted. Thus, by using each nucleotide in turn as the terminating unit, the distribution of each could be arrived at. A more rapid way of applying the same basic idea was developed as the "plus and minus" method.[6] A still more rapid and adaptable variation, which was the final one adopted, used 2′,3′-dideoxy nucleoside triphosphates to bring about the termination of synthesis of the growing chain.[7] These methods all require a single-stranded template which, with some DNA viruses, caused no difficulty. However, most DNAs are double-stranded, and the production of single strands required the cloning in a single-stranded virus of DNA fragments produced either by the use of sonication or restriction enzymes. In addition, increasingly efficient methods were required for resolving mixtures of oligonucleotides according to molecular weight, and the method ultimately adopted was ionophoresis on acrylamide gel.

Various techniques were developed with increasing degrees of success, and the dideoxy nucleotide method emerged as the climax of Sanger's work on sequencing polynucleotides. As discussed in connection with the work on insulin, progress in the work came by patient and persistent improvement and by the development of techniques to solve individual problems. The ultimate success is to be measured by the efficiency, speed, and general applicability of the dideoxy method. This could not have been achieved through following a predetermined experimental plan. Thus, the distinction of the award of the Nobel Prize was not only so well deserved for the ultimate outcome, but also as recognition of the importance to science of the careful resolution of problems as they present themselves, as opposed to the single flash of intellectual insight.

G. R. BARKER
*The Biochemical Society, London*
*(Deceased)*

---

[6]Sanger, F.; Coulson, A. R. "A Rapid Method for Determining Sequences in DNA by Primed Synthesis with DNA Polymerase"; *J. Mol. Biol.* **1975**, *94*, 441–448.

[7]Sanger, F.; Nicklen, S.; Coulson, A. R. "DNA Sequencing with Chain-Terminating Inhibitors"; *Proc. Natl. Acad. Sci. U.S.A.* **1977**, *74*, 5463–5467.

# Bibliography

Barley, K. "An Appreciation of the Scientific Contributions of Frederick Sanger"; *Chemistry and Industry* **1958,** *December 13 (50),* 1653–1654.

Sanger, F., "The Chemistry of Insulin"; In *Nobel Lectures, Chemistry, 1942–1962;* Elsevier: Amsterdam, The Netherlands, 1964; pp 544–556.

Fraenkel-Conrat, H. "The Chemistry of Proteins and Peptides"; *Ann Rev. Biochem.* **1956,** *25,* 291–330.

Judson, H. F. *The Eighth Day of Creation: Makers of the Revolution in Biology;* Simon & Schuster: New York, 1979.

Kolata, G. B. "The 1980 Nobel Prize in Chemistry"; *Science* **1980,** *210,* 887–889.

Sanger, F., "Sequences, Sequences and Sequences"; *Ann Rev. Biochem.* **1988,** *57,* 1–28.

Sanger, F.; Brownlee, G. G. "Methods for Determining Sequences in RNA"; *Biochem Soc. Symp.* **1970,** *30,* 183–197.

Sanger, F., "Nucleotide Sequences in DNA"; *Proc. Roy. Soc. London B.* **1975,** *191,* 317–333.

Sanger, F., "Determination of Nucleotide Sequences in DNA"; *Science* **1981,** *214,* 1205–1210.

Silverstein, A.; Silverstein, V. *Frederick Sanger; The Man Who Mapped out a Chemical of Life;* John Day: New York, 1969.

Sutton, C. "Genetic Engineers Sweep the Board"; *New Scientist* **1980,** *88,* 217–219.

# Jaroslav Heyrovský

## 1890–1967

Jaroslav Heyrovský was born in Prague, Czechoslovakia, on December 20, 1890. He received a B.S. degree from University College in London in 1913, a Ph.D. from Charles University in Prague in 1918, and a D.Sc. degree from the University of London in 1921. In 1926 Heyrovský married Marie Kořánová, and they had two children, Judith and Michael. His professional career was spent at Charles University and at the Polarographic Institute of the Czechoslovak Academy of Science. Heyrovský's fame is due to his invention of polarography and his development of the technique into a major method for chemical analysis. For these accomplishments he received the First State Prize of Czechoslovakia in 1951 and the Nobel Prize in chemistry in 1959. He was active in the Czechoslovak Chemical Society and made numerous contributions to the advancement of science in Czechoslovakia. He passed away on March 27, 1967, after a lengthy illness.

One of the most elegant methods of chemical analysis is the voltammetric method developed by Heyrovský. He was introduced to the dropping mercury electrode and electrochemical reactions at a mercury electrode during his doctoral studies. The development of polarography soon followed and a lifelong study ensued. Seldom has a chemist initiated an area of chemistry and then remained a central figure in the full development of that area, and seldom has a chemist devoted so much of a lifetime to one area of chemistry.

Jaroslav Heyrovský was the son of a professor of Roman law at what was then called the Charles-Ferdinand University in Prague. After developing an interest in science and mathematics in secondary school, the *Akademicke Gymnasium,* Heyrovský entered Charles University in 1909 at the age of 19. The work of Sir William Ramsay inspired him to transfer to University College in London. After receiving a B.S. degree in 1913, Heyrovský continued his studies under F. G. Donnan and initiated an investigation of the electrode potential of aluminum. During these studies, Donnan suggested that Heyrovský use a dropping mercury electrode to overcome the passivation problem that was encountered when using an aluminum electrode. Thus began the lifetime interest of the young scientist.

In 1914, during his postbaccalaureate studies, Heyrovský traveled to Prague for a holiday but was prevented from returning to London because of the outbreak of World War I. He obtained facilities for continuing his studies at Charles University in the lab of J. S. Štěrba-Böhm. However, Heyrovský was soon drafted into the army. Fortunately, he was assigned to work in a military hospital, where he served as a dispensing chemist and a roentgenologist. He was able to continue his studies during his free time there, and in 1918 he submitted his Ph.D. thesis entitled "The Electro-Affinity of Aluminum" to the university in Prague and was awarded a Ph.D degree.

Heyrovský became an assistant under Professor Bohuslav Brauner at Charles University and continued studies on aluminic acid and aluminates. This work was the basis for a habilitation thesis, which qualified him to become *Docent* of Physical Chemistry in 1920. As a result of three publications from this work, he was also awarded a Doctor of Science degree in 1921 from the University of London. In 1922 the paper describing the invention of polarography was published in the Czech journal *Chemicke Listy*.[1] With a strong record of scientific accomplishment, Heyrovský quickly advanced in position. In 1924 he became extraordinary professor and director of the newly established Institute of Physical Chemistry at Charles University. By 1928 he had become full professor. During the ensuing years, Heyrovský devoted all of his time to the advancement of polarography. Although many research laboratories were seriously disrupted by World War II, Heyrovský was able to continue his work at Prague owing to arrangements

---

[1]Heyrovský, J. "Elektrolysa se rtutovou kapkovou Kathodu"; *Chem. Listy* **1922**, *16*, 256–264.

made by the German scientist J. Böhm. In 1950 the Polarographic Institute was established with Heyrovský as director. In 1952 the Institute became part of the Czechoslovak Academy of Science. Heyrovský continued his research on polarographic methods and published his results regularly throughout his life.

The development of polarography resulted from the interplay of a number of fortunate events. In his thesis studies on the electrode potential of aluminum, Heyrovský was plagued by problems of the passivation of aluminum electrode surfaces because of the formation of an oxide coating. Donnan had recently used sodium amalgams to measure the activity of sodium, and he suggested that Heyrovský use this technique for aluminum also. The flow of mercury out of a glass capillary formed a mercury droplet. Periodically, the droplet fell and a fresh mercury droplet formed, providing a renewed mercury surface. A number of investigators were applying various methods to measure the electrocapillary curves of mercury, that is, the surface tension as a function of applied voltage. Bohumil Kučera was doing so by weighing mercury drops, since the surface tension was directly proportional to the weight of the drop.

As it turned out, Kučera was on Heyrovský's examination committee, and the examination turned into a discussion of the electrocapillary phenomenon. Kučera invited Heyrovský to work with a colleague of his, Dr. Šimůnek, in investigating some anomalous results that he had obtained. The investigation by Heyrovský using Kučera's method added nothing to the understanding of the anomalies. But to avoid the tedium of weighing drops, Heyrovský began a study of the current between the dropping mercury electrode and a reference pool of mercury at the bottom of the cell. Further investigations showed that characteristic stepped current versus voltage curves were obtained when reducible species were present. The position of the curve was characteristic of the particular species, and the magnitude of the current response was a function of the concentration of the electroactive species. In 1925 Heyrovský published a series of papers outlining the concepts of electrochemistry at the dropping mercury electrode, including general principles of the technique,[2] the instrument,[3] and the

[2]Heyrovský, J. "Researches with the Dropping Mercury Electrode. Part I. General Introduction"; Recl. Trav. Chim., Pays-Bas. 1925, 44, 488–495.

[3]Heyrovský, J.; Shikata, M. "Researches with the Dropping Mercury Electrode. Part II. The Polarograph"; ibid., 496–498.

theory of overpotential.[4] An apparatus that automatically recorded the current–voltage curves on photographic film was constructed through the assistance of Masuzo Shikata, a co-worker from Japan, and the term "polarograph" was coined. Worldwide development of the polarographic technique was the direct result of Heyrovský's development and promotion through publications and travel. The first major publications on the practice of polarography appeared in the mid-1930s. In 1933 Heyrovský spent six months as Carnegie Visiting Professor at the University of California at Berkeley, Stanford University, and the California Institute of Technology. He also visited several other American universities at the time. The following year he was invited to Leningrad (St. Petersburg) to present a lecture on polarography at the Mendeleev Centenary celebration. In addition, Professor W. Böttger, the noted German analyst, traveled to Prague for the purpose of acquiring the technique for use in his laboratory. Noted scientists from Japan (M. Shikata), Poland (W. Kemula), Australia (B. Breyer), Italy (G. Semierano), and the United States (O. H. Müller, I. M. Kolthoff, J. S. Lingane) visited the Prague laboratory and were influential in spreading the technique throughout the world.

The most common applications of polarography involve the qualitative identification and the quantitative determination of well-behaved (reversible) electroactive species in solution. Because metals usually exhibit multiple stable oxidation states, they are ideally suited to polarographic analysis. Polarography received great impetus during World War II, when its use in metallurgical analysis became extremely valuable. Because of the high overvoltage of hydrogen on mercury and the formation of stable alloys with mercury, even alkali metals can be analyzed polarographically. Multiple components of a sample can be analyzed simultaneously because of the separation of each polarographic wave. For example, iron, copper, or lead alloys can be analyzed for a number of components in a single procedure. Organic compounds usually undergo irreversible reductions, but polarographic analysis of these has still proved to be an area of great interest. When the electroactive species undergoes a chemical reaction during reduction, the current response is modified. Often the current response can be used to determine reaction rate constants. Other physical parameters such as equilibrium constants and diffusion coefficients can also be derived from polarographic analyses.

---

[4]Heyrovský, J. "Researches with the Dropping Mercury Electrode. Part III. Theory of Overpotential"; *ibid.*, 499–502.

A primary reason for the award of the Nobel Prize to Heyrovský was the importance of polarography in applied chemical analyses. Polarography has also found use in medical and pharmaceutical fields. It has been used to diagnose cancer and inflammatory diseases. One of the more notable tests is the Brdička filtrate test, in which the polarographic analysis of protein fractions of blood serum in the presence of divalent cobalt provides information about the pathological state of a patient. Polarography can provide rapid, sensitive analysis of the lead content of blood and has been used to ascertain the vitamin C content of fruits and vegetables.

Pulse polarography, oscillographic polarography, chronopotentiometry, AC polarography, cyclic voltammetry, and amperomeric titrations are all outgrowths of the original polarographic method. Heyrovský spent a great deal of time in developing the oscillographic technique. Although classical polarography has been supplanted by other electrochemical techniques in many applications, the dependence of these techniques on the fundamental voltammetric principles defined by Heyrovský cannot be overlooked.

Jaroslav Heyrovský was a prominent figure in Czechoslovak science. He lived by Faraday's motto "work, finish, publish", but he recognized the limitations facing Czech scientists in disseminating the results of their work. With his friend, Emil Voteček, he started the *Collection of Czechoslovak Chemical Communications,* which first appeared in 1929 under their coeditorship. The original publications appeared in French and English (usually translated by the editors) but later in Russian and German as well. Heyrovský was active in the early organization of the Czechoslovak Academy of Sciences and was the first director of the Polarographic Institute when it became part of the Academy. The staff of the Polarographic Institute and the Institute of Physical Chemistry was dominated by Heyrovský's co-workers for many years. Realizing the importance of documentation in the dissemination of scientific information, Heyrovský prepared bibliographies of the polarographic literature, a tradition that continued for years with the assistance of many colleagues, particularly O. H. Müller. As a result, literature in the area of polarography is much more accessible than that in most areas of chemistry.

That Heyrovský was an experimentalist is obvious from his work. He was quick to explain experimental facts but was happy to leave theoretical derivations to others. In addition, as editor of the *Collection of Czechoslovak Chemical Communications,* he refused to publish purely theoretical papers that contained no experimental verification. He was quick to point out that Einstein received the Nobel Prize for study of the photoelectric effect, not for the theory of relativity.

On a personal side, Heyrovský's dedication to the development of polarography was at the cost of significant personal sacrifice. In his early years, Heyrovský enjoyed tennis, soccer, mountaineering, and skiing. However, his commitment to polarography left no time for these activities. But he did find time for the music that he enjoyed so much. In addition, he loved to entertain guests with good cooking, red wine, good jokes, and, on occasion, satirical plays based on scientific life.

Heyrovský's professional pursuits became a family affair. In living up to Newton's motto as displayed at the Institute ("A man must resolve either to produce nothing new or to become a slave to defend it"), Heyrovský devoted all his energy to the advancement of polarography. In doing so he inspired both students and family members to do likewise. His wife Marie was an unofficial secretary and was deeply involved in the preparation of the bibliographical works. His son, Michael, yearned to be with his father in the laboratory, even at a very young age. When the time was right, he became a co-worker with his father and presently holds a position at the Jaroslav Heyrovský Institute of Physical Chemistry and Electrochemistry.

Heyrovský had an unusually strong influence on those around him through his example of dedication and purpose. His modest style, enthusiasm, patience, and personal warmth attracted many students to join him in his research. His associates came to sense his belief that life should be devoted to a higher purpose and that science is a particularly worthwhile purpose. These beliefs inspired his son to follow in his footsteps in spite of the warnings that "work in science consumes men physically and mentally for the price of only very rare flashes of pure happiness." Having achieved international prominence, Heyrovský was able to spread his personal influence worldwide. He passed on his sense of moral responsibilty in science and urged cooperation among scientists of all nations. People like Jaroslav Heyrovský, through their personalities and accomplishments, make the world a better place.

The helpful comments of Michael Heyrovský are gratefully acknowledged.

JERRY WALSH
*University of North Carolina at Greensboro*

# *Bibliography*

Belcher, R. "Obituaries: Professor Jaroslav Heyrovsky"; *Nature* **1967**, *214*, 953.

Brdička, R. "To the Sixtieth Birthday of Professor J. Heyrovský"; *Collect. Czech. Chem. Commun.* **1950,** *15,* 691–698.

Butler, J. A. V.; Zuman, P. "Jaroslav Heyrovský 1890–1967"; *Biogr. Mem. Fellows R. Soc.* **1967,** *13,* 167–191.

Heyrovský, J. "Trends in Polarography"; in *Les Prix Nobel* (1959); reprinted in *Science* **1960,** *132,* 123–130; and in *Nobel Lectures: Chemistry, 1942–1962;* Elsevier: New York, 1964.

Khás, L. *Jaroslav Heyrovský, Founder of Polarography;* Kavanova, J. R., Trans.; Orbis: Prague, Czechoslovakia, 1968.

Laitinen, H. A.; Ewing, G. W., Eds. *A History of Analytical Chemistry;* American Chemical Society: Washington, D.C., 1977.

*The Nobel Prize Winners: Chemistry;* Magill, F. N., Ed.; Salem: Pasadena, CA, 1990; Vol. 2.

Olander, A. "Jaroslav Heyrovský, Nobel Prize Winner"; *Review of Polarography* **1960,** *8,* 83–84.

*Progress in Polarography;* Zuman, P.; Meites, L., Eds.; Interscience: New York, 1962– 1972.

Teich, M. "Heyrovský, Jaroslav"; In *Dictionary of Scientific Biography;* Gillispie, C. C., Ed.; Charles Scribner's Sons: New York, 1972; Vol. 6, pp 370–376.

Zuman, P. "With the Drop of Mercury to the Nobel Prize"; In *Electrochemistry, Past and Present;* Stock, J. T.; Orna, M. V., Eds.; American Chemical Society: Washington, DC, 1989; pp 339–369.

Zuman, P.; Elving, P. J. "Jaroslav Heyrovský: Nobel Laureate"; *J. Chem. Educ.* **1960,** *37,* 562–567.

# Willard Libby

## 1908–1980

Willard Frank Libby was born on December 17, 1908, in Grand Valley, Colorado, and died on September 8, 1980. He was the son of a farmer, Ora Edward Libby, and Eva May Libby. He married Leonor Hickey in 1940 and was divorced in 1966, at which time he married Leona Woods Marshall. They were the parents of twin daughters, Janet Eva and Susan Charlotte. He earned degrees in chemistry from the University of California at Berkeley (B.S., 1931; Ph.D., 1933).

Libby's professional career began with a faculty appointment to the University of California at Berkeley (1933–1945). That was followed by a World War II assignment as a chemist at Columbia University in the War Research Division (1941–1945); then he became professor of chemistry at the Institute for Nuclear Studies and the Department of Chemistry, University of Chicago (1945–1954), followed by an appointment to the U.S. Atomic Energy Commission. He was director of the Institute for Geophysics and Planetary Physics, University of California at Los Angeles (1959) and also served as special visiting professor at the University of Colorado (1967). Libby received numerous awards. They included the Remsen Memorial Lecture Award (1955); Bicentennial Lecture Award, City College of New York (1956); Nuclear Applications in Chemistry Award (1956); Cresson Medal, Franklin Institute (1957); Willard Gibbs Medal of the American Chemical Society (1958); Priestley Memorial Award, Dickinson College (1959); Albert Einstein Medal (1959); Day Medal, Geological Society of America (1961); Gold Medal, American Institute of

Chemists (1970); and the Lehman Award, New York Academy of Sciences (1971). He was elected to the National Academy of Sciences in 1950.

The Nobel Prize was awarded for his work leading to radiocarbon ($^{14}$C) dating, which was developed from the knowledge that cosmic rays collide with air nuclei, generating a shower of secondary nuclei. These nuclei then collide with abundant isotopic $^{14}$N to form $^{14}$C. The $^{14}$C was found to spread throughout the atmosphere, mixing with ordinary carbon atoms that then entered into the carbon and carbon dioxide cycle and thus the food chain through animal and plant assimilation. Because $^{14}$C is a radioactive isotope of carbon, which is common to living plants and animals, a measurement of it enabled scientists to determine the absolute ages of various organic materials. Libby developed sensitive techniques to detect the amount of $^{14}$C activity in a sample. The detection was accomplished by the design, development, and construction of extremely sensitive Geiger counters. In order to reduce the effects of natural terrestrial radiation attributed to background, Libby enclosed the counters with 8-inch-thick iron shielding. To further reduce cosmic radiation, he constructed within the iron shielding around the detector, a permanent layer of Geiger counters in tangential contact with each other. These were connected in a manner such that they would turn themselves off during the thousandth of a second time interval that the passage of a cosmic ray was registered. This enabled him to relate the amount found to the constant amount when the substance was a living plant or animal, and thus the age of the archaeological sample could be determined. The accuracy of the technique was tested on objects of known age, such as the sequoia tree and various wooden artifacts from Egyptian tombs whose ages were quite well documented. His techniques later proved reliable for ages from about 5000 years, with a high degree of probability through 50,000 years, or as far back as the carbon-14 could be detected at the time. In later years, as the techniques and instrumentation were improved, radiocarbon dating provided insight into worldwide climatic changes, human evolutionary development, and other events, such as changes within the Earth.

During World War II Libby worked on the Manhattan project, focusing his research on developing the techniques of gaseous diffusion needed to separate the isotopes of uranium. While on assignment at Columbia he worked under Harold Urey. His contributions provided an essential step for the production of the atomic bomb. Libby's work on isotopes helped shift his interest toward nuclear physics. Libby's 1954 appointment to the U.S. Atomic Energy Commission was made by President Dwight Eisenhower. He served with the AEC until 1959, when he rejoined the faculty of the University of California at Los Angeles.

His early work (1946) on cosmic rays also led to the production of isotopic tritium ($^3$H). He found that trace amounts are always in the atmosphere and are thus present in water vapor found there. The further refinement of detecting and measurement techniques has led to the accurate quantitative determinations of tritium concentration that are useful in dating well water, wine, and other water-based solutions.

GEORGE F. WOLLASTON
*Clarion University of Pennsylvania*

# Bibliography

Berger, R. "Willard Frank Libby, 1908–1980"; In *$^{14}$C and Archaeology*; Mook, W. G.; Waterbolk, H. T., Eds.; Council of Europe: Strasbourg, 1983.

Berland, T. *The Scientific Life*; Coward-McCann: New York, 1962.

Burleigh, R. "W. F. Libby and the Development of Radiocarbon Dating"; *Antiquity* **1981**, *55*, 96–98.

Libby, W. F. "Chemistry and the Peaceful Uses of the Atom"; *Chem. Eng. News* **1957**, *35*, 14–17.

Libby, W. F. *The Publications of Willard Frank Libby*; Berger, R.; Libby, L. M., Eds.; GeoScience Analytical: Santa Monica, CA, 1982; 7 volumes.

Libby, W. F. *Radiocarbon Dating;* University of Chicago: Chicago, IL, 1952, 1955.

*Nobel Prize Winners* Wasson, T., Ed.; The H. W. Wilson Company: New York, 1987.

Renfrew, C. *Before Civilization: The Radiocarbon Revolution and Prehistoric Europe*; Alfred A. Knopf: New York, 1973.

# 1961

NOBEL LAUREATE

# Melvin Calvin

## 1911–

Melvin Calvin, winner of the 1961 Nobel Prize in chemistry for his research in
photosynthesis, is a native of St. Paul,
Minnesota, where he was born to Russian immigrant parents on April 8, 1911.
As a child, Calvin was fascinated with the
role of chemistry in his surroundings, and
this interest led him to major in chemistry at Michigan College of Mining and
Technology (now Michigan Technological
University), where he graduated with a
bachelor of science degree in 1931. In graduate school at the University of Minnesota,
Calvin studied aspects of the electron affin-
ity of halogens under the direction of Professor George A. Glockler.
On completing his Ph.D. thesis in 1935, Calvin traveled to the University of Manchester in England to study with Professor Michael
Polanyi; his stay was funded by a Rockefeller grant. At Manchester,
Calvin's interest in the chemistry of the photosynthetic process was
fostered. His research on coordination catalysis was specifically aimed
at understanding the electronic basis for the photochemical properties of the metalloporphyrins heme and chlorophyll and their synthetic
analogues.

At the invitation of Professor Gilbert N. Lewis, Calvin began his
career at the University of California at Berkeley in 1937 as an instructor of chemistry. In 1941 he became an assistant professor of chemistry
and in 1947 was promoted to full professor. Calvin was associated
with the Manhattan Project in 1944–1945, developing methods for the
purification of uranium and plutonium. From 1946 to 1980 Calvin was

director of the Bio-Organic Division of Lawrence Radiation Laboratory at Berkeley, which in 1960 became the Laboratory of Chemical Biodynamics. He currently maintains his position as University Professor of Chemistry.

Calvin's interest in theoretical aspects of the properties of organic molecules led to significant collaborations with colleagues at Berkeley, notably "The Color of Organic Substances" with Lewis and *The Theory of Organic Chemistry* with Professor G. E. K. Branch.

In the mid-1940s, with the conclusion of World War II, a long-lived radioactive isotope of carbon, carbon-14 ($^{14}C$), became readily available for research purposes. Calvin recognized that its availability made possible the study of one aspect of photosynthesis in a unique way: the sequence of reactions responsible for the reduction of carbon dioxide ($CO_2$) to carbohydrate using radioactive carbon-14-labeled $CO_2$ ($^{14}CO_2$). With the support of Ernest O. Lawrence, Director of the Radiation Laboratory at Berkeley, Calvin and his co-workers began the experiments that eventually led to the elucidation of the biochemical pathways responsible.

Prior to Calvin's work, some of the basic theoretical considerations that stimulated his plan to elucidate the reactions involved in photosynthetic $CO_2$ utilization were already known. From the early work of Antoine-Laurent Lavoisier, Nicholas de Saussure, Julius Sachs, and others, it was known that atmospheric $CO_2$ was the source of carbon utilized for the photosynthetic formation of carbohydrates in plants. More recently Robin Hill had shown that ferric iron could be utilized as an oxidant in place of $CO_2$, resulting in the generation of oxygen gas by isolated chloroplasts, and thus had demonstrated that the photochemical production of oxygen was an event distinct from the reduction of $CO_2$. This idea supported the earlier suppositions of Cornelius van Niel, F. F. Blackman, and Otto Warburg that the reduction of $CO_2$ to form carbohydrate was a reaction that was not necessarily directly dependent on the primary photochemical event.

Since it was known that atmospheric $CO_2$ enters the plant and that the carbon eventually appears in all of its organic constituents, Calvin's strategy for the elucidation of the pathways responsible was to treat a photosynthetic organism with $^{14}CO_2$ for defined and varied time intervals and then to abruptly terminate the metabolic reactions. Next the radioactively labeled cellular constituents were isolated and chemically analyzed to determine the fate of the radioactive carbon. The analysis of the time course for the appearance of the radioactivity in different metabolic chemical intermediates allowed the determination of the sequence of chemical reactions by which the $CO_2$ ultimately was converted to the major storage carbohydrates, starch and sucrose. Most of the initial experiments utilized a unicellular green photosynthetic alga, *Chlorella,* rather than a higher plant, because of the ease with which large quantities of it could be grown and experimentally manipulated. However, all of the important aspects of the research were eventually verified in a wide variety of photosynthetic organisms.

Some of the first $^{14}CO_2$ labeling experiments verified the hypothesis that the reduction of $CO_2$ to carbohydrates was not directly dependent on light, because after a period of illumination and exposure to $^{14}CO_2$, cells continued to incorporate some radioactivity into organic metabolic intermediates for a short period of time. The researchers observed a complex mixture of radiolabeled organic compounds. The separation of some of these components by ion-exchange chromatography, followed by chemical identification, provided important information, including the observation that many of the intermediates were anionic in nature and probably contained several negative charges per molecule. Calvin personally identified the three-carbon compound 3-phosphoglyceric acid as the predominant radioactive compound in the photosynthetic organism following a short (less than one second) exposure to the $^{14}CO_2$. This experiment was of primary importance, since it indicated that the first step in the formation of carbohydrate was the condensation of $CO_2$ with some organic intermediate to yield 3-phosphoglyceric acid.

It was apparent that classical methods for the isolation and identification of the large number of radiolabeled organic intermediates of the photosynthetic carbon pathway would be laborious and time-consuming and that large amounts of material would be necessary for the identification of some specific compounds. The necessity of a relatively rapid analytical method led to the development of a two-dimensional paper chromatographic map of the photosynthetic intermediates. The methodology became available with the advent of partition chromatography, developed by A. J. P. Martin and R. L. M. Synge, for which they received the Nobel Prize in 1952. Following exposure of photosynthetic organisms to $^{14}CO_2$, cell or tissue extracts were separated by paper chromatography, photographic film was layered next to the paper, the radioactive areas in the paper were identified by their exposure of the photographic film, and the radioactivity was quantitated. The chemical identification of the radioactive compounds that migrated charac-

teristically to unique positions on the chromatogram was the result of considerable effort, requiring both the determination of the
mobility of authentic standards and the elution and chemical analysis of
unknown components. With the development of the chromatographic
map of the photosynthetic intermediates, detailed kinetic analyses of
the incorporation of $^{14}CO_2$ into a variety of metabolic intermediates
became technically feasible. This work was completed in Calvin's lab
by many workers, notably Andrew A. Benson, a postdoctoral fellow,
and James A. Bassham, a graduate student with Calvin, who remained
in the laboratory as a postdoctoral fellow.

It later became apparent that four-, five-, six-, and seven-carbon
phosphorylated compounds were also rapidly labeled intermediates of
the photosynthetic carbon pathway. Initially, the roles of some of these
intermediates were unclear. The simplest explanation for the initial reaction in the scheme was that the $CO_2$ condensed with a two-carbon
intermediate to form the initial three-carbon product that had been
observed previously (3-phosphoglyceric acid). This erroneous supposition led to some confusion before it finally became apparent to Calvin
that the immediate precursor was actually a five-carbon intermediate
that condensed with $CO_2$ and then split to form two molecules of
a three-carbon product. With the realization that the five-carbon intermediate, ribulose-1,5-bisphosphate, was the acceptor molecule that
condensed with the $CO_2$, the other proposed reactions rapidly fell into
place in a complex cyclic series. This series of reactions is commonly
referred to as the Calvin cycle or the Calvin–Benson–Bassham cycle, in
honor of the researchers, or more descriptively as the reductive photosynthetic carbon cycle. In 1961 Melvin Calvin was awarded the Nobel
Prize in chemistry for this work on photosynthesis, which was carried
out between 1945 and 1961. The focus of the work is summarized in
the Nobel lecture he delivered on December 10, 1961, entitled "The
Path of Carbon in Photosynthesis".

Since the description of these reactions by Calvin and co-workers,
several closely related metabolic pathways have been elucidated. Sup-
plemental pathways for $CO_2$ utilization have been described that work
in concert with the Calvin cycle to allow certain types of plants to utilize
low levels of $CO_2$ very efficiently. A coexisting oxidative photosynthetic
carbon cycle, or photorespiratory pathway, has been described that is
initiated by a side oxidation reaction catalyzed by the same enzyme
that had originally been identified by Calvin and co-workers as the
enzyme that catalyzes the condensation of $CO_2$ with ribulose-1,5-
bisphosphate. Considerable research continues on the regulation of the

processes by which $CO_2$ is utilized and by which storage-carbohydrate forms are synthesized and translocated in photosynthetic organisms.

In recent years Calvin has focused his efforts on developing artificial systems to utilize solar energy in the photochemical generation of hydrocarbon fuel sources that might eventually be of practical use. In a somewhat unrelated area, he has also undertaken an interest in the mechanisms of carcinogenesis.

In addition to being awarded the Nobel Prize, Dr. Calvin has received six honorary degrees and numerous other honors and awards including the Sugar Research Foundation Prize (1950), the Flintoff Medal and Prize of the Chemical Society (1955), the Hales Award of the American Society of Plant Physiologists (1956), the Richards Medal (1956), Nichols Medal (1958), Priestley Medal (1978), and the Oesper Prize (1981) of the American Chemical Society, the Davy Medal of the Royal Society (1964), the Gold Medal of the American Institute of Chemists (1979), and the National Medal of Science (1989). Dr. Calvin was elected to the National Academy of Sciences in 1954 and is a member of the American Association for the Advancement of Science, the American Society of Biological Chemists, the American Society for Cell Biology, the American Society of Plant Physiologists (president 1963–1964), the Royal Society of London, the American Physical Society, and the American Chemical Society (president 1971).

H. DAVID HUSIC
*Lafayette College*

DIANE W. HUSIC
*East Stroudsburg University*

# Bibliography

Calvin, M. "Exploring the Path of Carbon in Photosynthesis"; In *Plant Physiology;* Salisbury, F. B.; Ross, C. W., Eds.; Wadsworth: Belmont, CA, pp 152–153.

Calvin, M. *Following the Trail of Light;* Profiles, Pathways, and Dreams Series; Seeman, J. I., Ed.; American Chemical Society: Washington, D.C., 1992.

Calvin, M. "The Path of Carbon in Photosynthesis"; In *Les Prix Nobel en 1961;* Liljestrand, M. G., Ed.; Imprimerie Royale P.A. Norstedt: Stockholm, Sweden, 1962; pp 156–183.

Calvin, M. "The Path of Carbon in Photosynthesis"; *Science* **1962**, *135*, 879–889.

Moritz, C., Ed.; *Current Biography Yearbook*; H. W. Wilson: New York, 1983.

Farber, E. *Nobel Prize Winners in Chemistry 1901–1961*; Abelard-Schuman: London, 1963; pp 301–308.

*McGraw-Hill Modern Men of Science;* Greene, J. E., Ed.; McGraw-Hill: New York, 1966; Vol. 1, p 85.

"Melvin Calvin"; In *Les Prix Nobel en 1961;* Liljestrand, M. G., Ed.; Imprimerie Royale P. A. Norstedt: Stockholm, Sweden, 1962; pp 100–102.

*Nobel Prize Winners;* Wasson, T., Ed.; H. W. Wilson: New York, 1987.

*The Nobel Prize Winners: Chemistry*; Magill, F. N., Ed.; Salem: Pasadena, CA, 1990; Vol. 2.

# John Kendrew

## 1917–

The Nobel Prizes of 1962 recognized scientific achievements in two broad segments of molecular biology: Francis Crick, James Watson, and Maurice Wilkins shared the award in physiology or medicine for their role in the elucidation of the structure of DNA, and Max Perutz and John Kendrew received the chemistry prize for their structural studies on the proteins hemoglobin and myoglobin. The Royal Swedish Academy of Sciences thereby emphasized the connection between nucleic acids, the genetic libraries of the cell, and proteins, the regulators of life processes. The two awards simultaneously highlighted the successes of a structural approach to biological questions, the mode of thought that Kendrew exemplified in his research on myoglobin. Born in the town of Oxford, England, on March 24, 1917, John Cowdery Kendrew has played a major part in the development of molecular biology.

Kendrew was reared in the academic environment of Oxford University, where his father (Wilfrid George) lectured on climatology. He received his early education at the Dragon School in Oxford and at Clifton College in Bristol before entering Cambridge University to pursue studies in chemistry. After graduation in 1939, he enlisted with the Ministry of Aircraft Production, where his wartime service dealt with the development of radar units for aviation. As a consequence of his defense work, he had frequent discussions with J. D. Bernal, a prescient physicist who was serving as an advisor to the British government. A pioneer in many fields, Bernal had undertaken the earliest crystallographic studies of proteins during the 1930s with Dorothy Crowfoot Hodgkin and later with Max Perutz. Convinced of the promise of the life sciences, Kendrew returned to Cambridge, where he had a fellowship remaining from before the war, and there became a doctoral candidate under W. H. Taylor, Reader in Crystallography at the Cavendish Laboratory. Also on the staff was Perutz, only three years older than Kendrew and a ready resource for the new graduate student, who would become a close colleague.

Kendrew's initial research involved a comparison of fetal and adult hemoglobins, a problem suggested by the physiologist Joseph Barcroft. Working with W. L. Bragg, an astute physicist who had received the Nobel Prize for his fundamental studies of X-ray diffraction, Kendrew and Perutz formulated a helical model for polypeptides that in many respects anticipated Linus Pauling's concept of protein structure.[1] After Kendrew received his doctorate in 1949, however, he shifted his attention to myoglobin, the molecule responsible for the storage of oxygen in muscle tissues. In retrospect, this study complemented Perutz's simultaneous work on hemoglobin, the oxygen-transport protein with a cooperative binding pattern that is absent in myoglobin. Nevertheless, Kendrew has emphasized that his choice of proteins was dictated by more mundane considerations. Myoglobin was a relatively small protein that would in principle be resolvable by existing techniques; it was available in large quantities; and most importantly, it could be obtained in crystalline form. Although myoglobin gave good X-ray photographs, the images proved impossible to interpret until Perutz recalled a strategy that his one-time mentor Bernal had proposed more than a decade earlier: the incorporation of heavy atoms as a frame of reference for obtaining the phase relationships of the diffracted waves. Kendrew and his co-workers examined several hundred such heavy atom derivatives of myoglobin before a suitable set of crystals was found. Using this technique, Kendrew was able in 1957 to sketch the rough outlines of myoglobin at 6 Å resolution. Additional X-ray photographs and more extensive calculations narrowed the focus two years later to 2 A, almost at the level of atomic resolution. The refinement of the X-ray analysis ultimately yielded a three-dimensional structural map of myoglobin in true atomic detail.

The solution of the myoglobin problem depended critically on technical innovations in several areas. The first was Perutz's

---

[1] Bragg, 1965.

introduction of isomorphous replacement into the domain of protein crystallography. Finally the processing of this information relied heavily on the electronic computing facilities at Cambridge. The University maintained a state-of-the-art computer laboratory, and Kendrew took full advantage of it in executing the mathematical manipulations required to solve the structure of myoglobin. Indeed, he pioneered the application of computers in the field of X-ray crystallography and actively promoted their use.

As the first protein to be viewed at high resolution, myoglobin provided the verification of many concepts that had been inferred from basic principles of peptide chemistry. The protein consists of eight helical segments varying in length from 7 to 24 amino acid units, with dimensions that followed exactly the theoretical predictions made by Pauling and Robert Corey. Although the amino acid sequence of myoglobin was largely unknown prior to Kendrew's structural studies, the X-ray map unexpectedly provided clues to the identity of many of the monomer units, and these assignments were confirmed by chemical analyses at the Rockefeller Institute in New York. The nonhelical strands in the protein often contained proline, which because of its ring structure should interfere with helix formation. The side chains of polar amino acids resided on the external surface of the myoglobin, whereas nonpolar residues with hydrophobic substituents were located in the interior of the structure, far away from the water molecules of the surrounding medium.

Nevertheless, the three-dimensional structure of myoglobin revealed a contorted and irregular molecule, lacking the aesthetic symmetry that characterized the double helix of DNA. Whereas an understanding of the complementary strands of DNA in 1953 led to broad generalizations about the mechanism of replication, the solution of the myoglobin problem did not permit any universal statements about the structure–function relationships of proteins. Instead, the significance of Kendrew's achievement rested on the fact that the structure could be determined at all. Prior to his studies on myoglobin, many scientists believed that the technical difficulties in solving such complex structures were too formidable, and this pessimistic group even included former practitioners such as Isidor Fankuchen. Kendrew himself referred to his and Perutz's research as a "mad pursuit".[2] The madness lay in the choice of large, complex molecules at a time when

---

[2]Olby, 1985.

the structures of many simpler compounds were still unknown. Such an endeavor required faith in the power of X-ray crystallography as well as perseverance to endure the years of painstaking effort necessary. The pursuit of protein crystallography contravened the conventional wisdom of choosing simple research problems that would lead to straightforward answers and rapid publication.

The laborious process by which the structure of myoglobin was solved offers a revealing contrast with the more abstract approach involved in the deduction of the DNA double helix. Kendrew has observed that molecular biology evolved as two distinct communities: the Conformational school (of which he and Perutz were members) and the Informational school (typified by the work of Max Delbrück and Salvador Luria). The latter group was concerned with the principles that governed the storage and inheritability of genetic traits, which Oswald Avery in 1944 had shown to be carried within DNA rather than proteins. The gulf between the two schools was not only intellectual but also geographical, with the Informationalists predominantly in the United States and the Conformationalists (aside from Pauling) in Britain. According to Kendrew, association between them was specifically hindered by Delbruck's disdain for chemistry and by his preference for quick, elegant experiments over the tedious approach of the X-ray crystallographers.[3] Watson, a product of Luria's group who was recruited by Kendrew to provide a biological perspective at Cambridge, formed a personal bridge between the communities, and the Watson–Crick model demonstrated in a dramatic way the relevance of three-dimensional structure to genetic information transfer. This successful interplay between structure and information has continued in the study of genetically altered proteins and in the characterization of proteins that regulate gene expression.

What the early crystallographers and the molecular geneticists had in common was a belief in the application of the physical sciences to the understanding of biological systems. Indeed, this was the goal of molecular biology in the sense that Warren Weaver of the Rockefeller Foundation had used the term. This philanthropic organization supported much of the early research in the field, including Max Delbrück's work at Caltech, William Astbury's research at Leeds University on fibrous proteins, and Perutz's early studies of hemoglobin. Beginning in 1947, however, Kendrew and Perutz

---

[3]Kendrew, 1967.

received their primary financial backing from a funding agency of the British government called the Medical Research Council (MRC). The MRC provided a secure base for the development of protein crystallography at Cambridge, and it created an institutional niche for the new specialty under the MRC Unit for Molecular Biology. Perutz served as the director from its beginning in 1947 as the Unit for the Study of the Molecular Structure of Biological Systems until the opening in 1962 of a new building known as the Laboratory for Molecular Biology. Kendrew also played an active administrative role as the deputy chairman of the Laboratory, which has provided a home for many of the most prolific molecular biologists, including Francis Crick, Sydney Brenner, Frederick Sanger, and Aaron Klug.

A significant manifestation of the growth of molecular biology was the establishment of specialized publications, the most important of which was the *Journal of Molecular Biology*. With the backing of publisher Kurt Jacoby, Kendrew founded this periodical in 1959 and remained the senior editor until 1987, publishing many of the seminal contributions to the field. The bias toward structural principles was apparent in the editor's stated intention to accept papers dealing with the "nature, production and replication of biological structure, and its relationship to function".[4] This definition of molecular biology in the new journal reflected the views of William Astbury, who was among the first protein crystallographers to popularize the term in Britain. A specific aim of Kendrew's periodical was the development of communication among the various disciplines within the emerging specialty, and to that end he encouraged the use of a common language that would be mutually intelligible. The *Journal of Molecular Biology* thus provided a vehicle for the unification of the divergent schools of molecular biology.

Kendrew has also promoted molecular biology through his many administrative positions; his extensive government service earned him a knighthood in 1974. As a member (1965–1972) of the British Council for Scientific Policy, he issued an influential report calling for greater teaching and research opportunities in the new field. He also served as Secretary-General (1970–1974) of the European Molecular Biology Conference, and in 1975 founded the European Molecular Biology Laboratory at Heidelberg, where he held the senior post of Director-General until 1982. He also sought to present the

---

[4]Kendrew, J. C. "Instructions to Authors"; *J. Mol. Biol.* **1959**,1 (*April*), inside cover.

principles of molecular biology to the general public through a series of televised lectures, which were subsequently published as a book.[5] Throughout his career, Kendrew has contributed to the vitality of molecular biology and guided its development as a subject rooted in the principles of physics and chemistry.

Building on the foundations laid by Bernal, Astbury, and W. L. Bragg, molecular biology in Britain reached maturity in the work of Kendrew and Perutz on protein crystallography, and it was their concentration on biological structure that also provided the intellectual environment in which Watson and Crick deduced the double helical model of DNA. Kendrew's elucidation of myoglobin revealed a new vista that encouraged his colleagues to chart other proteins in similar detail. X-ray crystallography thus became the primary method for viewing the molecular architecture of living systems in three dimensions, and it therefore broadened the perspective of the informational biologists who had previously confined themselves to one-dimensional blueprints. Kendrew, through his own research as well as in his editorial and administrative capacities, has stressed the necessary relationship between structure and information in the development of molecular biology. A historical analysis of the intellectual and social factors involved in the emergence of molecular biology will include a critical role for the contributions of Sir John Kendrew.

WILLIAM J. HAGAN, JR.
*College of Saint Rose*

# *Bibliography*

Bragg, W. L. "First Stages in the X-ray Analysis of Proteins"; *Reports on Progress in Physics* **1965**, *28*, 1–14.

Kendrew, J. C. "How Molecular Biology Got Started"; *Scientific American* **1967**, *216*, 141–143.

Kendrew, J. C. "Information and Conformation in Biology"; In *Structural Chemistry and Molecular Biology;* Rich, A.; Davidson, N., Eds.; W. H. Freeman: San Francisco, CA, 1968; pp 187–197.

Kendrew, J. C. "Myoglobin and the Structure of Proteins (Nobel Address)"; *Science* **1963**, *139*, 1259–1266.

---

[5]Kendrew, 1966.

Kendrew, J. C. "Some Remarks on the History of Molecular Biology"; *Biochemical Society Symposia* **1970**, *30*, 5–10.

Kendrew, J. C. *The Thread of Life;* Harvard University: Cambridge, MA, 1966.

Kendrew, J. C. "The Three-Dimensional Structure of a Protein Molecule"; *Scientific American* **1961**, *205*, 96–110.

Kendrew, J. C.; Bodo, G.; Dintzis, H. M.; Parrish, R. G.; Wyckoff, H.; Phillips, D. C. "A Three-Dimensional Model of the Myoglobin Molecule Obtained by X-ray Analysis"; *Nature* **1958**, *181*, 662–666.

Law, J. "The Development of Specialties in Science: The Case of X-Ray Protein Crystallography"; *Science Studies* **1973**, *3*, 275–303.

*Nobel Prize Winners*; Wasson, T., Ed.; H. W. Wilson: New York, 1987.

*The Nobel Prize Winners: Chemistry*; Magill, F. N., Ed., Salem: Pasadena, CA, 1990; Vol. 2.

Olby, R. C. "The 'Mad Pursuit': X-ray Crystallographers' Search for the Structure of Hemoglobin"; *History and Philosophy of the Life Sciences* **1985**, *7*, 171–193.

Perutz, M. "Origins of Molecular Biology"; *New Scientist* **1980**, *85*, 326–329.

Watson, J. D. *The Double Helix: A Personal Account of the Discovery of the Structure of DNA*; Atheneum: New York, 1968.

# 1962

## NOBEL LAUREATE

# Max Perutz

## 1914−

Max F. Perutz's long professional life has been devoted mainly to a single field: the structure and function of the oxygen-binding protein hemoglobin. Born in Vienna, Austria, on May 19, 1914, he has spent most of his academic life in Cambridge, England, where he first applied X-ray crystallography to the structure of this complex molecule. For his achievements, he shared the Nobel Prize in chemistry in 1962 with his colleague John C. Kendrew.

As a chemistry undergraduate at the University of Vienna, Perutz became interested in biochemistry and persuaded his father, a textile manufacturer, to allow him to pursue his graduate studies in England. Perutz was attracted to the Cambridge laboratory of Frederick Gowland Hopkins, who had won the 1929 Nobel Prize in physiology or medicine "for his discovery of growth-stimulating vitamins". Perutz asked his professor of physical chemistry, Herman Mark, to find him a place in Hopkins's laboratory on a visit to Cambridge, but Mark forgot. Mark had been excited by the X-ray diffraction photographs of the enzyme pepsin taken by John Desmond Bernal and his assistant Dorothy Crowfoot (Hodgkin) and suggested to Perutz that he join Bernal's laboratory instead.

Perutz began his graduate work in Cambridge with Bernal in September 1936. Trained under the elder Bragg at the Royal Institution, Bernal had applied X-ray methods to the structures of graphite, amino acids, sterols, and water, and was about to turn to the crystallographic analysis of viruses. Bernal was intensely interested in many other subjects, including the history and social function of science. Perutz writes "we called him Sage, because he knew everything and had original views on any subject from physics to the history of art".[1] Perutz had arrived hoping to pursue research on a biological problem, but Bernal had no suitable crystals and put him to work on the structure of minerals. He returned to Austria on vacation in the summer of 1937 and sought the advice of his cousin's husband, the biochemist Felix Haurowitz, who had worked on hemoglobin. Haurowitz suggested that Perutz should take up the X-ray analysis of hemoglobin and told him to contact the Cambridge physiologist G. S. Adair, who had been the first to determine the molecular weight of hemoglobin correctly. Adair gave Perutz crystals of horse methemoglobin.

Cambridge had a long tradition of hemoglobin research since the physiologist Joseph Barcroft had begun looking at its oxygen equilibrium curve in 1907 and A. V. Hill had given it a mathematical interpretation. The physiologist Francis J. W. Roughton continued Barcroft's work and the biologist David Keilin was deeply interested in all heme proteins. Keilin later gave Perutz bench space for preparing his hemoglobin crystals. Thanks to these great pioneers, Perutz enjoyed a wealth of expertise on the biochemistry and physiology of hemoglobin. Buoyed by Bernal's "visionary faith in the power of X-ray diffraction", he pursued his studies in protein crystallography.[2]

Adair's crystals of horse hemoglobin yielded very good X-ray diffraction photographs. Perutz also took X-ray pictures of crystals of the digestive enzyme chymotrypsin, which John Northrop of the Rockefeller Institute had given to Bernal, but the twinning of these crystals frustrated their analysis. He therefore abandoned chymotrypsin and concentrated on hemoglobin. Perutz determined the dimensions and symmetry of the unit cell, which showed hemoglobin to be a dimer made up of identical halves.

In the autumn of 1938 the "sage" moved to Birkbeck College in London. Perutz continued his work at Cambridge, but he had no job and supported himself on the remainder of the money his father had given him when he left Vienna. The new Cavendish Professor of Physics, Lawrence Bragg, had formulated the law that is named after him when he was only 22 years old and had solved the first crystal structures, beginning with common salt, in 1913. He was excited by

---

[1] Perutz, 1980.
[2] Perutz, 1985.

the prospect of applying X-ray techniques to molecules as large as proteins and secured a grant from the Rockefeller Foundation that enabled Perutz to continue his work and get his doctorate in 1940.

The structure of hemoglobin was not solved until 19 years later. Disruptions combined with the forbidding difficulties of the scientific problem that he had taken on in a rash mood of optimism: Perutz had to get his parents a residence permit in Britain after they fled Nazi-occupied Austria; he suffered a half-year internment as an enemy alien after Britain entered the war; and in 1943 he was called in to help build a gigantic aircraft carrier made of ice.

Perutz faced fundamental problems in the interpretation of his X-ray pictures of hemoglobin, because they contained only the intensities and not the phases of the diffracted rays. Without knowledge of the phases it was impossible to solve the structure. After many futile attacks on the phase problem, he finally drew on the method of isomorphous replacement with heavy atoms, first applied to the structure of the alums in W. L. Bragg's Manchester laboratory by the young American crystallographer J. M. Cork. In 1954 Perutz attached mercury and silver atoms to hemoglobin to determine the phases. This was the turning point in his own work, and it opened up protein crystallography to many others. In the next five years, Perutz collected X-ray diffraction data from six isomorphous derivatives with heavy atoms in different positions on hemoglobin. This procedure provided him with accurate phases and led to a three-dimensional map of hemoglobin at a resolution sufficient to distinguish the fold of its four polypeptide chains and the positions of its four hemes.[3] The fold of the chains resembled that shown in the model of myoglobin that Kendrew had constructed two years earlier. It was known from chemistry that each polypeptide chain contained one heme made up of an iron atom surrounded by the four nitrogens of a porphyrin and a fifth nitrogen from the imidazole of a histidine residue. The X-ray results showed the heme to be enclosed in pockets of the globin that are accessible to oxygen on one side.

The classic sigmoid shape of the oxygen equilibrium curve of hemoglobin had been attributed to heme–heme interaction, but the wide separation now found between the hemes was difficult to reconcile with any direct interaction between them. In 1938 Haurowitz had found that crystals of oxy- and deoxyhemoglobin have different symmetries, from which he inferred that hemoglobin changes its structure

---

[3]Perutz et al., 1960.

when it combines with and releases oxygen. When the wide separation between the hemes provided no clue to the nature of heme–heme interaction, Perutz decided that comparison of the structures of oxy- and deoxyhemoglobin might solve the puzzle.

In the early 1960s Jacques Monod, François Jacob, and Jean-Pierre Changeux at the Pasteur Institute in Paris noticed similarities between the cooperative binding of oxygen by hemoglobin and the catalytic behavior of certain biosynthetic enzymes. They began to realize that the sigmoid equilibrium curve was just one instance of a type of behavior exhibited by many proteins, termed allosteric. This led them to formulate a general theory that Monod, Wyman, and Changeux later extended and put into mathematical form.[4]

The theory postulates that cooperative behavior is exhibited by proteins that are made up of several subunits held together in two or more discrete, reversible arrangements: a relaxed (R) structure, in which the subunits are unconstrained, and a tense (T) structure, in which they are constrained by additional bonds between them.

Hilary Muirhead and Perutz found in 1962 that the change in crystal form discovered by Haurowitz was indeed due to a rearrangement of the α and β subunits during the binding and release of oxygen, but the mechanism that produced this change remained obscure until 1970, when the structures of both oxy- and deoxyhemoglobin had been solved at atomic resolution.

The constraints predicted by Monod in the T structure took the form of salt bridges between the subunits that are broken on transition to the R structure. The transition between the two structures is triggered by a movement of the iron atoms relative to the porphyrin. In deoxyhemoglobin the iron is displaced from the porphyrin plane and the porphyrin is domed, whereas in oxyhemoglobin the iron moves into the plane of a flat porphyrin ring. The movement is due to an electronic transition and a change in coordination of the iron. In deoxyhemoglobin the iron is high spin and five-coordinated, and the iron–nitrogen distances are longer than in oxyhemoglobin, where it is low spin and six-coordinated.[5]

Perutz later predicted that there should be a reciprocal relationship between the spin state of the iron and the two structures of hemoglobin. He proposed to test this idea in derivatives in which the

---

[4]Monod et al., 1965.
[5]Perutz et al., 1987.

iron is in a thermal equilibrium between two spin states, arguing that if a transition of hemoglobin from low to high spin changes the structure of the protein from R to T, then a change of the protein structure from R to T should induce a transition of the iron from low to high spin. An experiment done in collaboration with two Italian physicists and others and with Perutz's son Robin bore out this prediction.

Hemoglobin carries oxygen from the lungs to the tissues, and it facilitates the return transport of carbon dioxide from the tissues back to the lungs. It does so by taking up protons on the release of oxygen, which helps to convert poorly soluble carbon dioxide into soluble bicarbonate ion. In the lungs this process is reversed. Perutz and J. V. Kilmartin found that the uptake of protons is linked to the formation of salt bridges in the T structure and their rupture in the R structure. In the T structure some of the histidines and α-amino groups form bridges with acid groups, which makes them take up protons and become positively charged. In the R structure the bridges are broken and the protons are released.[6]

The elucidation of the three-dimensional structure of hemoglobin has also explained the molecular pathology of hemoglobin diseases. In sickle-cell anemia, the hemoglobin precipitates into long fibers on loss of oxygen; these fibers distort the red blood cells into a sickle shape and make them rigid. In 1956 Vernon Ingram, a protein chemist working in Perutz and Kendrew's group, found that sickle-cell hemoglobin differs from normal hemoglobin in the replacement of one pair of valine by a pair of glutamine acid residues. It was a surprise that the replacement of a single pair of amino acid residues among 287 other pairs should affect the properties of a protein so profoundly. Chemists recently asked Perutz to help them develop antisickling agents by using X-ray analysis to determine their binding sites. This study led Perutz to formulate general rules governing the interaction of drugs with proteins.[7]

The emergence of powerful new techniques of site-directed mutagenesis has made it possible to introduce specific changes into polynucleotide templates to prepare custom-made proteins with individual substitutions. Perutz and M. Brunori advanced a hypothesis to explain the *Root effect,* which is the anomalously low oxygen affinity at low pH observed in teleost (bony) fish hemoglobins; the low affinity allows fish to secrete oxygen into the swim bladder in response to the

---

[6]Perutz et al., 1969
[7]Perutz et al., 1986.

secretion of lactic acid into the blood. Perutz's colleague Kiyoshi Nagai tested the hypothesis by directed mutagenesis but found that it failed to explain the Root effect.[8] This humbling conclusion illustrated Perutz's maxim that "what is obvious is not necessarily true".[9] Protein engineering has made it possible to probe the mechanism of oxygen binding more profoundly than the natural mutants had allowed, and it has led to a modified human hemoglobin that could serve as a blood substitute.

In recent years Perutz has become concerned with the effect of science on society and has expressed his ideas in his 1989 book *Is Science Necessary?* This collection also contains some of his reviews of books about science and scientists and an account of his absurd wartime experiences. Max Perutz works in the Medical Research Council's Laboratory of Molecular Biology in Cambridge, England.

WILLIAM J. HAGAN, JR.
*College of Saint Rose*

## Bibliography

Monod, J.; Wyman, J.; Changeux, J.-P. "On the Nature of Allosteric Transitions: A Plausible Model"; *J. Mol. Biol.* **1965**, *12*, 88–118.

Nagai, K.; Perutz, M. F.; Poyart, C. "Oxygen Binding Properties of Human Mutant Hemoglobins Synthesized in *Escherichia coli*"; *Proc. Natl. Acad. Sci.* **1985**, *82*, 7252–7255.

Perutz, M. F. "Early Days of Protein Crystallography"; *Methods in Enzymology* **1985**, *114*, 3–18.

Perutz, M. F. "Hemoglobin Structure and Respiratory Transport"; *Scientific American* **1978**, *239 (December)*, 92–125.

Perutz, M. F. *Is Science Necessary? Essays on Science and Scientists*; Dutton: New York, 1989.

Perutz, M. F. "Origins of Molecular Biology"; *New Scientist* **1980**, *85*, 326–329.

Perutz, M. F. "Physics and the Riddle of Life"; In *Selected Topics in the History of Biochemistry: Personal Recollections, III*; Semenza, G.; Jaenicke, R., Eds.; Comprehensive Biochemistry, Vol. 37; Elsevier: Amsterdam, Netherlands, 1990; pp. 1–20.

---

[8]Nagai et al., 1985.
[9]Perutz, 1978.

Perutz, M. F. "The Significance of the Hydrogen Bond in Physiology"; In *The Chemical Bond: Structure and Dynamics*; Zewail, A. H., Ed.; Academic Press: San Diego, CA, 1992; pp 17–30.

Perutz, M. F. "X-Ray Analysis of Hemoglobin (Nobel Address)"; *Science* **1963,** *140,* 863–869.

Perutz, M. F.; Fermi, G.; Abraham, D. J.; Poyart, C.; Bursaux, E. "Hemoglobin as a Receptor of Drugs and Peptides: X-Ray Studies of the Stereochemistry of Binding"; *J. Am. Chem. Soc.* **1986,** *108,* 1064–1078.

Perutz, M. F.; Fermi, G.; Luisi, B.; Shaanan, B.; Liddington, R. C. "Stereochemistry of Cooperative Mechanisms in Hemoglobin"; *Acc. Chem. Res.* **1987,** *20,* 309–321.

Perutz, M. F.; Muirhead, H.; Mazzarella, L.; Crowther, R. A.; Greer, J.; Kilmartin, J. V. "Identification of the Residues Responsible for the Bohr Effect in Hemoglobin"; *Nature* **1969,** *222,* 1243–1246.

Perutz, M. F.; Rossmann, M. G.; Cullis, A. F.; Muirhead, H.; Will, G.; North, A. C. T. "Structure of Haemoglobin: A Three-Dimensional Fourier Synthesis at 55-Å Resolution, Obtained by X-ray Analysis"; *Nature* **1960,** *185,* 416–422.

Watson, J. D. *The Double Helix: A Personal Account of the Discovery of the Structure of DNA;* Antheneum: New York, 1968; reprinted W. W. Norton: New York, 1980.

# Giulio Natta

## 1903–1979

Giulio Natta, born on February 26, 1903, in Imperia, Liguria, Italy, had a distinguished career in chemical research, in teaching, and as director of the Milan Polytechnic Institute. He became well-known in Europe for several contributions to the development of commercially important chemical processes and achieved world fame through his discovery, in 1954, of stereospecific polymerization and isotactic polypropylene. This and subsequent discoveries that opened and then expanded the field of "tactic" polymers won him a share of the Nobel Prize in chemistry in 1963. Natta died on May 2, 1979, in Bergamo, Italy.

Natta's was an old Italian family whose male members traditionally followed the legal profession. His father, his uncles, and his cousins were magistrates. Giulio would doubtless have followed this tradition had he not "bumped into" chemistry at the age of 12 (as a result of a family essay assignment) and become fascinated with it. He started his scientific training in mathematics at the University of Genoa but moved to the Milan Polytechnic Institute and took his *Dottore* degree in chemical engineering at the early age of 21. This was followed by the *Libero Docente* degree at the same institute three years later.

Natta's professional career advanced with commensurate rapidity, beginning with an assistant professorship at Milan. In 1933 he moved to Pavia and became professor and director of the Institute of General Chemistry and soon thereafter (1935) moved to the University of Rome as professor and director of the Institute of Physical Chemistry. In

1937 he headed the Institute of Industrial Chemistry at Turin, and he returned to Milan in 1938 as Professor and Director of the Milan Institute of Industrial Chemistry.

His early research involved the use of X-rays and electron diffraction in studying the structure of solids, including catalysts and polymers. He also studied the kinetics of reactions such as the hydrogenation of unsaturated compounds and the synthesis of methanol and higher alcohols from hydrogen and carbon monoxide. His introduction to high polymers resulted from a contact with Hermann Staudinger in 1932.[1] The first work in his own laboratory directly related to the production of polymers had to do with the purification, by extractive distillation, of butadiene, essential for the manufacture of synthetic rubber.

As early as 1938, Natta was studying the polymerization of olefins, and when Karl Ziegler announced his *Aufbau* reaction, which converted ethylene to *linear* low polymers, Natta took an immediate interest. He was then a consultant to the Italian chemical company Montecatini, which sponsored his research, and he persuaded them to take a license on Ziegler's information. Under this agreement, research information was exchanged between the two laboratories, and men from Natta's laboratory spent considerable time in Ziegler's laboratory at Mülheim. Thus, when Ziegler later discovered that certain transition-metal catalysts would convert ethylene to a linear high polymer, Natta was promptly informed, and he immediately tried applying the same catalysts to propylene.

In striking contrast to previous efforts to polymerize propylene with Friedel-Crafts catalysts, which had produced only viscous liquids, Natta obtained a high-melting, crystalline polymer. He proceeded quickly to determine that it had a stereoregular structure, with all of the methyl side groups on the same side of the polymer chain, and a helical crystal pattern. These and subsequent analyses were greatly facilitated by Natta's familiarity with X-ray diffraction techniques.

Natta recognized that stereoregularity was responsible for the unexpected properties of this first solid polymer of propylene; polymer chains with all of the side groups on the same side would fit together sufficiently closely to form crystalline regions ("microcrystallites"), and this microcrystallinity gave the polymer its high melting point, strength, and the ability to be oriented by drawing (and thus to make

[1]Natta, G. "Macromolecular Chemistry"; *Science*, **1965**, *147*, 261.

fibers). He realized further that this finding should have much wider significance, since it should be possible to produce similar effects with other monomers. He immediately visualized three potential applications for the new class of polymers: new plastics, new fibers, and new synthetic rubbers, and all three applications eventually became very important commercially.

The laboratory was reorganized with these goals in mind; however, the research program was focused primarily on broadening the study of transition-metal catalysts and applying them to the polymerization of other monomers, particularly α-olefins and olefin-diolefin mixtures. Natta (with the aid of his wife, who was a semanticist) introduced the following terms to describe the various possible stereoregular (and irregular) configurations of polymer chains:

Isotactic—all side groups on one side; "All-D"

Syndiotactic—side groups alternating from one side to the other; "DLDLDL"

Ditactic—for polymers with two different side groups

Atactic—random configuration

The investigation of stereospecific polymerization was rapidly extended to include a great many other monomers. The crystal structures and morphologies of the resultant polymers were carefully and elegantly elucidated. In addition to making isotactic polymers, catalysts and conditions were found that would form polymers with the other steric configurations named above. Stereoregular polymers were prepared from all of the common α-olefins and from vinyl aromatics (styrene, substituted styrenes, vinyl naphthalene), cycloolefins, and numerous nonhydrocarbon monomers. Optically active polymers were synthesized with the aid of optically active catalysts, and a detailed theory was advanced regarding the structure of the active catalysts and the mechanism of stereospecific polymerizations.[2]

Natta returned to an earlier field of interest, synthetic elastomers, with the preparation and characterization of polymers of butadiene and copolymers of ethylene and propylene. The rubber cis-1,4-polybutadiene, which was also produced in other laboratories, is exceptionally elastic and has attained considerable commercial importance.

---

[2]Natta, G. "Some Remarks on the Mechanism of the Stereospecific Ionic Coordinate Polymerization of Hydrocarbon Monomers"; Pure and Applied Chemistry 1966, 12 (January 4), 155.

The ethylene–propylene copolymer rubbers, having no residual unsaturation, are highly resistant to degradation. Cross-linking (vulcanization) was first accomplished by reaction with peroxides, but further work in Natta's and other laboratories showed that the incorporation of small amounts of a (nonconjugated) diene as a third monomer offers the advantage that the polymer can then be vulcanized with conventional rubber technology (reaction with sulfur). Such rubbers are now in large-scale commercial use.

The work of Natta's laboratory was characterized by efficient organization, original ideas, and high productivity in conducting research and in promptly publishing the results. In one five-year period Natta and some 60 co-workers published no less than 170 papers.[3] All told, Natta's publications numbered some 450 scientific papers. This remarkable output and the rapidity with which it bore fruit in large-scale commercial developments owed much to the availability of a well-equipped and well-staffed laboratory and the support of Italy's largest chemical company, Montecatini. Primarily, however, to quote Sir Robert Robinson, "The successful initiation, prosecution, and completion of so much and so varied research is the result of his most unusual originality, drive, and power of sustained work".[4]

As a result of their numerous contacts, Giulio Natta and Karl Ziegler became good friends, but after the discovery of polypropylene a serious disagreement arose in regard to the true priorities and licensing rights in that field. An agreement was finally worked out whereby Natta and Montecatini became the licensing agents for polypropylene, with Ziegler retaining rights for polyethylene. Within a few years a great many licenses were issued, and a great many commercial plants were built around the world. Polypropylene became, and remains, one of the world's largest volume and most important plastics.

In recognition of his many achievements, Natta was awarded several honorary degrees and numerous other awards and medals. He was invited to lecture at many different locations, and on his visits he was pursued by those interested in obtaining information on, or license to use, his discoveries; this despite the fact that licensing negotiations were handled by Montecatini. His contributions to synthetic rubber were recognized in the award of the first international synthetic

---

[3] *Stereoregular Polymers and Stereospecific Polymerizations: The Contributions of Giulio Natta and His School to Polymer Chemistry*, Vols. I and II; Natta, G.; Danusso, F., Eds.; Pergamon: New York, 1967.

[4] International Synthetic Rubber Award citation, October 3, 1961.

rubber medal in 1961.[5] In 1963 he shared the Nobel Prize in chemistry with Karl Ziegler (Ziegler was recognized primarily for his discovery of linear polyethylene, and Natta for originating and elucidating stereospecific polymerization and stereoregular polymers).[6]

As impressive as the rapid-fire research achievements in Natta's group is the rapidity with which research findings were carried forward into commercial applications. Although his close association with Montecatini was a real advantage, the major factors were Natta's own abilities, personality, and *modus operandi*. His entire career was spent in research, yet he maintained a keen awareness of and interest in potentials for commercial applications and the importance of early patent protection. Natta himself gave much credit to his training as a chemical engineer. His colleagues speak repeatedly of his very strong intuition, in which he had justifiable confidence, and of the intensity and energy with which he attacked every problem. He spent a large fraction of each day in reviewing individual research workers' programs in detail and recommending the direction for further work, yet he was not an autocratic director and remained open to suggestions and ideas.

Despite the dedication and long hours he devoted to science, Natta did not lack for family interests or outside activities. True to the tradition that "all Ligurians are nature lovers", he enjoyed at various times spelunking, mountaineering, hunting, fishing, and skiing. He had a mountain lodge that he used as a retreat and often took long walks in the woods. During such expeditions he collected mushrooms, and he developed sufficient expertise on edible fungi to qualify as an expert mycologist.

In 1958 Natta developed the first symptoms of Parkinson's disease. The disease progressed slowly, and he continued resolutely with a full work load. His wife, Rosita, died in 1968 after a brief illness. Natta finally retired in 1978 and lived in seclusion with his daughter, Franca, until his own death in the following year.

Giulio Natta's primary legacy to science and the world at large is a new field of research and development, well-defined with its own special nomenclature and a substantial literature, which has spawned innumerable commercial activities and a sizable family of useful products. His impact continues to be felt through the numerous well-trained

[5] *Rubber and Plastics Age (London)* October 3, 1961.
[6] *Science* **1963,** *142,* 938.

students and colleagues who are still expanding the field he opened and applying his principles and methods in other fields and in other centers of science and technology.

**FRANK M. MCMILLAN**
*Calsec Consultants*

# *Bibliography*

Bamford, C. H. "Guilio Natta—an Appreciation"; *Chemistry in Britain* **1981**, *17*, 298–300.

*Biographical Encyclopedia of Scientists;* Daintith, J., et al., Eds.; Facts on File: New York, 1981; 2 vols.

Danusso, F. "Pioneers of Polymers: Giulio Natta"; *The Plastics and Rubber Institute (London)*, **1981**, *26*.

*Guilio Natta, Present Significance of His Scientific Contribution;* Carrà, S., et al., Eds.; Editrice di Chimica: Milan, 1982.

*McGraw-Hill Modern Scientists and Engineers;* McGraw-Hill: New York, 1980; 3 vols.

McMillan, F. *The Chain Straighteners: Fruitful Innovation; The Discovery of Linear and Stereoregular Synthetic Polymers:* Macmillan: London, 1979.

McMillan, F. *Interview with I. Pasquon, Director, Milan Polytechnic Institute;* Milan, Italy (June 5, 1972). Transcript on file at Beckman Center for the History of Chemistry, 3401 Walnut St, Philadelphia, PA 19104-6228.

Morris, P. J. T. *Polymer Pioneers: A Popular History of the Science and Technology of Large Molecules;* Center for History of Chemistry: Philadephia, PA, 1986.

Natta, G. "Evolution des recherches sur les hauts polymers a l'École Polytechnique de Milan"; *Chimie et Industrie* **1963**, 89, 545–557.

Natta, G. "Macromolecular Chemistry: From Stereospecific Polymerization to the Asymmetric Synthesis of Macromolecules"; Nobel Prize Lecture, *Science* **1965**, *147*, 261.

Natta, G. "Stereospecific Polymerization of Macromolecules"; In *Nobel Lectures: Chemistry, 1963–1970;* Elsevier: Amsterdam, 1973; pp. 27–60.

*Nobel Prize Winners:* Wasson, T., Ed.; H. W. Wilson: New York, 1987.

*The Nobel Prize Winners: Chemistry;* Magill, F. N., Ed.; Salem: Pasadena, CA, 1990; Vol 2.

Pasquon, Italo. "History of the Development of the Stereospecific Polymerization Processes"; *American Chemical Society Symposium;* New York, September 13, 1966.

"Signifcato di un Premio"; *La Chimica e L'Industria* **1963,** *45,* 1326A (in Italian, with photographs).

*Stereoregular Polymers and Stereospecific Polymerizations: The Contributions of Giulio Natta and His School to Polymer Chemistry,* Vols. I and II. Natta, G.; Danusso, F., Eds.; Pergamon: New York, 1967.

# 1963

# Karl Ziegler

## 1898–1973

Karl Ziegler was born on November 26, 1898, in Helsa, near Kassel, in Germany. The son of a minister, he became interested in chemistry at an early age and was conducting experiments in a home chemistry laboratory by age 11. He went to the University of Marburg in 1916 to study chemistry and received his doctorate under Karl von Auwers in 1920. He married Maria Kurtz in 1922 and officially began his academic career with a lectureship at the University of Frankfurt in 1925. In 1926 he joined the faculty at Heidelberg and spent 10 years there as a professor of chemistry, until he was offered a chairmanship at the University of Halle. He subsequently became director and professor of chemistry at the Halle/Saale University. He did not remain there long, as he accepted an offer to become director of the Kaiser Wilhelm Institute for Coal Research (later known as the Max Planck Institute) in Mülheim, Germany, in 1943. In late 1952, while studying the reactions between aluminum alkyls and ethylene, he made a discovery that led to an important chain of events culminating with a unique catalytic system enabling the low-pressure polymerization of ethylene to give linear polyethylene of high molecular weight. It was for this work that he shared the 1963 Nobel Prize in chemistry with Giulio Natta (who discovered that these catalysts can polymerize α-olefins such as propylene to high molecular weights *and* high stereoregularity). Karl Ziegler then spent much of his time guiding the Institute toward the further development of the now-famous Ziegler catalyst system. He retired from the directorship in 1969 but stayed on as a scientific member of the staff until his death on August 11, 1973, at the age of 74, in Mülheim. So great was the impact of his work that a week after his death, the American Chemical Society began to organize a memorial symposium in his honor; it was held at the national meeting the following spring in Los Angeles, California.

During his tenure at Heidelberg, Ziegler concentrated on three main areas of interest to him: the chemistry of three-valent carbon, the chemistry of many-membered rings, and the chemistry of organoalkali metal compounds. These three fields of interest overlapped considerably; together they helped to set the stage for the discoveries to come. (His interest in organoalkali metal compounds eventually led him to the study and development of various other organometallic compounds, particularly aluminum alkyls, which were later to become so important). He first became interested in the chemistry and stability of three-valent carbon free-radical compounds in 1923 and published the first of many publications in which he sought to identify the steric and electronic factors responsible for the dissociation of hexa-substituted ethane derivatives.[1] Working with B. Schnell, he devised a method for the preparation of reactive organoalkali metal compounds (by cleaving ethers with alkali metals) and found that these compounds could easily be converted to the hexa-substituted ethane derivatives. The nature of the substituent could be easily and systematically altered using this synthetic route by simply changing the identity of the ether starting material. The experimental results Ziegler obtained for the activation and dissociation energies agreed well with the theoretical values by Erich Hückel (who is well-known for his approach to molecular orbital theory).

Ziegler's work with many-membered ring compounds (cyclic organic molecules many carbon atoms in length) also utilized the reactive nature of alkali metal compounds. Ziegler used the lithium and sodium salts of amines, such as the strong bases lithium diethylamide and sodium methylanilide, to accomplish the cyclization of long-chain hydrocarbons possessing terminal cyano groups ($\alpha,\omega$-dinitriles). The initially formed ring compound was then converted to the desired macrocyclic ketone product. Interest in the synthesis of these compounds arose following work by Leopold Ružička on the preparation of the naturally occurring macrocyclic ketones that have the character-

---

[1] Wilke, G. "Karl Ziegler, in Memoriam"; In *Coordination Polymerization*; Chien, J. C. W., Ed.; Academic: New York, 1975, pp 1–13

istic musky odor. Ziegler's synthetic method, which included running reactions at high dilution to favor the intramolecular cyclization over competing intermolecular reactions, resulted in yields superior to those of existing procedures. Ziegler and co-workers published the first of their series of papers on the preparation of large ring systems in 1933. For his work in this area and in free-radical chemistry he was awarded the Liebig Memorial Medal in 1935.

Ziegler's true forte, however, was his work with organoalkali metal compounds. In 1927 he found that when the olefin stilbene was added to an ethyl ether solution of phenylisopropyl potassium (prepared using his ether cleavage reaction), an abrupt color change from red to yellow took place. What he had observed was the first addition of an organoalkali metal compound (or any other organometallic compound, for that matter) across a carbon–carbon double bond, a reaction that he had earlier proposed in the *Habilitation* thesis he submitted in 1923 while at Marburg. Further work showed that this addition was not necessarily limited to the 1:1 adduct; he found that he could successively add more and more of the olefinic hydrocarbon butadiene to a solution of phenylisopropyl potassium and obtain a long-chain hydrocarbon with the reactive organopotassium end still intact. Oligomers such as these were the forerunners of the so-called "living polymers", in which the anionic chain end of the polymer can be propagated by adding more monomer. A current example of this is the polymerization of butadiene by the reactive anionic initiator *n*-butyllithium. Ziegler and co-worker H. Colonius developed a convenient general preparation for this lithium alkyl complex in 1930.

Having a convenient method for preparing butyllithium (from butyl chloride and lithium chips), Ziegler began to extensively study the reactivity of this compound (and other lithium alkyls) with olefins. An important discovery was that ethylene behaved similarly to butadiene with regard to the addition reactions to lithium alkyls. Specifically, when ethylene was heated under pressure with ethyllithium (or butyllithium), a "growth reaction" occurred, as had been seen earlier. Because Ziegler was at the Max Planck Institute for Coal Research, where ethylene was readily available as a byproduct from coal gas, his experiments with this cheap feedstock were of interest to the coal industry and were encouraged. His goal was to synthesize polyethylene of high molecular weight, but his attempts were thwarted because a competing elimination reaction always occurred, limiting the carbon chain length to approximately 10 to 12 carbon atoms.

A major breakthrough occurred in 1949 when Ziegler and co-worker H. G. Gellert decided to try the ether-soluble lithium aluminum hydride, $LiAlH_4$, which had become available in 1947. Ziegler had earlier observed that butyllithium decomposed to lithium hydride (LiH) and 1-butene during attempts at distillation under high vacuum and had postulated that the reaction may have been reversible: The addition of the olefin under pressure to LiH might yield the addition product. The reaction did not work as stated, owing partly to the insolubility of LiH in the ether solvent. The reaction of ethylene with $LiAlH_4$ did work, however, and gave a product mixture containing $\alpha$-olefins from 4 to 14 carbon atoms in length. Further experiments showed that lithium triethylaluminum was formed and that the reaction actively took place on the triethylaluminum end, rather than the ethyllithium end. This meant that aluminum was more active than lithium! The Ziegler group continued working with lithium aluminum hydride and found that aluminum worked wonderfully for the same type of "growth reaction" (or *Aufbaureaktion*, as Ziegler called it) they had observed earlier.[2] Additionally, this growth reaction took place on all three aluminum–carbon bonds of the aluminum trialkyls.

Ziegler soon found that the reaction of triethylaluminum with ethylene under pressure yielded aluminum alkyls and that the alkyl groups incorporated as many as 100 ethylene units. The growth reaction proceeded favorably at the Al–C bond at temperatures of 100–120 ° C, but once again, a competing elimination reaction occurred that prevented higher molecular weight polyethylene from being obtained. Such was the state of affairs until late 1952, when Ziegler and his student E. Holzkamp, while repeating one of their growth reaction experiments, made an interesting and useful discovery.

Instead of the expected long-chain polymer, they obtained the smallest possible polymer of ethylene: the dimer 1-butene. They reasoned that a contaminant must have been present to cause the premature displacement reaction and after a strenuous search found this to be the case; the autoclave had been contaminated with traces of colloidal nickel, which had been used in a previous hydrogenation experiment.[3] Ziegler realized the significance of this finding; if one

---

[2] Ziegler, K. "A Forty Years' Stroll through the Realms of Organometallic Chemistry"; In *Advances in Organometallic Chemistry*; Stone, F. G. A.; West, R., Eds.; Academic: New York, 1968; p 4.

[3] Boor, J., Jr. *Ziegler–Natta Catalysts and Polymerizations*; Academic: New York, 1979, Chapter 2.

metal could cause a premature displacement reaction, then perhaps another metal might delay the displacement reaction. In other words, it was possible that the *Aufbaureaktion* could be modified by adding catalytic amounts of various transition-metal salts to the reaction. The big day (in 1953) came when, after finding that metals such as cobalt and platinum behaved like nickel, Ziegler and his student H. Breil tried tris(acetylacetanato)zirconium(III) and this time found in the reaction vessel not 1-butene but "a solid cake of snow-white polyethylene".[4] They had found a transition metal complex that, when used as a cocatalyst with triethylaluminum, allowed them to obtain high molecular weight polyethylene (MW > 30,000) and, most importantly, to do so at low ethylene pressures. The conventional process for the polymerization of ethylene at that time involved ethylene pressures on the order of thousands of atmospheres. Further work revealed that the chlorides of titanium were the most powerful catalysts for the system, and suddenly the Ziegler group had a polymerization procedure for ethylene superior to all existing processes.

Ziegler then disclosed his catalyst to the Montecatini Company in Italy, a company, which Ziegler had granted a license to, in 1952 and for which Giulio Natta was acting as a consultant. Natta, who was the first to denote this class of catalysts as "Ziegler catalysts", became extremely interested in their ability and potential to stereoregularly polymerize α-olefins such as propene (propylene). Ziegler, meanwhile, concentrated mainly on the large-scale production of polyethylene and copolymers of ethylene and propylene by what he began calling the "Mülheim atmospheric polyethylene process".

Soon the scientific community was informed of his discovery. Highly crystalline and stereoregular polymers that previously could not be prepared became synthetically feasible, and commercial plants began producing the new polymers (today in the billions of pounds). Many of the plastics in everyday materials that we often take for granted are produced by Ziegler–Natta catalysis: plastics such as high-density polyethylene, isotactic polypropylene, and various elastomers such as *cis*-1,4-polyisoprene ("natural" rubber). Even state-of-the-art polymers, such as *trans*-polyacetylene (which can be made to conduct electricity by treating or "doping" the polymer with small amounts

---

[4]Ziegler, K. "A Forty Year's Stroll through the Realms of Organometallic Chemistry" *Advances in Organometallic Chemistry*; Stone, F. G. A.; West, R., Eds.; Academic: New York, 1968; p 11

of iodine), are synthesized today using a Ziegler–Natta catalyst. For their work on the controlled polymerization of hydrocarbons through the use of these novel organometallic catalysts, Karl Ziegler and Giulio Natta shared the 1963 Nobel Prize in chemistry.

Karl Ziegler was a chemist's chemist, one who made *many* contributions in various areas of chemistry, only some of which have been discussed here. Ziegler was a pleasure to work with, and the loss at his death was felt by many. George Wilke, a former student, presents this homage: "The many students and scientists from all over the world who worked with Ziegler at various times were able to learn under his guidance to tackle a problem with enthusiasm and to seek the answer by systematic investigation. In addition it was one of Ziegler's principles to keep an eye open for unexpected developments and not to neglect new phenomena as irrelevant to the main project. However, not only was it possible to learn from his scientific approach, but also from his philosophy to life whereby he had a talent for finding simple solutions to complicated problems".[5] He also had many hobbies, among them collecting paintings and hiking in the mountains, where it has been said that he rarely "talked shop". Wilke has also suggested that his ability to enjoy himself outside of the lab "undoubtedly contributed to his success".

PETER V. BONNESEN
*Oak Ridge National Laboratory*

*Bibliography*

Bawn, C. E. H. "Karl Ziegler"; *Biographical Memoirs of Fellows of the Royal Society*; The Society: London, 1975; Vol. 21; pp 569–584.

Boor, J., Jr. *Ziegler–Natta Catalysts and Polymerizations*; Academic: New York, 1979.

Eisch, J. J. "Karl Ziegler: Master Advocate for the Unity of Pure and Applied Research"; *Journal of Chemical Education* **1983**, *60*, pp 1009–1014.

McMillan, F. M. *The Chain Straighteners*; Macmillan: London, 1979.

"Meet Karl Ziegler, Mulheim's Man of Many Parts"; *Chemical Week* **1958**, *May 31*.

---

[5]Wilke, G. "Karl Ziegler in Memoriam"; *Coordination Polymerization*; Chien, J. C. W., Ed.; Academic: New York, 1975; p 13

Overberger, C. G. "The 1963 Nobel Prizes—Chemistry"; *Science* **1963**, *142*, 938–939.

Wilke, G. "Karl Ziegler, in Memoriam"; In *Coordination Polymerization*; Chien, J. C. W., Ed.; Academic: New York, 1975; pp 1–13.

Ziegler, K. "Folgen und Werdegang einer Erfindung"; *Angewandte Chemie* **1964**, *76*, 545–553. (This is the text of Ziegler's Nobel lecture, delivered December 12, 1963.)

Ziegler, K. "A Forty Years' Stroll through the Realms of Organometallic Chemistry"; In *Advances in Organometallic Chemistry*; Stone, F. G. A., and West, R., Eds.; Academic Press: New York, 1968; pp 1–17.

Ziegler, K. "Organic Aluminum Compounds"; In *Organometallic Chemistry*; Ziess, H., Ed.; American Chemical Society Monograph 147; Reinhold: New York, 1960; pp 194–269.

# Dorothy Hodgkin

## 1910–

Dorothy Mary Crowfoot Hodgkin was the sole Nobel Laureate in chemistry in 1964, receiving the prize for her work on the determination by X-ray techniques of the structure of several important biochemical compounds, including vitamin $B_{12}$ and penicillin. She was the third woman, and the first Englishwoman, to receive a Nobel Prize in chemistry. Her work is considered classic among structural determinations.

She was born in Cairo, Egypt, of English parents on May 12, 1910. Her father, John Winter Crowfoot, an inspector for the Ministry of Education, was a distinguished classical scholar and archaeologist. Her mother, Grace Mary Hood Crowfoot, even though having little formal education beyond finishing school in Paris, was an excellent botanist, drawing flowers for *Flora of the Sudan*; she was also an authority on ancient textiles.

At age ten, while attending a small private school in Beccles, Suffolk, England, Crowfoot was introduced to elementary chemistry. It was here that she grew her first crystals and began learning crystallography. From the beginning she was intrigued by the elegance and beauty of their geometric shapes, with plane faces meeting at characteristic angles, and the orderly symmetry of their structures. In 1921 she was enrolled at Sir John Leman's School in Beccles, where she and another girl were "allowed" to join the boys for a chemistry class.

In 1923 Crowfoot visited her father in Khartoum for a six-month period. While there, she and her sister visited the Wellcome Laboratory, where Dr. A. F. Joseph, a good friend of her father's, was a soil chemist. To amuse the girls, the laboratory geologists threw a few grains of gold into a pile of sand and showed the girls how to pan away the sand. Trying this technique in the stream that ran through her parents' garden, she found a black substance and asked Dr. Joseph for help in analyzing it. It was ilmenite, a mixed oxide of iron and titanium. "Uncle" Joseph had to show her how to test for titanium, since this element was not included in her textbooks. Later, Dr. Joseph presented Crowfoot with a box like those used by surveyors for doing chemical analysis. This box became one item in the laboratory that she established in the attic of her home as soon as she returned to England.

In 1925 her mother gave her the book *Concerning the Nature of Things,* written by Sir William Henry Bragg, which intensified her interest in crystals. Sir William Henry Bragg and his son, Sir William Lawrence Bragg, had received the Nobel Prize in physics in 1915 for research on the structure of crystals by means of X-rays. At this time X-ray crystallography was still in its infancy. Crowfoot was particularly interested in Bragg's description of how X-rays have enabled scientists to 'see' individual atoms and molecules.

After finishing her studies at the Leman School in 1928, Crowfoot rejoined her parents, who were then in Palestine. She worked with them on the digs and showed great skill in drawing the mosaics of the ancient churches. Scientific history came close to being changed that summer, for her work with her father was so exciting and so enjoyable that she gave serious thought to leaving chemistry for archaeology. However, she entered Somerville College at Oxford in October 1928, where she read chemistry and physics.

During her first year at Oxford, she spent most of her spare time completing the drawings of the mosaics and carrying out a series of chemical analyses on colored-glass tesserae that her mother had found in Palestine. E. G. J. Hartley encouraged her work at Oxford, lending her his platinum crucible and a key to the laboratory so that she could work on her own, even during vacation times. Crowfoot picked up the reputation in her undergraduate days of knowing more about inorganic chemistry than most others of that period. After taking a special course in crystallography, she was encouraged to do a research project in X-ray crystallography. This led to H. M. Powell's selection of her as his research student. Unfortunately, Somerville College was not accustomed to having bright, energetic science students, and she was Powell's first research student. As a result, no one seemed to know what to do with her talent.

Dr. Joseph visited Oxford in 1932, the same year that Crowfoot received her bachelor's, and by chance happened to meet T. S. Lowry, professor of physical chemistry at Cambridge. Lowry suggested that she work with John D. Bernal at Cambridge. Bernal had just been appointed by Cambridge University to start research in X-ray analysis. Bernal had previously worked with W. H. Bragg at the Royal Institution from 1923 to 1927.

Dorothy Crowfoot left Oxford to spend two years at Cambridge, working with Bernal and Isidor Fankuchen on the structure of vitamin D and on several sex hormones. About this time she was awarded a small college research grant of £ 75, and, additionally, her aunt gave her about £ 200 to stay on for another year at Cambridge. In 1934 during the summer, the first photographs of a single crystal of the protein pepsin were taken.

Somerville College then offered her a two-year, full-time research fellowship. The first year was to be spent working with Bernal at Cambridge and the second year to be spent at Oxford. Later she was appointed fellow and tutor in chemistry at Oxford.

The research equipment at Oxford was meager and cumbersome. The X-ray apparatus that she first used was a small gas tube that took many hours to put together, take apart, clean, and maintain. At Cambridge she had used a new X-ray tube that worked with the turn of a switch. There was no real laboratory: Her work was scattered throughout the basement of the University Museum. With the help of Sir Robert Robinson, she was able to secure a small grant for some new laboratory equipment. She was able to buy two X-ray tubes, a Weissenberg camera, and two oscillation cameras. However, she felt "lonely scientifically", so she often returned to Cambridge during vacations to work with Bernal's group. She was given space in the mineralogy and crystallography laboratory by Professor H. L. Bowman.

In 1937 she was awarded her doctorate from Cambridge, having completed a thesis on the chemistry and crystallography of the sterols. During the spring, while in London to photograph insulin at the Royal Institution, she met Thomas L. Hodgkin. Hodgkin, an historian, was educated at Winchester and Oxford. He came from a remarkable family that includes Dr. Thomas Hodgkin (Thomas' great-uncle), who first described the neoplastic condition known as Hodgkin's disease. Thomas and Dorothy were married in December 1937. From this marriage came three children: Luke (1938), Elizabeth (1941), and Toby (1946).

Dorothy Hodgkin was appointed university lecturer and demonstrator at Oxford and chose first to work on the crystal structures of the halogen derivatives of cholesterol. In addition, new crystals kept coming to her from many people who had heard of her good results with X-ray photography. Among the first crystals to arrive was a small sample of zinc insulin from Professor Robinson. Although the insulin molecule was too large to study in detail then, her interest in it began. A few years later Ernst Chain, who had just completed successful tests of penicillin extracts, promised Hodgkin that someday he would have crystals for her to work with.

World War II created a large demand for penicillin, and late in the summer of 1942 Hodgkin began a study of penicillin hydrochloride. Structural information would be a great aid in the development of methods to synthesize and produce the large quantities needed. About halfway through the war, L. J. Comrie suggested that the calculations needed could be done in three dimensions on punch card machines. He arranged for the calculations of electron density maps to be made using an old IBM punch card machine located in an evacuated government building in Cirencester. The essential structure was determined by 1945. The three-dimensional structure of penicillin was published, with Charles Bunn and others, in 1949.

The large-scale production of penicillin by chemical synthesis proved to be very difficult, since the substance is sensitive to acid. A drug closely related to penicillin but having greater stability in acid, cephalosporin C, was worked on by Hodgkin's group. Cephalosporin C was able to wipe out many infections that resisted penicillin. During all of this work many new techniques were introduced that widened the range of what could be accomplished through crystallography.

In 1948 Hodgkin and her colleagues began work on the molecular structure of vitamin $B_{12}$, important in the understanding and control of pernicious anemia. The first photographs showed that the determination of the structure would be an almost impossible task. Vitamin $B_{12}$ had a structure that had not been seen in nature previously. Crystalline vitamin $B_{12}$ is a red coordination compound that contains cobalt, and it has a molecular weight (1355) larger than that of any other known vitamin. Dorothy Hodgkin coordinated the collection of data by several researchers for a six-year period. The same punch card machine used to determine the structure of penicillin was used at the beginning of the project, but in time electronic computers were developed. Ultimately, the structure of vitamin $B_{12}$ was found with the help of the Mark 1 at Manchester, Deuce at NPL (Natonal Physics Laboratory), and SWAC (National Bureau of Standards Western Automatic Computer) in Los Angeles. Vitamin $B_{12}$ was the largest and most complex organic molecule to have its structure determined in complete detail. The final paper was published in 1957. With the determination of the structure of vitamin $B_{12}$ came a better understanding of the blood-building processes in the body.

Dorothy Hodgkin's third major achievement was the determination of the structure of insulin. It came as a result of more than 30 years of research following the first X-ray photographs in 1935. Insulin contains 51 amino acids. Frederick Sanger determined the exact ordering of the chain, whereas Hodgkin was able to explain the unique three-dimensional arrangement of the amino acid side groups that lead to hydrogen bonds, disulfide bridges, and energetically favorable structures. By 1969 Hodgkin had shown that the long chains were folded into compact molecules that formed hexamers around two zinc atoms.

When Hodgkin's Nobel Prize was announced, she was Wolfson Research Professor of the Royal Society, a post she held from 1960 to 1977. The previous year Thomas Hodgkin's cousin Alan shared the Nobel Prize for physiology or medicine; therefore, the three Hodgkin children have a Nobel Laureate on each side of the family.

The achievements of Dorothy Hodgkin are many. She determined the structure of many very important and complicated molecules, including penicillin, vitamin $B_{12}$, and insulin. The knowledge of the way atoms are arranged in crystals has given a great deal of new data to chemistry. It brought about a general rewriting of inorganic chemistry in terms of ions. Hodgkin demonstrated that it is possible to determine such complicated structures by the use of X-ray diffraction. She also demonstrated that it is possible to think about molecular structure from a three-dimensional viewpoint. Her contributions were rare gifts, and as W. L. Bragg put it, her work exceeds the "sound barrier".

**NORMAN W. HUNTER**
*Western Kentucky University*

# *Bibliography*

Farago, P. "Interview With Dorothy Crowfoot Hodgkin"; *J. Chem. Ed.* **1977**, *54*, 214–215.

Haber, L. *Women Pioneers of Science*; Harcourt, Brace, Jovanovich: New York, 1979.

Ihde, A. J. *The Development of Modern Chemistry;* Harper & Row: New York, 1964.

Jeffrey, G. A. "Nobel Prize Awarded to Crystallographer"; *Science* **1964**, *146*, 748–749.

Julian, M. "Dorothy Crowfoot Hodgkin: Nobel Laureate"; *J. Chem. Ed.* **1982**, *59*, 124–125.

Leonardi, S. J. *Dangerous by Degrees. Women at Oxford and the Somerville College Novelists;* Rutgers Univ. Press: New Brunswick, NJ, 1989.

"Nobel Prize Given for Vitamin B$_{12}$ Studies"; *Chem. Eng. News* **1964**, *42* (*November 9*), 54.

# 1965

# Robert Woodward

## 1917–1979

Robert Burns Woodward, the foremost organic chemist of the twentieth century, accepted the Nobel Prize in Stockholm in 1965. He marked the event not with a recitation of previously described work but with the announcement of a new accomplishment he had raced to complete for the occasion: the total synthesis of a cephalosporin antibiotic. Even among the august ranks of Nobel laureates, Woodward displayed rare genius, and the Nobel Prize actually preceded his greatest achievements.

Born in Boston on April 10, 1917, the only child of Margaret Burns and Arthur Woodward, he sustained the loss of his father before age two. Difficult times ensued but a hardworking mother and a devoted son provided mutual support. Young Woodward obtained his primary and secondary education in the public school system of Quincy, outside Boston.[1] However, his intellectual life revolved around chemistry, both in readings, courtesy of the local public library, and in experiments conducted in his home laboratory. The gift of a chemistry set from his mother prompted this interest, but the experimental work became much more than child's play as Woodward methodically reproduced the repertoire of work typical of a university course. The scope of his precocious interests extended

---

[1] Arthur Woodward (no immediate relation), Director of Secondary Education for Quincy, MA, provided original records of Woodward's early education. Newspaper reports of poor conduct and effort were mistaken.

beyond chemistry to include mathematics and literature. His teachers clearly recognized the exceptional ability of their student; the record shows that Woodward skipped fourth grade, jumped to seventh grade halfway through sixth grade, and also skipped eleventh grade.

In the fall of 1933, at age 16, Woodward matriculated at the Massachusetts Institute of Technology.[2] In an unorthodox plan to maximize time for the library and laboratory, he did not attend class but simply took final examinations. His neglect of course work resulted in expulsion after $1\frac{1}{2}$ years, but he was readmitted for the third year. Undaunted, Woodward took and passed over two years' worth of courses during that year and graduated at age 19.

James F. Norris, chairman of the MIT chemistry department, guided Woodward and tolerated his nonconformist approach to the Institute because of the promise in Woodward's original ideas on organic synthesis. The laboratory synthesis of molecules found in nature provides proof of chemical structure and yields insights into how organic molecules react. Woodward proposed the construction of the natural steroid estrone from simple starting materials by a route that startled his mentors for its perceptiveness and sophistication. He stayed on at MIT and received the Ph.D. degree after a single year, even though he had not completed the estrone synthesis.

After spending the summer of 1937 at the University of Illinois, Woodward took a position at Harvard, where he remained for the rest of his life.[3] A brief assistantship with the chairman of the chemistry department, E. P. Kohler, led to an appointment as a Junior Fellow of the Harvard Society of Fellows in 1938. In 1941 he became an instructor in charge of the advanced undergraduate organic chemistry laboratory, attracting numerous students to implement his ideas. His productivity resulted in a rapid rise through the ranks with appointments as assistant professor in 1944, associate in 1946, and full professor in 1950, at the age of 33. When he was named to the Donner Professorship of Science in 1960, he became one of four Harvard faculty members exempted from formal teaching chores to concentrate solely on research. The scientific accomplishments that propelled his career cover almost all aspects of organic chemistry, including the synthesis of natural prod-

---

[2]Confirmed through sources at MIT.

[3]The Department of Chemistry at Harvard University provided Woodward's curriculum vitae of 1979 along with other biographical data.

ucts, organic structure determination, synthetic methods, biogenetic theory, and theoretical organic chemistry.[4]

R. B. Woodward's contributions to the total synthesis of natural products are legendary and unparalleled. Famed for the complexity of the objective, the subtlety in the strategic planning, the resourcefulness in the face of obstacles, and the deft stereochemical control at each step, the works unfold with an artistic perfection. Each synthesis tested the limits of chemical methodology as well as the chemist's ingenuity, and one after another of these previously unattainable objectives fell: quinine in 1944, patulin in 1950, cholesterol in 1951, cortisone in 1951, lanosterol in 1954, lysergic acid in 1954, strychnine in 1954, reserpine in 1956, chlorophyll in 1960, tetracyclines in 1962, colchicine in 1963, and, for the occasion of his Nobel lecture, cephalosporin C in 1965.

What can be measured of the man who accomplished such feats? To his raw intelligence, Woodward added an encyclopedic knowledge of the chemical literature, especially the old German literature. His total recall of obscure papers remains the subject of lore. A quest for singular recognition appeared to be a driving force for Woodward. Well-known were his perfectionism and pursuit of a distinctive style. The most notorious element of this style involved his preoccupation with the color blue: his invariant attire of dark-blue suit and light-blue tie, his blue-painted office, his blue Mercedes Benz sedan, and his parking space painted and repainted blue by his graduate students. More substantive though no less distinctive were the captivating 4- to 5-hour marathon lectures entitled "Recent Advances in the Chemistry of Natural Products", noted for blackboards covered by chemical structures hand-drawn with draftsmanlike precision. Woodward embraced a puritan work ethic: He frequently labored 16-hour days, functioned on 3 hours' sleep, and eschewed vacations. Woodward's boundless self-confidence, occasionally taken for arrogance, might lead him to undertake the most daunting projects. But beyond his intellect, knowledge, drive, effort, and fortitude, a deep conceptual divide separated Woodward from other practitioners of the art.

From the mid-1800s through the early part of the twentieth century, organic chemists focused on deciphering the structures of carbon compounds and the products derived from them under various

---

[4]Biographical data and citations for all 196 original works of R. B. Woodward may be found in Alexander Todd and John Cornforth's "Robert Burns Woodward", *Biographical Memoirs of Fellows of the Royal Society* **1981**, 27, pp 629–695.

transforming conditions. The synthesis of natural products during this period was an exceptionally arduous, heavily empirical task. Only in 1911 had Lord Rutherford experimentally defined the structure of the atom, and only over the next two decades had G. N. Lewis among others extended the Rutherford model to explain the chemical properties of ions, atoms, and the covalent bonds between atoms. A new, theoretical branch of organic chemistry began to develop particularly around the process of chemical reaction, that is, the reaction mechanism. Although not initially appreciated, by understanding the detailed mechanism of a reaction involving one type of compound, subtle predictions could be made about the likely result of a related reaction. It was this application of the insights of reaction mechanism to the individual reactions that make up an organic synthesis that allowed Woodward to anticipate how complex and then-unknown molecules would react. With a penchant for innovation Woodward embraced the newly developing spectroscopic methods for the information that they could provide about chemical structure, and in his laboratory the modern age of organic chemistry arrived. His revolution in methodology allowed him to construct multistep syntheses that were beyond the imagination of his teachers and peers. So radical was the conceptual advance that a college-level organic chemistry student could thereafter design a workable natural-product synthesis rivaling that of Woodward's great predecessors.

Woodward's analytic skill and mechanistically oriented approach together with an aggressive application of instrumental methods also allowed him to solve many of the great structural problems of his day. The breathtaking catalog of his natural-product structural determinations includes penicillin in 1945, strychnine in 1948, patulin in 1949, terramycin in 1952, aureomycin in 1952, cevine in 1954, magnamycin in 1956, gliotoxin in 1958, oleandomycin in 1960, streptonigrin in 1963, and tetrodotoxin in 1964.

Keen observations and an organizing intelligence led Woodward to several important generalizations. While a Junior Fellow at Harvard he took interest in the new method of ultraviolet spectroscopy for its potential to rapidly elucidate structure and obviate laborious degradative studies. $\alpha,\beta$-Unsaturated ketones are important in steroid and terpene chemistry, and in 1941 he published the first of four papers correlating their ultraviolet absorption spectra with their substitution pattern. Woodward's empirical correlations brought him international recognition at age 23.

Another Woodward generalization led to a new branch of organometallic chemistry. As a young Harvard faculty member, Geoffrey Wilkinson cast about for an important line of investigation, and Woodward suggested that he redetermine the just-published structure of iron biscyclopentadienyl. Woodward had surmised that the reported structure was incorrect and that this compound demonstrated a previously undescribed delocalized bonding. Woodward's supposition proved true; a joint publication resulted in 1952, and Wilkinson had taken his first step to a Nobel Prize.

A third generalization returned Woodward in 1961 to correlations of instrumental measurements to chemical structure. Compounds possessing a particular asymmetry distort polarized light in a phenomenon known as the Cotton effect. Woodward deduced the complex relationship between the structure of asymmetric ketones and the magnitude of the effect so that its measurement could be used to determine structure.

Woodward contributed to biogenetic theory, the study of how organisms synthesize organic molecules, in the areas of indole alkaloids (1948), steroids (1953), and macrolides (1956). Woodward's insight into steroid biogenesis illustrates his mechanistic orientation most vividly. Harvard colleague Konrad Bloch, reporting in 1952 that cholesterol biogenesis occurred through the intermediacy of squalene, suggested a wishful rearrangement of atoms that would immediately yield the product but by means of an unclear mechanism. Woodward countered with a mechanistically mandated rearrangement of atoms even though this hypothesis required additional reactions. Woodward's hypothesis proved correct; a joint publication with Bloch followed, and Bloch was another step closer to the Nobel Prize.

Finally, in the course of structural determination and synthesis of natural products Woodward contributed a number of synthetic methods. To that number would also be added his synthetic methods for polypeptides (1947) and peptides (1961).

The Nobel Prize was anticipated and in 1965 the Prize came. The Nobel citation recognized Woodward's "contributions to the art of organic synthesis" but his most heralded accomplishments were yet to come: the synthesis of vitamin $B_{12}$ and the conceptualization of the principle of orbital symmetry conservation.

The total synthesis of vitamin $B_{12}$, which surely was and arguably remains the most complicated molecule made by man, resulted from Woodward's collaboration with Albert Eschenmoser of Zurich and the effort of more than 100 postdoctoral fellows extending over a decade.

The formal synthesis was completed in 1972 with the synthesis of cobyric acid, because a sample of natural cobyric acid had previously been converted to the vitamin. Nonetheless, Woodward took his synthetic cobyric acid on to vitamin $B_{12}$ in 1976, perhaps just to enjoy the exquisite crimson crystals. Woodward distributed photomicrographs of the crystals on a blue background, and at the celebration of the completion of the work Woodward described the organic synthesis for all assembled, an activity so difficult that not a quest for results but only a love of the experimental details, the reactions, the crystals, could justify the effort.

Woodward could also supply the scientific justification that natural-product synthesis was a likely framework for investigations of chemical structure and reactivity. Early in the vitamin $B_{12}$ synthesis he noted an inexplicable finding: The thermal cyclization of an isomeric triene yielded only the less-favored of two possible isomeric products. Woodward exhaustively investigated this reaction and deduced that symmetry relationships between the electronic (orbital) structure of the reactant and product dictated the course of the reaction. He enlisted a young theoretician and Harvard Junior Fellow, Roald Hoffmann, in the development of these ideas, and in 1965 they produced the Woodward–Hoffmann orbital symmetry rules, which predict the relative ease and stereochemical outcome of concerted thermal and photochemical reactions. Hundreds of previous reports on reactions with curious stereochemical results were suddenly explained and thousands of papers followed confirming the myriad predictions of the theory. This most important theoretical advance in modern organic chemistry earned Hoffmann the Nobel Prize in 1981 and, were it not for the injunction against posthumous awards, would surely have yielded Woodward his second.

He undertook heroic efforts in effortless style and possessed an undeniable charismatic appeal. Woodward gradually evolved into legend. However, in his lab he inspired filial affection as well as glorious myth. For his sixty-first birthday his graduate students constructed an elaborate, canopied sedan chair, upholstered in the mandatory blue. As the party slowed past midnight, Woodward decided to take the chair for a spin to a local ice cream shop, and, hoisting him up, the students carried him through Harvard Yard making hapless undergraduates kneel before him while he assumed royal posture and prerogatives. Some months later Woodward had several of his graduate students transport the chair to the airport in Boston to surprise his secretary, who was leaving for her first trip to Europe. At his insistence she rode in the chair while he, walking beside, announced to all that Her Highness was embarking for Liechtenstein to reclaim her throne—chaos, and Woodward's good humor, reigned.

Through the course of his career, Woodward held various appointments in addition to his continuous tenure at Harvard: as a Member of Corporation at Massachusetts Institute of Technology, 1966–1971; to the Board of Governors of the Weizmann Institute of Science, 1968–1979; to the Board of Directors of Ciba-Geigy, Ltd., 1970–1979; and as the Alexander Todd Professor of Chemistry at Cambridge University, 1974–1975. In a unique honor Ciba, the Swiss pharmaceutical company, established the Woodward Institute in Basel under his directorship in 1963, and it was actually there that he conducted the synthesis of cephalosporin C. Over the years Woodward accepted 25 honorary degrees, dozens of prizes, and scores of honorary lectureships. He trained more than 400 chemists, and the leading ranks of academe and industry abound with his disciples.

In the mid-1970s he began the synthesis of the antibiotic erythromycin and in the late 1970s began a theoretical and synthetic effort to devise an organic superconductor. However, without warning, Woodward died of a heart attack at home in Cambridge on the morning of July 8, 1979. Twice divorced, he was survived by two daughters, Siiri and Jean, from his 1938 marriage to Irja Pullman, and a daughter, Crystal, and a son, Eric, from his 1946 marriage to Eudoxia Muller. The world of chemistry mourned the loss of R. B. Woodward, but more than one friend took solace in the fact that he passed without knowing the decline of his powers.

DONALD W. LANDRY*
*Columbia University*

---

*The author was the last graduate student accepted by R. B. Woodward to receive the doctoral degree from him.

# Bibliography

## Syntheses

Woodward, R. B. "The Total Synthesis of Colchicine"; In *The Harvey Lectures, 1965;* Academic Press: New York, 1965.

Woodward, R. B. "The Total Synthesis of a Tetracycline"; *Pure and Applied Chemistry* **1963,** *6,* 561.

Woodward, R. B. "The Total Synthesis of Vitamin $B_{12}$"; *Pure and Applied Chemistry* **1973,** *33,* 145.

Woodward, R. B.; Ayer, W. A.; Beaton, J. M.; Bickelhaupt, F.; Bonnett, R.; Buchschacher, P.; Closs, G. L.; Dutler, H.; Hannah, J.; Hauck, F. P.; Ito, S.; Langemann, A.; Le Goff, E.; Leimgruber, W.; Lwowski, W.; Sauer, J.; Valenta, Z.; Volz, H. "The Total Synthesis of Chlorophyll"; *J. Am. Chem. Soc.* **1960,** *82,* 3800.

Woodward, R. B.; Bader, F. E.; Bickel, H.; Frey, A. J.; Kierstead, R. W. "The Total Synthesis of Reserpine"; *J. Am. Chem. Soc.* **1956,** *78,* 2023.

Woodward, R. B.; Cava, M. P.; Ollis, W. D.; Hunger, A.; Daniken, H. U.; Schenker, K. "The Total Synthesis of Strychnine"; *J. Am. Chem. Soc.* **1954,** *76,* 4749.

Woodward, R. B.; Doering, W. E. "The Total Synthesis of Quinine"; *J. Am. Chem. Soc.* **1944,** *66,* 849.

Woodward, R. B.; Patchett, A. A.; Barton, D. H. R.; Ives, D. A.; Kelly, R. B. "The Synthesis of Lanosterol"; *J. Am. Chem. Soc.* **1954,** *76,* 2852.

Woodward, R. B.; Sondheimer, F.; Taub, D. "The Total Synthesis of Cholesterol"; *J. Am. Chem. Soc.* **1951,** *73,* 3548.

Woodward, R. B.; Sondheimer, F.; Taub, D. "The Total Synthesis of Cortisone"; *J. Am. Chem. Soc.* **1951,** *73,* 4057.

## Structures

Woodward, R. B.; Brehm, W. J.; Nelson, A. L. "The Structure of Strychnine"; *J. Am. Chem. Soc.* **1947,** *69,* 2250.

Woodward, R. B.; Gougoutas, J. Z. "The Structure of Tetrodotoxin"; *J. Am. Chem. Soc.* **1964,** *86,* 5030.

## Biogenetic Theory

Woodward, R. B. "Biogenesis of the Strychnos Alkaloids"; *Nature* **1948,** *162,* 155.

Woodward, R. B. "Struktur und Biogenese der Makrolide, eine neue Klasse von Naturstoffen"; *Angew. Chem.* **1957,** *69,* 50.

Woodward, R. B.; Bloch, K. "The Cyclization of Squalene in Cholesterol Synthesis"; *J. Am. Chem. Soc.* **1953,** *74,* 2023.

## General and Theoretical

Hoffmann, R.; Woodward, R. B. "Selection Rules for Concerted Cycloaddition Reactions"; *J. Am. Chem. Soc.* **1965,** *87,* 2046.

Moffitt, W.; Woodward, R. B.; Moscowitz, A.; Klyne, W.; Djerassi, C. "Structure and the Optical Rotatory Dispersion of Saturated Ketones"; *J. Am. Chem. Soc.* **1961,** *83,* 4013.

Wilkinson, G.; Rosenblum, M.; Whiting, M. C.; Woodward, R. B. "The Structure of Iron Bis-Cyclopentadienyl"; *J. Am. Chem. Soc.* **1952,** *74,* 2125.

Woodward, R. B. "Structure and Absorption Spectra. IV. Further Observations on α, β-Unsaturated Ketones"; *J. Am. Chem. Soc.* **1942,** *64,* 76.

Woodward, R. B.; Hoffmann, R. "The Conservation of Orbital Symmetry"; *Angew. Chem., Int. Ed.,* **1969,** *8,* 781.

Woodward, R. B.; Hoffmann, R. *The Conservation of Orbital Symmetry;* Verlag Chemie/Academic Press: New York, 1970.

Woodward, R. B.; Hoffmann, R. "Selection Rules for Sigmatropic Reactions"; *J. Am. Chem. Soc.* **1965,** *87,* 2511.

Woodward, R. B.; Hoffmann, R. "Stereochemistry of Electrocyclic Reactions"; *J. Am. Chem. Soc.* **1965,** *87,* 395.

Woodward, R. B.; Rosenblum, M.; Whiting, M. C. "A New Aromatic System"; *J. Am. Chem. Soc.* **1952,** *74,* 3458.


## On Woodward

Bartlett, P. D.; Westheimer, F. H. "Robert Burns Woodward, Nobel Prize in Chemistry for 1965"; *Science* **1965,** *150,* 585–587.

Bowden, M. E.; Benfey, O. T. *Robert Burns Woodward and the Art of Organic Synthesis;* Beckman Center for the History of Chemistry: Philadelphia, PA, 1992.

Wheeler, D. M. S. "R. B. Woodward und die moderne organische Chemie"; *Chemie in Unserer Zeit* **1984,** *18,* 109–119.

Woodward, C. "Art and Elegance in the Synthesis of Organic Compounds"; In *Creative People at Work: Twelve Cognitive Case Studies;* Wallace, D. B.; Gruber, H. E., Eds.; Oxford University Press: Oxford, 1989.

# Robert Mulliken

## 1896–1986

Robert Sanderson Mulliken received the 1966 Nobel Prize in chemistry for "fundamental work concerning chemical bonds and the electronic structure of molecules". He was born in Newburyport, Massachusetts, on June 7, 1896, and died on October 31, 1986, in Arlington, Virginia. Both his father (Samuel Parsons Mulliken, a professor of organic chemistry at Massachusetts Institute of Technology) and mother (Katherine Mulliken, an artist and music teacher prior to her marriage) descended from the same ancestor, a Robert Mullicken, believed to have sailed from Glasgow in 1680.

As a boy, Mulliken attended the local primary school; read the prescribed children's classics; attended the Unitarian Church with his mother; helped to proofread his father's four-volume treatise *A Method for Identification of Pure Organic Compounds;* accompanied his father on motorboat excursions and hikes in the summer; and helped him at times with laboratory experiments at MIT. Influenced by his father's academic interests and the extraordinary scientific ferment of the period, young Mulliken developed an early interest in science and pursued it with single-minded purpose.

In high school, motivated by both his interest in science and by the promise of financial support from the Wheelwright Fund, which "provided for young men from Newburyport to go to college, but especially MIT", he elected the scientific course, which included biology, physics, French, German, and English, to which he added one year of Latin. Fascinated by contemporary developments in science, he read books that "dealt with life and its origins as well as physics and chemistry". When he graduated from Newburyport High School in 1913, Mulliken was class salutatorian and delivered an essay entitled "Electrons: What They Are and What They Do".[1]

Mulliken entered MIT in the fall of 1913 to study chemistry. He enjoyed his course work, finding that he "loved molecules in general, and some molecules in particular". A research project undertaken in his sophomore year with his father's close friend Prof. Arthur Noyes produced only a "laundry list of molecules and ions", greatly disappointing the latter and dismaying his father. In his senior year (1916–1917), however, he "successfully carried out a small piece of research in organic chemistry with [J. F.] Norris", the results of which were published in 1920 in the *Journal of the American Chemical Society*.

When Mulliken graduated in 1917, the United States was at war. He accepted a job (which eventually came under the Chemical Warfare Service) working on poison gases at the American University in Washington, D.C., under the supervision of J. B. Conant. After he was mustered out of service as a private first class, he went to work as a technical assistant for the New Jersey Zinc Company, where he did routine analyses and some simple research on the compounding of rubber.

In the fall of 1919 he left New Jersey Zinc to work with William D. Harkins, a physical chemist at the University of Chicago. He chose Chicago because he "wanted to work on something really fundamental, namely atomic nuclei in which hardly anyone except Rutherford in England and Harkins in this country seemed interested." After some work on surface tension, another of Harkin's interests, he wrote his Ph.D. thesis on the partial separation of mercury isotopes by irreversible evaporation. During this period (1919–1922) he read the papers of G. N. Lewis on chemical bonding and valence and the related papers of Irving Langmuir; and he learned about the old quantum theory from Prof. Robert A. Millikan, who was a member of the physics faculty at the time (and with whom Mulliken is frequently confused).

From 1922 to 1923 he continued his work on mercury isotope separation as a National Research Council Fellow, building what he called an "isotope factory", which achieved a density change of many

[1]Mulliken, R. S. *Life of a Scientist*, Ransil, B. J., Ed.; Springer: New York, 1989. In that volume, birth date and place of death are incorrect.

parts per million. In the process he found that a dirty surface on the mercury acted as a diffusion membrane, anticipating by 35 years the barrier layer separation of uranium isotopes used for the Manhattan Project.

When he reapplied to the National Research Council to continue his work on isotopes, proposing to try out other methods of separation, the Council told him that he must do something different at another institution if he wanted to retain the Fellowship. He first proposed to go to work with Rutherford at Cambridge on beta ray spectra, but was told he did not have the necessary theoretical or experimental background. Suggestions from a friend, Norman Hillberry, led to an acceptable project: looking for isotope effects in band spectra of BN at Harvard's Jefferson Physical Laboratory under the sponsorship of E. A. Saunders. It was his introduction to the field of molecular spectra and to the beginning of a lifetime interest in the electronic spectra and structure of molecules.

What led Mulliken to propose the band spectra project were some photographs by Wilfred Jevons of King's College, London, of what Jevons believed to be the BN spectrum. Although Jevons had not reported bands of two isotopes, Mulliken's knowledge of isotopes led him to look for and find in the published photograph "a number of extra bandheads in the positions to be expected for the less abundant isotope of boron". Making new photographs of Jevons's BN bands on Harvard's spectrograph, Mulliken measured and analyzed them and discovered that there was a much better fit of data with theory if the spectra were ascribed to BO rather than BN and that "the vibrational energy comes in half quanta and is never less than one-half quantum". He communicated these results in a letter to *Nature* in 1924, anticipating the zero-point energy by a year or more. (Quantum mechanics was to show in 1925–1926 that half quanta are to be expected for *all* molecules.) Jevons responded with a rebuttal in the same journal. But instead of continuing the controversy in print, Mulliken met with Jevons in the summer of 1925 during a tour of European research centers, at which time they amicably resolved their differences in favor of the BO interpretation.

During his European visit in 1925 Mulliken met "nearly everybody in Europe who had been doing worthwhile research on band spectra, as well as ... other scientists who had been leaders in work on atomic spectra and quantum theory". Among those visited were Lord Rayleigh, Fowler, C. P. Snow, Gaydon, Aston, Mecke, Hertha Sponer, Sommerfeld, Bohr, and Born, but especially Friedrich Hund, who was Born's assistant, with whom Mulliken talked "extensively about what [he] had learned about molecular spectra".

Between 1926 and 1932, working independently but in periodic communication with each other's remarkably parallel thinking, Hund and Mulliken developed a molecular model based on the Bohr model of the atom in which electrons, characterized by individual quantum numbers, were distributed in molecular orbitals that permeated the molecule. Hund proposed the orbital symbols $\sigma$, $\pi$, $\delta$, ..., and state symbols $\Sigma$, $\Pi$, $\Delta$, ..., for diatomic molecules. This nomenclature, together with Mulliken's concept of "electron configuration" for ground and excited diatomic molecule states in 1928, and Lennard-Jones's introduction of linear combinations of atomic orbitals (LCAOs) in 1929 (by which molecular orbitals are built from atomic orbitals), expedited the development of the burgeoning field of molecular orbital (MO) theory.

During the early years of its development, the MO theory was jointly attributed to Hund and Mulliken, a fact that Mulliken often alluded to in his later years. In his account of the Nobel Prize, Mulliken recalls, "One of the happy consequences of . . . the Prize [was a] flood of congratulatory telegrams and letters. One which I especially cherished was a sympathetic and generous letter from Hund, with whom I would have been glad to share the Prize."

In 1926 Mulliken became assistant professor of physics at New York University, Washington Square, where he taught physics and supervised three graduate students in the measuring of diatomic spectra. Two years later he was recruited by A. H. Compton for the University of Chicago as associate professor in physics with the promise of quick promotion to full professor and, more importantly, a high-resolution spectrographic grating. However, once he got there, the grating was slow to materialize. While waiting for the grating to be ruled on Chicago's machines, his enthusiasm for measuring molecular spectra waned, and he found himself becoming more interested in their interpretation.

When Mulliken again visited Europe in 1927, Schrödinger introduced him to W. Heitler and F. London, whose paper applying the electron-pair theory (the Heitler–London valence bond method—which, eventually extended to a generalized form, became known as the Heitler–London–Slater–Pauling method) to the hydrogen molecule was soon to be published. In his autobiography, Mulliken recalls that "Chemists had long talked about valence bonds between atoms in molecules, but could not explain the forces involved. The paper of

Heitler and London for the first time seemed to provide a basic under-standing which could be extended to other molecules. Linus Pauling at the California Institute of Technology in Pasadena soon used the valence-bond method [to describe] the structure of molecules in general and also of many individual molecules. . . . Linus persuaded chemists all over the world to think of typical molecular structures in terms of the valence-bond method." Several decades were to elapse before molecular orbital theory became widely disseminated and generally appreciated.

Well-established in his career at the University of Chicago and at the age of 33, Mulliken married Mary Helen Noé, a trained artist like his mother, on Christmas Eve of 1929. Lucia was born in 1934 and Valerie in 1948. Mary Helen raised the children, subordinating her artistic interests to motherhood and to her career as faculty wife, enjoying the opportunities it afforded to meet visiting scientists, to entertain and to travel.

Mulliken spent World War II as an organizer and Director of the Information Division in the Manhattan Project at the University of Chicago with A. H. Compton as director and Norman Hillberry as associate director. The Division's main function was to supervise the production and distribution of classified reports. Two of his assistants were Eugene Rabinowitch and H. H. Goldsmith, whose spirited dis-cussions of the weighty ethical issues of atomic weaponry eventually evolved into the *Bulletin of the Atomic Scientists*. When the Manhattan Project drew to a close, Mulliken served as editor-in-chief of the Plu-tonium Project's historical volume and then returned to his electrons orbiting in molecules.

The immediate postwar years saw the rapid assembly of a group of students and research associates, the arrival of John Platt and Clemens C. J. Roothaan, the establishment of the Laboratory of Molecular Structure and Spectra (LMSS) in the Ryerson Physical Laboratory, the Department of Physics at the University of Chicago, and the publication of its annual Red Reports (1947–1970). Research emphasis was about equally divided between theoretical and experimental work on molecules; students came from the departments of chemistry and physics. In the late 1940s, Roothaan, a Ph.D. student of Mulliken's at Chicago, combined work on his thesis with teaching physics in Karl Herzfeld's department at the Catholic University of America in Washington, D.C. The thesis topic suggested by Mulliken, semiem-pirical MO calculations on substituted benzenes, entailed method-ological problems that led Roothaan to develop a genuine $N$-electron self-consistent-field model in which each MO, as in the semiempirical theory, is constructed as a linear combination of atomic orbitals (LCAO). Calling this the LCAO-SCF method, Roothaan proposed this model for the calculation of accurate *"ab initio"* atomic and molecular wave functions. This development, together with Roothaan's and Klaus Rudenberg's contributions to the solution of the difficult integrals problem, made it possible to undertake the systematic calculation on digital computer of wave functions for the first-row diatomic hydrides, homopolars, and selected heteropolars (the Diatomic Molecule Project), under the direction of Bernard J. Ransil (1956–1961) and Paul E. Cade (1961–1969). It led to a decade or more of accurate atomic and diatomic molecular calculations at LMSS under Roothaan's direction. In the basement of Ryerson, experimental work on diatomic and polyatomic spectra, performed by a small but dedicated cadre of research associates and students on a vacuum spectrograph, kept pace with the theoretical activities on the third floor. For over two decades LMSS was an important international center and magnet for scores of visiting scientists, research associates, and students from the United States, Europe, and Asia, who were interested in molecular science and molecular orbital theory.

Mulliken was a member of many scientific societies in the United States and abroad. He was a Guggenheim Fellow in 1930 and 1932 and a Fulbright research fellow at Oxford in 1952–1953. He served as science attaché to the American Embassy at London in 1955. Beginning in 1960 (at the age of 64) he received numerous medals and academic honors, culminating in the Nobel Prize for chemistry in 1966. In later years he was decorated by the Japanese government for his contribution to the training of Japanese spectroscopists (1984), was the recipient of the American Chemical Society's Priestley Medal (1983), and was inducted into the American Academy of Achievement (1983).

Mulliken's active scientific career spanned six decades, during which he published approximately 250 papers (the first in 1919, the last posthumously in 1987), over 70% of them as sole, and over 80% as first author. Whereas many of these papers proved seminal to many different communities of scientists (in 1980 at a colloquium given in his honor at the University of Chicago, six individuals—three of them Nobel Laureates—spoke of six key concepts introduced or influenced by Mulliken in as many fields), one of them, "Electronic Population Analysis on LCAO-MO Molecular Wave Functions, Part I", struck a resonant chord in enough readers to make it a Citation Classic. Between 1955 and 1985 this paper was cited in over 2000 publications.

Functioning neither as a "proper experimentalist nor a theorist, but as a middleman between experiment and theory, between chemistry and physics", Mulliken was equally at home at the faculty club's "Physics" and "Chemistry" tables, between which he oscillated with aperiodic frequency. He was the right man in the right place at the right time with the right mindset. Building on the work of Thomson, Bohr, Lewis, Sommerfeld, Heisenberg, Schrödinger, and Hund, he was quick to perceive relevant elements of each man's thinking to use in explaining his own knowledge of electronic behavior empirically derived from molecular spectra. He incorporated the elements into a model that "worked"; that is, a model that was readily visualized, amenable to mathematical modeling, and capable of "explaining" and predicting molecular phenomena.

Concepts and terms coined by Mulliken that have shaped and illuminated our understanding of molecular structure include orbital, molecular orbital, charge transfer process, electron population and population analysis, spin orbital, electron donor and acceptor, electron affinity, and electron promotion.

Although Mulliken will be rightly remembered in history as the codeveloper (with Hund) of the molecular orbital theory, his unique individual achievement was the devising of a conceptual framework and terminology with which to describe, explain, and predict chemical phenomena. This helped place chemistry, largely an empirical science until World War II, on a firmer theoretical foundation. Can anyone imagine "doing" chemistry without molecular orbitals?

BERNARD J. RANSIL
*Harvard Medical School*

# Bibliography

Berry, R. S. "Robert Sanderson Mulliken"; *American Philosophical Society Yearbook*; Philadelphia, PA, 1989, 187–194.

Longuet-Higgins, H. C. "Robert Sanderson Mulliken"; *Biographical Memoirs of Fellows of the Royal Society* **1989**, *35*, 327–354.

*Molecular Orbitals in Chemistry, Physics and Biology. A Tribute to R. S. Mulliken;* Löwdin, P.-O.; Pullman, B., Eds.; Academic Press: New York, 1964.

Mulliken, R. S. *Life of a Scientist;* Ransil, B. J., Ed.; Springer: New York, 1989.

Mulliken, R. S. "Molecular Orbitals"; In *Encyclopedic Dictionary of Physics 4;* Pergamon: New York, 1961.

Mulliken, R. S.; Ermler, W. C. *Diatomic Molecules: Results of* Ab Initio *Calculations;* Academic Press: New York, 1977.

Mulliken, R. S.; Ermler, W. C. *Polyatomic Molecules: Results of* Ab Initio *Calculations;* Academic Press: New York, 1981.

Mulliken, R. S.; Person, W. B. *Molecular Complexes* (a lecture and reprint volume); John Wiley & Sons: New York, 1969.

Nachtrieb, N. H. "Interview with Robert S. Mulliken"; *J. Chem. Ed.* **1975,** *52,* 560–564.

Platt, J. R. "1966 Nobel Laureate in Chemistry: Robert S. Mulliken"; *Science* **1966,** *154,* 745–747.

*Selected Papers of Robert S. Mulliken;* Ramsay, D. A.; Hinze, J., Eds.; University of Chicago Press: Chicago, IL, 1975.

# Ronald Norrish

## 1897–1978

Ronald George Wreyford Norrish, winner of the Nobel Prize in chemistry in 1967, was intimately and continuously associated with the City and University of Cambridge in England. He was born there on November 9, 1897, in a house close to the site of the present University Chemical Laboratory. He went to school in Cambridge, attended the university, and from the early 1930s until his death lived in a house located adjacent to Emmanuel College, of which he was a Fellow. He died on June 7, 1978, in Addenbrooke's Hospital, only a short distance from the Chemical Laboratory.

Norrish's father was a pharmaceutical chemist, as were other members of the family. In 1908 Norrish won a scholarship to the Perse School, a well-known independent school in Cambridge; he attended it as a day boy, and it was there that he developed his great interest in chemistry. With his father's encouragement he established a chemical laboratory in a garden shed, and the apparatus he used is still preserved at the Science Museum in South Kensington, London. In 1915 he won a Foundation Scholarship to Emmanuel College, Cambridge, but instead of at once taking up residence he enlisted in the Royal Field Artillery, serving as lieutenant on the Western Front. In 1918 he was captured and spent six months as a prisoner-of-war in Germany.

In 1919 he went into residence at Emmanuel College and first specialized in chemistry, physics, and botany for Part I of the Natural Sciences Tripos, obtaining first-class honors in 1920. For Part II of the Tripos he chose chemistry and again obtained first-class honors in 1921. As an undergraduate he was particularly influenced by Sir William Pope, the professor of chemistry, and by E. K. Rideal, then a fellow of Trinity Hall, who supervised some of his work. In 1921, having obtained his B. A. degree, he began research under the direction of Rideal, moving into the new Physical Chemistry Laboratory in 1922. Rideal had recently returned from an appointment at the University of Illinois, and Norrish found him to be "bubbling with ideas." Rideal was never strong on experimental techniques, unlike Norrish, who showed considerable experimental skill, and the two made a powerful combination. Their collaborative work concerned the photochemistry of potassium permanganate; the reaction between hydrogen and sulfur; and photochemical reactions involving hydrogen, oxygen, and chlorine. Together they published seven papers, and there is no doubt that this work stimulated Norrish's lifelong interest in photochemistry and the mechanisms of complex reactions.

Norrish obtained his Ph.D. in 1924 and was elected a research fellow of Emmanuel College, with which he remained associated in various positions until the end of his life. He at once embarked on independent research and soon became recognized as having exceptional ability. In 1926 he was appointed a university demonstrator in physical chemistry, and in that year married Anne Smith, who had been a lecturer in child psychology at University College, Cardiff; they had twin daughters. In 1936 T. M. Lowry, the professor of physical chemistry at Cambridge, died unexpectedly, and the following year Norrish was elected to succeed him. He occupied this position until 1965 and was at the same time professorial fellow of Emmanuel College; after 1965 he was a life fellow of the college.

Norrish's research covered a wide range of topics, mainly in the field of chemical kinetics. Much of his work was concerned with photochemical processes, and he had a particular interest in oxidation and polymerization reactions. His most noteworthy contributions were probably those on flash photolysis, initiated with the collaboration of George Porter with whom (and with Manfred Eigen) Norrish shared the Nobel Prize. However, even if the flash photolysis work had not been done, his other contributions would have marked him as outstanding.

The first research that Norrish initiated independently, in the 1920s, was on the mechanism of the photochemical reaction between hydrogen and chlorine, in the presence of various inhibitors. He also carried out some work on the photolysis of nitrogen dioxide and related processes. The latter work was important in leading to an understanding of the relationship between a photochemical threshold and the nature of the absorption spectrum.[1]

After about 1930, Norrish and his students, particularly C. H. Bamford, began to investigate the photochemistry of organic compounds, mainly those containing a carbonyl group. At that time the participation of organic free radicals as intermediates in reactions was only just beginning to be recognized, and Norrish's work was important in leading to an understanding of the types of free-radical reactions that occur. Shortly afterward he began the investigation of the kinetics and mechanism of the combustion of hydrocarbons and other organic substances; the investigation will be referred to later in this biography because it was continued after the flash photolysis technique had been introduced. In the 1930s he also began to investigate the kinetics of addition polymerization. In some early photochemical studies he had observed the deposit of viscous polymeric material, and he later made a careful study of the polymerization of formaldehyde and acetaldehyde;[2] the polymerization was found to be strongly catalyzed by formic acid. Other studies showed proton donors to be effective catalysts for certain types of polymerization, but most of the investigations in Norrish's laboratories were concerned with liquid-phase addition polymerization occurring by free-radical mechanisms.[3] Some of this work was done at high pressures.

Another significant investigation, carried out just before the outbreak of World War II, was on the quenching of the resonance radiation of sodium.[4] Few data had previously been available on quenching processes of this kind.

During the war Norrish directed a number of projects related to the war effort; problems worked on included the suppression of gun flash and the ignition and combustion of fuels. At the end of the war

[1]Norrish, R. G. W. "Photochemical Equilibrium in Nitrogen Dioxide II"; *J. Chem. Soc.* **1929**, 1158–1169; *J. Chem. Soc.* **1929**, "Photochemical Equilibrium in Nitrogen Dioxide III" 1604–1611, "Equilibrium in Nitrogen Peroxide IV"; *J. Chem. Soc.* **1929**, 1611–1621, "The Relationship of Fluorescence to Photolysis in Gaseous Systems"; *Trans. Faraday Soc.* **1939**, *35*, 21–28.

[2]Norrish, R. G. W.; Carruthers, J. E. "The Polymerization of Gaseous Formaldehyde and Acetaldehyde" *Trans. Faraday Soc.* **1936**, *32*, 195–208.

[3]Norrish, R. G. W.; Bengough, W. I. "The Mechanism and Kinetics of the Heterogeneous Polymerization of Vinyl Polymers I" *Proc. Roy. Soc.* **1949**, *A200*, 301–320, "The Mechanism and Kinetics of the Heterogeneous Polymerization of Vinyl Polymers II" *Proc. Roy. Soc.* **1953**, *A218*, 149–154 and "The Mechanism and Kinetics of the Heterogeneous Polymerization of Vinyl Polymers III" *Proc. Roy. Soc.* **1953**, *A218*, 155–163.

[4]Norrish, R. G. W.; Smith, W. M. "The Quenching of the Resonance of Radiation of Sodium" *Proc. Roy. Soc.* **1940**, *A176*, 295–312.

he resumed some of the work that had been interrupted and began work on a number of important new problems. The most important and far-reaching of these was flash photolysis, in which a flash of light of high intensity and extremely short duration brings about the formation of free radicals and other excited species in higher concentrations than had previously been possible. The idea of flash photolysis was first suggested to Norrish by T. M. Sugden, and an important factor that led to the development of this technique was the arrival in Cambridge, as a research student under Norrish, of George Porter (now Lord Porter). Then age 26, Porter had been a student at Leeds University, after which he served in the Royal Navy Volunteer Reserve as a radar officer. The electronic skills that he acquired in the navy allowed Porter to construct efficient flash photolysis equipment very suitable for the production and spectroscopic determination of short-lived reaction intermediates.[5]

During the next two decades Norrish and his students carried out many investigations on a wide variety of chemical reactions, with a view to identifying intermediates and studying their reactions. Some of the earlier work was on the photolysis of ketene and some aldehydes and ketones. In one investigation,[6] iodine was photodissociated and a study made on the kinetics of the recombination of the resulting iodine atoms in the presence of various added gases ("third bodies").

For many years Norrish had a great interest in combustion reactions, and particularly with F. S. (now Lord) Dainton and P. G. Ashmore made a series of comprehensive studies on the reaction between hydrogen and oxygen.[7] The investigations led to a further understanding of the mechanisms of the processes, particularly of the chain-ending steps involved. With the advent of flash photolysis Norrish applied the technique to oxidations; the first study concerned the hydrogen–oxygen system in the presence of nitrogen dioxide.[8] The

[5]Norrish, R. W. G.; Porter, G. "Chemical Reactions Produced by Very High Light Intensities" Nature 1949, 164, 658, and many later papers.

[6]Christie, M. I.; Norrish, R. G. W.; Porter, G. "The Recombination of Atoms I"; Proc. Roy. Soc. 1952, A216, 152–164.

[7]Norrish, R. G. W.; Dainton, F. S. "The Study of Sensitized Explosions V" Proc. Roy. Soc. 1941, A177, 393–410; "The Study of Sensitized Explosions VI"; 411–420; "The Study of Sensitized Explosions VII"; 421–447; Norrish, R. G. W.; Ashworth, P. G. "A Study of Sensitized Explosions VIII"; Proc. Roy. Sci. 1950, A203, 454–471; "A Study of Sensitized Explosions IX"; Proc. Roy. Sci. 1950, A203, 472–486; "A Study of Sensitized Explosions X"; Proc. Roy. Soc. 1950, A204, 34–50.; "A Phenomenon of Successive Ignitions"; Nature 1951; 167, 390–392.

[8]Norrish, R. G. W.; Porter, G. "Spectroscopic Studies of the Hydrogen–Oxygen Explosion Initiated by the Flash Photolysis of Nitrogen Dioxide"; Proc. Roy. Soc. 1952, A210, 439–460.

oxygen atoms produced in the photolysis initiated the chain reaction between hydrogen and oxygen. A technical improvement allowed the time resolution of the equipment to be reduced from 1 ms to about 50 μs in an investigation of acetylene–oxygen explosions photosensitized by $NO_2$.[9] This work led to valuable information about the formation and subsequent reactions of intermediates such as OH, CH, $C_2$, $C_3$, and CN.

Norrish had a great interest in oxidations of hydrocarbons, which show some features different from the hydrogen–oxygen reaction. Under certain conditions a mild type of explosion occurs, and since the temperatures are fairly low the expression "cool flame" is used. This behavior had been explained in 1930 by N. N. Semënov in terms of "degenerate chain branching", and in some of his early work Norrish had made contributions to the understanding of the mechanisms.[10] These investigations were resumed with the aid of flash photolysis, and special attention was paid to the presence of detonation waves in the gaseous mixtures.[11] This work was important in elucidating the function of lead tetraethyl, a widely used "antiknock" additive in fuels. The conclusion was that lead tetraethyl lengthens the induction period of a photosensitized ignition; the lead is present as PbO during the induction period but is reduced to free Pb as ignition occurs. Work was also carried out with other antiknock agents.[12] Another investigation led to a chain mechanism for the oxidation of phosphine.[13]

The work thus far was concerned mainly with the use of flash photolysis to identify free radicals and to study their reactions. Some investigations in Norrish's laboratory were also concerned with

---

[9]Norrish, R. G. W.; Porter, G.; Thrush, B. A. "Detection of Diatomic Radical Absorption Spectra During Combustion"; *Nature* **1952**, *169*, 582–583; "Studies of the Explosive Combustion of Hydrocarbons by Kinetic Spectroscopy I"; *Proc. Roy. Soc.* **1953**, *A216*, 165–182; "Studies of the Explosive Combustion of Hydrocarbons by Kinetic Spectroscopy II"; **1955**, *A227*, 423–433.

[10]Norrish, R. G. W. "A Theory of the Combustion of Hydrocarbons"; *Proc. Roy. Soc.* **1935**, *A150*, 36–57; Norrish, R. G. W.; Foord, S. G. "The Kinetics of the Combustion of Methane"; *Proc. Roy. Soc.* **1936**, *A157*, 503–525.

[11]Erhard, K. H. L.; Norrish, R. G. W. "Studies of Knock and Antiknock by Kinetic Spectroscopy"; *Proc. Roy. Soc.* **1956**, *A234*, 178–191; "Studies of Knock and Antiknock by Kinetic Spectroscopy II"; **1960**, *A259*, 297–393.

[12]Callear, A. B.; Norrish, R. G. W. "Mechanism of Antiknock"; *Nature* **1959**, *184*, 1704–1795; Ebhard, K. H. L.; Norrish, R. G. W. "Studies of Knock and Antiknock by Kinetic Spectroscopy II"; *Proc. Roy. Soc.* **1960**, *A259*, 297–303; Callear, A. B.; Norrish, R. G. W. "The Behaviour of Additives in Explosions and the Mechanism of Antiknock"; *Proc. Roy. Soc.* **1960**, *A259*, 304–324.

[13]Norrish, R. G. W.; Oldershaw, G. A. "The Flash Photolysis of Phosphine"; *Proc. Roy. Soc.* **1961**, *A263*, 1–9 and 10–18.

obtaining structural information about radicals. Work of this kind was done on sulfur monoxide, sulfur dioxide, and sulfur trioxide.[14] Structural studies were also made on the radicals ClO and the radical $NO_3$.[15]

In addition to free radicals, vibrationally excited molecules and their reactions were also investigated in Norrish's laboratories. In the photolysis of $ClO_2$ it was found, for example, that oxygen molecules were formed in highly excited vibrational states,[16] and excited oxygen was also found in the flash photolysis of ozone.[17] Other investigations were concerned with the mechanisms by which vibrationally excited species are produced in photochemical reactions. Such species are sometimes produced directly from molecules formed in repulsive states. They can also be produced by the collisional quenching of species formed in excited electronic states. Norrish's investigations with W. M. Smith of the quenching of excited sodium, published in 1940, have already been mentioned. Over 20 years later, flash photolysis was used to study the quenching of the metastable $Hg(^3P_0)$ species by a number of gases.[18]

In appearance and manner, Norrish was very far from what is popularly regarded as a typical professor or scientist. He was powerfully built, wore a closely cropped militarylike mustache, and was always neatly dressed. His personality was highly extroverted, and he was the most convivial and amusing of companions. His hospitality, in his college and at his home, knew no bounds, and he did everything he could to prevent his guests from leaving at the end of the evening. All who were entertained by him long remembered the experience, although some might have preferred to forget the effect of

[14]Norrish, R. G. W.; Oldershaw, G. A. "The Absorption Spectrum of SO and the Flash Photolysis of Sulphur Dioxide and Sulphur Trioxide"; *Proc. Roy. Soc.* **1959,** *A249,* 489–512.

[15]Lipscomb, F. J.; Norrish, R. G. W.; Porter, G. "Photolysis of Chlorine Dioxide and Absolute Rates of Chlorine Monoxide Reactions"; *Nature* **1954,** *174,* 785–786; Norrish, R. G. W.; Lipscomb, F. J.; Thrush, B. A. "The Study of Energy Transfer by Kinetic Spectroscopy I"; *Proc. Roy. Soc.* **1968,** *A233,* 455–465; Norrish, R. G. W.; Nicholas, J. E. "Some Reactions in the Chlorine and Oxygen System Studied by Flash Photolysis"; *Proc. Roy. Soc.* **1956,** *A307,* 391–397; Husain, R.; Norrish, R. G. W. "The Production of $NO_3$ in the Photolysis of Nitrogen Dioxide and of Nitric Acid Vapour Under Isothermal Conditions"; *Proc. Roy. Soc.* **1963,** *A273,* 165–179.

[16]Norrish, R. G. W.; Lipscomb, J. F.; Thrush, B. A. "The Study of Energy Transfer by Kinetic Spectroscopy I"; *Proc. Roy. Soc.* **1956,** *A233,* 455–465.

[17]McGrath, W. D.; Norrish, R. G. W. "The Flash Photolysis of Ozone"; *Proc. Roy. Soc.* **1957,** *A242,* 265–276; "Production of Vibrationally Excited Oxygen Molecules in the Flash Photolysis of Ozone"; *Nature* **1957,** *180,* 1272–1273; "Studies of the Reactions of Excited Oxygen Atoms and Molecules Produced in the Flash Photolysis of Ozone"; *Proc. Roy. Soc.* **1960,** *A254,* 317–326.

[18]Callear, A. B.; Norrish, R. G. W. "The Metastable Triplet State of Mercury, Hg $6(^3P_0)$: A Study of Its Reactions by Kinetic Spectroscopy"; *Proc. Roy. Soc.* **1962,** *A266,* 299–311.

his hospitality on their sobriety. The organizers of chemical lectures at Cambridge had to make sure that lecturers were not previously entertained by Norrish, since his lavish dispensing of beverages had sometimes led to more than usual incoherence. Although his research students sometimes found Norrish to be a little difficult, they retained their admiration and respect for him. He was remarkably successful as a director of research, and many of his students went on to do work of great distinction and to secure appointments of responsibility.

His style of research was highly intuitive. He had great persistence and experimental skill, with just enough theoretical background to be able to work extremely effectively in his own particular branch of kinetics. He did not have a deep understanding of thermodynamics, of valence theory, or of theories of chemical reaction rates, but knew enough for his own particular purposes. He once commented to the present writer that transition-state theory was "highfalutin' stuff", and it is unlikely that he had spent much time learning about it. His attitude to molecular orbital theory was very much the same. It is at first sight surprising that one who was so very successful in chemical kinetics and in certain aspects of spectroscopy should have been so uninterested in the theoretical treatments that lie behind those topics. The answer lies in his peculiarly intuitive approach to research. He had no objection to hypotheses based on a particular piece of research and framed many himself, but he regarded them as something to be shot at and perhaps shot down by subsequent work.

Norrish disliked administration and avoided it as far as possible. He was impatient with the kinds of arguments commonly brought up at committee meetings, and his impatience led him to be ineffective in influencing the opinions of others. He was not a good lecturer in the conventional sense. Many who heard his lectures on thermodynamics came away wondering whether he really understood the subject himself; it is not in fact likely that he had ever made a rigorous study of thermodynamics, preferring to rely on his intuition to gain a useful general idea of the subject in its practical aspects. His lectures on kinetics were inspiring to the better students but, again, did not convey to them a broad and comprehensive view of the subject.

Norrish wrote several review articles, particularly on his applications of flash photolysis, but he wrote no books. Here he is to be contrasted with his Cambridge colleague E. K. Rideal and his Oxford "arch rival" C. N. Hinshelwood. Both Rideal and Hinshelwood had a much more scholarly attitude toward science, and they wrote books that exerted considerable influence.

Norrish received many honors and awards. He was elected a Fellow of the Royal Society and was elected an honorary member of many societies in Britain and several European countries.

Three years before his death he suffered a stroke that left him considerably paralyzed and unable to speak, but a year later he was able to enjoy a luncheon given in his honor by Emmanuel College, which had awarded him a scholarship 63 years earlier.

KEITH J. LAIDLER
*University of Ottawa*

*The author is grateful to Lord Dainton, Lord Porter, and Dr. D. A. Ramsay for valuable suggestions relating to this biography.*

## *Bibliography*

Dainton, F.; Thrush, B. A. "Ronald George Wreyford Norrish"; *Biographical Memoirs of Fellows of the Royal Society* **1981,** 27, 379–424.

Norrish, R. G. W. "Chemistry and the Spectroscope"; *Advancement of Science* **1961,** 74, 1–12.

Norrish, R. G. W. "The Kinetics and Analysis of Very Fast Chemical Reactions"; *Chemistry in Britain* **1965,** 1, 289–311.

Norrish, R. G. W. "Kinetic Spectroscopy and Flash Photolysis"; *American Scientist* **1962,** 50, 131–157.

Norrish, R. G. W., "Liversidge Lecture; Some Isothermal Reactions of Free Radicals Studied by Kinetic Spectroscopy"; *Proc. Chem. Soc.* **1958,** 247–255.

Norrish, R. G. W. "Some Fast Reactions in Gases Studied by Kinetic Spectroscopy"; In *Les Prix Nobel 1967;* Norstedt and Söner: Stockholm, Sweden, 1967; pp 181–211.

Norrish, R. G. W. "The Study of Combustion by Kinetic Spectroscopy"; *Experientia, Supplementum* VII **1957,** 7, 87–112.

Norrish, R. G. W. "The Study of Energy Transfer in Atoms and Molecules by Photochemical Methods"; In *The Transfer of Energy in Gases;* Interscience: New York, 1964; pp 99–182.

Norrish, R. G. W.; Thrush, B. A. "Flash Photolysis and Kinetic Spectroscopy"; *Quart. Rev.* **1956,** 10, 149–168.

*Photochemistry and Reaction Kinetics* (dedicated to R. G. W. Norrish); Ashmore, P. G., Dainton, F. S., and Sugden, T. M., Eds.; Cambridge University: Cambridge, England, 1967.

# George Porter

## 1920—

George Porter, winner of a Nobel Prize in chemistry in 1967, was born in Stainforth, Yorkshire, on December 6, 1920. His father was a builder and a Methodist lay preacher. He won a scholarship to Thorne Grammar School, also in Yorkshire, and from 1938 to 1941 attended Leeds University as Ackroyd Scholar. In 1941 he enlisted in the Royal Naval Volunteer Reserve and until 1945 served as a radar officer in the Mediterranean and elsewhere.

From 1945 to 1948 he was a research student at Emmanuel College, Cambridge, and then served for three years as a university demonstrator in physical chemistry, at the same time carrying out research, particularly on flash photolysis, partly in collaboration with R. G. W. Norrish, with whom (along with Manfred Eigen) he shared the Nobel Prize. In 1951 he was appointed research fellow of Emmanuel College and assistant director of research in physical chemistry at Cambridge. After a year as assistant director of the British Rayon Research Association, he was appointed professor of physical chemistry at the University of Sheffield, becoming Firth professor and head of the Department of Chemistry in 1963.

He left the University of Sheffield in 1966 to become director of the Royal Institution of Great Britain; among his many illustrious predecessors were Sir Humphry Davy and Michael Faraday. As well as being director, he was Fullerian professor of chemistry and director of the Davy–Faraday Research Laboratory of the Royal Institution. He retired from the Royal Institution in 1987, and since then he has been

487

research professor and fellow of Imperial College, London, besides being President of the Royal Society from 1985 until 1990.

Porter's name is inextricably associated with flash photolysis, a technique that he developed at Cambridge in association with Norrish and that he greatly improved over a period of years. In this technique a light flash of high intensity and short duration brings about the formation of species such as atoms, radicals, and excited molecules, the structure and reactions of which can be studied by spectroscopic methods. In the early experiments at Cambridge in the late 1940s, the duration of the flash was on the order of a millisecond. Within a year or two, technical improvements allowed the duration to be reduced to about a microsecond, which is adequate for chemical reactions involving processes such as atom or radical combinations. For other processes, such as energy transfer, electron transfer, and primary steps in photosynthesis, flashes of shorter duration are required. They were achieved by important new techniques devised, in the 1960s and later, in Porter's laboratories and elsewhere.

The original work on flash photolysis was carried out between Porter's arrival at Cambridge in 1946 and 1949, when the first paper on the subject was published;[1] this was the second paper published by Porter, and it was quickly followed by others giving more details.[2] The first experiments were done with a bank of capacitors, supplied by the Royal Navy, allowed to discharge through a suitably designed lamp; energy of about 10,000 J was dissipated in a millisecond or so. At first the spectrum of the transients was observed by means of a continuous light source and electronic scan, but it was soon realized that it was a great improvement to employ a second flash, occurring after a time delay. This technique became known as the "pump and probe" or "pulse and probe" method. The delay between the two flashes was originally achieved by the use of a rotating sector, and this was the procedure used for several years. As the flash duration was reduced it became necessary to resort to electronic and other methods to introduce the necessary time delay. In earlier experiments the spectra were recorded photographically, which involved much labor in the identification of the absorption bands and in the study of their decay.

Two flash photolysis techniques have been used in Porter's laboratories. In one the spectrum is recorded photographically by means of

---

[1]Norrish, R. G. W.; Porter, G. *Nature* **1949**, *164*, 658.
[2]Porter, G. *Proc. Roy. Soc.* **1950**, *A200*, 284–290; *Disc. Faraday Soc.* **1950**, *9*, 60–69.

the second flash lamp. For a kinetic study the procedure is then repeated with various delays between the two flashes. Alternatively, a narrow wavelength interval may be selected for the second flash, and the kinetic behavior followed in a single experiment. The first method is convenient for the recording of spectra and a preliminary survey of the kinetics, whereas the second is more suitable for accurate and detailed kinetic studies. The first "pulse and probe" method was, however, essential for the much faster times that were used in later developments of the technique.

The first free radical to be studied by Porter in any detail was ClO, which was produced by the flash photolysis of mixtures of oxygen and chlorine.[3] The analysis of the spectra provided the dissociation energy and vibrational frequency of the radical both in its ground state and in an electronically excited state. It was also possible to obtain the rate constant for the formation of ClO by the process $2Cl + O_2 \rightarrow 2ClO$ and for the recombination of chlorine atoms with $N_2$ acting as a third body.

Aromatic free radicals were also studied in great detail in Porter's laboratories. A number of radicals, such as benzyl and anilino, were detected in rigid media,[4] and shortly afterward they were found to be produced in the flash photolysis of aromatic vapors.[5] The spectra of radicals such as phenoxy and anilino were found to vary considerably with the media in which they were present, and it was shown that this is due to their existence in protonated and unprotonated forms.[6] From the variation of the spectra with the media it was possible to determine the acid dissociation constants.

Much work was also done in Porter's laboratories on species existing in triplet states. It had been shown by G. N. Lewis and M. Kasha[7] that the phosphorescence of certain organic molecules in rigid media is the emission of light from excited species of triplet multiplicity, that is, having two unpaired electrons. Phosphorescence does not occur with the substances in the gas phase or in ordinary solution, because of the

---

[3]Porter, G. *Disc. Faraday Soc.* **1950**, *9*, 60–69; Porter, G.; Wright, F. J. *Disc. Faraday Soc.* **1953**, *14*, 23–34.

[4]Norman, I.; Porter, G. *Nature* **1954**, *174*, 508–509; *Proc. Roy. Soc.* **1955**, *A230*, 399–414.

[5]Porter, G.; Wright, F. J. *Trans. Faraday Soc.* **1955**, *51*, 1469–1474.

[6]Porter, G.; Ward, B. *J. Chim. Phys.* **1964**, *61*, 1517–1521; Land, E. J.; Porter, G.; Strachan, B. *Trans. Faraday Soc.* **1961**, *57*, 1885–1893; Land, E. J.; Porter, G. *Trans. Faraday Soc.* **1963**, *59*, 2027–2037.

[7]*J. Am. Chem. Soc.* **1944**, *66*, 2100–2116.

shorter lifetimes involved. Porter in the early 1950s recognized that flash photolysis might provide a means of identifying triplet states in ordinary solutions at room temperature, provided that the lifetimes were not shorter than a few microseconds, which corresponds to the duration of the flashes that could be produced in the early experiments.

The first work of this kind, done while Porter was still at Cambridge, was very successful.[8] For anthracene in hexane solution, for example, the half-life was on the order of a millisecond, and the flash photolysis technique provided an ideal way of observing the decay of the triplet state. A little later, similar work was done with the substance in the vapor phase, when the lifetimes were shorter but could still be measured.[9]

During subsequent years, at the University of Sheffield and at the Royal Institution, Porter and his colleagues carried out comprehensive studies of molecules in their triplet states and drew many conclusions of great significance. They found, for example, that under certain conditions the decay of the triplet state is a first-order process and that under other conditions a second-order process of triplet–triplet annihilation is predominant.[10] It was also shown that physical quenching of the triplet state is sometimes brought about by interaction with molecules in lower electronic states, to which energy is transferred.[11]

As well as using the flash photolysis technique to study the characteristics of free radicals and excited molecules, Porter devoted much attention to kinetic studies of chemical reactions of various types. The first reaction to which the technique was applied was the recombination of iodine atoms.[12] One surprising result was that $I_2$ was 1000 times more effective as a third body ("chaperon") than He, and that NO was 20 times more efficient still. This led to the suggestion that there may in some cases be a first a charge-transfer complex formed between an iodine atom and the chaperon molecule.

---

[8]Porter, G.; Windsor, M. W. *J. Chem. Phys.* **1953**, *21*, 2088; *Disc. Faraday Soc.* **1954**, *17*, 178–186; *Proc. Roy. Soc.* **1958**, *A245*, 238–258.

[9]Porter, G.; Wright, F. J. *Trans. Faraday Soc.* **1955**, *51*, 1205–1211.

[10]Porter, G.; Wright, M. R. *J. Chim. Phys.* **1958**, *55*, 705–711; *Disc. Faraday Soc.* **1959**, *27*, 18–27; Porter, G.; West, P., *Proc. Roy. Soc.* **1964**, *A279*, 302–312.

[11]Porter, G.; Wilkinson, F. *Proc. Roy. Soc.* **1961**, *A264*, 1–18.

[12]Christie, M. I.; Norrish, R. G. W.; Porter, G. *Proc. Roy. Soc.* **1952**, *A216*, 152–183; Christie, M. I.; Harrison, A. J.; Norrish, R. G. W.; Porter, G. *Proc. Roy. Soc.* **1955**, *A231*, 446–457; Porter, G.; Smith, J. A. *Proc. Roy. Soc.* **1961**, *A261*, 28–37; Porter, G. *Disc. Faraday Soc.* **1962**, *33*, 198–204.

A number of kinetic studies were also made of reactions in solution, using the flash photolysis technique. For example, some reactions involving quinones and dyes were investigated.[13] Measurements were also made of some protonation and deprotonation rate constants for species in ground and excited states.[14]

In about 1967, when Porter won his Nobel Prize, he began to introduce important improvements into the technique of flash photolysis, improvements that would greatly broaden the range of processes that could be investigated. Until that time flash photolysis had been limited to events that occur over a period of a few microseconds. It was capable of dealing with many processes of a chemical nature (involving the breaking of chemical bonds) and of physical processes such as the decay of certain species in triplet states. Many other physical processes were, however, excluded: for example, the decay of species in singlet states occurs in a matter of nanoseconds.

In the late 1960s Porter and his colleagues were successful in developing highly efficient laser flash systems that were capable of dealing with processes occurring in the nanosecond range.[15] Ruby lasers proved to be very effective, and by the use of a Q-switching technique it was possible to obtain a highly reproducible flash of a few nanoseconds' duration, a flash that had the energy of several joules, sufficient to produce the necessary excitation. The primary emission of the ruby laser is at 694 nm, at the red end of the spectrum, but the first harmonic is at 347 nm, which, because it is in the ultraviolet range, is absorbed by most organic substances.

Electronic methods are unreliable for controlling very short time delays between the primary flash and the monitoring flash, and resort was made to a device based on the speed of light. Light travels 30 cm in each nanosecond, so that a delay of about a nanosecond can be achieved by splitting the laser beam and allowing it to travel a convenient distance before reaching the reaction vessel. This type of equipment was used by Porter and his colleagues to study a number of rapid processes, including the radiative decay of singlet-excited species, both in the gas phase and in solution.

During the 1970s there were further developments in laser technology, in the form of mode-locked lasers, which allow the production

[13]Porter, G.; Windsor, M. W. *Proc. Roy. Soc.* **1958**, *A245*, 238–258.

[14]Godfrey, T. S.; Porter, G.; Suppan, P. *Disc. Faraday Soc.* **1965**, *39*, 194–199.

[15]Porter, G.; Steinfeld, J. I. *J. Chem. Phys.* **1966**, *45*, 3456–3457; Porter, G.; Topp, M. R. *Nature* **1968**, *220*, 1228–1229; *Proc. Roy. Soc.* **1970**, *A315*, 163–184.

of a pulse of only about a picosecond ($10^{-12}$ s) duration. Porter and his colleagues made many investigations in this field of picosecond flash photolysis, which allows the study of processes such as vibrational relaxation and redistribution, singlet-to-triplet transitions, and electron-transfer processes. For example, a study was made of the elementary processes involved in the reaction between excited thionine and ferrous ions.[16] In this well-known reaction the dye is reduced to a leuco (colorless) form and the process reverses in the dark. The reaction of excited dye with $Fe^{2+}$ occurs in the picosecond region and its rate is measured. Since the late 1960s much work has been done in Porter's laboratories on many of the elementary processes involved in photosynthesis and on related model systems. Early work included a study of energy transfer and light-harvesting mechanisms in model systems.[17] Numerous papers on topics related to photosynthesis have come from Porter's laboratories during the 1970s and 1980s, and in the early 1990s he began studies of the reaction centers of chloroplasts in the femtosecond time range.

Aside from his research work, Porter has made distinguished contributions to education and to scientific administration. He is an outstanding lecturer, able to present complex subjects with great clarity and to make them interesting and often entertaining. He has excelled not only as a lecturer to university students but as one who has given many lectures to persons with no particular scientific background (for example, his Royal Institution lectures). In the 1960s the British Broadcasting Corporation filmed a series of half-hour lectures on "The Laws of Disorder", covering in a very clear manner the basic ideas of thermodynamics and kinetics. Although intended for a lay audience, this series remains a valuable review of the subject even for university students.

Porter's direction of the research at the Royal Institution, over a period of 21 years, was highly effective, and led that distinguished organization into an entirely new era. In addition to his own contribution to photochemistry, his colleagues have carried out important research in photosynthesis and other fields. The Royal Institution has also had great influence in organizing conferences and in other ways.

The honors received by Porter include the Fellowship of the Royal Society, of which he was president from 1985 to 1990, and the Order

---

[16] Archer, M. D.; Ferreira, M. I. C.; Porter, G.; Tredwell, C. J. *Nouveau Journal de Chimie* **1976**, *1*, 9–12.

[17] Kelly, A. R.; Porter, G. *Proc. Roy. Soc.* **1970**, *A315*, 149–161.

of Merit. He was knighted in 1972 and was created Lord Porter of Luddenham in 1990. He is an honorary fellow or honorary member of many societies in the United Kingdom and elsewhere. He has received honorary degrees from over thirty universities in various parts of the world. He has presented named lectures at numerous universities and other institutions, and has received many medals for his contributions to research and education.

KEITH J. LAIDLER
*University of Ottawa*

# Bibliography

*Contemporary British Chemists;* Campbell, W. A.; Greenwood, N. N., Eds.; Taylor & Francis: London, 1971.

Doust, T. A. M.; Phillips, D.; Porter, G. "Picosecond Spectroscopy: Applications in Biochemistry"; *Biochem. Soc. Trans.* **1984,** *12,* 630–635.

Gore, B. L.; Doust, T. A. M.; Giorgi, L. B.; Klug, D. R.; Ide, J. P.; Crystal, B.; Porter, G. "The Design of a Picosecond Flash Spectroscope and Its Application to Photosynthesis"; *J. Chem. Soc. Faraday Trans.* **1986,** *2,* 82, 2111–2115.

Patterson, L.; Porter, G. "Lasers in Photochemical Kinetics"; *Chemistry in Britain* **1970,** *6,* 246–250.

Porter, G. "Flash Photolysis"; In *Techniques of Organic Chemistry;* Weissberger, A., Ed.; Interscience: London, 1963; Vol. 8, Part 2.

Porter, G. "Flash Photolysis"; In *Photochemistry and Reaction Kinetics;* Ashmore, P. G.; Dainton, F. S.; Sugden, T. M., Eds.; Cambridge University Press: Cambridge, England, 1967; pp 93–111.

Porter, G. "Flash Photolysis and Some of Its Applications"; In *Les Prix Nobel en 1967;* Norstedt and Sons: Stockholm, Sweden; pp 1–20; reprinted in *Science* **1968,** *160,* 1299–1307.

Porter, G. "Picosecond Chemical Kinetics"; In *Lasers in Chemistry;* West, M. A., Ed.; Elsevier: Amsterdam, The Netherlands, 1977; pp 288–298.

Porter, G. "Picosecond Chemistry and Biology"; In *Picosecond Chemistry and Biology;* Doust, T. A. M.; West, M. A., Eds.; Science Reviews: London, 1983.

Porter, G.; Topp, M. T. "Nanosecond Flash Photolysis"; *Proc. Roy. Soc.* **1970,** *A315,* 163–184.

# Manfred Eigen

## 1927–

Manfred Eigen was born in Bochum, Germany, on May 9, 1927, the son of chamber musician Ernest Eigen and his wife, Hedwig Eigen, née Feld. He began the study of physics and chemistry in the autumn of 1945 at the Georg-August University in Göttingen and received his doctorate in natural sciences in 1951. After two years as assistant lecturer in physical chemistry there, he transferred to the Max-Planck Institut für Physikalische Chemie, which had moved to Göttingen under the directorship of Karl Bonhoeffer. His scientific work focused on the study of fast reactions by various relaxation methods, for which he received the 1967 Nobel Prize in chemistry. Four areas have particularly attracted his interest: (1) ionic reactions, (2) proton transfer reactions, (3) multistage processes involving metal complexes, and, most recently, (4) complex biochemical processes. Among many other awards, he has received the Bodenstein Prize of the Deutsche Bunsen-Gesellschaft (1956), the Otto Hahn Prize (1962), the Kirkwood Medal (1963), and the Harrison Howe Award (1965) of the American Chemical Society. Among his honorary degrees are those from Harvard University, Washington University, and the University of Chicago. He holds memberships in numerous scientific societies both in Europe and abroad. He has written two books: *Hypercycles: Principles of the Self-Organization of Macromolecules* (with P. Schuster, Springer: New York, 1979) and *Laws of the Game: How the Principles of Nature Govern Chance* (with R. Winkler, translated by Robert and Rita Kimber, Knopf: New York, 1981). He is married to

Elfriede Eigen, née Muller. They have two children, Gerald, born in 1952, and Angela, born in 1960. He is a keen amateur musician and enjoys mountaineering as a holiday pastime.

The work of Manfred Eigen and his colleagues has opened up to experimental measurement the rates of reactions that go to completion in times less than a millisecond. The breakthrough came from a different way of dealing with the study of chemical kinetics. Previously, kinetics had been studied by mixing reagents and following the rate of the reaction that occurred during mixing. Various sampling procedures were used to determine the extent to which products were formed from the reagents. Direct chemical determination of concentrations was used when the reaction was slow enough. Otherwise, some physical property of one or more of the reagents, such as absorption spectrum, color, optical rotatory characteristics, or index of refraction, was monitored.

The limiting values for studying reactions following the mixing of reagents were reached by the rapid mixing technique invented by Hartridge and Roughton[1] in the 1920s and later developed by Britton Chance and his school. The limiting values so obtained were in the range of milliseconds. The method was particularly successful in the investigation of the kinetics of enzyme reactions.

However, acid–base neutralization reactions still proved to be too fast even for the rapid mixing technique to handle. It was possible to show on the basis of diffusion rate calculations that, for ions that react at every collision, reaction times could be as short as $10^{-11}$ seconds, some 7 or 8 magnitudes lower than the millisecond range accessible by rapid mixing techniques.

It became clear that a new approach was needed to deal with such rapid reactions. When a chemical reaction reaches "equilibrium", reaction does not cease. Rather, the rate at which products are formed from reactants is balanced by the rate at which reactants are reformed from products. Hence, what one needs to do is disturb the equilibrium of the system very rapidly and follow its return to equilibrium, that is, follow the "relaxation process". Then it is no longer necessary to mix the reaction partners. Rather it is necessary to transmit an inducing signal into the system and to follow the reaction signal. Obviously, both of these signals must be on a time scale much shorter than the millisecond range. The inducing signal may be mechanical or electrical.

---

[1] Hartridge, H.; Roughton, F. J. W. *Proc. Roy. Soc. London*, **1923**, *A104*, 376.

A mechanical signal, such as a pressure wave, propagates at the velocity of sound, about $10^{-5}$ cm/s in condensed phases. A traveling electric wave propagates about 100,000 times faster. Thus, in a cell of the dimensions 1–10 mm, a mechanical signal can achieve a resolution time of about $10^{-6}$ s, and an electrical signal, one of less than $10^{-10}$ s. Clearly, an optical or electrical signal must be used to follow the reaction.

The first breakthrough in the study of such rapid reactions came from the technical problem of measuring distances in seawater with ultrasonic probes. In certain frequency ranges, it was found that seawater absorbs sound even more strongly than distilled water. The magnesium sulfate in seawater was found to be responsible for this absorption. Two absorption peaks were found in the ultrasonic region, one at about $10^5$ Hz and the other at about $10^8$ Hz.[2] This result was best interpreted in terms of a stepwise substitution of water molecules bound to the coordination shells of the magnesium–aquo complexes by sulfate ions. The absorption continuum of the ion clouds was later discovered at higher frequencies.

From a consideration of the physics of energy absorption, it turns out that the conditions for observing relaxation phenomena most effectively exist when the rate of reestablishment of equilibrium is comparable to the rate of pressure change, that is, when the time constant for reestablishment of equilibrium is on the same order of magnitude as the period of the acoustic wave. In this case, a characteristic absorption maximum occurs, from which the relaxation time of the equilibrium state can be read off, for example, as the reciprocal of the frequency at the maximum absorption per wavelength. The hydrolysis of ammonia in aqueous solution follows this simple behavior. The situation for magnesium sulfate is more complicated because of the multistep process involved. It is, however, possible to obtain a system of linked differential equations, from which by appropriate mathematical treatment relaxation times are found as eigenvalues. The rate constant can then be determined from the relaxation time spectra, suitably transformed.

The physical theory of relaxation effects in a dissociating gas was worked out very early by Einstein.[3] In 1923 Walther Nernst unsuccessfully attempted to detect the effect in the $2NO_2 = N_2O_4$ system.

[2]Tamm, K.; Kurtze, G. Nature **1951**, *168*, 346; Acustica **1953**, *3*, 33.

[3]Einstein, A., Sitzber. Preuss. Akad. Wiss., Physik.-math. Kl. **1920**, 380.

At the time, the techniques of sound transmission were insufficiently developed. Forty years later the effect was observed in this system.[4] In the meantime, following the work of K. F. Herzfeld and H. O. Kneser (1928–1931), relaxation effects were studied in the establishment of equilibrium in energy transfer between the degrees of freedom of complex molecules. In general, however, the sound absorption method turned out be be too insensitive for studying chemical relaxation effects.

Variation of the electric field strength proved to be a more successful relaxation method. It was especially useful for the study of the kinetics of hydrogen bonding in nonpolar solvents and for the investigation of the kinetics of structural changes in polypeptides. The change in amplitude of the shift at equilibrium is plotted as a function of the pulse duration (relative to the relaxation time). From this dispersion curve, the relaxation time can be determined.

The chemical relaxation of the dissociation of $H_2O$, and thence the neutralization kinetics, was measured for the first time by Eigen, employing a single rectangular pulse produced by a double spark circuit. A stationary field method made possible the direct measurement of dissociation rates in ice crystals, thus enabling the kinetic parameters of the transport and neutralization of proton charges in hydrogen-bridged systems to be finally established.

Step-type disturbances are used almost exclusively today for studying complex, multistage processes. The relaxation spectrum can be followed directly by means of discrete steps on a logarithmic scale. The field effect of oxyhemoglobin was measured in this way by Ilgefritz with a time resolution of 50 ns, by means of a rectangular field pulse obtained in the form of a traveling wave by discharging a high-tension cable across two spark gaps.

Particularly useful in the study of fast reactions has been the temperature-jump method. It can be produced by either adiabatic compression or dilatation. In electrolytic systems, it can be produced by heating with electrical impulses. Electrical heating can be achieved in two ways: (1) with a current impulse in solutions having a finite electrolytic conductivity and (2) with microwave heating in the X-band range for any degree of conductivity. Fields on the order of 100 kV/cm are required in both cases. In the case of microwave heating, one needs

---

[4]Bauer, H. J.; Kneser, H. O.; Sittig, E. *Acustica* **1959**, *9*, 181; Sessler, G., *Acustica* **1960**, *10*, 44.

to produce radar impulses with powers on the order of megawatts. The chemical change can best be followed by optical methods. The temperature-jump method has a wide range of applications extending from inorganic to biological chemistry. Two improvements have extended the range of applications: (1) A flow arrangement is used instead of a static measurement cell; (2) The stationary-state reaction mixture flowing through an observation capillary can be heated periodically with the aid of repeated microwave impulses. The first of these improvements makes it possible to carry out relaxation measurements on systems that are not at equilibrium but that have reached a stationary state. This improvement is of decisive importance in the investigation of biochemical reaction mechanisms. With these methods, the time range has been covered from fractions of a nanosecond to several seconds, thus bridging the time gap between molecular spectroscopy and classical kinetics.

Using relaxation spectrometry, Eigen found that the reaction $H^+ + OH^- = H_2O$ occurs at every encounter between the solvated ions, with a rate constant $k = 1.43 \times 10^{-11}$ mole$^{-1}$sec$^{-1}$. Are all such charge-neutralization reactions diffusion-controlled? A study of inorganic metal ions shows that the rate constants for substitution in the inner coordination shell lie in the range from $10^3$ to $10^9$ sec$^{-1}$. The mechanism is a stepwise substitution of the negative ion for water molecules within the successive shells surrounding the metal ion; the slowest step is the replacement of water in the inner coordination shell. Specific effects of electronic structure show up only in the case of the transition metals.

In organic chemistry, the method has been even more successful in measuring the rates of proteolytic reactions. The kinetics of proton transfer were determined by Eigen for a large number of organic acids and bases. The reaction mechanism of acid–base catalysis has been elucidated, and the Brønsted relations extended and generalized. In particular, relaxation methods have been able to show that enol formation, like most proteolytic recombination processes, is entirely a diffusion-controlled reaction ($k \approx 10^{10}$ mole$^{-1}$sec$^{-1}$).

Perhaps the main uses for relaxation spectrometry today are in biochemistry and biology. Eigen developed methods for the study of complex sequences of biological reactions—as well as of single, specific, elementary steps. The "fine structure" of enzyme reaction mechanisms have become accessible to the biochemist. Two models for explaining the cooperative interaction by which enzymes bind their substrates have been developed. Each of these models employs three parameters and can account quantitatively for the sigmoid character of the curve for saturation of the enzyme as a function of substrate concentration. A decision can be made on the basis of relaxation measurements between these two models. Examples that fit each of these mechanisms have been found. The two models have been shown to represent limiting cases of a single, more general, reaction scheme. It is interesting to note that mechanisms of this kind explain properties that we meet nowhere else on the plane of molecules and that we have come to know only through the circuit and control elements of electronics and transistor technology. It is a new discovery that these properties are possessed by single—indeed, programmed—molecules and are not simply the consequence of complex or mutually coupled reactions. Eigen also found that the lifetime of a base pair is measured in fractions of a microsecond. Replication, with many correlated individual steps, takes place in fractions of a millisecond. It seems that we are even beginning to come to an understanding of how molecular reaction systems have found a way to "organize themselves".

**ERNEST G. SPITTLER, S.J.**
*John Carroll University*

# Bibliography

Eigen, M. *Das Spiel: Naturgesetze Steuern den Zufall* (with R. Winkler); München, 1975. *(Laws of the Game: How the Principles of Nature Govern Chance,* translation by Robert and Rita Kimber of *Das Spiel;* A. A. Knopf: New York, 1981.)

Eigen, M. *The Hypercycle: A Principle of Natural Self-Organization* (with P. Schuster); Springer: Berlin, 1979.

Eigen, M.; Diebler, H.; Czerlinski, B., "Relaxation Investigations in the Kinetics of Metal Complex Formation"; *Z. Physik. Chem. (Frankfurt)* **1959,** *19,* 246–249.

Eigen, M.; Hammes, G. G., "Kinematic Studies of Adenosine Triphosphate Reactions with the Temperature Jump Method"; *J. Am. Chem. Soc.* **1960,** *82,* 5951–5952.

Eigen, M.; Kurtze, G.; Tamm, K., "Reaction Mechanisms of Ultrasonic Absorption in Aqueous Electrolytic Solutions"; *Z. Elektrochem.* 57, 102–118.

Eigen, M.; Tamm, K., "Sound absorption in electrolytes as a consequence of chemical relaxation (I)"; "Relaxation theory of stepwise dissociation (II)"; *Z. Elektrochem.* **1962,** *66,* 93–107 and 107–121.

# Lars Onsager

## 1903–1976

Lars Onsager was born on November 27, 1903, in Oslo, Norway. He received the Nobel Prize in chemistry on December 10, 1968, for his discovery of the Onsager reciprocal relations, mathematical equations that are fundamental for the thermodynamic theory of irreversible processes. He married Margarethe Arledter in 1933; they had one daughter and three sons. In 1945 he became a citizen of the United States. He died on October 5, 1976, in Coral Gables, Florida. He worked with pencil and paper in the tradition of Svante Arrhenius (Nobel Laureate in chemistry, 1903) and Josiah Willard Gibbs, extending the methods of physical chemistry to elucidate mysterious, diverse, and fundamental phenomena.

His father was a lawyer and his mother was a teacher. His education in the Oslo schools was rich in literature, philosophy, music, and the fine arts. Lars showed an early aptitude and interest in mathematics. In 1920 he entered the Norges Tekniske Hogskole in Trondheim to study chemical engineering; he received the Ch.E. degree in 1925. His intellectual interests were not limited by the curriculum. He worked his way through the notoriously difficult problems in Edmund Taylor Whittaker and G. N. Watson's calculus text,[1] and he began to read beyond the textbooks into the research journals.

In 1887 Arrhenius had published a theory of electrically conducting solutions (such as solutions of acids, bases, and salts in water) based

---

[1] *A Course of Modern Analysis;* Cambridge University: Cambridge, England, 1915, 1920.

on the idea that the dissolved substances dissociate in water to produce electrically charged ions.[2] The idea of ionic solutions was revolutionary, leading to experimental investigations and theoretical speculations by many physical chemists. Because of mutual electrical attractions and repulsions, the interactions of ions are more complex than interactions of neutral molecules. In 1923 Peter Debye (Nobel Laureate in chemistry, 1936) and Erich Hückel published an impressive theoretical analysis of these ionic interactions. They imagined that each ion in solution was shielded by an electrostatic atmosphere produced by ions of the opposite charge.[3]

Onsager worked through the mathematics and physics of the publications of Debye and Hückel but was puzzled about certain implications of their theory. He found that the Debye–Hückel theory did not satisfy fundamental requirements of symmetry. Onsager recalled: "The relaxation effect ought to reduce the mobilities of anion and cation in equal proportion. Much to my surprise, the results of Debye and Hückel did not satisfy that relation, nor the requirement that whenever an ion of type A is 10 Å West of a B, then there is a B 10 Å East of that A. Clearly, something essential had been left out in the derivation of such unsymmetrical results".[4] Onsager was 22 in 1925, not yet graduated from Norges Tekniske Hogskole, when he went to Zurich to tell Debye that the Debye–Hückel theory was flawed. Debye listened to Onsager's criticism and then offered him a research assistantship at Eidgenössische Technische Hochschule in Zurich. With Debye's encouragement, he developed what became known as the Onsager limiting law for dilute solutions of electrolytes.[5] He continued as a graduate student at the Norges Tekniske Hogskole and was also a student at the Eidgenössische Technische Hochschule from 1926 to 1928.

Emigrating to the United States in 1928, Onsager accepted an appointment at Johns Hopkins University as an instructor in the freshman chemistry course, but by all accounts he did not communicate with the freshmen. After this brief stint at Johns Hopkins he became an associate in chemistry at Brown University, where he remained from

---

[2] Arrhenius, S. "Über die Dissociation der in Wasser gelösten Stoffe"; Z. Phys. Chem. Stöchiom. Verwandtschaftsl. **1887,** 1, 631–648.

[3] "Zur Theorie der Elektrolyte"; Phys. Z. **1923,** 24, 185–206, 305–325.

[4] "The Motion of Ions: Principles and Concepts"; In Les Prix Nobel en 1968; Norstedt: Stockholm, Sweden, 1969; pp 169–182.

[5] "Zur Theorie der Electrolyte", Phys. Z. **1926,** 27, 388–392; **1927,** 28, 277–298.

1928 to 1933. There he met Raymond Matthew Fuoss. Fuoss took his Ph.D. in 1932 with Onsager, was his first coauthor, and eventually became a colleague at Yale. It was also at Brown that Onsager developed the theory for which he received the Nobel Prize.[6] He submitted this work on his reciprocal relations to the Norges Tekniske Hogskole in Trondheim, but the faculty there judged it unacceptable for a doctorate. Thirty years later the Norges Tekniske Hogskole did award him the honorary degree Dr. Technicae.

Onsager's two 1931 papers presented a fundamental extension of chemical thermodynamics beyond its nineteenth-century formulation by Gibbs.[7] Gibbs developed the mathematics that relates properties of any chemical system *at equilibrium,* when no properties are changing. Onsager examined evolving systems in which two or more irreversible processes (such as flows of heat or electricity, diffusion of matter, or chemical reactions of substances) are taking place simultaneously, and developed the mathematics that relates the changing properties. Lord Kelvin, Hermann von Helmholtz, and others had examined specific processes; Onsager generalized their results in terms of symmetrical equations relating forces and flows within nonequilibrium systems. He added two new principles to the "laws" of thermodynamics: the principle of microscopic reversibility and the principle of least dissipation of energy.

Charles A. Kraus, an experimental chemist and chair of the Brown University Chemistry Department, often tried to convince Onsager to do some experiments to complement his theoretical work. Finally Onsager reported to Kraus that he had decided to separate isotopes by thermal diffusion. Kraus was especially pleased that the only equipment needed was a tube. Then Onsager revealed that the tube would be installed in a stairwell in the chemistry building, reaching from the basement to the third floor, and that the tube had to be made of platinum. With the Depression at hand, funds were not available for a multistory platinum tube. The experiment eventually was conducted successfully by others as part of the Manhattan Project for the atomic bomb.[8]

[6]"Reciprocal Relations in Irreversible Processes"; *Phys. Rev.* **1931**, Series 2, *37*, 405–426; **1931** *38*, 2265–2279.

[7]Gibbs, J. W. "On the Equilibrium of Heterogeneous Substances"; *Trans. Conn. Acad. Arts Sci.* **1876–1878**, *3*, 108–248, 343–524.

[8]Barry, J. F., Jr. "Lars Onsager: 'The greatest theoretical chemist'"; *Brown Alumni Monthly* **1976**, *77 (November)*, 7–8.

The deepening depression forced Brown to reduce its faculty, and Onsager lost his job in 1933. "In the summer of that year Lars was in Europe, and went to visit H. Falkenhagen, the Austrian electrochemist. Falkenhagen was unwell at the time and asked his sister Gretl (Margarethe Arledter) to entertain Lars. Gretl saw him coming up the stairs—a very handsome young man who her brother had told her was 'well ahead of his times'. They went out to dinner, but Lars was his usual reticent self. After dinner he fell asleep on the patio, and then woke up and asked: 'Are you romantically attached?' They became engaged 8 days later—Margarethe at 21 and Lars at 29—and got married on 7 September 1933."[9]

He received an appointment as a Sterling and Gibbs postdoctoral research fellow at Yale beginning in the fall of 1933. The Yale chemistry faculty, having awarded him a postdoctoral fellowship, were embarrassed to discover that he had no doctorate. They persuaded him to submit a thesis for a Yale Ph.D. Onsager did not want to resubmit the work on reciprocal relations that the Norges Tekniske Hogskole had rejected; instead he wrote on a mathematical subject that neither the Yale chemists nor physicists could evaluate. On the enthusiastic recommendation of Einar Hille of the Yale mathematics department, the chemistry department approved the thesis.[10]

There followed 39 years at Yale (1933–1972) in association with a distinguished group of experimentalists and theoreticians in the area of chemical solutions, thermodynamics, and statistical mechanics: Raymond Fuoss, Julian Munson Sturtevant, Herbert Spencer Harned, Benton Brooks Owen, and John Gamble Kirkwood. In 1945 Yale named him the J. Willard Gibbs Professor of Theoretical Chemistry, appropriately recognizing this towering intellect of the twentieth century as a successor to Yale theoretician Gibbs, whose work in thermodynamics and statistical mechanics is probably the United States's outstanding nineteenth-century contribution to science.

Onsager told an interviewer that "my field of interests is as unspecified as anybody's in the physical sciences".[11] Certainly his interests were never narrow. Onsager's major accomplishments ranged throughout theoretical chemistry and physics, including the dipole theory of dielectrics, disordered solids, phase transitions, properties of liquid

---

[9]Quotation from Longuet-Higgins and Fisher, 1978, p 451.

[10]"Solutions of the Mathieu Equation of Period $4\pi$ and Certain Related Functions", Ph.D. dissertation, Yale University, 1935.

[11]Saar, R. "Lars Onsager"; *The Yale Scientific Magazine* **1971,** *96 (November),* 5–7.

helium, magnetic properties of ice, quantization of vortex motion in superfluids, cooperative phenomena, electron correlation in metals, and protonic semiconductors.

"Perhaps what he is best at," commented Yale inorganic chemist Andrew Patterson, Jr., "is making subtle mathematical substitutions which will collapse a formidable problem into a solvable one. This is real insight. In a time before computers became popular, he was able to make mathematical approximations which allowed problems to be solved that couldn't have been solved in any other way".[12]

His office was a windowless cubbyhole in the medieval-looking Sterling Chemistry laboratory. Joseph Hubbard was appointed a post-doctoral fellow with Onsager in 1971. "On first setting eyes on his new postdoc, Lars embraced him in Russian style and took him to his office to show him a reprint. There was chaos on every surface, including the floor. Suddenly Lars disappeared, and Hubbard found him underneath the desk, where he had located the reprint (which turned out to be a 400-page thesis) and a two-month-old paycheck. Observing Onsager's contortions, Hubbard thought to himself: 'Here's a fellow who scratches his left ear by reaching round the back of his head with his right hand. I wonder how he ties his shoes!' ".[13]

His courses were challenging and his lectures were obscure. Yale wisely confined his teaching duties to graduate-level courses. His statistical mechanics courses were unofficially and variously known as "Advanced Norwegian I and II", or as "Sadistical Mechanics". He had no ambition to build a large research group. He enjoyed working alone, and almost all his research publications were written either by himself or with a single collaborator. Yet he was a warm and genial man who was a fellow of Yale's Trumbull College and often ate meals there with undergraduates.

Onsager was elected to the National Academy of Sciences in 1947. He received the Rumford Gold and Silver Medals of the American Academy of Arts and Sciences (1953), the Gilbert Newton Lewis Medal of the Berkeley Section of the American Chemical Society (1962), the Willard Gibbs Medal of the Chicago ACS Section (1962), the Theodore William Richards Medal of the Northeastern ACS Section (1964), the Peter J. W. Debye Award in Physical Chemistry (1965), the Belfer award in Pure Science of Yeshiva University (1966), and the U.S. National Medal of Science (1968). He was the first recipient of the John

---

[12]Onsager papers, Sterling Memorial Library, Yale University.
[13]Quotation from Longuet-Higgins and Fisher, **1978**.

Gamble Kirkwood Medal of the New Haven ACS Section (1962), and in 1958 became the sixth recipient of the Lorentz Medal (first awarded in 1927) of the Royal Netherlands Academy of Sciences.

Upon his retirement from Yale, Onsager accepted an appointment at the Center for Theoretical Studies at the University of Miami at Coral Gables, where he turned his attention to biophysics. He never underestimated the challenge of neuroscience. He remarked that using an electroencephalogram to investigate the brain is like trying to discover how a telephone system works by measuring the fluctuations in the electric power used by the telephone company. The Center honored him on his seventieth birthday with a conference on "Quantum Statistical Mechanics in the Life Sciences" at which he gave a paper ("Life in the Early Days") on the origin of life on Earth. He remained an active scientist until the day of his death.

Yale physical chemist Philip A. Lyons wrote that "for the experimentalist, Onsager is a pure delight. His analyses of phenomena invariably define the practical scope of an experimental study and, what is more important, the direction of the study. Never has one of his theoretical proposals led to anything but substantial and fruitful research.... Lars Onsager's contributions are those of a scientist with a marvelous blend of intuition and sophistication, zest and discipline. Though they differ widely in content, his papers have one thing in common: they offer a fresh look, usually at a vista."

**GEORGE FLECK**
*Smith College*

# Bibliography

Cole, R. H. "Obituary: Lars Onsager"; *J. Non-Equilib. Thermodyn.* **1977**, *2*, 129–131.

Fuoss, R. M. "Nobel Laureates in Physics and Chemistry. 2. Chemistry"; *Science* **1968**, *162*, 646–648.

Ihde, A. J. *The Development of Modern Chemistry,* Harper & Row: New York, 1964.

Kirkwood, J. G. "The Scientific Work of Lars Onsager"; *Proc. Am. Acad. Arts Sci.* **1953**, *82*, 298–300.

Longuet-Higgins, H. C.; Fisher, M. E., "Lars Onsager, 27 November 1903–5 October 1976"; *Biog. Mem. Fellows R. Soc.* **1978**, *24*, 443–471. Contains bibliography.

Margenau, H. "Lars Onsager (1903–1976)"; *Foundations of Physics* **1977**, *7*, 783–784.

Miller, D. G. "Thermodynamics of Irreversible Processes: The Experimental Verification of the Onsager Reciprocal Relations"; *Chem. Rev.* **1960**, *60*, 15–37.

"Onsager, Lars"; *Modern Men of Science;* McGraw-Hill: New York, 1966; pp 357–359; an autobiography.

# Derek Harold Richard Barton

## 1918–

The 1969 Nobel Prize in chemistry was awarded to two scientists: Odd Hassel, a physical chemist who applied diffraction techniques to organic molecules, and Derek H. R. Barton, a synthetic organic chemist who understood the importance of geometry in the chemical behavior of organic compounds. The prize was given to these men "for developing and applying the principles of conformation in chemistry", that is, how the preferred shapes of molecules determine their physical and chemical properties. Barton, the organic chemist, was born in Gravesend, Kent, England, on September 8, 1918, to William Thomas Barton and Maude Lukes Barton. He attended Gillingham Technical College and received his B.Sc. with First-Class Honors at Imperial College, University of London, in 1940, and his Ph.D. at the same institution in 1942. He did his graduate research under Ian Heilbron and E. R. H. Jones. In his professional career, Barton has worked at several institutions and is presently Distinguished Professor of Chemistry at Texas A&M University.

Sometimes the Nobel Prize is awarded for a lifetime of contributions to a particular area of chemistry, and sometimes it is awarded for a single breakthrough effort. The latter is true for Barton's award, which honored a four-page paper that appeared in the Swiss journal

*Experientia* in 1950, entitled "The Conformation of the Steroid Nucleus".[1] Although this journal had limited circulation, the importance of the work was immediately recognized, and the nature of organic chemical research was changed dramatically. Barton's analysis of the shape of steroids was based on the structure of the six-membered ring of carbon atoms (cyclohexane), which is the fundamental unit not only of steroids, but of many naturally occurring compounds, such as terpenes, alkaloids, and carbohydrates, as well as synthetic compounds produced in the laboratory. The concept of conformational analysis developed by Barton was quickly utilized by practical organic chemists in structure determination and synthesis of complex molecules and at the same time by teachers who included his work in beginning courses on organic chemistry. Indeed, one of the advantages of Barton's work is its inherent ease of comprehension, something appearing so obvious and rational that one wonders why it was not discovered long before 1950.

The detailed shape of the cyclohexane molecule was a subject of speculation for many years. It was assumed by Baeyer (Nobel Prize in chemistry, 1905) that the cycloalkanes were planar rings and that bond angles in these rings would deviate from the normal "tetrahedral" bond angle of carbon, $109°28'$. The greater the deviation from the normal angle, the greater the strain in the molecule. A planar cyclohexane would have bond angles of $120°$ and thus would be a strained molecule. In 1890, Hermann Sachse, a relatively unknown 28-year-old assistant at the Charlottenburg Institute, published alternate structures for cyclohexane that used tetrahedral carbon atoms and were "strain-free".[2] However, the models proposed by Sachse were ignored or discounted, or chemists believed that "strain-free" cyclohexanes rapidly interconverted and that the molecule could be approximated by a planar structure. Certainly, the geometric (cis-trans) isomers could be accommodated by an average planar ring. Ernst Mohr, professor of chemistry at Heidelberg, resurrected the strain-free cyclohexane ring in 1918 and drew pictures similar to the structures used by modern chemists. Rather than a planar structure for cyclohexane, Mohr used two non-planar structures. In one of these, referred to as the *chair,* the carbon atoms are alternately arranged above and below the average molecular plane (i.e., atoms 1,3, and 5 are above the plane and atoms 2, 4, and

---

[1]Barton, D. H. R. *Experientia* **1950**, *6*, 316.

[2]Sachse, H. "On the Geometrical Isomers of Hexamethylene Derivatives"; *Chem. Ber.* **1890**, *23*, 1363.

6 are below the plane). In the other structure, called the *boat*, two opposite carbons (atoms 1 and 4) are above the plane that passes through the other four atoms. Mohr recognized that cyclohexane itself could be flexible and that chair and boat forms could interconvert by rotation around the carbon–carbon bonds in the ring. Although several known experiments could be rationalized in terms of Mohr's model, there was no direct physical evidence that confirmed it until Hassel's electron-diffraction studies of cyclohexane derivatives, reported in 1943.[3] The details of Hassel's work can be found in the next chapter, but the major result was a confirmation of Mohr's chair as the more stable shape, or *conformation,* of cyclohexane. Further, hydrogens or other atoms bonded to the carbons of the rings must occupy specific positions with respect to the ring. One position was above the ring (attached to carbons above the average molecular plane) or below the ring, with the bond between the carbon and substituent perpendicular to the average molecular plane; Hassel called this the "standing" position. The other position was more or less near the molecular plane and directed to the side opposite the "standing" substituent: This was the "reclining" position, according to Hassel. Hassel's studies showed that a substituent attached to a "reclining" position was thermodynamically more stable than one attached to the "standing" position. Hassel's experiments were necessary to advance the concepts of conformational analysis, but it remained for Barton to show how the structures of substituted cyclohexanes were of fundamental importance to chemists.

Barton became interested in the chemistry of steroids and related compounds when he was an undergraduate, because of their complexity and their importance in medicine. He spent a short time working in this area as a graduate student, and then, at the suggestion of E. R. H. Jones, investigated the molecular rotations of steroids and triterpenes. After he received his Ph.D., Barton spent two years at Imperial College at a military intelligence laboratory and a year in the chemical industry with Albright and Wilson in Birmingham. During these years (1942–1945), Barton continued reading the literature of steroids and triterpenes in his spare time, awaiting the chance to actually carry out research with these compounds. He returned to Imperial College after the war, but no position was available in organic chemistry at the time. He spent four years teaching inorganic and physical chemistry and carrying out research in various areas of physical and organic chemistry.

---

[3]Hassel, O. "The Cyclohexane Problem"; *Tidsskr. Kjemi Bergvesen og Met.* **1943,** *3,* 32.

He became aware of Hassel's work and realized the importance of the shape of cyclohexane to his studies of steroids. He designed a set of molecular models that accurately represented the actual geometry of steroids and had them built in 1948. These models enabled Barton to examine the elaborate stereochemistry of these complex molecules in a way that was far beyond the understanding of other chemists at the time.

In 1949 Barton was invited to Harvard University as a visiting lecturer, replacing R. B. Woodward for his sabbatical year. Louis Fieser, professor of organic chemistry at Harvard, had made many contributions to steroid chemistry, and his textbook, *Natural Products Related to Phenanthrene*, had been the standard reference work since its first edition in 1936. The third edition (written with Fieser's wife and fellow chemist, Mary) appeared in 1949, within a month of the announcement of the use of cortisone in therapy. Barton had arrived at an important center of steroid research at the moment of the most intense interest in the field. During a seminar lecture by Fieser, in which he discussed unsolved problems in the hydrolysis of certain esters of steroids, Barton realized that the results could be explained by the precise shapes of the molecules that he knew from his models. He added other examples and formulated a short paper, which became the stimulus for countless investigations in practical and theoretical chemistry.

In his paper, Barton first stated what he meant by "conformation": differing strainless arrangements in space of bonded atoms. He cited Hassel's results on cyclohexane and used the nomenclature of Kenneth Pitzer for the substituent positions: Hassel's "standing" became "polar", and "reclining" became "equatorial".[4] Subsequently "polar" became "axial", and this term will be used here. Barton emphasized that in cyclohexane, although equatorial substituents were thermodynamically more stable than axial substituents, if the barrier for interconversion was small, some molecules would have axial substituents. After illustrations of the conformations of several disubstituted cyclohexanes and decalin (two fused cyclohexane rings) derivatives, Barton showed the portion of the steroid nucleus that contains three fused cyclohexane rings in detail: Each bond was clearly labeled for the system with rings A and B in a *cis*-fusion and in a *trans*-fusion. He then cited experimental evidence to show how the conformation of the steroid accounted for its chemical behavior.

---

[4]Beckett, C. W.; Pitzer, K. S.; Spitzer, R. "The Thermodynamic Properties and Molecular Structure of Cyclohexane"; *J. Am. Chem. Soc.* **1947,** *69,* 2488.

First, equilibration of alcohols at various carbon atoms in the cholestane and coprostane series always showed the equatorial alcohol to be more stable. Second, elimination reactions, such as dehydration and dehydrohalogenation, require that the eliminated atoms and the carbons to which they are bonded must be coplanar. This can only be achieved if both eliminated atoms are axial substituents. Third, axial substituents are crowded ("sterically hindered"), so that reactions such as esterification or ester hydrolysis are faster for the less-hindered equatorial substituent. Finally, Barton showed that conformational analysis allowed the assignment of structures for compounds that contained fused cyclohexane structures, such as di- and triterpenoids, as well as steroids.

Barton's paper immediately opened the third dimension to organic chemists, who now had to consider the conformations of molecules as well as their traditional structural formulas. The next edition of Fieser's text (now called *Steroids*) was published in 1959, and the most striking difference can be seen in the inclusion of steroid conformations, which, of course, were absent in the 1949 version. Conformational analysis was applied to all types of compounds and extended to other cyclic and acyclic systems. Theoreticians applied themselves to studying rotational barriers and their origins, hoping to develop a fundamental understanding of conformation. The reader is directed to any beginning organic chemistry textbook or any chemistry journal to see examples of the extent of the consequences of Barton's paper.

Barton returned to England in 1950 and accepted an appointment as reader at Birkbeck College, University of London. He was promoted to professor in 1953 and left to become regius professor at Glasgow in 1955. During this period he continued his work on conformations of natural products, as well as on synthesis. At Birkbeck he developed a synthesis of usnic acid, the main constituent of lichens, which involved free-radical coupling of phenols. This work led later to a general theory of alkaloid biosynthesis through phenolic coupling, a process that occurs in morphine and many other alkaloids. Barton explored this theory by classical laboratory synthesis of the alkaloids and by isolating compounds from plants that had been fed with radioactively labeled precursors.

At Glasgow, Barton turned his attention to organic photochemistry, especially the complex rearrangements that the compound santonin undergoes when irradiated. This work preceded the numerous papers on organic photochemistry that appeared in the early 1960s and was often cited by others as seminal. During vacations at Glasgow,

Barton returned to the United States and worked at the Research Institute for Medicine and Chemistry in Cambridge, Massachusetts. There he developed a photochemical procedure that came to be known as the Barton reaction.[5] This reaction brought together two of Barton's interests, photochemistry and stereochemistry. An alcohol is converted to its corresponding nitrite ester, and the nitrite subsequently photolyzed. Photolysis converts the nitrite back to the alcohol and nitrosates a methyl group that is in the precise position required to accept the NO group. The nitrosomethyl group tautomerizes to the oxime, which is hydrolyzed to the final product, an aldehyde. This is one of the few known methods for effecting substitution on an angular methyl group in a steroid and was used by Barton to achieve a two-step synthesis of the steroid aldosterone.[6] At the time, it became possible for 40 to 50g of aldosterone to be obtained by this synthesis. Formerly the total world supply was only 10 mg.

Barton returned to Imperial College as professor in 1957 and continued his work in the synthesis of natural products and the development of new synthetic reactions. At the time of the award of the Nobel Prize, he had published 300 papers. In 1978 he became director of the Institut de Chimie des Substances Naturelles in Gif-sur-Yvette, France, and in 1986 accepted his present position at Texas A & M University. Since his appointment at Harvard in 1949, Barton returned many times to the United States to lecture, and he has consulted extensively with companies in the chemical industry. He continues his interest in the design of new reactions, especially high-yielding radical chain reactions and highly selective oxidations using elements in unusual oxidation states.

Derek Barton is an extremely hard-working chemist who has devoted many hours each day to his study of a wide variety of organic chemical problems. His contributions have been recognized by universities and scientific societies in many nations, and his appreciation of the importance of conformation in chemistry has profoundly influenced the development of the science.

MARTIN R. FELDMAN
*Howard University*

[5]Barton, D. H. R. "The Use of Photochemical Reactions in Organic Synthesis"; *Pure Appl. Chem.* **1968,** *16,* 1.
[6]Barton, D. H. R.; Beaton, J. M. "A Synthesis of Aldosterone Acetate"; *J. Am. Chem. Soc.* **1961,** *83,* 4083.

# Bibliography

Barton, D. H. R. *Some Recollections of Gap Jumping;* Profiles, Pathways, and Dreams Series; Seeman, J. I., Ed.; American Chemical Society: Washington, DC, 1991.

Barton, D. H. R. "Molecular Models for Conformational Analysis"; *Chem. Ind.* **1956,** 1136.

Barton, D. H. R. "The Conformation of the Steroid Nucleus"; *Topics in Stereochem.* **1972,** *6,* 1–10. (This is a more accessible reprint of the paper in *Experientia.*)

Campbell, W. A.; Greenwood, N. N. *Contemporary British Chemists;* Barnes & Noble: New York, 1971.

Eliel, E. L. "Nobel Prize in Chemistry"; *Science* **1969,** *166,* 718–720.

Farago, P. "Interview with Derek Barton"; *J. Chem. Educ.* **1973,** *50,* 234–237.

Fieser, L. F.; Fieser, M. *Steroids;* Reinhold: New York, 1959.

*McGraw-Hill Modern Scientists and Engineers.* McGraw-Hill: New York, 1980; Vol. 1.

*Nobel Prize Winners;* Wasson, T., Ed.; H. W. Wilson: New York, 1987.

*The Nobel Prize Winners: Chemistry;* Magill, F. N., Ed.; Salem: Pasadena, Calif. 1990; Vol 3.

Ramsay, O. B. "The Early History and Development of Conformational Analysis"; In *Essays on the History of Organic Chemistry;* Traynham, J. G., Ed.; Louisiana State University: Baton Rouge, Louisiana, 1987; pp 54–77.

Russell, C. A. "The Origins of Conformational Analysis"; In *Van't Hoff–Le Bel Centennial;* ACS Symposium Series 12; Ramsay, O. B., Ed.; American Chemical Society: Washington, DC, 1975; pp 159–178.

# Odd Hassel

## 1897–1981

In one way, the development of chemistry can be seen as the improvement of our understanding of molecular structure. From the early stages of knowing only which elements in what mass ratio make up a compound, chemists have progressed to knowledge of how many atoms of each element are involved, the order in which those atoms are linked together, and how they are arranged in space. A major step along this path of increasing insight into molecular architechture was provided by the work of Norwegian chemist Odd Hassel. Born in Kristiania (now Oslo) on May 17, 1897, to Mathilde (Klaveness) and Ernst August Hassel, a physician, Hassel shared the 1969 Nobel Prize in chemistry with Englishman Derek Barton for contributions to the concept of conformation and its application in chemistry. Hassel died in Oslo on May 15, 1981.

When Hassel was eight years old, his father died. Hassel, his sister, and three brothers (one, Leif, being his twin) were raised by their mother. Although not an outstanding student before he entered Oslo University (Universitas Regia Fredericiana), he earned the *Candidatus Realium* degree when 23 years old with a thesis on the reaction kinetics of the reduction of nitro compounds with stannous chloride. After graduation Hassel spent five years in France, Italy, and Germany. He studied French, and then theoretical physics, to which he was strongly attracted, before settling down to chemistry in 1922. His discovery of adsorption indicators, a new class of indicators for use in analytical chemistry, while working in Kasimir Fajan's laboratory in Munich, gained him wide recognition as a chemist.[1]

Leaving Munich after seven months because of difficulties in his relationship with Fajans, Hassel went to the Kaiser Wilhelm Institute in Berlin, where he began work in X-ray crystallography. At that time, the Kaiser Wilhelm Institute was one of the two preeminent centers for work with this new technique. This was the start of what proved to be a lifelong preoccupation with questions of molecular structure. At Berlin he took part in the determination of a number of crystal structures of inorganic substances, bismuth and graphite among them.

Hassel received a doctorate from the University of Berlin in 1924 and, on the nomination of Fritz Haber, a Rockefeller Fellowship for 1924–1925. He returned to Norway in 1925 to teach at Oslo University, where he remained until his retirement. In 1934 he was named to the chair of physical chemistry, the first in Norway, and headed the newly created department of physical chemistry.

Faced with meager resources for research on his return to Oslo, Hassel nevertheless instituted a research program of X-ray crystallography and designed some of the necessary equipment himself. With his students, he studied a wide variety of Werner complexes (coordination compounds). His book on crystal chemistry achieved wide popularity and was translated from its original German into English (1935) and Russian (1936).[2]

By 1930 Hassel had abandoned crystallography for the use of electrical dipole moment measurements as a means to gain information about molecular structure. X-ray crystallography can only be applied to solids, and the structure of a molecule may be perturbed by intermolecular forces arising from the way the molecules are packed together in the crystal. Moreover, the location of hydrogen atoms, a major component of organic molecules, cannot be determined directly. However, a consideration of dipole moments can, at best, only serve to limit the number of possible structures for a molecule. In 1938 he began to study the electron diffraction of molecules in the gas phase. The rotating sector apparatus that he used was designed and built with the help of C. Finbak, who had first suggested the rotating sector technique in 1937. It was with this technique that he succeeded over

[1]Fajans, K.; Hassel, O. Z. Elektrochem. **1923**, *29*, 495–500.

[2]Hassel, O. *Kristallchemie;* T. Steinkopf: Dresden and Leipzig, Germany, 1934.

the next five years in deciphering the structure of cyclohexane and its derivatives, the work for which he was awarded the Nobel Prize some 30 years later.

It was known that when a carbon atom forms four single bonds, the angles between any pair of the bonds is approximately $109°28'$, the so-called tetrahedral bond angle. It was also known that in the cyclohexane molecule, $C_6H_{12}$, the carbon atoms are linked in a six-membered ring with two hydrogen atoms bonded to each carbon, so that each carbon forms four single bonds. These two facts are incompatible with the assumption that the ring lies flat with all six carbons in the same plane. However, by twisting the ring, two nonplanar structures are possible in which tetrahedral bond angles can be maintained. In one of these shapes, or conformers, one carbon is raised above the plane and a second, opposite to it in the ring, is lowered below the plane. This is the "chair" conformer. In the other conformer, two carbons opposite each other in the ring are both raised above the plane, forming the "boat" conformer. Only one kind of cyclohexane is known and before Hassel's work, no one could say which of the two conformers it was, or whether it was an inseparable mixture of the two.

Cyclohexane had been an object of Hassel's research as far back as 1930. In that year he had shown, by means of X-ray crystallography, that the cyclohexane molecules were in the "chair" form in the crystalline solid.[3] But it was only the use of the electron diffraction of the vapor that allowed him to determine in 1942 that the chair form was the preferred form in the gas phase.[4] This was followed by a fuller discussion of the structure of cyclohexane in what is considered to be the seminal paper in the field.[5] In this paper Hassel summarized his experimental work and showed that in the chair form there are two different kinds of carbon-hydrogen bonds: those that project out more or less perpendicular to the approximate plane of the ring and those that project out more or less parallel to the approximate plane of the ring. Although Hassel was not the first to point out that

[3] Hassel, O.; Kringstad, H. Tidsskr. Kjemi Bergves. 1930, 10, 128–130.

[4] Hassel, O.; Ottar, B. Arch. Math. Naturvidenskab 1942, 45, No. 10, 1–7.

[5] Hassel, O. Tidsskr. Kjemi. Bergves. Metall. 1943, 3, 32–34; English translation by K. Hedberg in Topics in Stereochemistry; Allinger, N. L.; Eliel, E. L., Eds.; Wiley–Interscience: New York, 1971; pp 11–17.

the geometry of the puckered ring led to this situation, he was the first to give experimental confirmation of the fact.

Hassel called these two kinds of bonds $\epsilon$ and $\kappa$ bonds, respectively, from the Greek *estekos* ("standing") and *keimenos* ("reclining"). After a period of confusion attributable to competing systems of nomenclature, workers in the field agreed to call them "axial" and "equatorial" bonds, as proposed in a letter to *Nature* and to *Science*.[6] Each carbon in the cyclohexane ring forms one C-H bond of each kind. Hassel showed that in the boat form, hydrogen atoms on one side of the ring would be crowded close together, reducing the stability of the molecule relative to the chair form.

For example, when one of the hydrogen atoms is replaced by a chlorine atom, the carbon–chlorine bond may be either axial or equatorial. Hassel showed that chlorocylohexane adopts the chair arrangement with the chlorine in an equatorial position.[7] With the information Hassel presented, it became possible to describe the preferred arrangement (or conformation) of almost any molecule containing cyclohexane rings. The English organic chemist Derek Barton became aware of Hassel's work on cyclohexane through a 1946 English review paper and in 1950 showed how this structural information could be used to correlate a number of facts concerning the reactivity of molecules containing rings and the products they gave in various types of reactions. This was the start of what is now called "conformational analysis".

Unfortunately, the German occupation of Norway in 1940 hindered Hassel's work and limited the audience he was able to reach. He stopped publishing in German journals, but his only alternatives were relatively obscure Norwegian journals, which had only small circulation and none outside occupied Europe. Moreover, Norwegian was not a language many chemists were familiar with. *Chemical Abstracts* did not publish an abstract of the 1942 paper until the middle of 1944, and then the abstract was taken from the German abstract journal *Chemisches Zentralblatt* for 1942. The 1943 paper on "The Cyclohexane Problem" was not abstracted until the middle of 1945. Shortly after publication of that 1943 paper, the German occupation authorities closed Oslo Uni-

---

[6]Barton, D.; Hassel, O.; Pitzer, K.; Prelog, V.; "Nomenclature of Cyclohexane Bonds"; *Nature* **1953**, *172*, 1096–1097; *Science* **1954**, *119*, 49.

[7]Hassel, O.; Viervoll, H. *Tidsskr. Kjemi. Bergves. Metall.* **1943**, *3*, 35–36.

versity and imprisoned Hassel, along with many others of the faculty. Released toward the end of 1944, he was unable to continue work until after the war.

Although the circumstances in postwar Europe were difficult, Hassel returned to work on the cyclohexane problem, publishing the results of new studies on the structures of cyclohexane derivatives. Hassel was the Norwegian editor of the *Acta Chemica Scandinavica* from its founding in 1947 until 1956 and published the bulk of his postwar work in it (more than 65 papers, in English). It is noteworthy that before the war, Hassel published most of his work in German in German journals, but that after 1940, almost all of his papers were in Norwegian or English. Only one paper appeared in German after 1943, and that was in 1954 in the Austrian journal, *Monatshefte der Chemie*.

By the middle 1950s, Hassel had turned his attention away from the cyclohexane problem and electron diffraction to the crystal structures of charge-transfer complexes. These addition compounds between ethers, ketones, and amines on the one hand and halogens and interhalogens, on the other, had been studied in solution but not in the solid state. Because they are not stable enough to exist in the gas phase, they could not be studied by means of electron diffraction. Hassel found that structures predicted on theoretical grounds were not always those actually existing in the solids. He pursued this line of research with a series of students at Oslo University into the early 1970s, well after his official retirement in 1964, and it was the subject of his Nobel Prize address.

Hassel received honorary degrees from the Universities of Copenhagen (1950) and Stockholm (1960) and was an honorary Fellow of the Norwegian Chemical Society and of the Chemical Society, London. In 1964 he received the Gunnerus medal from the Royal Norwegian Academy of Sciences and the Guldberg–Waage Medal from the Norwegian Chemical Society. His Nobel Prize, which he received in 1969, was the first awarded to a Norwegian scientist for work done almost completely in Norway.

RUSSELL F. TRIMBLE
*Southern Illinois University*

# Bibliography

Bastiansen, O. "Odd Hassel 1897–1981"; *Chem. Britain* **1982** *18*, 442.

Bykov, G. V. *Istoriya Stereokhimii Organicheskikh Soedinënii*; Nauka: Moscow, **1966**; pp 299–309.

Eliel, E. L. "Nobel Prize in Chemistry"; *Science* **1969**, *166*, 718–720.

*J. C. Poggendorffs biographisch-literarisches Handwörterbuch;* Stobbe, H., Ed.; Verlag Chemie: Berlin, Germany, 1937; Vol. 6, Pt. 2, p 1044; Salie, H., Ed.; Akademie Verlag: Berlin, Germany, 1970; Vol. 7B, No. 3, Pt. 4, pp 1881–1884.

Johnson, R. "Barton, Hassel Win Chemistry Nobels"; *Chem. Eng. News* **1969**, *47*, 11.

*McGraw-Hill Modern Scientists and Engineers;* Parker, S. P., Ed.; McGraw-Hill: New York, 1980; Vol. 2, pp 26–27. This biographical sketch was approved by Hassel.

Hassel, O. "Structural Aspects of Interatomic Charge-Transfer Bonding"; *Nobel Lectures: Chemistry, 1963–1970*. Elsevier: New York, 1972; pp 295–331.

Nobel Lecture delivered on June 9, 1970; also in *Science* **1970**, *170*, 497–502.

Ramsay, O. B. *Stereochemistry;* Nobel Prize Topics in Chemistry; Heyden: London, 1982.

Russell, C. A. "The Origins of Conformational Analysis"; In *Van't Hoff–Le Bel Centennial*; ACS Symposium Series 12; Ramsay, O. B.; Ed.; American Chemical Society: Washington, DC, 1975; pp 159–178.

# Luis Leloir

## 1906–1987

Copyright Nobel Foundation

Luis Leloir received the Nobel Prize in chemistry in 1970 for his achievements in clarifying the biochemical pathways and the intermediates involved in the synthesis of sucrose, lactose, trehalose, and most importantly the energy reservoir in mammalian cells, glycogen. Leloir was born in Paris of Argentinean parents on September 6, 1906, but from the age of two spent most of his life in Buenos Aires. He died there on December 4, 1987. The magnitude of his discoveries is remarkable considering the primitive conditions under which his early work was conducted in the 1940s and the early 1950s, particularly that of isolating and identifying a cofactor essential for interconverting glucose and galactose.

Leloir received a medical degree from the University of Buenos Aires in 1932, but his interests became focused on basic research while working at the Institute of Physiology with Bernardo Houssay on the effects of the adrenal gland on carbohydrate metabolism. In 1936 he went to the Biochemical Laboratory of Sir Frederick Gowland Hopkins, at Cambridge, England, where he worked with Malcolm Dixon and David E. Green. On returning to Buenos Aires he studied fatty acid oxidation in livers with J. Munoz and later worked with E. Menendez and co-workers on renal hypertension. In 1943 he married Amelia Zuberbuhler and left Argentina for the United States when Houssay was dismissed by Peron. A daughter, Amelia, resulted from this marriage. In the United States he worked with Carl and Gerty Cori at Washington University in St. Louis and then with David Green at Columbia

University. When Houssay obtained private funds to establish an institute for biochemical research, Leloir returned to Argentina in 1945 to direct it. It was there that he and his co-workers R. Caputto, C. Cardini, R. Trucco, and A. Palladini initiated their seminal studies on galactose metabolism in yeast.

Their initial studies were concerned with verifying the thesis that glycogen was transformed in the mammary gland to lactose. When it was realized that this was not the route, they turned their attention to the metabolism of lactose in the yeast *Saccharomyces fragilis*. They showed that on hydrolysis of this disaccharide to glucose and galactose the latter sugar was phosphorylated by a specific enzyme, galactokinase, to galactose-1-phosphate (Gal-1-P).[1] Although evidence for the presence of this compound had been presented earlier by others, this was the first demonstration of its direct formation. From this point on, Leloir's contributions had a major impact on the biochemical community, particularly with the exciting discovery that Gal-1-P was transformed in yeast extracts to glucose-6-phosphate (Glc-6-P) through the mediation of two thermostable cofactors isolated from heat-denatured yeast extracts. The first of these was shown to be α-glucose-1,6-diphosphate, which aided in catalyzing the conversion of glucose-1-phosphate (Glc-1-P) to Glc-6-P by the enzyme phosphoglucomutase[2] The second factor, which appeared to be directly involved in the conversion of glucose to galactose, absorbed ultraviolet light at 260 nm and possessed a spectrum similar to that of the recently described nucleoside uridine. After a brilliant series of deductions based on the results of rudimentary chemical analyses, the factor was proposed to be uridine diphosphate α-D-glucose (UDPG),[3] which was verified some five years later by its chemical synthesis. Considering the rather primitive isolation techniques employed at that time and the minute amounts of material available for analysis, this was indeed a major coup. The conversion of galactose to glucose was then shown by Leloir and his group to involve two enzyme reactions, with the first (a) resulting in an exchange of Gal-1-P for the Glc-1-P of UDPG to yield UDP-α-D-galactose (UDPGal), and the second (b) effecting an inversion of the hydroxyl on the 4-position of galactose (termed

---

[1] Trucco et al., 1948.

[2] Caputto et al., 1948.

[3] Caputto et al., 1950.

an *epimerization*) to yield a glucose moiety attached to UDP. The net result (c) of these reactions is the conversion of Gal-1-P to Glc-1-P,

a) Gal-1-P + UDPG → Glc-1-P + UDPGal
b)          UDPGal → UDPG
_____
c)          Gal-1-P → Glc-1-P

with the latter converted by phosphoglucomutase to Glc-6-P, a reaction medicated by α-glucose-1,6-diphosphate, as discussed above. The nucleotide cofactor, UDPG, would prove to be the first in a series of nucleotide diphosphate sugars that are involved in the biosynthesis of disaccharides, polysaccharides, and the glycoproteins.

Thus, it was first shown by Leloir and co-workers that UDPG was the sugar donor involved in the synthesis of the yeast disaccharide, trehalose,[4] and in the transfer of glucose to fructose in wheat germ extracts to form sucrose.[5] As a testimonial to the impressive intuition of Leloir and his group, they soon discovered other sugar nucleotide-transferring vehicles in the guise of UDP-*N*-acetylglucosamine (UDP-GlcNAc), guanosine diphosphate-mannose (GDPM), and UDP-*N*-acetylgalactosamine (UDP-GalNAc), and each was shown subsequently by others to be involved in one way or another in the biosynthesis of bacterial cell walls, lipopolysaccharides, plant polysaccharides, eukaryotic polysaccharides, chitin and the mucoproteins.

One of the ultimate consequences of the discovery of UDPG is that it led to refutation of one of the accepted dogmas of biochemistry: that glycogen is formed as a result of the phosphorolysis of glucose-1-P by the enzyme phosphorylase A. In another series of definitive studies, Leloir and Cardini were able to clearly demonstrate in 1957 that glycogen was formed by the transfer of glucose from UDPG to glycogen[6] and to put to rest, once and for all, the belief that phosphorylase was responsible for the synthesis of glycogen. By extending this work to starch, the plant equivalent of glycogen, these investigators were able to demonstrate that adenosine diphosphate glucose was a much better donor than UDPG.[7]

After receiving the Nobel Prize for these incisive discoveries, Leloir continued to contribute to the field of glycosyl transfer reac-

---

[4]Leloir and Calib, 1953.
[5]Cardini et al., 1955.
[6]Leloir and Cardini, 1957.
[7]Recondo and Leloir, 1961.

tions, particularly those concerned with mammalian glycoprotein synthesis. From the studies on bacterial cell wall synthesis, which was shown by others to involve the lipid intermediate,[8] undecaprenol pyrophosphate, as a mediator of *en bloc* polysaccharide transfer, Leloir and colleagues proposed that a similar process might occur in eukaryotic systems. This thesis was based on the finding of a polyprenol (termed *dolichol*) in mammalian tissues, similar to, but not identical to that described for bacterial sugar transport.[9] It was soon found that liver microsomes contained enzymes that transferred the corresponding sugars from UDPG, UDP-GlcNAc, and GDPM to dolichol-P to form a lipid oligosaccharide that could contain as many as 20 sugar residues. The oligosaccharide portion was then transferred as a unit to specific asparagine residues in proteins to yield glycoproteins.[10]

Of all of Leloir's contributions, the discovery of UDPG and other nucleotide sugars and their involvement in the biosynthesis of various disaccharides and polysaccharides cleared the way for progress to be made in elucidating the biosynthesis of bacterial cell walls and their counterparts in eukaryotes, the glycoproteins. Considering that his laboratory was not located in a mecca of modern biochemical research, it more than made up for its deficiencies by providing the scientific world with some of its most important discoveries.

It has been stated that early in Leloir's career he had to make a choice between polo and science. Fortunately for all of us, he decided on the latter.

FRANK MALEY
*New York State Department of Health*

## Bibliography

Cardini, C. E.; Leloir, L. F.; Chiriboga, J. "The Biosynthesis of Sucrose"; *J. Biol. Chem.* **1955**, *214*, 149–155.

Caputto, R.; Leloir, L. F.; Cardini, C. E.; Paladini, A. C. "Isolation of the Coenzyme of the Galactose Phosphate–Glucose Phosphate Transformation"; *J. Biol. Chem.* **1950**, *184*, 333–350.

---

[8]Wright et al., 1967.
[9]Hemming, 1969.
[10]Parodi et al., 1972.

Caputto, R; Leloir, L. F.; Trucco, R. E.; Cardini, C. E.; Paladini, A. C. "A Coenzyme for Phosphoglucomutase"; *Archives of Biochemistry* **1948**, *18*, 201–203.

Hemming, F. W. "Polyprenols"; *Biochemical Journal* **1969**, *113*, 12P–25P.

Leloir, L. F. "Two Decades of Research on the Biosynthesis of Saccharides", *Science* **1971**, *172*, 1299–1303.

Leloir, L. F.; Calib, E. "The Enzymic Synthesis of Trehalose Phosphate"; *J. Am. Chem. Soc.* **1953**, *75*, 5445–5446.

Leloir, L. F.; Cardini, C. E. "Biosynthesis of Glycogen from Uridine Diphosphate Glucose"; *J. Am. Chem. Soc.* **1957**, *79*, 6340–6341.

Parodi, A. J.; Behrens, N. H.; Leloir, L. F.; Carminatti, H. "The Role of Polyprenol-Bound Saccharides as Intermediates in Glycoprotein Synthesis in Liver"; *Proceedings of the National Academy of Sciences USA* **1972**, *69*, 3268–3272.

Recondo, E.; Leloir, L. F. "Adenosine Diphosphate Glucose and Starch Synthesis"; *Biochemical and Biophysical Research Communications* **1961**, *6*, 85–92.

Trucco, R. E.; Caputto, R.; Leloir, L. F.; Mittleman, N. "Galactokinase"; *Archives of Biochemistry* **1948**, *18*, 137–146.

Wright, A.; Dankert, M.; Fennesey, P.; Robbins, P. W. "Characterization of a Polyisoprenoid Compound Functional in O-Antigen Biosynthesis"; *Proceedings of the National Academy of Sciences USA* **1967**, *57*, 1798–1803.

# 1971

# Gerhard Herzberg

## 1904–

Gerhard Herzberg received the 1971 Nobel Prize in chemistry for "contributions to the knowledge of electronic structure and geometry of molecules, particularly free radicals." Born on December 25, 1904, in Hamburg, Germany, where his father was the co-manager of a small shipping company, Herzberg received most of his early education in that city. His early interest in astronomy was supported by inspiring teachers in physics and mathematics, and their stories of research conducted by friends intrigued him. Dissuaded from seeking advanced training in astronomy by the Director of the Hamburg Observatory, Herzberg entered the Technische Universität at Darmstadt in 1924, where he found that basic physics was as interesting as astronomy. Studying with Hans Rau, a student of Wilhelm Wien, Herzberg received his doctorate in engineering physics in 1928. According to Herzberg, it was his "good fortune to start scientific work in the period between 1926 and 1928, which coincided closely with the high point in the development of the subject of molecular spectroscopy. This fortunate accident made it possible for me to learn the subject while it was being developed."

Herzberg was indeed a witness to the birth of modern quantum physics, for it was during this period that Werner Heisenberg and Erwin Schrödinger presented their first solutions to the puzzles of atomic structure that grew out of the earlier theory of Niels Bohr. Excited by these developments and the explosion of results from that period, Herzberg subsequently spent more than 50 years on the study of molecular spectra (both in emission and absorption) to determine the electronic, vibrational, and rotational energy levels of diatomic and polyatomic molecules and to derive from these observed energy levels information about the geometrical structure of the molecules. After leaving Darmstadt in 1928, Herzberg spent the next year at the University of Göttingen, dividing his time equally between the experimentalist James Franck and the theoretician Max Born. Collaborating with Walter Heitler, who had just developed the valence theory of molecular bonds with Fritz London, Herzberg used the Raman spectrum of $N_2$ measured by Franco Rasetti to show that the nitrogen nuclei follow Bose statistics rather than Fermi statistics as was generally accepted at that time.[1] This result, derived before the discovery of the neutron, was the first indication that electrons are not present in the atomic nucleus.

Before leaving Göttingen, Herzberg made two more important spectroscopic contributions. With Professor G. Scheibe, he observed for the first time the vacuum ultraviolet spectra of the methyl halides, his first introduction to the spectroscopy of polyatomic molecules.[2] After studying the 1928 paper of E. Wigner and E. E. Witmer on the correlation of diatomic molecular states to those of the separated atoms, Herzberg correctly concluded that the then accepted value for the dissociation energy of $O_2$ must be modified.[3] Later, he determined the new value with precise measurements of new spectra.

After listening to Herzberg's series of lectures on introductory spectroscopy presented at the First Physical Institute at Göttingen, Scheibe suggested that they be collected in book form. Finding that his introduction to atomic spectra had reached the publisher's limit of 160 pages, Herzberg published that part in 1936, with an English translation appearing a year later. A timeless classic, it is still in print.[4]

---

[1]Heitler, W.; Herzberg, G. "Gehorchen die Stickstoffkerne der Boseschen Statistik?"; *Naturwissenschaften* **1929**, *17*, 673.

[2]Herzberg, G.; Scheibe, G. "The Absorption Spectra of Methyl Halides and Some Other Methyl Compounds in the Ultraviolet and the Schuman Region"; *Trans. Far. Soc.* **1929**, *25*, 716–717; "Ueber die Absorptionsspektra der dampfförmigen Methylhalogenide und einiger anderer Methyl-verbindungen im Ultraviolett und im Schumann-Gebiet"; *Z. Phys. Chem.* **1930**, *7*, 390–406.

[3]Herzberg, G. "Die Dissoziationsarbeit von Sauerstoff"; *Z. Phys. Chem.* **1929**, *4*, 223–226.

[4]*Atomic Spectra and Atomic Structure;* 1st ed.; Prentice-Hall: New York, 1937; 2nd ed.; Dover: New York, 1944.

As he continued to chronicle the developments in molecular spectroscopy, Herzberg found that his early attempt to summarize a series of introductory lectures continued to expand and required additional subdivision into separate books. The result was a remarkable series of four volumes, *Molecular Spectra and Molecular Structure*, spanning four decades (1939–1979).

At the invitation of J. E. Lennard-Jones, who was interested in his theoretical work on molecular orbital theory, Herzberg spent a second postdoctoral year at the University of Bristol, where he found that the physics department was very well equipped and even better supported financially than the department at Göttingen. Arriving in time for the General Discussion on Molecular Spectra and Molecular Structure of the Faraday Society, Herzberg heard C. V. Raman, who would receive a Nobel Prize in physics the next year for his spectroscopic work, and met many of the leading spectroscopists of the time. In spite of some initial difficulties with new surroundings and an unfamiliar language, Herzberg continued his research and writing, determining and examining both the spectrum of diatomic phosphorus[5] and the vacuum ultraviolet spectra of other polyatomic molecules such as formaldehyde and acetylene.

After submitting his theoretical work on molecular orbitals as a thesis (beyond the Ph.D. level) and presenting a test lecture to the faculty (a procedure called *habilitation*), Herzberg was appointed *Privatdozent* at Darmstadt in 1930. This position did not carry with it a salary. He did earn a very modest salary by assisting in the supervision of the practical laboratories and by advising students working on advanced degrees.

For the next five years there was intense experimentation primarily related to predissociation and forbidden transitions in such diatomic molecules as $P_2$, $N_2$, CP, PN, BeO, NO, and $O_2$. To facilitate these measurements new equipment was constructed, including a new 3-m grating spectrograph, a 12.5-m absorption tube, and a mirror system to allow measurement of the solar spectrum. Herzberg also collaborated with Edward Teller in the development of selection rules governing polyatomic electronic transitions.[6] Shortly after the discovery of deuterium by Harold Urey in 1932, Herzberg made his own

[5]Herzberg, G. "Bandenspektrum Prädissoziation und Struktur des $P_2$-Moleküles"; *Ann. Phys.* **1932**, *15*, 677–706.

[6]Herzberg, G.; Teller, E. "Schwingungsstruktur der Elektronenübergänge bei methratomigen Molekülen"; *Z. Phys. Chem.* **1933**, *B21*, 410–446.

"heavy" water and synthesized simple deuterium-labeled compounds to measure their spectra.[7]

Among the new laws established by the Nazis in the early 1930s was one that prohibited faculty members with Jewish wives from teaching at a university. By the end of 1933 Herzberg was seeking a position outside of Germany, but it was not until August 1935 that he and his wife left Hamburg for New York. J. W. T. Spinks, a young physical chemist from the University of Saskatchewan, had spent the 1933–1934 academic year with Herzberg, and after his return to Canada he managed to convince the university's president to offer Herzberg a guest professorship, using funds provided by the Carnegie Foundation to the British Commonwealth for the purpose of short-term hiring of refugees. Within three months, after the resignation of a staff member, Herzberg was offered a permanent position in the physics department, and the uncertainties of a two-year appointment were removed.

Saskatchewan was teaching-oriented, and research equipment was sparse by European standards. Nevertheless, Herzberg continued spectroscopic measurements with the available Hilger spectrograph, and through a grant from the American Philosophical Society was able to buy a grating and construct a high-caliber spectrograph by 1939.

Using the new instrument, A. E. Douglas, a Herzberg student, was the first to measure the spectrum of $B_2$ and determine the molecule's structure.[8] In the summer of 1941 Herzberg was able to produce and measure the spectrum of $CH^+$, matching it exactly with an interstellar spectrum of previously unknown origin, thereby confirming the presence of this molecular ion in interstellar space.[9] Herzberg was also concerned with the spectra of comets and attempted to identify a feature of the spectrum known as the 4050 group, whose origin had not been identified. Speculating that the $CH_2$ radical was responsible, Herzberg was able to produce the 4050 group in the laboratory but unable to establish that $CH_2$ was responsible. Later A. E. Douglas showed it to be due to $C_3$.[10]

[7]Herzberg, G.; Patat, F.; Spinks, J. W. T. "Rotationsschwingungsspektren im photographischen Ultrarot von Molekülen, die das Wasserstoffisotop der Masse 2 enthalten I. Das $C_2HD$-Spektrum und der C–C– und C–H– Abstand im Acetylen"; *Z. Phys.* **1934**, *92*, 87–99.

[8]Douglas, A. E.; Herzberg, G. "Spectroscopic Evidence of the $B_2$ Molecule and Determination of Its Structure"; *Can. J. Res.* **1940**, *18A*, 165–174.

[9]Douglas, A. E.; Herzberg, G. "$CH^+$ in Interstellar Space and in the Laboratory"; *Astrophys. J.* **1941**, *94*, 381.

[10]Douglas, A. E. "Laboratory Studies of the λ4050 Group of Cometary Spectra"; *Ap. J.* **1951**, *114*, 466.

Herzberg's achievements in astronomical applications of molecular spectra led to an offer in 1943 of a position from the Yerkes Observatory to set up a spectroscopic laboratory. Herzberg moved to the Observatory in 1945 when wartime restrictions had been lifted. With the aid of highly skilled technicians he quickly constructed another spectrograph and a high-quality absorption tube whose mirrors gave a path length that was 250 times the actual tube length.

Photographic infrared and ultraviolet spectra of a number of simple molecules were measured. The ultraviolet spectrum of $O_2$, significant to the understanding of upper atmosphere physics, was studied. Perhaps the most important discovery was that of the quadrupole spectrum of molecular hydrogen, because it would subsequently play an important part in the identification of hydrogen in planetary and stellar atmospheres and interstellar space.

Unable to obtain funding in the United States to support the Yerkes work, in 1947 Herzberg accepted an offer from the National Research Council (NRC) in Canada. Originally appointed as Principal Research Officer, Herzberg became Director of the Physics Division in 1949, a position he held for the next 20 years. The NRC had several attractive features. The original offer was specifically designed for the establishment of a spectroscopic laboratory, and it would not be necessary to seek research funds from outside agencies since sufficient funding and support staff were part of the research operation. In addition, Dr. E. W. R. Steacie, Director of the Chemistry Division and later NRC president, was a firm believer in giving his research staff complete freedom in problem selection and operating conditions as they responded to the creative challenge. Herzberg was quick to adopt a similar attitude in the Physics Division.

The NRC appointed Herzberg its first "Distinguished Research Scientist" in 1969, allowing him to continue his research activities virtually indefinitely without administrative responsibilities. In 1974 the NRC established the Herzberg Institute of Astrophysics, where the astronomical activities of the Council would continue under a new administrative title and where Herzberg would continue to find the opportunity to engage in scientific work.

Herzberg and his colleagues at the NRC have described their experimental results and interpretations in hundreds of papers. In addition to studying the spectra of diatomic and polyatomic systems, the Herzberg group continued to seek applications of such spectra in the fields of planetary atmospheres, interstellar space, and the composition of comets. Among the more important discoveries are the identification

GERHARD HERZBERG

of $H_2O^+$ in the tail of the comet Kohoutek[11] and the observation of a spectrum of neutral $H_3$, proving its existence in a number of excited states.[12] Both in the observed states of $H_3$ and in the ground state of $H_3^+$, discovered shortly afterward in Herzberg's laboratory by T. Oka, the three protons are at the corners of an equilateral triangle. Herzberg and his associates developed spectroscopic techniques for examining very short-lived free-radical species. The structures of over 30 of these species have now been elucidated.

Because of the experimental skill of its members and their theoretical insights the Herzberg laboratory was cited by the Nobel Committee as having "attained a unique position as the foremost center for molecular spectroscopy in the world." Because his work has had such an important bearing on the structure and properties of molecules, Herzberg became one of the few physicists to receive the Nobel Prize in chemistry because his "ideas and discoveries have stimulated the whole modern development [of chemistry] from chemical kinetics to cosmochemistry."

JAMES J. BOHNING
*Beckman Center for the History of Chemistry*

[11]Herzberg, G.; Lew, H. "Tentative Identification of the Water(+) Ion in Comet Kohoutek"; *Astronomy and Astrophysics* **1974**, *31*, 123–124.

[12]Herzberg, G. "A Spectrum of Triatomic Hydrogen"; *J.Chem.Phys.* **1979**, *70*, 4806–4807.

# Bibliography

Herzberg, G. Interview by Christine King in Ottawa, Ontario on May 5, 1986; Beckman Center for the History of Chemistry; Philadelphia, PA; Oral History Transcript 23.

Herzberg, G. *Molecular Spectra and Molecular Structure.*

   a.   *Spectra of Diatomic Molecules;* 1st ed.; Prentice-Hall: New York, 1939; 2nd ed.; Van Nostrand: New York, 1950.

   b.   *Infrared and Raman Spectra of Polyatomic Molecules;* Van Nostrand: New York, 1945.

   c.   *Electronic Spectra and Electronic Structure of Polyatomic Molecules;* Van Nostrand: New York, 1966.

   d.   *Constants of Diatomic Molecules* (with K. P. Huber); Van Nostrand: New York, 1979.

Herzberg, G. "Molecular Spectroscopy: A Personal History"; *Ann. Rev. Phys. Chem.* **1985**, *36*, 1–30.

Herzberg, G. "Spectra of Molecular Ions"; *Proc. Indian Natl. Sci. Acad.* **1985**, *51A*, 495–521.

Herzberg, G. *The Spectra and Structures of Simple Free Radicals: An Introduction to Molecular Spectroscopy;* Cornell University Press: Ithaca, NY, 1971.

Herzberg, G. "Spectroscopic Studies of Molecular Structure"; *Science* **1972**, *177*, 123–138.

Herzberg, G. "Spectroscopy and Molecular Structure"; In *Physics Fifty Years Later, 14th General Assembly, International Union of Pure and Applied Physics, 1972;* Brown, S. C., Ed.; NASA: Washington, DC, 1973; pp 101–162.

Stokes, L. D. "Canada and an Academic Refugee from Nazi Germany: The Case of Gerhard Herzberg"; *Canadian Historical Review* **1976**, *57*, 150–170.

# *Christian Anfinsen*

### 1916–

Christian Boehmer Anfinsen was a recipient of the 1972 Nobel Prize in chemistry for his work on the relationship between the structural properties of proteins and their biological functions, specifically with regard to the enzyme ribonuclease. Anfinsen, the son of a Norwegian engineer, was born on March 26, 1916, in the western Pennsylvania town of Monessen. He attended Swarthmore College, near Philadelphia, obtaining a Bachelor of Arts degree in 1937. He then began graduate studies at the University of Pennsylvania and in 1939 received a Master of Science degree in organic chemistry.

Anfinsen was the recipient of a Rockefeller Public Service Award to spend the following year as a visiting investigator at the Carlsberg Laboratory in Copenhagen, where he worked in the laboratory of Kaj Linderstrøm-Lang. In this research environment, he became exposed to many aspects of the chemistry of protein molecules, an interest that was central to the theme of his research in the following years. Subsequently, Anfinsen attended graduate school at Harvard Medical School and received a Ph.D. in biological chemistry in 1943. He remained at Harvard as instructor of biochemistry (1943–1945), assistant professor of biochemistry (1945-1947), and then associate professor of biochemistry (1948-1950). In 1947–1948 he spent one year as a senior fellow of the American Cancer Society, working at the Medical Nobel Institute in Sweden with Hugo Theorell, who was noted for his research in flavoproteins.

Anfinsen was appointed chief of the Laboratory of Cellular Physiology at the National Heart Institute of the National Institutes of Health (NIH) in 1950, and in 1952 he became chief of the Laboratory of Cellular Physiology and Metabolism. It was during those years at NIH that Anfinsen began studying the relationship between the structure and the catalytic function of ribonuclease. In 1959 Anfinsen wrote the book *The Molecular Basis of Evolution*, a treatise that described his interest in the relationships between protein chemistry and genetics and the promise that the combination of these areas held for the understanding of evolution. He spent 1962–1963 as a professor of biological chemistry at Harvard, but then returned to the NIH, where he became chief of the Laboratory of Chemical Biology at the National Institutes of Arthritis, Metabolic and Digestive Diseases. After a year as visiting professor of biochemistry at the Weizmann Institute of Science in Israel from 1981 to 1982, Anfinsen became a professor of biology at Johns Hopkins University, a position he currently holds.

The telegram to Anfinsen from the Royal Swedish Academy of Sciences notifying him of the 1972 Nobel Prize stated his accomplishments as "studies on ribonuclease, in particular the relationship between the amino acid sequence and the biologically active conformation". In his Nobel lecture, delivered on December 11, 1972, entitled "Principles That Govern the Folding of Protein Chains", Anfinsen summarized the research on the processes involved in the folding of protein molecules into their unique three-dimensional conformations. Anfinsen was a corecipient of the award with Stanford Moore and William Stein of Rockefeller University. Their complementary research, also using the enzyme ribonuclease as a model, involved the elucidation of the composition and sequence of amino acid residues in the enzyme.

In addition to receiving the Nobel Prize in chemistry in 1972, Anfinsen has received the Public Service Award of the Rockefeller Foundation (1954) and honorary degrees from Swarthmore College (1965), Georgetown University (1967), New York Medical College (1969), the University of Pennsylvania (1973), Gustavus Adolphus College (1975), Brandeis University (1976), Providence College (1978), University of Naples (1982), Adelphi University (1987), Bar Ilan University (1990), and University of Las Palmas, Gran Cananias (1993). He is a member of the American Society of Biochemistry and Molecular Biology (in which he served as president in 1971–1972) and is an honorary fellow and member of the board of governors for the Weizmann Institute of

Science. Anfinsen was elected to the American Academy of Arts and Sciences in 1958, to the U.S. National Academy of Sciences in 1963, to the Royal Danish Academy in 1964, and the Pontifical Academy of Science in 1981.

Protein molecules are composed of various combinations of 20 types of amino acid units bound covalently to adjacent amino acid residues by peptide bonds, forming linear polypeptide chains. The sequence of the various amino acids within a chain is unique to each type of protein. In addition, a linear chain may fold on itself in a characteristic way that results in a three-dimensional structure unique to that protein and essential for the exhibition of the specific biological function of the protein. The focus of Anfinsen's research was the events involved in the folding of polypeptide chains. Most of the experiments were carried out with enzymes that catalyze the hydrolysis of nucleic acids, specifically, ribonuclease from the bovine pancreas and an extracellular nuclease from *Staphylococcus aureus*. Important principles were developed from his studies with these enzymes; the principles provide a general model for the processes governing the folding of protein molecules into their biologically active conformations.

Important early observations carried out by Anfinsen and others were those that described the denaturation of the enzyme ribonuclease by the disruption of the noncovalent forces responsible for maintaining the three-dimensional structure. This denaturation, which leaves the covalently bonded amino acid backbone intact but results in a loss of the biological activity of the protein, was carried out by subjecting the protein to conditions such as high temperature or exposure to chemical denaturants such as urea. However, it was discovered that with the carefully controlled removal of the denaturing conditions, the protein could often renature into its native conformation, with a concomitant restoration of the biological activity of the protein.

Anfinsen's research was designed to understand how a protein folds to obtain a characteristic three-dimensional structure. As a polypeptide chain folds on itself, each chemical group of each amino acid residue of the chain may potentially interact in some way with each chemical group of the other amino acid residues of the protein. There are an astounding number of possibilities for the interactions that can occur. For example, Anfinsen noted in his Nobel lecture that a protein molecule like the *Staphylococcus* nuclease, which contains 149 amino acid units, may theoretically exist in at least one of any $4^{149}$ to $9^{149}$ conformations in solution if it is assumed that there are two rotatable bonds per amino acid residue and three favored orientations about each of these rotatable bonds. Theoretical estimates of the average time required to obtain a unique conformation by the random folding of the protein indicate that this could not occur in a reasonable time frame in vivo. Anfinsen and his co-workers eventually described the basis for this folding process in terms of a "thermodynamic hypothesis".[1] In his Nobel lecture, Anfinsen explains, "The three-dimensional structure of a native protein molecule in its normal physiological milieu (solvent, pH, ionic strength, presence of other components such as metal ions or prosthetic groups, temperature, and other) is the one in which the Gibbs free energy of the whole system is lowest; that is, that the native conformation is determined by the totality of interatomic interactions and hence by the amino acid sequence, in a given environment."

Much of Anfinsen's research to substantiate the thermodynamic hypothesis focused on one specific type of interaction as a model, namely, the interaction of the sulfhydryl group of the amino acid cysteine with another cysteine sulfhydryl to form a stable covalent disulfide linkage within the protein chain. In the case of the bovine pancreatic ribonuclease, the enzyme is a chain of 124 amino acid residues, eight of which are cysteines. It can be calculated that there are 105 possible pairings of these cysteines to form four disulfide linkages, but Anfinsen's work indicated that only one of these combinations results in a catalytically competent enzyme.[2] In a key experiment, Haber and Anfinsen treated the isolated catalytically active enzyme first with urea to chemically denature it and then with a sulfhydryl reducing reagent to break the disulfide linkages.[3] After removal of the sulfhydryl reducing reagent, a mixture of the 105 possible disulfide-bonded isomeric forms developed if the enzyme remained in the urea denaturant. However, if the urea was removed prior to the sulfhydryl reducing reagent treatment to allow the renaturation of the protein into its native conformation, only the four specific disulfide linkages that resulted in the catalytic activity were formed.

---

[1] Epstein, C. J.; Goldberger, R. F.; Anfinsen, C. B. "The Genetic Control of Tertiary Protein Structure: Studies with Model Systems"; *Cold Spring Harbor Symposia of Quantitative Biology* **1963**, *28*, 439–449.

[2] Sela, M.; White, F. H., Jr.; Anfinsen, C. B. "Reductive Cleavage of Disulfide Bridges of Ribonuclease"; *Science* **1957**, *125*, 691–692.

[3] Haber, E.; Anfinsen, C. B. "Side-Chain Interactions Governing the Pairing of Half-Cystine Residues in Ribonuclease"; *J. Biol. Chem.* **1962**, *237*, 1839–1844.

The authors concluded that only the single most thermodynamically stable three-dimensional structure of the protein would result on renaturation and that this would bring specific cysteines close enough to one another to promote the formation of the necessary disulfide bonds. In the presence of the urea denaturant, the probability of the protein existing in this conformation by random folding was very low because the urea disrupted the amino acid and side-chain interactions necessary for the protein to obtain its most stable conformation. Although this renaturation process occurred only slowly over a period of several hours, enzymes were identified in the endoplasmic reticulum of the cell that accelerated the rate at which the cysteine disulfide bonds were formed and exchanged. When these enzymes were included during renaturation of the enzyme, the form with the lowest free energy (the native form of the enzyme) was attained in less than a few minutes.[4] Under these conditions, the time required for the rapid folding of the enzyme into an active conformation was comparable to that expected to occur in vivo following the synthesis of a nascent polypeptide chain.

Anfinsen's kinetic studies of polypeptide chain folding led to the conclusion that the process was positively cooperative, that is, that the initial interactions between portions of the polypeptide chain enhanced the rates of formation of the other important interactions necessary for the native conformation to be obtained. It was postulated that the rapidity of the refolding makes it essential that the process take place along a limited number of pathways and that the process likely begins with only a limited number of initiation or nucleation events, which begin the cooperative process involved in the folding.

Anfinsen's hypotheses imply that the information required to confer biological activity on a protein is resident within the sequence of the 20 different amino acid residues in the polypeptide chain and that these amino acid residues interact in the most energetically favorable manner to produce a unique three-dimensional structure characteristic of that protein. This work provided some of the theoretical foundation for experiments by R. B. Merrifield and Ralph Hirschmann that showed that a catalytically active enzyme could be chemically synthesized from its amino acid precursors.

---

[4]Goldberger, R. F.; Epstein, C. J.; Anfinsen, C. B. "Acceleration of Reactivation of Reduced Bovine Pancreatic Ribonuclease by a Microsomal System from Rat Liver"; *J. Biol. Chem.* **1963**, *238*, 628–635.

## CHRISTIAN ANFINSEN

Anfinsen's current research relates to the structure and stability of enzymes of hyperthermophilic bacteria at the Johns Hopkins University in Baltimore, Maryland, where he resides with his wife, Libby Esther.

**H. DAVID HUSIC**
*Lafayette College*

## *Bibliography*

Anfinsen, C. B. *The Molecular Basis of Evolution*; John Wiley & Sons: New York, 1959.

Anfinsen, C. B. "Principles That Govern the Folding of Protein Chains"; In *Les Prix Nobel en 1972*; Odelberg, W., Ed.; Norstedt & Sons: Stockholm, Sweden, 1973; pp 103–119.

Anfinsen, C. B. "Principles That Govern the Folding of Protein Chains"; *Science* **1973**, *181*, 223–230.

"Anfinsen, Christian Boehmer"; *McGraw-Hill Modern Men of Science;* Crouse, W. H., Ed.; McGraw-Hill: New York, 1968; Vol. 2, pp 10–11.

"Christian B. Anfinsen"; *Les Prix Nobel en 1972;* Odelberg, W., Ed.; Norsted & Sons: Stockholm, Sweden, 1973; pp 100–102.

# Stanford Moore

## 1913–1982

Copyright Nobel Foundation

Stanford Moore was a corecipient of the 1972 Nobel Prize in chemistry for pioneering research into the development of methodology for the elucidation of the structural properties of protein molecules. He shared the prize with his colleague William H. Stein and with Christian B. Anfinsen of the National Institutes of Health.

Born in Chicago on September 4, 1913, Moore spent his early childhood in Illinois, Florida, and Georgia, before his family settled in Nashville, Tennessee. He died on August 23, 1982. His father, John Howard M. Moore, was a law professor at Vanderbilt University, and his mother, Ruth (Fowler) Moore, was a graduate of Stanford University. Moore was an excellent student, earning all A's at Peabody Demonstration School, where he first became interested in the sciences. In 1931 he entered Vanderbilt University, where he considered a major in aeronautical engineering but made the decision to major in chemistry after completing organic chemistry. Moore graduated summa cum laude with a bachelor of arts degree in chemistry in 1935 and received the Founder's Award, given to the premier student in his graduating class.

Funded by a Wisconsin Alumni Research Foundation Fellowship, Moore began graduate studies in organic chemistry at the University of Wisconsin in 1935. Moore's research mentor was Karl Paul Link, and in the early part of his graduate career, Moore learned some of the microanalytical methods for elemental analysis that Link had learned

538

previously from Fritz Pregl (Nobel Prize in chemistry, 1923) in Austria. Moore's graduate research primarily involved the synthesis and characterization of crystalline benzimidazole derivatives formed by the reaction of o-phenylenediamine with monosaccharides. This work led to the development of novel methods for the analysis and quantification of certain monosaccharides and culminated in the completion of his Ph.D. thesis in 1938, which was entitled "Identification of Carbohydrates as Benzimidazole Derivatives".

Immediately after receiving his Ph.D., Moore accepted an invitation to join the research group of Max Bergmann, a friend of Link's who had moved his laboratories from Germany to the Rockefeller Institute for Medical Research in New York in 1934. Moore chose the research position in Bergmann's laboratory in favor of a four-year fellowship at Harvard Medical School. Moore spent three years in Bergmann's lab developing methods for the quantitative determination of amino acids from protein hydrolysates by the gravimetric analysis of selectively precipitated amino acid salts. Bergmann suggested that Moore begin a collaborative project with William Stein, a postdoctoral associate who had been in Bergmann's research group since 1937. Together they developed novel methods for the quantification of amino acids by precipitation with aromatic sulfonic acids. These procedures yielded salts that were not as insoluble as those normally considered desirable for the quantification of materials by gravimetric methods, but, by utilizing selective precipitants and accounting for the solubility products of the derivatives, quantitative determinations of the amino acids were obtained. This collaborative research was the beginning of a close working relationship between the two scientists, which continued for many years and would result in many important discoveries.

The onset of World War II interrupted this research on amino analysis in Bergmann's laboratory; many laboratories devoted scientific expertise to projects of value to national defense. Bergmann's research group undertook several projects for the Office of Scientific Research and Development related to the physiological effects of mustard gas and related chemical compounds. Their goal was the development of potential treatments for exposed individuals. During this time Moore left the laboratory and joined the National Defense Research Committee of the Office of Scientific Research and Development in Washington in 1942. His responsibilities involved the coordination of the academic and corporate research into the biological effects of blister agents. During 1944 and 1945 he served with the Chemical Warfare Service, initially in

Washington as a member of the Project Coordination Staff, and later in Hawaii with the Operational Research Section.

Following Bergmann's death in 1944, the director of the Rockefeller Institute, Dr. Herbert Gasser, gave both Stein and Moore the opportunity to establish a research program at Rockefeller in the laboratories that had been occupied by Bergmann's group. Both accepted the positions and began their development of sensitive methods for the separation and quantification of amino acids. Their research was soon productive, and they were given permanent positions. Moore was named an associate member in 1949 and in 1952 was promoted to member, a title that was changed to professor when the Rockefeller Institute became the Rockefeller University in 1955. Moore was named the John D. Rockefeller Professor at the university in 1981.

Moore left Rockefeller temporarily in 1950 to hold the Francqui Chair at the University of Brussels, where he set up a laboratory of amino acid analysis in the School of Medicine. While in Europe he spent time at the University of Cambridge in 1951, sharing laboratory space with Frederick Sanger (Nobel Prize in chemistry, 1958 and 1980) who at that time was carrying out his landmark studies on the determination of the structure of the peptide hormone insulin. When Moore returned to New York, he and Stein began researching their new methodology for amino acid analysis to determine the amino acid sequence of ribonuclease. Moore also left Rockefeller in 1968 to spend a year as a visiting professor of health sciences at Vanderbilt University School of Medicine.

Some of the achievements of the long collaboration between Stanford Moore and his colleague William Stein are summarized below, particularly those relating to the techniques they developed that were instrumental in deciphering the complex chemical structures of protein molecules, and, specifically, the enzyme ribonuclease. Because it was such a close collaboration, assigning individual accomplishments to one or the other of these outstanding scientists is impossible. In 1972 Stanford Moore and William Stein were jointly awarded the Nobel Prize in chemistry for their work on the structure of ribonuclease. They shared the prize with Christian B. Anfinsen of the National Institutes of Health, who independently completed complementary studies on different aspects of the structure of ribonuclease.

Moore never married and concentrated his energies almost exclusively on his research and on projects of benefit to the scientific community. He lived modestly, and neither he nor Stein received personal financial rewards for their discoveries, which ultimately had great commercial value. Moore was stricken with amyotrophic lateral sclerosis (Lou Gehrig's disease) in the middle 1970s but remained actively involved with research at the Rockefeller University. He died on August 23, 1982, at the age of 68. Moore willed his estate to Rockefeller to provide funding for an investigator in biochemistry.

In addition to the Nobel Prize, Moore received honorary doctorates from the Faculty of Medicine at the University of Brussels (1954), the University of Paris (1964), and the University of Wisconsin (1974), and he was an honorary member of the Belgian Biological Society and the Belgian Royal Academy of Medicine. He also shared several awards with William Stein, including the American Chemical Society Award in Chromatography and Electrophoresis (1964), the Richards Medal of the American Chemical Society (1972), and the Linderstrøm-Lang Medal (1972). Moore was elected a member of the National Academy of Sciences in 1960 and served as chairman of the Section of Biochemistry from 1969 to 1972. He was a member of the American Society of Biological Chemists, serving as treasurer from 1956 to 1959, as president in 1966–1967, and as member of the editorial board of the *Journal of Biological Chemistry* from 1950 to 1960. In addition Moore was president of the Federation of American Societies for Experimental Biology in 1970–1971.

Moore and Stein's collaborative efforts in Max Bergmann's laboratory at Rockefeller initiated a lifelong association. With their backgrounds in microanalysis and chemical derivatization, their research led to a method for the rapid and simultaneous quantitative analysis of each of the 20 common amino acids that comprise protein molecules. With these new and sensitive methodologies at hand, the technical means were available to complete the sequencing of amino acids in a large protein. Their choice was ribonuclease from the bovine pancreas, a protein about twice the size of the insulin molecule that had been sequenced previously by Frederick Sanger at Cambridge University. They were successful in the determination of the complete sequence of the amino acids within the ribonuclease molecule, the first enzyme for which the amino acid sequence had been determined. This accomplishment was the first step both in their eventual understanding of the three-dimensional arrangement of portions of the polypeptide chain and in their research into the catalytic roles and spatial orientations of individual amino acids directly involved in the catalytic process. A summary of Moore and Stein's research is provided in the combined texts of their

Nobel lectures, delivered on December 11, 1972, entitled "Chemical Structures of Pancreatic Ribonuclease and Deoxyribonuclease".

Pancreatic ribonuclease is responsible for the hydrolysis of ribonucleic acids. In order to study the structural details of the enzyme, it was first necessary to purify the protein to homogeneity. Ribonuclease had been partially purified by earlier workers, and Moore and Stein's first postdoctoral associate, Werner Hirs, examined the elution characteristics of the ribonuclease catalytic activity after subjecting the preparation to ion-exchange chromatography on a polymethacrylate resin. These were among the earliest studies in the utilization of such materials for the chromatographic resolution of proteins. These experiments revealed that the enzyme preparations contained two separate forms of the enzyme, designated ribonucleases A and B, which could be resolved chromatographically. The major component of the mixture, ribonuclease A, was the species utilized for further investigations. The subsequent elucidation of the primary structure of ribonuclease A would have been virtually impossible without the availability of a homogenous form of the enzyme.

The first step in the structural analysis of ribonuclease A was to determine the composition of the amino acids within it. At the time, the only available methods for analysis were those that used selective precipitations of amino acid salts, along with their subsequent quantification by gravimetric methods. These methods were familiar to Moore and Stein since they had collaborated in the development of new methodologies for gravimetric amino acid analysis as research associates in Bergmann's laboratory. Furthermore, Stein utilized similar methods for his Ph.D. thesis research to determine the amino acid composition of elastin.

In 1945, research by A. J. P. Martin and R. L. M. Synge (who shared the Nobel Prize in chemistry, 1952) into the development of partition chromatography and Lyman Craig's invention of the countercurrent distribution apparatus resulted in the establishment of new high-resolution methods for the separation of molecules. Synge and co-workers had been successful in the separation of amino acids using potato starch as an absorbent chromatographic matrix, with the elution of the amino acids from the columns with organic acid–alcohol–water mixtures. Moore and Stein decided to develop this method with the hope of eventually achieving resolution of all 20 amino acids in protein hydrolysates.

It soon became apparent that the desired amino acid analyses would require time-consuming chromatographic steps. Initially, the eluted fractions from the starch columns were collected manually; however, Moore and Stein quickly developed the drop-counting fraction collector to collect fractions of the desired size automatically. Similar collection instruments eventually were manufactured commercially and exist today as routine apparatuses in many scientific laboratories.

To quantify the amino acids separated and eluted from the chromatographic columns, Moore and Stein developed a method that was technically simpler and more sensitive than the standard gravimetric methods. Many years earlier it had been discovered that ninhydrin reacted with amino acids to form colored products. Stein and Moore developed procedures to utilize this reagent for the sensitive and reproducible spectrophotometric quantification of amino acids eluted from the chromatographic columns. The important modifications they used were the incorporation of a reducing agent into the reaction mixture to reduce the concentration of interfering oxygen in the reaction and the addition of a water-soluble organic solvent to increase the solubility of the colored reaction products.

With these technical tools, Moore and Stein succeeded in the development of a method to separate and quantify all of the amino acids in protein hydrolysates utilizing three different starch columns and sequentially eluting the amino acids with different solvent combinations. They applied their method initially to the determination of the amino acid compositions of β-lactoglobulin and albumin. These procedures required considerably less protein than the gravimetric methods of analysis and had a high degree of accuracy. Although this new methodology was a remarkable accomplishment for the time, the complete analysis still required two weeks, and the starch columns could be used for only a single sample. Thus, Moore and Stein began to investigate the use of sulfonated polystyrene ion-exchange resins in the separations. Eventually they were successful in the development of procedures for the resolution of all 20 amino acids on one column in a single six-hour chromatographic separation, by washing the column with a step-wise gradient of buffers of increasing pH. This methodology was the basis for the commercially available amino acid analyzers that became essential tools in protein chemistry research laboratories in the 1960s. Throughout the development of these important methodologies, Moore and Stein were noted for their willingness to share the details of their experimental protocols and instrumental configurations so that others might use them in their laboratories.

Having developed procedures for the rapid quantification of the amino acids, Moore, Stein, and their co-workers began to study the

124-amino acid sequence of ribonuclease A. The initial step was to prepare peptide fragments of the enzyme that were small enough so that their individual amino acid sequences could be determined. To accomplish this, the enzyme was first oxidized with performic acid to break the covalent disulfide bonds that joined pairs of cysteine residues to one another, and then it was treated with proteolytic enzymes with known specificities to partially hydrolyze the polypeptide chain. These methods were similar to those used by Sanger for his elucidation of the amino acid sequence of insulin. The peptide fragments were separated by ion-exchange chromatography, and the amino acid sequences of the individual peptides were determined. Eventually they were able to determine the order of the sequenced peptide fragments in the original polypeptide chain by the comparison of overlapping sequences. The sequencing of the peptide fragments was greatly facilitated by the sequential chemical degradation of amino acids from the amino-terminal end of the peptides using phenylisothiocyanate, as had been developed by Pehr Edman in Sweden. This work led to the determination of the entire sequence for ribonuclease A, published in 1963. Throughout these studies, there was ongoing discussion with Christian Anfinsen at the National Institutes of Health, who was performing complementary studies with ribonuclease that focused on the disulfide bond formation between pairs of cysteine residues.

Moore and Stein's next goal was to study the role of individual amino acids in the catalytic reaction for which the enzyme is responsible. Their approach was to utilize chemical reagents that specifically modified certain amino acids and then to examine the effect of the modifications on the catalytic properties of the enzyme. Their initial studies involved the modification of ribonuclease with iodoacetate, a reagent that caused catalytic inactivation of the enzyme. Although iodoacetate was known to modify sulfhydryl groups, Anfinsen had earlier shown that ribonuclease contained no free sulfhydryl groups; all eight cysteine residues in the enzyme were disulfide-bonded. Utilizing analyses of the derivatized enzyme, Moore and Stein found that treatment of the enzyme with iodoacetate resulted in the carboxymethylation of methionine, histidine, or lysine residues and that the amino acids that were modified depended on the pH during the reaction. Utilizing such information, they were able to infer that the catalytic site of the enzyme contained two histidine residues and a lysine residue, all of which could be alkylated by iodoacetate. In addition, it was possible to estimate the distances between these three amino acids. Their conclusions were verified in 1967 when the X-ray crystallographic structural determination of the enzyme was completed in other laboratories and the involvement of these amino acids in the catalytic mechanism was confirmed.

In the years after these discoveries, Moore and Stein began extensive studies of a similar nature on a similar enzyme, bovine pancreatic deoxyribonuclease, as well as with a number of proteolytic enzymes. In addition, they continued the study of various chemical modifications of proteins.

The accomplishments of Moore and Stein's extensive research collaboration resulted in the development of new methodologies for the structural determination of protein molecules, and their work marked some of the earliest studies to define clearly the functional roles of specific amino acids within protein molecules. Many of the principles and technologies they developed remain today as the basis of modern techniques of the study of proteins.

H. DAVID HUSIC
*Lafayette College*

# Bibliography

Desmond, P. "Stanford Moore"; In *The Annual Obituary 1982;* Podell, J., Ed.; St. Martin's Press: New York, 1983; pp 400–402.

Hirs, C. H. W. "Stanford Moore: Some Personal Recollections of His Life and Times"; *Anal. Biochem.* **1984**, *136*, 3–6.

Moore, S.; Stein, W. H. "Chemical Structures of Pancreatic Ribonuclease and Deoxyribonuclease"; *Science* **1973**, *180*, 458–464.

Moore, S.; Stein, W. H., "Chemical Structures of Pancreatic Ribonuclease and Deoxyribonuclease"; in *Les Prix Nobel en 1972;* Odelberg, W., Ed.; Norstedt: Stockholm, Sweden, 1973; pp 128–143.

*Nobel Prize Winners;* Wasson, T., Ed.; H. W. Wilson: New York, 1987.

*The Nobel Prize Winners: Chemistry;* Magill, F. N., Ed.; Salem: Pasadena, CA, 1990; Vol. 3.

Smith, E. L.; Hirs, C. H. W. "Stanford Moore"; In *Biographical Memoirs of the National Academy of Sciences* **1987**, *56*, 355–386.

"Stanford Moore"; in *Les Prix Nobel en 1972;* Odelberg, W., Ed.; Norstedt: Stockholm, Sweden, 1973; pp 120–122.

# William Stein

## 1911–1980

William Howard Stein was a corecipient of the 1972 Nobel Prize in chemistry with his colleague Stanford Moore, and with Christian B. Anfinsen of the National Institutes of Health, for research into the structural properties of protein molecules, and in particular, of the enzyme ribonuclease.

Stein was born in New York City on June 25, 1911, the second of three children born to Fred M. S. and Beatrice S. (Borg) Stein. He died on February 2, 1980. His parents were actively involved in causes aimed at improving public health and the well-being of disadvantaged children in New York City, and their social commitment and concern was an inspiration for Stein. In his autobiography, Stein wrote, "During my childhood, I received much encouragement from both of my parents to enter into medicine or a fundamental science."

Stein's secondary education began at the Lincoln School of the Teacher's College of Columbia University in New York City, where he was exposed to a diverse and progressive curriculum that included the sciences. At the age of 16, he transferred to Phillips Exeter Academy, a private preparatory school in New Hampshire, where he was challenged by a more rigid and demanding curriculum than he had experienced in New York.

In 1929 Stein became a student at Harvard University, as had his father and older brother. He enjoyed his undergraduate career at Harvard, and although by his own admission did not distinguish himself academically, he completed the requirements of the Bachelor of Science degree in chemistry in 1933. Stein remained at Harvard for graduate study in organic chemistry. He struggled in his first year of graduate school and contemplated alternatives to a career in science but instead transferred to the Department of Biochemistry in the College of Physicians and Surgeons at Columbia University.

At Columbia Stein worked under the research direction of Professor E. G. Miller. His thesis project was to determine the amino acid composition of elastin, an extensively cross-linked protein that contributes to both the elastic properties and mechanical strength of connective tissues. For these analyses, Stein utilized gravimetric methods for the quantification of the insoluble amino acid salts formed by the treatment of protein hydrolysates, making use of techniques he learned from Erwin Brand at Columbia that had been developed by Max Bergmann in Germany. In 1934 Bergmann had moved to the Rockefeller Institute for Medical Research in New York, where his laboratory became involved in both the determination of the amino acid composition of a number of proteins and the substrate specificity of several proteolytic enzymes. Stein earned his Ph.D. from Columbia in 1937, with a doctoral dissertation entitled "The Composition of Elastin" and immediately joined Bergmann's prospering laboratory at the Rockefeller Institute.

Stein concentrated his efforts in Bergmann's laboratory on improving gravimetric methods for the analysis of amino acids. In 1939 Stanford Moore joined Bergmann's laboratory, having recently completed the requirements for his Ph.D. at the University of Wisconsin. Soon after, Stein and Moore began a lifelong collaboration, which started with the development of gravimetric methods for the analysis of glycine and leucine by preparing aromatic sulfonic acid salts of the amino acids. These methods were applied to the analysis of these two amino acids in several proteins. The tedious nature and limited sensitivity of the gravimetric methods of analysis would eventually yield to simpler microscale methods, which Stein and Moore jointly developed in future years and for which they received great recognition.

The research on amino analysis at the Rockefeller Institute was interrupted with the beginning of World War II; many scientific laboratories devoted their research expertise during these years to the national defense effort. The scientists in Bergmann's laboratory undertook several projects under contract to the Office of Scientific Research and Development related to the physiological effects of mustard gas and related compounds with the idea of developing potential treatments for individuals exposed to these gasses. Stein was specifically involved in the characterization of the products formed by the reactions of these gasses with the functional groups of some amino acids within proteins. During this time, Stanford Moore had left the laboratory to work for the Chemical Warfare Service.

Following Bergmann's death in 1944, Dr. Herbert Gasser, the Director of the Rockefeller Institute, offered both Stein and Moore positions at Rockefeller in the laboratories that had been occupied by Bergmann's group. Their initial research, related to the development of sensitive methods for the separation and quantification of amino acids, was productive right away, and their positions became permanent. Stein was named an associate member in 1949 and was promoted to member in 1952. In 1955 the Rockefeller Institute became the Rockefeller University, and Stein's title was changed to professor of biochemistry.

Stein and Moore were coauthors of the great majority of their contributions to the scientific literature. In a biographical memoir of Stein's life, Moore wrote, "During the early years of our cooperation, Stein and I worked out a system of collaboration that lasted for a lifetime. Stein combined an inventive mind and a deep dedication to science with great generosity. Over a period of 40 years, we approached problems with somewhat different perspectives and then focused our thoughts on the common aim. If I did not think of something, he was likely to, and vice versa, and this process of frequent interchange of ideas accelerated progress in research."

A summary of the collaborative research carried out by Stein and Moore is provided in the article on Stanford Moore in this volume. Briefly, the major emphasis of this work was the development of rapid and sensitive chromatographic methods for the separation and quantification of the amino acids present in protein hydrolysates. These methods were utilized for the determination of the amino acid sequence of the enzyme ribonuclease A, the first enzyme molecule for which the primary structure was determined. Subsequent research identified the amino acid residues within ribonuclease that are critical for catalysis, and similar studies were also carried out with a number of other enzymes.

In 1972 the significance of their research was recognized by the Royal Swedish Academy of Sciences, and Stein and Moore were awarded the Nobel Prize in chemistry. Their Nobel lectures were delivered on December 11, 1972, and the combined text of their lectures was entitled "Chemical Structures of Pancreatic Ribonuclease and De-oxyribonuclease". The award was shared with Christian B. Anfinsen of the National Institutes of Health, whose complementary research involved aspects of the three-dimensional structure of the ribonuclease molecule and the study of the role of the cysteine sulfhydryl groups on the catalytic activity of the enzyme.

Dr. Stein left Rockefeller to serve as a visiting professor at the University of Chicago in 1960, and as a visiting lecturer at Harvard University in 1963. He served his profession and community in a number of important capacities. He was a member of the Medical Advisory Board of Hebrew University–Hadassah Medical School in Israel (1957–1970), trustee of Montefiore Hospital in New York (1947–1974), member of the council of the Institute of Neurological Diseases and Blindness at the National Institutes of Health (1961–1966), vice chairman (1962–1963) and chairman (1965–1968) of the U.S. National Committee for the International Union of Biochemistry, and chairman of the publications committee for the Sixth International Congress of Biochemistry (1964). Beginning in 1955, Stein served as an editor for the *Journal of Biological Chemistry,* was chairman of the Editorial Committee from 1958 to 1961, a member of the Editorial Board from 1962 to 1964, became an associate editor of the *Journal* in 1964, and was appointed editor-in-chief in 1968. However, he was able to serve in that capacity only until 1969, when health problems required that he resign.

In 1969, while attending a symposium in Copenhagen, Stein was stricken with an illness eventually diagnosed as Guillain-Barré syndrome. Paralysis resulted, and it became necessary for him to remain confined to a wheelchair for the remainder of his life. Despite this physical handicap, for the next decade he continued to provide valuable advice and criticism to his colleagues and contributed to discussions with Moore and others in the laboratory at Rockefeller until his death at the age of 68 on February 2, 1980, from heart failure. Stein was survived by his wife Phoebe (Hockstadter) Stein, whom he had married in 1936, and their three sons, William, David, and Robert.

In addition to the Nobel Prize, Dr. Stein received other prizes and awards jointly with Stanford Moore, including the Chromatography and Electrophoresis Award (1964) and the Richards Medal of the American Chemical Society (1972), and the Kaj Linderstrøm-Lang Gold Medal and Prize (1972). Stein was elected to membership in the National Academy of Sciences and to the American Academy of Arts and Sciences in 1960, and he was awarded honorary doctorate degrees from Columbia University (1973) and the Albert Einstein College of Medicine of Yeshiva University (1973).

The cooperative research accomplishments of Stein and Moore led to the establishment of new methods for the analysis of protein structure and the relationship between protein structural properties and biological function. These methodologies and principles remain as the basis for many modern techniques used for the study of protein structure and function.

H. DAVID HUSIC
*Lafayette College*

## *Bibliography*

Moore, Stanford. "William H. Stein"; *Biographical Memoirs of the National Academy of Sciences* **1987,** *56,* 415–442.

Moore, Stanford; Stein, William H. "Chemical Structures of Pancreatic Ribonuclease and Deoxyribonuclease"; In *Les Prix Nobel en 1972;* Odelberg, W., Ed.; Norstedt: Stockholm, Sweden, 1973; pp 128–143.

Podell, Janet; Scharf, Betty. "William (Howard) Stein"; In *The Annual Obituary 1982;* Turner, R., Ed.; St. Martin's Press: New York, 1983; pp 75–77.

Richards, Frederic M. "The 1972 Nobel Prize for Chemistry"; *Science* **1972,** *178,* 492–493.

Stein, William H. "William H. Stein"; In *Les Prix Nobel en 1972;* Odelberg, W., Ed.; Norstedt: Stockholm, Sweden; pp 125–127.

# Ernst Otto Fischer

## 1918–

The 1973 Nobel Prize in chemistry was awarded jointly to Ernst Otto Fischer and Geoffrey Wilkinson for their pioneering work in organometallic chemistry, with particular emphasis on their independent efforts toward developing the chemistry of, and recognizing the new type of bonding in, the transition metal "sandwich" compounds known as metallocenes. Fischer was born on November 10, 1918, in Solln, Germany (near Munich), the son of Professor Karl T. Fischer of the Physics Institute of the Technische Hochschule in Munich. He followed in his father's scientific footsteps and studied chemistry at the Technische Hochschule; however, the bulk of his formal education in chemistry was obtained after World War II, on account of his military service. He received the Ph.D. degree in the field of organometallic chemistry in 1952, under the supervision of Professor Walter Hieber, who has been called "the father of modern carbonyl chemistry".[1] He then remained at the Technische Hochschule as a scientific assistant, and, with his first students, he began his research on the chemistry and structure of the metallocenes. In 1957 he moved to the University of Munich, where he became a professor at the Institute for Inorganic Chemistry. In 1964 Professor Hieber retired as director of the Inorganic Chemistry Laboratory at the Technische Hochschule, and Fischer returned there to succeed him.

---

[1]Seyferth, D.; Davison, A. "The 1973 Nobel Prize in Chemistry"; *Science* **1973**, *182*, 699.

The Inorganic Chemistry Laboratory at the Technische Hochschule (now Technische Universität) has since "become one of the leading centers in the world for research in organometallic chemistry".[2]

As Fischer was beginning his scientific career at the Technische Hochschule in early 1952, two papers concerning a new, and apparently quite stable, complex between iron and cyclopentadiene appeared in the literature. Both papers described the new compound, $C_{10}H_{10}Fe$ (iron bis-cyclopentadienyl), as an air-stable sublimable orange solid that melted at about $173°C$ without decomposition and was soluble in various organic solvents but insoluble in water. The structural description given by both papers showed two molecules of cyclopentadienide anion ($Cp^-$) coordinated to one molecule of divalent iron ($Fe^{2+}$); the nature of the bonding of each $Cp^-$ ring to the iron center was explained using simple sigma bonds, stabilized by ionic-covalent resonance.[3] Fischer (as did Geoffrey Wilkinson) realized that sigma bonding could not account for the unusual thermal stability and chemical properties observed for this compound. As Wilkinson and his colleague R. B. Woodward (at Harvard University) set out to prove the correct structure of iron bis-cyclopentadienyl using chemical, physical, and spectroscopic methods, Fischer and co-workers independently tried to prove the structure by using X-ray crystallography. Fischer and his co-worker W. Pfab submitted their first paper in late June 1952. (They were unaware of Wilkinson's first paper, submitted about three months earlier, until after they had finished their manuscript.) Fischer's paper corroborated Wilkinson's evidence that the structure was a "sandwich", with the iron atom centered between the two planar pentagonal cyclopentadienide rings. Fischer's structure also showed that the two $Cp^-$ rings were not mutually coincident but were staggered in conformation, such that the entire structure formed a pentagonal antiprism, a geometry that Wilkinson had seriously considered in his paper, but could not prove without a crystal structure. The sandwich structure of iron bis-cyclopentadienyl signifies that the bonding between the $Cp^-$ rings and iron is due to good orbital overlap between the *pi-electrons in the p-orbitals* of the $Cp^-$ rings and the d-orbitals on iron, the pi-complexation being responsible for the high stability of the compound. This type of

[2] Seyferth, D.; Davison, A. "The 1973 Nobel Prize in Chemistry"; *Science* **1973**, *182*, 699.

[3] Kealy, T. J.; Paulson, P. L, "A New Type of Organo–Iron Compound"; *Nature* **1951**, *168*, 1040; Miller, S. A.; Teboth, J. A.; Tremaine, J. F. "Dicyclopentadienyliron"; *J. Chem. Soc.* **1952**, 632–635.

bonding, while suspected for many years, had not been formally identified until now. In fact, with the discovery and recognition of this new type of bonding between metals and organic unsaturated molecules, "the classical period of organometallic chemistry drew to a close and the modern era began".[4]

X-ray diffraction studies on iron *bis*-cyclopentadienyl from other laboratories around the world followed the Fischer paper, but Fischer was always a frontrunner in the field of what was now called metallocene chemistry. (The name arose from the name "ferrocene" given by a member of Woodward's group to iron *bis*-cyclopentadienyl, owing to the compound's similar chemical reactivity to aromatic molecules such as benzene.) A series of papers by Fischer and his co-workers on the preparation and crystal structure analysis of ferrocene analogues (such as "cobaltocene" and "nickelocene") soon followed, demonstrating that this type of structure was not unique to iron. Fischer further extended the concept of organic aromatic molecules pi-bonded to transition metals by correctly predicting that dibenzenechromium(0) was also a sandwich compound. Although organometallics resembling pi-arene compounds had been known since 1919, their structure was not recognized until Fischer published an X-ray crystal structure of $Cp(C_6H_6)_2$, showing it to have a neutral chromium atom sandwiched between two neutral benzene molecules. Fischer and Walter Hafner developed a high-yield synthesis for this compound (a reductive-ligation procedure now known as the Fischer–Hafner method) using aluminum powder and aluminum halides to activate a transition metal halide (in this case $CrCl_3$) toward reaction with the neutral, and not particularly reactive, benzene molecules.

Fischer continued his ground-breaking work in arene–metal chemistry before entering a highly significant new area: transition metal–carbene and –carbyne chemistry. A carbene (such as :CRR′) is a short-lived and highly reactive organic species in which the central carbon atom has two substituents, represented here by R and R′, and a lone pair of electrons that is capable of bonding to one or two other species. An alkene can be formed by the combination of two carbenes; the two lone pairs form a double bond between the two carbons to give R′RC=CRR′. If instead, one *in principle* reacts a transition metal M containing at least two d-electrons with a carbene, one would obtain a transition metal carbene such as M=CRR′, in which there is a metal-

[4]Haiduc, I.; Zuckerman, J. J. *Basic Organometallic Chemistry;* Walter deGruyter: Berlin, 1985; p 5.

to-carbon double bond. Similarly, a transition metal carbyne is the metal analogue of an alkyne: a compound of the type M≡CR, possessing a metal-to-carbon triple bond. These compounds were believed to be implicated in various transition metal–catalyzed organic transformations, and even though they were presumed to be highly unstable, chemists nonetheless searched for stable examples of these complexes.

In a short communication published in 1964, Fischer and Maasböl reported for the first time the preparation and isolation of stable carbene complexes. In this paper they described the reaction between tungsten hexacarbonyl and phenyl- or methyllithium in ether to form the lithium acylpentacarbonyltungstates $[(OC)_5WC(O)(C_6H_5)]^-Li^+$ and $[(OC)_5WC(O)(CH_3)]^-Li^+$, respectively (by attack of the carbanions $C_6H_5^-$ or $CH_3^-$ to the partially positive carbon of one of the carbonyl ligands). These acylpentacarbonyltungstates (after conversion to the tetramethylammonium salts) were then protonated by reaction with acid to give the corresponding pentacarbonyl[hydroxy(organo)carbene]tungsten(0) complexes, which were highly unstable. Although these complexes were not isolable, they could be "successfully converted to the substantially more stable methoxycarbene compounds by treatment with diazomethane" to give $(OC)_5W=C(OC_6H_5)(CH3)$ and $(OC)_5W=C(OCH_3)(CH_3)$, respectively.[5] Further work led them to a superior and general synthesis that involved the direct alkylation of lithium acylcarbonylmetalates using alkyloxonium tetrafluoroborate reagents, such as $[(CH_3)_3O]^+BF_4^-$, a good source of methyl cation. Fischer and co-workers paved the way in transition metal–carbene chemistry by synthesizing many new types of carbenes of other transition metals; characterizing them using chemical, spectroscopic, and X-ray diffraction methods; and by discussing the bonding involved in these complexes. The transition metal carbenes of the type M=C(ER)(R'), where E is a heteroatom such as O, S, or NR", have since become commonly known as "Fischer carbenes" in the organometallic literature, in honor of their founder and developer.

These transition metal carbenes have a very interesting chemistry all their own. Fischer and co-worker Weiss found that the alkoxycarbenes could react with primary and secondary amines, in a manner similar to that of esters, to give the metal–carbene analogue of an amide, for example, M=C(NHR)(R'). Fischer writes, "This observa-

---

[5]E. O. Fischer, "On the Way to Carbene and Carbyne Complexes"; In *Advances in Organometallic Chemistry*; Stone, F. G. A.; West, R., Eds. Academic: Orlando, FL, 1976; p 3.

tion led us into peptide chemistry along a path that proved to be quite surprising to a coordination chemist. We could show that the alkoxy group of alkoxy(organo)carbene complexes can be substituted not only by mono- or dialkylamino residues but also by free amino groups of amino acid and peptide esters".[6] The amino acid or peptide could later be removed from the metal by treatment with a strong acid such as trifluoroacetic acid. The use of transition metal–carbene residues as amino-protective groups for amino acids and peptides opened up many interesting possibilities to synthetic organic chemists and biochemists. Today, Fischer carbene complexes, their derivatives, and immediate precursors, find many applications, and new ones are still being developed.

The first transition metal–carbyne complexes were also prepared by Fischer, quite serendipitously, during attempts to exchange the methoxy group of a methoxycarbene with a halogen by reaction of the carbene with boron trihalides at low temperature. Instead of the expected substitution product, they obtained a compound that spectroscopically could only be explained as having four CO ligands (instead of the previous five), as well as a halogen atom and a CR group bonded to the metal in a *trans* orientation. In 1973 Fischer and his co-workers published the first of their papers detailing the synthesis and properties of transition metal–carbyne complexes of the type *trans*-$X(OC)_4M{\equiv}CR$ (where for M = W, X = Cl, Br, or I, and R = methyl or phenyl, and for M = Cr and Mo, X = Br, and R = phenyl). In this paper they reported the crystal structure of $I(OC)_4W{\equiv}CC_6H_5$, which showed a very short tungsten–carbon bond distance (for the carbyne moiety), indicative that there is indeed a triple bond between these atoms.[7]

Fischer's collective work dealing with the different types of bonding between transition metals and organic groups has revolutionized organometallic chemistry. His work has not gone without notice, as he has received numerous awards during his career in addition to the Nobel Prize. He has been described by his colleagues as being a warm and congenial man who, in addition to being a dedicated and imaginative scientist, is also a superb teacher and lecturer who has "inspired his

---

[6]E. O. Fischer, "On the Way to Carbene and Carbyne Complexes"; In *Advances in Organometallic Chemistry;* Stone, F. G. A.; West, R., Eds.; Academic: Orlando, FL, 1976; p 11.

[7]Fischer, E. O; Kreiter, C. G.; Müller, J.; Huttner, G.; Lorenz, H. "*trans*-Halogeno-[alkyl(aryl)carbyne]tetracarbonyl Complexes of Chromium, Molybdenum, and Tungsten—A New Class of Compounds Having a Transition Metal–Carbon Triple Bond"; *Angewandte Chemie, International Edition in English,* **1973,** 12, 564–565.

students to give the best of themselves".[8] Ernst Otto Fischer is both a model scientist and person, and organometallic chemists have been extremely fortunate to have such a man to emulate.

PETER V. BONNESEN
Oak Ridge National Laboratory

## Bibliography

Fischer, E. O. "On the Way to Carbene and Carbyne Complexes"; In *Advances in Organometallic Chemistry*; Stone, F. G. A.; West, R., Eds.; Academic: Orlando FL, 1976; vol. 14, pp 1–32.

Fischer, E. O.; Kreiter, C. G.; Müller, J.; Huttner, G.; Lorenz, H. "*trans*-Halogeno-[alkyl(aryl)carbyne]tetracarbonyl Complexes of Chromium, Molybdenum, and Tungsten—A New Class of Compounds Having a Transition Metal–Carbon Triple Bond"; *Angew. Chem., International Edition in English* **1973**, *12*, 564–565.

Fischer, E. O.; Maasböl, A. "On the Existence of a Tungsten Carbonyl Carbene Complex"; *Angew. Chem., International Edition in English* **1964**, *3*, 580–581.

Fischer, E. O.; Pfab, W. "Cyclopentadien Metallkomplexe, ein neuer Typ metallorganischer Verbindungen"; *Z. Naturforsch.* **1952**, *76*, 377–379. The X-ray crystallographic study of iron *bis*-cyclopentadienyl ("ferrocene").

Haiduc, Ionel; Zuckerman, J. J. *Basic Organometallic Chemistry;* Walter deGruyter: Berlin, Germany, 1985; Chapter 1.

Kealy, T. J.; Pauson, P. L. "A New Type of Organo Iron Compound"; *Nature* **1951**, *168*, 1039–1040.

Miller, S. A.; Tebboth, J. A.; Tremaine, J. F. "Dicyclopentadienyliron"; *J. Chem. Soc.* **1952**, 632–635.

Seyferth, D.; Davison, A. "The 1973 Nobel Prize in Chemistry"; *Science* **1973**, *182*, 699–701.

Weiss, E.; Fischer, E. O. "Zur Kristallstruktur und Molekulgestalt des Dibenzol-chrom (0)"; *Z. Anorg. Allg. Chem.* **1956**, *286 (July)*, 142–145.

---

[8]Seyferth, D.; Davison, A. "The 1973 Nobel Prize in Chemistry"; *Science,* **1973**, *182,* 700.

# Geoffrey Wilkinson

### 1921–

Sir Geoffrey Wilkinson and Ernst Otto Fischer shared the 1973 Nobel Prize in chemistry for their pioneering work in the field of organometallic chemistry. In particular, they were cited for their independent efforts in developing the chemistry, and deducing the correct structure, of the transition metal "sandwich" compounds known as metallocenes. Wilkinson was born on July 14, 1921, in Todmorden, England, and received his B.Sc. degree from the Imperial College of Science and Technology in London in 1941. He served as a junior scientific officer in the UK–US–Canadian atomic energy project in Canada during World War II from January 1943 to June 1946. He received his Ph.D. degree from Imperial College in 1946. From January 1946 until June 1950, he was with Professor Glenn T. Seaborg's group at the Lawrence Radiation Laboratory of the University of California in Berkeley studying the radioisotopes of the lanthanide group metals. In 1950 he became a research associate at the Massachusetts Institute of Technology, where his focus shifted from nuclear to inorganic chemistry, working initially with Professor J. W. Irvine, Jr., on $Ni(PCl_3)_4$. In September 1951 he arrived at Harvard University as a new assistant professor of chemistry, and it was there that the initial Nobel Prize winning work on "sandwich compounds" was performed. He returned to England in 1956 to the Chair of Inorganic Chemistry at Imperial College. He retired in 1988 and continues to make remarkable contributions to inorganic and organometallic chemistry as professor emeritus and senior research fellow.

As an assistant professor at Harvard, Wilkinson had the job of teaching courses in inorganic chemistry and nuclear chemistry, the former being something of a new experience for him. He recalls using, among other texts, Sidgwick's *The Chemical Elements and Their Compounds* (which was then considered the "inorganic bible"), and Eméleus and Anderson's *Modern Aspects of Inorganic Chemistry*, to aid him in teaching the class.[1] (As a point of interest, what is today considered by many to be the "inorganic bible" is the book coauthored by Wilkinson and his former student F. Albert Cotton, entitled *Advanced Inorganic Chemistry: A Comprehensive Text*, now in its fifth edition.) While reading these texts, he learned about several complexes between transition metals and olefins (unsaturated hydrocarbons), most notably Zeise's salt, $K^+[(C_2H_4)PtCl_3]^-$ (a platinum complex containing ethylene), and butadieneiron tricarbonyl, a compound prepared by Reihlen in 1930. The precise nature of the bonding of the olefins to the metals in these complexes was not well established at the time, and Wilkinson, intrigued, began to look more closely at reactions between transition metal carbonyls and unsaturated compounds. It was Wilkinson's belief, which was later shown to be correct, that the ethylene in Zeise's salt was "side-bound" to the platinum and that the butadiene in Reihlen's compound was coordinated in a *cis*-chelate fashion to iron, again with the individual alkene portions perpendicular or side-bound to iron.

On the afternoon of January 30, 1952, Wilkinson went to the department library to read the current literature, as he usually did, and noticed in the December 15, 1951, issue of *Nature* a note by T. J. Kealy and P. L. Pauson describing a new and apparently quite stable complex between iron(II) and two molecules of cyclopentadienide anion (biscyclopentadienyliron). Structurally, the cyclopentadienide ion has five carbon–hydrogen (CH) groups in a five-membered symmetrical ring, with six $\pi$-electrons delocalized throughout the five carbon corners of the ring. The anion is isoelectronic with the neutral molecule benzene, in which the six CH groups comprising the ring form a regular hexagon. The neutral molecule cyclopentadiene is electronically similar to butadiene, in that it possesses a pair of adjacent *cis*-oriented double bonds. The structure Kealy and Pauson gave for biscyclopentadienyliron showed a central iron atom with a single ($\sigma$) bond on either side attaching it in each case to only one of the five carbon atoms of

---

[1]Wilkinson, G. "The Iron Sandwich: A Recollection of the First Four Months"; *J. Organomet. Chem.* **1975,** *100,* 273.

the flat, planar cyclopentadienide (Cp) ring. A second structural representation (a resonance structure of the first) showed the iron as a dication, with the two $Cp^-$ anions balancing the charge on either side ($Cp^-Fe^{2+}Cp^-$), to give an overall effect of a high degree of ionicity to the compound. An elemental analysis had been performed on the molecule, showing that the $C_{10}H_{10}Fe$ molecular formula corresponding to the structural formula was correct. The compound was described as an orange solid that melted without decomposition at 173–174 °C, sublimed readily above 100 °C, and was "insoluble in, and apparently unattacked by, water".[2] Wilkinson immediately recognized that the structure depicted could not account for the unusual stability of this compound: A simple σ bond between the Cp carbon and iron would be too unstable (metal alkyls were known to be unstable with regard to air oxidation), and the ionic saltlike interaction would not be consistent with the observed volatility. The molecule had to possess unusually strong bonding between the Cp ligands and iron. It was the electronic similarity between cyclopentadiene and butadiene that first occurred to Wilkinson, and he drew a structure showing the Cp ligands coordinated to iron by a *cis*-chelate interaction, with one Cp "above" the iron and the other "below". Since the Cp ligands were anions here, each with five electronically equivalent carbons, he quickly realized that all five carbons in each Cp ring must contribute equally in their mode of bonding to iron. He then drew a "sandwich", in which the iron atom was centered in the middle of two slices of Cp "bread". The bonding was strong owing to the good orbital overlap between the iron metal d orbitals and (the π electrons in) the p orbitals of the Cp ligands. What really excited Wilkinson at the time was "the thought that if iron did this, the other transition metals must also form sandwich compounds".[3]

It turned out that Harvard colleague R. B. Woodward, a premier organic chemist (who won the 1965 Nobel Prize in chemistry) had also seen the Kealy and Pauson paper that Friday, had discussed the paper with members of his research group, and had arrived at the same conclusion concerning the structure. The following Monday, Wilkinson and Woodward agreed to collaborate in proving their proposed

[2]Kealy, T. J.; Pauson, P. L. "A New Type of Organo-Iron Compound"; *Nature* **1951**, *168*, 1040.
[3]Wilkinson, G. "The Iron Sandwich: A Recollection of the First Four Months"; *J. Organomet. Chem.* **1975**, *100*, 273.

structure. One of Woodward's graduate students, Mike Rosenblum, prepared some of the compound following the synthetic preparation described in the Kealy and Pauson paper. Wilkinson and members of Woodward's group then proceeded to characterize the compound by infrared and ultraviolet spectroscopy, Gouy magnetic susceptibility, and by measuring the dipole moment. The infrared spectrum showed that there was only one type of C–H stretch (meaning only one type of C–H bond) in the molecule, which was consistent with the proposed structure. The magnetic susceptibility experiment revealed the compound to be diamagnetic (no unpaired electrons), and the dipole moment was found to be "effectively zero" (0.05 Debye), which, together with the infrared data, showed that the π-complexed sandwich structure had to be correct.[4] Wilkinson found that he could readily oxidize the iron center from the +2 oxidation state to the +3 oxidation state and prepared a number of derivatives of the resultant blue cation of the type $[Fe(C_5H_5)_2]^+X^-$, where $X^-$ is a large anion, such as $GaCl_4^-$. The results were quickly written up by Wilkinson and Woodward and submitted in mid-March to the *Journal of the American Chemical Society* for publication. Kealy and Pauson were not the first to make $(C_5H_5)_2Fe$; independent work by Miller, Tebboth, and Tremaine appeared in February 1952 in the *Journal of the Chemical Society*. This paper gave an alternate synthetic route to biscyclopentadienyliron, but nevertheless depicted the structure in the same (incorrect) manner as the Kealy and Pauson paper.

A series of papers (one by Woodward's group and four by Wilkinson) concerning biscyclopentadienyliron and analogues were submitted for publication during the month of June. Woodward wondered if the aromatic nature of the Cp rings was similar enough to that of benzene to allow classical electrophilic aromatic substitution reactions, such as Friedel–Crafts reactions, to be performed on the Cp ring as they are on the benzene ring. The Woodward group found that they could indeed perform Friedel–Crafts chemistry on both Cp rings, prompting one of Woodward's postdoctoral fellows, Mark Whiting, to propose the name "ferrocene" for the compound. Subsequently, the entire class of transition metal biscyclopentadienyliron compounds became collectively known as the "metallocenes." Wilkinson, eager to try other metals, prepared the ruthenium analogue "ruthenocene" and the cationic

[4]Wilkinson, G.; Rosenblum, M.; Whiting, M. C.; Woodward, R. B. "The Structure of Iron *bis*-Cyclopentadienyl"; *J. Am. Chem. Soc.* **1952,** *74,* 2125.

cobalt analogue (the "cobalticenium" ion $Cp_2Co^+$), using an improved synthetic route that he devised.

Wilkinson soon found that he was not alone in the field of metallocene chemistry. Independently, and at virtually the same time, another chemist, Professor Ernst Otto Fischer of the Technische Hochschule in Munich, Germany, was also working on the structure of biscyclopentadienyliron. In collaboration with W. Pfab, he published the results of an X-ray diffraction study (of the unit cell of) biscyclopentadienyliron, which clearly showed the compound to be just as Wilkinson (and Woodward) had predicted: a $\pi$-complexed sandwich structure.

While at Harvard, Wilkinson continued to devote much of his energy toward the further development of transition metal cyclopentadienyl chemistry and, in collaboration with his students, he published many papers on cyclopentadienyl metal carbonyls, nitrosyls, and the first organometallics characterized by nuclear magnetic resonance (NMR) spectroscopy, $Cp_2ReH$ and the cation $Cp_2ReH_2^+$. One compound, $Cp(CO)_2FeCp$ (possessing both a $\pi$- and a $\sigma$-bound Cp group), had an infrared (IR) spectrum that seemed to conflict with its NMR spectrum. Wilkinson's keen analytical mind was quick to arrive at a logical answer; he correctly interpreted the phenomenon as attributable to a rapid intramolecular rearrangement of the $\sigma$-bound Cp ring known as a 1,2 shift or "ring whizzing". The shift was slow on the IR time scale, causing the molecular motion to appear in one case as being "frozen out", but fast on the NMR time scale, causing the molecular motion to appear in this case as being "blurred". This was the first molecule recognized as being fluxional; it was a decade before another, similar one was recognized.

After returning to England in 1956, Wilkinson continued his research on transition metal complexes and forged ahead with many new and exciting accomplishments. His work with metal hydrido complexes is especially noteworthy, as it led (in the mid-1960s) to the development of the versatile homogeneous catalyst $Rh(Cl)(PPh_3)_3$, better known as "Wilkinson's catalyst", for the hydrogenation of many olefinic compounds under mild conditions. The catalyst is especially versatile in that it can hydrogenate olefins without disturbing other functional groups such as esters, nitriles, or ketones, present within the same molecule. Wilkinson's pioneering and continued efforts in homogeneous catalysis have paved the way for the remarkable advances in this area, particularly in olefin isomerization, hydroformylation, hydrogenation, and hydrosilylation reactions.

The discovery, recognition, and development of the new π mode of bonding between metals and unsaturated molecules by Wilkinson and Fischer revolutionized the field of organometallic chemistry and was undoubtedly responsible for the explosive growth that has taken place in the field since 1952. Wilkinson has continued as a pacesetter in the fields of inorganic and organometallic chemistry and is also a prominent editor or coeditor of major works and journals in these fields. He has received many honors, among them being knighted by H.R.H. Queen Elizabeth II of England, and has received many awards, such as the American Chemical Society Award in Inorganic Chemistry (1965), the Lavoisier Medal of the French Chemical Society (1968), and the Chemical Society Award in Transition Metal Chemistry (1971). He has also been described as a warm and witty man, whose creativity and imagination, together with the confidence he has instilled in his students, has made his research group "an exciting and enjoyable one to be associated with".[5]

PETER V. BONNESEN
*Oak Ridge National Laboratory*

## Bibliography

Jardine, F. H. "Chlorotris(triphenylphosphine)rhodium(I): Its Chemical and Catalytic Reactions"; In *Progress in Inorganic Chemistry;* Lippard, S. J., Ed.; John Wiley & Sons: New York, 1981; Vol. 28, pp 63–202.

Kealy, T. J.; Pauson, P. L. "A New Type of Organo-Iron Compound"; *Nature* **1951,** *168,* 1039–1040.

Miller, S. A.; Tebboth, J. A.; Tremaine, J. F. "Dicyclopentadienyliron"; *J. Chem. Soc.* **1952,** 632–635.

Osborn, J. A.; Jardine, F. H.; Young, J. F.; Wilkinson, G. "The Preparation and Properties of Tris(triphenylphosphine)halogenorhodium (I) and Some Reactions thereof including Catalytic Homogeneous Hydrogenation of Olefins and Acetylenes and their Derivatives"; *J. Chem. Soc., Section A* **1966,** 1711–1729.

Powell, P. *Principles of Organometallic Chemistry,* 2nd ed.; Chapman & Hall: London, 1988.

---

[5] Seyferth, D.; Davison, A. "The 1973 Nobel Prize for Chemistry"; *Science* **1973,** *182,* 701.

Seyferth, D.; Davison, A. "The 1973 Nobel Prize for Chemistry"; *Science* **1973,** *182,* 699–701.

Wilkinson, G. "Die lange Suche nach stabilen Alkyl-Übergangsmetall-Verbindungen" ("The Long Search for Stable Alkyl-Transition Metal Compounds"); *Angew. Chem.* **1974,** *86,* 664–667 (Nobel lecture, in German; Published in English in *Les Prix Nobel,* 1973, by the Nobel Foundation, Stockholm, Sweden).

Wilkinson, G. "The Iron Sandwich: A Recollection of the First Four Months"; *J. Organomet. Chem.* **1975,** *100,* 273–278 (an autobiographical sketch including references for the early papers on ferrocene).

Wilkinson, G.; Rosenblum, M.; Whiting, M. C.; Woodward, R. B. "The Structure of Iron *bis*-Cyclopentadienyl"; *J. Am. Chem. Soc.* **1952,** *74,* 2125–2126 (first note on the structure of ferrocene).

Wilkinson, G.; Cotton, F. A. "Cyclopentadienyl and Arene Metal Compounds"; In *Progress in Inorganic Chemistry;* Cotton, F. A., Ed.; Interscience: New York, 1959; Vol. 1, pp 1–124.

# Paul Flory

## 1910–1985

Paul John Flory was awarded the Nobel Prize in chemistry in 1974 "for his fundamental achievements, both theoretical and experimental, in the physical chemistry of macromolecules." Flory was born on July 19, 1910, in Sterling, Illinois. His parents were Ezra Flory, a clergyman–educator, and Martha Brumbaugh Flory, who had been a schoolteacher. He died on September 8, 1985, at his home in Big Sur, California, while preparing a paper for presentation at the National Meeting of the American Chemical Society. His interest in chemistry was inspired by Carl W. Holl, Professor of Chemistry at Manchester College, a liberal arts college in Indiana, where Flory received his undergraduate education. After graduation in 1931, Flory entered graduate school at Ohio State University, where he completed the requirements for the Ph.D. in 1934. His dissertation research in physical chemistry was carried out under the guidance of Professor Herrick L. Johnston.

On completion of his Ph.D., Flory joined the Central Research Department of the DuPont Company in Wilmington, Delaware. He was assigned to the small research group headed by the great Wallace H. Carothers, the inventor of nylon and other synthetic polymers, whose name must be placed alongside that of Hermann Staudinger (Nobel Laureate in chemistry, 1953) as a founder of macromolecular chemistry. Carothers introduced Flory to polymers and convinced him that they were valid objects of scientific inquiry. It was an ideal time to become involved in polymer research. The work of Staudinger and

Carothers had only recently convinced the scientific community that macromolecules were linked by ordinary covalent bonds rather than by mysterious "secondary linkages".[1] Carothers's research group was beginning to make significant progress in the synthesis of polymers. In the process they were raising a number of important questions concerning the rates of polymerization reactions.

Flory's doctoral research on the photochemical dissociation of nitric oxide had provided him with valuable experience in chemical kinetics, the study of the rates and mechanisms of reactions. He was able to put this knowledge to immediate use in deciphering the rates of formation of nylon and other condensation polymers. The formation of a simple condensation polymer like nylon involves the repetition of an elementary chemical reaction to form an amide linkage. As the reaction proceeds, larger and larger molecules are formed. It was assumed at the time that because larger molecules move more slowly in solution, reactions between them would be more sluggish than those between the original small molecules. Flory, in a theoretical investigation of the distribution of sizes of molecules formed in the polymerization reaction, took the contrary view that reactivity is independent of molecular size.[2] Using this assumption and straightforward statistical methods, he was able to calculate the distribution of molecular sizes, a result ordinarily called the "most probable distribution". In addition, Flory performed careful kinetic experiments to validate his hypothesis of equal reactivity.[3] The most probable distribution does describe, at least approximately, the distribution of products in a real polymerization reaction. Its derivation was Flory's first major contribution to polymer physical chemistry.

In this early work Paul Flory exhibited the four characteristics that made him a truly outstanding scientist. The first was physical intuition, the ability to formulate a mental model of the system that incorporates the most important physical and chemical processes that determine its behavior. The hypothesis of equal reactivity was the insight needed to solve the problem, and it was Flory who was able to ignore conventional wisdom and see that the size of the reacting molecule was

---

[1] Herbert Morawetz, *Polymers: The Origins and Growth of a Science;* J. Wiley & Sons, New York, 1985; Chapter 10.

[2] Flory, P. J. "Molecular Size Distribution in Linear Condensation Polymers"; *J. Am. Chem. Soc.* **1936,** *58* (October), 1877–1885.

[3] Flory, P. J. "Kinetics of Polyesterification: A Study of the Effects of Molecular Weight and Viscosity on Reaction Rate"; *J. Am. Chem. Soc.* **1939,** *61* (December), 3334–3340.

irrelevant and only the local environment was important. Flory was able to see to the heart of the problem and to state the essential physics and chemistry in terms of a simple model that is amenable to calculation and that produces results useful to the theorist and experimentalist alike. The second characteristic was confidence in his own intuition. The hypothesis of equal reactivity was controversial, but Flory stood his ground, producing a convincing conceptual model and, ultimately, experimental data to support his position. Flory showed the same insight and tenacity throughout his career. On those occasions when his ideas were challenged, Flory persisted and usually was proved correct. The third characteristic was mathematical ability. Flory was always able to state the problem mathematically and then find a solution. The final characteristic, of course, was experimental skill. Flory designed and carried out many of the important experiments to test his own and other theories.

After the untimely death of Carothers in 1937, Flory moved to the University of Cincinnati. The urgency of war-related research on the development of synthetic rubber lured him back to industry, first at the Esso (now Exxon) Laboratories (1940–1943) and then at the Research Laboratories of the Goodyear Tire and Rubber Company (1943–1948). Liberal policies on basic research and publication at these two companies allowed him to continue his scientific career. Following his work on the most probable distribution, Flory made more important contributions to the understanding of polymerization, particularly the introduction of the concept of "chain transfer", the process by which the reactive center on one molecule is transferred to another molecule.[4] This process controls the ultimate size of the products in many polymerization reactions and is, therefore, of tremendous industrial importance. He also applied his statistical theory of polymerization to nonlinear polymers and produced a theory of gel formation, wherein a three-dimensional molecular network is formed from smaller molecules.[5] Gel formation is important in such diverse areas as the making of gelatin desserts and jams in the kitchen and the production of rubber tires.

During the war years Flory also developed a theory of polymer solution thermodynamics, the famous Flory–Huggins theory (M. L.

---

[4]Flory, P. J. "The Mechanism of Vinyl Polymerizations"; *J. Am. Chem. Soc.* **1937,** *59* (February), 241–255.

[5]Flory, P. J. "Molecular Size Distributions in Three-Dimensional Polymers. I. Gelation"; *J. Am. Chem. Soc.* **1941,** *63* (November), 3083–3090.

Huggins produced essentially the same theory independently), a theory that is still the basis for much of our understanding of the solution properties of polymers.[6] In addition, he began his studies of the phenomenon of rubber elasticity, a subject on which he continued to work until the last days of his life.

In 1948 Flory was invited to deliver the George Fisher Baker Lectures at Cornell University. He subsequently accepted a faculty appointment in the chemistry department at Cornell and moved there in the fall of 1948. The Baker Lectures were ultimately expanded into the book *Principles of Polymer Chemistry,* published by Cornell University Press in 1953. This book was one of the definitive contributions to pedagogy in the field. It has been the "bible" for generations of students and remains in print and in wide usage today. It is written in Flory's characteristically elegant prose with clear expositions of both the basic physical and chemical principles and the mathematical details.

While he was preparing the Baker Lectures, Flory thought of a way to treat the so-called excluded-volume problem in polymers.[7] This is probably his most remarkable contribution. Chain polymers are long flexible objects that, in solution, adopt a random configuration, like a strand of cooked spaghetti in a pot of water. In order to understand the motions of the molecule in solution, it is necessary to know the volume occupied by the molecule and how that volume depends on the chain length. The simplest way to treat the problem is to assume that the conformation, the way the chain is arranged in space, can be modeled as a random flight and to ignore the possibility that two portions of the chain might come into contact with each other. A random flight is a series of independent random steps in which any point in space can be visited any number of times. This random flight problem had been solved originally by Lord Rayleigh at the turn of the century and had been applied to the problem of polymer conformations by Werner Kuhn in 1934. It provides a quantitative prediction of the relationship between the volume occupied by the molecule and the chain length. Real polymers, however, cannot overlap and therefore should occupy a larger volume than that predicted by the random flight model for a given chain length. A proper treatment of this effect,

[6]Flory, P. J. "Thermodynamics of High Polymer Solutions"; *J. Chem. Phys.* **1942,** *10* (January), 51–61.

[7]Flory, P. J. "The Configuration of Real Polymer Chains"; *J. Chem. Phys.* **1949,** *17* (March), 303–310.

termed the "excluded-volume" problem, was first worked out by Paul Flory. He used a mean field treatment in which the correlations between chain segments were calculated using a simple approximation to show that the excluded-volume effect caused a significant expansion of the volume occupied by the coil. In fact, the perturbation of the coil size becomes larger as the chain length increases, and Flory was able to calculate the magnitude of this effect accurately with his simple model. This was a completely unexpected result, but subsequent work, both experimental and theoretical, has confirmed the essential correctness of the original Flory result. Flory also showed that the effect of the excluded volume depends on both the solvent and the temperature and that there will be a particular temperature at which the effect of excluded volume will vanish and the coil dimensions will be accurately described by the simple ideal random flight model. Flory named this condition the "theta point". He went on to explore the application of these ideas to the study of the hydrodynamic, or flow, properties of polymer solutions.

There are many topics in the physical chemistry of macro-molecules to which Flory has made significant contributions, including the conformations of proteins, the elasticity of fibrous proteins, the structure of the interface between polymer crystals and amorphous polymers, and the conformation of polymer chains in the melt.

Flory left Cornell in 1957 to become executive director of the Mellon Institute in Pittsburgh and four years later moved to Stanford University, where in 1966 he was appointed to the J. G. Jackson–C. J. Wood Professorship in Chemistry. During his years at Stanford he focused on the calculation of the detailed conformational properties of chain molecules, using rigorous mathematical and computational methods. This work was summarized in his second book, *Statistical Mechanics of Chain Molecules,* published in 1969. In his later years Flory returned to the study of rubber elasticity and also worked extensively on the theory of liquid crystalline substances.

Paul Flory was clearly the most important figure in the early development of the physical chemistry of polymers. His contributions have, in a real sense, defined the field. His tremendous insight, coupled with remarkable theoretical and experimental skills, helped reveal many aspects of the fascinating behavior of macromolecules. In addition, he worked tirelessly to promote the view that polymer science is a legitimate subject for basic research and not just a branch of technology. He argued that material on macromolecules should be a greater part of the usual undergraduate chemistry curriculum. He was highly regarded as both a teacher and a person by his students and colleagues. His strong sense of honor led him to become a courageous advocate for the cause of human rights in all countries of the world.

JEFFREY KOVAC
*University of Tennessee, Knoxville*

# Bibliography

Chayut, M. *Historical Studies in the Physical and Biological Sciences* **1992**, *23(2)*, 193–218.

*Current Biography Yearbook 1975;* Moritz, C., Ed.; The H. W. Wilson Company: New York, 1975–1976; pp 127–130.

"Flory, P. J."; *McGraw-Hill Modern Scientists and Engineers;* McGraw-Hill: New York, 1980; Vol. 1.

"Flory, P. J."; In *Macromolecular Science: Retrospect and Prospect;* Vol. 1 of *Contemporary Topics in Polymer Science;* Ulrich, R.D., Ed.; Plenum: New York, 1978; pp 69–97.

Flory, P. J. *Principles of Polymer Chemistry;* Cornell University Press: Ithaca, NY, 1953.

Flory, P. J. *Statistical Mechanics of Chain Molecules;* Interscience: New York, 1969.

Hounsell, D. A.; Smith, J. K., Jr. *Science and Corporate Strategy: Du Pont R&D, 1902–1980;* Cambridge University: Cambridge, England, 1988.

Mandelkern, L.; Mark, J. E.; Suter, U. W.; Yoon, D. Y. *Selected Works of Paul J. Flory;* 3 Volumes; Stanford University Press: Stanford, CA, 1985; includes an autobiographical sketch, Flory's Nobel lecture, and all the original papers cited in the text.

Mark, H. "An Architect of Polymer Science"; *Journal of Polymer Science* **1976**, *54*, 1–2.

McMillan, F. M. *The Chain Straighteners;* Macmillan: New York, 1979.

Morawetz, H. *Polymers: The Origins and Growth of a Science;* John Wiley & Sons: New York, 1985.

Morris, P. J. T. *Polymer Pioneers: A Popular History of the Science and Technology of Large Molecules;* Beckman Center for the History of Chemistry: Philadelphia, PA, 1986.

*The New York Times,* October 16, 1974; November 22, 1974; September 12, 1985.

*Nobel Prize Winners;* Wasson, T., Ed.; H. W. Wilson: New York, 1987.

*The Nobel Prize Winners; Chemistry;* Magill, F. N., Ed.; Salem Press: Pasadena, CA, 1990; Vol. 3.

Scheraga, H. A., "Paul J. Flory on His 70th Birthday;" *Macromolecules* **1980**, *13*, 8A–10A.

Seltzer, R. J. "Paul Flory: A Giant Who Excelled in Many Roles"; *Chemical and Engineering News* **1985**, *63(51)*, 27–30.

Stockmayer, W. H. "The 1974 Nobel Prize in Chemistry"; *Science* **1974,** *106* (November 22), 724–726.

Paul Flory's scientific papers are in the archives of the Chemical Heritage Foundation's Othmer Library in Philadelphia, Pennsylvania. His books and journals are housed in the Paul J. Flory collection at the Institute of Polymer Science, University of Akron, Akron, Ohio.

# John Cornforth

## 1917–

John Warcup Cornforth was born in Sydney, Australia, on September 7, 1917. He received his secondary education at Sydney High School and then attended the University of Sydney, where he received his B.Sc. degree in 1937 and was awarded the University Medal in Organic Chemistry. He received his M.Sc. degree a year later. At Sydney, Cornforth developed his interest in the chemistry of natural products and worked with J. C. Earl on chemical compounds isolated from Australian plants. He was clearly an exceptional student and received one of the scholarships established from proceeds of the 1851 Exhibition in London to enable overseas students to study in Britain. For the rest of his professional career Cornforth studied and worked in the United Kingdom in a variety of areas in organic chemistry and bioorganic chemistry. On October 17, 1975, the announcement came from Stockholm that John Warcup Cornforth had been awarded the Nobel Prize in chemistry for 1975, jointly with Vladimir Prelog, for "an outstanding intellectual achievement" in his work on the stereochemistry of (in other words, the spatial relationships involved in) reactions catalyzed by enzymes.

Cornforth left Australia and came to Oxford University in 1939 to work with the celebrated organic chemist Robert Robinson, himself a Nobel Laureate in chemistry in 1947. Cornforth's work with Robinson included approaches to the laboratory synthesis of steroids, a leading problem in the organic chemistry of the period. The steroids are a group of complex organic molecules that function in several ways in living systems. Many sex hormones are steroids, but the commonest of the steroids, cholesterol, is an essential component of the membranes that surround the cells of the body.

World War II broke out in Europe in September 1939, and after Cornforth completed his D.Phil. degree at Oxford in 1941 he joined the team of scientists working on problems related to the structure and production of the then-novel antibiotic penicillin. He correctly guessed the structure of a major portion of the penicillin molecule, namely the amino acid D-penicillamine, currently used as an antidote in some cases of heavy metal poisoning, and he was the first to synthesize it.

When the war ended, Cornforth joined the staff of the Medical Research Council at the National Institute for Medical Research in London in 1946. He continued his collaboration with Robert Robinson and, with the aid of a critical reaction that Cornforth had discovered in 1941, they published the first total synthesis of a nonaromatic steroid in 1951. An alternate route was published at the same time by R. B. Woodward of Harvard University. But at the National Institute for Medical Research many of Cornforth's colleagues were biochemists; he began to investigate problems of chemistry that were very closely related to biochemistry. He worked for 16 years, from 1946 to 1962, at the National Institute for Medical Research, and during that time made contributions to heterocyclic chemistry; to the chemical synthesis of biologically important natural products including cortisone, sialic acid, and squalene; to the chemotherapy of tuberculosis and leprosy; and to elucidation of the pattern in which acetic acid molecules are assembled by enzymes in the liver into squalene and then cholesterol.

His attention had been caught by a short but significant note written by Alexander Ogston in 1948 about how a particular approach to stereochemistry could clarify thinking about certain biochemical reactions. As Cornforth himself said in his Nobel Prize lecture, given in Stockholm on December 12, 1975, "Up to that time I had, as an organic chemist interested in the synthesis of natural products, the same kind of feeling for stereochemistry that a motorist might have for a system of one-way streets—a set of rules forming one more obstacle on the way to a destination".[1] This new outlook stimulated what was to become a major part of his research program, the study of the stereochemistry of reactions catalyzed by enzymes.

---

[1]Cornforth, J. W. "Asymmetry and Enzyme Action"; *Science* **1976**, *193*, 121–125.

An understanding of the way that enzymes work is essential to understanding the chemistry of biological processes. Molecules, like other three-dimensional objects, can be divided into two groups: those that are identical with their mirror images (a familiar example might be a sock, because for an ordinary sock it doesn't matter whether you put it on your right or on your left foot) and those that are not identical with their mirror images (now think of a shoe: It certainly does matter whether you put it on your right or on your left foot). Many of the molecules of importance in living systems, including small molecules like glucose, sucrose, or the amino acids and large molecules like proteins, enzymes, and nucleic acids (DNA and RNA), have a characteristic that turns out to be essential for the way they function: They are not identical with their mirror images. Such molecules are given a particular description by chemists; they are called chiral. In general, only one of the mirror-image forms of a chiral molecule is found in living systems and only that form can function in its role in the system. The other mirror-image form is either much less reactive in the system or may even interfere.

Ogston had pointed out in his 1948 note that some apparently nonchiral molecules might appear to be chiral when they were bound to a chiral enzyme. This could explain how enzymes worked to make possible the synthesis of only one mirror-image form of a molecule inside a cell. Cornforth and his collaborators decided to study how this actually happened by using a technique of labeling atoms by replacing them with an isotopic species. Cornforth's principal collaborators when he started this study included his wife, Rita (née Harradence), whom he married in 1941. She was also Australian and also an organic chemist. Biochemical support and expertise came from George Popják, at that time also a member of the staff of the National Institute for Medical Research, with whom Cornforth collaborated until 1967.

The main isotopes that Cornforth's group used in their work were the isotopes of hydrogen. Hydrogen, the lightest of the elements, has three isotopes that have atoms of different masses but are otherwise very similar in their chemical properties. The first of the isotopes is ordinary hydrogen, of mass 1 on the normal atomic mass scale; this isotope is given the symbol H. A second isotope is the naturally occurring deuterium, of mass 2, symbolized D; and the third is the artificially prepared, and radioactive, tritium, of mass 3, given the symbol T. As a typical example of the application of these isotopes, consider the class of enzymes called the dehydrogenases that operate by reversibly transferring one hydrogen atom from a coenzyme (a nicotinamide nucleotide) to a substrate. This hydrogen atom comes from a methylene ($CH_2$) group in the coenzyme molecule. Ogston's hypothesis would predict that enzymes can discriminate between the two hydrogen atoms of the methylene group, and this was verified experimentally. Popják and his collaborators, then working at Hammersmith Hospital, induced an enzyme to introduce deuterium, instead of ordinary hydrogen, into the methylene group. This produced a CHD group that, if the transfer were under stereochemical control, would be chiral in a particular sense (either right- or left-handed). Cornforth's group then degraded the labeled coenzyme chemically and separated this chiral center as part of a small molecule, succinic acid. They then chemically synthesized a sample of succinic acid containing a CHD group of known chirality and compared it with the degradation product by using a newly developed instrument that could measure very low levels of optical activity. In this way they could directly deduce the stereochemistry of hydrogen transfer by the enzyme they used and analogously by all enzymes that use nicotinamide coenzymes. Information of this kind cannot prove the mechanism of an enzymic reaction, but it does limit the number of acceptable mechanisms, since they have to agree with the stereochemical findings.

In 1962 Cornforth and Popják were appointed joint directors of the Shell Company's Milstead Laboratory of Chemical Enzymology, at Sittingbourne in Kent. The basic direction of Cornforth's work was unchanged by this move. He worked out the chemistry and stereochemistry of the biochemical reactions of a relatively simple starting material, called mevalonic acid, which contains six carbon atoms in each molecule. The importance of mevalonic acid is that it serves as starting material for two quite different kinds of biologically important classes of compounds. One of these is the steroids, mentioned earlier. Mevalonic acid is the parent of most steroids via an intermediate known as squalene, which contains 30 carbon atoms per molecule. Mevalonic acid also is the parent of the terpenes, a whole group of compounds of different molecular complexities that constitute many of the flavor and odor components of plants. Cornforth's work elucidated the spatial details of how the simple precursor mevalonic acid is converted by a sequence of reactions catalyzed by enzymes either into steroids or into terpenes.[2] This is the type of work that

---

[2]Retey, J.; Robinson, J. A. *Stereospecificity in Organic Chemistry and Enzymology;* Verlag Chemie: Weinheim, Germany, 1982.

the Nobel Committee recognized as "an outstanding intellectual achievement".

In 1968 Popják left the Milstead Laboratory and Cornforth became its sole director. He continued his stereochemical studies and found another biochemical collaborator in Professor Herman Eggerer at Munich. Together they solved the problem of the chiral methyl group. The methyl group, $CH_3$, is present in most biological molecules, and it is modified or generated by many enzymic reactions. When a molecule $X-CH_3$ is transformed into, or is formed from, a molecule $X-CH_2-Y$, the stereochemistry of the change cannot be deduced from simple isotopic labeling. However, if a methyl group contains one atom of each of the isotopes of hydrogen, the resulting CHDT group is chiral. If CHDT groups of known chirality could be made and distinguished from their mirror images, then the stereochemistry of the generation and transformation of methyl groups could be deduced. Cornforth and his co-workers solved the problem of making methyl groups of known chirality by purely chemical methods, and Eggerer and his co-workers devised an enzymic assay that distinguished between the mirror-image forms and measured the proportions of each. The key substance for both the synthetic method and the enzymic assay was acetic acid, the sour principle of vinegar and a basic building block of biochemistry. Chiral methyl groups have been used to trace the stereochemistry of dozens of enzymic reactions.[3]

Cornforth was a visiting professor at the University of Warwick from 1965 to 1971. From 1971 to 1975 he was visiting professor at the University of Sussex, and in 1975 he left Milstead and took up the position of Royal Society research professor at the University of Sussex, where he stayed until 1982, when he retired and became emeritus professor. Cornforth's contributions to chemistry have been recognized by his colleagues worldwide, and he has received many honors from them. He was elected to the Royal Society in 1953 and was made a foreign associate of both the American Academy of Sciences and the Royal Netherlands Academy of Science in the same year—1978. His medals and prizes include the Corday-Morgan Medal of the Chemical Society of London, in 1953; the CIBA Medal of the Biochemical Society, jointly with Popják, in 1965; the Stouffer Prize in 1967; the Ernest Guenther Award of the American Chemical Society in 1969;

[3]Cornforth, J. W. "The Chiral Methyl Group—Its Biochemical Significance"; *Chemistry in Britain* **1970**, *9*, 431–436.

the Davy Medal of the Royal Society, jointly with Popják, in 1968; the Prix Roussel in 1972; the Royal Medal of the Royal Society in 1976; and the Copley Medal of the Royal Society in 1982. He has received honorary degrees in Switzerland, England, Scotland, Ireland, and Australia, and was knighted, as Sir John Cornforth, by Queen Elizabeth II in 1977. In 1991 his native country awarded him its highest honor: Companion of the Order of Australia.

No account of Cornforth's career can be complete without a mention of the fact that he is a handicapped scientist. He started to lose his hearing as a boy and became totally deaf by his mid-twenties. His wife Rita has been a major channel of communication between Cornforth and his colleagues, and it is obvious from his achievements that his handicap has not prevented him from moving to the forefront of the scientists of his time. Chemistry and biochemistry are not his only recreations, however. He was a first-rate lawn tennis player and enjoys gardening, but chess has occupied a significant part of his leisure time. He writes, "I have been close to master strength at that game for about 50 years, and still hold the Australian record for blindfold chess (12 games simultaneously against sighted opponents; 8 wins, 2 draws, 2 losses) set up in 1937. As you know, I have been totally deaf for about the same period, and chess (where the sense of hearing is more often a drawback than an advantage!) was a logical choice".[4]

Cornforth's work in developing our understanding of how enzymes spatially direct biochemical reactions has led to a greatly increased depth of knowledge about vital processes; the tools he fashioned for his research have been used and adapted by many other chemists and biochemists, and in their hands have led to steady advances in the comprehension of the subtle details of how life processes work.

HAROLD GOLDWHITE
*California State University, Los Angeles*

---

[4] Cornforth, J. W., private communication, 1987.

# *Bibliography*

Callow, R. K. "Cornforth's Prize"; *Nature* **1975,** *258,* 5.

Choate, R. "Cornforth Wins Nobel Prize"; *The Times of London* **1975,** 3.

"Sir John (Warcup) Cornforth"; In *McGraw-Hill Modern Scientists and Engineers*; Greene, J. E., Ed.; McGraw-Hill: New York, 1980; Vol. 2, pp 233–234.

"Cornforth, Sir John (Warcup)"; In *Who's Who*; St. Martin's Press: New York, 1986; p 372.

Eliel, E. L.; Mosher, H. S. "The 1975 Nobel Prize for Chemistry"; *Science* **1975,** *190,* 772–774.

# 1975

NOBEL LAUREATE

# Vladimir Prelog

## 1906–

Copyright Nobel Foundation

Vladimir Prelog was born on July 23, 1906, in Sarajevo, Bosnia–Herzegovina. His mother tongue is Croatian. He attended a college preparatory school in Zagreb and, for two years, in Osijek. His father, Milan, was a teacher of history and later a professor at the University of Zagreb. From 1924 to 1929 Prelog studied chemistry at the Institute of Technology, Prague, where his mentor Rudolf Lukes, a graduate assistant, was able to kindle his enthusiasm for organic chemistry. He received his diploma in chemical engineering in 1928 and his doctorate in 1929.

No academic positions were available because of the Depression, so he worked in Prague from 1929 to 1935 as a chemist in the Laboratory of G. J. Driza, preparing chemicals not available commercially. There he also supervised his first doctoral candidate, his boss Driza, who graduated summa cum laude. In 1933 he married Kamila Vitek.

From 1935 to 1942 he was lecturer and associate professor at the Technical Faculty of Zagreb University. In 1941 he emigrated to Switzerland, where he was lecturer and associate professor at the Swiss Federal Institute of Technology (ETH) until 1950, when he became full professor. In 1957 he succeeded Leopold Ružička as head of the Laboratory of Organic Chemistry. Because he disliked administrative tasks—and to keep the other seven full professors of organic chemistry from leaving the ETH—in 1965 he introduced a rotating chairmanship and came to be known as "Der Dorfälteste" (the oldest in the village).

In 1975 Prelog was awarded the Nobel Prize in chemistry. He retired in 1976.

Prelog developed an early interest in chemistry in Osijek under the influence of a very good teacher; he published his first paper at age 15.[1] In Prague he worked on his doctoral thesis on the constitution of rhamnoconvolvulinolic acid (3,12-dihydroxypalmitic acid), which is an aglycone of a glycoside, under Emil Votocek, a well-known sugar chemist and pupil of the noted Bernhard Tollens.[2] Natural products became one of the two main interests of his career, the other being stereochemistry. The two fields are related because to understand the structure of natural products, one needs stereochemistry.

He fostered this interest in Prague, where he ran the lab in the daytime and did research in the evening. He maintained contact with Lukes, whose fascination with the "bizarre" structures of alkaloids was contagious. Because he wanted to do something of social importance, he chose to study the *Cinchona* alkaloids, that is, quinine, which was at that time still the most important antimalarial substance. The constitution of quinine was known but not its configuration, (or three-dimensional structure). His work on quinine and compounds related to its quinuclidin moiety was continued in Zagreb and Zurich.

At the University of Zagreb, where the pay was poor and research facilities were modest, he was fortunate to have the financial assistance of a small, prosperous pharmaceutical company in building up a research laboratory. With the help of his enthusiastic group he continued work on *Cinchona* alkaloids.[3] In Zagreb he synthesized adamantane, which brought him international recognition.[4] The technical synthesis of Streptazol (sulfanilamide) brought him some financial gain and allowed him in 1937 to spend several months in the laboratories of Ružička at the ETH in Zurich, a visit that became crucial for his future career.

When in 1941 Yugoslavia was occupied by the German army, Prelog emigrated to Switzerland. Because Richard Kuhn, who was president of the German Chemical Society, had invited him to lecture

---

[1] "Eine Titriervorrichtung"; *Chem-Ztg.* **1921,** *45,* 736.

[2] Votocek, E.; Prelog, V. "Sur l'acide 3,12-dioxylpalmitique composant de l'acide rhamnoconvolvulique"; *Collect. Czech. Chem. Commun.* **1929,** *1,* 55.

[3] Prelog, V.; Seiwerth, R.; Hahn, V.; Cerkovnikov, E. "Synthetische Versuche in der Reihe der Chinaalkaloide I"; *Ber. deutsch. chem. Gesell.* **1939,** *72B,* 1325-1333.

[4] Prelog, V.; Seiwerth, R. "Über die Synthese des Adamantans"; *Ber. deutsch. chem. Gesell.* **1941,** 74B, 1644–1648.

in Germany, he was able to get passports; Ružička helped Prelog to obtain Swiss visas for himself and his wife. Prelog came to Zurich at the right time, because a number of Ružička's experienced, Jewish co-workers had just emigrated to the United States and he could help fill the vacuum created by their departure. Ružička asked him to continue work on an abandoned project dealing with the isolation of steroids from organ extracts. Although the results of the work were disappointing, it was a good exercise in natural products chemistry.

Prelog's main interest was still focused on alkaloids. He found an ideal topic in the elucidation of the structure of solanidin; he continued his work on *Cinchona* alkaloids and started to investigate strychnine. He showed that Robert Robinson's formula for strychnine was not correct.[5] Although the formula he proposed was also not the right one, the discovery increased his international prestige. Later he worked on elucidating the structures of aromatic *Erythrina* alkaloids and (together with D. R. H. Barton, O. Jeger, and R. B. Woodward) of cevin, which was his last project in alkaloid chemistry.

At mid-century, the instrumental revolution (that is, the introduction into organic chemistry of the physical methods of X-ray analysis, molecular spectroscopy, and chromatography) necessitated a new approach to structural elucidation. Purely chemical methods had become outdated and had lost some of their intellectual appeal. Recognizing the growing importance of microbial metabolites, Prelog started working on these compounds, which possess unusual structures and interesting biological properties. It led him into antibiotics, and he subsequently elucidated the structures of such compounds as nonactin, boromycin, ferrioxamins, and rifamycins (used against tuberculosis and leprosy). For Prelog, natural products represented more than a chemical challenge: He considered them a record of billions of years of evolution.

In 1944 at the ETH, he managed to separate enantiomers with "asymmetric" trivalent nitrogen by column chromatography at a time when this method was still in its infancy.[6] His work on medium-sized alicyclic and heterocyclic rings established him as a pioneer in stereochemistry and conformational theory and brought an invitation to give the first Centenary Lecture of the Chemical Society in London in 1949.

---

[5]Prelog, V.; Szpilfogel, S. "Die Konstitution des Strychnins"; *Experientia* **1945**, *1*, 197–198.

[6]Prelog V.; Wieland, P. "Über die Spaltung der Tröger'schen Base in optische Antipoden, ein Beitrag zur Stereochemie des dreiwertigen Stickstoffs"; *Helv. Chim. Acta* **1944**, *27*, 1127.

He synthesized medium-sized ring compounds with 8 to 12 members from dicarboxylic acid esters by acyloin condensation and explained their unusual chemical reactivity by a "nonclassical" strain because of energetically unfavorable conformations. He also contributed to the understanding of Bredt's rule (in bridged-ring systems, a double bond cannot start from a bridgehead) by showing that a double bond may occur at the bridgehead if the ring is large enough. "I believe that the investigations which we have carried out show clearly that a medium-size ring effect exists which was not predicted by classical chemistry".[7]

In his research on asymmetric syntheses, Prelog studied enantio-selective reactions and established rules for the relationship between the configuration of educts and products.[8] From the stereoselectivity of asymmetric syntheses to a study of enzymatic reactions was only a small step, because "the mystery of enzymic activity and specificity will not be elucidated without a knowledge of the intricate stereochemical details of enzymic reactions".[9]

From Prelog's researches into the stereospecificity of microbio-logical reductions of alicyclic ketones and the enzymic oxidation of alcohols, he contributed not only to the knowledge of the mechanism of stereospecificity of enzymic reactions in general but also to the structure of the active site of the enzyme.

Specifying the growing number of stereoisomers of organic compounds became for Prelog one of his important aims. In 1954 he joined R. S. Cahn and Sir Christopher Ingold in their efforts to build a system for specifying a particular stereoisomer by simple and unambiguous descriptors that could be easily assigned and deciphered: The CIP (Cahn-Ingold-Prelog) system was developed for defining absolute configuration using "sequence rules". Together they published two papers.[10] After Cahn and Ingold died, Prelog published a third paper on the topic.[11] The CIP system was accepted by Beilstein's *Handbuch der organischen Chemie* (1956), which served as a first test for its validity, and into

---

[7]Prelog, V. "Newer Developments of the Chemistry of Many-Membered Ring Compounds"; *J. Chem. Soc.* January **1950**, 420–428.

[8]Prelog, V. "Über den sterischen Verlauf der Reaktion von α-Ketosaureestern optisch aktiver Alkohole mit Grignard'schen Verbindungen"; *Helv. Chim. Acta* **1953**, *36*, 308.

[9]Prelog, V. "Chirality in Chemistry"; Nobel lecture reprint, *Science* **1976**, *193*, 24.

[10]Cahn, R.S.; Ingold, C.K.; Prelog, V. "The Specification of Asymmetric Configuration in Organic Chemistry"; *Experientia* **1956**, *12*, 81, and "Specification of Molecular Chirality"; *Angew. Chem. Int. Ed.* **1966**, *5*, 385.

[11]Prelog, V.; Helmchen, G. "Basic Principles of the CIP-System and Proposals for a Revision"; *Angew. Chem. Int. Ed.* **1982**, *21*, 567–583.

the Chemical Abstracts on-line service. Its rules can be used on every "stereogenic" unit and also to define conformations, including helices. Prelog's later activities include the separation of enantiomers by partition between liquid phases in the studies of simple systems that may serve as models for complex biological membranes.[12]

As director of the Organic Laboratory at the ETH, Prelog was a direct successor in a line of Nobel Laureates: Willstätter, Staudinger, Kuhn, Ružička. Among his other predecessors were the distinguished stereochemists Johannes Wislicenus, Victor Meyer, and Arthur Hantzsch. Prelog has published some 400 papers; has lectured extensively in Europe, the United States, India, Australia, Japan, and Israel; and has received many honors. He is considered the leading figure in stereochemistry of our era and one of the important natural products chemists.

TONJA A. KOEPPEL

*Bibliography*

Eliel, E. L.; Mosher, H. S. "The 1975 Nobel Prize for Chemistry"; *Science* **1975**, *190*, 772–774.

Jeger, O; Prelog, V. "Steroid Alkaloids: The Solanum Group" In *The Alkaloids;* Manske, R. H. F., Ed.; Academic Press: New York, 1960; Vol. 7, p 343–361.

Prelog, V. "Chiral Ionophores"; *Pure Appl. Chem.* **1978**, *50*, 893–904.

Prelog, V. "From Configurational Notation of Stereoisomers to the Conceptual Basis of Stereochemistry";  *van't Hoff-Le Bel Centennial;* Ramsay, O. B., Ed.; ACS Symposium Series 12; American Chemical Society: Washington, DC, 1975; pp 179–188.

Prelog V. "Conformation and Reactivity of Medium Sized Ring Compounds"; *Pure Appl.Chem.* **1963**, *6*, 545–560.

Prelog, V. "Conformational Analysis—Scope and Present Limitations"; *Pure Appl. Chem.* **1971**, *25*, 465–468.

Prelog, V. "The Constitution of Rifamycins"; *Pure Appl. Chem.* **1963**, *7*, 551–564.

Prelog, V. "Influence sterique dans la synthese asymmetrique"; *Bull. Soc. Chim. Fr.* **1956**, 987–995.

Prelog, V. Interview by Tonja A. Koeppel at the Swiss Federal Institute of Technology on January 17, 1984; Beckman Center for the History of Chemistry: Philadelphia, PA, 1984; Transcript 38.

---

[12]Prelog, V.; Stojanac, Z.; Kovacevic, K. "Über die Enantiomerentrennung durch Verteilung zwischen flüssigen Phasen"; *Helv. Chim. Acta* **1982**, *65*, 377–384.

Prelog, V. *My 132 Semesters of Studies of Chemistry*; Profiles, Pathways, and Dreams; Seeman, J. I., Ed.; American Chemical Society: Washington, DC, 1991.

Prelog V. "The Role of Certain Microbial Metabolites as Specific Complexing Agents"; *Pure Appl. Chem.* **1971,** *25,* 197–210.

Prelog V.; Traynham J. B. "Transannular Hydride Shifts"; In *Molecular Rearrangements;* De Mayo, P., Ed.; Interscience–Wiley: New York, 1963; Vol. 1, pp 593–615.

# *William N. Lipscomb, Jr.*

1919–

Copyright Nobel Foundation

William Nunn Lipscomb, Jr. won the 1976 Nobel Prize in chemistry in recognition of his outstanding contribution to boron hydride (borane) chemistry, with particular emphasis on "his studies on the structure of boranes illuminating problems of chemical bonding". Lipscomb was born on December 9, 1919, in Cleveland, Ohio, and attended the University of Kentucky, where he graduated with a B.S. degree in 1941. He then went to the California Institute of Technology for his graduate work, where he was strongly influenced by Professor Linus Pauling (Nobel Laureate in chemistry, 1954) and Pauling's ideas concerning molecular structure and chemical bonding. After receiving the Ph.D. degree under Pauling in 1946, he became an assistant professor of physical chemistry at the University of Minnesota. In 1959 he accepted a professorship at Harvard University and later became Abbott and William James Lawrence Professor of Chemistry.

Lipscomb first became interested in X-ray and electron diffraction methods while a graduate student at Caltech. At the time, the field was in its infancy, and its power as an indispensable analytical tool for uncovering the secrets of molecular structure and bonding was only just beginning to be realized. While at Caltech, Lipscomb published two electron diffraction studies, one concerning the structure of vanadium tetrachloride and the other concerning the structure of dimethylketene dimer. In the former paper, he correctly showed that the unpaired electron in vanadium tetrachloride is not involved in bond formation and that vanadium tetrachloride has a regular tetrahedral structure within the limits of accuracy then available.[1]

At the time, the correlation between molecular structure and bonding was not clearly understood for many molecules, and this was especially the case for the boron hydrides. Although a number of the neutral boron hydrides, including diborane ($B_2H_6$, the simplest stable borane) and such higher homologes as $B_4H_{10}$, $B_5H_9$, $B_5H_{11}$, and $B_{10}H_{14}$, were first prepared and characterized in the early 1920s by German chemist Alfred Stock, their definitive structures were not known until much later, as the bonding in boron hydrides could not necessarily be extrapolated from the simple Lewis theory of bonding that was used to describe the bonding in the carbon hydrides (hydrocarbons). Boron combines four valence orbitals with only three valence electrons when forming bonds, whereas carbon has the full complement of four of each. This apparent "electron deficiency" of boron caused much frustration and confusion among theoretical chemists, because the chain and ring structures that so typified the bonding in hydrocarbons could not be readily applied to the boranes and in fact seemed to be ruled out entirely. A radically different approach and explanation was thus necessary, and Lipscomb, while a young assistant professor at the University of Minnesota, was intrigued and was among those who searched for a solution. Clues were first uncovered in 1947 when the correct structure of diborane was established experimentally by W. C. Price and then in 1948, when Kasper, Lucht, and Harker (a man very influential in the development of modern X-ray crystallography) published preliminary results concerning the structure of $B_{10}H_{14}$. Diborane was found to have a structure consisting of two $BH_2$ groups connected to each other by two bridging hydrogens, and, although the B-H-B bridge was later described as a three-center two-electron bond by H. C. Longuet-Higgins, the overall structure had been correctly predicted as far back as 1921 by W. Dilthey. However, $B_{10}H_{14}$ was found to "have a basket-like arrangement of boron atoms" that was totally unlike the hydrocarbons and that had not been remotely

[1]Lipscomb, W. N.; Whittaker, A. G. "An Electron Diffraction Investigation of Vanadium Tetrachloride"; *J. Am. Chem. Soc.* **1945**, *67*, 2019–2021.

predicted by anyone.[2] Lipscomb, realizing the potential of X-ray and electron diffraction methods, set out to solve the structures of the remaining Stock boranes by employing these techniques. It proved to be a difficult task, not only because X-ray diffraction was a long and time-consuming procedure at that time, but also because the extreme volatility and pyrophoricity (tendency to inflame spontaneously in air) of these boranes demanded that they be handled by special vacuum-line techniques and that the X-ray data be collected at low temperatures. Nonetheless, he successfully solved the structures for all the Stock boranes, as well as for many other boranes. Lipscomb, however, was not content to simply report the bewildering assortment of polyhedra and cages that he had uncovered as the structures for these boranes; he also wanted to establish some general rule that could relate the given structure of a particular borane with its molecular formula.

In 1954, in collaboration with W. H. Eberhardt and Bryce Crawford, Jr., Lipscomb published a landmark paper that set forth his topological approach to boron hydride structure: the *styx* numbers. Each letter in the "styx" formula stood for one type of "building block" or structural linkage of the four possible for the boron hydrides. One of these four had not been formally proposed prior to this paper: the three-center, two-electron B–B–B bond, a crucial extension of the B–H–B bond described above. The four possible types of linkages, then, are three-center B–H–B bonds (labeled $s$), three-center B–B–B bonds (labeled $t$), two-center B–B bonds (labeled $y$), and finally, two-center terminal B–H bonds (labeled $x$, where $x$ equals the number of $BH_2$ units). Several relationships were put forth between the molecular formula of a given boron hydride, $B_pH_q$, and the *styx* numbers. For example, for three-center orbital balance, $4p = (q-s)+2s+3t+2y$, and for electron balance, $3p = (q-s)+s+2(t+y)$ (these have since been extended and simplified by Lipscomb and others).[3] Taken together with a formula for hydrogen balance (and with certain other constraints), these equations of balance describe the relationship between the formula of the boron hydride and the number and kinds of bonds in the molecule. Thus, the topology of any boron hydride can be described in terms of the resultant *styx* number. Diborane, for example, can be described by the *styx* number 2002, which means there are two linkages

[2]Grimes, R. N. "The 1976 Nobel Prize for Chemistry"; *Science* **1976**, *194*, 709.

[3]Purcell, K. F.; Kotz, J. C. *Inorganic Chemistry;* W. B. Saunders: Philadelphia, PA, 1977, Chapter 18.

of type $s$, none of types $t$ and $y$, and two of type $x$, which precisely describes the actual bonding.[4] This topological theory of bonding not only provided a means of explaining the known structures of boron hydrides, but also greatly aided in the discovery of new boron hydride compounds. The theory also clearly showed that the three-dimensional cage structures that the boron hydrides adopted were nature's way of getting around the "electron deficiency" of boron.

Lipscomb and his group published many papers on the boron hydrides over the next several years, dealing not only with further developments and extensions of his topological theory, but also with geometric structure, chemical reactivity, and charge distribution. By the early 1960s, the boron hydrides had become the second largest class of molecular hydrides (behind the hydrocarbons), and although many other scientists were working in the area, Lipscomb and his group were responsible for the development of nearly all the general theory.[5] Lipscomb also published many papers on the crystal structures of various inorganic and organic compounds.

Following his move to Harvard in 1959, Lipscomb expanded his theoretical and X-ray studies of the boranes, looked for ways to prepare new boranes, and started up a second research group to study protein structure and bonding by X-ray crystallography. Other research groups, building on the groundwork Lipscomb had provided, came forth with new developments, such as the carboranes (boron cage complexes in which varying numbers of carbon atoms have been substituted for boron within the framework), and large polyhedral dianions. Again, Lipscomb and his group played a key role in the further development of these areas.

In 1963 he published his first major book, *Boron Hydrides*. It became a landmark and indispensable text for all those working in the field. He also was a pioneer in the field of boron-11 nuclear magnetic resonance (NMR) spectroscopy and eventually coauthored a book with Gareth Eaton (entitled *NMR Studies of Boron Hydrides and Related Compounds*) in 1969. As he became more interested in biochemistry through his work on proteins, his thoughts turned to the biological applications of boron chemistry, such as uses in radiation therapy for certain types of cancer.

---

[4] Eberhardt, W. H.; Crawford, B., Jr.; Lipscomb, W. N. "The Valence Structure of the Boron Hydrides"; *J. Chem. Phys.* **1954**, *22*, 989–1001.

[5] Grimes, R. N. "The 1976 Nobel Prize for Chemistry"; *Science* **1976**, *194*, 709–710.

As time progressed, Lipscomb began to focus his energies more on his work with proteins and made many significant contributions in this area as well. X-ray crystallography has become an extremely powerful technique for studying the three-dimensional structure and function of enzymes, and Lipscomb has become one of the world leaders in the field. Many scientists who are currently prominent in the field of protein structure determination have studied under Lipscomb. It should be noted that it is a quantum leap in complexity in going from solving a small borane molecule of perhaps two dozen atoms to solving the structure of the metalloenzyme carboxypeptidase A, which has thousands of atoms. Carboxypeptidase A (CPA) is an important zinc-containing digestive enzyme found in the intestine. Lipscomb's group determined the structure and probed the functioning of CPA and suggested a mechanism for the enzymatic activity, which included the role of the metal cofactor zinc. Carboxypeptidase A was the first metalloenzyme for which both the high-resolution structure and the sequence were determined.[6] Other enzymes Lipscomb and his group determined the structures for include the allosteric regulatory enzyme aspartate carbamoyltransferase (also called aspartate transcarbamylase), found in the intestinal bacteria *Escherichia coli,* which catalyzes the first unique step in the biosynthesis of pyrimidine nucleotides (which are building blocks for DNA).[7]

Lipscomb's work in molecular structure determination is awe-inspiring and has redefined molecular and bonding theory. Yet Lipscomb still finds time for many personal pursuits. He is an accomplished clarinetist of professional caliber and is extremely well read, having a particular fancy for Sir Arthur Conan Doyle's Sherlock Holmes detective stories (he is a member of the "Baker Street Irregulars"). One can easily say that all the fine scientific detective work Lipscomb has used to uncover scientific truths would have made Sherlock Holmes proud.

PETER V. BONNESEN
*Oak Ridge National Laborotory*

---

[6]Lipscomb, W. N. "Structure and Mechanism in the Enzymatic Activity of Carboxypeptidase A and Relations to Chemical Sequence"; *Acc. Chem. Res.* **1970,** *3,* 81–89.

[7]Monago, H. L; Crawford, J. L.; Lipscomb, W. N. "Three-Dimensional Structures of Aspartate Carbamoyltransferase from *Escherichia coli* and Its Complex with Cytidine Triphosphate"; *Proc. Natl. Acad. Sci.* **1978,** *75,* 5276–5280.

# Bibliography

Eberhardt, W. H.; Crawford, B., Jr.; Lipscomb, W. N. "The Valence Structure of the Boron Hydrides"; *J. Chem. Phys.* **1954,** *22,* 989–1001.

Grimes, R. N. "The 1976 Nobel Prize for Chemistry"; *Science* **1976,** *194,* 709–710.

Lipscomb, W. N. "Acceleration of Reactions by Enzymes"; *Acc. Chem. Res.* **1982,** *15,* 232–238.

Lipscomb, W. N. "Bonding in Boron Hydrides"; *Pure Appl. Chem.* **1972,** *29,* 493–511. This paper presents a summary of the theoretical work done by Lipscomb and his students and was one of the main lectures presented at the International Meeting on Boron Compounds, held at Castle Liblice near Prague, Czechoslovakia, June 21–25, 1971.

Lipscomb, W. N. "Framework Rearrangement in Boranes and Carbonanes"; *Science,* **1966,** *153,* 373–378.

Lipscomb, W. N. "Structure and Mechanism in the Enzymatic Activity of Carboxypeptidase A and Relations to Chemical Sequence"; *Acc. Chem. Res.* **1970,** *3,* 81–89.

Lipscomb, W. N. "Three-Center Bonds in Electron Deficient Compounds. The Localized Molecular Orbital Approach"; *Acc. Chem. Res.* **1973,** *6,* 257–262.

Lipscomb, W. N.; Whittaker, A. G. "An Electron Diffraction Investigation of Vanadium Tetrachloride; *J. Am. Chem. Soc.* **1945,** *67,* 2019–2021.

Monago, H. L.; Crawford, J. L.; Lipscomb, W. N. "Three Dimensional Structures of Aspartate Carbamoyltransferase from *Escherichia coli* and its Complex with Cytidine Triphosphate"; *Proc. Natl. Acad. Sci.* **1978,** *75,* 5276–5280.

Purcell, K. F.; Kotz, J. C. *Inorganic Chemistry;* W. B. Saunders: Philadelphia, PA, 1977; Chapter 18.

# 1977

NOBEL LAUREATE

# *Ilya Prigogine*

## 1917—

Ilya Prigogine was awarded the Nobel Prize in chemistry in 1977 for "his contributions to nonequilibrium thermodynamics, particularly the theory of dissipative structures." Prigogine was born in Moscow, Russia, on January 25, 1917, just nine months before the outbreak of the Bolshevik revolution. He was the second son of Roman Prigogine, a chemical engineer and factory owner, and Julia Wichman Prigogine, who had been a student at the Conservatory of Music in Moscow. In 1921 his family left Moscow to escape the restrictions on private enterprise imposed by the new government. They spent a year in Lithuania and several years in Berlin before settling in Brussels, Belgium. Prigogine received his secondary school education at the Athenée in Brussels, a public school noted for its rigorous classical curriculum. He obtained both his undergraduate and graduate education in chemistry at the Université Libre de Bruxelles, receiving the M.S. degree in 1939 and the Ph.D. in 1941. In 1947 he was appointed professor in the Université Libre, a position he has held ever since. In 1962 he was named director of the Instituts Internationaux de Physique et de Chimie, Solvay, also in Brussels. In 1967 he founded the Center for Statistical Mechanics (renamed the Ilya Prigogine Center for Studies in Statistical Mechanics in 1977) at the University of Texas, Austin, and in 1979 he was appointed Regental Professor of Physics and Chemical Engineering at Texas. He currently divides his time between Brussels and Austin.

Despite the disruptive political and economic situation of the times, Julia Prigogine attempted to give her sons the kind of cultural education that she had enjoyed. Ilya Prigogine could read music before he could read words and, in his high school years, dreamed of becoming a classical pianist. He read widely in the classics as well as books on archaeology, literature, and philosophy. He was particularly influenced by the thinking of Henri Bergson, who emphasized the differences between the concept of time used in science and the time of ordinary experience. Prigogine's interest in chemistry developed in a rather unusual way. He and his parents had agreed that he would pursue a legal career, and he decided that the best way to begin would be to understand the mind of the criminal. While looking for books on criminal psychology, he discovered a volume on the chemical composition of the brain. His plans to study law soon gave way to a passionate interest in chemistry, and he entered the university with chemistry as his prospective major.

At the Université Libre, Prigogine was strongly influenced by two professors: Théophile De Donder and Jean Timmermans. De Donder was interested in the applications of thermodynamics to nonequilibrium situations following the pioneering work of Pierre Duhem. Duhem's early efforts had not received much attention, but the field had been reinvigorated by the work of Lars Onsager (Nobel Laureate in chemistry, 1968) on the reciprocal relations. Timmermans was interested in the applications of equilibrium thermodynamics to the study of solutions and other complex systems. These became the two themes of Prigogine's research career.

Prigogine's work on the thermodynamics of solutions resulted in the development and application of the averaged potential method for cell models of liquids. In this method the complicated interactions between molecules in a dense liquid are treated by using a simple model in which each molecule is confined to a particular region of space (the cell). This model is amenable to quantitative calculation. Prigogine was able to predict new phenomena, most notably the phase separation of mixtures of isotopes of helium. In a phase-separation process, a homogenous liquid mixture divides into two immiscible layers, like oil and water, as the temperature is changed. This is a common phenomenon in mixtures of molecules with significantly different shapes and intermolecular interactions, but the phase separation of isotopic mixtures is rather surprising because the molecules differ only in mass and not in their intermolecular interactions. These methods were also successfully applied to polymer solutions. Extensions of Prigogine's original work on polymers are in use today.[1]

Prigogine's major interest, however, has been the study of nonequilibrium processes. Since his college days he has been fascinated by the role of time in scientific theory. In our everyday lives we experience a flow of time. Tomorrow differs from yesterday. In microscopic physical theories, classical and quantum mechanics, which deal with the motions of single particles, time enters only as a directionless parameter. The equations are the same whether time goes forward or backward. Prigogine found this to be strange, and much of his research has been an attempt to resolve this paradox.

In science, the distinction between past and future only appears strongly in the second law of thermodynamics, a macroscopic theory that considers systems containing a large number of particles. The second law states that a system isolated from the rest of the universe will eventually run down to a state of equilibrium in which all of its properties, such as temperature, pressure, and composition, do not vary with time, and there are no flows of matter or energy in the system or at its boundaries. Conventional thermodynamics allows us to study the properties of such equilibrium systems. Linear nonequilibrium thermodynamics, which originated with the work of Onsager, provides a way to study the flows of matter and energy that bring a system into an equilibrium state. In his early career Prigogine worked extensively on linear nonequilibrium thermodynamics, developing the formalism and proving the important theorem of minimum entropy production.[2]

Entropy is the important quantity in the theory of nonequilibrium processes. Roughly, it is a measure of the disorder in the system. A liquid has a higher entropy than a solid because the molecules of the liquid are at random positions, whereas in the solid they are arranged in a regular array. The second law of thermodynamics states that the entropy of an isolated system increases as it moves toward equilibrium. The theory of nonequilibrium thermodynamics provides equations to calculate the entropy increase in terms of the rates of the various irreversible processes and the magnitudes of the forces that drive them. If

---

[1]Prigogine, I.; Bellemans, A.; Mathot, V. *The Molecular Theory of Solutions;* North Holland: Amsterdam, the Netherlands, 1957.

[2]Prigogine, I. *Introduction to Thermodynamics of Irreversible Processes;* Interscience: New York, 1962.

a system is provided with energy or matter from an external source, it is possible to maintain it in a nonequilibrium steady state. In such a state the properties of the system are independent of time but, unlike in an equilibrium system, there are currents of matter or energy within the system and at the boundaries. An example of such a system is a section of a smoothly flowing river. The level of water, the temperature, and other properties are constant, but there is a continuous flow of water through that section. Prigogine was able to show that the stable steady states of a system near equilibrium are states in which there is a minimum increase in entropy per unit time.

Prigogine's most remarkable contributions have come in his studies of irreversible processes that are far from equilibrium. The phenomenon that kindled his interest in this area was the Bénard instability. When a fluid is heated from below, it becomes unstable because the warm liquid on the bottom is less dense than the cooler liquid on the top and hence will try to rise. As the temperature difference between the bottom and the top of the liquid is increased, a spectacular change in the behavior of the system can occur. At small values of the temperature difference, the movement of the liquid is chaotic. At a certain critical value of the temperature difference, the system spontaneously forms a regular hexagonal array of convection cells, with the liquid in each one circulating from bottom to top and back again. An ordered nonequilibrium stationary state has formed. Prigogine named these states "dissipative structures" to emphasize that they are the result of irreversible processes in which entropy is continually being produced but that they do have a definite structure in space and, perhaps, in time.

Prigogine and his collaborators have studied a variety of simple mechanisms that bring about spontaneous organization as the system is driven far from equilibrium. The most interesting among these are the oscillating chemical reactions. In most chemical reactions the concentrations of the starting materials decrease steadily toward a final equilibrium value while the concentrations of the products increase similarly. In an oscillating reaction the concentrations cycle between two values, increasing and decreasing in a regular pattern in time, like a chemical clock. Some examples even show a spectacular pattern of oscillation in space, giving rise to alternating bands of color. Prigogine and his collaborators G. Nicolis and R. Lefever devised a mathematical model, nicknamed the "Brusselator", that shows this oscillatory behavior. The Brusselator and similar mathematical models describe the behavior of real oscillating reactions, such as the Belousov–Zhabotinski reaction, that display complex structures in space and time when driven far from equilibrium.[3]

In collaboration with P. Glansdorff, Prigogine extended thermodynamic theory to include the possibility of the creation of order far from equilibrium.[4] They developed a thermodynamic stability theory that spans the whole range of equilibrium and nonequilibrium phenomena. The stability theory considers the response of a thermodynamic system to a fluctuation in one of its properties. In a steady state or equilibrium situation, the values of the properties are, on the average, constant in time. Because the molecules in the system are always moving and have a distribution of velocities, however, there are instantaneous deviations of the system properties from their average values. These deviations are called fluctuations. The Glansdorff–Prigogine stability theory calculates the effect of such fluctuations on the state of the system. Near equilibrium the fluctuations always disappear and the equilibrium or nonequilibrium steady state is stable. Far from equilibrium it is possible for a fluctuation to grow and carry the system into a new state. This is called a bifurcation and is exactly what happens in the Bénard instability when the hexagonal convection cells form. Prigogine has called the range of behavior where fluctuations die out the "thermodynamic branch" and the range of behavior where the fluctuations are magnified to create nonequilibrium, ordered structures the "dissipative structures branch".

This research has profound consequences for the study of biological systems. The large degree of order exhibited by biological organisms is maintained only through a large flux of matter and energy from the environment.[5] The theory provides a model with which to study the formation of both individual organisms and species as examples of dissipative structures. The formalism can also be used to study such diverse areas as economics, population dynamics, meteorology, and chemical engineering.

In addition to his work on nonequilibrium thermodynamics, Prigogine has always been interested in the microscopic theory of ir-

[3]Prigogine, I. *From Being to Becoming*; W. H. Freeman: San Francisco, California, 1980.

[4]Glansdorff, P.; Prigogine, I. *Thermodynamic Theory of Structure, Stability and Fluctuations*; Wiley-Interscience: New York, 1971.

[5]Nicolis, G.; Prigogine, I. *Self-organization in Non-equilibrium Systems;* John Wiley & Sons: New York, 1977.

reversibility.[6] His work is a continuation of the classic work of Boltzmann and leads to a non-Markovian master equation that is valid in
situations other than dilute gases. Prigogine has always felt that the resolution of the classic paradox of macroscopic irreversibility and microscopic reversibility would come from the study of unstable dynamical
systems. Recently, he has made considerable progress in this area, and
it now appears possible to formulate a classical and quantum theory
for unstable systems that includes irreversibility at the basic dynamical
level.

Prigogine has also been concerned with the broader philosophical
issues raised by his work. In the nineteenth century the discovery of
the second law of thermodynamics, which predicts a relentless movement of the universe toward a state of maximum entropy, generated
a pessimistic attitude about nature and science. Prigogine feels that his
discovery of self-organizing systems gives a more optimistic interpretation of the consequences of thermodynamics. In addition, his work
leads to a new view of the role of time in the physical sciences. In order
to open a dialog with the lay public about the intellectual consequences
of his research, he has written, in collaboration with Isabelle Stengers,
a popular exposition. This book, published in French under the title
*La Nouvelle Alliance* and in English as *Order Out of Chaos,* not only
explains the scientific discoveries of Prigogine and his collaborators
in nontechnical language but also attempts to place them in a broad
historical and philosophical context.

The main theme of the scientific work of Ilya Prigogine has been
a better understanding of the role of time in the physical sciences and
in biology. He has contributed significantly to the understanding of irreversible processes, particularly in systems far from equilibrium. The
results of his work on dissipative structures have stimulated many scientists throughout the world and may have profound consequences for
our understanding of biological systems. He has also recognized that
scientific research can have an important influence on society and he
has made a significant attempt to establish a dialog with the intellectual
community and with the lay public.

JEFFREY KOVAC
*University of Tennessee, Knoxville*

---

[6]Prigogine, I. *Non-equilibrium Statistical Mechanics*; Interscience: New York, 1962.

# *Bibliography*

Cousins, N. *Nobel Prize Conversations with Sir John Eccles, Roger Sperry, Ilya Prigogine, Brian Josephson;* Saybrook: San Francisco, Calif., 1985.

*For Ilya Prigogine;* Rice, S. A., Ed.; Wiley: New York, 1978.

Hershey, D. *Must We Grow Old: From Pauling to Prigogine to Toynbee;* Basal Books: Cincinnati, OH, 1984.

*The Omni Interviews;* Weintraub, P., Ed.; Ticknor and Fields: New York, 1984; pp 333–349.

*Physics and the Ultimate Significance of Time: Bohm, Prigogine, and Process Philosophy;* Griffin, D. R., Ed.; State University of New York Press: Albany, NY, 1986.

Prigogine, I. *From Being to Becoming;* W. H. Freeman: San Francisco, CA, 1980.

"Prigogine, Ilya"; *Current Biography* **1987,** *48,* 35–38.

Prigogine, I. "Prigogine, Ilya"; In *McGraw-Hill Modern Scientists and Engineers;* Parker, S. P., Ed.; McGraw-Hill: New York, 1980; pp 440–441.

Prigogine, I.; Stengers, I. *Order Out of Chaos;* Bantam Books: New York, 1984.

Procaccia, I.; Ross, J. "The 1977 Nobel Prize in Chemistry"; *Science* **1977,** *198,* 717–718.

Weber, R. *Dialogues with Scientists and Sages;* Routledge and Kegan Paul: London, 1986; pp 181–197.

# 1978

NOBEL LAUREATE

# *Peter Mitchell*

## 1920–1992

Copyright Nobel Foundation

Peter Dennis Mitchell, winner of the 1978 Nobel Prize in chemistry, was born in Mitcham, Surrey, England, on September 29, 1920, and died on April 10, 1992. He was the son of Christopher Gibbs Mitchell, a civil servant, and Kate Beatrice Dorothy née Taplin. Mitchell completed his secondary education at Queens College in Taunton, England, and shortly after the beginning of World War II in 1939 he entered Jesus College at Cambridge University. His studies at Cambridge focused on the sciences, and he became interested in the application of chemical principles to the study of biological phenomena.

In 1942 Mitchell began graduate studies at Cambridge in the Department of Biochemistry and was the first doctoral student of Professor James F. Danielli, who introduced Mitchell to the study of the structure and permeability properties of biological membranes, an area of interest that was continued and expanded throughout Mitchell's research career. As a graduate student at Cambridge, Mitchell's research addressed the mechanisms for the bactericidal action of penicillin and also involved the development of a dithioglycerol glucoside to be used as an antidote to arsenic poisoning. This work led to the investigation of the countertransport of arsenate and phosphate ions across the bacterial cell membrane. Mitchell's Ph.D. dissertation was entitled "The Rates of Synthesis and Proportions by Weight of the Nucleic Acid Components of Micrococcus during Growth in Normal and Penicillin-Containing Media, with Reference to the Bactericidal Action of Penicillin". As a result of these studies, Mitchell became familiar with mechanisms involved in the facilitated movement of ions across cellular membranes and acquired an understanding of the regulation and energetics associated with such processes.

While at Cambridge, Mitchell met David Keilin, Director of the Moltento Institute of Biology and Parasitology. Keilin was a distinguished scientist who had made important advances in the area of bioenergetics and was influential to Mitchell both personally and scientifically. Among Keilin's pioneering studies were the identification of the cytochromes, membrane-associated metalloproteins that participate in biological electron transfer through the oxidation–reduction reactions that occur at their metallic centers. His discoveries, and others', led to the formulation of the respiratory chain concept for the transfer of electrons through a series of membrane-associated electron and hydrogen carriers, resulting in the production of chemical energy in the form of adenosine 5'- triphosphate (ATP), which could be used by a cell to perform work. In the manuscript of his lecture delivered on receipt of the Sir Hans Krebs Medal in 1978, Mitchell described Keilin as "[my] . . . unofficial benefactor . . . who thoughtfully provided me with a mirror that could be used to reflect upon, and clearly see, the gorgonian transport aspect of redox chain metabolism without danger of petrifaction." The general theory of vectorial metabolism and chemiosmotic systems that Mitchell developed in his research career provided first a demarcated hypothesis and subsequently a well-researched explanation for how the transfer of electrons through the membrane-associated electron carriers of Keilin's electron transport chain was coupled to the synthesis of ATP.

Mitchell received his Ph.D. from Cambridge University in 1951 and remained there until 1955 as a demonstrator in the Department of Biochemistry. In 1955 he relocated to Edinburgh University in Scotland, where he coordinated the establishment of the Chemical Biology Unit in the Department of Biology. At Edinburgh he formally described the critical points of his chemiosmotic hypothesis. Mitchell became a senior lecturer in 1961 and in 1962 was appointed to the position of readership. However, in 1963 he was stricken with gastric ulcers that caused him to take leave from his position and, later, to resign.

For several years after leaving Edinburgh, Mitchell devoted his time to the establishment of the Glynn Research Laboratories within a large mansion in Bodmin, Cornwall, England. The independent research laboratory was founded with funds provided by Mitchell and his brother, Christopher, and Mitchell directed the restoration of the facility at the Glynn House and assumed the role of research director of the laboratory. In collaboration with Jennifer Moyle, a former colleague from Cambridge, and a small research group, Mitchell began the research necessary to test his chemiosmotic hypothesis experimentally. This work successfully substantiated many of the predictions arising from his hypothesis and was important in its eventual acceptance by the scientific community.

In addition to receiving the Nobel Prize in chemistry in 1978, Mitchell received numerous honors and awards, including honorary Doctor of Science degrees from the University of Liverpool and Exeter University in the United Kingdom (1977) and the University of Chicago (1978). He had also received the CIBA Medal and Prize of the British Biochemical Society (1973), the Warren Triennial Prize (1974, jointly with Efraim Racker), the Louis and Bert Freedman Foundation Award of the New York Academy of Sciences (1974), the Wilhelm Feldberg Foundation Prize (1976), the Lewis S. Rosensteil Award of Brandeis University (1977), the Sir Hans Krebs Lecture and Medal of the Federation of European Biochemical Societies (1978), and the Copley Medal of the Royal Society (1981). Mitchell held the positions of fellowship in the Royal Society (1974), honorary member of the American Society of Biological Chemists (1974), foreign honorary member of the American Academy of Arts and Sciences (1975), and foreign associate of the U.S. National Academy of Sciences (1977).

Peter Mitchell was awarded the Nobel Prize in chemistry in 1978 and was cited by the Royal Swedish Academy of Science for " . . . his contribution to the understanding of biological energy transfer through the formation of the chemiosmotic theory." Mitchell summarized this theory in his Nobel address, delivered on December 8, 1978, entitled "Keilin's Respiratory Chain Concept and Its Chemiosmotic Consequences" and in reviews, a few of which are listed in the bibliography.

The background for Mitchell's application of his general chemiosmotic theory to oxidative phosphorylation was provided by earlier studies by other investigators of respiratory metabolism. David Keilin had demonstrated in the 1920s the involvement of several membrane-associated proteins in biological oxidation–reduction reactions, and he coined the term "cytochrome" to describe these colored proteins. The cytochromes directly participated in oxidation–reduction reactions at a transition-metal center of the protein molecule. Other relevant studies were those of Otto Warburg in Germany, who was studying soluble enzymes that catalyze the reduction of molecular oxygen, and research by Heinrich Wieland, who was investigating enzymes that catalyze the transfer of electrons from cellular metabolites to specific electron acceptors. The result of these discoveries, and others, was the development of the concept of the respiratory chain. The respiratory chain hypothesis described the transfer of hydrogen-atom-derived electrons from soluble intracellular metabolic substrates to several oxidized enzyme cofactors, including nicotinamide adenine dinucleotide ($NAD^+$) and flavin adenine dinucleotide (FAD), and the transfer of the hydrogen-atom-derived electrons from the reduced forms of these cofactors, NADH and $FADH_2$, to the membrane-associated cytochrome system. The hydrogen-atom-derived electrons are passed from one membrane-bound carrier to the next, in an energetically favorable manner, and eventually to molecular oxygen, the terminal electron acceptor in the process. Kalckar and others realized that energy derived from the sequential reduction and oxidation of the carriers of the respiratory chain was coupled in some way to provide energy for the otherwise energetically unfavorable synthesis of ATP from adenosine 5′-diphosphate (ADP) and inorganic phosphate ($P_i$). These principles were eventually applied universally to energy-transducing membrane systems within mitochondria, chloroplasts, and in bacteria.

Although the mechanism by which energy from the reactions within the respiratory chain became coupled to ATP synthesis was not understood, most researchers believed that there were several steps along the chain where sufficient free energy was released as a result of an oxidation–reduction reaction to promote the formation of an energy-rich chemical intermediate, which could then provide sufficient energy to promote the synthesis of ATP. However, despite the search for evidence of such an intermediate in many laboratories over a number of years, none could be demonstrated unequivocally. The inability to identify an intermediate led some workers to suggest that the transfer of energy to drive ATP synthesis might involve the energy associated within another type of chemical interaction, perhaps of a conformational, osmotic, or electrical nature.

Mitchell integrated the experimental results and ideas from many laboratories working in the area of bioenergetics, and in 1961 he proposed a hypothesis to explain how energy derived from the oxidation – reduction reactions in energy-transducing membrane systems was coupled to the synthesis of ATP from ADP and $P_i$. The hypothesis did not require the existence of an energy-rich intermediate but instead considered the significance of the localization of electron carriers within the membrane systems; the necessity of an intact closed-membrane system to serve as an electrical and osmotic barrier; and the role of the vectorial movement of electrons, hydrogen atoms, protons, and other ligands in the generation of an electrochemical-potential difference of protons across the membrane.

The essential points of Mitchell's chemiosmotic hypothesis can be briefly summarized as follows. As hydrogen atoms (i.e., electrons and protons) derived from soluble cytoplasmic substrates are transferred to the membrane-associated electron and hydrogen carriers of the respiratory chain, electrons are transferred through the series of carriers. Protons are both utilized and generated during this process in an asymmetric manner with respect to the two sides of the membrane, giving a net result of the vectorial translocation of protons across the membrane. Thus, a protonic potential difference across the membrane is generated that consists of both an electrical potential arising from the asymmetric charge distribution and a thermodynamic potential arising from the proton-concentration differential between the two sides of the membrane. Mitchell proposed that the potential energy difference and corresponding force associated with this asymmetric distribution of protons was sufficient to promote the otherwise energetically unfavorable formation of ATP from ADP and $P_i$. The process is mediated by the membrane-bound enzyme, ATP synthase (also called ATPase, because of the ATP-hydrolyzing activity associated with the isolated enzyme), which serves to catalyze the formation of ATP by using the energy and force associated with the electrochemical potential gradient as protons flow back through a pore in the ATP synthase molecule. In his Nobel lecture, Mitchell made the analogy between the movement of electrons (electricity) and what he refers to as "proticity": "Separate protonmotive redox (or photoredox) and reversible protonmotive ATPase complexes are conceived as being plugged through a topologically closed insulating membrane between two proton-conducting aqueous media at different protonic potential. Thus coupling may occur, not by direct chemical or physical contact between the redox and reversible ATPase systems, but by the flow of proticity around an aqueous circuit connecting them."

Initially, Mitchell's chemiosmotic hypothesis was not accepted, and vigorously opposed, by many individuals working in the area of bioenergetics. However, over the years, experiments by Mitchell's group and others have substantiated most of the predictions made in his original hypothesis. Many of these studies focused on the stoichiometries for the flow of electrons through the chain, the numbers of protons translocated across the membrane, and the numbers of molecules of ATP synthesized from ADP and $P_i$. This work also established the link between the electron-transport-mediated establishment of a protonic and electrical potential and the transport of specific metabolites across the membrane barrier. The research resulted in the recognition of the critical role of the membrane as a physical barrier to the movement of protons, hydroxide ions, and other small ionic species.

Today, the chemiosmotic theory is almost universally accepted, even by most of those who initially were not in agreement. This acceptance was particularly satisfying to Mitchell, and in his Nobel lecture he remarked, "The fact that what began as the chemiosmotic hypothesis has now been acclaimed as the chemiosmotic theory—at the physiological level, even if not at the biochemical level—has therefore astonished and delighted me, particularly because those who were formerly my most capable opponents are still in the prime of their scientific lives."

H. DAVID HUSIC
*Lafayette College*

# Bibliography

*Chemiosmotic Proton Circuits in Biological Membranes.* Skulachev, V. P.; Hinkle, P. C., Eds.; Addison-Wesley: Reading, Mass., 1981. (Volume published in honor of Mitchell's 60th birthday)

Huszagh, V. A.; Infante, J. P. "The Hypothetical Way of Progress"; *Nature* **1989**, *338*, 109.

Mitchell, P. "Bioenergetic Aspects of Unity in Biochemistry: Evolution of the Concept of Ligand Conduction and Chemical, Osmotic, and Chemiosmotic Reaction Mechanisms"; In *Of Oxygen, Fuels, and Living Matter;* Semenza, G., Ed.; John Wiley & Sons: New York, 1981; Pt. 1, pp 1–160.

Mitchell, P. "Compartmentation and Communication in Living Systems. Ligand Conduction: A General Catalytic Principle in Chemical, Osmotic, and Chemiosmotic Reaction Systems"; *Eur. J. Biochem.* **1979**, *95*, 1–20.

Mitchell, P. "Coupling of Phosphorylation to Electron and Hydrogen Transfer by a Chemi-Osmotic Type of Mechanism"; *Nature* **1961**, *191*, 144–148.

Mitchell, P. "David Keilin's Respiratory Chain Concept and Its Chemiosmotic Consequences"; In *Les Prix Nobel 1978;* Odelberg W., Ed.; Almquist and Wiksell International: Stockholm, Sweden, 1979; pp 137–172.

Mitchell, P. "Keilin's Respiratory Chain Concept and Its Chemiosmotic Consequences"; *Science* **1979,** *206,* 1148–1159.

Mitchell, P. "Osmochemistry of Solute Translocation"; *Res. Microbiol.* **1990,** *141,* 277–400.

Mitchell, P. "Osmoenzymology: The Study of Molecular Machines"; In *Cell Function and Differentiation:* Akoyunoglu, G.; Evangelopoulus, A. E.; Georgastos, G.; Palaiologos, G.; Trakatellis, A.; Tsiganos, C. P., Eds.; Alan R. Liss: New York, 1982; pp 399–408.

Mitchell, P. "Possible Protonmotive Osmochemistry in Cytochrome Oxidase"; In *Cytochrome Oxidase: Structure, Function, and Physipathology;* Annals of the New York Academy of Sciences, Vol. 550 (December 31, 1988); pp 185–198.

Mitchell, P. "Realistic Models of Transport Processes"; In *Integration and Control of Metabolic Processes: Pure and Applied Aspects;* Kon, O. L., Ed.; ICSU Press: Miami, FL, 1987; pp 232–245.

"Peter Mitchell"; In *Les Prix Nobel 1978;* Odelberg, W., Ed.; Almquist and Wiksell International: Stockholm, Sweden, 1979; pp 134–136.

# 1979

# Herbert Charles Brown

## 1912–

Herbert Charles Brown, usually called H.C. Brown, was born on May 22, 1912, in London, England. In 1914 his family moved to Chicago, Illinois. In 1926 Brown's father died, so he dropped out of school for three years to help run the family hardware store. Brown enrolled in Crane Junior College in February 1933, intending to become an electrical engineer. However, at Crane he took his first chemistry course. Fascinated by the subject, H. C. Brown committed himself to chemistry as his future career. In 1933 the school closed because of a lack of funds, but Brown was one of a few fortunate students invited by one of his professors at Crane, Dr. Nicholas Cheronis, to do experimental work in his small commercial laboratory, "Synthetical Laboratories," operated in the converted garage of Cheronis's home. Sarah Baylen was another. Wright Junior College opened in 1934, with Cheronis in charge of chemistry, and both Sarah and Herbert enrolled there. They graduated in 1935; according to Brown, Sarah autographed his yearbook, "To a future Nobel Laureate!" Brown won a partial scholarship to the University of Chicago in 1935 and received his B.S. in 1936 by taking ten courses per quarter (at no extra cost). He then worked with H. I. Schlesinger at the University of Chicago, where he entered graduate school in 1936. Brown married Sarah Baylen on February 6, 1937. They have one son, Charles A. Brown, also a chemist.

Brown received his Ph.D. in inorganic chemistry in 1938, then did postdoctoral study with Morris Kharasch at Chicago, and served as instructor of chemistry there until 1943. He moved to Wayne University in Detroit in 1943 to become an assistant professor. In 1947 he made a final move, to Purdue University, first as full professor and then as R. B. Wetherill Professor from 1959 (R. B. Wetherill Research Professor in 1960) until his "retirement" in 1978. He has been the R. B. Wetherill Research Professor Emeritus since 1978 and remains active in research. Brown shared the Nobel Prize in chemistry with the German organic chemist Georg Wittig in 1979. He has written and coauthored over 1100 papers and five books.

H. C. Brown's parents, Charles and Pearl (Gorinstein) Brovarnik had fled Ukraine in 1908 to escape persecution. They settled for five years in Britain in housing provided by the Rothschilds for Jewish refugees of Czarist persecution. When the family moved to Chicago in 1914 their name was "anglicized" to Brown. After attempting to work as a carpenter, Brown's father, who had been a cabinetmaker, went into the hardware business in 1920. The future Nobel Laureate attended public school in Chicago, advancing several times and entering high school at age 12. After graduating from high school in 1930, he held a series of temporary jobs (the hardware store was sold in 1930) before he enrolled in Crane Junior College.

Sarah Baylen did more to influence H. C. Brown's future than just offer encouragement. As a graduation present, she bought him a copy of Albert Stock's book *Hydrides of Boron and Silicon,* evidently because it was the least expensive chemistry book at the University of Chicago's bookstore.[1] Choosing Stock's book proved fortuitous in two ways: It introduced Brown to the type of compounds on which most of the rest of his work in chemistry and his Nobel Prize would rest, and it led him to do graduate work with H. I. Schlesinger at the University of Chicago. Schlesinger and Stock, at the Technische Hochschule in Karlsruhe, Germany, were the only chemists who were working with boron hydrides at the time.

Brown married Sarah Baylen in 1937, against the advice of the renowned University of Chicago chemist Julius Stieglitz, who had encouraged Brown to go on to graduate school and to delay marriage until his education was complete. However, since his graduate assistantship only provided $400 a year and tuition was $300, Sarah's work as a medical chemist at Billings Hospital kept the young couple from

[1]Brown, H. C. "Adventures in Research"; *Chemical and Engineering News* **1981**, *59*, 24–29.

extreme poverty and permitted him to give full attention to his graduate work.

At that time the simplest hydride of hydrogen and boron, diborane, was rare and difficult to make. Diborane is a colorless gas that readily burns in air with a green flame. As his doctoral research, Brown began his lifelong work with the "Green Lady", his chief love after Sarah.[2] For his thesis, Brown studied the reactions of diborane with two common, related classes of organic compounds, aldehydes and ketones. He discovered that diborane readily reacts with these compounds, both of which have a carbon-to-oxygen double bond, and converts them to alcohols. His thesis results were quickly published.[3] However, the method that he had discovered was not adopted by chemists, though conversion of aldehydes and ketones to alcohols was a valuable process. There were several reasons, including the great rarity and expense of diborane, the difficulty of working with a gaseous reagent, and the fact that there were simpler alternatives available. In his later research, Brown overcame every one of these disadvantages by the development of new reagents and methods. He did this so well that the Nobel citation listed one of these very useful new reagents, sodium borohydride, as part of the reason for his being awarded the Nobel Prize.

His professors at Chicago recognized Brown's outstanding abilities. They offered him a postdoctoral fellowship when he could not find industrial employment. About nine months before his thesis work was published, he had already published some work that he had done independently.[4] Brown worked for Morris Kharasch, who first assigned the problem of isolating the active principle in pituitary glands. The assignment was a departure from all his previous work, but Brown has stated, "We were much less specialized then and it was expected that we could adapt to research in any chemical area."[5] Negotiations for a gift of several thousand dollars' worth of pituitary glands proceeded very slowly, so Brown started on other work until the negotiations

[2]Brewster, J. H. "To Live for Wisdom—An Appreciation of Herbert C. Brown"; *Aldrichimica Acta* **1987**, *20* (1), 3–8.

[3]Brown, H. C.; Schlesinger, H. I.; Burg, A. B. "Hydrides of Boron. XI. The Reaction of Diborane with Organic Compounds Containing a Carbonyl Group"; *J. Am. Chem. Soc.* **1939**, *61*, 673–680.

[4]Brown, H. C. "A Convenient Preparation of Volatile Acid Chlorides"; *J. Am. Chem. Soc.* **1938**, *60*, 1325–1328.

[5]Brown, H. C. "Adventures in Research."

were completed. The pituitary project was never carried out by him; he stated that the person who did try it spent three fruitless years of work on it. Brown's other work, dealing with the reactions of certain chlorine-containing compounds in light, convinced Kharasch of Brown's excellence as a chemist, even persuading Kharasch to raise his salary.

After a year of work with Kharasch, Brown was offered an instructorship/research assistantship at Chicago by Schlesinger on the departure of his current assistant, Anton Burg, for the University of Southern California in 1939. He accepted. After beginning some research on solid derivatives of gaseous diborane, known as borohydrides, Brown and Schlesinger became involved in the early stages of government research on methods to purify uranium and separate its isotopes in late 1940. They were asked to prepare volatile uranium compounds without the corrosive properties of $UF_6$ and eventually made uranium borohydride. World War II was in progress, and although their uranium compound never was used, the intermediate, sodium borohydride, proved useful for the generation of hydrogen in portable units on battlefields. To make sodium borohydride in quantity, they were forced to develop improved methods for its production that did not require the difficult-to-make diborane as a starting point. They also discovered that sodium borohydride was unexpectedly stable in water.

In attempts to purify the crude sodium borohydride their reaction produced, Brown and Schlesinger tried acetone, a good solvent that is also a ketone. The sodium borohydride dissolved but could not be recovered from the solvent. They then discovered that the two had reacted, with the ketone converted to an alcohol. Brown had discovered the first of his Nobel Prize-winning reactions, the reduction of aldehydes and ketones with the mild but effective solid reagent sodium borohydride. The reaction can be done in solution with ordinary equipment, uses a stable solid reagent, occurs quickly, and does not produce significant byproducts. This surpassed the diborane reaction of his doctoral days and bettered most competing methods known at the time of reducing aldehydes and ketones to alcohols.

Because the professors were very busy with war research, Brown was allowed to have his own graduate students, despite his instructor status. He began to explore with them the ways in which the spatial arrangement of the atoms in chemical species determines how they will react or even if they can react. At the time, such "steric" explanations were in very low repute. Brown is fond of quoting a popular textbook of the time that stated that "steric hindrance is the last refuge of puzzled organic chemists".[6] With a series of studies over the next decades, Brown enhanced the credibility of reasoning based on steric interactions between reagents.

When Brown determined that he would not be tenured at Chicago, he moved in 1943 to Wayne University, where a friend of Kharasch's, Neil Gordon, had recently become chairman of the chemistry department. Wayne (later Wayne State) was then operated by the city of Detroit and only offered M.S. degrees in chemistry. Little research of any kind went on, but Brown built on his work at Chicago to establish an effective research group that studied steric effects in reactions of a variety of compounds. It was in connection with these studies that Brown later entered some of his greatest conflicts, over what became known as "the nonclassical ion problem". Some chemists explained the various changes in reaction rates observed in processes much like those Brown had studied by exclusively "electronic" explanations, in which the charges produced during these reactions were modified or redistributed over several atoms, or "delocalized". Brown felt that these explanations were not necessary and that they ignored the effects of steric strains in the original molecules that also could enhance the rates of reaction. Frequently citing Occam's razor, he carried out many reaction studies and publicly argued in papers and lectures that complex electronic effects were not needed to explain many observations. The jury is still out on this matter, but most chemists today acknowledge that Brown has done a service by raising important questions about nonclassical ions, and some of the more exuberant advocates have moderated their claims. It should be noted, however, that the Nobel Committee did not mention the matter in its description of Brown's work.

Brown moved to Purdue University in 1947 to expand his research program in an environment more suited to such efforts. He remained there throughout his career, despite offers from other institutions. At Purdue, Brown's research program resumed studies with boron hydrides and their derivatives. He systematically studied the various compounds intermediate in reactivity between his own sodium borohydride and the much stronger reagent later discovered by Schlesinger and Finholt, lithium aluminum hydride. He also investigated the abil-

---

[6]Ray, F. E. *Organic Chemistry*; Lippincott: Philadelphia, PA, 1941, p 522; cited by Brown in "Adventures in Research".

ity of diborane and a related aluminum compound, alane, to reduce organic substances. It was during a study of the reactions of diborane that Brown made his other Nobel discovery.

Brown has been described as "almost religious in resisting facile conclusions and has thereby succeeded in avoiding erroneous ones. . . . 'The main thing,' Brown is fond of saying, 'is to find out exactly what is going on'."[7] This ethic is reflected in the way "hydroboration" was discovered: A graduate student was following Brown's usual exhaustive method of studying a problem; he was using diborane to react with many different compounds of the class known as esters. They are similar to aldehydes and ketones in that they also have carbon-to-oxygen double bonds, but other features of their molecules are different. As with aldehydes and ketones, the double bond of esters is also reduced by diborane, but the student, B. C. Subba Rao, observed an anomaly. One ester used excess diborane. Subba Rao wanted to discard that result, but Brown pressed for an explanation. They determined that the ester in question, which also had a carbon-to-carbon double bond, reacted even more readily at that site. Brown had discovered a new way to form bonds easily from boron to carbon, dubbed hydroboration. Brown soon discovered that the boron-to-carbon bond formed during hydroboration could be quantitatively replaced with a bond from carbon to oxygen, nitrogen, another carbon atom, or other elements. Although the full breadth of the process is still being explored 30 years after the basic reaction's discovery, it has proven to be one of the most useful tools in the synthetic organic chemist's bag of tricks. An important factor that makes this use of diborane much more appealing than Brown's 1939 work is that he developed methods and equipment to make the use of diborane much simpler and safer. Now used in ether solutions, rather than as gas, the transfer of diborane solutions with ordinary syringes is easy.

Brown continues to explore the vast domains of organoboron chemistry that he first entered 50 years ago. He is sometimes compared to an inventor in the mold of Edison, tirelessly exploring every possible way of using what he has discovered. His discoveries have been of immense usefulness to the work of organic chemists, both in terms of practical application and in theory. He has left a lasting im-

---

[7]Brewster, J. H.; Negishi, E.-I. "Brown: Passes Through the Mountains" *Science* **1980**, *207*, 44–46.

pression on more than 300 graduate and postgraduate students, over 50 of whom now hold faculty positions around the world.[8]

Herbert C. Brown has been honored by chemical and scientific societies from all over the world, especially Israel, Japan, Taiwan, Thailand, India, Britain, and, of course, the United States, where he has won almost every award and medal given by the American Chemical Society and other organizations. He has been one of the strongest influences on organic chemistry in this century. His major leisure interest is travel abroad, usually as part of lecture trips, and visits with his son Charles and Charles's family: Dalia, Tamar, and Ronni.

LEROY C. KROLL
*Taylor University*

# *Bibliography*

*Aspects of Mechanism and Organometallic Chemistry*; Brewster, J. H., Ed.; Plenum Press: New York, 1978.

Brown, H. C. *Boranes in Organic Chemistry;* Cornell University Press: Ithaca, NY, 1972.

Brown, H. C. *Herbert C. Brown, A Life in Chemistry*; Department of Chemistry, Purdue University: West Lafayette, IN, 1980.

Brown, H. C. *Herbert Brown's Remarks*; Colonial Society of Massachusetts: Boston, MA, 1975.

Brown, H. C. "The Nonclassical Ion Problem—Twenty Years Later"; *Pure Appl. Chem.* **1982,** *54 (October)*, 1783–1796.

Davenport, D. A. "On Opinion in Good Men: An Oblique Tribute to H. C. Brown"; *Aldrichimica Acta* **1987**, *20(1)*, 25.

Negishi, E.-I. "Profile of Professor H. C. Brown"; *Heterocycles* **1982,** *18 (Special Issue)*, 7–11.

Pelter, A.; Smith, K.; Brown, H. C. *Borane Reagents;* Academic Press: London, 1988.

Worthy, W. "Herbert Brown Wins 1980 Priestley Medal"; *Chem. Eng. News* **1980,** *58 (July 7)*, 21–22.

---

[8]Negishi, E.-I. "Scientific Contributions of Professor Herbert C. Brown", *Journal of Organometallic Chemistry* **1978**, *156* (1), xi–xv.

# 1979

# Georg Wittig

## 1897–1987

Georg Wittig was born on June 16, 1897, in Berlin, Germany, and died on August 26, 1987, in Heidelberg. He was the son of Professor of Fine Arts Gustav Wittig and Martha (Dombrowski) Wittig and showed a lifelong interest in the fine arts, especially piano music. After wavering for a time between chemistry and a career as a classical pianist, Georg Wittig entered the University of Tübingen in 1916 to study chemistry. However, World War I prevented the completion of his studies. He was captured by the British during the war and was held in a prisoner-of-war camp in the United Kingdom. At the war's end, he returned to Germany and resumed his studies, this time under Karl von Auwers at the University of Marburg/Lahn. His Ph.D. was granted in 1923. From 1923 until 1932, Wittig was an instructor (*Dozent*) at the University of Marburg's Chemical Institute. From 1932 to 1937, he served as a special (*ausserplanmässiger*) professor at the Technical Institute in Braunschweig. In 1937 Wittig began seven years as associate professor at the University of Freiburg-Breisgau. He became a full professor of chemistry at Tübingen in 1944 and remained there until a final move in 1956 to Heidelberg, where he served as Director of the Chemical Institute and professor until retirement in 1965. During the first two years of retirement, Wittig still served as codirector of the Chemical Institute. He was no longer very active in chemistry when the announcement came in 1979 that he had been jointly awarded the Nobel Prize in chemistry with fellow synthetic organic chemist Herbert C. Brown of the United States.

Wittig received more than 20 medals and prizes for scientific distinction from chemical and scientific societies in many nations of Europe and the Americas. He was made an honorary fellow of British, Swiss, French, Peruvian, Argentine, and other chemical and scientific societies. Wittig wrote a *Textbook on Stereochemistry* in 1930 and published important research steadily for almost 50 years. In his retirement Wittig returned to his interests in the arts, especially piano and painting, and continued to engage occasionally in mountaineering, another longtime avocation. Wittig married Waltraut Ernst, who also held a Ph.D. in chemistry, in 1930; she died in 1978. They had two daughters.

The Nobel citation of the Swedish Academy described Georg Wittig's researches as having "provided many significant contributions in organic chemistry." Wittig's career spanned a period in which organic chemistry was enriched by many new techniques. Georg Wittig was at the heart of the development of several of these techniques, with the so-called Wittig reaction preeminent among them. The history of his discovery of that reaction and its variations illustrates the underlying methodology and flexibility that he displayed so often in his half-century in chemistry. In an interview reflecting his mountaineering interests, Wittig explained this methodology when his Nobel Prize was announced: "The paths of research rarely lead in straightforward fashion from starting point to desired goal. . . . Although intention predisposes the route, chance or occurrences along the way often enforce a change of course. . . . Along the way, we come upon various points of interest which invite us to linger awhile. Ours, like all such rambling tours, possess that special attraction that comes from knowing that the landscape spread out before us will be opened to view, not by intention, but by chance and surprise".[1]

The general direction that Georg Wittig traveled during his years of research was to lead to a new chemical territory via the route of theoretical interests. One "change of course" followed another as he encountered "various points of interest". Some of Wittig's earliest work, on highly strained molecules, led him along one such route. Wittig had an interest in the extent to which certain structural strains could exist in a molecule without causing it to fall apart. This theoretical interest led him to attempt to make compounds known as diradicals. Some thought these were the probable products when an overly strained

---

[1] "Two Organic Chemists Share Nobel Prize"; *Chemical and Engineering News* **1979**, *57*, 6–7.

molecule fell apart. The pair of electrons that make up a normal bond that broke probably would divide, one electron going to each formerly bonded atom. A series of syntheses of strained molecules performed in the late 1920s did not give stable diradicals (or "diyls" as Wittig called them). Wittig demonstrated his continued interest in this area by reporting the synthesis of another related compound over 40 years later!

These events seem to have led Wittig into a box canyon, but they provided a resource that enabled him later to scale some of his greatest peaks. This resource was the reagent phenyllithium. Benzene ($C_6H_6$) is a simple compound consisting of a planar hexagon of carbon atoms, with one hydrogen atom on each carbon atom. When one of the hydrogen atoms is replaced with a lithium atom, the product is called phenyllithium. This replacement is not usually done in one step; instead, it is done by the replacement of a hydrogen atom with a halogen atom (bromine, chlorine, or iodine). A lithium atom then replaces that new atom by reaction with lithium metal. Phenyllithium was first used by Wittig in 1931; he called it his "divining rod," according to British chemist Robert Shaw.[2] Phenyllithium has two often-related properties that organic chemists prize and that Wittig used very effectively. It is very basic, so basic that it cannot be used in water or related solvents and must instead be used in nearly inert hydrocarbon or ether solvents. Phenyllithium is a good nucleophile, meaning that it tends to react with any somewhat positively charged region in another molecule.

Wittig first used phenyllithium in attempts to synthesize other compounds needed for his pursuit of diradicals from strained molecules. His idea was to use the nucleophilic properties of phenyllithium to displace a bromine atom and form a new carbon-to-carbon bond. Instead he found that phenyllithium acted as a base, removing a hydrocarbon atom from the other molecule, not the bromine atom. Use of chlorine or iodine for bromine did not alter the result, but when he used the remaining halogen, fluorine, he opened a new door. In contrast to its chemical cousins, fluorine was quickly removed. Studies of the products and related substances led Wittig to the conclusion that he had formed what would be a far *more* strained molecule than he had originally intended in the earlier, planned work and one that also acted something like a diradical. The fluorine atom left, and so did an adjacent hydrogen atom. This new compound was called "de-

---

[2]Shaw, R. A. "Georg Wittig—Virtuoso of Chemical Synthesis"; *Nature* **1979**, *282*, 231–232.

hydrobenzene" by Wittig and "benzyne" by others.[3] Recent studies agree that the actual dehydrobenzene molecule is not very different in shape from benzene's regular hexagon, but some distortions do occur. The diradical character of dehydrobenzene is not very evident in most of its reactions; however, the best description of its structure uses a diradical model. The name *benzyne* implies that there is a triple bond between the two reacted carbons of the ring; this fits the reactions of the molecule but not its structure.

More important than the diradical or triple bond character of Wittig's dehydrobenzene is that this very reactive, easily formed, simple molecule set off a whole series of explorations into uncharted territories by other chemists. What other unusual molecules awaited discovery and study? Since then, many other simple but reactive molecules have been thought of, made, and studied by chemists trying to better discern the limits of their science.

The key discovery made by Wittig, for which he was honored by chemists when they named a reaction after him and by the Nobel committee when they gave him the prize, began with Wittig's study of an idea of exclusively theoretical interest. Yet the Wittig reaction and its close relatives may be one of the most useful synthetic conversions ever developed. The discovery is an interesting mixture of planning, accident, use of past discoveries, and fortunate ignorance of other past discoveries.

With the divining rod of phenyllithium, Wittig sought to prepare compounds having five atoms bonded to nitrogen. To do this, he reacted phenyllithium with salts in which a positively charged nitrogen was bound to four carbon atoms. He hypothesized that phenyllithium's strongly nucleophilic nature would overcome the known resistance of nitrogen to forming five bonds and that its negative phenyl group would attach to the positive nitrogen. What he discovered was that phenyllithium's other major property, its strongly basic character, caused it to remove a hydrogen atom from one of the four carbon atoms already attached to nitrogen, removing it as a positive hydrogen ion. This produced a new type of compound that Wittig termed an "ylide" because it had a dual bonding nature. A shared-electron-pair bond existed between nitrogen and carbon, as between a pair of "-yl" groups. The compound also had a second bonding force between the

[3] Wittig, G. *Naturwissenschaften* **1942**, *30*, 696–703.

same nitrogen and carbon. This force was due to the positive charge on nitrogen and a negative charge on the carbon, produced when phenyllithium had removed a hydrogen ion. This second force is of the type found in salts and other ionic compounds, named with an "-ide" suffix. Combining the two suffixes for each type of bonding gives "ylide". Ylides of nitrogen were of some interest theoretically but were hard to form and not very stable. When Wittig turned his attention to the analogues made using the element phosphorus, he found that they were much easier to form and were more stable. Phosphorus ylides are more stable because the ability of phosphorus to form five bonds reduces the destabilizing ionic character of the second (-ide) force. Nitrogen steadfastly refuses to do any similar stabilization. The positive charge on phosphorus and the negative charge on the attached carbon atom are reduced to more moderate levels, because the two electrons that make up the negative charge on carbon form a second shared-electron bond from carbon to phosphorus.

Further studies with the phosphorus ylides led Wittig to the discovery that when these ylides react with aldehydes or ketones, common organic substances having carbon-to-oxygen double bonds, the carbon–oxygen double bond of the aldehyde or ketone is replaced by a new carbon–carbon double bond.[4] In effect, the phosphorus–carbon double bond of the ylide switches partners with the carbon–oxygen double bond of the aldehyde or ketone. The phosphorus–oxygen and carbon–carbon doubly-bonded compounds thereby formed are more stable than the initial phosphorus–carbon and carbon–oxygen compounds. This reaction is the Wittig reaction.

The Wittig reaction proved to be of immense practicality to organic chemists seeking to produce a variety of compounds containing carbon–carbon double bonds. This type of bond is common in nature but had previously been difficult to form simply. Now chemists had a tool that allowed them to convert readily available aldehydes and ketones into new compounds or duplicates of rare natural compounds. Since its discovery, chemists have used the Wittig reaction and several closely related processes to make β-carotene, progesterone and other steroids, juvenile hormones and pheromones of many insects, some prostaglandins, vitamin $D_2$, vitamin $A_1$ (done on a multiple-ton-per-year scale by BASF using the Wittig reaction), and many other substances.

---

[4] Wittig, G.; Schöllkopf, U. *Chemische Berichte* **1954**, *87*, 1318–1330.

The balance of factors that led Wittig to the Wittig reaction is interesting.[5] Wittig made his discovery at the right time. Phosphorus ylides were actually discovered in 1919 by students of Hermann Staudinger. They also observed the reaction with aldehydes and ketones but did not see that reaction as useful and published little about it. Wittig wrote an interesting article on his having followed (unknowingly) in Staudinger's footsteps.[6] Kröhnke discovered the basic reaction of ylides with aldehydes again but described it only in a dissertation of 1937. Finally, Carl Marvel of the University of Illinois, who was aware of the work of Staudinger, had also reacted phenyllithium with positively charged nitrogen and phosphorus compounds in the hope of forming five bonds to either atom in the mid-1920s. It is likely that his results would have directed Wittig away from the reactions that he later used had Wittig known about them, because Marvel did not report success using this method with the same kind of compounds. Alternate substances that could have been tried would not have formed ylides, but they would have formed the five-bonded phosphorus compounds that Wittig originally sought. Fortunately, Wittig did not know of Marvel's work, and he did discover ylides, spending much of the next decade exploring their usefulness as reagents. In the years after World War II a rapid expansion of synthetic organic chemistry occurred, and Wittig's reagents became an important part of that expansion. As is so often the case, the right mixture of timing, knowledge, ignorance, and thoughtful analysis of unexpected results produced a significant discovery.

Wittig engaged in studies of theoretical aspects of chemistry through the remaining years of his active research, studying the reactive substances known as carbenes (again synthesized with phenyllithium). Carbenes are akin to dehydrobenzene in that they are another example of simple compounds that previous experience would not have predicted. In carbenes, as with dehydrobenzene, there are two fewer atoms on the molecule than expected. However, carbenes are simpler in that the two fewer atoms are missing from the same carbon atom. Many carbenes contain only one carbon atom and two other atoms. They are highly reactive and like dehydrobenzene are detectable most often from the products they form by reacting immedi-

[5]Vedejs, E. "Wittig: Fortune Favors the Prepared Mind"; *Science* **1980**, *207*, 42–44.

[6]Wittig, G. "Variations on a Theme of Staudinger: A Contribution to the Story of Organophosphorus Carbonyl-Olefination"; *Pure and Applied Chemistry* **1964**, *9*, 245–254.

ately with substances present in the same solution. Wittig also synthesized his originally intended five-bonded phosphorus compounds. He extended this exploration to analogous compounds. Studies of substances formed by making additional bonds to elements like boron led Wittig to one of the few contacts that he had directly with the research of his fellow prize winner, H.C. Brown. In a 1948 paper, "On Boron-Alkalimetallo-Organic Complex Compounds", Wittig refers to the postdoctoral work done by Brown in Chicago and published in 1940. [7]

Wittig's students and colleagues held him in high regard. Robert Shaw ended his *Nature* article on Wittig with a reflection on the working atmosphere at Wittig's Chemical Institute at Heidelberg: "I well recall the kindness with which he later received me and many other young scientists at his Institute in Heidelberg and his and his wife's generous hospitality. His Institute mirrored his personality, courteous and always kind, and was not one where continuous internecine warfare was current: his kindness made him an exceptionally effective teacher to those fortunate enough to have been his pupils."

LEROY C. KROLL
*Taylor University*

# Bibliography

Bestman, H. J. *Wittig Chemistry*; Springer-Verlag: Berlin, 1983.

*Organophosphorus Reagents in Organic Synthesis;* Cadogan, J. I. G., Ed.; Academic Press: Orlando, FL, 1980.

Furhop, J.; Penzlin, G. "Formation of Alkenes and Alkynes"; In *Organic Synthesis: Concepts, Methods, Starting Materials;* Verlag Chemie: Deerfield Beach, Florida, 1983; pp 26–35.

Hoffman, R. *Dehydrobenzene and Cycloalkynes;* Academic Press: Orlando, FL, 1967.

Shaw, Robert. "Georg Wittig: Virtuoso of Chemical Synthesis"; *Nature* 1979, *282*, 231–232.

Trippett, S. "The Wittig Reaction"; *Pure App. Chem.* 1974, *9*, 255–269.

Vedejs, E. "Wittig: Fortune Favors the Prepared Mind"; *Science* 1980, *207*, 42–44.

Wittig, G. "From Diyls over Ylides to My Idyll"; *Accounts of Chemical Research* 1974, *7*, 6–14.

---

[7]Wittig, G.; Rückert, A.; Keicher, G.; Raff, P. *Justus Liebigs Annalen der Chemie* 1949, *563*, 110–126.

# Paul Berg

## 1926–

Paul Berg, one of today's pioneering leaders in the molecular biology of nucleic acids, shared the 1980 Nobel Prize in chemistry (see also Nobel Laureates Walter Gilbert and Frederick Sanger) "for his fundamental studies of the biochemistry of nucleic acids, with particular regard to recombinant-DNA." Berg was born on June 30, 1926, in New York City, the son of Harry Berg and Sarah (Brodsky) Berg. He is currently the Willson Professor of Biochemistry in the Department of Biochemistry at Stanford University School of Medicine, and he is the Director of Stanford's Arnold and Mabel Beckman Center for Molecular and Genetic Medicine. Berg is perhaps most popularly recognized as the father of gene-splicing techniques, which, in conjunction with other important methodologies such as cloning and DNA sequencing, have formed the basis for the development during the 1980s of the new technology of genetic engineering (sometimes called recombinant DNA technology, gene splicing, or biotechnology). Of equal significance are Berg's numerous accomplishments in biochemistry and molecular biology, which, since the 1950s, have consistently been at the leading edge of advances in the field of nucleic acid research.

Berg majored in biochemistry as an undergraduate at Pennsylvania State University. After graduation in 1948 he attended graduate school at Western Reserve University (now Case Western Reserve University), where he received his doctorate in biochemistry in 1952 for studies on the metabolism of one-carbon compounds in the folate coenzyme system. In 1952 he was awarded a two-year U.S. Public Health Service postdoctoral fellowship. During the first year of his fellowship, Berg studied in H. Kalckar's laboratory at the Institute of Cytophysiology in Copenhagen. There, in collaboration with W. K. Joklik, he discovered a new enzyme (nucleoside diphosphokinase) involved in nucleotide metabolism. It is now recognized that both of these enzyme systems, studied by Berg during the early stages of his career, play important roles in the complex biosynthesis of the nucleotide subunits, which become the building blocks for the subsequent synthesis of ribonucleic acid (RNA) and deoxyribonucleic acid (DNA).

In 1953, the same year James D. Watson and Francis H. Crick proposed their double-helix model of DNA, Berg continued his postdoctoral studies as a research fellow in Arthur Kornberg's laboratory in the department of microbiology at Washington University in St. Louis. Previously Kornberg had been studying several nucleotide-containing coenzymes and had discovered that their biosynthesis required a chemical activation step that involved nucleoside triphosphate intermediates (the same general type of polyphosphate nucleosides produced by the enzyme Berg discovered in Copenhagen). Kornberg had begun to consider the intriguing question of how cells synthesize the complex biomolecules, such as DNA and RNA, from their smaller subunit molecules. He reasoned that chemically active precursor molecules such as the nucleoside triphosphates might be combined with each other under the influence of an appropriate enzyme to form the long chains of DNA.

While Kornberg was initiating his experiments on the cell-free synthesis of DNA, Berg began to investigate the mechanism associated with the biosynthesis of another nucleotide coenzyme, acetyl coenzyme A (acetyl CoA). During his investigations, Berg made the important discovery that acetic acid reacts with ATP so that its carboxyl group becomes chemically linked by way of an enzyme-bound intermediate to the nucleotide adenosine monophosphate. This enzyme-bound acyl adenylate intermediate then reacts with an acceptor molecule, CoA, to form acetyl CoA. In other words, the "unreactive" acid can be chemically "activated" by ATP in an enzymatic reaction that transfers or connects the carboxyl group of the acid to CoA (its biological "acceptor" molecule). This mechanism was later shown by a number of investigators to be the first step involved in the metabolic utilization of all fatty acids. Most significantly, a similar mechanism was found to be involved in the biochemical activation of amino acids for protein biosynthesis.

Guided by his success in elucidating the enzymatic mechanism of acyl group activation, Berg demonstrated that other enzymes exist that chemically activate the carboxyl group of amino acids (the building blocks of cellular proteins) by a similar mechanism. Berg reasoned that some type of biological "acceptor" molecule should therefore exist that would be attached to the activated amino acid, just as the CoA acceptor became attached to the activated acetic acid.

In the mid-1950s, while searching for this amino acid "acceptor" molecule, Berg independently discovered the existence of specialized RNA molecules to which the amino acids become attached.[1] These molecules, now known as transfer RNA, were later shown to play an integral role in the processing, or "translation", of the genetic messages contained in the specific nucleic acid sequences (the four-letter nucleic acid alphabet) of DNA and RNA into the corresponding amino acid sequences (the 20-letter protein alphabet) of cellular proteins.

Berg remained at Washington University as an assistant and then associate professor of microbiology until 1959 when he moved to Stanford University as a professor of biochemistry. That same year he received the ACS Award in Biological Chemistry (sponsored by Eli Lilly) for his work on the activation of fatty acids and of amino acids for protein biosynthesis. Following these studies Berg continued to make other important contributions to the understanding of protein synthesis, which included insights into the process of "transcription", the mechanism by which DNA directs RNA synthesis. He demonstrated that this process was controlled by an enzyme, an RNA polymerase, which uses the base sequence in one of the two strands of DNA to direct the order of assembly of RNA nucleotide sequences in "messenger" RNA.[2] This step is fundamental in gene expression because it is the RNA, not the DNA, that is responsible for carrying the genetic message coded by DNA to the cellular ribosomes, where the proteins are synthesized.

During the mid-1960s, Berg focused his studies on understanding the processes of transcription and translation and, in particular, the problem of how these two processes might be regulated. He helped to clarify the important role of transfer RNA as the "adaptor" molecule

---

[1] Berg, P.; Ofengand, E. J. "An Enzymatic Mechanism for Linking Amino Acids to RNA"; *Proc. Natl. Acad. Sci. U.S.* **1958**, *44*, 78–86.

[2] Chamberlin, M.; Berg, P. "Deoxyribonucleic Acid-Directed Synthesis of Ribonucleic Acid by an Enzyme from *Escherichia coli*"; *Proc. Natl. Acad. Sci. U.S.* **1962**, *48*, 81–94.

in protein synthesis. Crick and others proposed an adaptor hypothesis that assumed that amino acids did not interact directly with messenger RNA but were brought together during protein synthesis by another adaptor molecule. Figuratively stated, one hand of the adaptor molecule would shake hands with a specific nucleotide sequence on RNA, while the other hand would deliver the correct amino acid to the growing polymer chain of the protein. Berg helped to clarify the specificity with which the different amino acids are activated and attached to their corresponding transfer RNA molecules. The specificity is of particular importance because it ensures the accuracy with which the genetic information is translated during protein synthesis.

After spending a sabbatical leave (1967–1968) in Renato Dulbecco's laboratory at the Salk Institute, Berg began exploring the possibility that animal tumor viruses could be used as probes to study the organization, expression, and regulation of mammalian genomes. His ideas were inspired by Dulbecco's pioneering discoveries on the interaction between tumor viruses and animal cells, discoveries that demonstrated that certain viruses induce a cancerous state in an infected cell by taking control over the expression of genetic information in the host cell for their own reproduction. The transformation of normal cells, in which growth is controlled, to cancerous cells, which continually grow and divide, had been shown to be a direct result of the incorporation of the virus's DNA into the host cell's DNA. Berg began to wonder if it might be possible to artificially insert foreign genes into the virus DNA and then use the virus as a vehicle, or vector, to incorporate the foreign DNA into cells. In this manner, the mechanisms governing the expression and control of the foreign genes could be studied.

Berg approached this problem by first attempting to artificially combine, or splice, genes from one virus into another. Using a cell-free system, he discovered that DNA from an animal tumor virus could be covalently joined to DNA from the bacterial lambda virus.[3] This remarkable experiment pointed the way to a general methodology that would allow the joining of DNA from any source with that of another, to form hybrid DNA genes (now referred to as recombinant DNA).

Let us consider this landmark gene-splicing experiment in more detail. First, the circular pieces of the simian virus (SV40) DNA were

---

[3]Jackson, D. A.; Symons, R. H.; Berg, P. "Biochemical Method for Inserting New Genetic Information into DNA of Simian Virus 40: Circular SV40 DNA Molecules Containing Lambda Phage Genes and the Galactose Operon of *Escherichia coli*"; *Proc. Natl. Acad. Sci. U.S.* **1972**, *69*, 2904-2909.

isolated from infected cells and were opened by a special "restriction" enzyme, which cuts DNA at specific sites. The enzyme Berg used (*Eco* R1) was discovered in Herbert Boyer's laboratory at the University of California—San Francisco and shown by Berg to cut the SV40 DNA in exactly one place. DNA molecules containing the lambda phage genes were snipped with the same enzyme scissors used on the SV40. The ends of each DNA were modified so that the two DNAs could be combined, or "annealed". Such joined molecules could then be filled in and covalently sealed by a "ligase" enzyme discovered simultaneously by C. Richardson and H. G. Khorana and also by I. R. Lehman. Later it was found that DNA ends created by *Eco* R1 are naturally cohesive, greatly simplifying the problem of making recombinant DNAs.

Although it had been demonstrated previously that DNA molecules could be synthetically joined, Berg and his colleagues were the first to synthetically introduce foreign genes into a mammalian virus, which could then be used as a transforming carrier for introducing the foreign genes into animal cells. The process of gene splicing occurs quite commonly in nature, but Berg was able to demonstrate that this process is a relatively simple and easy procedure to reproduce in the laboratory. In conjunction with the development of cloning techniques pioneered by Stanley Cohen, Herbert Boyer and their colleagues, it became the first of a number of essential steps necessary for the development of the remarkable technology of genetic engineering. From the point of view of basic research, gene splicing opened the door to the many tremendously exciting and unexpected findings about the organization, function, and control of DNA in animal and plant cells that have been made over the past several years by Berg and many other scientists in biochemistry and molecular biology. These advancements have begun the new era of "molecular" medicine: the diagnosis, treatment, and possibly the eventual cure of genetic diseases using recombinant DNA methods, in which defective genes might someday be replaced with their normal counterparts.

The science and applications of recombinant DNA have also opened a Pandora's box of social, legal, and ethical questions, which began in the 1970s, continue today, and most likely will continue for decades to come. Berg and other distinguished scientists took a leadership role in addressing questions of possible hazards in recombinant DNA research at its earliest stages by calling for a voluntary moratorium on such experiments until potential risks were

PAUL BERG

assessed.[4] As chairman of the National Academy of Science's Committee on Recombinant DNA Molecules, Berg was instrumental in organizing an international conference held in Asilomar, California, in 1975. The committee's recommendations from this conference became the basis for strict guidelines governing gene-splicing research, which were published by the National Institutes of Health in 1976. Since then, the NIH guidelines have been revised a number of times, easing the safety restrictions that were initially imposed on researchers in the field. Today, many of the scientists who participated in the Asilomar meeting have different views on both the conference and its impact but would agree that any potential hazards are far less likely than were originally thought possible.[5]

Clearly, Paul Berg's contributions to the development of nucleic acid biochemistry and molecular biology have been, and continue to be, of tremendous importance. His latest experiments are designed to explore the chemistry and biology of mammalian, including human, chromosomes in the hope of providing basic knowledge for

[4]Committee on Recombinant DNA Molecules; "Potential Biohazards of Recombinant DNA Molecules" *Proc. Natl. Acad. Sci. U.S.* **1974**, *71*, 2593–2594.

[5]Miller, J. A. "Lessons from Asilomar"; *Science News* **1985**, *127*, 122–123.

the prevention, management, and cure of hereditary diseases. He is a leader in the scientific community as evidenced not only by the 1980 Nobel Prize but also by the numerous other awards and distinctions granted him by learned organizations and professional societies. He is a distinguished educator, highly regarded by his students and colleagues, and twice recognized with the Kaiser Award for his outstanding teaching at Stanford. As a concerned scientist, educator, and public citizen, he has worked intensely to promote the view that responsible laypersons as well as scientists should become educated regarding the substance of controversial issues that relate not only to the extraordinary science of recombinant DNA but also to other potentially controversial new technologies. Then and only then can decisions be made in the absence of distortion or personal prejudice, decisions that will allow new advances in science to benefit society as a whole.

CHRISTIAN G. REINHARDT
*Rochester Institute of Technology*

# *Bibliography*

Berg, P. "Dissections and Reconstructions of Genes and Chromosomes"; *Science* **1981,** *213,* 296–303; Paul Berg's Nobel lecture, given December 8, 1980, in Stockholm, Sweden.

"Berg, P." In *McGraw-Hill Modern Scientists and Engineers;* Parker, S. P., Ed.; McGraw-Hill: New York, 1980; Vol. 1, pp 82–83.

Berg, P.; Singer, M. *Dealing with Genes: The Language of Heredity;* University Science Books: Mill Valley, CA, 1992.

Freifelder, D. *Readings from Scientific American: Recombinant DNA;* W. H. Freeman: New York, 1978; includes a discussion of the recombinant DNA debate and references.

Kolata, G. B. "The 1980 Nobel Prize in Chemistry"; *Science* **1980,** *210, 887–889.*

Stevens, P.; Price, B. *Genetic Engineering: Prospects for the Future.* Edited and produced by Peter Cochran. HRM Science Division of Human Relations Media: Pleasantville, New York, 1981; a nontechnical audio-visual introduction for students.

Wade, N. *The Ultimate Experiment;* Walker: New York, 1977.

Watson, J. D.; Tooze, J.; Kurtz, D. T. *Recombinant DNA: A Short Course;* W. H. Freeman: New York, 1983; an excellent introduction to recombinant DNA.

Wright, S. "Recombinant DNA Technology and Its Social Transformation, 1972–1982"; *Osiris* second series **1986**, *2*, 303–360.

# 1980

NOBEL LAUREATE

# *Walter Gilbert*

## 1932–

Walter Gilbert received the Nobel Prize in chemistry in 1980, together with Paul Berg and Frederick Sanger. Gilbert and Sanger were honored for their diverse and complementary contributions of methods for the sequencing of DNA. Gilbert's methods, developed in collaboration with Allan Maxam, used well-known chemical reactions to determine the precise sequence of bases in the DNA strands in a known gene (the *lac* operon gene). These methods greatly advanced the understanding of the molecular mechanisms of genetic regulation and evolution.

Born on March 21, 1932, in Boston, Massachusetts, Gilbert graduated from Harvard College in chemistry and physics in 1953. He went for his Ph.D. to Cambridge University on a National Science Foundation (NSF) predoctoral fellowship. There he completed a dissertation in mathematical physics in 1957 with Abdus Salam (Nobel Laureate in physics, 1979). Gilbert returned to the United States on an NSF postdoctoral fellowship with Julian Schwinger (Nobel Laureate in physics, 1965) at Harvard. Following a stint as a junior faculty member in theoretical physics (1957–1964), during which he coauthored eight papers and began research in molecular biology, Gilbert fully switched to molecular biology in 1964 as an associate professor of biophysics. Except for the period 1981–1984, he served as Chief Executive Officer of Biogen N.V. and Chairman of its Supervisory Board of Directors, Gilbert has spent his entire career at Harvard. In 1986 he became

H. H. Timken Professor of Science in the Department of Cellular and Developmental Biology. Gilbert was elected to membership in the U.S. National Academy of Sciences in 1976; holds honorary doctorates of science from the University of Chicago and Columbia, Rochester, and Yeshiva Universities; and is the recipient of prizes from several scientific societies and academies.

Gilbert's first work in molecular biology, at the instigation of his Harvard colleague James D. Watson, included participation in a team collaborating on experiments designed to prove the existence of messenger RNA by methods of pulse labeling. Parallel work at Caltech by Sidney Brenner, François Jacob, and Mathew Meselson produced the conclusive experimental evidence in favor of the concept of messenger RNA.

After three years of work on transfer RNA, protein synthesis, the genetic code, and especially the way that messenger RNA was read by ribosomes, Gilbert began pursuing, together with Benno Muller-Hill, a postdoctoral fellow from Germany, the ambitious goal of isolating the *lac* repressor. Gilbert had been fascinated by the isolation problem ever since he heard Monod and Jacob's elegant theory of the cellular regulation of genetic expression at the Cold Spring Harbor Symposium for Quantitative Biology in 1961. Much like the neutrino in particle physics in 1932, the repressor in molecular biology in 1967 was a theoretical inference that turned into an experimental fact owing to the perseverance and methodological ingenuity of Gilbert and Muller-Hill. Their method combined genetic and physicochemical techniques: They isolated mutants that made a repressor that bound the inducer more tightly than normal. Then they further doubled the normal number of regulator genes in the cell by bacterial mating. They kept the ground bacteria that contained the repressor in a cellophane bag immersed in water containing a radioactive inducer while allowing for equilibrium to be achieved. If the inducer bound to the repressor, the radioactivity within the bag would increase.

It was, however, Gilbert's experience with the *lac* repressor that led him almost by chance to start the new line of work on chemical DNA sequencing, for which he was awarded the Nobel Prize. As he told the story in his Nobel lecture, the repressor acted by preventing the RNA polymerase from copying the *lac* operon genes into RNA. Because the *lac* repressor bound to DNA at a specific region, the operator, he and his colleagues could isolate the DNA of this region and work out the sequence of this small fragment (25 base pairs). Gilbert and his main collaborator for the DNA sequencing work, Allan Maxam, did the DNA sequencing by copying the fragment into short fragments of RNA and sequencing those by using methods developed by Sanger and his colleagues. As part of this work, they also sequenced operator constitutive mutations to determine the polymerase binding site and other elements in the *lac* gene control. The focus of the work became the interaction of various control factors with DNA.

At that time, in early 1975, Gilbert was urged by a visitor to his lab, the Soviet Andrey Mirzabekov, to do experiments on the selective methylation of DNA with the *lac* repressor bound to it. The agent they used, dimethyl sulfate, methylated the guanines uniquely at the N-7 position, which is exposed in the major groove of the DNA double helix, and the agent methylated the adenines in the N-3 position, which is exposed in the minor groove. At Mirzabekov's suggestion, Gilbert used this groove specificity to explore the interaction of the *lac* repressor with the *lac* operator. There was no way to attack the problem directly, so Gilbert and Maxam devised an indirect sequencing method based on the following idea. They conjectured that if they labeled one end of one strand of the DNA fragment with radioactive phosphate they might determine the point of methylation by measuring the distance between the labeled end and the point of breakage (obtained by the hydrolysis of the backbone sugar, relieved of its methylated base by heat). The labeled fragment of DNA was prepared by using restriction enzymes so that only the 5′ end of one of the DNA strands of the fragment bearing the operator carried a radioactive label. If the experiment was done in the presence of the *lac* repressor protein bound to the DNA fragment and if the repressor lay close to the N-7 of a guanine, the "protected" DNA would not be modified at that base and the corresponding fragment would not appear in the analytical pattern. This method, when coupled with differences in base reaction during heating, allowed for an unambiguous discrimination between bands corresponding to fragments differing by one base.

They realized that the technique could determine the position of adenines and guanines along DNA for distance on the order of 40 nucleotides. The next step was to find a reaction that would determine the position of the two other pyrimidine bases, the cytosines and the thymines. First they tried hydrazine, benzaldehyde, and alkali, but the DNA broke without discrimination. Eventually, Maxam found that salt suppressed the reactivity of the thymines only. With this finding, their method of sequencing was conceptually completed. Better results were obtained when hydrazine was substituted with the primary amine aniline, which released the phosphate from the 3' position on the sugar only. The logic behind their chemical method was to divide the attack into two steps. First, a specificity-carrying reagent was used to target one base only. In the second stage, the complete cleavage of the DNA was ensured by nonspecific means. Later modifications of these techniques enabled the reading out of sequences between 200 and 400 bases from the point of labeling. The important part is the linearity of the method, enabling one to construct the restriction map of the gene as one goes, rather than to sequence randomly.

The new techniques were used by a Gilbert student to sequence the gene for the *lac* repressor, a process that further corrected the protein sequence. Other immediate implications of the DNA-sequencing method included a molecular explanation, a transition to a thymine, for the existence of "hot spots", or sites of much higher mutation frequency, in the *lac* repressor gene. Moreover, the use of thymine in DNA was explained by its capacity to suppress the effects of the natural rate of deamination. The accuracy and unambiguous quality of DNA sequencing was further tested by another student, who studied the gene for the ampicillin resistance of *E. coli,* a gene that is carried on a variety of plasmids. Both strands of this 4362 base pair-long plasmid were sequenced, particularly to cope with the problem of compression during mobility of longer DNA fragments on the gel.

The sequencing of DNA revealed an expected gene structure for bacteria, that is, a contiguous series of codons lying on the DNA between an initiation signal and one of the terminator signals. The bacterial gene could thus be understood in terms of a binding site for RNA polymerase and further binding sites for repressor and activator proteins around and under the polymerase. However, DNA sequencing revealed an unexpected structure for vertebrate genes, a revelation that has broad evolutionary implications. Vertebrate genes, for example, those for globin, immunoglobulin, and ovalbumin, which were transferred into bacteria by recombinant DNA techniques and sequenced, consisted of series of codons interrupted by noncoding DNA. The exons, regions of DNA that will be expressed in the mature message, are separated from each other by introns, regions of DNA that lie within the genetic element but whose transcripts will be spliced out of

WALTER GILBERT

the message. Thus, the original transcript of a gene (now thought of as a transcription unit) will undergo a series of splices before being able to function as a mature message in the cytoplasm. Vertebrate genes can have many exons, which are short coding sequences separated by long noncoding sequences. It further appears that the presence of introns correlates with evolutionary position; for example, the simplest eukaryotes, such as yeast, have few introns. Further up the evolutionary ladder the genes are more broken up. The question persists whether the emergence of the intron–exon structure rising to ever greater degrees of complexity reflects higher evolutionary options or rather the loss of preexisting structure as we move down the evolutionary ladder. The possible role of splicing as an adaptation mechanism with a regulatory role has also been considered. It appears that the decisive element of intron function is not their sequence (which drifts rapidly by mutation) but their length, that is, their capacity to move the exons apart along the chromosome. A consequence of the separation of exons by long introns is that the recombination frequency between exons, both legitimate and illegitimate (a recombination event that leads to the fusing of two DNA sequences at a point where there is no matching of sequence), will be higher. This will increase the rate, over evolutionary time, at which the exons, representing parts of the protein structures, will be shuffled and reassorted to make new combinations, resulting in a faster way for evolution to form the final, new (multiple-length) gene. Moreover, recombination within introns provides an immediate way to build polymeric structures out of simpler units. Indeed, the prediction that polymeric structures will be found to have genes in which the intron–exon structure of the primitive unit is repeated, separated again by introns, was borne out.

According to Gilbert, the most striking prediction of the above molecular evolutionary view is that separate elements defined by the exons have some functional significance, that is, that known proteins constitute gene products that were assembled out of previously achieved solutions of the structure–function problem. The far-reaching implications of rapid-sequencing methods and molecular cloning are that the gene product cannot be read directly from the chromosomes by DNA sequencing alone, but investigation must be made on the sequence of the actual protein to learn the intron–exon structure of the gene. And conversely, the DNA sequence in the chromosome will reveal not only the primary structure of proteins but also their functional structure.

PNINA G. ABIR-AM
*Johns Hopkins University*

# Bibliography

Gilbert, W. "Broken Symmetries and Massless Particles"; *Phys. Rev.* **1964,** *12,* 713.

Gilbert, W. "DNA Sequencing and Gene Structure"; *Science* **1981,** *214,* 1305–1312. Nobel lecture delivered on December 8, 1980.

Gilbert, W. "Genes-in-Pieces Revisited"; *Science* **1985,** *228,* 823–824.

Gilbert, W. "Molecular and Biological Characterization of Messenger RNA"; *Cold Spring Harbor Symposia* **1961,** *26,* 111–132.

Gilbert, W. "Why Genes in Pieces?"; *Nature* **1978,** *271,* 4973.

Gilbert, W.; Church, G. "Genomic Sequencing"; *Proc. Natl. Acad. Sci.* **1984,** *81,* 1991–1995.

Gilbert, W.; Maxam, A. "The Nucleotide Sequence of the *lac* Operator"; *Proc. Natl. Acad. Sci.* **1973,** *70,* 3581.

Gilbert, W.; Muller-Hill, B. "Isolation of the *lac*-Repressor"; *Proc. Natl. Acad. Sci.* **1966,** *56,* 1891–1898.

Gilbert, W. Muller-Hill, B. "The *lac*-Operator Is DNA"; *Proc. Natl. Acad. Sci.,* **1967,** *58,* 2415–2421.

Gilbert, W., et al. "Unstable Ribonucleic Acid Revealed by Pulse Labelling of *E. coli*"; *Nature* **1961,** *190,* 581–585.

Hall, S. S. "Biologist in the Boardroom"; *Science* **1985,** *Jan.–Feb.,* 42–50.

Judson, H. F. *The Eighth Day of Creation: The Makers of the Revolution in Biology;* Simon & Schuster: New York, 1979; pp 62, 216, 442, 443, 447, 459, 460, 500, 585–590.

Kolata, G. B. "The 1980 Nobel Prize in Chemistry"; *Science* **1980,** *210,* 887–889.

# 1980

NOBEL LAUREATE

# Frederick Sanger

## 1918—

Frederick Sanger was born on August 13, 1918, in Gloucestershire, England, and after education at Bryanston School he studied at St. John's College, Cambridge, obtaining his B.A. degree in natural sciences in 1939. Next he followed an advanced course in biochemistry, in which he obtained a first-class degree, and then he began research for the Ph.D. degree in the department of biochemistry under the supervision of A. Neuberger. His work leading to the award of the degree in 1943 was concerned with lysine metabolism and proved to be a sound basis for his subsequent work on insulin, for which he was awarded the Nobel Prize in chemistry in 1958. From 1944 to 1951 he held a Beit Memorial Fellowship for Medical Research, and from 1951 until his retirement in 1983 he was a member of the staff of the Medical Research Council. In 1962 the Council opened the new Laboratory for Molecular Biology in Cambridge, and Sanger moved there to continue his studies on nucleic acids, which he had begun following the award of the Nobel Prize for his work on insulin. These studies were brought to successful conclusions, and in 1980 he was again awarded a Nobel Prize in chemistry (together with Paul Berg and Walter Gilbert). He is one of the very few scientists who have been awarded two Nobel Prizes.

Sanger began his work on insulin at an exciting time in the study of protein chemistry. The work had been made possible partly by the preparation of proteins in pure form such that they could be expected to have a unique structure. The protein insulin had become readily available in crystalline form, and A. C. Chibnall, who had recently taken up the chair of biochemistry at Cambridge, suggested to Sanger that he should study the free amino groups in insulin. Chibnall's work had shown that the number of free $\alpha$-amino groups in insulin could be equated with the number of polypeptide chains in the protein molecule, and it had been shown that one such terminal group was phenylalanine.[1] Sanger first developed an improved method for characterizing amino-terminal residues using the reagent 1,2,4-fluorodinitrobenzene, which interacts with free amino groups under mild alkaline conditions to give a relatively stable derivative.[2] Using this method, he confirmed the presence of a free $\alpha$-amino residue attached to phenylalanine and also showed the presence of some attached to glycine. Thus it appeared that insulin had two peptide chains of which phenylalanine and glycine were the respective N-terminal residues. Work in other laboratories had shown considerable variation in values for the molecular weight of insulin, and it was only during the course of Sanger's work that a firm value was obtained corresponding to the presence of two polypeptide chains.

Any attempt to determine the sequence of amino acid residues in the insulin molecule as a whole clearly required the separation and purification of the two polypeptide strands. Because insulin is rich in cystine residues, union of the strands through disulfide bridges seemed the most likely. The fission of the disulfide residues by reduction proved inappropriate because of the reversibility of the process and the formation of polymeric products. Sanger introduced a new technique whereby the disulfide residues were oxidized by performic acid, producing a cysteic acid residue at each side of the bridge.[3] This technique allowed the production of pure phenylalanine and glycine chains in the oxidized form and gave material suitable for meaningful experiments on the sequence of amino acid residues.

The interaction of a polypeptide with dinitrofluorobenzene followed by hydrolysis with acid gives only the nature of the amino-(N)-terminal residue. In order to determine sequences of residues beyond this, peptides were required resulting from partial hydrolysis. A mild

[1]Jensen, H.; Evans, E. A. "Studies on Crystalline Insulin XVIII: The Nature of the Free Amino Groups in Insulin on the Isolation of Phenylalanine and Proline from Crystalline Insulin"; *J. Biol. Chem.* **1935,** 108, 1–9.

[2]Sanger, F. "The Free Amino Groups of Insulin"; *Biochem J.* **1945,** *39,* 507.

[3]Sanger, F. "Fractionation of Oxidized Insulin"; *Biochem. J.* **1949,** *44,* 126.

acid hydrolysis of the dinitrophenyl (DNP) peptide resulted in the isolation of products containing two, three, and a few more amino acid residues, thus giving the initial sequence, but a lack of suitable separation methods prevented this approach from giving more than very limited information. The technique of paper chromatography had been introduced in 1944, and it proved eminently adaptable to Sanger's problem. (The yellow color of DNP derivatives was an added advantage.) The hydrolysis of the polypeptides from insulin by enzymes yielded peptides of varying length. By using acid and enzymes to cause fission at different locations, the determination of the sequences of the resulting small polypeptides was carried out, and, by matching one sequence against another, longer and longer lengths of total polypeptide could be sequenced. This systematic approach led to the full sequences of both the phenylalanine and glycine chains.

The determination of the positions of the cystine $-S-S-$ linkages in the insulin molecule proved to be a difficult task. First, one $-S-S-$ linkage occurs between two cysteine residues in the glycine chain, and conditions for the fission of peptide linkages between these two residues proved difficult to find. The other two $-S-S-$ linkages join pairs of cysteine residues in the two chains, thereby linking them together. An unforeseen difficulty that had to be overcome was that the interchange of cystine-bridging links takes place during the acid hydrolysis of unoxidized insulin. However, by using $-SH$ inhibitors this difficulty was overcome, and the precise location of the $-S-S-$ linkages, including the two that link the two chains, completed the elucidation of the structure of the protein.

These researches differed from many equally noteworthy discoveries in which a single observation, or reaction by hypothesis, led to outstanding results. In the case of insulin, by virtue of the fact that it led to the first elucidation of the structure of a protein, many previously unexpected experimental problems arose. The outstanding success of the achievement depended on a clear enunciation of problems as they arose and on persistence in finding solutions. It showed for the first time that the determination of the chemical structure of a protein is an attainable objective. Because of the peculiar nature of the insulin molecule, it did not lay down methods of general applicability. It did, however, encourage others to tackle other proteins and to solve the problems intrinsic to the elucidation of their structures. Sanger's work was carried out using bovine insulin. Subsequent work showed small differences among insulins from different species that might result in immunological reactions in diabetic patients under treatment with the bovine hormone. This was an incentive very much later to find a way of producing a human insulin using the methods of genetic engineering. It must be pointed out at this stage that the achievement of that objective depended, in part, on Sanger's later researches for which he received his second Nobel Prize in 1980.

The beginning of Sanger's work on the sequencing of nucleic acids coincided approximately with his move to the newly established Laboratory of Molecular Biology. The problems associated with sequencing nucleic acids are different from those met with proteins, because only four monomer units are involved, necessitating longer units to be sequenced so that overlapping segments can be matched. This, in turn, increases the difficulty of separating closely similar oligonucleotides, and Sanger saw that radioactive labeling with $^{32}P$ might hold the answer to this problem. To achieve the high degree of resolution required, high levels of labeling were needed together with the improved separation techniques, including two-dimensional ionophoresis.[4] Nucleases cutting RNA at specific positions were already available, and ribonuclease $T_1$ cutting after guanine nucleotides was used extensively. These methods led to the sequencing of the largest RNA up to that time, namely, 5S RNA.[5] However, before the general approach could be extended to determining the genetic code in m-RNAs, this had been achieved by quite different methods not involving determinations of sequences.

The application to DNA of the experimental methods of radioactive labeling and the ionophoresis of oligonucleotides presented greater difficulties because of the larger molecular weights and, hence, greater complexity of mixtures to be resolved. Experiments designed to overcome these difficulties led, almost accidentally, to what was to be perhaps the biggest step forward in the approach to sequencing nucleic acids.

Not only were efficient methods of separation required but also specific degradative enzymes analogous to ribonuclease $T_1$, used in the work on RNA. Moreover, the naturally occurring available single-stranded polydeoxyribonucleotides of unique structure were too large for the development of the necessary techniques. During the course of attempts to prepare suitable ogliomers synthetically, using DNA polymerase with the single-stranded ØX174 as template, high spe-

---

[4]Sanger, F.; Brownlee, G. G.; Barrell, B. G. "Two-Dimensional Procedure for Radioactive Nucleotides"; *J. Mol. Biol.* **1965,** *13,* 373–398.

[5]Brownlee, G. G.; Sanger, F.; Barrell, B. G. "The Sequence of 5S Ribosomal Ribonucleic Acid"; *J. Mol. Biol.* **1968,** *34,* 379–412.

cific radioactivity in the product was sought by using low concentrations of the highly labeled monomers. This resulted in the cessation of polymerization when all the labeled monomer was used up. Thus oligomers were produced, terminating in the nucleotide *preceding* the one that had been labeled and had been exhausted. Thus, by using each nucleotide in turn as the terminating unit, the distribution of each could be arrived at. A more rapid way of applying the same basic idea was developed as the "plus and minus" method.[6] A still more rapid and adaptable variation, which was the final one adopted, used 2′, 3′-dideoxy nucleoside triphosphates to bring about the termination of synthesis of the growing chain.[7] These methods all require a single-stranded template which, with some DNA viruses, caused no difficulty. However, most DNAs are double-stranded, and the production of single strands required the cloning in a single-stranded virus of DNA fragments produced either by the use of sonication or restriction enzymes. In addition, increasingly efficient methods were required for resolving mixtures of oligonucleotides according to molecular weight, and the method ultimately adopted was ionophoresis on acrylamide gel.

Various techniques were developed with increasing degrees of success, and the dideoxy nucleotide method emerged as the climax of Sanger's work on sequencing polynucleotides. As discussed in connection with the work on insulin, progress in the work came by patient and persistent improvement and by the development of techniques to solve individual problems. The ultimate success is to be measured by the efficiency, speed, and general applicability of the dideoxy method. This could not have been achieved through following a predetermined experimental plan. Thus, the distinction of the award of the Nobel Prize was not only so well deserved for the ultimate outcome, but also as recognition of the importance to science of the careful resolution of problems as they present themselves, as opposed to the single flash of intellectual insight.

**G. R. BARKER**
*The Biochemical Society, London*
*(Deceased)*

---

[6]Sanger, F.; Coulson, A. R. "A Rapid Method for Determining Sequences in DNA by Primed Synthesis with DNA Polymerase"; *J. Mol. Biol.* **1975,** *94,* 441–448.

[7]Sanger, F.; Nicklen, S.; Coulson, A. R. "DNA Sequencing with Chain-Terminating Inhibitors"; *Proc. Natl. Acad. Sci. U.S.A.* **1977,** *74,* 5463–5467.

# Bibliography

Barley, K. "An Appreciation of the Scientific Contributions of Frederick Sanger"; *Chemistry and Industry* **1958,** *December 13 (50),* 1653–1654.

Sanger, F., "The Chemistry of Insulin"; In *Nobel Lectures, Chemistry, 1942–1962;* Elsevier: Amsterdam, The Netherlands, 1964; pp 544–556.

Fraenkel-Conrat, H. "The Chemistry of Proteins and Peptides"; *Ann Rev. Biochem.* **1956,** *25,* 291–330.

Judson, H. F. *The Eighth Day of Creation: Makers of the Revolution in Biology;* Simon & Schuster: New York, 1979.

Kolata, G. B. "The 1980 Nobel Prize in Chemistry"; *Science* **1980,** *210,* 887–889.

Sanger, F., "Sequences, Sequences and Sequences"; *Ann Rev. Biochem.* **1988,** *57,* 1–28.

Sanger, F.; Brownlee, G. G. "Methods for Determining Sequences in RNA"; *Biochem Soc. Symp.* **1970,** *30,* 183–197.

Sanger, F., "Nucleotide Sequences in DNA"; *Proc. Roy. Soc. London B.* **1975,** *191,* 317–333.

Sanger, F., "Determination of Nucleotide Sequences in DNA"; *Science* **1981,** *214,* 1205–1210.

Silverstein, A.; Silverstein, V. *Frederick Sanger; The Man Who Mapped out a Chemical of Life;* John Day: New York, 1969.

Sutton, C. "Genetic Engineers Sweep the Board"; *New Scientist* **1980,** *88,* 217–219.

# *Kenichi Fukui*

## 1918–

Kenichi Fukui, corecipient of the 1981 Nobel Prize in chemistry for "his frontier orbital theory of chemical reactivity", was born on October 4, 1918, in Nara, Japan, the eldest of three sons of Ryokichi Fukui, a foreign trade merchant and factory manager, and Chie Fukui. After receiving a B.A. degree in engineering from the Department of Industrial Chemistry of Kyoto Imperial University in 1941, Fukui took a position with the Japanese Army Fuel Laboratory, where he was involved in the development of synthetic fuels. In 1943 he returned to Kyoto University as a lecturer in the Engineering School's Department of Hydrocarbon Chemistry. There he began work on a doctoral degree, while simultaneously advancing through the academic ranks, receiving his Ph.D. in chemical engineering in 1948 and the rank of full professor in 1951. Fukui remained at Kyoto University for the next 30 years, becoming professor emeritus in 1982 and president of the Kyoto Institute of Technology the same year.

Despite the fact that he had been trained as an engineer and had begun his career as an experimentalist doing applied fuel research, Fukui had developed an interest in quantum chemistry while still a student. Although he was, as a consequence, largely self-taught as a theoretician, he began to cultivate this interest actively after returning to the university. By the early 1950s he had built up a small but active group of theoreticians within the largely experimentally oriented Department of Hydrocarbon Chemistry. In the next three decades Fukui's group of experimentalists and theoreticians would produce more than 450 papers, dealing with such diverse topics as reaction engineering and catalysis, the statistical theory of gelation, the use of inorganic salts in organic synthesis and, most importantly from the standpoint of the Nobel Prize, the electronic theory of organic reactions.

The central thread in Fukui's work on the theory of organic reactivity, like so much in the history of modern organic chemistry, can be traced back to the benzene molecule. Just as the elucidation of the structure of benzene became a central problem for the development of structural organic chemistry in the last half of the nineteenth century, so the understanding of its reactivity would become the central problem for the newly developing discipline of theoretical organic chemistry in the first half of this century. The basic problem was to explain why an attacking molecule or ion (usually called the reagent) chose to react more frequently at certain carbon atoms of an organic molecule (usually called the substrate) than at others and why this frequency pattern changed when the substrate was modified or the attacking reagent was changed.

One of the first to approach this problem using the newly emerging electrical theory of matter, developed by physicists during the last decade of the nineteenth century and the early decades of the twentieth century, was H. S. Fry of the University of Cincinnati. By using a polar bonding model derived from the atomic models of J. J. Thomson and assigning net positive and negative charges to alternate atoms in a manner similar to the modern assignment of formal oxidation numbers, Fry was able, as early as 1908, to rationalize qualitatively the reactivity of various monosubstituted benzene species studied a decade earlier by the British team of A. Crum Brown and J. Gibson. The preferred site of attack in the monosubstituted benzene species was determined electrically, with positive reagents favoring the negative carbon sites (meta to a positive substituent and ortho–para to a negative substituent) and negative reagents favoring the positive carbon sites (with the opposite distributions).

However, the ionic models of Fry and other early pioneers of the electronic theory of organic reactivity, such as K. G. Falk, J. M. Nelson, J. Stieglitz, and A. A. Noyes, were not in keeping with the physical properties of organic molecules. In the 1920s most of Fry's results were translated into the covalent or shared-electron pair bonding model of G. N. Lewis. Particularly prominent in this translation process were the American chemist H. J. Lucas, who showed that the evidence for the alternate polarity hypothesis central to the ionic model was faulty, and the English chemists T. Lowry, R. Robinson, and C. K. Ingold. According to the new approach, differential reactivity was still a function of differential electrical charges on the various atoms of a molecule, though these were now "partial" charges, attributable either to an inherent inequality in the sharing of the bonding electrons (the static effect) or to their induction by the attacking reagent (the dynamic effect). By the early 1930s this had led to a qualitative electronic classification of reagents as radical (odd electron), electrophilic (electron-pair acceptor), and nucleophilic (electron-pair donor), which is still used today, and to a classification of substrate electronic-shift mechanisms as inductive, inductometric, mesomeric, and electrometic.

The 1930s and 1940s saw yet a further modification of these concepts under the impact of ideas imported from the newly developing field of quantum mechanics. Initially, the most important of these ideas was the concept of resonance, developed by L. Pauling and extensively applied to problems of organic reactivity by G. W. Wheland of the University of Chicago. The "English School" of organic chemistry had already recognized that many organic molecules departed from the localized electron-pair models of Lewis by delocalizing, or spreading out, certain of their bonding electrons over more than two adjacent atoms, an effect related to the mesomeric and electromeric electron-shift mechanisms. Pauling pointed out that an analogue to this process could be found in a quantum mechanical calculation procedure known as the valence bond (VB) method and that this analogy implied that so-called mesomeric delocalization—or "resonance", as Pauling called it—should lead to an increase in the stability of the molecule. Pauling and Wheland proceeded to develop a series of rules for qualitatively estimating the relative stabilities of resonance-stabilized species based on the superposition of limiting-case localized electron-pair structures.

When applied to the problem of organic reactivity, resonance theory implied that one should compare the resonance stabilization of each of the various hypothetical transient "supermolecules" or "transition states" produced by weakly attaching the attacking reagent to each of the various kinds of carbon atoms in the substrate. The transition state displaying the greatest amount of resonance stabilization should correspond to the position showing the greatest frequency of attack. This procedure implied the use of a dynamic reactivity index, though in practice, it was usually applied to the isolated substrate itself and was quite successful in qualitatively rationalizing the static electrical partial charges used by the English school to explain the reactivity of the monosubstituted benzene derivatives originally rationalized by Fry. Ultimately, however, the most significant impact of quantum mechanics lay in the introduction of methods for quantitatively calculating the relative reactivity of organic molecules. In this area a second computational method, known as molecular orbital (MO) theory, quickly supplanted the more difficult procedures of VB theory. This method was first extensively applied to benzene and other conjugated hydrocarbons by the German theoretician E. Hückel in 1931. Although he had to use a large number of oversimplifications and estimations, Hückel was able to show that the method was capable of quantitatively mimicking the static partial charge distribution patterns in the monosubstituted benzene derivatives demanded by the older qualitative rationales, a result also confirmed by Pauling and Wheland a few years later.

Up to this point, virtually all of the work on organic reactivity had centered on substituted hydrocarbon substrates containing at least one atom other than carbon or hydrogen and especially on the monosubstituted derivatives of benzene. This "hetero" atom acted as a key or starting point relative to which the partial charges of the carbon atoms in the substrate could be determined. However, with the introduction of VB, MO, and resonance theory, procedures were developed for predicting the relative reactivity of neutral conjugated hydrocarbons containing carbon atoms that were nonequivalent by virtue of symmetry rather than bonding to a hetero atom of some kind. Because simple Hückel MO theory predicted that all of the carbon atoms in these species had the same total electron density, this required the development of relative reactivity indices based on something other than static partial charges. As early as 1938 Pauling had used qualitative resonance theory to predict both enhanced electrophilic and nucleophilic reactivity at the alpha carbons in naphthalene. In the late 1940s and early 1950s French theoreticians used the theory to develop a series of quantitative static reactivity indices for conjugated hydrocarbons based on the use of so-called $\pi$ bond orders and free valence numbers. About the same time, the English theoretician C. A. Coulson showed that similar static indices could be calculated on the basis of MO theory, as well as a new dynamic index called the self-atom polarizability.

Even earlier, in 1942, Wheland had introduced a picture of the transition state for aromatic substitution reactions known as the localization model, in which the resonance energy of the isolated substrate was compared to that of various alternative transition states in which the reacting carbon atom, as well as two $\pi$ electrons (in the case of electrophilic attack), were deleted from the conjugated system. In the case of nucleophilic attack, two additional electrons were added to the shortened conjugated system instead. The localized transition state showing the least resonance destabilization corresponded to the site with the greatest reactivity. By the early 1950s, various modifications of this procedure had added a number of dynamic localization indices for predicting reactivity to the array of already competing static indices.

This was the state of affairs when Fukui published his first paper on organic reactivity in 1952. In it he introduced yet another static reactivity index for predicting the favored site of electrophilic attack on aromatic hydrocarbons. This index was based on an examination of the substrate molecular orbital containing the electrons with the greatest energy (later called the highest occupied molecular orbital, or HOMO). Fukui referred to these as the species' "frontier electrons" and postulated that electrophilic attack would occur at the carbon atom having the greatest frontier electron density, rather than at the atom with the greatest total electron density (which, as we have seen, was the same at all carbon atoms). Although this index worked as well, if not better, than the other competing indices, Fukui could not provide a good theoretical rationale for its use and, as a result, the index was criticized by a number of theoreticians.

Luckily, R. S. Mulliken, one of the founders of MO theory, began to publish a series of papers about this time dealing with the quantum mechanical treatment of charge-transfer complexes or weakly-bonded Lewis acid–base adducts. Lewis acids and bases are essentially the inorganic chemist's equivalent of electrophilic and nucleophilic reagents, and Mulliken was able to show that the structures of the resulting complexes (which could be viewed as weak electrophilic–nucleophilic addition reactions) were the result of maximizing the overlap between the highest occupied MO of the base (the donor orbital) and the lowest unoccupied MO of the acid (the acceptor orbital). This result became known as the "overlap-orientation principle" and inspired Fukui to continue his exploration of the frontier electron index.

In 1954 he and his co-workers produced two more important papers. In the first of these, Fukui extended his method to include nucleophilic and free-radical attack on conjugated hydrocarbons, showing that they could be correlated, respectively, with the substrate position having the greatest virtual electron density for lowest unoccupied MO (or LUMO, as it was later called) and the position having the greatest combined HOMO–LUMO densities. He also extended his substrates and reaction types to include substitution reactions for heteroaromatics and addition reactions for conjugated alkenes and began to speak of frontier orbitals as well as frontier electrons. In the second paper, Fukui used perturbation theory to provide the missing theoretical rationale for his approach. Perturbation theory indicated that, when two species began to react with or perturb each other, stabilizing interactions could result only when the occupied MOs of one species interacted with the unoccupied MOs of the other species. Moreover, the extent of this stabilization depended directly on the overlap between the orbitals and inversely on their energy separation. Because the HOMO and LUMO had the smallest energy separation of all the orbitals on the substrate and reagent, their importance in determining reactivity was explained. The criterion of maximum overlap explained why the favored position of attack corresponded to the maximum in the frontier orbital density.

Throughout the late 1950s and early 1960s, Fukui and his co-workers continued to refine and extend the frontier orbital method. They used various combinations of frontier electron densities to generate new reactivity indices (known as the "delocalizability" and "superdelocalizability" indices), explored new theoretical derivations and, most importantly, continued to apply the method to new classes of reactions and substrates, including polycentric cycloaddition reactions, substitution reactions for saturated alkanes, three-species interactions, catalysis, and photochemical reactions. This work was eventually summarized in a monograph written in 1975, but before its publication two important events happened that would bring Fukui's work to the attention of a much broader range of practicing organic and inorganic chemists.

The first of these events resulted from the publication of a series of short communications by R. B. Woodward and R. Hoffmann in 1965. In these notes Woodward and Hoffmann showed how an increasing number of previously unclassified reactions could be organized into three major groups, which they called electrocyclic, cycloaddition, and sigmatropic. Common to all three classes were the existence of a cyclic transition state and a set of simple rules for predicting the stereochemistry and photothermal "allowedness" of the reactions as a function of the π-electron counts. Even more importantly, Woodward and Hoffmann showed how these rules could be rationalized by using the criterion of positive overlap between various lobes of the species' frontier orbitals. The response of the organic chemistry community to the resulting "principle of the conservation of orbital symmetry", as it became known, was immediate. A deluge of extensions and alternative theoretical derivations appeared in the literature. Most prominent among them were the orbital correlation diagram approach of H. C. Longuet-Higgins and E. W. Abrahamson (also extensively used by Woodward and Hoffmann), the concept of aromatic and antiaromatic transition states developed by M. J. S. Dewar, the concept of Hückel and Möbius transition states developed by H. E. Zimmerman, and the work of R. F. W. Bader and its elaboration by L. Salem and R. G. Pearson. However, it eventually became clear that, in terms of both simplicity and breadth of application (though not necessarily rigor), the frontier orbital method was by far the most attractive.

The second event of importance was the publication of a paper in 1968 by G. Klopman of Case Western Reserve University. Klopman showed that a more extended perturbational treatment of chemical reactions actually resulted in a two-term reactivity index. The first term, which Klopman called the orbital-control term, corresponded to Fukui's approach and pointed to the importance of maximizing the overlap between the frontier orbitals of the reactants. The second term, which Klopman called the charge-control term, pointed to the importance of net partial charges in determining reactivity. The two indices frequently predicted two different points of attack, though a simple consideration of the relative energies of the frontier orbitals of the reagent and substrate allowed one to determine which index was the most important for the reaction in question, with a small energy separation resulting in orbital control and a large separation in charge control. This result could be linked to the observation that two alternative competing reactivity sites were frequently present in organic substrates, corresponding to thermodynamic- versus kinetic-controlled reactivity, and that preference for them varied with the nature of the attacking reagent. Even more importantly, Klopman was able to link his result to the concept of hard and soft Lewis acids and bases (the HSAB principle), which had been developed in the 1960s by R. G. Pearson, and so direct the attention of most inorganic chemists to the concept of frontier orbitals and their implications for chemical reactivity.

A final and more subtle result also came out of the Woodward–Hoffmann–Klopman results. Although the computational superiority of MO theory had been apparent to theoreticians since the early 1950s, most of the early results discussed above, including Fukui's, had little impact outside a small circle of practicing theoretical organic chemists.

Textbooks and most papers by synthetic organic chemists continued to use qualitative resonance theory to explain reactivity. This happened in part because it was not only qualitative and easy to use, but because it had a simple and effective symbolism involving the use of curved arrows (to indicate electron shifts) and conventional Lewis structures. The delocalized orbitals of MO theory, on the other hand, appeared to be too difficult to draw and the computational results of Fukui and others too laborious for a nonspecialist to calculate. The development of orbital symmetry, however, focused attention on a qualitative consideration of the orbital nodal signs and sizes in order to evaluate the relative extent of positive orbital overlap. This, coupled with the introduction of the convention of representing MOs in terms of combinations of their constituent atomic orbitals, by superimposing p-type atomic orbital lobes on the conventional line formula for the underlying sigma bonding framework, seemed to provide the missing qualitative symbolism for MO theory. The further simple identification of the conventional donor electron pair with the HOMO frontier orbital and the conventional acceptor site with the LUMO frontier orbital appears to have completed this translation process. Since the 1970s, MO theory has increasingly become a part of every chemist's conceptual repertoire. This popularity has, in turn, greatly facilitated the acceptance of Fukui's frontier orbital theory of reactivity, which is rapidly attaining a similar status, if one is to judge from the number of both specialist monographs and introductory textbook treatments appearing in recent years.

WILLIAM B. JENSEN
*University of Cincinnati*

# Bibliography

Bykov, G. V. "Historical Sketch of the Electron Theories of Organic Chemistry"; *Chymia* **1965**, *10*, 199–253.

Fleming, I. *Frontier Orbitals and Organic Chemical Reactions;* Wiley-Interscience: New York, 1976.

Fry, H. S. *The Electronic Conception of Valence and the Constitution of Benzene*; Longman & Green: London, 1921.

Fukui, K. "The Role of Frontier Orbitals in Chemical Reactions (Nobel Lecture)"; *Angewandte Chemie: International Edition* **1982**, *21*, 801–809.

Fukui, K. *Theory of Orientation and Stereoselection;* Springer: New York, 1975.

Ingold, C. K. *Structure and Mechanism in Organic Chemistry;* Cornell University Press: Ithaca, NY, 1953.

Mulliken, R. S.; Person, W. B. *Molecular Complexes: A Lecture and Reprint Volume;* Wiley–Interscience: New York, 1969.

Tedder, J. M.; Nechvatal, A. *Pictorial Orbital Theory;* Pitman: London, 1985.

Wheland, G. W. *The Theory of Resonance and Its Application to Organic Chemistry;* Wiley: New York, 1944.

Woodward, R. W.; Hoffmann, R. *The Conservation of Orbital Symmetry;* Verlag Chemie: Weinheim, Germany, 1971.

# Roald Hoffmann

## 1937—

The part of the world where Roald Hoffmann was born in 1937 and where he spent his earliest years was in that part of Poland absorbed by the Soviet Union in the aftermath of the Second World War. Hoffmann, of Jewish parentage, was fortunate to survive those terrible times. He left Poland in 1946 and after an eventful journey arrived in the United States in 1949. He went to Stuyvesant High School in New York City, became a U.S. citizen after graduation, and went to Columbia College, where he majored in chemistry, graduating with a B.A. in 1958. During the summer vacations he worked at the National Bureau of Standards and at Brookhaven National Laboratory. His first two scientific publications date from that time: One deals with the heat of formation of cements and the other with an efficient low-level counting system for carbon isotopes. There was not yet an inkling of the interest in the area of theoretical chemistry that would make him famous and lead to the 1981 Nobel Prize in chemistry.

Hoffmann went to Harvard in 1958 and obtained an M.A. in physics in 1960 and a Ph.D. in chemical physics with Martin Gouterman and William Lipscomb in 1962. The importance of symmetry, which he learned from Gouterman (Hoffmann was his first student), and the beauty of molecular orbital ideas, which came from Lipscomb, strongly molded the way he looked at chemical problems. The initials of the last names of the group of Lipscombites at the time (Hoffmann, Lohr, Stevens, Pitzer) recalled an acronym from an earlier age of molecular quantum mechanics, as did the first letters of the header (Harvard Laboratory for Structural Properties) of their computer output. During this time at Harvard he also did his last piece of experimental research. His synthesis of tetraphenylporphyrin using a bomb went awry, and purple dye spattered all over the basement of the new Conant Laboratory. As Martin Gouterman later remarked, this event "was a gain for the world of theoretical chemistry".

Hoffmann attended the Löwdin summer school in Sweden, where he met his wife, Eva, and also spent nine months with A. S. Davidov at Moscow University, where he worked on the application of second quantization methods to excitons in helical polymers. These were formative years for Hoffmann. At Harvard he worked on the development of "an extended Hückel method" with which to evaluate the orbital energies of molecules. Fellow graduate student L. L. Lohr's interests were in the application of one-electron methods to transition metal systems, and Hoffmann put his efforts into main group and organic compounds. His work included studies of the molecular orbital theory of polyhedral boranes, molecules that Lipscomb had been studying experimentally. The molecular orbital method would later prove to be very profitable for building global theories of the structures and reactivity of molecules.

Encouraged by E. J. Corey, he began to view some of the mechanistic problems of organic chemistry. After finishing his thesis he became a junior fellow at Harvard, an enviable position that allowed him to work on any problem he wanted. At this time his two children, Hillel and Ingrid, were born, and he began his legendary collaboration with R. B. Woodward on the theory of concerted reactions. He learned from Woodward about the vital stimulus that experimentation has on theory. The one-electron ideas with which he had experimented a few years earlier provided just the right kind of theoretical model with which to launch the concept of orbital symmetry conservation.[1] The new concept provided a breathtaking connection between theory and experiment in an area that was ripe for rationalization. In a series of five papers in 1965, before he was 30 years old, the foundations of what we now know as the Woodward–Hoffmann rules were presented to the chemical world.[2] Their impact was so immediate that in the

---

[1]Woodward, R. B.; Hoffmann, R. *The Conservation of Orbital Symmetry;* Verlag Chemie: Weinheim, Germany, 1970.

[2]Woodward R. B.; Hoffmann, R. "Stereochemistry of Electrocyclic Reactions"; *J. Am. Chem. Soc.* **1965**, *87*, 395.

following year they appeared in the undergraduate curriculum at the University of Cambridge, where this biographer was a student.

The general idea was as follows. During a concerted chemical process the movement of the atoms as they change from reactants to products engenders changes in the energy of the molecular orbitals associated with the system. How these orbitals move in energy is strongly controlled by symmetry properties of the reaction pathway. If an occupied reactant orbital is forced to correlate with a high-energy orbital of the product, the process will be of high energy and thus will be forbidden. The attraction of the ideas was that they were so general. The central theme could be applied to a wide spectrum of seemingly unrelated processes.

By 1965 Hoffmann had a university job as an associate professor at Cornell, where he continued to work on organic problems. He looked at the electronic structures of important organic intermediates, such as carbenes, benzynes, carbonium ions, and diradicals. Several important ideas and techniques of great use to chemistry came from this period: Through bond coupling, the use of perturbation theory, and frontier orbital concepts were but three of them. Hoffmann was promoted to professor in 1968; in 1974 he became the John A. Newman Professor of Physical Science. With interruptions for sabbaticals, he has lived in Ithaca, New York, ever since. In 1972 he was made a member of the National Academy of Sciences and in 1981 he won the Nobel Prize in chemistry, jointly with Kenichi Fukui, for his efforts in what he likes to call "applied theoretical chemistry". The title of his address[3] is typical of his philosophy. If one believes in the unity of nature then the division of chemistry into inorganic, organic, organometallic, and solid-state subdisciplines must be artificial. Hoffmann is continually telling us not to be parochial in our endeavors but to look for a broader picture, to build links between different areas of chemistry.

In the early 1970s Earl Muetterties moved to Cornell from DuPont and was instrumental in getting Hoffmann interested in inorganic and organometallic problems. Here was a greater challenge than before. The sheer variety of molecules and structural nuances in this part of chemistry is enormous. What followed was Hoffmann's second great contribution to chemistry: a seminal series of papers that organized in a rigorous theoretical way an enormous variety of structural prob-

---

[3]Hoffmann, R. "Building Bridges between Inorganic and Organic Chemistry"; *Angewandte Chemie (Int. Ed.)* **1982**, *21*, 711.

lems involving the inorganic and organometallic type coordination complexes. Some of these papers united existing ideas with a modern viewpoint. Inorganic chemists have always been more interested in the mysteries of structure and bonding than their organic colleagues, and there was a patchy theoretical inorganic literature, which was usually a mix of crystal field and molecular orbital theory. Hoffmann was a great admirer of the work of Leslie Orgel, a pioneer in inorganic theory, and the influence of Orgel's work may often be seen in papers dating from Hoffmann's earlier "inorganic" days. There is one long, splendid paper written in 1973 and 1974 that set the stage for much of today's inorganic and organometallic theory. The paper was initially submitted to the *Journal of the American Chemical Society*, which declined to publish it on account of its lengthiness. It eventually found a home in *Inorganic Chemistry* and has since been a well-used source paper for inorganic and organometallic chemists alike.[4] What makes these papers quite incredible reading is that they are written by a person who clearly has a comprehensive knowledge of the whole area at his fingertips, in addition to having a real feel for the important problems in the field. Given that Hoffmann only became an "inorganic" chemist in the early 1970s, his achievement is remarkable. His work in this area was recognized in 1982 by the Inorganic Chemistry Award of the American Chemical Society. There is a nice story concerning a visit to England, dating from this time, when his hosts expected to receive an "organic" title for his seminar. There was some consternation when an obviously "inorganic" title was forthcoming, but it was a good way to get publicity for his change in research direction.

One very striking feature of these papers is the style of presentation. Hoffmann started his scientific career at a particularly good time in the development of computers. They had then grown to be sufficiently big and fast for him to do exactly what he needed. How he treated the results, however, was very different from almost all of his contemporaries. His publications are, of course, the result of much number crunching by computer and the generation of tables of eigenvalues and lists of eigenvectors. Few of these numbers are ever published. Instead, his papers — organic, inorganic, and solid-state — are full of extremely useful little pictures and figures, well-conceived and executed, that convey the essence of the orbital ideas behind the

---

[4]Elian, M.; Hoffmann, R. "The Bonding Capabilities of Transition Metal Carbonyl Fragments"; *Inorg. Chem.* **1975**, *14*, 1085.

problem.[5] The eigenvectors have been turned into little shaded balloons, which have now become, because of his efforts, the accepted hieroglyphs of chemistry. One very important observation is that in Hoffmann's case, the use of the computer has never been a substitute for thinking.[6] When he gave the Stieglitz Lecture at the University of Chicago a week before the announcement of his Nobel Prize, Robert Mulliken dubbed him the "Picasso of Chemistry". After reading many of these papers one often finds oneself saying "how simple" or "how obvious", so effective are they at conveying both the outline and the detail of the problem at hand. It was Hoffmann who exploited the use of the fragment approach in these studies to simplify complex chemical problems and to pare them down to their essentials so that we can understand them. When one reads these papers one appreciates Roald Hoffmann's perhaps greatest role, that of a teacher.

Some of the papers are full of very detailed explanations concerning why energy levels go up or down during a bend, twist, or some other deformation. Many of the explanations deal with complex problems, which are sometimes heavy going for the nonspecialist. Nevertheless, from these many "bread and butter", but complete, explorations of the chemistry of series of compounds, Hoffmann was able to derive a much larger picture. It came in the late 1970s and is known as the Isolobal Analogy, a term coined by Joe Lauher. It has its roots in Halpern's observations concerning the similarities in chemistry between main group and transition metal compounds, Dahl's structural observations on clusters, and Hawthorne's analogy between $C_2B_9H_{11}^{2-}$ and $C_5H_5^-$. In its completeness it owes much to the fragment orbital ideas developed earlier in Hoffmann's group and to the polyhedral ideas of Wade and Mingos. (One may note that it is a little surprising, given his talent for immediately appreciating the significance of a wide range of experimental data, that during his thesis research he had missed the electron-counting algorithm for polyhedral species. It was up to Wade some 10 years later to sort this out.) The importance of the Isolobal Analogy lies in its broad application. Hoffmann provided the ability to relate the structures of diverse classes and groups of molecules from all areas of chemistry and to anticipate some of the chemistry of predicted molecules by simple consideration of the orbital structure of the fragments from which they were made.

---

[5]See, for example, Rossi, A. R.; Hoffmann, R. "Transition Metal Pentacoordination"; *Inorg. Chem.* **1975**, *14*, 365.

[6]Hoffmann, R. "Theory in Chemistry"; *Chem. & Eng. News* **1974**, *52(30)*, 32.

In more recent years Hoffmann has started to look at solid-state problems. This is an area more complex than molecular organic or inorganic chemistry and is one of which most chemists are somewhat afraid. The solid-state physicist has long used tight-binding theory, the equivalent of simple Hückel theory, to study a variety of problems, and once again Hoffmann's extended Hückel ideas have come into their own. We are beginning to see in this area the carefully executed studies that have characterized his earlier research, and we can anticipate even greater things to come.

Hoffmann's work, like that of many who are creative, can be difficult to accept for one reason or another. One of the established practices at his group seminars at Cornell was the reading of the juicier, negative referee comments, which seemed to arrive on a regular basis. He felt that the problems associated with the dissemination of new ideas should be shared with his younger colleagues. It is therefore appropriate to close this biography with an extract from a letter written to Hoffmann by his mentor W. N. Lipscomb on the occasion of the award of Hoffmann's Nobel Prize. "You know that to feel the great beauty of Nature, one need not fear either inexactness or criticism. Science moves too fast and as it does, the wonderful feeling of discovery, and the pain of error, drive the braver ones to new pathways and to new syntheses of ideas."

JEREMY K. BURDETT
*University of Chicago*

## *Bibliography*

"Cornell's Roald Hoffmann Wins ACS's Highest Award in Chemistry"; *Chem & Eng. News* **1989,** *June 5.*

"Five Win Nobels for Chemistry, Physics"; *Chem. & Eng. News* **1981,** *Oct. 26.*

Ruthen, R. "Profile: Modest Maverick"; *Scientific American* **1990,** *July,* 32–35.

# Aaron Klug

## 1926–

Aaron Klug was born on August 11, 1926, and his early days were spent in Durban, South Africa. After education at Durban High School he moved to the University of Witwatersand in Johannesburg, where he graduated in 1945 in science with first-class honors. He then moved to Cape Town, where he obtained the degree of M.Sc. in 1946. It was probably this period, followed by two years as junior lecturer, that shaped the outstanding career that was to come. At Cape Town he was particularly influenced by R. W. James, who had formerly been a member of the School of Crystallography at Manchester, England, with Sir Lawrence Bragg (he had graduated at Cambridge in the same year as Bragg). James is still remembered at Manchester by a few contemporaries, not only as a key member of the Bragg team, but as a man of wide interests, having participated as a physicist in the Shackleton Antarctic expedition. He is remembered as a man who would be likely to stimulate young scientists, and it is interesting that Klug's interest in X-ray crystallography has links with the early days of that discipline. It was to be some years before this influence bore fruit, but James's interest in "imaging" may well have initiated the importance, brought out in Klug's later work, of understanding at a deep level the physical basis of a technique. During a period at the Cavendish Laboratory, in Cambridge, England, with Hartree, Klug obtained the Ph.D. degree in 1949 for a study on the phase changes in the solidification of molten steel involving the austenite–pearlite transition. Some of the ideas involved here later had a resonance in Klug's work on tobacco mosaic virus, in which he divided the assembly of the virus into two stages, nucleation (or initiation) and growth (or extension).

Klug's move to Birkbeck College, London, in 1954 brought him in contact with a number of scientists applying physical techniques to biological problems, including Rosalind Franklin, whose work that contributed to the double helical structure of DNA was to be posthumously recognized. Klug remained at Birkbeck College until the opening of the Medical Research Council's Laboratory of Molecular Biology at Cambridge in 1962, where he was to carry out his major work leading to the award of a Nobel Prize in chemistry in 1982 "for his development of crystallographic electron microscopy and his structural elucidation of biologically important nucleic acid–protein complexes."

It is not possible in a brief résumé of his work to describe in detail Klug's influence on the development of knowledge of biological materials: His contributions have used intricate techniques, and he has enunciated theoretical concepts, all of which have focused on the elucidation of very complex structures. Historically, the earliest chemical studies of biological materials were confined to isolated molecules that did not necessarily represent structures present in the live material. These compounds were relatively stable and held together by covalent bonds. Other less stable bonding was first recognized in the proteins and later in the nucleic acids in the structures of which hydrogen bonding plays important roles. Even at that stage, the criticism was often justified that the materials studied in isolation did not properly represent structures present in the living cell. With the advent of differential centrifugation and other techniques of separation, it became possible to isolate relatively uniform preparations of subcellular particles: ribosomes, mitochondria, nuclei, and chromosomes, to name a few. These materials posed problems for a structural investigation of a different magnitude, and Klug made outstanding advances in understanding their structures by bringing to bear on them a variety of experimental methods, notably X-ray diffraction and electron microscopy, backed up by the appropriate biochemistry. He also produced new theoretical ideas based on geometry and on more sophisticated principles concerned with the packing together of macromolecules. His collaborators have been numerous and from diverse backgrounds, but, in addition to his coauthors, he has influenced and stimulated many others who have benefited from his fund of ideas.

Klug's interest in materials of biological origin appears to date from his beginnings as a medical student but was only realized after his move to Birkbeck College in 1954. Early publications were concerned with the structure of plant viruses in which he collaborated with Rosalind Franklin.[1] Later, X-ray diffraction analysis of turnip yellow mosaic virus showed it to possess icosahedral symmetry, as had been shown to be the case earlier by Caspar in tomato bushy stunt virus. In collaboration with Crick, Klug made a notable advance in the Fourier analysis of diffraction patterns by working out the detailed theory of helical diffraction, which has been widely applied.[2] A notable example was the deduction of the helical path of the RNA in the tobacco mosaic virus, on which subject there had been previous disagreement.[3] Klug's move to Cambridge in 1962 coincided with his application of electron microscopy to other viruses, including bacteriophages, as well as the tobacco mosaic virus. In addition to pushing the use of X-ray diffraction and electron microscopy to the limits of their potential, the translation of electron microscopic images quantitatively into three-dimensional molecular models was one of his major contributions. It had formerly not been possible to interpret these images objectively in terms of three-dimensional models, because an image is essentially a two-dimensional projection of the three-dimensional structure. However, as a result of Klug's theoretical approach, it became possible to avoid the time-consuming assumption and testing of structures until agreement was reached.

Klug also made an important contribution to the principles of virus structure. Watson and Crick had proposed that spherical viruses should have the symmetry of one of the cubic point groups and so could have not more than 60 units arranged symmetrically on the surface. Klug and Caspar found a greater number of units and developed the principle of quasi-equivalence to explain the discrepancy.[4] This principle was put forward to help interpret the structures of spherical viruses, but the underlying concept of self-assembly was believed to be of general application to macromolecular complexes. Whether this is so will become clear as different types of complexes are investigated. For example, one of the possible structures for muscle myosin does not agree with the rule of quasi-equivalence, but the true answer remains to be found. Actin bundles in nonmuscle cells do, however, appear to use the principle.

---

[1] Franklin, R. E.; Klug, A. *Biochim. Biophys. Acta* **1956**, *19*, 403.

[2] Klug, A.; Crick, F. H. C.; Wyckoff, H. W. *Acta Cryst.* **1958**, *11*, 199.

[3] Klug, A. Caspar, D. L. D. *Advances in Virus Research* **1960**, *7*, 223.

[4] Caspar, D. L. D.; Klug, A. *Cold Spring Harbor Symp. Quant. Biol.* **1962**, *27*, 1–24.

The phenomenon of self-assembly has been studied by Klug, both experimentally and theoretically, in the case of the tobacco mosaic virus. The protein subunits of this virus are able to assemble themselves in a two-layer disk structure, in which the packing of the subunits differs somewhat from the arrangement in the complete virus containing both RNA and protein. The disk is an obligatory intermediate in the assembly of the virus, which simultaneously fulfills the physical requirement for nucleating the growth of the virus particle and the biological requirement for specific recognition of the viral RNA. Klug's achievement in this problem was to elucidate the process of self-assembly of such partial structures to give the complete virus. This process, together with the solution of the structure of the protein disk by X-ray analysis, yielded a full description of the protein and RNA interaction.[5]

Klug's interest moved more completely to nucleic acid–containing structures, and one of his successes was to solve the structure of phenylalanine tRNA. That this structure agreed with that published almost simultaneously by Alexander Rich's team did not detract from the value of the work, but confirmed the reliability of the methods used in both laboratories. Mammalian mitochondrial tRNA differs from the majority of RNA molecules in containing unusual bases formed after transcription of the DNA. As a result, this RNA contains nonstandard base pairs, and Klug has deduced that unusual physical configurations are present.[6] Moreover it is only in tRNA that stable G–U wobble pairs have been demonstrated by Klug and others.[7]

During recent years a major part of Klug's work has centered on the structures present in chromatin, the material of which the chromosomes are made. This is undoubtedly the biggest challenge he has set himself. The elucidation of the structure of chromatin is far from complete, but Klug has contributed considerably to present knowledge. Some of this work postdates the award of the Nobel Prize.

The detailed structure of chromatin depends on the fine structure of its components, namely DNA and histone proteins. DNA in chromatin adopts the B-configuration, and Klug and others have studied this in detail by using X-ray diffraction patterns. The nucleosome is

[5]Bloomer, A. C.; Champness, J. N.; Bricogne, G.; Staden, R.; Klug, A. *Nature (London)* **1978**, *276*, 362–368.

[6]de Bruijn, M. H. L.; Klug, A. *EMBO J.* **1983**, *2*, 1309–1321.

[7]Ladner, J. E.; Jack, A.; Robertus, J. D.; Brown, R. S.; Rhodes D.; Clark, B. F. C.; Klug, A. *Proc. Nat. Acad. Sci. USA* **1875**, *72*, 4414–4418.

the primary repeating unit of DNA organization in chromatin and was first described by Roger Kornberg.[8] Extensive digestion with micrococcal nuclease releases a nucleosome core consisting of 146 base pairs of DNA surrounding an octamer of histones. Klug and his colleagues crystallized the nucleosome core and, using high-resolution X-ray crystallography, have shown that the DNA forms two superhelical turns surrounding the histone core, allowing numerous interactions with the histones within.[9] Earlier, Klug had examined the DNA-free histone octamer by using electron microscopy. By using the method of image reconstruction referred to above, a three-dimensional density map was produced, from which the general nucleosome structure could be deduced.

Klug has also studied the higher-order structure of chromatin, namely, how the chain of nucleosomes is further folded, by the fifth histone, to produce further compaction required in chromosomes. He and his colleagues have shown that the nucleosome filament is coiled into yet another helix, the "solenoid," thus establishing another level in the hierarchy of DNA packaging.

In recent years Klug has discovered a novel class of protein transcription factors, the "zinc finger" proteins, that consist of tandem repeated minidomains, each stablized by a zinc ion. They appear to act as reading heads for specific recognition of DNA sequences. They occur widely throughout nature.[10]

Klug's work has ranged over a variety of complex natural cellular materials, the most complex being chromatin; although this latter problem is by no means completely settled, his contributions toward its solution are undoubtedly his most noteworthy achievement. His success has come, in every case, by bringing together results using a variety of techniques and, above all, through his ability to match results of experimental work with insight revealing the crucial core of a problem and developing the theoretical basis essential for its resolution.

G. R. BARKER
*The Biochemical Society, London*
*(Deceased)*

---

[8]Kornberg, R. D. "Structure of Chromatin"; *Ann. Rev. Biochem.* **1977,** *46,* 931–954.

[9]Richmond, T. J.; Finch, J. T.; Rushton, B.; Rhodes, D.; Klug, A. *Nature (London)* **1984,** *311,* 532–537.

[10]Klug, A.; Rhodes, D. *Trends Biochem. Sci.* **1987,** *12,* 464–469.

# Bibliography

Caspar, D. L. D.; DeRosier, D. J. "The 1982 Nobel Prize in Chemistry"; *Science* **1982,** *218,* 653–655.

Holmes, K.C. "Aaron Klug — Nobel Prize for Chemistry"; *Trends in Biochemical Sciences* **1983,** *8,* 3–5.

Igo-Kemenes, T.; Horz, W.; Zachau, H. G. "Chromatin"; *Ann. Rev. Biochem.* **1982,** *51,* 89–121.

Singer, B.; Kuśmierck, J. T. "Chemical Mutagenesis"; *Ann. Rev. Biochem.* **1982,** *51,* 655–693.

Stryer, L. *Biochemistry;* W. H. Freeman: New York, 1981; pp 690–694.

Zimmerman. "The Three-Dimensional Structure of DNA"; *Ann. Rev. Biochem.* **1982,** *51,* 395–427.

# Henry Taube

## 1915–

Henry Taube was born on November 30, 1915, in Neudorf, Saskatchewan, Canada. He received his education at the University of Saskatchewan (B.S., 1935 and M.S., 1937) and at the University of California at Berkeley (Ph.D., 1940). He became a naturalized U.S. citizen in 1941. He served as an instructor at Berkeley from 1940 to 1941; as an instructor and assistant professor at Cornell University from 1941 to 1946; as assistant, associate, and full professor at the University of Chicago from 1946 to 1961; and as professor of chemistry at Stanford University from 1961 to the present. He served as department chairman at Chicago from 1955 to 1959 and as chairman at Stanford from 1972 to 1974. Taube's major awards include the Nobel Prize in chemistry in 1983 and the Priestley Medal, the highest award of the American Chemical Society, in 1985. Taube is an inorganic chemist with interests in substitution and electron transfer reactions of coordination compounds and redox reactions of inorganic compounds.

When faced with the task of preparing an undergraduate course on coordination chemistry, Henry Taube was "bored silly."[1] (That was in the post-Werner–pre-Taube era.) He realized how little was known about the differing rates of reactions for various coordination complexes. Was this the event that inspired the 1983 Nobel Prize winner to

---

[1]Milgrom, L.; Anderson, I. "Understanding the Electron"; *New Scientist* **1983**, October 27, 253–254.

lead the resurgence of inorganic chemistry? Henry Taube has been the forerunner in many aspects of modern coordination chemistry, particularly reaction chemistry. His monumental research accomplishments are a fitting complement to the modest, unpretentious scientist.

Henry Taube grew up in the agricultural environment of rural Saskatchewan, where education was highly valued but opportunities did not always come easily. Starting from a small elementary school of only about a dozen students, Taube found an opportunity to attend college and he grasped it eagerly. He received a B.S. degree at the University of Saskatchewan and went on to obtain an M.S. degree under J. W. T. Spinks at the same institution. His master's research involved studies of the photodecomposition of chlorine dioxide in solution. In 1937 Taube traveled to the University of California at Berkeley and began his doctoral studies under Professor William C. Bray. His doctoral research involved a study of redox chemistry of main group compounds, particularly oxidation reactions of ozone and hydrogen peroxide, in aqueous solution.

Taube's professional career began at Cornell in the years from 1941 to 1946. During that time, his research continued in the same areas as during his graduate student investigations, that is, redox reactions of oxygen- and halogen-containing oxidizing agents. (Interestingly, during this period and up until 1950, Taube had very few publications concerning coordination chemistry. However, that would soon change.) In studies of aqueous solutions of chlorate, Taube applied oxygen-atom-labeling studies, using oxygen-18 and radioactive chlorine to study the mechanisms of the redox reactions of oxochlorine species and related molecules in aqueous solutions. In 1955 Taube received the American Chemical Society Award for Nuclear Applications in Chemistry in recognition of his perceptive applications of isotopic labeling techniques.

In the area of the inorganic chemistry of metal ions, the contributions of Alfred Werner, leading to an understanding of the composition and structure of coordination complexes, are well-known. Although a great deal was known about the reactivity of coordination complexes by 1950, it remained for Henry Taube to place reactivity into a unified concept based on the electronic nature of the metal and on the influence of the ligands around the metal center. In many ways, Taube is the leading figure in modern coordination chemistry, because he laid the groundwork for many conceptual advancements and probed various aspects of chemical reactivity with new techniques.

Concepts and techniques developed during Taube's early professional work on main group chemistry extended to his investigations in coordination chemistry as he drifted into that area in the early 1950s. Taube was the first to determine the number of water molecules in the inner coordination sphere of aqueous metal cations by applying oxygen isotopic labeling,[2] NMR chemical shift measurements, and NMR paramagnetic shift techniques. Such studies are fundamental in determining the precise nature of metal ions in solution and in understanding pathways available for reactions.

Henry Taube has made multiple contributions to the concepts of ligand substitution reactions of coordination complexes. In a classic paper in 1952,[3] Taube pointed out a correlation between ligand substitution rates and the electronic configuration of the metal. In that paper, the slow substitution rate of metal complexes with $d^3$ or low-spin $d^4$, $d^5$, or $d^6$ configurations was noted. The classification of metal complexes as inert or labile, for slow or fast ligand exchange systems, respectively, was introduced, which served to highlight the consequences in chemical behavior of the differences in lability to substitution. The importance of electronic configuration in determining transition metal reactivity was an idea promoted by Taube and is a fundamental aspect of any discussion of substitution or redox reaction.

In 1953 Taube published the first of his many papers that define the concepts of the inner sphere and the outer sphere mechanisms of electron transfer between transition metal complexes.[4] This work serves as the classic textbook example defining the concept of inner-sphere electron transfer. The choice of labile chromium(II) and inert pentaamminechlorocobalt(III) as electron-transfer partners was crucial. Simply the formation of pentaaquochlorochromium(III) defined the reaction as inner-sphere. The chloro ligand in the inert cobalt reactant could not have left the cobalt coordination sphere before electron transfer. Neither could it have entered the coordination sphere of the inert chromium(III) after electron transfer. Thus, the only logical conclusion is that the chloride was part of both the cobalt *and* the chromium co-

[2]Hunt, J. P.; Taube, H. "The Exchange of Water between Aqueous Chromic Ion and Solvent"; *J. Chem. Phys.* **1950**, *18*, 757–758.

[3]Taube, H. "Rates and Mechanisms of Substitution in Inorganic Complexes in Solution"; *Chem. Rev.* **1952**, *50*, 69–126

[4]Myers, H.; Rich, R.; Taube, H. "Observations on the Mechanism of Electron Exchange in Solution"; *J. Am. Chem. Soc.* **1953**, *74*, 4118–4119.

ordination spheres at the moment of electron transfer. Thus, the class of electron-transfer reactions occurring in a two-step process involving, first, formation of a ligand bridge between two metal centers and, second, electron transfer, was defined. Henry Taube was instrumental in defining criteria that can be used to distinguish between inner-sphere and outer-sphere reactions.[5] In addition, he critically probed the role of the bridging ligand in inner-sphere electron-transfer reactions.

As analogs of the ligand-bridged intermediates that exist during inner-sphere electron transfer, mixed-valence complexes, in which a ligand permanently binds two metal centers together, provide valuable models for study of electron-transfer reactions. Three classes of mixed valence species were defined by Robin and Day: (a) Class I species, in which valence-isolated metal centers exhibit the characteristic properties of corresponding mononuclear metal centers; (b) Class III species, in which total electron delocalization occurs so that no characteristics of localized valence state metal centers remain; and (c) Class II species, in which localized valence states of the two metals are slightly perturbed by electron delocalization. The Creutz and Taube ion, $\mu$-pyrazinebis[(pentaammine)ruthenium](5+), was the first in a series of iron, ruthenium, and osmium complexes that were synthesized and characterized by Taube's group to probe the theoretical models of the electronic nature of mixed-valence ions.

A marriage of electron-transfer and ligand-substitution concepts exists in several of Taube's studies. Taube reported an oxidative substitution reaction in which chlorine oxidized an iodide ion that was coordinated to a cobalt metal center. In the process, the iodine was released while chloride coordinated to the cobalt center. Several studies by Taube defined the role of a labile metal ion in promoting substitution at an inert metal center through a combination of electron transfer and ligand substitution reactions. These include Pt(II)–Pt(III) and Pt(III)–Pt(IV), Au(II)–Au(III), and Ru(II)–Ru(III) systems.

Taube's research has paved the way not only for our present understanding of the structure and reactivity of classical coordination complexes but has also been the basis of discussion for the behavior of metal ions in biological systems. The binding of biologically active compounds to metal centers and the redox reactions of such species have been studied by Taube, and he has been active in developing

---

[5]Taube, H. *Electron Transfer Reactions of Complex Ions in Solution;* Academic Press: New York, 1970.

a strategy for studying intramolecular electron transfer between metal ions, which is widely applied in biological systems.[6].

Another area where Henry Taube has made significant contribution is molecular nitrogen complexes. Taube was the first to report formation of a nitrogen complex in aqueous solution by direct action of molecular nitrogen with pentaammineaquoruthenium(II). He also reported the first complex containing a bridging dinitrogen. The significance of Taube's contribution in this area goes beyond the report of these firsts. He placed the metal–dinitrogen bonding interaction in perspective. These species were just a few of a whole series in which Taube defined $\pi$-backbonding interaction within classical coordination complexes. The concept was commonplace in organometallic chemistry but largely uncharacterized in traditional complexes before Taube's work.

The wide-ranging contributions of Henry Taube to the advancement of chemistry are apparent from just a casual survey of the areas in which he has published. In addition to major contributions in classical substitution and redox chemistry of transition metal complexes of Cr, Co, Ru, and Os, Taube's publications include papers on (1) main group nonmetal ions and molecules, particularly oxygen compounds and chlorine oxide compounds; (2) organic acids and amino acids; (3) rare earth compounds; (4) main group metal complexes; (5) sulfur ligands in coordination complexes; (6) organometallic compounds of transition metals. The impressive aspect of these studies is the way in which Taube universally applies the fundamental concepts and techniques that he so masterfully developed and refined.

When, just after learning that he had been awarded the Nobel Prize, he was asked by a reporter if he could explain his work in lay terms in a few words, Taube replied "Nope," but commented that when considering that question earlier, he found himself "writing the lectures for the first year in general chemistry."[7] This work serves as the classic textbook example The comment should not be construed as an overinflated view of the value of his own work. Indeed that would be totally out of character for Henry Taube. Rather it reflects the fundamental nature of his contributions to reaction chemistry as well as his approach to understanding chemical systems. Taube would surely advise one to take a fundamental approach, perform the right experiment, and observe carefully both what happens and what

[6]Taube, H.; Isied, S. S. "Rates of Intramolecular Electron Transfer"; *J. Am. Chem. Soc.* **1973**, *95*, 8198–8200.

[7]Milgrom, L.; Anderson, I. "Understanding the Electron"; *J. Am. Chem. Soc.* **1953**, *74*.

does not happen. This concept of scientific investigation was instilled in him by Bray, his Ph.D. mentor. Add to that concept Taube's brilliance and insight, and it is understandable how the current knowledge of inorganic reaction chemistry could owe so much to one man.

The greatness of a scientist is not measured only by the number (over 300) and significance of his publications. An important parameter in the measure of a scientist is the overall impact of his "school" on the scientific community. Taube associates (students, postdocs, visiting faculty, and coauthors) make an impressive list of scientists who have made tremendous contributions in their own right. Of nearly 200 associates, nearly half have served as faculty on university campuses throughout the United States and the world. The Taube school is represented at places such as Cornell, Iowa State, Georgetown, the State University of New York at Stony Brook, Texas Tech, Georgia Tech, Michigan State, Boston, North Carolina, Indiana, MIT, Rice, and Pittsburgh, to name a few. Over a dozen Taube associates are prominent scientists in foreign countries. A similar number hold positions at national research laboratories such as Los Alamos, Brookhaven, and Argonne. Add to this the number of associates in major industrial laboratories and the influence of Henry Taube on the scientific community is indeed significant.

Prior to the conferment of major awards (the Nobel Prize and the Priestley Medal), Henry Taube's work was celebrated and honored in a symposium at a national meeting of the American Chemical Society (in Las Vegas, Nevada, 1982) and in Volume 30 of *Progress in Inorganic Chemistry,* entitled "An Appreciation of Henry Taube". The impact of Henry Taube is reflected in the prominence of the science and the scientists who contributed to these celebrations.

On a personal level, Henry Taube has been characterized as a modest, unpretentious scientist who generously credits others for inspiration and accomplishments. He is greatly respected by his students and colleagues for his guidance and inspiration to approach chemical problems from a critical but rational viewpoint. The success of the Taube school is due not so much to their training in chemistry as to the infectious enthusiasm and creative approach to chemistry that Taube has passed on to his colleagues. Taube associates transmit a sense of high esteem and appreciation for Taube in all references to their professional and personal associations with him.

A reflection of Taube's generosity and respect for others may be best represented by his choice of topics for his Priestley Medal address. On that occasion, he honored William C. Bray, his Ph.D. mentor, four decades after Bray's death. The warmth, inspiration, and generosity of Bray was obviously a strong influence on Taube, and he is surely gratified to earn similar praise from his students.

Taube remains an active scientist at Stanford, continuing to publish on electron-transfer reactions, organometallic chemistry, and various aspects of redox chemistry, including oxygen-containing species similar to those he studied at the start of his career. His immediate family consists of his wife, Mary, two daughters, Linda and Marianna, and two sons, Heinrich and Karl. He enjoys listening to old opera records from his vast collection and is said to have a passion for gardening and sour mash whiskey.

The scientific contributions of Henry Taube have inspired revolutionary advancements in inorganic reaction mechanisms. His direct achievements have laid much of the groundwork, and his associates have built on that to define much of what we know as inorganic reaction chemistry. A simple statement from the nomination papers for the Nobel Prize may be the most appropriate way to end. "Henry Taube founded the modern study of inorganic reaction mechanisms."

JERRY WALSH
*University of North Carolina at Greensboro*

# Bibliography

"An Appreciation of Henry Taube"; Lippard, S. J., Ed.; *Prog. Inorg. Chem.* **1983**, *30*.

Gray, H. B.; Collman, J. P. "The 1983 Nobel Prize in Chemistry"; *Science* **1983**, *222*, 986–987.

Taube, H. "Electron Transfer between Metal Complexes: Retrospective"; *Science* **1984**, *226*, 1028–1036.

Taube, H. *Electron Transfer Reactions of Complex Ions in Solution;* Academic Press: New York, 1970.

Taube, H. "Rates and Mechanisms of Substitution in Inorganic Complexes in Solution"; *Chem. Rev.* **1952**, *50*, 69–126.

# 1984

# Robert Bruce Merrifield

## 1921—

Robert Bruce Merrifield was born on July 15, 1921, in Fort Worth, Texas. He is the son of George E. Merrifield, a furniture salesman, and Lorene (Lucas) Merrifield. Although the family moved frequently, Bruce Merrifield essentially grew up in California. During a five-year residence in Montebello, California, he decided on a career in chemistry as the result of a high school essay assignment on careers. He attended Pasadena Junior College, then completed his bachelor's degree in chemistry at the University of California—Los Angeles (UCLA) in 1943.

After a year of biochemistry work at the Philip R. Park Research Foundation, Merrifield returned to UCLA for graduate work in biochemistry with professor Max Dunn. Merrifield received his Ph.D. in June 1949; his thesis dealt with microbiological assays of certain components of nucleic acids. In the same week he married Elizabeth Furlong. They have six children: five girls and a boy. Immediately after marriage Merrifield went to the Rockefeller Institute for Medical Research (today, Rockefeller University), where he remained throughout his career. He began a long association with Dr. Dilworth Wayne Woolley at Rockefeller, serving first as an assistant biochemist and then as research associate in 1953. Merrifield became an associate professor in 1957, a full professor in 1966, and John D. Rockefeller, Jr., Professor in 1983. He has received many awards and honorary doctorates, including the American Chemical Society Award for Creative Work in Synthetic Organic Chemistry in 1972. He received the Nobel Prize in chemistry in 1984.

Bruce Merrifield is an example of a scientist thoroughly immersed in one field of study who revolutionizes that field by his practical approach to a knotty problem. Merrifield began from his earliest days as a chemist to study some key substances of life: the proteins and nucleic acids. His earliest work as a student at UCLA led him to doctoral studies on methods to find the amount of various nucleic acid building blocks present in biological samples. His work with these same materials continued when he began research with D. W. Woolley at Rockefeller.

Woolley had discovered several vitamins and "antimetabolites", compounds similar to vitamins or other essential substances that block normal metabolism because they are enough like the natural compounds to fool biological systems into trying to use them. At first, Merrifield worked on the isolation and study of some yeast nucleic acid fragments.[1]

In the early 1950s, Merrifield began to work with proteins under Woolley's direction. Proteins are polymers made up of about 20 different amino acids bonded into long chains in a specific order. The average protein contains several hundred amino acid units, and the absence or replacement of even one of them can make the protein nonfunctional. Because most enzymes, many hormones, and many structural and transport molecules in living things are proteins, the consequences of the production of nonfunctional protein can be catastrophic. Most genetic diseases produce their effects owing to the synthesis of one or more such nonfunctional proteins by the organism. Merrifield's first publication in protein chemistry dealt with the similarities of natural protein fragments and a simple synthetic polypeptide, a polymer of amino acids too small to be called a protein.[2] This article shows some of the difficulties plaguing protein chemistry in this period, difficulties later reduced greatly by Merrifield's Nobel Prize-winning research. Because of the difficulty in carefully separating, analyzing, and synthesizing even simple polypeptides, all that Merrifield and Woolley could say about the materials they were comparing was that they must have some structural similarities, but they were not identical in structure, and that certain general features were necessary for the biological activity that they observed.

The following few years were occupied with the painstaking synthesis of several polypeptides by Merrifield using the best techniques of

[1] Merrifield, R. B.; Woolley, D. W. "The Structure and Microbiological Activity of Some Dinucleotides Isolated from Yeast Ribonucleic Acid"; *J. Biol. Chem.* **1952,** *197*, 521–537.

[2] Merrifield, R. B.; Wooley, D. W. "High Strepogenin Potency of Synthetic Oxytocin and of Purified Vasopressin"; *J. Am. Chem. Soc.* **1954,** *76,* 316

the time. The reason for such tedium in the process is the highly specific ordering of amino acids. To combine amino acids together into polypeptides and ultimately proteins, it is necessary to form specific bonds between pairs of amino acids. Each amino acid has an alkaline portion called an amino group and an acidic portion called a carboxylic acid group. To form a bond between two amino acid molecules, the amino group of one molecule must react with the carboxylic acid group of the other amino acid molecule. There are two major complications: the wrong amino group may react with the wrong carboxylic acid group, and one amino acid may react with other molecules of its own type. For example, what if we react some amino acid A molecules with a similar number of amino acid B molecules? Four possible combinations would result, each with about the same abundance (where — indicates the new bond formed during reaction):

(amino A acid) + (amino B acid)→
                    (amino A acid)—(amino B acid)
             and    (amino B acid)—(amino A acid)
             and    (amino A acid)—(amino A acid)
             and    (amino B acid)—(amino B acid)

Clearly this would be a waste of the original amino acids. In addition, the three undesired products would be hard to separate from the one that was desired. By the time of Merrifield's research in the early 1950s, chemists had determined ways to reduce this problem.

The original amino acids would first be "blocked" by reacting the amino group of one amino acid with a removeable reagent that made the amino group unreactive. The carboxylic acid group of the second amino acid would be similarly blocked with a different reagent. Now only one combination is possible (brackets show blocked groups):

([amino] A acid) + (amino B [acid]) →
                    ([amino] A acid)—(amino B [acid])

If the blocking groups can be removed selectively, one can now repeat the sequence, and a series of specific amino acids can be bonded to form a polypeptide with a known, planned sequence of amino acids:

(amino A acid)—(amino B [acid]) + ([amino] C acid) →
             ([amino] C acid)—(amino A acid)—(amino B [acid])

This much had been done by Merrifield's time, but the process of making a synthetic polypeptide was still very slow and gave a low final yield of product. Each amino acid's addition to the growing chain had to be followed by the removal of its blocking group. Isolation of the new polypeptide from the reagents used to promote the attachment of the amino acid followed. It has been estimated that, when carried out under the typical circumstances for such reactions, a simple peptide of 100 amino acids could only be produced in 0.003% yield.[3] Nevertheless, Bruce Merrifield prepared several polypeptides of interest to himself and to those he worked with at Rockefeller. Much like Geiger, whose tedious counting of radioactivity with a visual scintillation counter prompted him to develop the Geiger–Müller counter, the tedium of repeated polypeptide syntheses prompted Merrifield to develop a more automated method to carry out such syntheses.

On May 26, 1959, Merrifield had noted in his laboratory notebook: "There is a need for a rapid, quantitative, automatic method for the synthesis of long chain peptides. A possible approach may be the use of chromatographic columns where the peptide is attached to the polymeric packing and added to by an activated amino acid, followed by the removal of the protecting group and with repetition of the process until the desired peptide is built up. Finally the peptide must be removed from the supporting medium."[4]

In the casual atmosphere of Rockefeller University, Merrifield could suggest his idea for peptide synthesis to Woolley, get the older man's encouragement, and carry out what turned out to be three years of work until he finally developed a workable system. Merrifield developed the basic technique almost entirely using his original idea as a foundation. The most difficult aspect was to develop the necessary supporting medium. Merrifield finally arrived at tiny porous beads of polystyrene, which he made reactive toward amino acids and to which he attached the first amino acid via its carboxylic acid group, with the amino group still blocked, or "protected". The protecting group could then be removed, exposing the amino group to react with a second amino acid's carboxylic acid group, followed by the deprotec-

---

[3]"Nobel Prizes: Merrifield Wins Chemistry Award"; *Chem. Eng. News* **1984,** *62* (October 6), 6.

[4]Merrifield, R. B. Notebook page reproduced in Henahan, J. F. "R. Bruce Merrifield, Designer of Protein-Making Machine"; *Chem. Eng. News* **1971,** *49* (August 2), 22–26.

tion of that second amino acid's amino group and the reaction of it with a third amino acid, and so on. There were two major advantages to this method: the attachment of the growing amino acid chain to a supporting medium that kept it from washing away during purification steps and the removal of other substances present—unreacted amino acids, removed protecting groups, and reagents—by washing the insoluble polystyrene beads. After the addition of the last desired amino acid, the new polypeptide chain could be removed by the use of proper reagents. The finished polypeptide was released in the absence of contaminating smaller peptide fragments.

Through extensive research, Merrifield determined the best conditions to produce nearly perfect attachment of each successive amino acid, to avoid premature detachment from the polystyrene support, and remove all contaminants from the polystyrene beads during the process. He reported his first successes in 1962, publishing them in 1963.[5] The new method was an incredible improvement on the previous techniques, especially for combining more amino acids into a longer chain. For example, Dr. John Stewart, a fellow researcher at Rockefeller, made synthetic variations on a natural nonapeptide (a peptide of nine amino acids) using both Merrifield's and the conventional methods. In the time that he had made three analogues by the older method, he made 50 using Merrifield's method! The yield estimated for a 100 amino acid polypeptide made by Merrifield's method is 61%, compared to the earlier quoted 0.003% for the conventional method.

Having established the method, Merrifield and Stewart then automated the process by 1965. John Stewart, a ham radio enthusiast, proved integral to the development of the automated aspect of the method Merrifield had pioneered, because he had the electrical expertise to help build the equipment. The order of amino acids added, the sequence of reagent and solvent additions, and the final release of the polypeptide became programmable steps. The early machines used a large perforated drum into which pegs could be inserted. As the drum rotated, the pegs would contact a row of electrical switches. Each switch controlled one reagent's addition. By placing a peg in a specific row of perforations, addition of one reagent would occur at a specific time. Newer commercial models have microcomputer controls.

---

[5]Merrifield, R. B. "Solid Phase Synthesis. I. The Synthesis of a Tetrapeptide"; *J. Am. Chem. Soc.* **1963**, *85*, 2149–2154

Since the automated version of the method became available, Merrifield has continued to do research on how to improve the yield and product purity obtained from the solid-phase synthetic technique. In his Nobel lecture, he cited at least three major improvements made over the years and how to avoid some problems associated with incomplete reactions.[6]

Simple variations of the Merrifield method are used to make other biological polymeric substances, including carbohydrates and nucleic acids. Many important biochemical peptides and their analogues have been synthesized by the Merrifield solid-phase technique, including hormones such as vasopressin, insulin, and epidermal growth factor; enzymes such as ribonuclease; and a variety of other materials, such as interferon. Merrifield's 1968 ribonuclease synthesis (achieved independently by a team led by Ralph Hirschmann and Robert Denkewalter of the Merck Sharp & Dohme Research Laboratories) is especially notable because it was the first total synthesis of any enzyme. Even though it will be increasingly possible in the future to make many such substances via recombinant DNA techniques, the synthesis and study of nonnatural analogues and of entirely new compounds will still require that the method developed by Merrifield be used widely.

Bruce Merrifield is often described as modest and self-effacing. In his hometown of Cresskill, New Jersey, he is known as a good neighbor and a Boy Scout leader. His leisure activities of tennis, camping, and hiking combine with an evident love of home and family. Elizabeth Merrifield worked with her husband in his later research on the synthesis of interferon. He has a strongly practical problem-solving nature. When his method was criticized on introduction because of uncertainties in product purity, he said, "The pessimist says that a pure product cannot be isolated and that, even if it were, its purity could not be demonstrated. Since this attitude produces no progress, we prefer the pragmatic approach of being aware of the problem of purity and simply using the best methods currently available during synthesis, isolation, and characterization of the products . . . what cannot be achieved today may seem simple tomorrow."[7] This confidence has been borne out, as shown by this description of the results of Merrifield's technique in *Scientific American:* "The solid phase technique has enabled investigators to elucidate the effect of structure on biological function of many pro-

[6]Merrifield, R. B. "Solid Phase Synthesis"; *Science* **1986**, *232*, 341–347.
[7]Kaiser, E. T. "The 1984 Nobel Prize in Chemistry"; *Science* **1984**, *226*, 1151–1153.

teins, including enzymes, hormones, and antibodies... helped in the development of monoclonal antibodies... [and may help to develop] synthetic vaccines against such viral diseases as influenza, rabies, and hepatitis."[8]

LEROY C. KROLL
*Taylor University*

# Bibliography

Henahan, J. F. "R. Bruce Merrifield, Designer of Protein Making Machine"; *Chem. Eng. News* **1971,** *49 (August 2),* 22–26.

Kaiser, E. T. "The 1984 Nobel Prize in Chemistry"; *Science* **1984,** *226,* 1151–1153.

Merrifield, R. B. "The Automatic Synthesis of Proteins"; *Sci. Am.* **1968,** *218 (March),* 56–74.

Merrifield, R. B. *The Concept and Development of Solid-Phase Peptide Synthesis;* Profiles, Pathways, and Dreams; Seeman, J.I., Ed.; American Chemical Society: Washington, DC, 1993.

Merrifield, R. B. "Solid Phase Synthesis"; *Science* **1986,** *232,* 341–347.

Merrifield, R. B. "Solid Phase Synthesis. I. The Synthesis of a Tetrapeptide"; *J. Am. Chem. Soc.* **1963,** *85,* 2149–2154.

---

[8]"Nobel Prizes: Chemistry"; *Scientific American* **1984,** *251 (December),* 71–72.

# Herbert Aaron Hauptman

## 1917–

Herbert Aaron Hauptman was the corecipient with Jerome Karle of the 1985 Nobel Prize in chemistry for "the development of direct methods for the determination of crystal structures." Born on February 14, 1917, in New York City, Hauptman received his early education there, graduating from Townsend Harris High School. His interest in science and especially mathematics began at an early age, virtually as soon as he had learned to read. Encouraged by his parents to pursue his own interests, Hauptman received a B.S. degree in mathematics from the City College of New York in 1937 (in the same class as Jerome Karle) and a M.S. degree in mathematics from Columbia University in 1939.

From 1940 to 1942 Hauptman was a statistician with the U.S. Census Bureau. After three years of military service as a weather officer and two years as an electronics instructor, Hauptman decided to obtain an advanced degree and seek a research career. In 1947 he joined the staff of the Naval Research Laboratory (NRL) as a physicist–mathematician, and shortly thereafter began his collaboration with Jerome Karle, who had joined the Physical Optics Division of the NRL in the previous year. At the same time, he started in the Ph.D. program at the University of Maryland.

Hauptman and Karle were the first explicitly to apply the nonnegativity criterion of the electron density function to X-ray diffraction data for the elucidation of electron density distributions around free atoms.[1] Hauptman and Karle then applied the nonnegativity criterion to the X-ray diffraction technique for the structure determination of a many-atom system in the unit cell of a crystal.[2] In 1953 they published a monograph that contained "for the first time a set of probabilistic formulas and measures for attacking the phase problem."[3] Over the next eight years Hauptman and Karle developed in a number of other papers their methodology for the determination of crystal structures, but it was more than a decade before crystallographers began to accept the concepts. (For a more complete discussion, see the biographical entry on Jerome Karle.)

Hauptman's collaboration with Karle on the phase problem also suggested the topic for his doctoral dissertation, "An N-Dimensional Euclidean Algorithm", and he received his Ph.D. degree in mathematics in 1955. Hauptman remained at NRL until 1970, where he was head, Mathematical Physics Branch, Solid State Division (1965–1967); acting superintendent, Mathematics and Information Sciences Division (1967–1968); head, Applied Mathematics Branch, Mathematics and Information Sciences Division (1968–1969); and head, Mathematics Staff, Optical Sciences Division (1969–1970). During that time he continued the development of direct methods, including the concepts of structure invariants and seminvariants.[4]

In 1968 Hauptman began a collaboration with Dr. Dorita Norton of the Medical Foundation of Buffalo (MFB) on the application of direct methods to steroids.[5] In 1970 he left the NRL to become the head of the Mathematical Biophysics Laboratory at the MFB, an

---

[1]Hauptman, H.; Karle, J. "The Structure of Atoms from Diffraction Studies"; *Phys. Rev.* **1950**, *77*, 491–499.

[2]Karle, J.; Hauptman, H. "The Phases and Magnitudes of the Structure Factors"; *Acta Crystallogr.* **1950**, *3*, 181–187.

[3]Hauptman, H.; Karle, J. *Solution of the Phase Problem. I. The Centrosymmetric Crystal*; American Crystallographic Association: New York, 1953.

[4]Hauptman, H. "Some New Relationships among the Structure Invariants"; *Acta Crystallogr.* **1963**, *16*, 792–795; Hauptman, H. "On the Equivalence of Structure Invariants"; *Acta Crystallogr.* **1966**, *21*, 816–819.

[5]Cooper, A.; Norton, D. A.; Hauptman, H. "Estrogenic Steroids. III. The Crystal and Molecular Structure of Estriol"; *Acta Crystallogr.* **1969**, *B35*, 814–828.

independent research institute established by a private donation in 1956 to further research in endocrinology and hormone-related disorders. In 1972 he became Research Director and Executive Vice President of the Foundation and was appointed the Foundation's President in 1986. Hauptman has also been a research professor of biophysical sciences at the State University of New York at Buffalo since his move there in 1970.

At Buffalo, Hauptman established a molecular structure group that was primarily concerned with molecules of biological interest, especially steroids.[6] Since 1972 the MFB has conducted a number of international workshops on the practical application of direct methods to crystal structure determination.[7]

Hauptman's major contributions continue to be in the mathematical procedures of the direct methods and include the following: quartet and quintet invariants,[8] the neighborhood concept,[9] and isomorphous replacement and anomalous dispersion.[10] More recently Hauptman has introduced a new concept, the so-called minimal principle, in the attempt to strengthen the traditional techniques of direct methods.[11]

Hauptman is the first mathematician to win a Nobel Prize, and the methodology of Hauptman and Karle, which have revolutionized

[6]Duax, W. L.; Hauptman, H. "The Crystal and Molecular Structure of Aldosterone"; *J. Am. Chem. Soc.* **1972**, *94*, 5467–5471.

[7]Langs, D. A.; Hauptman, H. A. "The Buffalo System for Direct Methods"; In *Proceedings of the International Summer School on Crystallographic Computing, Twente, the Netherlands, July–August 1978;* Schenk, H.; Olthof-Hazekamp, R.; van Konigsveld, H.; Bassi, G. C., Eds.; Delft University Press: Delft, the Netherlands, 1978.

[8]DeTitta, G. T.; Edmonds, J. W.; Langs, D. A.; Hauptman, H. "Use of Negative Quartet Cosine Invariants as a Phasing Figure of Merit: NQUEST"; *Acta Crystallogr.* **1975**, *A31*, 472–479; Hauptman, H. "A New Method in the Probabilistic Theory of the Structure Invariants"; *Acta Crystallogr.* **1975**, *A31*, 680–687; Fortier S.; Hauptman, H. "Quintets: A Joint Probability Distribution of Fifteen Structure Factors"; *Acta Crystallogr.* **1977**, *A33*, 572–575.

[9]Hauptman, H. "A Heuristic Study of the Neighborhoods of the Structure Seminvariants in the Space Group P1"; *Acta Crystallogr.* **1976**, *A32*, 934–940.

[10]Hauptman, H. "On Integrating the Techniques of Direct Methods and Isomorphous Replacement. I. The Theoretical Basis"; *Acta Crystallogr.* **1982**, *A31*, 289–294; Hauptman, H. "On Integrating the Techniques of Direct Methods with Anomalous Dispersion. I. The Theoretical Basis"; *Acta Crystallogr.* **1982**, *A38*, 632–641.

[11]Hauptman, H. "A Minimal Principle and Its Role in Direct Methods"; American Crystallographic Assn. Meeting, 1988, Philadelphia, PA, Abstract R4 and Video Monograph: Nobel Laureates Symposium, Ohio: Polycrystal; Hauptman, H. "A Minimal Principle in the Phase Problem"; *Crystal Computing 5: From Chemistry to Biology;* Moras, D.; Podjarny, A. D.; Thierry, J. D., Eds; Oxford University Press, New York, 1991; pp 324–332; Miller, R.; DeTitta, G.; Jones, R.; Langs, D.; Weeks, C.; Hauptman, H. "On the Application of the Minimal Principle to Solve Unknown Structures"; *Science* **1993**, *259*, 1430–1433.

the analysis of molecular structure, have had an impact throughout all branches of chemistry. In his introduction of Hauptman and Karle at the award ceremonies, Professor Ingvar Lindqvist noted that "the imagination and ingenuity of the laureates have made it unnecessary to exercise these qualities in normal structure determinations. On the other hand, they have increased the possibilities for the chemists to use their imagination and their ingenuity."

JAMES J. BOHNING
*Beckman Center for the History of Chemistry*

# Bibliography

*Direct Methods in Crystallography, Proceedings of the 1976 Intercongress Symposium*; Hauptman, H., Ed.; Medical Foundation of Buffalo: Buffalo, NY, 1978.

Hauptman, H. "The Direct Methods of X-ray Crystallography"; *Science* **1986**, *233*, 178–183.

Hauptman, H. "Direct Methods and Anomalous Dispersion"; *Chemica Scripta* **1986, 26**, 277–286.

Hauptman, H. "The Phase Problem During the Seventies"; In *Crystallography in North America*; McLachlan, D., Jr. and Glusker, J. P., Eds.; American Crystallographic Association: New York, 1983.

Hauptman, H. *Crystal Structure Determination: The Role of the Cosine Seminvariants;* Plenum Press: New York, 1972.

# *Jerome Karle*

## 1918–

Jerome Karle was the corecipient with Herbert A. Hauptman of the 1985 Nobel Prize in chemistry for "their outstanding achievements in the development of direct methods for the determination of crystal structures." Born on June 18, 1918, in New York City, Karle grew up in the Ocean Parkway and Coney Island sections of that city, attending the local public schools. Encouraged by his mother to become a professional pianist, Karle acquiesced partially to her wishes, but he found a much stronger attraction to science. The broad interest generated by reading the popularized accounts of Sir James Jeans and by visits to the *New York Daily News* Science Museum became more focused when Karle entered Abraham Lincoln High School. Challenged by the chemistry and physics courses, he decided to pursue a career in science. (Two other Nobel Laureates, Arthur Kornberg (physiology or medicine, 1959) and Paul Berg (chemistry, 1980) also graduated from Abraham Lincoln.)

Karle's intense interest in a diversity of subjects led to the completion of the high school graduation requirements before his fifteenth birthday, and in the fall of 1933 he began a subway commute from Coney Island to the City College of New York, where the only fee was $0.50 per semester for a library card. Immersing himself in a full complement of chemistry and biology courses, Karle took extra courses in physics and mathematics while meeting the liberal arts component required of all students.

After graduation in 1937 Karle went to Harvard, where he received an M.S. in biology the next year. He then went to work for the New York State Health Department in Albany and developed what later became the standard method for the determination of the amount of fluorine in water supplies. The Albany position enabled Karle to save money for graduate school, and in 1940 he entered the University of Michigan.

In the physical chemistry courses at Michigan the students were assigned laboratory benches alphabetically, and Karle found himself next to Isabella Lugoski. They were married in 1942 and completed the work for their Ph.D.s in 1943 under the direction of Professor Lawrence O. Brockway. Attracted by Brockway's charismatic personality and his research in electron diffraction by gaseous molecules, the Karles investigated several molecular structures while assisting in the improvement of the electron diffraction unit that Brockway designed and assembled with his own hands.[1]

In October of 1943 Karle joined the Manhattan Project, working at the Metallurgical Laboratory at the University of Chicago on the reduction of uranium and plutonium oxides. Isabella joined the group in January of 1944 and worked on the synthesis of plutonium chloride. The Karles returned to the University of Michigan in July of 1944, where Jerome worked on a defense-related Naval Research Laboratory (NRL) project and Isabella was an instructor in the Chemistry Department.

During this time Jerome investigated the structures of monolayer films related to the boundary lubrication of metallic surfaces,[2] and derived a theory that explained the electron diffraction patterns of oriented monolayers.[3]

After World War II ended, the Karles, seeking scientifically and geographically similar employment, found a unique opportunity in the Physical Optics Division of the NRL in Washington, D.C., which they joined in 1946. In the postwar era the NRL was changing rapidly, and the newly developing emphasis on basic research provided the

---

[1]Brockway, L. O.; Karle, J. "An Electron Diffraction Investigation of the Monomers and Dimers of Formic, Acetic, and Trifluoroacetic Acids and the Dimers of Deuterium Acetate"; *J. Am. Chem. Soc.* **1944,** *66,* 574–584.

[2]Brockway, L. O.; Karle, J. "Electron Diffraction Study of Oleophobic Films on Copper, Iron and Aluminum"; *J. Colloid Sci.* **1947,** *2,* 277–287.

[3]Karle, J. "The Scattering of Electrons by Hydrocarbon Films"; *J. Chem. Phys.* **1946,** *14,* 297–305.

ideal environment for the Karles and their research interests. Taking advantage of the NRL's extensive facilities, the Karles constructed a more sophisticated electron diffraction apparatus in order to increase accuracy and observe internal motion, and continued their work on the structure of gaseous molecules.

When a beam of electrons is passed through a gas it is diffracted into a series of concentric rings whose diameters and intensities are dependent on the arrangement and movement of the atoms in the molecule. Obtaining quantitative information about the molecular geometry involved a special mathematical treatment of the experimental data that made use of Fourier transform theory. In order to separate the weaker molecular scattering from the total diffracted intensity, the Karles adjusted for the background scattering by establishing that the Fourier transform had to be a nonnegative function. This nonnegativity criterion led quickly to increased accuracy in molecular parameters that were verified by independent methods, thus establishing the reliability of the procedure.[4]

As the electron diffraction work progressed, Herbert Hauptman joined the Karle group at the NRL. The initial intention was to develop the area of slow electron diffraction. Although suitable apparatus was designed and built, this never came to pass, because it was interrupted by an intense interest in applying the nonnegativity criterion to other areas of structural research. At the same time that a method was being derived for the elucidation of electron density distributions around free atoms,[5] the implications of the nonnegativity criterion for crystal structure analysis, that is, the determination of atomic arrangements in the unit cells of crystals, was investigated, resulting in the foundation mathematics for all the formulas used in the direct methods of crystal structure determination.[6] The result was an infinite set of inequalities, which are the necessary and sufficient conditions for a nonnegative Fourier series. The elements are the structure factors, the coefficients of the Fourier series. This was a development of the work of Otto Toeplitz.[7]

---

[4]Karle, I. L.; Karle, J. "Internal Motion and Molecular Structure Studies by Electron Diffraction"; *J. Chem. Phys.* **1949**, *17*, 1052–1058.

[5]Hauptman, H.; Karle, J. "The Structure of Atoms from Diffraction Studies"; *Phys. Rev.* **1950**, *77*, 491–499.

[6]Karle, J.; Hauptman, H. "The Phases and Magnitudes of the Structure Factors"; *Acta Crystallogr.* **1950**, *3*, 181–187.

[7]"Über die Fouriersche Entwicklung positiver Funktionen"; *Rend. Circ. Mat. Palermo* **1911**, 191–192.

The determinantal inequalities were preceded by the work of David Harker and John Kasper[8] who obtained useful inequalities among the Fourier coefficients (structure factors) by application of Schwarz and Cauchy inequalities. The inequality relationships permit, in favorable circumstances, the evaluation of angles (phases) associated with structure factor magnitudes. The latter magnitudes are obtained from the experimental measurement of the scattered X-ray intensities. Once the phases and magnitudes are known, the Fourier series can be computed to obtain the electron density distribution in the unit cell of a crystal, revealing the arrangement of the atoms.

The inequalities, as they were applied, were limited to very simple structures. When certain measured X-ray intensities were large enough, phase information could be obtained that was certain to be correct. The number of sufficiently large intensities decreases very rapidly, however, as the complexity of the crystals increases, thus affording little or no information except in rather simple cases. It was recognized that the applicability of the phase-determining formulas could be considerably extended by viewing them in a probabilistic context rather than as a definitive source of phase information, and the development of appropriate mathematical tools was begun soon after the determinantal inequalities were obtained. The first attempt involved the use of the theory of the random walk. It did not prove suitable, but certain features of the latter mathematics of the random walk were found to be quite useful in developing appropriate probabilistic measures in the form of joint probability distributions. Probabilistic measures for centrosymmetric crystals and a methodology for using them were published in 1953.[9] By 1956 the main probabilistic measures and auxiliary tools for origin and enantiomorph specification were also in place for noncentrosymmetric crystals.[10]

One early motivation for pursuing probabilistic methods arose from a visit Jerome Karle made to David Harker's laboratory at the

[8]"Phases of Fourier Coefficients Directly from Crystal Diffraction Data"; *Acta Crystallogr.* **1948**, *1*, 70–75.

[9]Hauptman, H.; Karle, J. *Solution of the Phase Problem. I. The Centrosymmetric Crystal*; American Crystallographic Association: New York, 1953.

[10]Cochran, W. "Relations between the Phases of Structure Factors"; *Acta Crystallogr.* **1955**, *8*, 473–478; Karle, J.; Hauptman, H. "A Theory of Phase Determination for the Four Types of Non-Centrosymmetric Space Groups 1P222, 2P22, 3P₁2, 3P₂2"; *Acta Crystallogr.* **1956**, *9*, 635–651; Hauptman, H.; Karle, J. "Structure Invariants and Seminvariants for Non-Centrosymmetric Space Groups"; *Acta Crystallogr.* **1956**, *9*, 45–55.

Brooklyn Polytechnic Institute in 1948. Harker told Karle that a scaling oversight made the intensities too large by a factor of two in a structural investigation being carried out by Joseph Gillis in his laboratory. However, it was found that correct phase values were being obtained even though the inequality relations would not have been satisfied by use of the correct, smaller values for the intensities. It was apparent that the inequality relationships would give correct phase values in a probabilistic context even when the X-ray scattering intensities were not quite large enough to satisfy the relationships.

The determinantal inequalities published by Karle and Hauptman in 1950 contained probabilistic features. Some of them were discussed by Georges Tsoucaris.[11] Soon after, J. Karle, in a paper that generalized the so-called tangent formula,[12] pointed out that a general relation for bounding any element (structure factor) in a determinant, which was expressed as an inequality in the 1950 paper with H. Hauptman, could be readily put in the form of a probability distribution for the element because the bound was given in terms of an expected value and a measure of the variance, the two quantities required for the application of the central limit theorem. Karle used this property in many subsequent theoretical investigations. In the early work, as noted, the joint probability distribution had been used for obtaining probabilistic information. Because the determinantal inequalities contain not only the phase relationships but also appropriate probability measures, they may be regarded as providing the foundation mathematics for phase determination.

Theoretical studies were continued over the years, but by 1956 enough theory was in place to provide a considerable basis for the development of generally practical procedures that still required several more years.

In 1956 Jerome Karle asked his wife, Isabella Karle, if she would be interested in developing a facility for carrying out X-ray diffraction experiments. Thus began her major contributions toward "bridging the gap" between the probabilistic mathematics and the actualities of analyzing experimental data. After a number of years the experience of applying the methodology of the 1953 monograph, modify-

---

[11]"A New Method of Phase Determination. The Maximum Determinant Rule'"; *Acta Crystallogr.* **1970**, *A26* 492–499.

[12]"A Generalization of the Tangent Formula"; *Acta Crystallogr.* **1971**, *B27* 2063–2065.

ing the procedures, and overcoming many problems in extending the methodology to noncentrosymmetric crystals produced the "symbolic addition procedure", the first general method for solving both centrosymmetric and, for the first time, noncentrosymmetric crystal structures. The first application concerned a synthetic peptide having four different conformers in the unit cell with 98 nonhydrogen atoms to be located.[13] The second application concerned a noncentrosymmetric structure, arginine dihydrate.[14] It was found that phase-determining procedures often produce partial structures, particularly with noncentrosymmetric crystals. In order to facilitate completion of the structure determination, Karle developed a widely used calculation for producing a completed structure from a partial structure.[15]

The general importance of structure determination has been widely demonstrated in the Karles' laboratory. They have used structural investigations to clarify the nature of the products obtained in photochemical rearrangements when the reaction produced totally unexpected substances: to establish structural formulas for natural products and products of chemical reaction; to determine folding, preferred conformations, and changes in conformation on complexation or hydration; to establish the stereoconfiguration and, at times, the absolute configuration of compounds having numerous asymmetric centers; and to obtain structural information concerning precursors, reaction intermediates, and final products that could facilitate synthesis and provide insight into reaction mechanisms.

A large variety of substances have been investigated by the Karle group at the NRL, including radiation-damaged genetic material; heart, antimalarial, and antiradiation drugs; cytostatic agents; antibiotics; natural products; toxins; membrane-active substances; reversible oxygen carriers; energetic materials; and strained compounds, such as may be found in cage compounds and significantly nonplanar ethylene derivatives. Research programs in the Laboratory for the Structure of Matter have also included structural investigations concerning substances in the gaseous state, amorphous solids, surface films, and macromolecules.

---

[13]Karle, I. L.; Karle, J. "An Application of a New Phase Determination Procedure to the Structure of Cyclo-(Hexaglycyl) Hemihydrate"; *Acta Crystallogr.* **1963**, *16*, 969–975.

[14]Karle, I. L; Karle, J. "An Application of the Symbolic Addition Method to the Structure of L-Arginine Dihydrate"; *Acta Crystallogr.* **1964**, *17*, 835–841.

[15]Karle, J. "Partial Structural Information Combined with the Tangent Formula"; *Acta Crystallogr.* **1968**, *B24* 182–186.

Numerous theoretical investigations by Karle have concerned such subjects as the use of phase values to specify an origin in a crystal; the analysis of neutron diffraction data when the atomic scattering factors are both positive and negative; higher order phase invariants; the analysis of amorphous materials; theory for macromolecular crystallography; protein refinement; internal rotation in molecules; and the effect of excitation in gas electron diffraction. Members of Karle's group have recently been developing theoretical and analytical procedures for determining the structures of extremely small solid samples, a few angstroms in width and about 50 Å thick, by use of diffraction data obtained with a fine electron probe (nanodiffraction).

Before the advent of a practical procedure for phase determination, which had its origins in the application of the nonnegativity constraint to X-ray structure analysis, crystal structures devoid of disproportionately heavy atoms required months or years of laborious analysis to elucidate, and noncentrosymmetric crystals were virtually beyond solving. By the use of direct methods and modern computers, research centers around the world are determining new structures in only a few days. As a result, at least 10,000 new structures are being determined every year.

Since 1968 Jerome Karle has held the Chair of Science as chief scientist of the NRL's Laboratory for the Structure of Matter. He continues his theoretical work in crystal structure analysis, seeking to improve existing techniques while searching for new methods. He also collaborates occasionally with Isabella, who continues her own award-winning experimental work. Among his interests has been the algebraic analysis of multiple-wavelength anomalous dispersion;[16] the result given in this reference is a set of exact, linear simultaneous equations for phase determination that are straightforward to apply and easy to solve. The analysis has begun to be applied with increasing frequency and success.[17]

The analysis of very large molecules with complex structures has eluded the power of the direct methods. Newer mathematical methods now being developed by Karle and others will be tested by experimentalists using the latest X-ray sources and computers. Success

---

[16]Karle, J. "Some Developments in Anomalous Dispersion for the Structural Investigation of Macromolecular Systems in Biology"; *Int. J. Quantum Chem.: Quantum Biology Symposium,* **1980**, 7, 357–367.

[17]Karle, J. "Macromolecular Structure from Anomalous Dispersion"; *Phys. Today* **1989**, *June,* 22–29.

in improving the speed and reliability of the structure determination of such substances as proteins and viruses will dramatically illustrate the validity of the Nobel citation to Karle and Hauptman, " . . . in order to understand the nature of chemical bonds, the function of molecules in biological contexts, and the mechanism and dynamics of reactions, knowledge of the exact molecular structure is absolutely essential."

JAMES J. BOHNING
*Beckman Center for the History of Chemistry*

# *Bibliography*

Julian, M. M.; Festa, R. R. "Isabella L. Karle and a New Mathematical Breakthrough in Crystallography"; *J. Chem. Ed.* **1986,** 63, 66-67.

Karle, I.; Karle, J. Interview by James J. Bohning and David van Keuren at the Naval Research Laboratory on February 26, June 15, and September 9, 1987; The Beckman Center for the History of Chemistry, Philadelphia, PA; Oral History Transcript #0066.

Karle, I. L.; Karle, J. "Recollections and Reflections"; In *Crystallography in North America;* McLachlan, D., Jr., Glusker, J. P., Eds.; American Crystallographic Association: New York, 1983; pp 277-283.

Karle, J. "Recovering Phase Information from Intensity Data"; *Chemica Scripta* **1986,** 26, 261–276. Nobel lecture.

Karle, J. "Structure Determination of Crystalline Substances by Diffraction Methods: Philosophic Concepts and Their Implementation (A Review)"; *Proc. Nat. Acad. Sci. U.S.A.* **1978,** 75, 3540–3547.

Karle, J. "Structures That Matter"; *Report of NRL Progress* **1973,** (July), 86–89.

Karle, J; Karle, I. "Electron Diffraction at the Naval Research Laboratory"; In *Fifty Years of Electron Diffraction;* Goodman, P., Ed; D. Reidel: Boston, MA, 1981; pp 243–253.

Karle, J.; Karle, I. "The Symbolic Addition Procedure for Phase Determination for Centrosymmetric and Noncentrosymmetric Crystals"; *Acta Crystallogr.* **1966,** 21, 849–859.

# 1986

NOBEL LAUREATE

# Dudley R. Herschbach

## 1932—

"The atoms move in the void and catching each other up jostle together, and some recoil in any direction that may chance, and others become entangled with one another in various degrees according to the symmetry of their shapes and sizes and positions and order, and they remain together and thus the coming into being of composite things is effected."

Dudley Robert Herschbach must have pondered this quotation from the sixth-century writings of Simplicius as he contemplated the challenge of unraveling the molecular motions that lead to chemical reaction. As corecipient of the 1986 Nobel Prize in chemistry, he was cited for "providing a much more detailed understanding of how chemical reactions take place." When he was interviewed by National Public Radio following the award announcement, his description of his own contributions was clearly influenced by the fact that his beloved Boston Red Sox were competing in the baseball World Series. Herschbach said that observing the contents of a conventional reaction vessel would be like watching a baseball game in which "zillions of pitchers throw zillions of balls at zillions of batters at the same time; it would be very hard to tell what is going on." The development of the field of molecular-reaction dynamics, especially the contributions that molecular-beam reactive scattering has made, allows an observer to watch a baseball game in which "one pitcher throws one ball to one batter". Guided by the words of Simplicius, Herschbach has spent the last 30 years of his career devising ingenious techniques to watch atoms and molecules jostle each other.

686

Dudley Herschbach was born in San Jose, California, on June 18, 1932, the eldest of six children. Growing up in what was then a rural area near San Jose, Herschbach's interests were those of a typical boy: the outdoors, scouting, and sports. He recalls that at the age of nine, he read an article about astronomy in *National Geographic* and was immediately captivated by the mysteries of the stars and planets. He often climbed into a locust tree in the backyard, star map in hand, trying to identify constellations.

As a high school student, Herschbach read voraciously and delved into the challenges of science and mathematics, but his interests in sports, particularly football, equaled his devotion to his studies. As a star right end for the Campbell High School football team, Herschbach learned about collisions and scattering firsthand, but it was not until he entered college that he chose molecular collisions over those occurring on the gridiron.

Offered both academic and athletic scholarships to attend Stanford, Herschbach pursued a degree in mathematics, taking enough courses to complete degrees in chemistry and physics as well. He was introduced to chemical kinetics shortly after he arrived at Stanford: His freshman advisor was Professor Harold S. Johnston, who was just beginning a research program to measure the rates of a number of fast gas-phase reactions and to interpret the results through applications of transition-state theory. Recognizing the talent in Herschbach, Johnston hired him as a summer research assistant after his sophomore year and put him to work on the construction of a vessel to study high-temperature rate processes. As a senior, Herschbach took Johnston's chemical kinetics course, an experience that launched him on a study of transition-state theory and ultimately led to a master's thesis, in which he devised careful calculations of Arrhenius A-factors for several simple gas-phase reactions, incorporating a consistent treatment for the internal motions of all reaction complexes.[1] By that time, Herschbach had been seduced by chemical kinetics, but he was convinced that to really study reaction rates carefully he had to learn about molecular structure first. Thus, he pursued Ph.D. studies at Harvard under the direction of Professor E. Bright Wilson, Jr. Following up on the internal rotation problem that had been so important in his work with Johnston, Herschbach employed microwave spectroscopy to explore the tunneling splittings in molecules with barriers to internal rotation.

---

[1]Johnston, H. S. *Gas Phase Reaction Rate Theory*; Ronald Press: New York, 1966.

Herschbach completed his Ph.D. in 1958 and remained at Harvard for one year as a junior fellow in the Society of Fellows.

Before Herschbach entered Harvard, he had learned of molecular beams and Otto Stern's tests of the Maxwell–Boltzmann distribution of molecular speeds, and he recalls the thrill of contemplating the power of the technique: "It was love at first sight. I remember a flush of excitement at the thought that this was *the way* to study elementary reactions, the unequivocal way to know a reaction is elementary and to study directly its properties." Herschbach had clearly taken to heart Johnston's admonition that critical tests of theory could only be performed on chemical reactions known to be elementary, single-step processes. As Herschbach visited several universities during the spring of 1958 as a faculty candidate, his work on microwave spectroscopy was enthusiastically received, but his plans to initiate molecular-beam studies of elementary chemical processes received mixed reviews. Many of the established scientists of the day firmly believed that "collisions do not occur in crossed molecular beams". Nonetheless, Herschbach was offered an appointment as assistant professor of chemistry at the University of California at Berkeley and in the fall of 1959 began to construct a crossed-beam instrument designed to study reactive collisions of alkali atoms with a wide variety of molecular partners.

Herschbach began his work at Berkeley with a style he characterized as "evangelical fervor". With two graduate students, George Kwei and James Norris, he began work on "Big Bertha", a crossed-beam machine with very large vacuum-chamber dimensions. Although the pioneering experiments in 1955 by Ellison Taylor and Sheldon Datz at Oak Ridge on the K + HBr system, as well as important but unheralded experiments by Bull and Moon in 1954 on the Cs + CCl$_4$ reaction at Birmingham, England, provided hope that the experiments would be successful, Herschbach hedged his bets and built a machine large enough to be converted to a molecular beam electron resonance spectrometer if the reactive scattering experiments were to fail. Although this particular option was never necessary, it is interesting to note that a decade later, Herschbach, then at Harvard, teamed up with William Klemperer to measure internal state distributions for the CsF product of the reaction Cs + SF$_6$, by using molecular-beam electron resonance methods. The initial Berkeley experiments on K + CH$_3$I were quite successful and provided the first detailed view of an elementary collision, demonstrating that the reaction occurred on a time scale comparable to the time for the reagents to pass by one another. The KI product recoiled in the backward direction relative to the direction of the incoming K atom beam, providing an example of a *direct* reaction, termed a *rebound* process. A number of studies of alkali atom collisions with alkyl halides followed, and rudimentary kinematic analysis showed that much of the energy available to the reaction products from initial translational energy and the reaction exothermicity appeared in product internal excitation. The recoil that Simplicius described fourteen centuries before had been observed, although with much greater directionality that he had predicted!

The role of serendipity in science was demonstrated well in the next phase of Herschbach's research. Following up on some diffusion flame experiments performed by Michael Polanyi (father of John Polanyi, corecipient of the 1986 Nobel Prize in chemistry with Herschbach and Yuan Tseh Lee) some 30 years before, Herschbach and his group attempted to explore the dynamics of the K + Br$_2$ reaction. Unexpectedly they discovered that their hot-wire surface ionization detector behaved unreliably, preventing the work from proceeding. It was only after learning of some work by Touw and Trischka that they realized the importance of the prior use of the tungsten or platinum filament: The surface was readily poisoned by contaminants, and only when it was pretreated properly could reliable results be obtained. In the detecting mode, achieved by a pretreatment of the surface with O$_2$, alkali atoms and alkali-containing compounds could be detected. In the nondetecting mode, achieved by a pretreatment with a hydrocarbon, only alkali atoms could be detected. In Herschbach's initial experiments on K + CH$_3$I, the CH$_3$I reagent had provided adequate treatment of the filament to give it the proper characteristics required for reactive scattering. In K + Br$_2$, however, adequate pretreatment was a matter of serendipity, with a stable carbide coating arising from contaminants such as vacuum pump oil. Herschbach likened their discovery to the famous Stern–Gerlach experiment: For some time, the photographic plate detectors had only shown the famous splittings of the beam when Stern viewed the plates. Only after some head-scratching did they realize that the high sulfur content of Stern's cheap cigars had been responsible for the development of an image on the plates as Stern breathed on them!

The K + Br$_2$ reaction soon yielded to the now-reliable differential surface ionization method, and new diversity in reaction dynamics emerged: The KBr product was formed with high internal excitation in a direct reaction but was scattered forward relative to the incident K atom beam, an example of a *stripping* reaction. Soon a number of examples of direct rebound and stripping reactions became known, and it became possible to make correlations of reaction dynamics with the electronic structure of reactants and products. This relationship developed as a result of a visit from Michael Polanyi, who described the reaction dynamics quite picturesquely as a "harpooning" process: The relatively low ionization potential of the alkali atom allowed the valence electron to "jump" to the electronegative halogen on an alkyl halide or diatomic halogen. The nascent alkali ion could then "haul in" the halide ion with the Coulomb force, creating products.

In 1963 Herschbach uprooted his Berkeley operation and returned to Harvard as professor of chemistry. Many of the ideas initiated at Berkeley reached a new level of sophistication in the first few years at Harvard, and the "Alkali Age" of molecular-beam reaction dynamics flourished. However, a particularly important discovery in late 1966 made the words of Simplicius even more prescient. Working with two graduate students, Sanford Safron and Walter Miller, on reactions of alkali atoms with alkali halides, Herschbach discovered that the exchange reaction appeared to proceed through a persistent complex that lasted many rotational periods. Under such conditions, the products *did* recoil "in any direction that may chance". Moreover, analysis of the disposal of rotational and orbital angular momentum demonstrated that the product angular distributions reflected the degree of reagent entanglement according to the oblate or prolate symmetric top nature of the intermediate complex, precisely as Simplicius speculated!

By 1967 the Alkali Age was in a state of maturity, and Herschbach realized that a qualitative step forward into the complexities of covalent chemistry required a much more sophisticated experimental approach. At this time, Yuan Tseh Lee, fresh from graduate work and a brief postdoctoral stint with Bruce Mahan at Berkeley, joined the Herschbach group as a postdoctoral fellow. In just 10 months, Lee and Herschbach, working together with graduate students Doug McDonald and Pierre LeBreton, constructed the first "supermachine", employing supersonic nozzle and effusive sources, mass spectrometric detection, extensive differential pumping in the detector and beam sources, as well as computerized time-of-flight product velocity analysis. The first experiments were on the $Cl + Br_2$ system and were followed by hydrogen–halogen and halogen–hydrogen halide experiments such as the $H + Cl_2$ and $Cl + HI$ reactions. These latter experiments were particularly important because John Polanyi at the University of Toronto had observed infrared emission from the products of these same reactions a few years before, by using the method of "arrested relaxation". In concert with the product state resolved measurements, the beam experiments provided another dimension in the emerging, detailed picture of the anatomy of elementary chemical reactions. The era of molecular-beam chemistry that followed was termed the "Organic Age" by Herschbach, and a number of interesting halogen substitution reactions with vinyl and allyl halides illustrated the concepts of atom and bond migration. During the same period, studies of van der Waals molecule structure by colleague William Klemperer led Herschbach to examine such chemical reactions, including the system $Br_2 + (Cl_2)_2 \rightarrow 2BrCl + Cl_2$, in which covalent bonds cleave, as well as $Xe + Ar_2 \rightarrow XeAr + Ar$, in which van der Waals bonds are both made and broken.

An abiding interest in Herschbach's laboratory has been the role that angular momentum, particularly its vector properties, plays in chemical reaction dynamics. Herschbach has been a pioneer in the measurement and theoretical interpretation of correlations that exist among the vector properties of reaction dynamics, including relative velocities and rotational and orbital angular momenta. The new field of "molecular stereodynamics" rests on the firm foundation of this work.

In recent years Herschbach's research has moved toward theoretical descriptions of electronic structure based on interpolation between exact solutions of the Schrödinger equation in the one- and infinite-dimensional limits. Such methods hold great promise for electronic structure computations, particularly the vexing electron-correlation problem.

The Nobel Prize recognizes Herschbach's prowess in research, but he is also a dedicated teacher, presenting a very popular freshman chemistry course at Harvard. He and his wife, Georgene, also a chemist, have served as Co-Masters of Currier House, a Harvard residence hall.

Herschbach has received numerous awards and lectureships in recognition of his contributions to chemistry. He was a visiting professor at the Göttingen University in 1963, a Guggenheim fellow at Freiburg in 1968, a visiting fellow at the Joint Institute for Laboratory Astrophysics in 1969, and a Sherman Fairchild Scholar at the California Institute of Technology in 1976. He was selected to the American Academy of Arts and Sciences in 1964 and to the U.S. National Academy of Sciences in 1967. He received the American Chemical Society's Award in Pure Chemistry in 1965, the Linus Pauling Medal in 1978, the Michael Polanyi Medal in 1981, and the Irving Langmuir Prize in Chemical Physics in 1983. He was also awarded the D.Sc. degree *honoris causa* from the University of Toronto in 1977.

The broad spectrum of Herschbach's work, both in experiment and theory, reflects the deep creativity that he has brought to scientific

## DUDLEY R. HERSCHBACH

research, which he describes as "having conversations with Nature". Surely the scientific community is richer for the efforts of this man, whose career has been built on developing a dictionary to help us understand the language of molecular structure and chemical reactions.

JAMES M. FARRAR
*University of Rochester*

# Bibliography

Herschbach, D. R. *Chemical Reactions Atom by Atom* (video recording); International Merchandising: New York, 1990.

Herschbach, D. R. "Molecular Dynamics of Elementary Chemical Reactions"; *Agnew. Chem. Int. Ed. Engl.* **1987,** *26,* 1221–1243.

Herschbach, D. R. "New Dimensions in Reaction Dynamics and Electronic Structure"; *Faraday Discuss. Chem. Soc.* **1987,** *84,* 465.

Herschbach, D. R. "Reactive Scattering in Molecular Beams"; *Adv. Chem. Phys.* **1966,** *10,* 319–393.

Kim, S. K.; Herschbach, D. R . "Angular Momentum Disposal in Atom Exchange Reactions"; *Faraday Discuss. Chem. Soc.* **1987,** *84,* 159.

Lee, Y. T.; McDonald, J. D.; LeBreton, P. R .; Herschbach, D. R . "Molecular Beam Reactive Scattering Apparatus with Electron Bombardment Detector"; *Rev. Sci. Instrum.* **1969,** *40,* 1402–1408.

Levine, R. D.; Bernstein, R. B. *Molecular Reaction Dynamics and Chemical Reactivity;* Oxford University Press: New York, 1987.

# 1986

NOBEL LAUREATE

# Yuan Tseh Lee

## 1936–

When the Nobel Prizes in chemistry were announced in October 1986, excited talk of baseball and the World Series was in the media. How appropriate, then, that one of the recipients of the award, Professor Yuan Tseh Lee of the University of California at Berkeley, when interviewed by the press, described the research that led to this recognition for elucidating the collision dynamics of elementary chemical reactions through analogies with his favorite sport. Lee likened the nature of collisions leading to energy transfer and chemical reaction to the way in which a batter strikes a baseball: The nature of the "impact parameter", or aiming error, with which the bat strikes the ball determines the trajectory of the baseball. Lee's research over the past 25 years has been devoted to developing sophisticated probes of molecular fastballs, curves, sliders, and knuckleballs. His incisive spectroscopic studies of cluster ions and solvation have even allowed him to probe the characteristics of molecular spitballs!

Yuan Tseh Lee was born on November 19, 1936, in Hsinchu, Taiwan. His father was an accomplished artist and his mother was a school teacher. Lee's early education began during the Japanese occupation of Taiwan but was interrupted when his family left the city and moved to the safety of the mountains to avoid bombing during World War II.

The perseverance that characterizes Lee no doubt was sharpened by the adversity that he and his family experienced during wartime. When the war ended and Taiwan was reunited with China, Lee resumed his formal schooling. An important part of his education was

693

the development of language skills, and in addition to his knowledge of Chinese, Japanese, and English, Lee acquired proficiency in Russian and German, the latter taught to him by Catholic missionaries in Taiwan. In elementary school, Lee filled his free time with sports and served as the second baseman for his school's baseball team as well as a member of a championship ping-pong team. In high school he played tennis and was a trombone player in the marching band, in which he became an aficionado of the music of John Philip Sousa. During that time he developed his interest in science, and it was primarily his reading of Eve Curie's biography of her mother, Marie Curie—a biography that describes the idealism, altruism, and scientific and personal integrity exemplified in her life—that convinced him to pursue a career in scientific research and education.

Lee entered National Taiwan University in 1955, and because of his excellent academic performance in high school, he was not required to take the usual entrance examinations. Lee was captivated by the excitement of learning and, through the stimulus of the excellent teaching of many of his professors, quickly chose to become a student of chemistry. He was introduced to research during college and conducted an investigation of the electrophoretic separation of the alkaline earths Sr and Ba under the direction of Professor Hua-sheng Cheng. He followed his undergraduate education with a master's degree from the Institute for Nuclear Studies at Tsinghua University and remained as a research assistant under the direction of Professor C. H. Wong, carrying out X-ray crystallographic studies of the structure of tricyclopentadienyl samarium.

In 1962 Lee entered the graduate program at the University of California at Berkeley. Although his path crossed that of his future collaborator and Nobel Prize corecipient Dudley Herschbach, a young faculty member at the time, Herschbach had no room in his newly established molecular beam group, and Lee joined the research group of Professor Bruce Mahan. His thesis work on the chemiionization reactions of electronically excited alkali atoms led to the surprising result that the bond energy of $Cs_2^+$ is greater than that of $Cs_2$, despite the fact that the ion is formed by removal of a bonding electron. This result, in conflict with the predictions of elementary molecular orbital theory, was met with skepticism by some members of his Ph.D. thesis committee, but the experimental results were unequivocal, and more sophisticated quantum mechanical calculations ultimately confirmed the observations. Following the receipt of his Ph.D. in 1965,

Lee remained in Mahan's group as a postdoctoral fellow. It was during that time, when scattering phenomena were defining an important new discipline within chemical physics, that Lee developed his interest in ion–molecule reactions and collision mechanics. With Mahan's encouragement and the collaboration of graduate student W. Ronald Gentry, Lee constructed a versatile and sophisticated ion beam instrument that yielded a detailed picture of ion–molecule reaction dynamics by determining the product flux energy and angular distributions in velocity space.

In January 1967 Lee began a postdoctoral appointment in Professor Dudley Herschbach's laboratory at Harvard, where he quickly used his mastery of the art of designing and building molecular beam instruments to construct a technologically advanced "universal" machine with an electron bombardment ionizer and mass spectrometer detector that moved the field of molecular beam kinetics from the "Alkali Age" to a new era of chemical variety in the study of reaction dynamics. Working with two able graduate students, J. Douglas McDonald and Pierre LeBreton, Lee and his co-workers developed a sophisticated ultrahigh vacuum detector that had the sensitivity and low background required to detect chemical reaction products corresponding to partial pressures of $10^{-16}$ torr. Their first success occurred on the system $Cl + Br_2 \rightarrow ClBr + Br$ and was followed by a number of additional studies. Lee then moved on to his first academic position, assistant professor in the James Franck Institute and the Department of Chemistry at the University of Chicago, in October 1968.

During the Chicago years, Lee and his students continued to develop the technology of the universal machine. Coupled with supersonic beam sources, time-of-flight velocity analysis, and improved pumping in the detector region, this instrument and its successors set the standard for molecular beam instrumentation designed for reactive collision studies. Lee and his group measured differential cross sections for rare gas pairs, allowing them to extract reliable potentials, especially in the region of the attractive minimum. The group also began studies on He* metastable atom scattering with various atomic and molecular collision partners as well as studies of the dissociation of alkali halide molecules induced by collision with hyperthermal Xe atoms.

Perhaps the most significant research accomplished during this period was the reactive scattering work on fluorine atoms with a variety of molecules. By crossing beams of fluorine atoms and $D_2$ molecules with narrow velocity distributions, Lee and his group were able to observe individual vibrational quantum states of the DF product molecules through their characteristic scattering into different regions of velocity space. This work provided important insight into the functioning of the DF chemical laser and served as one of the key stimuli for the currently active field of state-to-state chemistry. In addition to this work, the group began a series of studies in which fluorine atoms attacked various organic substrates. The chemically activated radicals thereby formed decayed by cleaving carbon–hydrogen, carbon–carbon, or carbon–halogen bonds. By measuring product angular and kinetic energy distributions, one could probe the lifetimes of the intermediate complexes and energy partitioning in the products, ultimately probing the fundamental nature of intramolecular vibrational relaxation.

The work at Chicago established Lee as a clear leader in the field of molecular-reaction dynamics, and his laboratory rapidly became the North American capital of molecular-beam studies of collision processes. Lee rose quickly through the academic ranks at Chicago, being promoted to associate professor in October 1971 and to professor of chemistry in February 1973. During that period Lee developed his very personal, interactive style of working with his graduate students and postdoctoral fellows. Molecular-beam experiments require patience, attention to detail, and perseverance. Many a student nursing a long experimental run on the "graveyard shift" would be joined by Lee, who would provide encouragement, support, and an occasional instrumental adjustment to ensure the success of the run. Such manipulations were often followed by the proclamation "Should be O.K.", a statement widely regarded by his students as a harbinger of experimental success. Occasionally, Lee would make a modification to one of the instruments late at night, after all the students had gone home. In one such case, the gas load from one of the molecular beams was creating a high level of interfering background. Lee opened the vacuum chamber and installed an auxiliary beam trap, termed a "beam catcher". His note to the student simply stated that he had installed a "Johnny Bench" in the machine. The student, who was quite new in the group and fairly ignorant of baseball (and beam catchers), had to wait until Lee came in the next day to discover the nature of the modification.

In August 1974 Lee moved his research operation from Chicago to the University of California at Berkeley, returning as professor of chemistry and principal investigator at the Lawrence Berkeley Laboratory. Also in 1974 Lee became a United States citizen.

The Berkeley phase of Lee's career ushered in a new era of experimentation in which lasers and molecular beams were combined to understand energy partitioning in molecular photodissociation as well as the role of internal excitation of reagents in promoting chemical reactions. Among the first such experiments was a study of infrared multiphoton dissociation (IRMPD), which was new at the time. The molecular-beam experiments carried out in Lee's laboratory demonstrated unequivocally that IRMPD is a statistical process, and that rapid intramolecular vibrational relaxation results in the dominance of the lowest energy dissociation channel. Other applications of lasers to the collisionless environment of the molecular beam led to studies of the role of alkali atom electronic excitation in promoting chemical reactions, experiments that could be termed the "Alkali Age Revisited". The development of an ion trap instrument and a corona discharge and supersonic beam ion source resulted in studies of the infrared spectra of ionic clusters and carbocations. More refined experiments on the $F + H_2/D_2$ reactions led to the observation of dynamical resonances. Supersonic molecular-beam techniques in conjunction with double-resonance methods allowed the group to observe highly resolved spectra of the local C–H stretching modes of benzene. The ultraviolet and infrared multiphoton chemistry of cyclohexene and 1,4-cyclohexadiene provided initial opportunities for investigation in the rich field of organic photochemistry.

Although laser photochemistry on a variety of molecules ranging from ionic clusters to van der Waals molecules to complex organic structures clearly has been a continuing focus in Lee's group, studies of reactive collisions have also maintained a prominent place in his group's repertoire. Crossed-beam studies of free-radical reactions with molecular targets, such as the reactions between methyl radicals and iodoalkanes, and endoergic substitution reactions of Br atoms with chlorinated aromatic species are representative examples of Lee's interest in testing fundamental concepts of organic chemistry. Reactive scattering studies of oxygen atoms with hydrocarbons such as acetylene and ethylene have allowed the primary reaction products of these important combustion reactions to be unambiguously identified. At the most fundamental level, reactive collision studies on the dynamics of the simplest chemical reaction, $D + H_2 \rightarrow HD + H$, have been undertaken. In conjunction with the complementary experimental efforts of Gentry, Valentini, and Zare, these studies are already providing critical tests of exact quantum theories of reactive scattering.

In addition to the Nobel Prize, Professor Lee's many contributions to science have been recognized with numerous awards. He was a recipient of the Department of Energy's Ernest O. Lawrence Award in 1981 and received the National Medal of Science in 1986. The American Chemical Society awarded him the Peter Debye Award in 1986, and the Rochester Section of the ACS presented him with the Harrison Howe Award in 1983. Elected a member of the National Academy of Sciences in 1979, Lee is also a member of the Academia Sinica of Taiwan, a fellow of the American Academy of Arts and Sciences, and a Corresponding member of the Göttingen Academy of Sciences. He is an honorary professor at several universities in the Peoples' Republic of China, and is honorary director of the National Laboratory for Chemical Kinetics and Reaction Dynamics at Dalien and Beijing. Honorary degrees have been bestowed on him from the University of Waterloo (1986), the Chinese Academy of Sciences (1986), the Chinese University of Hong Kong (1989), Arizona State University (1990), and the University of Rome (1992).

Lee's family has always been an important part of his life. He and his wife, Bernice Wu, whom he has known since childhood, have a daughter, Charlotte (born in 1969), and two sons, Sidney (born in 1966) and Ted (born in 1963).

Through Lee's ability to identify important research problems, his mastery of the construction of sophisticated instrumentation to address those problems, and his ability to attract and motivate outstanding collaborators, he has firmly established the molecular-beam technique in modern chemical physics. His influence pervades scientific research worldwide. Professor Lee is particularly proud of the fact that nearly 20 of his former associates hold professorships in major universities, and many others are making important scientific contributions in national laboratories and in industry. Through the work of his own group, which continues at a rapid pace, the work of his former students and postdoctorals and those influenced by his research, the legacy of Yuan Tseh Lee in modern chemical physics will live for many years to come.

JAMES M. FARRAR
*University of Rochester*

# Bibliography

Bernstein, R. B. *Chemical Dynamics via Molecular Beam and Laser Techniques;* Oxford University Press: New York, 1982.

Gentry, W. R.; Gislason, E. A.; Lee, Y.; Mahan, B. H.; Tsao, C. "Product Energy and Angular Distributions from the Reaction of $N_2^+$ with Isotopic Hydrogen Molecules"; *Discussions Faraday Soc.* **1967**, *44*, 137.

Lee, Y. T. "Molecular Beam Studies of Elementary Chemical Processes"; *Science* **1987**, *236*, 793.

Lee, Y. T.; McDonald, J. D.; Lebreton, P. R.; Herschbach, D. R. "Molecular Beam Reactive Scattering Apparatus with Electron Bombardment Detector"; *Rev. Sci. Instrum.* **1969**, *40*, 1402.

Levine, R. D.; Bernstein, R. B. *Molecular Reaction Dynamics and Chemical Reactivity;* Oxford University Press: New York, 1987.

Schulz, P. A.; Sudbø, Aa. S.; Krajnovich, D. J.; Kwok, H. S.; Shen, Y. R.; Lee, Y. T. "Multiphoton Dissociation of Polyatomic Molecules"; *Ann. Rev. Phys. Chem.* **1979**, *30*, 379.

# 1986

NOBEL LAUREATE

# John C. Polanyi

## 1929–

John Charles Polanyi, winner of a 1986 Nobel Prize in chemistry, was born on January, 23, 1929, in Berlin, where his father, Michael Polanyi, was professor of physical chemistry. Michael Polanyi, who did distinguished work both in chemical kinetics and in philosophy, was born in Budapest. In 1933, when his son John was aged 4, he left Berlin for England to become professor of chemistry at the University of Manchester. In Berlin and in Manchester he did pioneering work on the mechanisms of elementary reactions. One of his more significant contributions was to construct, with the American chemist Henry Eyring, the first potential-energy surface for any chemical reaction. Another was to formulate, in 1935, transition-state theory in collaboration with M. G. Evans; their work was done independently of that of Eyring, whose paper giving essentially the same treatment appeared a few months earlier.

John Polanyi was educated at the Manchester Grammar School and at the University of Manchester, where he obtained his Ph.D. degree in 1952. His research supervisor was Ernest Warhurst, who had been a research student with Michael Polanyi. From 1952 to 1954 John Polanyi was a postdoctoral fellow at the National Research Council of Canada in Ottawa, working mainly on transition-state theory in the kinetics laboratories of E. W. R. Steacie. For a few months he worked in the spectroscopic laboratories of Gerhard Herzberg, where he assembled infrared spectroscopic equipment for observing vibrational and rotational excitation in molecules. Some of this equipment he took in 1954 to Princeton, where he was a research associate for two years.

In 1956 Polanyi was appointed to the University of Toronto, where he has remained in various positions ever since; in 1974 he was given the special title of University Professor. In 1958 he married Anne Davidson, always known as Sue, who is an accomplished musician. They have one son and one daughter.

At the University of Toronto, Polanyi at once embarked on a comprehensive program of research in the rapidly developing field of reaction dynamics, which is concerned with details of the elementary acts that occur during the course of chemical change. His work has been particularly effective in combining both experimental and theoretical approaches to the problem. He has been concerned with understanding the intramolecular and intermolecular motions that occur during chemical change and with determining the quantum states of reactant and product molecules. His investigations made it possible to draw conclusions about the transition species that lie between the initial and final states, and in the early 1980s he was able to obtain more direct information about these species.

The investigations first carried out by Polanyi and his students were mainly concerned with the energy distributions among the products of an elementary chemical reaction. One of the first reactions they studied was the exothermic process $H + Cl_2 \rightarrow HCl + Cl$. Atomic hydrogen produced in an electric discharge was introduced into a stream of chlorine gas at low pressure, and the products were observed by means of an infrared spectrometer. Although the temperature of the reaction system remained low, the hydrogen chloride produced was in excited vibrational states corresponding to an effective temperature of thousands of degrees. This *infrared chemiluminescence* was first described in "Infrared Chemiluminescence from the Gaseous Reaction Atomic $H+Cl_2$",[1] written with one of his first graduate students, J. K. Cashion, and this was followed by many subsequent publications. Some of the later work was concerned with overcoming effects of vibrational and rotational relaxation, in order to determine the states of the newly formed products before their energy had become dissipated by collisions. This was done either by measuring the relaxation or by arresting it through cryopumping.

---

[1] *J. Chem. Phys.* **1958**, *29*, 455–456.

An important result of these investigations was that a widely variable fraction of the energy released passes into vibrational energy of the product. To explain these different findings, Polanyi and his coworkers constructed a graded series of potential-energy surfaces for an exothermic reaction.[2] At one extreme was a so-called attractive surface in which the energy barrier was at an early stage of the reaction path; at the other extreme was a repulsive surface in which the barrier was at a late stage. It was shown that for moderately attractive surfaces the energy tends to pass into vibration of the products, whereas for a repulsive surface less energy does so. It was noted that this type of behavior was observed particularly for reactions of the type designated **L + HH**, where **L** stands for a light atom and **H** for a heavy one (the $H+Cl_2$ reaction being an example). Polanyi and his students also considered different mass combinations and explored the effect on the dynamics of the reactions.

Polanyi's early work on this type of problem had important practical applications to the construction of lasers, in which a population inversion, corresponding to a negative temperature, is in general produced in some medium. Polanyi, in a paper that appeared in 1959,[3] showed how there could be such a population inversion with respect to vibrational states. A little later he showed that for vibrational lasing one can have a *positive* temperature, with vibration hotter than rotation.[4] This type of population inversion is produced in partially cooled gases and many chemical reactions. The implication of this is that vibrational and chemical lasers should be readily realizable. Four years later J. V. V. Kasper and G. C. Pimentel[5] followed up this suggestion. By using the $H + Cl_2$ reaction proposed by Polanyi, they were able to construct the first operating laser based on a chemical reaction. Polanyi's 1961 papers had one year earlier been submitted to *Physical Review Letters* but had been rejected, just as was Theodore Maiman's paper announcing the construction of the first operating laser! Polanyi asks sponsors of research, who are apt to require practical applications, whether they would have had the foresight to realize that this basic reaction, involv-

[2]Kuntz, P. J.; Nemeth, F. M.; Polanyi, J. C.; Rosner, D. S.; Young, C. E.; *J. Chem. Phys.* **1966,** *44,* 1168–1184.

[3]"Energy Distribution among Reagents and Products of Atomic Reactions"; *J. Chem. Phys.* **1959,** *31,* 1338–1351.

[4]"Proposal for an Infrared Maser Dependent on Vibrational Excitation"; *J. Chem. Phys.* **1961,** *34,* 347–348.

[5]Kasper, J. V. V.; Pimentel, G. "HCl Chemical Laser"; *Phys. Rev. Lett.* **1965,** *14,* 352–354.

ing feeble infrared luminescence, would lead to the development of extremely powerful lasers.

An important achievement of the work in Polanyi's laboratory was the development of "triangle plots", which give the product energy distribution, obtained from infrared chemiluminescence, over vibration, rotation, and translation. This detailed information was not available from any source, and it provided the point of departure for many theoretical studies in Polanyi's laboratory and elsewhere, leading to an improved understanding of the correlation between the forces operating in a reactive encounter and the resulting molecular motions. Thus, many covalent reactions are largely "repulsive", and the repulsion in a bent transition state gives rotation. In addition, reactant often "flirts" with one end of a molecule under attack before "migrating" to the other end.

In the early 1970s Polanyi began to shift his attention to a related problem, the selective enhancement of a chemical reaction. Suppose, for example, that a reaction is endothermic: What type of energy— translational, vibrational, or rotational—is most effective in causing the system to surmount the energy barrier? In view of the principle of microscopic reversibility, this kind of problem could be related to the reverse problem of the disposal of excess energy. For endothermic reactions, which generally have a late barrier crest, Polanyi and his coworkers showed that vibrational energy is most effective in leading to reaction.[6]

Polanyi also directed his attention to the question of what happens to energy that is in excess of that required to surmount the energy barrier. Experimental and theoretical studies in his laboratories led to the conclusion that there is a tendency for excess translational energy to appear as translational energy in the products and for excess vibrational energy to appear as vibrational energy. In other words, there is a tendency for this energy to exhibit *adiabaticity*.[7]

In the 1970s in a number of laboratories, work was initiated in state-to-state kinetics, which involves a still-more-detailed probing of molecular events in a chemical reaction. The reactants are put into

---

[6]Perry, D. S.; Polanyi, J. C.; Wilson, C. W., "Location of Energy Barriers. VI. The Dynamics of Endothermic Reactions, AB+C"; *Chem. Phys.* **1974**, *3*, 317–331.

[7]Cowley, L. T.; Horne, D. S., Polanyi, J. C. "Infrared Chemiluminescence Study of the Reaction Cl+HI→HCl+I at Enhanced Collision Energies"; *Chem. Phys. Lett.* esc **1971**, *12*, 144–149; Parr, C. A.; Polanyi, J. C.; Wong, W. H. "Distribution of Reaction Products (Theory). VIII. Cl+HI, Cl+DI"; *J. Chem. Phys.* **1973**, *58*, 5–20; and many later papers.

particular vibrational and rotational states, and a study is made of the vibrational and rotational states of the products. In some laboratories, such as those of P. R. Brooks, Y. T. Lee, and D. R. Herschbach, the molecular beam method was employed, and laser excitation was used to put the reactants into desired quantum states. An alternative technique, which involved bulk experiments, was introduced by Polanyi.[8] Reactants were put into particular vibrational and rotational states by producing them in an exothermic pre-reaction. The course of the subsequent reactions was followed by two methods, one of which was to monitor the infrared chemiluminescence of the products. The second method involved following the decrease in the luminescence of the reactant molecules; it is known as the method of *chemiluminescence depletion*. Polanyi's methods have the advantage that the reactant molecules can be formed in a much wider range of vibrational and rotational states than through laser excitation. It has the disadvantage of allowing no variation of translational energy, as is possible with the molecular beam technique. Polanyi and his co-workers studied a number of reactions by these methods and were able to interpret the results in terms of the shapes of the potential energy surfaces.

An important line of investigation in which Polanyi was the first to be successful was the detection by spectroscopic means of molecular species having configurations between those of the reactants and products of an elementary reaction. Such entities can be called *transition species*, a special case of them being the *activated complexes*, which are defined as existing in an arbitrarily small region of phase space. A transition species was first detected experimentally in 1979 by Polanyi and co-workers for the reaction.[9]

$$F + Na_2 \rightarrow F \cdots Na \cdots Na \rightarrow NaF + Na^*$$

The reaction was caused to occur in crossed uncollimated molecular beams. The product $Na^*$ was in an excited electronic state and emitted the familiar yellow D-line. On both sides of the line there was "wing" emission, and this as well as later evidence from Polanyi's laboratory

---

[8]Douglas, D. J.; Polanyi, J. C.; Sloan, J. J. "Effect of Reagent Vibrational Excitation on the Rate of a Substantially Endothermic Reaction; $HCl(v' = 1–4) + Br \rightarrow \begin{smallmatrix} Cl + HBr; \\ | \\ HBr'' \end{smallmatrix}$" *J. Chem. Phys.* **1973**, *59*, 6679–6680, and later papers.

[9]Arrowsmith, P.; Bartoszek, F. E.; Bly, S. H. P.; Carrington, T.; Charters, P. E.; Polanyi, J. C. "Chemiluminescence During the Course of a Reaction Encounter; $F + Na_2 \rightarrow FNaNa \rightarrow NaF + Na^*$"; *J. Chem. Phys.* **1980**, *73*, 5895–5897.

indicated that it was due to the transition species $F \cdots Na \cdots Na$. This general approach, with important variants, has been extended to other reactions and is now being actively pursued in several laboratories. More recently the efforts of Polanyi and his collaborators have been directed toward the new field of *surface aligned photochemistry*. The work is concerned with the interaction of light with molecules adsorbed on surfaces and with exploring the details of the processes that occur. One objective is to determine the effects of surface forces on the dynamics of photodissociation processes. Another is to study photoinduced reactions between species that are coadsorbed on a surface, with the aim of positioning the adsorbed molecules so that they can be caused to react in a predetermined way. The first investigations in this field involved the ultraviolet photodissociation and photodesorption of methyl bromide adsorbed on the 001 face of lithium fluoride.[10] A substantial number of other systems have been investigated more recently, from both the experimental and theoretical points of view.

Aside from his research, Polanyi has exerted a wide influence in many ways. A large number of undergraduates, graduate students, and postdoctoral fellows have come under his influence, and he has always been unsparing in his help to others. He has given many named lectures, and other lectures, in various parts of the world. In 1970 he prepared a film that expounds in a very clear way the principles of reaction dynamics.

Polanyi has also played an important role in explaining to politicians and the general public the importance of science in our culture and economy. He has been an eloquent and forceful advocate of high-quality science, carried out without regard to its apparent relevance to practical matters. In this connection he has served on Canada's National Advisory Board on Science and Technology, the chairman of which is the Prime Minister.

His influence has been by no means confined to scientific matters. He has a highly developed social conscience and urges his fellow scientists to increase their participation in public affairs. He himself has been tireless in pursuing this objective and has written many articles and given many speeches on matters of public concern. In particular, he has devoted much attention to world peace in this age of nuclear weapons. He is an outspoken critic of the arms race among the superpowers and

---

[10]Bourdon, G. B. D.; Cowin, J. P.; Harrison, I.; Polanyi, J. C.; Segner, J.; Stanners, C. D.; Young, P. A. "UV Photodissociation and Photodesorption of Adsorbed Molecules. I. $CH_3Br$ on LiF(001)"; *J. Phys. Chem.* **1984,** *88,* 6100–6103.

JOHN C. POLANYI
Photo courtesy of Richard Palmer, Calgary.

has been a powerful advocate of nuclear disarmament. Ronald Reagan's Strategic Defense Initiative (Star Wars) frequently came under his attack. In 1960 he became the founding chairman of the Canadian Pugwash Group, and in 1978 he chaired an international symposium entitled "The Dangers of Nuclear War" and coedited a book based on the symposium.

Polanyi's influence over others is greatly enhanced by his personal appearance and style. Many who have met him for the first time, having previously known him by reputation, have found it difficult

JOHN C. POLANYI
Photo courtesy of Dave Sideway, *The Montreal Gazette.*

to relate this youthful-looking man of diffident manner and disarming modesty to their preconception of a man of such outstanding qualifications and achievements.

Polanyi has received many honors, including honorary degrees from a number of universities. He is a Fellow of the Royal Society of Canada, receiving its Henry Marshall Tory Medal in 1977, and is a founding member of its Committee on Scholarly Freedom. He is a Fellow of the Royal Society of London, receiving its Royal Medal in 1989, and is a Foreign Associate of the U.S. National Academy of Sciences. He received the Isaac Walton Killiam Memorial Prize in 1988 and is a Companion of the Order of Canada.

KEITH J. LAIDLER
*University of Ottawa*

# Bibliography

Carrington, T.; Polanyi, J. C.; Wolf, R . J. "Probing the Transition State in Reactive Collisions"; in *Physics of Electronic and Atomic Collisions*; Datz, S., Ed.; North Holland: Amsterdam, Netherlands, 1982; pp 393–401.

Ding, A. M. C.; Kirsch, L. J.; Perry, D. S.; Polanyi, J. C.; Schreiber, J. L. "Effect of Changing Reagent Energy on Reaction Probability and Product Energy Distribution"; *Faraday Disc. Chem. Soc.* **1973**, *55*, 252–276.

Harrison, I.; Polanyi, J. C.; Young, P. A. "Photochemistry of Adsorbed Molecules. IV. Photodissociation of $H_2S$ on LiF (001)"; *J. Chem. Phys.* **1988**, *89*, 1498–1523.

Polanyi, J. C. "Dynamics of Chemical Reactions"; *Disc. Faraday Soc.* **1967**, *44*, 293–307.

Polanyi, J. C. "Energy Distribution among Reaction Products and Infrared Chemiluminescence"; *Chem. Britain* **1966**, *2*, 151–158.

Polanyi, J. C. "Infrared Chemiluminescence"; *Quant. Spectros. Radiat. Transfer* **1963**, *3*, 471–496.

Polanyi, J. C. "Some Concepts in Chemical Dynamics"; *Acc. Chem. Res.* **1972**, *5*, 161–168.

Polanyi, J. C. "Some Concepts in Reaction Dynamics"; *Science* **1987**, *236*, 680–690.

Polanyi, J. C. "The Transition State"; In *The Chemical Bond: Structure and Dynamics;* Zewail, A. H., Ed.; Academic: San Diego, CA, 1992; pp 149–173.

Polanyi, J. C. "Vibrational Rotational Population Inversion"; *Applied Optics. Supp. 2 of Chemical Lasers* **1965**, 109–127.

# 1987

NOBEL LAUREATE

# *Donald J. Cram*

1919–

The third edition of a widely used textbook in physical organic chemistry, published in 1987, lists in its general index the names of a few of the most prominent contributors to the field. Included in this select group is Donald J. Cram, professor of chemistry at the University of California at Los Angeles and corecipient of the 1987 Nobel Prize in chemistry. One might be surprised, however, that there is no mention in the textbook of the work for which he received the prize. Instead, it deals only with his seminal contributions to the structure and characteristics of phenonium ions, the stereochemistry of carbanions as studied by proton-exchange processes, and the stereochemistry of carbonyl addition reactions, these being some of the areas of research in which Professor Cram was engaged for approximately the first half of his illustrious scientific career. It was not until midcareer, and as the result of a conscious, bold, and prescient decision, that his work took on the character that has so fascinated the chemical public and that, 20 years after its inception, captured the attention of the Nobel committee.

Donald J. Cram was born on April 22, 1919, in Chester, Vermont, the son of a Scottish father and a German mother who had migrated from Ontario, Canada, with their three girls just prior to his birth. Two years later the family moved to Brattleboro, Vermont, where young Donald spent the rest of his youth, a time that was generally a happy one though made difficult by the death of his father when Donald was only four years old. According to his siblings, Donald was a precocious child, always curious, and frequently causing mischief. He was first an early reader, devouring children's books by the age of four, and then a voracious reader, making his way through Dickens, Kipling, Scott, Shaw, Shakespeare, and many other authors. But by no means was he a bookish lad. Endowed by nature with a big and strong physique, he also "learned how to run up spring-fed brooks and jump from one glacier-polished stone to another". And, to help the precarious finances of a mother rearing five children in a fatherless family, he carried firewood, emptied ashes, shoveled snow, picked apples, raked leaves, farmed, delivered newspapers, and mowed the lawn of a dentist who paid him by filling his cavities and pulling his teeth. By the age of 16 he had learned how to adapt to 18 different employers. Also he had become a good enough athlete to play high school varsity tennis, football, and ice hockey.

Family rearrangements led to a change of scene, and young Donald finished his high school work at Winwood, a small private school on Long Island, where he took a course in chemistry, taught himself solid geometry, and won a $6000 four-year National Rollins College Honor Scholarship. At Rollins College in Florida he worked diligently at chemistry, dabbled in philosophy, extended his reading to Dostoyevsky, Spengler, and Tolstoy, sang in the choir and in a barbershop quartet, obtained an airplane pilot's license, acted in plays, and produced and announced a radio program. It was an unusually wide-ranging set of activities presaging the diversity of a professional career that lay ahead.

Summers during his college years, 1938–1941, were spent at the National Biscuit Company in New York City, initially as a salesman, which brought him face to face with the poverty of the ghetto. The experience, though educational, was physically taxing, reducing his normally robust 195 pounds to a gangly 155 pounds. In contrast, the next summers were spent in the laboratory doing the physically less demanding job of analyzing cheeses for moisture and fat content. These summer jobs had a profound effect on the young man, teaching him self-discipline, enlightening him as to the differences among people, and convincing him of his dislike for repetitive activities. Back at school after his first summer in New York, he concluded that research had a magic aura about it and that to pursue it was the most enticing of all. To quote him directly, "Research became my god and the conducting of it my act of prayer, a thought that has sustained me from that time to the present."

In contrast to today, when departments of chemistry seek graduate students with an intensity rivaling the recruitment of football players, teaching assistantships in 1941 were scarce. Of the 17 applications that the young college graduate Cram submitted for entry into a graduate program in chemistry, only three resulted in offers. He accepted the one from the University of Nebraska, and a year later, working under the supervision of Professor Norman O. Cromwell, he received an M.S. degree. World War II was underway by that time, so he joined Merck & Co., Inc. to work on the penicillin program. At the conclusion of the war in 1945 his mentor at Merck, the great Max Tishler, arranged for him to attend Harvard and work for Professor Louis F. Fieser. Eighteen months later, during which time he had been supported by a National Research Council Fellowship, Mr. Cram became Dr. Cram. Having accepted an academic position from the University of California at Los Angeles and with four months to spare, but lacking funds for the cross country trip, he accepted a three month postdoctoral position with Professor John D. Roberts at M.I.T.

At UCLA, assistant professor Cram became a colleague of Saul Winstein, one of the pioneers in the field of physical organic chemistry and who, but for his premature death in 1969 at the age of 57, might himself have won a Nobel Prize. Cram's early work was clearly influenced by this association, as it was also influenced by his contact with Professors Paul D. Bartlett and Robert B. Woodward at Harvard and Professor John D. Roberts, then at M.I.T. and later at the California Institute of Technology. Delving fearlessly into the thicket of neighboring group effects, a phenomenon actively studied by Winstein at the time, Cram carried out the solvolysis of L-*threo* and L-*erythro* isomers of 3-phenyl-2-butyl tosylate in acetic acid and interpreted the results in terms of an intermediate that he called the "phenonium ion", demonstrating at this early stage in his career a talent for coining catchy names for chemical structures and phenomena. Over the next 20 years, extending these early studies of the phenonium ion and also initiating experiments in a variety of these areas, with particular emphasis on carbanion chemistry, he established himself as a major figure in the area of physical organic chemistry, rising quickly through the ranks at UCLA to become a full professor in 1955. In 1965 he wrote *Fundamentals of Carbanion Chemistry*, a book summarizing the state of knowledge in the field to which he was a major contributor. Missing from the book, however, was any discussion of his other contributions during those first 20 years, including "Cram's rule for asymmetric induction", his exploration of a class of compounds to which he gave the name "cyclo-phanes", and his studies of the stereochemistry of sulfur compounds. But all of this proved a mere prologue for what was to follow. In 1967 Charles Pedersen of the DuPont Company published a paper describing the synthesis of a new class of macrocyclic polyethers, which he named "crown ethers". Pedersen showed them to be capable of forming organic solvent soluble complexes with $K^+$, $Na^+$, and various other metal ions. Cram was quick to perceive a use for these metal-complexing agents in his studies of electrophilic substitution at saturated carbon centers, an application that he discussed in June 1969 at a national organic chemistry symposium in Salt Lake City entitled "Invisible and Revealed Reaction Mechanisms of Carbanions". He soon realized that the crown ethers had possibilities far beyond this limited application, and in 1973 he published a short communication in which he stated that "the pioneering 'Pedersen Papers' on the crown ether chemistry and the remarkable properties of Simmons and Park's out–in bicyclic amines and Lehn's cryptates stimulated us to turn from the study of effects of crown ethers on ion-pair phenomena to synthesizing multiheteromacrocycles with cavities shaped to complex selected species." Starting first with a few postdoctoral associates and graduate students, this embryonic group quickly expanded and became the dominant part of Cram's research program. From the outset Cram focused on the construction of chiral crown ethers, realizing the critical role that optically active compounds had played in much of his earlier work. Also from the outset great reliance was placed on the use of space-filling molecular models, by means of which the complementary relationships between guest and host molecules can be readily perceived. Aided by an army of talented graduate students and postdoctoral associates; provided with the fine facilities of one of the country's greatest academic institutions; and working with a high degree of insight, imagination, and tenacity, Cram succeeded in making the crown ethers do his bidding. He published a body of work over the next 14 years for which the 1987 Nobel Prize was awarded.

In the years immediately following Cram's conversion in the early 1970s to this area of investigation, which he named "host–guest chemistry", a number of short reports of results were published, as well as a longer article in *Science*. Not until 1977, however, did the first major paper appear in what was to become a long series of publications that was still growing at the time of the Nobel Prize award. This paper described in exhaustive detail the synthesis and testing of a family of optically active crown ethers capable of selectively complexing one of the mirror images (enantiomers) of a mirror image pair (a racemic mixture) of guest molecules, thus mimicking the enzymes' remarkable ability to perform this feat. However, these first chiral crown ethers paled in comparison to the enzymes with respect to tightness of complexation and the degree of enantiomeric selectivity. Cram's attention was therefore directed to designing other kinds of compounds containing oxygen functions in a cyclic array. This proved to be a remarkably productive endeavor and resulted in the synthesis of several new classes of macrocyclic compounds, to which Cram gave such descriptive names as spherands, hemispherands, cryptaspherands, and carcerands. With these new compounds the goal was expanded to mimic not only the ability of enzymes to discriminate among enantiomers but also their ability to increase the rates of reactions by many orders of magnitude. The boldest and most challenging of these syntheses was a 30-step sequence that yielded a macrocyclic, basket-shaped molecule carrying hydroxyl and imidazole groups designed to mimic the enzyme chymotrypsin, which catalyzes the hydrolysis of a variety of substances in living systems. It worked! It showed enantioselectivity, and it accelerated the rate of acetyl transfer from the guest molecule. Although the absolute magnitude of the rate enhancement shown by the Cram chymotrypsin mimic depends on one's choice of a reference system, the ability to make and test this compound provided a striking demonstration of how far crown ether chemistry had come from Pedersen's discovery of 18-crown-6 in the 1960s. Cram's focus on the detailed structures of a wide variety of host molecules and their complementary relationship to guest molecules revealed the great importance of preorganization of the host, adding insight to the riddle of host–guest and enzyme–substrate interactions. His work, notable for its exquisite attention to detail, its careful execution, and its thorough and comprehensive consideration of the many facets of the problem, was instrumental in bringing about this transformation. The extraordinary expansion and proliferation of host–guest chemistry in the 1980s gave credence to Cram's bold move of changing the direction of his research in midcareer.

Teaching and research, though generally carried on concurrently by the academic scientist, are often perceived to be in competition with one another. In the case of Donald Cram, however, they were truly complementary activities. To a degree remarkable for a scientist of great distinction in research, Cram addressed the problems of the undergraduate classroom and became one of the great chemistry teachers in the United States. Three of his seven books were organic chemistry textbooks, used by the 10,000 students who took his courses at UCLA as well as by many more thousands at colleges and universities elsewhere throughout the world. One or more of his textbooks were translated into 12 different languages. Popular as a lecturer because of his clear and lucid style, he is also remembered by his students as an accomplished musician who ended the semesters, to great applause, by singing folk songs while accompanying himself on the guitar.

What makes a distinguished scientist? Donald Cram places great significance in his case on the early death of his father, a circumstance that deprived him of an immediate role model and required him, as he states, "to construct a model composed of pieces taken from many different individuals, some being people I studied and others lifted from books". Dominant among those whom he studied was Professor Saul Winstein, his colleague, competitor, and friend for many years at UCLA, whose name adorns the endowed professorship to which Cram was appointed in 1985. But also important to Cram's success was his ability to choose realizable goals and to work toward them with single-minded attention. Skiing, surfboarding, playing tennis, and tending his garden were leisure-time enjoyments, but professional activities always took precedence. Although childless in two marriages, the first to Jean Turner Cram from 1940 to 1968 and the second to Jane Maxwell Cram in 1969, he considers his more than 200 graduate students and post-doctoral associates to be his extended family.

Professor Cram's many contributions to science have been recognized by numerous awards in addition to the Nobel Prize. From the American Chemical Society he received the Award for Creative Work in Synthetic Organic Chemistry (1965), the Arthur C. Cope Award for Distinguished Achievement in Organic Chemistry (1974), and the Roger Adams Award in Organic Chemistry (1985). Local sections of the ACS have awarded him the Willard Gibbs Medal and the Tolman Medal. In 1961 he was elected to the National Academy of Sciences, in 1974 he was California Scientist of the Year, and in 1976 he was Chemistry Lecturer and Medalist of the Royal Institute of Chemistry (U.K.). He received the National Academy of Science Award in the Chemical Sciences in 1992. He has also received honorary doctorates from Uppsala University (1977), the University of Southern California (1983), Rollins College (1988), the University of Nebraska (1989), the University of Western Ontario (1989), and the University of Sheffield

(U. K., 1991). Although 68 years old at the time of his Nobel Prize, Professor Cram's research pace did not slacken, for he viewed the award simply as a means to facilitate and promote the continued exploration of the field of chemistry he has been so instrumental in establishing. His 52-year-old research career has resulted to date in more than 400 papers.

C. DAVID GUTSCHE

*Texas Christian University*

# Bibliography

Cram, D. J. "The Design of Molecular Hosts, Guests, and Their Complexes"; *J. Inclusion Phenomena* **1988,** *6,* 397–413; *Angewandte Chemie, International Edition in English* **1988,** *27,* 1009–1112.

Cram, D. J. "Designed Host–Guest Relationships"; In *Design and Synthesis of Organic Molecules Based on Molecular Recognition;* von Binst, G., Ed.; Springer-Verlag: New York, 1986; pp 153–172.

Cram, D. J. *From Design to Discovery;* Profiles, Pathways, and Dreams Series; Seeman, J. I., Ed.; American Chemical Society: Washington, DC, 1990.

Cram, D. J. *Fundamentals of Carbanion Chemistry;* Academic Press: Orlando, FL, 1965.

Cram, D. J. "Molecular Cells, Their Guests, Portals, and Behavior"; *CHEMTECH* **1987,** 120–125.

Cram, D. J. "Preorganization—From Solvents to Spherands"; *Angewandte Chemie, International Edition in English* **1986,** *25,* 1039–1134.

Cram, D. J.; Cram, J. M. "Host–Guest Chemistry"; *Science* **1974,** *183,* 803–809.

*Cyclophanes;* Volgtle, F., Ed.; Springer-Verlag: New York, 1983.

*International Who's Who: 1991–1992,* 55th ed.; Europa: London, 1991; p 349.

Kyba, E. P.; Helgeson, R. C.; Madan, K.; Gokel, G. W.; Tarnowski, T. L.; Moore, S. S.; Cram, D. J. "Host–Guest Complexation. 1. Concept and Illustration; *J. Am. Chem. Soc.* **1977,** *99,* 2564–2571.

# Jean-Marie Lehn

## 1939–

Jean-Marie Lehn was awarded the Nobel Prize in chemistry in 1987, jointly with Donald J. Cram and Charles J. Pedersen, "for their development and use of molecules with structure-specific interactions of high selectivity". Lehn was born on September 30, 1939, in Rosheim, a small city of Alsace in France. His parents were Pierre Lehn, who had been a baker and later the organist of the city, and Marie Lehn, who kept the house and brought up the four boys of the family. In 1965 Jean-Marie Lehn married Sylvie Lederer, who is teaching mathematics at the University Louis Pasteur in Strasbourg; they have two sons, David and Mathias.

He attended high school from 1950 to 1957 and studied classics, but he oriented himself to science when he entered the University of Strasbourg. In 1960 he obtained the degree of *Licencié-ès-Sciences* (bachelor's). The same year, to work toward a Ph.D. degree, he joined, as a junior member of the Centre National de la Recherche Scientifique, the Laboratoire de Chimie des Substances Naturelles, which was headed by Professor Guy Ourisson, one of the most dynamic young chemists in France. Lehn practiced organic synthesis and was in charge of the laboratory's first NMR spectrometer; he was rapidly considered an expert in NMR spectrometry. These years had a great influence on his subsequent career, mainly because he acquired high-level training in both organic and physical chemistry. His *Doctorat-ès-Sciences* (Ph. D.) thesis at Strasbourg in 1963 dealt with the NMR of the triterpenes. Lehn was considered at that time a highly gifted young scientist: By the year of his doctorate, after only three years of research, he already had his name on 16 publications. After receiving his Ph.D., he spent a year in the laboratory of the great Robert B. Woodward at Harvard University, where he participated in the total synthesis of vitamin $B_{12}$.

After his return, Lehn developed an interest in several fields, mainly in physical organic chemistry. Topics studied by his group included nitrogen inversion, conformational rate processes, quadrupolar relaxation, molecular motions, and liquid structure. To achieve a better understanding of some of these phenomena, he expanded his investigations to quantum chemical calculations on nitrogen inversion, electronic structure, and stereoelectronic effects. In 1970 Lehn wrote a review, "Nitrogen Inversion", that collected a large number of results from his group and demonstrated his important contributions to this field.[1]

Concerning this period (1965–1970) of his scientific activity, it is important to stress one aspect of his approach to physical organic chemistry: Because of his dual training, he pursued the synthesis and the physicochemical studies of the desired compounds. Many molecules were synthesized in order to reveal a given property or to check the validity of a relationship between a property and the structure of the chemical species. This way to operate was simply and already that of designing a compound for a given purpose, an approach that would be of central importance later. In 1966 Lehn was appointed assistant professor at the chemistry department of the University of Strasbourg and promoted to full professor in 1970.

Although the various projects and results at his laboratory were developing very well, Lehn was anxious to initiate a new field. Because of his interests in neurochemistry, he paid close attention to the progress made by biochemists in the understanding of certain fundamental processes, in particular, the electrical events in nerves. Because the phenomena rest on changes in the distribution of sodium and potassium ions across the membrane, he focused his attention on these cations. Strangely, but for several obvious reasons, those involved at that time in coordination chemistry did not consider the complexation of the cations of group I (and II) to be an important target of research. The discovery that an antibiotic, valinomycin, was able to mediate potassium transport in mitochondria attracted Lehn's attention, and when Pedersen published his first paper on the subject in 1967, on macro-

---

[1] Topics in Current Chemistry 1970, 15, 311.

cyclic polyethers, he realized immediately the full significance of the described results. He realized also that macrobicyclic systems with their three-dimensional cavity, rather than macrocyclic ones, would achieve a much stronger binding of the cations. This idea led to the design of 4,7,13,16,21,24-hexaoxa-1,10-diazabicyclo[8,8,8]hexacosane. The synthesis of this compound was accomplished in the fall of 1968, and its strong complexation ability toward the potassium cation was established very rapidly thereafter by several methods. Crystal structure determination demonstrated unambiguously the location of the cation inside the macrobicyclic cavity. Lehn proposed the names *cryptand* for the free ligand and *cryptate* for its complex.

The first synthesis was followed by the modification of the cavity size by altering the length of the bridging chains. This led to a series of ligands exhibiting high selectivity of complexation toward almost all representatives of groups I and II. Macrobicyclic ligands incorporating other donor atoms (e.g., sulfur, nitrogen) in place of oxygen were also synthesized and studied.

In 1973 Lehn wrote a broad article, "Design of Organic Complexing Agents. Strategies towards Properties"[2], which gave the complete story of the newborn field. This paper is a masterpiece; it contains the entire philosophy of the molecular design. Twenty years after its publication, this article is still of great value to read.

The strong and selective complexation by a ligand of alkali and alkaline earth cations is the simplest recognition process (spherical recognition). Once this goal was achieved, Lehn developed systems able to bind efficiently a tetrahedral species, the ammonium cation, and its substituted derivatives, the primary ammonium ions $R-NH_3^+$. A receptor for the planar guanidinium cation was also synthesized. In all these examples the binding of the cation rests on several $N^+-H\cdots O$ hydrogen bonds.

A further step in complexity was accomplished by the design of ligands capable of complexing more than one cation or of substrates bearing several functional groups able to interact with the binding sites of the receptor. This design requires a discrete arrangement of the ligand interacting atoms, that is, the formation of two or more binding subunits, each able to complex one cation or one functional group. The compounds of this class were named polytopic receptors. More precisely, some of such systems can form dinuclear or trinuclear metal complexes. Other receptors of this type can bind linear diammonium

---

[2] *Structure and Bonding* **1973**, *16,1.*

substrates, $H_3N^+ - (CH_2)_n - NH_3^+$, with high selectivity. This specificity was obtained by an appropriate separation depending on the length of the substrate (determined by $n$), thus achieving of the two binding subunits of the receptor, a linear recognition of the substrate.

In the mid-1970s Lehn realized that the tools (macrocyclic and macrobicyclic structures) used in cation complexation could be extended to the unexplored area of anion coordination chemistry, provided that the ligand bears anion-binding sites (guanidinium or ammonium units, for example). Many anion-complexing ligands were synthesized by his group; they display efficient binding of halides, carboxylates, phosphates, and the like.

All the results described above concern the recognition of a species by a specifically designed receptor. The research has yielded numerous applications: the separation of cations (including isotopic enrichment and the removal of harmful metals) by extraction or by polymer-bound ligands, and the formation of alkali anions, enabled by cryptands. In addition, the good solubility of many cryptates in an apolar medium involves a high reactivity of the associated anion, which allows cryptates to be used in anion activation, in phase-transfer catalysis, or in the formation of a highly basic medium.

For each of these applications the ligand first tested (for example, a diaza-polyoxa-macrobicycle) can be modified slightly to optimize the behavior of its corresponding complex. This approach is elegantly exemplified in the design of a *carrier* molecule; the first series of cryptands form very stable complexes with the alkali cations, but they are poor carriers. By suitable structural changes (lipophilic or hydrophilic balance and number of binding sites), the receptor molecule can be transformed into a genuine carrier.

A higher level of sophistication was achieved by Lehn when he designed ligands that, in addition to recognition ability, display also chemical reactivity. Thus a ligand bearing an appropriate reactive group can bind a substrate and react with it. Ligands of this class have been designed for both cationic and anionic substrates, and their efficiency has been demonstrated by several examples. A more elaborate system allowed the formation of a bond between two substrates bound by the ligand, thus performing cocatalysis. Such reagents also represent approaches to enzyme models or so-called artificial enzymes.

In the course of his investigations Lehn suspected that the original character of the new field needed to be more clearly specified. In 1978, he proposed the name *supramolecular chemistry* for the field, and for the complex entity (receptor+substrate) he suggested the name *su-*

*permolecule*. Thus he defined "supramolecular chemistry as chemistry beyond the molecule, bearing on the organized entities of higher complexity that result from the association of two or more chemical species held together by intermolecular forces"; he added that "one may say that supermolecules are to molecules and the intermolecular bond what molecules are to atoms and the covalent bond." These two terms should not just be considered two new names; they are of great conceptual importance. A supermolecule has to be seen as a whole, as a new species with its own properties. Therefore, the design of the receptor is directed toward several very precise targets; that is, the designer defines all the specifications of the projected supermolecule. The supermolecule concept is structuring the thoughts of the chemist involved in this type of chemistry. Lehn gave a detailed account on supramolecular chemistry in the formal *Leçon Inaugurale* when he was elected professor at the Collège de France in 1979.

Many other aspects of supramolecular chemistry have been investigated by Lehn. They include attempts to design an artificial cation channel based on the superposition of macrocyclic units, the photochemistry of several photoactive cryptates, and the molecular self-assembling of linear ligands and suitable metal ions leading to metal complexes of double helix structure. The summary of Lehn's contribution to supramolecular chemistry can be found in his Nobel lecture. The more recent developments of supramolecular chemistry concern mainly self-assembly processes and the design of programmed molecular systems[3] as devices for the storage and manipulation of molecular information.

It is not necessary to say that Lehn is a hard-working man, but certainly more important, he is totally relaxed. This explains how he is able to conduct all the various aspects of the laboratory organization simultaneously, spanning from the creativity in projects, to the smallest technical problem, to chatting with collaborators on nonscientific matters. He has a tremendous imagination and finds, apparently without any difficulty, numerous new subjects. A detailed look at his publications reveals that he stays on a project until the very important facts have been demonstrated and then he stops and turns to a new subject.

This short biography on Jean-Marie Lehn would not be complete without an account of his extraprofessional activities and interests. As a legacy from his father he likes music and plays the piano and the organ.

---

[3] *Angew. Chem.* **1990**, *29*, 1304.

He is an amateur practitioner of arts (painting and sculpture), chiefly in their modern expressions. He takes advantage of opportunities during his numerous professional travels to visit museums. He considers these short breaks as his holidays and never takes vacations.

By 1993, at the age of fifty-four, Lehn had published about 400 publications and reviews; he has presented a number of distinguished invited lectures and received many honors. At present he has a triple life: in Paris, as professor at the Collège de France, director of the Laboratoire de Chimie des Interactions Moléculaires; in Strasbourg, as director of the Laboratoire de Chimie Supramoléculaire and science advisor for chemistry to the Rhône-Poulenc Company. Despite all these obligations his influence and creativity in chemistry will continue for a long time.

BERNARD DIETRICH
*Université Louis Pasteur*

# Bibliography

Colquhoun, H.; Stoddart, F.; Williams, D. "Chemistry Beyond The Molecule"; *New Scientist* **1986**, *110* (May 1), 44–48.

Dietrich, B. "Cryptate Complexes"; In *Inclusion Compounds;* Atwood, J. L.; Davies, J. E. D.; MacNicol, D. D., Eds.; Academic Press: London, 1984; p 337.

Lehn, J.-M. "Cryptates: The Chemistry of Macropolycyclic Inclusion Complexes"; *Acc. Chem. Res.* **1978**, *11,* 49.

Lehn, J.-M. "Dinuclear Cryptates: Dimetallic Macropolycyclic Inclusion Complexes. Concepts—Design—Prospects"; *Pure & Appl. Chem.* **1980**, *52,* 2441.

Lehn, J.-M. "Macrocyclic Receptor Molecules: Aspects of chemical reactivity. Investigations into molecular catalysis and transport processes"; *Pure & Appl. Chem.* **1979**, *51,* 979.

Lehn, J.-M. "Perspectives in supramolecular chemistry—From molecular recognition towards molecular information processing and self-organization"; *Angew. Chem. Int. Ed. Engl.* **1990**, *29,* 1304.

Lehn, J.-M. "Supramolecular Chemistry: Receptors, Catalysts and Carriers"; *Science* **1985**, *227,* 849.

Lehn, J.-M.; Popov, A. I. "Physicochemical Studies of Crown and Cryptate Complexes"; In *Coordination Chemistry of Macrocyclic Compounds;* Plenum: New York, 1979; p 5.

Potvin, P. G.; Lehn, J.-M. "Design of Cation and Anion Receptors, Catalysts, and Carriers"; In *Synthesis of Macrocycles: The Design of Selective Complexing Agents;* Izatt, R. M.; Christensen, J. J., Eds.; John Wiley & Sons: New York, 1987; p 167.

Stoddart, J. F. "Host-Guest Chemistry"; *Royal Society of Chemistry, Annual Reports B* **1988.** *85,* 353.

Vögtle, F. *Supramolekulare Chemie;* B. G. Teubner: Stuttgart, Germany, 1989.

# 1987

NOBEL LAUREATE

# Charles J. Pedersen

## 1904–1989

Charles John Pedersen was awarded the Nobel Prize in chemistry in 1987 with Donald J. Cram and Jean-Marie Lehn "for their development and use of molecules with structure-specific interactions of high selectivity." Pedersen was born October 3, 1904, in Fusan, Korea, and died at his home in Salem, New Jersey, on October 27, 1989. His parents were Brede Pedersen, a Norwegian engineer, and Takino Yasui, a Japanese woman from Kyushu whose family had emigrated to Korea to trade in soybeans and silkworms. An early interest in science was sparked by his life at the Unsan gold mines in northwestern Korea, where his father worked. After primary education at boarding schools in Japan, two years in Nagasaki and seven at St. Joseph College, a French preparatory school in Yokohama, he came to America in 1922. He received a degree in chemical engineering in 1926 from the University of Dayton, Ohio, and an M.S. in organic chemistry in 1927 at the Massachusetts Institute of Technology. His professor, James F. Norris, recognized the ability of this unassuming young man and tried to persuade him to seek a doctorate, but Pedersen no longer wished to burden his father financially and chose to go to work.

With Norris's help he obtained a position with the DuPont Company, where he spent his entire career of 42 years, mostly at Jackson Laboratory, their most diversified applied research organization. After a brief indoctrination in analytic work, he gained the support of two mentors, his director W. S. Calcott and associate director A. S. Carter, who did all they could to guide and encourage him while at the same time becoming close friends. Calcott sensed Pedersen's unusual quality and chose to keep him in research rather than send him to manufacturing as was usual for those with no doctoral degree. They also fostered some of his lifelong interests: love of the natural world, stamp collecting, and fishing, which was originally inspired by his father's love for the sea. A voracious reader, Pedersen rapidly broadened his knowledge of science and his hobbies.

Pedersen soon learned the nature of research and relevant DuPont technology and also developed a personal investigative style, intuitive and unfettered by scientific dogma. He was a hands-on chemist, an experimentalist rather than theoretician, a keen observer of what was happening. He had an uncanny ability to ignore confusing details and find a direct route to his goal, usually through the simplest of experiments. Although scientific in his methodology, he was very much an industrial chemist. Above all he wanted his research to be useful, to have practical application.

After five years of success on a variety of problems typical of a diversified chemical business, Pedersen had established his self-confidence and his reputation as a fine chemist. Then in 1932 he was confronted with a situation in which the yield of tetraethyl lead in the plant was less than that indicated by the consumption of sodium and lead alloy. By simple experiments he showed that the missing product was adsorbed on finely divided lead sludge and could be recovered by the addition of a nonfoaming wetting agent before the steam distillation.[1] Thus he increased the yield of a high-volume process substantially and made what was probably his most commercially profitable discovery.

In 1935 he made another key discovery: the first deactivators of metal contaminants (especially copper), which catalyze the oxidation of oils, gasoline, and rubber. Clues in a search for hydrocarbon-soluble precipitants or chelating agents for copper led him to prepare disalicylal propylenediimine. It proved to be very effective in countering degradative effects of copper and other heavy metals.[2] It was soon used in almost all gasoline and much rubber and is still important today. It

---

[1] Pedersen, C. J. et al., "Tetraalkyl Lead"; U.S. Patent 2,004,160, July 11, 1935.

[2] Downing, F. B.; Pedersen, C. J. "Stabilization of Organic Substances"; U.S. Patents 2,181,121, and 2,181,122, November 28, 1939.

was a source of great personal satisfaction for him to see the name of his metal deactivator on all gasoline pumps.

The metal deactivator discovery reshaped Pedersen's career. It turned his attention toward the catalytic effects of metal ions and especially to the behavior of metals in complexes. In exquisite detail he varied the structures of ligands to determine those features that led to the strongest complexes and the most effective deactivation of the transition metals.[3] He also discovered an unusual synergistic effect of metal deactivators in improving antioxidant efficacy.[4] These studies led him to peroxide and radical behavior and to the chemistry of oxidation processes. As a result he received about 30 patents covering antioxidants, stabilizers, and inhibitors, particularly chelating compounds for transition metals.[5] Even outside DuPont he was now famous as an outstanding researcher in the field of stabilization and degradation.

Although Pedersen carried a Norwegian passport, he had never been to Norway to activate it. Throughout World War II he was officially classified as a stateless alien. This caused no real problem. As a supervisor he continued his work on nonclassified problems in the petroleum field. As an acknowledged expert in the stabilization of petroleum products he was of value to the government and the armed forces and served as a wartime consultant. He was very highly regarded.

In 1947 Pedersen married Susan J. Ault. Sue and Charlie became devoted companions until her death in 1983. Concurrently, in recognition of his ability in research he was appointed research associate, the top scientific position in DuPont. This freed him of supervisory responsibilities, which he found onerous, and allowed him to practice research as he wished: independently, with complete responsibility for his programs. Delighted with his new freedom, he returned to studies of the prooxidant catalytic activity of metal chelates, of oxidation mechanisms, and of the decomposition products of peroxides.

In this capacity he came to work with his friend H. E. Schroeder, who encouraged him to broaden his chemical horizons. New develop-

[3]Pedersen, C. J. "The Suppression of Metal Catalysis in Gasoline Gum Formation"; *Oil and Gas Journal* **1939,** July 27, 97.

[4]Downing, F. B.; Pedersen, C. J. "Stabilization of Organic Substances"; U.S. Patent 2,336,598, December 14, 1943.

[5]Pedersen, C. J. "Inhibition of Deterioration of Cracked Gasoline during Storage"; *Industrial and Engineering Chemistry* **1949,** 41 (May), 824.

ments soon appeared. An interest in the chemistry of macrocycles was engendered by puzzling contrasts in azaporphyrin chemistry. Iron and cobalt phthalocyanines showed reversible redox behavior not shared by copper, nickel, or metal-free types. Intrigued by such differences, Pedersen soon found that the purportedly stable types were easily oxidized and formed solvent-soluble adducts with peroxides and hypochlorites. Subsequent chemical or photochemical reduction quickly restored the original pigments. These reactions could be used to produce brilliant blue and green colors on textiles or to generate images.[6]

Pedersen's curiosity as to the fate of $N,N'$-diaryl $p$-phenylenediamine antioxidants during use led to the discovery of new p-quinonediimine $N,N'$-dioxides. These interesting substances function as both polymerization inhibitors and antioxidants, in which capacities they react with hydrocarbon radicals rather than with peroxy intermediates. Further, they decompose photochemically into $p$-quinoneimine $N$-oxides and azo compounds.[7]

In 1951 Kealy and Pauson's note in Nature on the unusual iron complex later called ferrocene caught Pedersen's eye. He felt that the authors had not fully grasped the nature of the substance or its significance (his conception was later proved to be correct). He soon found that ferrocene was a fine combustion-control agent of value because it was both hydrocarbon-soluble and volatile. More importantly, at very low concentration in gasoline it was an antiknock agent, the first substance to be highly effective since tetraethyllead, yet it was not so toxic as the lead compound. In view of potential military importance, his March 1952 patent application was immediately placed under a U.S. government secrecy order. That was rescinded in 1960, but by then engine tests had shown that ferric oxide formed on combustion and caused excessive wear.[8] Nevertheless, patent rights were traded for rights of considerable value. He also discovered that ferrocene was a very effective combustion catalyst. Small amounts stop the formation of soot or smoke in a hydrocarbon flame. When hydrocarbon fuels containing ferrocene were mixed with fuming nitric acid, ignition

---

[6]Pedersen, C. J. "Reversible Oxidation of Phthalocyanines"; *J. Org. Chem.* **1957**, *22*, 127.

[7](Pedersen, C. J. *J. Am. Chem. Soc.* **1957**, *79*, 2295; 5014.

[8]Pedersen, C. J. "Liquid Hydrocarbon Fuels"; U.S. Patents 2,867,516, 1959; 3,341,311, September 2, 1967.

was spontaneous. Such hypergolic mixtures were of potential value as rocket fuels.[9]

In 1957 Pedersen chose to join a new DuPont department dedicated to elastomeric polymers because he welcomed a new challenge and because his friends Carter and Schroeder would be the research management. With his strong focus on the utility of research he undertook a series of very fruitful studies of polymerization catalysis, polymer stability, and new polymers. Then in 1961, with Schroeder's encouragement, he left polymer research and returned to his beloved coordination chemistry. He hoped to unravel the effects that various ligands have on the catalytic behavior of vanadium in oxidation and polymerization reactions. He chose to synthesize new phenolic ligands with multiple oxygen-bonding sites, starting with ethanoxy structures derived from polyhydric phenols.

His most famous discovery occurred in July 1962 when he tried to make a five-toothed ligand from a monoether of catechol and dichlorodiethyl ether and isolated some fibrous crystals in 0.4% yield. It was a neutral substance with the ability to bind with sodium and potassium ions rendering them soluble in organic media. This behavior was so unusual that he pursued its identity with unrelenting zeal. He found that it had a toroidal 18-membered ring structure and resulted from a reaction involving catechol contaminant. This was the first of what he later named "crown ethers", dibenzo-18-crown-6.

Intuitively and almost immediately Pedersen saw the significance of his find, and he devoted all his energy to the study of these novel substances. He first described his work in 1967 at a conference and in his monumental 1967 note and paper describing about 50 crown ethers with from 9- to 60-membered oligoether rings of different hole sizes with the ability to accept ions of various diameters. This discovery was an amazing accomplishment for a single chemist assisted by one technician.[10] He tells of events leading to the discovery in "Macrocyclic Polyethers for Complexing Metals".[11] Reaction to the work was fast and enthusiastic.

Knowing full well the unusual nature of his discovery, Pedersen spent the rest of his career with DuPont on the chemistry of the crown

---

[9]Pedersen, C. J. "Fuels and Process for Burning Them"; U.S. Patents 3,038,299–300, June 12, 1962.

[10]Pedersen, C. "Cyclic Polyethers and Their Complexes with Metal Salts"; *J. Am. Chem. Soc.* **1967,** *89,* 2495, 7017.

[11]*Aldrichimica Acta* **1971, 4,** 1.

ethers. His studies, which are described in 10 papers and 10 patents, broadened the scope greatly and included syntheses of new crown ethers plus products with sulfur and nitrogen in the rings, extensions of the preparation and characterization of crystalline salt complexes, ionic complexes, and unusual thiourea complexes. He made many suggestions for uses of the crown ethers that have proved prophetic. He was fully aware of the potential of these substances for carrying various metal ions into organic media to effect reactions.[12]

After he retired in 1969, he collaborated with Professor Mary Truter (University College, London). Their work on the crystal structures of crown ethers and their metal derivatives confirmed his view on the planarity of polyethermacrocycle complexes.[13]

The major impact of Pedersen's discovery is shown by the frequency of citation of his crown ether papers and by the many new directions of research that the discovery has opened up involving many groups, for example: cryptates, J.-M. Lehn; host–guest chemistry, D. J. Cram; naked-ion chemistry and phase-transfer catalysis, C. J. Liotta; solubilization of alkali metals, J. L. Dye; ion pairing in nonpolar solvents, J. Smid; X-ray structure of complexes, M. J. Truter; and crown ethers as ionophores, G. Eisenman.

It is worth noting that, despite his influence, accomplishments, and contributions, Pedersen did not receive a major award or prize, except for the American Chemical Society's Delaware Section award, until he won a share of the 1987 Nobel Prize in chemistry. Fellow Nobel Laureate Donald Cram wrote the following in a January 20, 1975, letter to H. K. Frensdorff supporting the nomination of Pedersen for one award:

> Pedersen's discoveries illustrate organic synthesis and the organic chemist at their best. Using a simple Williamson ether synthetic procedure and simple starting materials he synthesized in good yields a new family of cyclic polyethers. He noted that these "crown compounds" complex and lipophilize metal guest cations whose diameters are similar to the diameters of the host crown compounds. He attributed the good yields of the crown compounds in part to templating effects during the ring closures. He varied the ring sizes over a wide range by appropriate selection of starting materials and the orders in which the molecular parts are assembled. The anions of salts lipophilized by the crown compounds have in many cases never been

---

[12]Pedersen, C. J.; Frensdorff, H. K. "Macrocyclic Polyethers and Their Complexes"; *Angew. Chemie, International Edition* **1972**, *119* (January), 16–25.

[13]Truter, M. R.; Pedersen, C. J. *Endeavor* **1971**, *30*, 142.

brought into non-polar organic media before, and show greatly enhanced reactivity over what they have shown in polar media . . . .
Pedersen's discoveries were not the usual result of the efforts of a group of competitors that he led or beat into print. Nor are they those of a large group of scientists under a director; they were his alone . . . he was alone — he was original — he did something important — he knew it was important — and he knew what to do with his discovery!

The man who did all this was not after riches, glory, or status. He loved the thrills and problems of research, and in his heart he most wanted his work to be worthwhile and so recognized. As a Nobel Laureate his modest and unassuming nature never changed.

HERMAN E. SCHROEDER

# Bibliography

Dietrich, B.; Sauvage, J.-P. "In 1987, the Nobel Award goes to Crown-Ethers, Cryptands, and other Host–Guest Molecular Systems"; *Nouv. J. Chem.* **1988**, *12*, 725–728.

Hiraoka, M. *Crown Compounds: Their Characteristics and Applications;* Elsevier: New York, 1982.

Izatt, R.; Christiansen, J. J. *Synthesis of Macrocycles, Design of Selective Complexing Agents, Progress in Macrocyclic Chemistry;* John Wiley & Sons: New York, 1987; Vol. 3.

Lewin, R. "Chemistry in the Image of Biology"; *Science* **1987**, *238*, 611–612.

"Nobel Prizes: U.S., French Chemists Share Award"; *Chemical & Engineering News,* **1987,** 65(Oct. 19), 4–5.

Pedersen, C. J. "Antioxidants"; In *Encyclopedia Brittanica,* 1953.

Pedersen, C. J. "Oil Soluble Chelating Agents as Metal Deactivators"; In *Advances in Chelate Chemistry;* New York, 1954; pp 113–123.

Pedersen, C. J. "Synthetic Multidentate Macrocyclic Compounds"; In *Synthetic Multidentate Macrocyclic Compounds;* Izatt, R. M.; Christiansen, J. J., Eds.; Academic Press: New York, 1978; pp 1–53.

Schroeder, H. "The Productive Scientific Career of Charles J. Pedersen"; *Pure and Applied Chemistry* **1988**, *80 (April)*, 445-451. (See also *Current Topics in Macrocyclic Chemistry in Japan*; Kimura, E., Ed.; Hiroshima University School of Medicine: Hiroshima, Japan; p 9–14, for a version with biographical details and for the proceedings of a symposium honoring C. J. Pedersen for his discovery of the crown ethers; also see the Pedersen Memorial Issue of the *J. Inclusion Phen. Mol. Recognition Chem.* **1992**, *12*, 11–21.

# 1988

NOBEL LAUREATE

# Hartmut Michel

1948–

# Johann Deisenhofer

1943–

# Robert Huber

1937–

| Michel | Deisenhofer | Huber |

In 1988, the Nobel Prize in chemistry was awarded to Hartmut Michel, Johann Deisenhofer, and Robert Huber for the crystallization and structural elucidation of a membrane protein, the reaction center protein complex from the purple photosynthetic bacterium *Rhodopseudomonas viridis*.

Hartmut Michel was born on July 18, 1948, in Ludwigsburg, Württemberg, Germany. He holds a degree in biochemistry from the University of Tübingen (1975) and a Ph.D. degree from the University of Würzburg (1977), where he studied physics at the Technical University of Munich, from which he graduated in 1971. Deisenhofer obtained a doctoral degree from the Max Planck Institute for Biochemistry and the Technical University of Munich under the supervision of Robert Huber in 1974. His thesis work involved high-resolution crystallographic studies of the basic pancreatic trypsin inhibitor.

Robert Huber was born on February 20, 1937, in Munich. He received a university degree (1960) and a Ph.D. (1963) in chemistry from the Technical University of Munich, working with the crystallographer Walter Hoppe and the biochemist P. Karlson. The title of his dissertation was "Three-Dimensional Crystal and Molecular Structure of a Diazote (The Solution of a Classical Stereochemical Problem)". Michel, Deisenhofer, and Huber broke new ground in the fields of membrane protein structure and photosynthesis.

The energy necessary for the sustenance of life on Earth comes from the sun and is trapped by plants, algae, and certain bacteria in the process of photosynthesis. In photosynthesis, light is first absorbed by an array of pigments, which transfer their excitation energy to a specialized chlorophyll–protein complex called the reaction center. In plants and algae a series of very fast electron-transfer reactions with participation of other membrane protein complexes in the reaction center leads to the production of NADPH and ATP. These highly energetic molecules are then used in a series of "dark" reactions, that reduce $CO_2$ to carbohydrates.

The mechanism of the primary photochemical events of photosynthesis cannot be determined without a picture of the three-dimensional disposition of the electron donors and acceptors embedded in the protein milieu. Filling this void in our knowledge entails analyzing the X-ray diffraction patterns from highly ordered crystals of reaction center proteins—a gargantuan undertaking. One of the primary difficulties is the fact that the photosynthetic reaction center lies inside a membrane. Diffraction-quality crystals of membrane proteins are so difficult to obtain that, until recently, biochemists had deemed the task impossible. Even if a good reaction center crystal can be made, accurate amino acid sequence data and very sophisticated mathematical techniques are necessary to fit the diffraction data, which originate from thousands of atoms.

With an impressive combination of ingenuity and perseverance, Michel, Deisenhofer, and Huber solved the crystal structure of the reaction center of *Rhodopseudomonas viridis*. The collaboration among these three scientists was successful because their scientific backgrounds were uniquely complementary. Michel brought to the project his biochemical expertise, and Deisenhofer and Huber possessed extensive experience in solving crystal structures of complex biological assemblies.

Michel's scientific career began in 1969 when, after approximately 18 months of obligatory military service, he enrolled at the University of Tübingen to study biochemistry. During this period, he did research at the University of Munich, the Max Planck Institute for Biochemistry, and the Friedrich Miescher Laboratory of the Max Planck Society in Tübingen.

While completing his work in biochemistry at the University of Tübingen, Michel began what would become a long-term collaboration with Dieter Oesterhelt. After graduating from Tübingen in 1975, Michel became a graduate student in Oesterhelt's group, which had moved from Tübingen to the University of Würzburg. Oesterhelt had become a very influential figure in the field of bioenergetics with his discovery, along with Walter Stockenius, that halobacteria contained a membrane protein, bacteriorhodopsin, which can act as a light-driven proton pump. While performing experiments on bacteriorhodopsin, Michel became interested in developing procedures for crystallizing membrane proteins.

Membrane proteins had eluded crystallization because they are largely hydrophobic. Michel's approach consisted of solubilizing the hydrophobic domains of the proteins by first allowing them to interact with amphophilic molecules, that is, molecules that contain both hydrophobic and hydrophilic parts. In a relatively short period of time, Michel was able to produce small crystals of bacteriorhodopsin,[1]

---

[1]Michel, H.; Oesterhelt, D. "Three-Dimensional Crystals of Membrane Proteins: Bacteriorhodopsin"; *Proc. Natl. Acad. Sci. U.S.A.* **1980,** 77, 338–342.

which, unfortunately, never yielded diffraction patterns amenable to analysis.

During this period, Oesterhelt moved to the Max Planck Institute for Biochemistry at Martinsried. A bit frustrated, Michel began working on the crystallization of other membrane proteins. His previous interest in light-driven processes led him to study the reaction center proteins of the purple nonsulfur photosynthetic bacterium *Rhodopseudomonas viridis*. First indications of success came in 1981: in July, Michel obtained the first reaction center crystals and, in September, he and his colleague Wolfram Bode determined that the crystals could produce high-quality diffraction patterns. Michel wrote a report of his exciting results and submitted it to the prestigious British journal *Nature,* which rejected the manuscript, claiming that a structural analysis of the crystals was needed to warrant publication. Michel eventually published his results in the *Journal of Molecular Biology.*[2] By the beginning of 1982, Michel himself was collecting massive amounts of X-ray data; the time had come to seek collaborators who were well versed in crystallographic data analysis.

The institute at Martinsried housed excellent facilities for protein crystallography. The director of the crystallography group, Robert Huber, became interested in the reaction center project after hearing Michel give a seminar on the subject. Michel and Huber agreed that Johann Deisenhofer was the best-qualified member of Huber's group to perform the analysis of the X-ray data.

Johann Deisenhofer came to Martinsried as a graduate student and remained there after his Ph.D. dissertation was completed. He worked closely with Robert Huber, both as a postdoctoral fellow and as a research associate. His tenure in Huber's group gave him extensive experience in biological crystallography. The impressive array of systems he studied included the trypsin–BPTI complex, human immunoglobulin Ko1, the Fab fragment of the immunoglobulin Ko1, human Fab fragment, the complex of the Fc fragment with fragment B of protein A from *Staphylococcus aureus,* C3a anaphylatoxin, leghemoglobin, citrate synthase, and the alpha 1 protease inhibitor.

When presented with the X-ray data from the reaction center of *Rhodopseudomonas viridis,* Deisenhofer embraced the task of analyzing them with vigor. The work was facilitated by collaborations with

---

[2]Michel, H. "Three-Dimensional Crystals of a Membrane Protein Complex. The Photosynthetic Reaction Centre from *Rhodopseudomonas viridis*"; *J. Mol. Biol.* **1982,** *158,* 562–567.

Kunio Miki, Otto Epp, and Michel, who had become a close friend. Of course, the project would not have succeeded without Robert Huber's insights and suggestions.

Robert Huber had been at the Max Planck Institute since 1963. His first efforts were directed toward developing crystallographic methods for the study of complex organic molecules. In time, his interests shifted toward structural problems in protein chemistry. He implemented protein crystallography at the Max Planck Institute, which thus became the first German laboratory capable of carrying out such work. In 1972, he became a director at the Max Planck Institute for Biochemistry at Martinsried. His group made significant advances in the study of proteases, protease inhibitors, and immunoglobulins. For excellence in these areas, Huber received the E.K. Frey Medal (1971), the Otto Warburg Medal (1977), the Emil von Behring Medal (1982), and the E.K. Frey–E. Werle Memorial Medal (1989).

The analysis of the X-ray data from crystals of *Rhodopseudomonas viridis* was a long process. The work could not be completed until Michel finished the amino acid sequence of the protein subunits. In this task, Michel received help from Karl A. Weyer and Heidi Gruenberg of Oesterhelt's laboratory.

Eventually, the first reports of the structure of the reaction center appeared.[3] The group's results showed not only the folding pattern of the four-subunit protein, but also details of the positions of the cofactors, which act as electron carriers. Schematically, the reaction center chromophores can be represented as shown below:

Cytoplasmic side

| (Q) | $Fe^{2+}$ | Q |
| Bpheo | | Bpheo |
| Bchl | | Bchl |
| | $Bchl_2$ | |
| | Cyt | |

Outside

---

[3]Deisenhofer, J.; Epp, O.; Miki, K.; Huber, R.; Michel, H. "X-Ray Structure Analysis of a Membrane Protein Complex: Electron Density Map at 3 Å Resolution and a Model of the Chromophores of the Photosynthetic Reaction Center from *Rhodopseudomonas viridis*"; *J. Mol. Biol.* **1984**, *180*, 385–398; and Deisenhofer, J.; Epp, O.; Miki, K.; Huber, R.; Michel, H. "Structure of the Protein Subunits in the Photosynthetic Reaction Centre of *Rhodopseudomonas viridis* at 3 Å Resolution"; *Nature*, **1985**, *318*, 618–624.

Closest to the inside of the membrane are four heme groups belonging to a cytochrome subunit. Immediately above the cytochrome is a dimer of bacteriochlorophylls (the so-called "special pair"). On each side of the special pair and directly above it, another bacteriochlorophyll and then a bacteriopheophytin molecule are found (pheophytins are chlorophyll molecules lacking the central magnesium ion). The crystal structure revealed a quinone molecule above one of the pheophytins. An empty quinone site lies across from the first bound quinone and above the other bacteriopheophytin. This second quinone is lost during the isolation and crystallization procedures. Between the two quinone sites lies an iron ion.

The crystallographic work spawned attempts to correlate the structure of the reaction center to its function. The sequence of electron-transfer events taking place in the reaction center has been determined by time-resolved spectroscopic methods.[4] Excitation energy enters the reaction center and generates the excited singlet state of $Bchl_2$. The excited special pair transfers an electron to bacteriopheophytin within 3 picoseconds. The reduced Bpheo then transfers an electron to quinone in 200 picoseconds. The oxidized special pair is reduced by the cytochrome.

The reaction center is indeed constructed to carry out vectorial electron transfer from the cytochrome to quinones. However, some features of the structure seem a little puzzling. First, there is a considerable degree of symmetry between the two branches of the reaction center; an axis of symmetry is defined by the $Bchl_2$-Fe vector. It is thought, however, that electron transfer occurs through only one branch of this structure. Second, it is thought that pheophytin is the immediate electron acceptor, yet a bacteriochlorophyll seems to be better positioned to accept electrons from the special pair.

In short, the crystal structure of the reaction center from *R. viridis* helped explain previous spectroscopic results, but also revealed gaps in our understanding of photosynthetic electron transfer. Nonetheless, the knowledge already in hand provided insight into the workings of reaction centers from algae and higher plants. A thorough understanding of bacterial reaction centers can also assist in the development of a complete theory of electron transfer and of effective artificial

---

[4]See the review by Budil, D. E.; Gast, P.; Chang, C.-H.; Schiffer, M.; Norris, J. R. "Three Dimensional X-Ray Crystallography of Membrane Proteins: Insights into Electron Transfer"; *Ann. Rev. Phys. Chem.* **1987**, *38*, 561–583.

photosynthesis. Hence, the crystallographic results from bacterial re-action centers showed potential applicability to a number of important problems in physics, chemistry, and biology. Realizing this fact, the scientific community has rewarded Michel, Deisenhofer, and Huber for their contributions. Michel and Deisenhofer shared the Biological Physics Prize of the American Physical Society (1986) and the Otto Bayer Prize (1988). Huber received the Keilin Medal from the Biochemical Society of London (1987) and the Richard Kuhn Medal from the German Chemical Society (1987). The string of awards culminated with the 1988 Nobel Prize in chemistry, which was shared by Michel, Deisenhofer, and Huber.

Before receiving the Nobel Prize, Michel and Deisenhofer accepted new positions. In October of 1987, Michel moved with wife Ilona, daughter Andrea, and son Robert Joachim to Frankfurt. He is department head and director at the Max Planck Institute for Biophysics and also holds a joint appointment at the University of Frankfurt. Deisenhofer and wife Kirsten Fischer Lindahl are now in the United States, where he is an investigator at the Howard Hughes Medical Institute and a professor of biochemistry at the University of Texas Southwestern Medical Center in Dallas, Texas. Huber remains a director at the Max Planck Institute at Martinsried and teaches at the Technical University of Munich. He is married to Christa Huber and has four children: Ulrike, Robert, Martin, and Julia.

JULIO C. DE PAULA
*Haverford College*

# Bibliography

Budil, D. E.; Gast, P.; Chang, C.-H.; Schiffer, M.; Norris, J. R. "Three-Dimensional X-Ray Crystallography of Membrane Proteins: Insights into Electron Transfer"; *Ann. Rev. Phys. Chem.* **1987,** *38,* 561–583.

Clarke, M. "1988 Nobel Prizes Announced for Physics and for Chemistry"; *Nature* **1988,** *335,* 752–753.

Clayton, R. K. *Photosynthesis: Physical Mechanisms and Chemical Patterns*; Cambridge University Press: Cambridge, England, **1980.**

Dagani, R.; Stinson, S. "Nobel Prizes: Photosynthesis, Drug Studies Honored"; *Chem. Eng. News* **1988,** *66 (October 24),* 4.

Deisenhofer, J.; Epp, O.; Miki, K.; Huber, R.; Michel, H. "X-Ray Structure Analysis of a Membrane Protein Complex: Electron Density Map at 3 Å Resolution and a Model of the Chromophores of the Photosynthetic Reaction Center from *Rhodopseudomonas viridis*"; *J. Mol. Bio.* **1984,** *180,* 385–398.

Deisenhofer, J.; Epp, O.; Miki, K.; Huber, R.; Michel, H. "Structure of the Protein Subunits in the Photosynthetic Reaction Centre of *Rhodopseudomonas viridis* at 3 Å Resolution"; *Nature* **1985,** *318,* 618–624.

Lewin, R. "Membrane Protein Holds Photosynthetic Secrets"; *Science* **1988,** *242,* 672–673.

Michel, H. "Three-Dimensional Crystals of a Membrane Protein Complex. The Photosynthetic Reaction Centre from *Rhodopseudomonas viridis*"; *J. Mol. Biol.* **1982,** *158,* 562–567.

Reigchel, D. "Johann Deisenhofer: Can a Nobel Prize Winner Find Happiness in Dallas? Naturlich, He Says"; *Dallas Morning News,* June 11, **1989.**

# Sidney Altman

## 1939–

Sidney Altman was awarded the Nobel Prize in chemistry in 1989 for a fundamental biochemical discovery. He showed that the catalytic agent for a vital cellular reaction was a distinct, specific molecule of RNA. Until this discovery and the related work of Thomas Cech, specific biochemical catalysis was thought to be affected only by proteins. Altman was born in Montreal, Quebec, on May 8, 1939. He has lived in the United States since 1956, and received U.S. citizenship in 1984. Both of his parents were immigrants to Canada in the early 1920s. His father, born in 1900, came from Cherny Ostrow, a village in the Ukraine, moving from a collective farm in the Soviet Union to the farm of a Canadian organization that sponsored his immigration. Altman's mother, the second oldest of 11 children living in Bialystok, near the Russian–Polish border, came to Canada with her older sister at age 18. She promptly enrolled in elementary school to learn English. With earnings from factory work, she and her sister were able to bring the rest of their family to Canada. Altman's father supported the family, including an older brother, adequately by means of a small grocery store in Montreal. Montreal had a vigorous Jewish community that was somewhat excluded from the British- and French-oriented communities. Learning and education were encouraged within that community; Einstein was a hero.

Altman was a bookish child. At age 12 he read Selig Hecht's *Explaining the Atom* (Viking: New York, 1949), his interest having been aroused at age 6 by the first American nuclear weapons. The book is a popular account of nuclear physics to 1947, not written for children. He was impressed by the beauty and simplicity of the periodic table.

Altman enrolled at MIT as a physics major. He played ice hockey on the MIT team, although he had been unable to meet higher Canadian standards for his high school team. His academic achievement at MIT was good but not outstanding. Nevertheless, his interest in science matured and he was eager to be initiated into laboratory research. When he asked some of the faculty about the possibility of a senior research project, one physics professor told him that his mediocre course grades showed that he was not receptive to new ideas. A project was arranged under the sponsorship of Lee Grodzins, a younger faculty member. The experience remains clear in the memory of both; Grodzins' encouragement was a major stimulus for Altman's career in science. Grodzins had a small laboratory in a "temporary" building erected during World War II. He proposed that Altman reexamine certain electron scattering experiments published in 1928 that should have been interpreted as showing the nonconservation of parity—a concept that was not established until 1956 with the work of Yang and Lee. Altman and Grodzins redesigned the earlier apparatus to accommodate more modern instrumentation and waited while the MIT machinists built the equipment. Grodzins worked closely with his undergraduate students, more so than with his graduate students. The physics seniors met weekly to describe and discuss their progress on thesis problems. To Altman this helped develop a clear sense of scientific reality. Grodzins recalls that Altman was a wonderful student, always questioning and working hard. The results of the project were reported by Grodzins in his historical memoir including extensive quotations from Altman's thesis: "Sidney's thesis is a model: logical, analytical, thorough. His lengthy introduction is followed by a novel calculation of the intensity of the asymmetry expected from an unpolarized source. The apparatus is described in terms of design problems and instrumental uncertainties".[1] Altman succeeded in improving and reproducing the older experiments except for an inversion of sign, + for − and vice versa. It appears that an error was made in the 1928 publication; not in Altman's thesis. During his senior year, Altman also took professor Cyrus Levinthal's

---

[1] Grodzins, L. "A Comment on the History of Double Scattering of Beta Rays"; In *Adventures in Experimental Physics; Maglich, B., Ed., Gamma volume. Princeton: World Science Education, 1973; pp 154–160.

course in molecular genetics. Later, he was particularly stimulated by the elegance of the paper by Crick, Barnett, Brenner, and Watts-Tobin (1961, see also Barnett et al., 1967) that demonstrated the translation of the nucleotide code into amino acids in the groups of three in living cells. He graduated with a B.S. in physics in 1960.

Altman spent about a year beginning graduate work in physics at Columbia University, but the death by cancer of a close friend and the seemingly endless cares of students at Columbia left him depressed and unable to continue. The work for beginning graduate students consisted entirely of classroom courses, no laboratory work. For a time he served as a science and poetry editor for Collier Publishing Company, selecting previously published books for reprinting in paperback editions, including a book of poetry by Robert Frost and a small book of essays by Heisenberg, Born, Schrödinger, and Auger. He was offered a job by the National Film Board of Canada as a screenwriter when he left New York, but lost interest when his assigned task was to prepare a script to train military pilots. In Montreal, he started to translate a French-Canadian novel, but decided that the original was not worth the effort. His interest in literature continues, and he continues to read widely in many fields.

From New York, Altman went to Boulder, Colorado, for a position as a science writer at the National Center for Atmospheric Research, but the Center was still in its formative stages, without real need for a writer. Encouraged by George Gamow, then in the physics department at the University of Colorado, and remembering Levinthal's course, he visited Theodore Puck's department of biophysics in the Medical School in Denver, and spoke with the faculty. He seemed to me, as one of the interviewers, to be an intelligent, highly motivated student with a background that had well served other distinguished scientists in molecular biology. It was clear that he was seriously interested in science in the broader sense. We had few enough graduate students in biophysics that the program was effectively tutorial, with the faculty, staff, postdoctoral researchers, medical associates, and students in other departments serving as our students' community. Altman was strongly attracted by the possibility of embarking immediately on laboratory research. He said that I was thought to be tough, uncompromising on students, but understanding. Sidney became virtually a member of my family and a good friend of my children, for whom he sometimes wrote stories. Our laboratory group included Rose Litman, who gave Sidney untiring help, friendship, and encouragement. Sidney accompanied me to Nashville to complete his dissertation research when I joined the department of molecular biology at Vanderbilt University. There, he was also encouraged by Gisela Mosig, among others.

Altman's thesis project was traceable to the suggestions I had received at the Cambridge Medical Research Council Laboratory of Molecular Biology (MRC) lab in 1960 to explore the biological mechanism of acridine mutagenesis. He undertook to see whether mutations could be attributable to anomalies in DNA synthesis in the T4 phage, and indeed found profound disturbances in the lengths of DNA molecules produced after infection in the presence of 9-aminoacridine, but the mechanism remained obscure. However, his kinetic analysis of the chain of intermediates of T4 DNA synthesis after infection without the mutagen clearly supported a then-novel mechanism, and received more attention. Our mathematical formulation of the intracellular kinetics, conventional among chemists, fell on blind eyes among molecular biologists. Altman felt particular satisfaction on two occasions—devising a valuable but very simple laboratory technique, and deriving biological insight from the formal kinetic analysis.

His two postdoctoral years in the laboratory of Matthew Meselson at Harvard were an introduction to the role of specific nucleases in the intracellular development of viral nucleic acid molecules—in that laboratory, DNA of the T4 genome. The Harvard atmosphere was quite different. The postdoctoral researchers worked independently and were buffeted by competitive pressures. There was little respect for Altman's Colorado degree.

From Harvard, Altman went to Cambridge to pursue tRNA studies. He settled comfortably into English life with friends in Barley, Hertfordshire, where he rented a 16th-century house. He played soccer with the Barley and the MRC lab's teams. It was unusual for a North American to play acceptably well. He regarded the MRC Laboratory scientists, Francis Crick, Sydney Brenner, Frederick Sanger, and others as "giants"; he noted that "they worked like hell". From his first conversation, nearly an hour, with Sydney Brenner, Altman was impressed by his originality, the flow and blossoming of ideas, and his enormous erudition.

The Cambridge laboratory was competitive, but on a friendlier basis among a particularly select group of young people. Given the opportunity to work independently in that stimulating environment, he took the first steps on the path that led to the RNA enzyme.

Working with John Smith, he found an RNA transcript containing the genomic sequence for tRNA-Tyr that was considerably longer than the mature tRNA sequence in a mutant of *E. coli*. That mutant was examined because it is unable to compensate for certain amino acid coding errors.[2] The transcript was not further processed along the chain of reactions by which transcripts would normally be converted into functional tRNA. Altman and Smith showed also that an extract of normal cells of *E. coli* could trim the excessively long, raw transcript obtained from the mutant cell to a correct end point for mature tRNA.

After two years in Cambridge, he arranged as many university speaking invitations in the United States as could be accommodated in one or two trips over the Atlantic. A tour during the previous year, before he identified the immature tRNA and its processing, had been fruitless. Of 14 universities visited in the second trip, in 1971, four had positions available. Altman chose Yale because of its excellent research faculty and the prospect of recruiting interested, capable graduate students. He has remained at Yale since that time.

In 1972, after taking an assistant professorship at Yale, Altman married Ann Korner from Bristol, UK, whom he had met when he arrived at New Haven. They have two children.

His continued studies on tRNA processing at Yale focused on the question of how the original transcript was cut at a particular base along its 5' end. The agent of the cleavage was named RNase P. The cleavage of other tRNAs showed that there was not simply a unique base sequence at the cleavage site, analogous to the specific sites in DNA for restriction endonucleases. Investigation of the nature of the enzyme and how it could select a unique site within diverse sequences led to the realization that RNase P consisted of separable protein and RNA parts. No enzyme previously known carried an RNA component. The report was met with skepticism, and acceptance of the essential role of an RNA component in the enzyme remained an uphill struggle. One publication was delayed by its editor long enough for a previous critic to conduct laboratory studies and publish similar claims simultaneously. This research topic was not deemed to be a high priority (even after the need for the RNA subunit was shown), and its NIH funding was interrupted for part of 1979. Testing each component separately led to the discovery that the catalytically active component of RNase P was the RNA molecule alone.[3] The RNA moiety of ribonuclease P is the catalytic subunit of the enzyme.

---

[2] Altman, S. "Isolation of tRNA Precursor Molecules"; *Nature New Biology* **1971**, *19*, p 229.

[3] Guerrier-Takada, C.; Gardiner, K.; Marsh, T.; Pace, N.; Altman, S. "The RNA Moiety of Ribonuclease Is the Catalytic Subunit of the Enzyme"; *Cell* **1983**, *35*, pp 849–857.

Altman served as chairman of the Department of Biology at Yale from 1983 to 1985, and as dean of Yale College from 1985 to 1989. He was respected and effective among the humanities departments in the college as well as the science departments. However, he was concerned that there was asymmetry in the educational breadth of the undergraduate curriculum. The faculty, under his leadership, strengthened the science requirements in the core curriculum. He established a tutorial program in science and mathematics. He was regarded as independent and courageous in upholding academic standards and responsibilities. His laboratory remained active and productive during his tenure as dean. Before he was awarded the Nobel Prize, Altman was given the Rosenstiel Award for Basic Biomedical Research in 1989.

All known life on earth relies on a basic set of a few thousand chemical reactions common to all cells and organisms, plus thousands more that are more narrowly distributed, more or less characteristic of the particular kind of cell or organism in which they occur. Each substance involved in any particular cellular reaction might also, in principle, participate as actively in dozens of other reactions. However, every cell is a tightly integrated system; it depends critically on the correct and efficient use of every constituent in a well-defined reaction pathway. The selection of which reactions occur and their progress at appropriate rates depend on the presence of a large number of highly specific catalysts, called enzymes (although that term was originally introduced to refer only to the agents of fermentation), together with an elaborate set of control systems that regulate the amount of each enzyme present and sometimes regulate the rate of catalysis. Up to the early 1980s, it was accepted that every enzyme could be identified with a particular and unique type of protein molecule, the last doubts on this question having been resolved half a century earlier. Widely used biochemistry textbooks stated: "Enzymes are proteins specialized to catalyze biological functions" (Lehninger, 1975). "Chemical reactions in biological systems rarely occur in the absence of a catalyst. These catalysts are specific proteins called enzymes" (Stryer, 1981).

Like many other principles that seemed to have withstood time's tests, this solid principle awaited revision. When Altman and his associates reported in a series of papers beginning in 1983 that at least one well-characterized RNA molecule exhibited all of the properties that define an enzyme—biological origin, catalysis in the technical sense, sharply limited specificity, and a distinct role for its catalytic function in the economy of the cell—the principle was violated. Catalysis by a number of other RNA types has been demonstrated since then. In the previous year, Cech reported a surprising intramolecular reaction in a particular large RNA molecule: two remote parts of the polymer chain become linked with the complete removal of an intervening segment of the chain. Cech showed that no protein or any other separate macromolecule was needed, thus demonstrating the possibility of covalent bond interactions between different components of RNA. The reaction implied precise specificity because it occurred only at the boundaries of a defined sequence. Cech's observations were entirely independent of Altman's studies; they provided a hint that catalysis might be involved.

Altman's discovery was revolutionary because it conflicted with conventional wisdom and opened previously unimagined paths for attack on difficult biological problems. It implied no logical contradiction of otherwise coherent theory. Rather, it represented a major and astonishing extension of our perception of the mechanisms of nature. To chemists and biologists, it was a new mechanism with strong implications for the origin of life. As far as we know, life throughout biological time has maintained a sharp division between the storage of hereditary information in chemically inert molecules, DNA, and fulfilling vital functions with chemically active molecules, proteins, that cannot convey instructions for their own synthesis to the next generation. With the recognition of catalysis by RNA, it appears that RNA can serve in both roles, and it suggests a plausible explanation for the earliest events when life began.

Instead of the excitement that might be expected when a well-supported discovery that radically revises fundamental textbook wisdom is announced, Altman's paper attributing a catalytic role to an RNA molecule was received with near indifference. Its impact can be estimated from the frequency with which the original publication is cited in the bibliography of subsequent research papers and by examination of the context within which each citation occurred. Among publications from 1984 through 1988 there were citations relevant to RNA catalysis in only 32 American papers reporting original research and in only 13 in papers of foreign origin. Some of the American papers originated from colleagues in the same department. Only three of the citations appeared in journals published by the American Chemical Society. The total of 202 citations through 1989 includes a large number of reviews, bibliographic collections, or commentaries as well as further papers by Altman, Cech, or their collaborators. The bandwagon that usually follows a surprising advance remained stuck for at least five years.

L. S. LERMAN
*Massachusetts Institute of Technology*

# Bibliography

Altman, S. "Ribonuclease P; An Enzyme with a Catalytic RNA Subunit"; *Advances in Enzymology* **1989** *62*, 1–36.

Barnett, L.; Brenner, S.; Crick, F. H.C.; Shulman, R.; Watts-Tobin, R.; "Phase Shift and Other Mutants in the First Part of the rII Cistron of Bacteriophage T4"; *Phil. Trans. Roy. Soc. B* **1967**, *252*, 487.

Cech, T. R.; Bass, B. L. "Biological Catalysis by RNA"; *Ann. Rev. of Biochem.* **1986, *55*,** 599–629.

Corelli, R. "Chemistry of Life: A Native of Montreal Shares a Major Award"; *Maclean's* Oct. 23, **1989**, *102*, 58.

Crick, F. H. C.; Barnett, L.; Brenner, S.; Watts-Tobin, R. J. "General Nature of the Genetic Code for Proteins"; *Nature* **1961**, *192*, 1227–1232.

Guerrier-Takada, C.; Lumelsky, N.; Altman, S. "Specific Interactions in RNA Enzyme–Substrate Complexes"; *Science*, **1989**, *246*, 1578–1584.

Haseloff, J.; Gerlach, W. L. "Simple RNA Enzymes with New and Highly Specific Endoribonuclease Activities"; *Nature* **1988**, *334*, 585–591.

Joyce, G. F. "RNA Evolution and the Origins of Life"; *Nature* **1989**, *338*, 217–224.

Lehninger, A. *Biochemistry: The Molecular Basis of Cell Structure and Function*; Worth Publishers: New York, 1975.

Pace, N. R.; Smith, D. K.; Olsen, G. J.; James, B. D. "Phylogenetic Comparative Analysis and the Secondary Structure of Ribonuclease P RNA—a Review"; *Gene* **1989**, *82*, 65–75.

Pendlebury, D. "The New Nobelists: A Look at Their Citation Histories"; *The Scientist*, Nov. 13, **1989**, *3*.

Orgel, L. E. "Evolution of the Genetic Apparatus"; *Cold Spring Harbor Symp. Quant. Biol.* **1987**, *52*, 9–16.

Stark, B. C.; Kole, R.; Bowman, E. J.; Altman, S. "Ribonuclease P: An Enzyme with an Essential RNA Component"; *Proc. Natl. Acad. Sci.* **1978**, *75*, 3717–3721.

Stryer, L. *Biochemistry*, 2nd ed.; W. H. Freeman: San Francisco, CA, 1981.

Vioque, A.; Arnez, J.; Altman, S. "Protein–RNA Interactions in the RNase P Holoenzyme from *Escherichia coli*"; *J. Mol. Biol.* **1988**, *202*, 835–848.

# Thomas R. Cech

## 1947–

Thomas R. Cech was born in Chicago, Illinois, on December 8, 1947. He was co-winner of the 1989 Nobel Prize in chemistry and delivered his Nobel Lecture on his 42nd birthday.

Cech was brought up in Iowa City, Iowa, where his father, Robert, was a physician and his mother, Annette, was a homemaker. Both sides of the family immigrated to the United States from Czechoslovakia around 1900. The sheltered environment of a midwestern university town was an ideal incubator for Cech's interest in science. Encouraged by his father's broad scientific interests, he was attracted to many aspects of the world around him, but especially to rocks and minerals.

From 1966 to 1970, Cech attended Grinnell College in his home state of Iowa. This small, highly acclaimed four-year college prides itself on innovative teaching and a broad-based liberal arts curriculum. During this period, Cech's interest in science gravitated primarily to physical chemistry. Two undergraduate research experiences at Argonne National Laboratory and Lawrence Berkeley Laboratory provided Cech with his first glimpses of science at the international level.

At Grinnell, Cech met Carol Martinson, a fellow chemistry major from Cedar Rapids, Iowa. They were married in 1970. Carol Cech's academic career in biochemistry has closely paralleled Thomas's. She is currently working in the biotechnology industry. They have two daughters, Allison and Jennifer.

In 1970 Thomas and Carol Cech entered the Ph.D. program in chemistry at Berkeley. Both gravitated to the biophysical chemistry group. Four faculty had assembled a tightly knit group of about 35 students and postdoctoral researchers who were enthusiastically taking different types of physical measurements of biological macromolecules. The warm, easygoing scientific and social environment contrasted sharply with the political strife that had dominated life on the campus in the preceding years. Cech chose John Hearst as a thesis advisor. Hearst's major interests at the time were the characterization of the sequence organization of eukaryotic DNA by a variety of physical methods, and the structure of eukaryotic chromosomes. Cech soon developed an enthusiasm for these areas that remains in place today.

Cech's thesis research focused on the properties and organization of DNA in the mouse genome. In these precloning days, the major methods of fractionating DNA molecules depended heavily on their physical properties: the temperature at which the strands dissociated, the rate at which strands reassociated, and the small differences in buoyant density in cesium chloride density gradient centrifugation. After a DNA fragment was purified, visualization by electron microscopy was a useful analytical tool. Cech used all of these methods to study the various types of mouse DNA that renatured rapidly. These included certain highly repeated sequences and long stretches of palindromic sequences ("snapback DNA") of still undetermined function. Cech's Ph.D. thesis, completed in 1975, resulted in six research papers in first-rate journals, excellent work for a graduate student. As is typical of papers on this topic from the early 1970s, the results were soon superseded by those using molecular cloning methods. As a result, this fine early work is rarely referred to.

In 1975, Cech moved to MIT for a postdoctoral position in Mary Lou Pardue's laboratory in the biology department. Although Pardue had a similar interest in chromosome structure and function, her laboratory was more biologically oriented, providing Cech the opportunity to round out his background. Cech's postdoctoral research focused on using psoralen derivatives to provide information about DNA and chromatin structure. Trimethyl-psoralen is capable of entering living cells and causing DNA crosslinks upon ultraviolet irradiation. Subsequent examination of the distribution of crosslinks within the DNA provides information about the structure of chromosomes inside cells, thereby avoiding potential artifacts obtained during chromosome isolation.

An important outcome of Cech's postdoctoral period was his growing interest in the ciliated protozoan *Tetrahymena* as a potential experimental organism. *Tetrahymena* is a free-living, single-celled pond organism that is easily cultured in the laboratory. The major attraction of this system for someone interested in chromosomes is that genes for ribosomal RNA are contained on separate small chromosomes and that these minichromosomes are amplified to about $10^4$ copies per cell. This offered the possibility of studying not only the isolated gene, but the associated structural proteins and enzymes. The relatively small, friendly group of ciliate biologists, including J. Gall (Yale), J. Engberg (Copenhagen), and M. Gorovsky (Rochester), welcomed Cech to their ranks and were instrumental in getting him started. Cech's choice of this somewhat unusual system was an inspired one. As is often the case, some of the most interesting biochemical principles have emerged from organisms outside of the mainstream.

With ten papers in prestigious journals, Cech was a strong candidate for several academic positions. He chose the chemistry department at the University of Colorado not only for the youth and enthusiasm of the faculty but because of the strength of the nearby department of molecular, cellular, and developmental biology. He arrived at Colorado in January 1978, and in less than three years he was performing the experiments that would lead to the Nobel Prize. This astonishing fact is especially remarkable because this work was not in the fields of eukaryotic DNA organization or chromosome structure that Cech had trained in, but in RNA biochemistry, a quite separate field.

In order to characterize the function of the *Tetrahymena* minichromosomes, Cech examined the synthesis of the gene product, ribosomal RNA, by incubating isolated *Tetrahymena* nuclei with radioactive precursors. Analysis of the reaction revealed not only the expected ribosomal RNA product, but a fragment of RNA corresponding to the intron present in the gene. This result was very exciting because it was only the second example of RNA splicing occurring in vitro that had been found (second to yeast transfer RNA splicing in John Abelson's lab). In addition, *Tetrahymena* promised to contain high concentrations of the enzymes necessary to splice the RNA produced from the $10^4$ copies of the gene. Cech immediately entered the field of biochemistry called RNA processing and started to try to purify the splicing apparatus. The confidence needed to make such a drastic change in scientific direction as a starting assistant professor is quite remarkable.

The crucial experiment, with research associate Arthur Zaug and graduate student Paula Grabowski, soon followed. When unspliced precursor ribosomal RNA was mixed with an extract of *Tetrahymena* nuclei, the expected in vitro splicing occurred. However, splicing also occurred in the control tube in which no extract was added. Thus, it appeared that the RNA was either "preactivated" for splicing or could splice itself. In April 1981, Cech first presented his results to the RNA processing community. The "failed control" was so strange that it was only briefly mentioned at the end of his talk. The response was very favorable, although the possibility of self-splicing was met with understandable skepticism.

It soon became clear that Cech was at least as skeptical about his own data as everyone else. The series of experiments demonstrating that RNA was a catalytic macromolecule and establishing the reaction pathway (the first RNA splicing mechanism correctly determined) occupied the next year and was a model of careful, conservative science. It was not until 1982 that he was willing to propose without caveats that self-splicing RNA was a reality.

The 1980s saw the discovery of several other RNA molecules with catalytic activity and a general flowering of the field of RNA biochemistry. Although Cech and his co-workers are focusing on the structure of the RNA and the mechanism of the reaction, he has not neglected his first love, chromosome structure. Several members of his research group study telomeres, the unusual DNA structure and DNA–protein complex at the termini of linear chromosomes.

As a result of the revolutionary nature of his basic discovery and his subsequent work, recognition of Cech's accomplishments came rapidly. He has received numerous national and international awards, including the Pfizer Award in Enzyme Chemistry, the Heineken Prize, and the Lasker Award. He was elected to the U.S. National Academy of Sciences in 1987 and received an honorary D.Sc. degree from Grinnell College the same year.

OLKE C. UHLENBECK
*University of Colorado, Boulder*

## Bibliography

Cech, T. R. "RNA as an Enzyme"; *Scientific American* **1986**, *255(5)*, 64–75.

Cech, T. R. "Self-Splicing of Group I Introns"; *Ann. Rev. Biochem.* **1990**, *59*, 543–568.

Cech, T. R.; Bass, B. L. "Biological Catalysis by RNA"; *Ann. Rev. Biochem.* **1986**, *55*, 599–629.

Cech, T. R.; Zaug, A. J.; Grabowski, P. J. "*In vitro* Splicing of the Ribosomal RNA Genes of *Tetrahymena*: Involvement of a Guanosine Nucleotide in the Excision of the Intervening Sequence"; *Cell* **1981,** *27,* 487–496.

Zaug, A. J.; Cech, T. R. "*In vitro* Splicing of the Ribosomal RNA Precursor in Nuclei of *Tetrahymena*"; *Cell* **1980,** *19,* 331–338.

# *Elias J. Corey*

## 1928–

As Elias James Corey was walking to work at Harvard University on the morning of October 17, 1990, a walk he makes almost every day, a graduate student approached him and said, "Congratulations". Although he was slightly puzzled by the comment, Corey politely smiled, thanked the student, and continued on his way. Corey had been unaware for the moment that he had just been named as the recipient of the 1990 Nobel Prize in chemistry. The announcement may have surprised Corey, but it surprised few others, since many of his colleagues had deemed him as destined to be awarded the Prize since as early as the mid-1970s. The Nobel Prize, often touted as the ultimate award, was actually the culmination of a series of prestigious honors, the latest, for Corey, being the Wolf Prize, the National Medal of Science (1988), the Japan Prize in Science (1989), and the Order of the Rising Sun, Gold and Silver Star (Government of Japan, 1989).

Corey, who was born on July 12, 1928, in Methuen, Massachusetts, developed a philosophy of dedication to a creative life in his youth during the Depression. His research portfolio carries the mark of this attitude, with nearly 100 syntheses and almost 800 scientific publications to his credit. Even though his syntheses are impressive in number, each one serves as a fascinating illustration of his capacity for finding simple solutions to difficult and complex problems. Corey has earned a reputation for a large number of remarkable syntheses, exemplified by longifolene, maytansine, bilobalide, and ginkolides A and B, he pioneered in the chemical synthesis of the large family of eicosanoids, including the prostaglandins and leukotrienes. His crowning achievements have involved his contributions to the transformation of organic synthesis from a form of "magic" into a precise and logical science. Indeed, his development of "retrosynthetic analysis", as he refers to it, was one of the bases for which the Royal Swedish Academy of Science awarded him the Nobel Prize.

Elias James Corey was born the fourth child of Elias Corey, a successful and respected businessman, and of Fatina (née Hasham) Corey. Although he had originally been named William, his mother changed his name to Elias shortly after his father's death when he was only 18 months old. Through hard work and determination his family conquered the adversities of his father's death, the Depression, and World War II. The ability to cope with adversity was not alien to his family, as his grandparents on both sides had survived the domination of the Ottoman empire in Lebanon to emigrate to the United States. Consequently, the legacy of his grandparents and father and the determination that his mother displayed during difficult times have served to inspire and motivate Corey throughout life.

Shortly after the death of his father, Corey's family grew to include both his mother's sister, Naciby, and her husband, John Saba. It was from his aunt Naciby that Corey learned to appreciate the value of hard work and attention to detail. Yet, like a typical child, he still preferred sports and outdoor activity over work. His uncle John died in 1957, while Corey was teaching at the University of Illinois, and his aunt died in 1960. It was during the period following the death of his uncle that Corey first conceived of the idea of retrosynthetic analysis, an idea that was to change the way synthesis was done. Fortunately, his mother would live long enough to witness more of her son's brilliant career. She died in 1970.

Corey received his earliest education at the Saint Laurence O'Toole elementary school, a Catholic school located in the nearby town of Lawrence. Later, while attending Lawrence Public High School, he developed a liking for languages, especially Latin, and for mathematics. Following high school graduation Corey enrolled almost immediately at the Massachusetts Institute of Technology (M.I.T.). Although his preparation had been very thorough, or most likely because it had been thorough, his interests were broad and he was unsure of what to study. He enjoyed mathematics most of all, but had been warned early on that he couldn't make a living in that field. Electrical engineering seemed an interesting direction to him. Nevertheless, like all first-year students at M.I.T., Corey started by taking a broad range of courses, including calculus, physics, chemistry, literature, military science, and mechanical drawing. By the end of his first year he began to gravitate toward chemistry, a move which he attributes largely to the enjoyment of doing experiments and to the enthusiasm displayed by his teachers. This is easily understood, because the chemistry faculty at the time included Arthur C. Cope, John C. Sheehan, John D. Roberts, and Charles Gardner Swain. Eventually his fascination focused further to organic chemistry because of its applicability to human health and welfare.

Corey finished his undergraduate work after three years and continued at M.I.T. as a graduate student in the research group of John C. Sheehan, where pioneering work was being done on penicillin synthesis. After two years he had completed the essentials of his thesis work and had received an offer for a teaching position at the University of Illinois. Because the offer stipulated that he be ready in six weeks, he stopped all lab work and began his "write-up". Thus, in January 1951, with his thesis miraculously completed in four weeks, he began his teaching career as instructor of chemistry under the supervision of Roger Adams and Carl S. Marvel.

University of Illinois policy at the time prohibited instructors from taking on graduate students, so Corey initially carried out his own research with occasional help from undergraduates. Some of his early work had a distinctive physical-organic flavor. One of his most important contributions in this area was his study of molecular orbital interactions in transition states of reactions involving cyclic ketones and cycloalkenes. One of his efforts was to prove his "stereoelectronic" hypothesis that axial bonds adjacent to the $\pi$-bond groups would be easier to break than equatorial bonds because of the possibility of delocalization into the $\pi$-system. This was a pioneering use of stereoelectronic arguments to make predictions about reactions and reaction products.

Since Corey's interests spanned virtually all aspects of chemistry, from pure theory to biological applications, he had decided early on to keep his studies as broad as possible. At first, this was not practical because of the restrictions placed on him as an instructor. Fortunately he was promoted to assistant professor in 1954 and starting with a research group with three graduate students, he was then able to transform his numerous ideas into experimental reality. The success of his group's research in just a few years led to his appointment as professor of chemistry in 1956, at the young age of 27. This allowed further expansion of his research into areas of natural-product synthesis, enantioselective synthesis, metal complexes, enzyme chemistry, and new synthetic methodology.

Corey took his first sabbatical leave in the fall of 1957, which he divided between Harvard at the invitation of R. B. Woodward, and Europe. His uncle John's death in September brought deep sadness but also a period of intense concentration on his work. During this period he developed a new way of thinking about synthesis and sketched out some specific synthetic plans. As a guest in the Harvard labs, he took the opportunity to show one of his ideas for the synthesis of longifolene to Woodward, who greeted the ideas with enthusiasm. Corey continued his sabbatical in Switzerland, London, and finally Lund, Sweden, where he visited with Karl Sune Bergstrom, with whom he had a joint research project on steroid biosynthesis. While at Lund, Corey first developed an interest in the prostaglandins, a group of biologically important compounds that would become the focal point of some of Corey's best work.

It was in the spring of 1959 that Corey received an offer of a professorship at Harvard University. Without delay, he accepted the offer. In joining the Chemistry Department at Harvard, Corey became a member of one of the most powerful departments in organic chemistry, with an opportunity to work alongside several giants of chemistry, including Paul D. Bartlett, Konrad Bloch, Louis F. Fieser, Frank H. Westheimer, and R. B. Woodward.

Once at Harvard, Corey's group grew to include about 16 coworkers. Developments in structure determination technology in the 1950s—NMR, IR, and X-ray diffraction techniques—coupled with Corey's newly developed insightful and logical approach to synthesis allowed his group to attempt some of the most challenging syntheses of the time. He applied his rapidly improving retrosynthetic techniques to sesquiterpenes of unprecedented structure, which include longifolene[1] and the caryophyllenes.[2] In addition to this research, Corey was maintaining his broad base of study and delving into the fields of transition-element-mediated synthesis and enantioselective synthesis of

---

[1]Corey, E. J.; Ohno, M.; Mitra, R. B.; Vatakencherry "Total Synthesis of Lonifolene"; *J. Am. Chem. Soc.* **1964,** *86,* 478–485.

[2]Corey, E. J.; Mitra, R. B.; Uda, H. "Total Synthesis of d,1-Caryophyllene and d,1-Isocaryophyllene"; *J. Am. Chem. Soc.* **1964,** *86,* 485–492.

amino acids. In 1965, as a result of the fantastic progress his group was making, Corey was appointed to the positions of Sheldon Emery Professor of Organic Chemistry and Chairman of the Department.

At this point, with a wealth of information gathered from the synthesis of the sesquiterpenes, Corey decided that it was time to formalize his "logical" approach to synthesis. In the past, organic synthesis had relied heavily on trial-and-error methodology, often based on the "mysterious" insights of the chemist. Often the chemist would base his synthetic strategy on a previously available compound that was an obvious subunit of the target structure. Corey realized the limitation of this process when he began studying the sesquiterpenes, which lack recognizable subunits. He therefore formulated a general set of principles that all chemists could understand and follow, rules that were drawn in part from synthetic ideas previously demonstrated.

In using retrosynthetic analysis, the chemist begins with the structure of the "target" molecule and, using several strategies either singly or concurrently, identifies key bond disconnections that will lead to a set of simple, cheap, and readily available starting compounds. A key concept is the process of recognition of a "retron", which is a structural feature of a molecule that results when it is produced by a particular chemical reaction. For instance, a typical Diels–Alder reaction of 1,3-butadiene and ethene yields cyclohexene. Thus, the basic retron for a Diels–Alder reaction is a six-membered ring with a double bond in it. Because there are countless ways to synthesize any given compound, the process of retrosynthetic analysis generates an enormous "tree" of potential intermediates, reactions, and starting materials, from which the best synthetic route can be selected.[3]

With the development of the appropriate computer technology and especially computer graphics in the 1960s, Corey decided that retrosynthetic analysis was ideally suited for translation into a computerized format. One of his students, Todd Wipke, agreed to attempt the programming of this project, and in 1969 they published an article describing their early progress.[4] The program, now named LHASA (Logic and Heuristics Applied to Synthetic Analysis), though limited in scope at the outset, was capable of taking a molecular structure,

---

[3]Corey, E. J.; Cheng, X.-M. *The Logic of Chemical Synthesis;* John Wiley & Sons: New York, 1989.

[4]Corey, E. J.; Wipke, T. W. "Computer-Assisted Design of Complex Organic Synthesis"; *Science* **1969,** *166,* 178–192.

input by the chemist using a light-pen, or a mouse and, by applying the retrosynthetic strategies and accessing a database of chemical reactions, converting it step by step into simple starting materials. As the technology has improved and the process has been refined, the program has grown significantly in power. It is now possible for LHASA to design syntheses that duplicate or improve upon what are currently regarded as the "best" syntheses.

Although at first this technique drew a great deal of skepticism, even ridicule from colleagues who believed natural products were too complex for this type of analysis, Corey's persistence has paid off. His concept of retrosynthetic analysis, which he taught to all of his students, is now being taught by them at colleges and universities throughout the world and is discussed in numerous introductory organic chemistry textbooks, educating the next generation of chemists in the use of this powerful tool. The usefulness of retrosynthetic analysis is almost startling. Within the three months of a course Corey teaches to first-year graduate students at Harvard, the students acquire the ability to design possible syntheses of virtually any compound. Corey's pioneering application of computer graphics now pervades organic chemistry.

During his studies on the sesquiterpenes, Corey became interested in the synthesis of prostaglandins. These compounds, now known to be involved in many bodily functions—including the regulation of immune response, blood pressure, blood coagulation, and contractions during childbirth—were only available in milligram quantities at the time, because the body produces them only in minute quantities. In 1968 he completed the first prostaglandin synthesis[5] and, by mid-1969, routes were available to all known prostaglandins. The syntheses were scaled-up for medicinal studies and pharmaceutical production, permitting much of fundamental medicinal research. In the 1970s Corey extended his studies to include the leukotrienes, another group of biologically active fatty acids, which, together with the prostaglandins, make up the eicosanoids, a term Corey coined in the late 1970s.

Throughout the 1970s and 1980s research in Corey's group continued to push the boundaries of synthesis, constantly generating new synthetic methods, some of which are in such common use that few

---

[5]Corey, E. J.; Andersen, N. H.; Carlson, R. M.; Paust, J.; Vedejs, E.; Vlattas, I.; Winter, R. E. K. "Total Synthesis of Prostaglandins. Synthesis of the Pure d, 1 − $E_1$ − $F_{1\alpha}$, −$F_{1\beta}$, −$A_1$, and −$B_1$ Hormones"; *J. Am. Chem. Soc.* **1968**, *90*, 3245–3248.

chemists realize that they are Corey's creations. One of his most impressive accomplishments was completed in 1988: the synthesis of ginkolide B. This chemical, the active substance found in extracts of the *ginko biloba* tree, has been used as a Chinese folk medicine since ancient times. It is now manufactured by pharmaceutical companies, to treat blood circulation problems and asthma, with over $500 million a year in sales.[6]

One area that Corey believes to be an important direction of organic chemistry is the design of "chemzymes": small, manufactured molecules capable of catalyzing reactions selectively like the much larger enzymes. For a catalyst to qualify as a chemzyme, it must function like a molecular robot and be able to assemble the reagents in the proper three-dimensional orientation, convert them to products, and then release these products, freeing the chemzyme for another catalytic cycle. One of the first chemzymes of his design was the CBS catalyst for the enantioselective reduction of ketones to alcohols.[7] More recently he has been developing catalysts for enantioselective aldol and Diels–Alder reactions, which are quickly becoming some of the most powerful synthetic tools available.[8]

E. J. Corey currently lives near the Harvard campus in Cambridge, Massachusetts, with his wife Claire Higham, a graduate of the University of Illinois, whom he married in September 1961. They have three children, all three of whom are actively pursuing their interests with the same vigor as their father. Their oldest son, David, a graduate of Harvard (A.B., 1985) and the University of California at Berkeley (Ph.D., 1990), is presently an assistant professor of chemistry/molecular biology at the University of Texas Medical School in Dallas. Their second son, John, also a Harvard graduate (A.B., 1987), is involved in classical music composition at the Paris Conservatory of Music. Their daughter, Susan, is also a graduate of Harvard (A.B.,

---

[6]Corey, E. J.; Kang, M.-C.; Desai, M. C.; Ghosh, A. K.; Houpis, I. N. "Total Synthesis of (±)-Ginkolide B"; *J. Am. Chem. Soc.* **1988,** *110,* 649.

[7]Corey, E. J.; Bakshi, R. K.; Shibata, S. "Highly Enantioselective Borane Reduction of Ketones Catalyzed by Chiral Oxazaborolidines. Mechanism and Synthetic Implications"; *J. Am. Chem. Soc.* **1987,** *109,* 5551–5553; Corey, E. J.; Bakshi, R. K.; Shibata S. "An Efficient and Catalytically Enantioselective Route to (S)–(–)–Phenyloxirane"; *J. Org. Chem.* **1988,** *53,* 2861–2863.

[8]Corey, E. J.; Imwinkelried, R.; Pikul, S.; Xiang, Y. B.; "Practical Enantioselective Diels–Alder and Aldol Reactions Using a New Chiral Controller System"; *J. Am. Chem. Soc.* **1989,** *111,* 5493–5495; Corey, E. J.; Kim, S. S. "Versatile Chiral Reagent for the Highly Enantioselective Synthesis of Either Anti or Syn Ester Aldols"; *J. Am. Chem. Soc.* **1990,** *112,* 4976–4977.

ELIAS J. COREY

1990) and Columbia (M.A., 1992) with majors in anthropology and education. Corey's leisure interests include virtually all outdoor activities, including his morning jog, and music.

Undoubtedly many great things will continue to evolve from Corey's laboratories. His research group's activity has shown no signs of decreasing, and he has begun to investigate several new, interesting areas of research. Additionally, although his brilliant research highlights his career, Corey has always placed a great deal of importance on teaching and transmitting his knowledge and enthusiam to the young. He continues to teach first-year graduate students and maintains an open-door policy for those who need assistance. He obviously feels an obligation to provide his knowledge to those students who seek it. But Corey also understands the value of allowing students to work on their own so that they may develop their own individual talents in science. Furthermore, he encourages them to approach research in the same manner that he does: to try to achieve more than they think they can. Because each chemist to pass through the Corey group becomes a carrier of this philosophy, they have become some of the most sought-after organic chemists in both industry and academics.

E. J. Corey's influence on chemistry is astounding. He has advised over 400 graduate students and postdoctoral fellows, who are employed at universities and pharmaceutical research laboratories all over the world. Over 100 of these occupy university positions in Europe, Asia, and the Americas. His resume lists 37 of the most prestigious awards in chemistry, of which the 1990 Nobel Prize in chemistry is the latest, and he holds honorary degrees from 14 institutions.

Additionally, in receiving the 1990 Nobel Prize, he continues an impressive tradition by becoming the thirty-third Nobel Laureate from Harvard (the sixth in chemistry). There is no doubt that he has had a profound impact on the science of organic chemistry. Perhaps his colleague Dudley Herschbach (Nobel Laureate in chemistry, 1986) said it best: "E. J. has changed the whole way modern chemistry is done."

CHARLES R. ALLERSON
*Harvard University*

# Bibliography

*American Men & Women of Science: 1992–93,* 18th ed.; R. R. Bowker: New York, 1992; Vol. 2, p 409.

*Chemical & Engineering News* **1990,** *68(43),* 4.

Corey, E. J. "E. J. Corey"; (Autobiographical sketch); In *Les Prix Nobel 1990;* Almqvist & Wiksell: Stockholm, Sweden, 1992; pp 171–175.

Corey, E. J.; Cheng, X.-M. *The Logic of Chemical Synthesis;* John Wiley & Sons: New York, 1989.

Corey, E. J.; Green, R. A. "Stereoelectronic Control in Enolization and Ketonization Reactions"; *J. Am. Chem. Soc.* **1956,** *78,* 6269–6278.

Corey, E. J.; Wipke, T. W. "Computer-Assisted Design of Complex Organic Synthesis"; *Science* **1969,** *166,* 178–192.

*The International Who's Who: 1991–92,* 55th ed.; Europa: London, 1991; p 349.

Mitra, A. *The Synthesis of Prostaglandins;* Wiley–Interscience: New York, 1977.

Sheehan, J. C.; Buhle, E. L.; Corey, E. J.; Laubach, G. D.; Ryan, J. J. "The Total Synthesis of a 5-Phenyl Penicillin"; *J. Am. Chem. Soc.* **1950,** *72,* 3828–3829.

Thompson, K. "Elias J. Corey - 20th Century Folk Hero"; *Chemistry in Britain* **1989,** *25(2),* 113–114.

# Richard R. Ernst

## 1933–

Richard R. Ernst was awarded the Nobel Prize in chemistry in 1991 "for his contributions to the development of the methodology of high-resolution nuclear magnetic resonance (NMR) spectroscopy." Ernst was born on August 14, 1933, in Winterthur, Switzerland. His father, Robert Ernst, taught architecture at the local technical high school.

Ernst's interest in chemistry was sparked by his experimentation with a case of chemicals found in the attic of his home. The chemicals had belonged to a deceased uncle who had been a metallurgical engineer. These early fascinating experiments encouraged Ernst to read as much as possible about chemistry and, after high school, he studied chemistry professionally at the Eidgenössische Technische Hochschule (ETH) in Zurich. He earned his diploma in 1956 from the ETH, and completed his Ph.D. in physical chemistry, also at the ETH, under the direction of Professor Hans Primas in 1962. His thesis work involved instrumentation and a theoretical study of nuclear magnetic resonance with random excitation.

The discovery of nuclear magnetic resonance (NMR) was recognized in 1952 by the award of the Nobel Prize in physics to Felix Bloch of Stanford and Edward Purcell of Harvard. Nuclear magnetic resonance can be observed when a material in which nuclei have a magnetic dipole moment is placed in a magnetic field. The most notable example is the nucleus of the hydrogen atom, which is a single proton. Following the laws of quantum mechanics, the nuclei can occupy only a limited number of energy states in a magnetic field corresponding to different orientations of their spin angular momentum. Transitions between these levels can be induced by an oscillating magnetic field of the appropriate frequency. In typical experiments, the frequency is usually in the range of tens to hundreds of megahertz, using magnets that can produce fields of 10 Tesla or more. The resonance frequency of nuclear spins is modified slightly by their environment. In a liquid sample, the major influence is exerted by the magnetic shielding effects of the electrons in a molecule. The application of NMR to chemical analysis stems from the observation that similar nuclei (e.g., hydrogen nuclei) in different positions in a molecule experience slightly different magnetic fields or chemical shifts and hence have different resonance frequencies. The branch of NMR in which such effects can be detected is known as high-resolution NMR and requires magnets of extremely high uniformity.

By the late 1950s, when Ernst began his graduate work, many of the more subtle aspects of high-resolution NMR were being studied and exploited. Interactions between different nuclei in a molecule led to splitting of resonance lines, which made spectra more complex but provided useful information. The removal of these interactions by the introduction of additional excitation provided insight into the source of the interaction for a more detailed understanding of the molecular structure. For his graduate thesis, Ernst studied the effects of excitation with a stochastic (random) signal that would excite a wide range of frequencies simultaneously. Although this work seemed somewhat esoteric at the time of its publication, it clearly indicated Ernst's understanding of the mathematical behavior of nuclear spin systems and was a precursor to the work for which he was awarded the Nobel Prize.

The work of Felix Bloch was continued by a number of his students at a company in California founded by Russel and Sigurd Varian. In 1963, following his graduate work in Switzerland, Ernst joined Varian Associates to work with one of those brilliant students, Weston Anderson. A significant problem with NMR was its very low sensitivity compared with other spectroscopic techniques. Inspired by ideas from Russel Varian, Anderson was working on methods to improve the NMR sensitivity by simultaneously exciting all resonance lines in a spectrum rather than sweeping through them sequentially, as was the standard practice. Out of Ernst and Anderson's collaboration came the first of the improvements in NMR methodology that were cited in the awarding of the Nobel Prize. They realized that excitation of a material by a strong pulse stimulates all the resonances of a particular nuclear species at once and that the subsequent response could be stored in a computer and converted to a familiar spectrum by a Fourier transformation. For typical spectra of protons, this new method could increase the sensitivity by a factor of 30 or more. The method was not immediately adopted in commercial instruments because the computation necessary for the Fourier transformation was costly and time-consuming. In his earliest experiments, Ernst stored the signal response on punched paper tapes, had them converted to punched cards at IBM San Jose, and brought them to a computer center, where they waited in line for computer analysis. The results were usually ready the next week. The development of minicomputers that could be dedicated to a spectrometer and the introduction of improved algorithms for the calculation of Fourier transforms revolutionized the process. Before the end of the 1960s, commercial NMR spectrometers that operated in the Fourier transform mode were available. At Varian, Ernst made other significant contributions based upon his earlier work with Hans Primas. He developed the concept of *noise decoupling*, in which all of the couplings of one nuclear species to another can be eliminated with a source of broadband excitation.

In 1968, Ernst left Varian to return to Switzerland, where he took a position as *Privatdozent* in physical chemistry at the ETH. He has remained at the ETH since that time, continuing his work in liquid–state NMR and, more recently, in solid–state NMR and electron paramagnetic resonance (EPR). He became a full professor in 1976.

In parallel with Reinhold Kaiser from New Brunswick and after correspondence with Anderson, he developed the idea of using noise excitation in an analogous way to pulse excitation for the simultaneous collection of spectral data. This latter idea has not received the wide acceptance of the pulsed Fourier transform technique but is still the subject of research in a few laboratories today.

The introduction of pulse excitation and Fourier transformation had a profound impact on the application of high–resolution NMR spectroscopy. The spectroscopy of a number of nuclear species, which previously had been severely limited by poor sensitivity, became a practical reality. The most notable example was $^{13}C$. The low natural abundance (1.1%) of $^{13}C$ contributes to the low sensitivity but also results in relatively simple spectra from which molecular structures can be determined. In addition, the greater dispersion of $^{13}C$ spectra compared with the spectra of protons means that larger molecules can be studied. The use of pulsed excitation also allowed the study of the transient behavior of the nuclear spins and provided additional information through the measurement of relaxation times of individual lines of a spectrum. Pulse experiments in the time domain led to an unprecedented unification of liquid state and solid state NMR methodology.

In 1971, at Ampère International Summer School, Basko Polje, Yugoslavia, Jean Jeener of Belgium proposed a novel experiment in which new information could be extracted by applying two separated pulses to a sample and analyzing the two-dimensional data set by a double Fourier transformation. This proposal did not generate great excitement, partially because of the limited computing power available and the perceived complexity of double Fourier transformations. In 1975, Ernst published the first experimental results extending Jeener's suggestion and began, with his students, a new series of developments that opened the application of NMR to increasingly larger molecules. In the two-dimensional method, nuclear spins are first excited by one or more pulses. The response of the spins is allowed to evolve for a certain period of time (the evolution time), after which the spins are perturbed again by the mixing process, and the subsequent response is recorded. Fourier transformation of this last response results in a spectrum with features that depend on the evolution time. The time evolution of these spectral features can be followed by repeating the experiments with different evolution times. A second Fourier transformation translates these time-varying features into frequencies, and the final result is displayed as a two-dimensional spectrum. Oscillations that are not affected by the mixing process appear along the diagonal of such a two-dimensional spectrum. New information is provided by peaks that lie off the diagonal.

The power of two-dimensional spectroscopy lies in the visual representation of the mutual relations between nuclei in molecules in terms of their proximity and their spatial arrangement in the chemical bonding network. This forms the basis for the determination of molecular geometry. Moreover, the information displayed can be manipulated by the selection of the sequence of pulses or other perturbations used prior to the detection of the spin response. Experiments may be designed to separate different interactions, such as chemical shifts and spin–spin couplings, to correlate transitions of coupled spin systems, or to study dynamic processes such as chemical exchange. The creation of new pulse sequences became a major area of NMR research following Ernst's pioneering work. More recent developments have expanded the two-dimensional experiments to three or more dimensions. The introduction of these multidimensional techniques, coupled with the technological development of superconducting magnets with high fields and sufficient field homogeneity, allows NMR to challenge and in some cases surpass X-ray diffraction methods for the determination of the three-dimensional structure of large molecules, in particular for proteins and nucleic acids. In addition to his work on experimental methods, Ernst and his students have had a strong influence on the theoretical understanding of the behavior of nuclear spins and their response to various perturbations.

In 1972, a completely new application of NMR was initiated when Paul Lauterbur at the State University of New York at Stony Brook described a method of producing cross-sectional images of medical and other objects with NMR by using magnetic field gradients to correlate spatial position with frequency dispersion of the NMR response. A similar idea was independently proposed by Peter Mansfield at Nottingham, with particular application to the study of solid samples. A number of groups soon developed medical imaging methods with a variety of techniques. In 1975, Ernst and his students published a new method for collecting image information. It was an extension of the method of two-dimensional spectroscopy in which the sample was excited by a pulse and the spin response allowed to evolve for a certain time while a field gradient was applied. The gradient was then switched to an orthogonal direction and the spin response recorded during its subsequent evolution. By repeating the experiment a number of times, each time increasing the duration for which the first gradient was applied, a series of spin responses was collected that could be translated by a double Fourier transformation into a cross-sectional image of the object. The method was readily extended to three dimensions by the inclusion of a third orthogonal gradient. A modification of Ernst's original idea by James Hutchison and his colleagues at Aberdeen, in which the amplitude of the gradient is varied during a fixed evolution time, has become a major feature of most magnetic resonance imaging (MRI) techniques. Although Ernst has not worked directly in the area of MRI, his pioneering proposal has had a significant impact, and his continuing developments in NMR spectroscopy have been applied as well to MRI methods.

Richard Ernst's contribution to NMR has been recognized by numerous honors, beginning in 1962 with the award of the Silver Medal

Ernst in his function as president of the research council of ETH-Zürich behind a pile of research proposals.

of the ETH for his Ph.D. thesis. In 1991, he and Alex Pines of Berkeley shared the prestigious Wolf Prize for chemistry conferred by the Wolf Foundation in Israel. In the same year, he shared the Louisa Gross Horwitz Prize for Biology or Biochemistry with Kurt Wüthrich of ETH Zurich. He continues to inspire workers in the field of magnetic resonance with his new ideas for further applications in chemistry and other fields of science.

Besides his research in magnetic resonance, Ernst is much involved in science education and research administration in Switzerland. His spare time is occupied by his love for classical music and, in partic-

Ernst loading a sample into the superconducting NMR magnet.

ular, by his fascination for the incredibly rich Tibetan art, culture, and philosophy. Since 1963, he has been married to Magdalena Kielholz, an elementary school teacher, and is the father of three children; Anna Magdelena (kindergarten teacher), Katharina Elisabeth (elementary school teacher), and Hans-Martin Walter (student in psychology).

HOWARD HILL
*Varian Associates*
*Palo Alto, California*

# Bibliography

*American Men and Women of Science 1992–93*, 18th ed.; R. R. Bowker: New York, 1992; Vol. 2, p 984.

Aue, W. P.; Bartholdi, E.; Ernst, R. R. "Two-dimensional Spectroscopy-Application to Nuclear Magnetic Resonance"; J. Chem. Phys. **1976**, *64*, 2229–2246.

"Nobel Chemistry Prize: Ernst Honored for NMR Achievement"; *Chem. & Eng. News,* **1991**, Oct. 21, pp 4–5.

Ernst, R. R. "Sensitivity Enhancement in Magnetic Resonance"; *Adv. Magn. Resonance* **1966**, *2*, 1–135.

Ernst, R. R.; Anderson, W. A. "Applications of Fourier Transform Spectroscopy to Magnetic Resonance"; *Rev. Sci. Instrum.* **1966**, *37*, 93–102.

Ernst, R. R.; Bodenhausen, G.; Wokaun, A. *Principles of Magnetic Resonance in One and Two Dimensions*; Clarendon Press: Oxford, 1987.

Griesinger, C.; Sørensen, O. W.; Ernst, R. R. "Three-dimensional Fourier Spectroscopy Application to High Resolution NMR"; *J. Magn. Reson.* **1989**, *84*, 14–63.

Kumar, A.; Welti, D.; Ernst, R. R. "NMR Fourier Zeugmatography"; *J. Magn. Reson.* **1975**, *18*, 69–83.

Lauterbur, P. C. "Image Formation by Induced Local Interactions: Examples Employing Nuclear Magnetic Resonance"; *Nature* **1973**, *242*, 190–191.

Mansfield, P.; Morris, P. G. *NMR Imaging in Biomedicine*; Academic Press: New York, 1982.

*Who's Who in the World 1991:1992*, 10th ed., Marquis Who's Who: Wilmette, IL, 1990; p 311.

# Rudolph A. Marcus

## 1923-

Rudolph A. Marcus received the 1992 Nobel Prize in chemistry for "his contributions to the theory of electron-transfer reactions in chemical systems".

Born on July 21, 1923, in Montreal, Canada, Marcus received most of his education in that city, except for a six-year stay in Detroit beginning when he was three years old. He was encouraged by both parents, but his mother exerted more intellectual influence. She even told him, as she pushed his baby carriage around the campus, that he would some day attend McGill University. Although he had an early fascination with mathematics, Marcus majored in chemistry when he entered McGill in 1941 because it was the field in which he felt "most comfortable". There he found C. A. Winkler's course in kinetics "intriguing", but a future in theoretical chemistry never occurred to him. According to Marcus, "I always loved school. It was natural to go on to graduate school. It wasn't as though I was going to make terrific discoveries. That didn't play any part."

With his graduate research dictated by the war effort, Marcus worked on the Canadian RDX project with Winkler, receiving his Ph.D. in 1946. Recruited by a former Winkler student and continuing in his "passive frame of mind", he spent several years as a postdoctoral fellow with E. W. R. Steacie at the National Research Council of Canada in Ottawa. It was his first exposure to "front-line research", as he moved from solution kinetics to the gas-phase kinetics of free-radical reactions.

Marcus enjoyed working with his hands, but in building the glass apparatus necessary for gas-phase reactions, his impatience led to "a number of breakages" and a "fair amount of wasted effort". Dissatisfied with these results, Marcus began reading current theoretical papers by Henry Eyring and others on reaction theory, primarily related to the calculation of activation energies. Steacie was generally skeptical of theory, but he was delighted when Marcus showed him how to use a "simple algebraic method" to extract the activation energy of a methyl radical attacking a hydrocarbon without knowing the entire mechanism of the overall process.

Realizing that "it might be possible to learn and do theory", Marcus wrote to six theoretical chemists and fortunately obtained a postdoctoral position with Oscar K. Rice at the University of North Carolina that lasted from 1949 to 1951. It was a turning point in his life. For Marcus, everything after that was "sheer joyousness". After three intense months of reading more theoretical papers, he responded to a suggestion by Rice that he examine unimolecular reactions. Unclear about which direction to follow, Marcus concentrated on the RRK (Rice–Ramsperger–Kassel) theory proposed in the 1920s, before quantum mechanics. The result, achieved by slowly "putting little pieces together", eventually became known as the RRKM theory, after lying dormant for a number of years.[1]

Although he got no positive responses to 35 letters of inquiry for a faculty position, Marcus managed to secure a position at the Polytechnic Institute of Brooklyn in 1951. It was through the initial (and unknown to him) recommendation of a former student at "Poly" who had been in one of his North Carolina graduate courses. At Brooklyn, the lack of experimental data on pressure effects in unimolecular reactions prompted Marcus to leave the theoretical area temporarily and return to doing experiments in that field and others. When a student's question on polyelectrolytes in his statistical mechanics course led Marcus to read everything that was available in electrostatics, he subsequently returned to calculations that others would later find useful.[2] More importantly, it "proved useful" when he "happened to chance on a 1952 symposium journal issue on electron-transfer reactions."

---

[1] Marcus, R. A. "Unimolecular Dissociations and Free Radical Recombination Reactions"; *J. Chem. Phys.*, **1952**, *20*, 359–364.

[2] Marcus, R. A., "Calculation of Thermodynamic Properties of Polyelectrolytes"; *J. Chem. Phys.* **1955**, *23*, 1057–1068.

Marcus was drawn to an electrostatic calculation by Willard F. Libby to explain the small rate constants found in the electron-transfer reactions of simple ions such as the ferric–ferrous and ceric–cerous systems. His immersion in the literature of electrostatics caused him to sense that something was wrong with Libby's calculation, although he was unsure what the error was. Within a month Marcus had the answer. He recognized that Libby's calculation was not appropriate for thermal electron transfers, and he then found a way to calculate the free energy of the ions in solution as a function of a nonequilibrium dielectric polarization occurring in these reactions.[3]

Marcus recalls that "it was unquestionably the most exciting time of my life—one of sheer elation." In an intense period of activity over the next nine years, he formed the basis of what became the "Marcus theory", extending his original concept to include intramolecular vibrational effects, numerically calculated rates of self-exchange and "cross-reactions", electrochemical electron-transfer reactions, chemiluminescent electron transfers, the relation between nonequilibrium and equilibrium solvation free energies for arbitrary geometries, and spectral charge-transfer processes. Marcus's work attracted immediate attention, serving as the basis for extension and application that he and many others still continue to explore.

The ultimate test of any theory is its experimental verification. Marcus admits that making predictions without experimental data is "a gamble", but if "the physical picture is okay, the equations are just a vehicle, and the key is the physics of the problem." In a 1962 conversation with Norman Sutin at Brookhaven National Laboratory, Marcus found experimental results that, to his delight, showed that the cross-relation did work.[4] But it took 25 years for a verification of his 1960 prediction of an "inverted region". In this region, the rate constants of a series of related reactions should at first increase with an increase in the driving force of the reaction (the negative value of the standard free energy), pass through a maximum, and then decrease. John R. Miller, Gerhard L. Closs, and Lydia Calcaterra provided the first experimental verification of this unusual prediction.

---

[3]Marcus, R. A. "On the Theory of Oxidation–Reduction Reactions Involving Electron Transfer. I."; *J. Chem. Phys.* **1956**, *24*, 966–978; Marcus, R. A. "Electrostatic Free Energy and Other Properties of States Having Nonequilibrium Polarization. I."; *J. Chem. Phys.* **1956**, *24*, 979–989.

[4]Marcus, R. A. "On the Theory of Oxidation–Reduction Reactions Involving Electron-Transfer. V. Comparison and Properties of Electrochemical Rate Constants"; *J. Chem. Phys.* **1963**, *67*, 853–857, 2889.

In his 13 years at Brooklyn Poly, Marcus had only one student who was interested in theoretical aspects of chemistry. In addition to his pioneering work in electron transfer, he continued some experimental research, especially in the field of flash photolysis. However, his move to the University of Illinois in 1964 marked an end to his experimental investigations. Beginning in 1960, when he spent a year and a half at the Courant Institute of Mathematical Sciences, he began to reduce and ultimately stop his experimental work. At Illinois he had no difficulty in attracting theoretical students. Besides continuing with the electron-transfer studies, he began work at Illinois on the development of analytical dynamics of vibrationally near-adiabatic chemical reactions and the semiclassical theory to bridge the gap between classical and quantum calculations of collision dynamics, an effort that continues today.[5]

New experimental results and techniques, especially those from molecular beam experiments, induced Marcus to return to the RRKM theory after he moved to the California Institute of Technology in 1978. He began by extending it to treat the more complicated vibrational-molecular rotational motion that determines the transition states.[6] Remarkably, more than half of Marcus's list of publications have appeared since he has been at Caltech as the Arthur Amos Noyes Professor of Chemistry.

Throughout his career Marcus has rejected all invitations to assume administrative positions, stating clearly that he has never been interested and "never tempted, even for a moment", because he has not wanted anything to interfere with his love for problems and research. This passionate affair with theoretical chemistry is symbolized by some of the "miscellaneous" areas he considers "fun to study": reactions of solvated electrons, a two-site behavior in photosynthesis, a unified approach to the electrochemical hydrogen evolution reaction, microcanonical transition-state theory and its consequences, Lie mechanics,

---

[5]Marcus, R. A. "On the Analytical Mechanics of Chemical Reactions. Quantum Mechanics of Linear Collisions"; *J. Chem. Phys.* **1966**, *45*, 4493–4499; Marcus, R. A. "On the Analytical Mechanics of Chemical Reactions. Classical Mechanics of Linear Collisions"; *J. Chem. Phys.* **1966**, *45*, 4500–4504; Marcus, R. A. "Extension of the WKB Method to Wave Functions and Transition Probability Amplitudes (S-Matrix) for Inelastic Collisions"; *Chem. Phys. Lett.* **1970**, *7*, 525–532.

[6]Wardlaw, D. M.; Marcus, R. A. "RRKM Reaction Rate Theory for Transition States of Any Looseness" *Chem. Phys. Lett.* **1984**, *110*, 230–234; Klippenstein, S. J.; Marcus, R. A. "Application of Unimolecular Reaction Rate Theory for Highly Flexible Transition States to the Dissociation of $CH_2CO$ into $CH_2$ and CO. II. Photofragment Excitation Spectra for Vibrationally-Excited Fragments"; *J. Chem. Phys.* **1990**, *93*, 2418–2424.

microwave transients, complex isotopic exchange reactions, conformal electronic structures, and vibrational nonadiabaticity and curvilinearity in transition–state theory.

Theories developed by Marcus have been highly influential and are in extensive use by other scientists. The electron-transfer work for which he was recognized by the Nobel committee is now being applied in various areas such as photosynthesis, electrically conducting polymers, chemiluminescence, and corrosion. When the first experimental measurement of the rate constant of an electron transfer across the liquid–liquid interface was recently made, Marcus recognized there was a need for further theoretical work on that process. He extended his existing theory to include the phenomenon and then compared it with experimental results, one of the first times that experiment preceded his theory.[7] He recently investigated the effect of solvents in slowing down electron transfers, and he is currently investigating electron transfers across organic bridges of varying length, electron transfers across organic monolayers on electrodes, long-range electron transfers in proteins, and scanning tunneling microscopy, as well as a number of problems on intramolecular dynamics and unimolecular reactions.

In reflecting on his accomplishments, Marcus emphasizes that "one of the most enjoyable aspects has been the interplay between experiment and theory." Confessing that he gets "all keyed up about it", he admits that when he was making the break from experiment to theory he "questioned all that wasted effort doing those experiments." But he has realized that the time was not wasted, for "it put the focus on experiment, and certainly one thing that's characterized our work has been the tremendous amount of interaction between theory and experiment."

JAMES J. BOHNING
*Beckman Center for the History of Chemistry*

---

[7]Marcus, R. A. "On the Theory of Electron-Transfer Rates Across Liquid–Liquid Interfaces"; *J. Phys. Chem.* **1990**, *94*, 4152, 7742; Marcus, R. A. "Theory of Electron-Transfer Rates Across Liquid–Liquid Interfaces. II. Relationships and Applications"; *J. Phys. Chem.* **1991**, *95*, 2010–2013.

# Bibliography

Bernstein, R. B.; Zewail, A. H. "Chemical Reaction Dynamics and Marcus Contributions"; *J. Phys. Chem.* **1986,** *90,* 3467–3469.

Connor, J. N. L. "Marcus' Contributions to the Semiclassical Theory of Collisions and Bound States"; *J. Phys. Chem.* **1986,** *90,* 3466.

Marcus, R. A. Interview by James J. Bohning at the California Institute of Technology on June 20, 1991; Beckman Center for the History of Chemistry: Philadelphia, PA; Oral History Transcript 0097.

Marcus, R. A. "Theory, Experiment, and Reaction Rates. A Personal View"; *J. Phys. Chem.* **1986,** *90,* 3460–3465.

Marcus, R. A.; Sutin, N. "Electron Transfers in Chemistry and Biology"; *Biochim. Biophys. Acta* **1985,** *811,* 265–322.

Sutin, N. "Marcus' Contributions to Electron-Transfer Theory"; *J. Phys. Chem.* **1986,** *90,* 3465–3466.

# The Nobel Prize in Chemistry
## A Bibliographical Guide

# The Nobel Prize in Chemistry
## A Bibliographical Guide

## On Alfred Nobel

Bergengren, E. *Alfred Nobel, The Man and His Work*; With a supplement on the Nobel Institutions and the Nobel Prizes by Nils K. Ståhle; Blair, A., Trans.; T. Nelson and Sons, Ltd.: London/New York, 1962.

Halasz, N. *Nobel: A Biography of Alfred Nobel*; Orion Press: New York, 1959.

Kant, H. *Alfred Nobel*; Biographien hervorragender Naturwissenschaftler, Techniker, und Mediziner, Bd. 63; BSB B. G. Teubner: Leipzig, Germany, 1983.

Schück, H.; Sohlman, R. *The Life of Alfred Nobel*; Lunn, B. and B., Trans.; Heinemann: London, 1929.

Sohlman, R. *The Legacy of Alfred Nobel: The Story Behind the Nobel Prizes*; Schubert, E. H., Trans.; Bodley Head: London, 1983.

## The History and Sociology of the Nobel Prizes

Crawford, E. *The Beginnings of the Nobel Institution: The Science Prizes, 1901–1915*; Cambridge University Press: Cambridge, England/New York, 1984.
    Based on extensive research in the Nobel Archives, Crawford's study details the workings of the Nobel selection process and analyzes the evolution of the Nobel Prizes in the context of disciplines, discoveries, and the international scientific community at the turn of the century.

Crawford, E. *Nationalism and Internationalism in Science. 1880–1939: Four Studies of the Nobel Population*; Cambridge University Press: Cambridge, England/New York, 1992.
    Crawford uses the population of Nobel candidates and prize winners from 1901–1939 as the starting point for examinations of the role of nationalism and internationalism in science during and after World War I, the relations of center and periphery in Central European science, the use of the Nobel Prizes in the Kaiser-Wilhelm Gesellschaft, and the role of the Nobel elite in American science.

Friedman, R. M. "The Nobel Physics Prize in Perspective"; *Nature* **1981**, *292*, 793–798.

Küppers, G.; Ulitzka, N.; Weingart, P. "Factors Determining the Award of Nobel Prizes in Physics and Chemistry, 1901–1929"; *Endeavour* **1983**, 7, 203–206.

Küppers, G.; Ulitzka, N.; Weingart, P. *Die Nobelpreise in Physik und Chemie, 1901–1929: Materialien zum Nominierungsprozess*; Report Wissenschaftsforschung, 23; B. Kleine Verlag: Bielefeld, Germany, 1982.

*The Nobel Population, 1901–1937: A Census of the Nominators and Nominees for the Prizes in Physics and Chemistry*; Crawford, E.; Heilbron, J. L.; Ullrich, R., Eds.; Berkeley Papers in History of Science, 11; Uppsala Papers in History of Science, 4; Office for History of Science and Technology: University of California, Berkeley; Office for History and Science: Uppsala University, Sweden, 1987.

Nobelstiftelsen; *Nobel, The Man and His Prizes*, 3rd ed.; Nobel Foundation; Odelberg, W., Eds.; 3rd ed.; Elsevier: Amsterdam/London/New York: 1972.
    This official guide to the Nobel Prizes and their history first appeared in Swedish in 1950 to commemorate the 50th anniversary of the Nobel Prize. The book includes chapters on Alfred Nobel (by Henrik Schück), on Nobel and the Nobel Foundation (by Ragnar Sohlman), on the administration and finances of the Nobel Foundation (by Nils Ståhle), and on the scientific work of the chemistry laureates (by Arne Westgren).

*Science, Technology, and Society in the Time of Alfred Nobel*; Bernhard, C. G.; Crawford, E.; Sörbom, P., Eds.; Nobel Symposium 52; Published for the Nobel Foundation by Pergamon Press: Oxford/New York, 1982.
    Of particular interest for understanding the context of the Nobel Prizes in chemistry before 1930 are the essays by John Heilbron on *"Fin-de-siècle* Physics" (pp 51–73); Joseph S. Fruton on "The Interplay of Chemistry and Biology at the Turn of the Century" (pp 74–96); Erwin N. Hiebert on "Developments in Physical Chemistry at the Turn of the Century" (pp 97–115); and L. F. Haber on "Chemical Innovation in Peace and War" (pp 271–282).

*The Nobel-Prize Awards in Science As a Measure of National Strength in Science*; U.S. House of Representatives, Committee on Science and Technology, Task Force on Science Policy; Science Policy Study Background Report No. 3; Report Prepared by the Congressional Research Service, Library of Congress, Ninety-Ninth Congress, Second Session, Serial S; U.S. Government Printing Office: Washington, DC, September 1986.
    An intriguing statistical study of trends in Nobel Prize awards as indicators of the comparative standing of national scientific enterprises.

Wilhelm, P. *The Nobel Prize*; Springwood Books: London, 1983.
    This illustrated volume contains a brief biography of Alfred Nobel, an account of the Nobel Foundation and their selection processes, and an interesting chapter on Kenichi Fukui's visit to Stockholm to receive the 1981 Nobel Prize in chemistry (pp 53–71).

Zuckerman, H. *Scientific Elite: Nobel Laureates in the United States*; Free Press: New York, 1977.
    A thorough sociological study of the 92 Nobel laureates who conducted their prize-winning research in the United States between 1901 and 1972. Zuckerman analyzes the education, recruitment, and funding of this elite group in the context of social stratification and hierarchy within the American scientific community.

## The Chemistry Laureates and their Lectures

Nobelstiftelsen; *Les Prix Nobel*; P. A. Norstedt & Söner: Stockholm, Sweden, 1901–1977; Almqvist & Wiksell International, 1978–  .
The official record of the annual award of the Nobel Prizes, including presentation speeches, laureates' biographies, and the Nobel Prize lectures (in their original languages). For the text of each chemistry lecture in English, see the next entry.

Nobelstiftelsen; *Nobel Lectures: Chemistry*; Elsevier: Amsterdam/London/New York, 1964–1972.
This four-volume set, published under the auspices of the Nobel Foundation, includes presentation speeches, laureates' biographies, and the text of each Nobel laureate's acceptance speech from *Les Prix Nobel*. The first volume covers 1901–1921; subsequent volumes cover 1922–1941, 1942–1962, and 1963–1970.

*Nobel Lectures in Chemistry, 1971–1980*; Forsen, S., Ed.; World Scientific Publishing: River Edge, NJ, 1993.

*Nobel Lectures in Chemistry, 1981–1990*; Malmstrom, B., Ed.; World Scientific Publishing: River Edge, NJ, 1993.

## Biographical Sources on Nobel Prize Winners in Chemistry

An indispensable resource for reliable biographical information on most Nobel laureates in chemistry is the *Dictionary of Scientific Biography*; Gillispie, C. C., Ed.; Charles Scribner's Sons: New York, 1970–1980; 16 vols. The biographical essays in the DSB include annotated bibliographies of primary and secondary sources. The 1990 *Supplement II*, edited by Frederic L. Holmes, adds essays on scientists who died after 1970.

Authoritative memoirs of many Nobel laureates, usually including complete bibliographies of their scientific writings, will also be found in the *Biographical Memoirs of Fellows of the Royal Society* (Royal Society: London, 1932–  ; the series was titled *Obituary Notices* prior to 1955) and in the *Biographical Memoirs of the National Academy of Sciences* (National Academy of Sciences: Washington, DC, 1877–  ).

Another helpful source is *J. C. Poggendorff biographisch-literarisches Handwörterbuch der exakten Naturwissenschaften*; Sächsischen Akademie der Wissenschaften zu Leipzig et al., Eds; Akademie-Verlag: Berlin, Germany, 1956–  ; Vols. VIIa, VIIb. As the title suggests, *Poggendorf* provides detailed bio-bibliographies for the scientists covered. These volumes focus on items from the years 1932 through 1962, but entries (particularly in Vol. VIIb) routinely include references to more recent secondary literature and monographs. Previous editions of *Poggendorf* (first published in 1863) are also useful for tracking the work of the earlier Nobel Prize winners in chemistry.

In addition to *Chemical Abstracts*, which provides useful coverage of historical items in the chemical literature, the single best source for following current scholarship on the history of Nobel laureates in chemistry is the annual *ISIS Critical Bibliography*, edited by John Neu for the History of Science Society. See also the *ISIS Cumulative Bibliography: A Bibliography of the History of Science Formed from ISIS Critical Bibliographies 1–90, 1913–1965*; Whitrow, M., Ed.; Mansell: London, 1971–1984; Vol. 1, Part I and Vol. 2, Part I, "Personalities"; Vol. 2, Part II, "Institutions"; Vol. 3, "Subjects"; Vols. 4 and 5, "Civilizations and Periods"; Vol. 6, "Author Index". The *ISIS Cumulative Bibliography, 1966–1975*; Neu, J., Ed.; Mansell: London, 1980–1985; 2 vols.; and *ISIS Cumulative Bibliography, 1976–1985*; Neu, J., Ed.; G. K. Hall: Boston, MA, 1989; 2 vols.; are continuations of Whitrow's compilation.

Bertsch, S. McG. *Nobel Prize Women in Science: Their Lives, Struggles, and Momentous Discoveries*; Birch Lane Press: New York, 1993.
   Includes lively biographies of Marie Curie (pp 11–36), Irène Joliot-Curie (pp 117–143), and Dorothy Crowfoot Hodgkin (pp 225–254), with useful notes on additional sources.

Farber, E. *Nobel Prize Winners in Chemistry, 1901–1961*, revised edition; The Life of Science Library, No. 41; Abelard-Schuman: London/New York, 1963.
   Farber includes a biography of each laureate, a description of their prize-winning work (including excerpts from the Nobel lectures and other writings by the laureate), and an assessment of the consequences of their research for chemical theory and practice. A useful bibliography of additional sources is also included (pp 309–328).

Farber, E. *Nobel Prize Winners in Chemistry, 1901–1950*; The Life of Science Library, No. 31; H. Schuman: New York, 1953.

*German Nobel Prizewinners*, 2nd ed.; Hermann, A., Ed.; Inter-Nationes: Bonn-Bad Godesberg, Germany, 1978.
   Nobel Prize-winners in chemistry are covered on pages 113–160, with a guide to additional sources on pages 168–169.

*Great Chemists*; Farber, E., Ed.; Interscience: New York, 1961.

*The Nobel Prize Winners: Chemistry*; Magill, F. N., Ed.; Salem Press: Pasadena, CA/ Englewood Cliffs, NJ, 1990; 3 vols. (1901–1937, 1938–1968, 1969–1989).
   Each entry includes an abstract of the presentation speech, the laureate's Nobel lecture and critical reaction to the award, a biography of the laureate including an assessment of his or her scientific career, and a bibliography of relevant primary and secondary sources. Robert J. Paradowski's informative introductory essay on "The Nobel Prize in Chemistry: History and Overview" in Vol. I (pp 1–42) concludes with an annotated bibliography of selected readings.

*Nobel Prize Winners: An H. W. Wilson Biographical Dictionary*; Wasson, T., Ed.; H. W. Wilson: New York, 1987.
   This collection includes narrative biographies with portraits of each laureate and brief bibliographies of selected works by and about them. The volume also includes a brief biography of Alfred Nobel (by Alden Whitman, pp xxiii–xxviii) and an essay by Carl Gustaf Bernhard on "The Nobel Prizes and Nobel Institutions" (pp xxix–xxxiii).

*Nobel Prize Winners, Supplement 1987–1991: An H. W. Wilson Biographical Dictionary*; McGuire P., Ed.; H. W. Wilson: New York, 1992.
   The first of a planned series of supplements to H. W. Wilson's *Nobel Prize Winners* (1987), this volume continues coverage through 1991, including indexes of Nobel Prize winners by prize, year, and country of residence, 1901–1991, and a table of Nobel laureates who died between 1986 and 1991.

Opfell, O. S. *The Lady Laureates: Women Who Have Won the Nobel Prize*, 2nd ed.; Scarecrow Press: Methuchen, NJ, 1986.

Thompson, L. *America's Nobel Laureates in Medicine, Physiology, and Chemistry: A Tribute to America's Living Nobel Prize Winners*, 2nd ed.; Friends of the National Library of Medicine: Washington, DC, 1988.

*The Who's Who of Nobel Prize Winners, 1901–1990*, 2nd ed.; Schlessinger, B. S.; Schlessinger, J. H., Eds.; Oryx Press: Phoenix, AZ, 1991.
    This edition of the *Who's Who of Nobel Prize Winners* (first published in 1986) includes new entries for 1986–1990, an introductory chapter by W. Odelberg on "A. Nobel: The Man and His Prizes", and indexes to the laureates by name, educational institution, nationality/citizenry, and religion.

## *Related Publications on the Nobel Prizes in Physics and Physiology or Medicine*

Heathcote, N. H. de V. *Nobel Prize Winners in Physics, 1900–1950*; The Life of Science Library, No. 30; H. Schuman: New York, 1953.

*Nobel Laureates in Physiology or Medicine: A Biographical Dictionary*; Fox, D. M., Ed.; Garland Publishing: New York, 1990.

*Nobel Lectures in Molecular Biology, 1933–1975*; with a foreword by D. Baltimore; Elsevier-North Holland: New York, 1977.

*Nobel Lectures in Physics, 1971–1980*; Lundqvist, S., Ed.; World Scientific Publishing: River Edge, NJ, 1992.

*Nobel Lectures in Physics, 1981–1990*; Ekspong, G., Ed.; World Scientific Publishing: River Edge, NJ, 1993.

*Nobel Lectures in Physiology or Medicine, 1971–1980*; Lindsten, J., Ed.; World Scientific Publishing: River Edge, NJ, 1992.

*Nobel Lectures in Physiology or Medicine, 1981–1990*; Lindsten, J., Ed.; World Scientific Publishing: River Edge, NJ, 1993.

Nobelstiftelsen; *Nobel Lectures: Physics*; Elsevier: Amsterdam/London/New York, 1964–1972; 4 vols.

Nobelstiftelsen; *Nobel Lectures: Physiology or Medicine*; Elsevier: Amsterdam/London/New York, 1964–1972; 4 vols.

Sourkes, T. L. *Nobel Prize Winners in Medicine or Physiology, 1901–1966*; The Life of Science Library, No. 45; Abelard-Schuman: New York, 1967.

Weber, R. L. *Pioneers of Science: Nobel Prize Winners in Physics*; Lenihan, J. M. A., Ed.; Institute of Physics: Bristol/London, 1980.

Jeffrey L. Sturchio<br>
Merck & Co., Inc.

# Index

# Index

## A

Abel, John Jacob, 381
Abelson, John, 747
Abelson, Philip, 340, 345
Abrahamson, E. W., 645
Accelerators, 341
Accessory food factors, 184
Actinide elements, 347
Actinium, 276
 discovery, 77
Activated complexes, 704
Acyl group activation, 619–620
Adair, G. S., 161, 436
Adams, Roger, 263, 752
Adiabatic demagnetization process,
 324–325, 326
Adiabaticity, 703
Adsorption indicators, discovery, 514
Affinity, 62, 63, 89, 94, 130
Alcohol dehydrogenase, 47
Alcohol fermentation, 43, 182–183
Alder, Kurt, 328, 335
Alder–Stein rules, 329
Alicyclic compounds, 69
Alkali Age, 690, 695
Alkaloids, synthesis, 306, 310–311
Alpha globulin, 319
Alpha helix, 373
Alpha-particle scattering, 57
Alpha rays, study, 55
Altman, Sidney, 737
Americium, 347
Amino acids, 10–11
 separation, 353–354
 sulfur-containing, 382

Ammonia
 oxidation, 200, 201
 synthesis, 114, 118–119
Anderson, Carl, 278
Anderson, Weston, 760
Androsterone
 isolation, 253, 255, 257
 study, 262
Anesthesia, 375
Anfinsen, Christian, 532, 538, 546, 548
Anisotropies, 160
Anson, Mortimer, 295
Antibody structure, 372
Antimetabolites, 668–670
Antivitamins, 251
Applied theoretical chemistry, 650
Argon, discovery, 25–26
Aromatic sextet, 312
Arrested relaxation, 690–691
Arrhenius, Svante, 4, 15, 27, 63–64, 65,
 80, 116, 125, 126, 128, 176, 393, 500
Artificial radioactivity, 217, 218, 223,
 225, 278
Ascorbic acid, 239–240
Astbury, William, 431, 432
Aston, Francis William, 140
Aston dark space, 141
Atomic energy, 53
Atomic explanation, 79
Atomic hydrogen welding, invention,
 206
Atomic number, 105
Atomic structure, 49, 56–57
 study, 140
Atomic weights, universality, 100,
 104–105

Atoms, 66
  existence, 153
ATP synthesis, 600–601
Autocatalytic reaction, 296, 297
Avogadro's law, 128
Azocarboxylic ester, 328


**B**

B-complex vitamins, 251
Back-scattering effect, 57
Bacteriophages, 296
Bader, R. F. W., 645
Baeyer, Adolf von, 8, 11, 30, 109,
  261, 508
Baeyer strain theory, 33
Bamford, C. H., 481
Band spectra research, 473
Barbier, Philippe A., 84, 86
Barcroft, Joseph, 429, 436
Bartlett, Paul D., 710
Barton, Derek Harold Richard, 507
Barton reaction, 512
Bassham, James A., 425
Bauer, Alexander Anton Emil, 152
Bawden, Frederick, 296, 303
Becquerel, Henri, 50, 54, 75, 76, 78
Beijerninck, Martinus, 301
Belousov–Zhabotinski reaction, 594
Bémont, Gustave, 77
Bendikt, Rudolf, 152
Benson, Andrew A., 425
Benzaldehyde, 71
Benzene, 614
Berg, Paul, 48, 406, 618, 627
Bergius, Friedrich, 192
Bergmann, Max, 362, 381, 539, 547
Bergstrom, Karl Sune, 753
Berkelium, 347
Bernal, John D., 166, 303, 428, 458
Berthelot, Marcellin, 3, 71, 93, 130
Berthollet, Claude-Louis, 62, 127
Berzelius, Jöns Jacob, 65
Beta globulin, 319
Bifurcation, 594
Biilman, Einar, 117
Bile acids, 164, 165–166
Bilirubin, 189
Billwiller, John, 194
Biochemical catalysis, 737
Biochemical processes, 494

Biochemistry, 20
  development, 175
  origins, 6
Biogenetic isoprene rule, 263
Biogenetic theory, 466
Biological oxidation, 167
Biophysical chemistry, 158
Biophysics, 318
Biotin, 383
Bischof, Georg, 24
Bjerrum, Niels, 179
Blackman, F. F., 423
Blaise, E. E., 85
Bloch, Felix, 759, 760
Bloch, Konrad, 263, 466
Blodgett, Katharine, 208
Blomstrand, C. W., 94
Bodenstein, E. A., 65
Bodenstein, Max, 388
Bohr, Niels, 57, 211, 267, 279
Boltwood, Bertram, 51, 56, 273, 275
Boltzmann, Ludwig, 126, 128, 129
Bonhoeffer, Karl, 122, 494
Boron
  bonding, 585
  electron deficiency, 585, 587
Boron hydride chemistry, 584, 585–587
*Boron Hydrides* (Lipscomb), 587
Borsche, Walther, 401
Bosch, Carl, 195, 198
Bouveault, Louis, 84
Boyer, Herbert, 622
Bragg, Lawrence, 436
Bragg, William Henry, 457
Bragg, William Lawrence, 429, 457
Branch, G. E. K., 423
Branly, Edouard, 80
Brauer, Ernst, 65
Brauner, Bohuslav, 413
Brazelein, 310
Bredig, Georg, 159
Bredt's rule, 581
Brenner, Sydney, 740
Brevium, 277
Brickwedde, Ferdinand, 212
Brooks, P. R., 704
Brown, A. Crum, 640
Brown, Harrison, 215
Brown, Herbert C., 604, 611
Brownian motion, 24, 155, 158, 159
Brusselator, 593
Buchner, Eduard, 33, 42, 176
Bunsen, Robert, 23, 115

Bunte, Christian, 259
Bunte, Hans, 116
Burg, Anton, 607
Bush Report, 372
Butenandt, Adolf, 249, 253, 260, 262
Butter, storage, 286

## C

Cacodylic compounds, 31
Cade, Paul E., 476
Calcaterra, Lydia, 768
Calcott, W. S., 722
Calvin, Melvin, 422
Calvin–Benson–Bassham cycle, 425
Cameron, Ewan, 376
Camphors, 71
Carbanion chemistry, 708, 710
Carbohydrates, 8, 9,
    10–11, 236
Carbonyl addition reactions, 708
Cardiac poisons, 170
Carius method, 148
Carotene, 110
Carotenoids, 249, 250
Carothers, Wallace H., 564
Carter, A. S., 722
Cashion, J. K., 701
Castle, William B., 372
Catalysis, 61, 88, 91
Catalysts, 65
Cathodic potential, control of, 116–117
Cavendish, Henry, 24
Cech, Thomas, 737, 745
Cell diffusion, 295
Cell-free fermentation, 42
Cellulose conversion, 196
Celtium, 269
Cephalosporin antibiotic, synthesis,
    462, 464–466
Chadwick, James, 57, 219, 234, 277
Chain, Ernst, 309, 310
Chain reactions, 394, 395–396
Chain transfer, 566
Changeaux, Jean-Pierre, 438
Charge-control term, 645
Chaulmoogric oil, 301
Chemical bonding, 368, 471, 472, 584
    Lewis–Langmuir theory, 208
Chemical high-pressure methods, 192,
    193–194, 195, 198, 203

Chemical kinetics, 1, 3, 387, 480–481, 485,
    565, 687
    application to combustion and
        explosions, 393, 396
    *Chemical Kinetics and Chain Reactions*
        (Semenov), 392, 395, 396
Chemical paleogenetics, 375
Chemical reactions
    chain theory, 394, 395–396
    energy distributions, 701–702
    explanation, 686, 688–689
    gas-phase, 393, 395
    liquid-phase, 395
    rates, 61, 62
    selective enhancement, 703–704
Chemical reactivity
    coordination complexes, 661
    orbital theory, 639, 644–646
Chemical solutions, 15
Chemical stereochemistry, 1
Chemical thermodynamics, 4, 125
Chemical transformation, 393
Chemical warfare, 121, 123, 294, 472,
    539, 548
Chemiluminescence depletion, 704
Chemiosmotic theory, 598, 599–602
Chemisorption, 91, 397
Chemistry curricula, 64–65
Chemotherapy, 170
Chemzymes, 756
Cheronis, Nicholas, 604
Chibnall, A. C., 407, 634
Chiral molecule, 573, 574, 575
Chlorophyll, 187, 189
    isolation, 110, 111
Cholesterol, 170, 171, 255, 262, 572
    oxidation, 333
Choline, 382–383
Christiansen, Johann, 388, 394
Chromatin, structure, 657–658, 746
Chromatography, 110, 250, 261, 320, 352,
    353–354
CIP (Cahn–Ingold–Prelog) system, 581
Clausius, Rudolf, 8, 17
Cleve, Per Theodor, 16, 94
Cloning, 618, 622
Closs, Gerhard L., 768
Co-ferment, 182–183
Coal hydrogenation, 202
Coal liquefaction, 194–195
Cobyric acid, synthesis, 467
Cocaine, structure, 109
Cockcroft, John, 58

Cohen, Stanley, 622
Collisional activation, theory, 387
Colloid chemistry, 158–159, 291, 316
Colloid Chemistry, ACS Division, 161
Colloidal gold, 153, 154, 155
Colloids, 151, 153, 154–155, 365
Colors, theory, 66
Common-ion effect, 127
Comrie, L. J., 459
Conant, J. B., 102, 472
Conductivity, 16
Conformation, 507, 509, 510–511, 514
Conformational analysis, origins, 517
Conservation of energy, law, 78
Conservation of orbital symmetry, 645
Cooke, Josiah Parsons, 100, 101, 103
Coordination chemistry, 98
Coordination complexes, structure,
    661–663
Coordination theory of crystal
    structures, 370
Coprosterol, 171
Corey, E. J., 649, 750
Corey, Robert, 430
Cornforth, John, 263, 571
Cortisone, 257
Cotton effect, 466
Coulson, A. C., 642
Covalent bond, 371
Cox, John, 51
Cozymase, 177, 178
Craig, Lyman, 542
Cram, Donald J., 708, 715, 722
Cram's rule for asymmetric induction, 710
Crawford, Bryce, Jr., 586
Crick, Francis, 374, 428, 619, 740
Cromwell, Norman, 710
Crookes, William, 26, 53, 54, 141
Crookes's dark space, 141
Crown ethers, 711, 726–727
Crystal structures, determination, 674,
    675–676, 678, 683–684, 717
Crystallographic electron microscopy, 655
Crystalloids, 151
Cunningham, Burris B., 346
Curie, Marie, 75, 86, 217, 218, 267
Curie, Pierre, 52, 75, 76, 79, 218
Curie's law, 76
Curium, 347
Curly arrow, use, 312
Curtius, Theodor, 10, 43, 45
Cyclohexane, 515–516
Cyclophanes, 711

D

Danielli, James F., 597
Daniels, Farrington, 102
Dark-field illumination, 155
Dark space, 141
Datz, Sheldon, 688
Davy, Humphry, 37
De Donder, Théophile, 591
de Vries, Hugo, 4
Debierne, André, 77
Debye, Peter, 129, 179, 228, 501
Debye–Hückel theory, 501
Debye–Scherrer powder diffraction
    method, 231
Decalcomania, 61
Decay series research, 55
Degenerate chain branching, 483
Dehydrobenzene, 613–614
Dehydrogenation, 261, 262, 334
Deisenhofer, Johann, 729
Delbrück, Max, 431
Denkewalter, Robert, 536
Destructive distillation, 194
Deuterium, discovery, 212–213,
    269, 527
Dewar, M. J. S., 645
Diamonds, synthesi 'f artificial,
    38–39, 40
Diatomic Molecule Project, 476
Diborane, 606
Dickinson, Roscoe, 369
Diels, Otto, 170, 328, 329, 332
Diels–Alder reaction, 329, 330–331, 332,
    335–336
Diels hydrocarbon, 334
Diene synthesis, 329, 330–331, 332,
    335–336
Differential reactivity, 641
Diffraction, applied to organic
    molecules, 507
Digitalis, 170
Diradicals, 612
Disaccharides, 239
Disintegration theory, 135
Dissipative structures, theory, 590,
    593–594
Dissociation, 64, 129, 193, 267
Diterpenes, 262
DNA
    and chromatin structure, 746
    study, 374, 428
DNA sequencing, 618, 627, 628–631

Doisy, Edward, 254
Dorn, Friedrich, 27
Douglas, A. E., 528
Droplet nucleation, 208
Duisberg, Carl, 33
Dulbecco, Renato, 621
Dumas, Jean-Baptiste, 102
Dumas method, 148
Du Vigneaud, Vincent, 380

**E**

Earl, J. C., 571
Eberhardt, W. H., 586
Edlund, Eric, 16
Eigen, Manfred, 494
Einhorn, Alfred, 109
Einstein, Albert, 81, 131, 155, 159, 221, 230, 496
Einsteinium, 348
*Electric Furnace, The* (Moissan), 40
Electrochemistry, 116
Electrode potential, 117
Electrolyte solutions, thermo-
    dynamics, 1
Electrolytic dissociation, 15, 18–19
Electrometric effect, 312
Electron configuration, 474
Electron diffraction, 228, 231, 370
Electron microscope, 303
Electron microscopy, 656
Electron-pair bond, 207–208
Electron-transfer reactions, 766, 768–769, 770
Electronegativity, 371
Electronic theory of organic chemistry, 308–309
Electrons, 18
Electrophoresis, 315, 317, 318–319
*Elektrochemische Gesellschaft*, 65
Elements
    disintegration, 49
    No. 72, 268–269
    No. 93, 340, 345
    No. 94, 340, 346
    No. 95, 346
    No. 96, 347
    No. 99, 348
    No. 100, 348
    No. 101, 348
    No. 102, 348
Elimination reactions, 511

Emanation, 135
Embden–Meyerhof pathway, 47
Endle, R., 360
Engberg, J., 747
Engelmann, Wilhelm, 64
Engler, Carl, 259, 360
Enriched hydrogen, 213–214
Entropy, 130, 322–323, 592
    law of increasing, 20
Enzyme replication theory, 297
Enzymes, 10, 44, 112
    crystallization, 288, 292, 295–296
    reactions catalyzed, 571
    role, in biochemical reactions, 571, 573–575, 576
Enzymology, 47
Equal reactivity, 565, 566
Equilibrium, 3–4, 16, 19, 61, 64, 130–131, 495, 497, 502
Ergosterol, 171, 172
Ernst, Richard R., 759
Eschenmoser, Albert, 466
Esson, William, 3
Estrogen, 257
    isolation, 253, 254–255, 257
Ettingshausen, Albert von, 128
*Etudes de dynamique chimique*
    (van't Hoff), 3, 4, 18
Euler-Chelpin, Hans von, 46, 175, 181
Europium, 347
Evans, M. G., 700
Evolution, 627
Eyring, Henry, 700, 767

**F**

Fajans, Kasimir, 137, 275, 515
Fankuchen, Isidor, 303, 458
Faraday, Michael, 37
Fermentation
    alcoholic, 43, 182–183
    cell-free, 42
    sugar, 177
Ferments, 45
Fermi, Enrico, 278, 345
Fermium, 348
Ferrocene, 725
Fertilizers, 200–201, 204
Fieser, Louis, 510, 710
Fischer, Emil, 8, 31, 32, 146, 169, 187, 238, 274, 276, 332
Fischer, Ernst Otto, 561
Fischer, Franz, 12

Fischer, Hans, 187
Fischer, Hermann, 8
Fischer, Otto, 9, 33, 181
Fischer carbenes, 554
Fischer–Hafner method, 553
Fischer–Tropsch synthesis of
  hydrocarbons, 12
Fission, 220, 225, 272, 279–280,
  339–340, 346
Fittig, Rudolf, 23
Flash photolysis, 480–482, 483–484,
  487–488, 490–492, 769
Flashpoint, 396–397
Fleck, Alexander, 137
Floderus, M. M., 15
Florey, Howard, 309, 310
Flory, Paul, 564, 565
Flory–Huggins theory, 566–567
Fluorescein, 32
Fluorine, 36, 37–38
*Fluorine and Its Compounds* (Moissan), 40
Fodder, preservation, 282, 285–286
Folin, Otto, 288
Fourier transform theory, 680
Fractional distillation, 352–353
Frank, Otto, 188
Frankland, Edward, 26, 84
Free energy, 130, 323
Free-energy values, 117
Free radicals, 395, 398, 451, 489, 525
Frémy, Edmond, 37
Freundlich, Herbert, 122
Friedrich, Walter, 231
Frontier electrons, 643, 644
Fry, H. S., 640
Fukui, Kenichi, 639
*Fundamentals of Carbanion Chemistry*
  (Cram), 710
Fuoss, Raymond Matthew, 502

**G**

Gall, J., 747
Gamma globulin, 319
Gamow, George, 57–58, 739
Gas adsorption and heterogeneous
  catalysis, theory, 206
Gas-filled electric lamp, invention, 206
Gas-liquid partition chromatography, 352,
  354, 357
Gas-phase reactions, 393, 395
Gaseous ionization, theory, 50, 51
Gasser, Herbert, 540, 548

Geiger, Hans, 55, 56, 267
Gel formation, theory, 566
Gellert, H. G., 452
Gene splicing, 618, 621–622
General displacement law, 137
Genetic engineering, 48, 618
Gentry, W. Ronald, 695
Geometry in the behavior of organic
  compounds, 507
German Chemical Society, 32
Giauque, William Francis, 321
Gibbs, Josiah Willard, 64, 65, 130,
  500, 502
Gibbs–Helmholtz equation, 131
Gibson, J., 640
Gilbert, Walter, 406, 618, 627
Gillis, Joseph, 682
Glansdorff, P., 594
Glansdorff–Prigogine stability theory, 594
Glass electrode, development, 117
Glucose, conversion to galactose,
  521–522
Glycogen, 520, 521–522
Glycolysis, 47
Glycosyl-transfer reactions, 522–523
Gold
  colloids, 153, 154, 155
  harvesting from the ocean, 121, 123
Goldberg, Moses, 263
Gordon, Neil, 608
Gorovsky, G., 747
Gouterman, Martin, 648
Grabowski, Paula, 747
Graham, Thomas, 151
Greenhouse effect, 19–20
Grignard, Victor, 83, 335
Grignard reaction, 83, 335
Grignard reagent, 85
Grodzins, Lee, 738
Guldberg, Cato Maximilian, 3, 62, 63
György, Paul, 383

**H**

Haber, Fritz, 114, 200, 259, 276, 277
Haber–Bosch process, 204
*Habilitation*, 527
Hafner, Walter, 553
Hafnium, 266
  discovery, 269
Hahn, Otto, 12, 51, 136, 225, 249, 272,
  339, 345
Halban, Hans, 219

Half-life, 52, 54
Hall, Norris Folger, 103
Happiness, theory, 66
Harcourt, A. G. Vernon, 3
Harden, Arthur, 46, 176, 181
Harker, David, 681
Harkins, William D., 207, 472
Harned, Herbert S., 5
Hassel, Odd, 514
Hauptman, Herbert A., 674, 678, 682
Haurowitz, Fritz, 436, 437
Haworth, Walter, 236
Heat theorem, 118, 131–132
Heat transfer, 207
Heavy hydrogen, 213–214
Heitler, Walter, 526
Heitler–London valence bond method, 474
Helium
  boiling point, 322
  discovery, 26
Helmholtz, Hermann von, 126, 127, 130
Hemin, synthesis, 187, 188–189
Hemoglobin, 428, 435, 437–439
Henderson, Lawrence, 289
Heredity, biochemistry, 257
Herriot, Roger, 296
Herschbach, D. R., 686, 694, 704
Herty, Charles, 95
Herzberg, Gerhard, 525, 700
Herzfeld, Karl, 388, 497
Herzog, Reginald, 362
Hess, Alfred, 170
Hess, K., 362
Heuer, Werner, 364
Hevesy, George de, 137, 178, 266
Heyrovsky, Jaroslav, 412
High-molecular-weight compounds,
  362, 363
Hilbert, David, 230, 231
Hill, A. V., 436
Hill, Robin, 423
Hillebrand, William, 26
Hinshelwood, Cyril, 386, 393, 485
Hirschmann, Ralph, 536
Histamine, 170
Hodgkin, Dorothy, 310, 429, 456
Hoffmann, Roald, 467, 644, 648
Hofmann, August von, 8, 9, 115
Hollemann, Arnold, 33
Host–guest chemistry, 711
Hubbard, Joseph, 504
Huber, Robert, 729
Hückel, Erich, 129, 450, 501, 642

Huggins, M. L., 567
Hund, Friedrich, 473
Hutchinson, James, 763
Hydrazine, 9
Hydroboration, 609
Hydrocarbons, 116
Hydrogenation, 88, 90–91

I

IG Farben, 202, 203
Immersion ultramicroscope, 156
Immunochemistry, 372
Immunoglobulin, 319
Indigo, synthesis, 30, 32
Indigo problem, 31
Indole, 9, 31
Influenza vaccine, 303–304
Infrared chemiluminescence, 701
Infrared multiphoton dissociation
  (IRMPD), 697
Ingold, C. K., 312
Inner-sphere electron transfer, 662
Inorganic and organometallic theory, 651
Inorganic chemistry, resurgence, 661
Insulin, 381
  structure, 460
  study, 406, 407–409, 633–636
Intercalation compounds, 96
Interionic attraction model, 232
Interpretation of Radium, The
  (Soddy), 136
Investigations on Chlorophyll
  (Willstätter), 111
Ionic reactions, 494
Ionic solutions, 501
Ionists, 17, 64, 176
Ionium, discovery, 275
Iron metabolism, study, 270
Irreversibility, theory, 594–595
Is Science Necessary? (Perutz), 440
Isochore, 130–131
Isolobal analogy, 652
Isomers, 2
Isoprene rule, 72–73, 261, 263
Isotactic polypropylene, discovery, 442,
  445, 449, 453
Isotope, 54
Isotope factory, 472
Isotopes, 100, 104, 134, 137, 142, 212,
  213, 266
  separation, 214
Isotopic dilution, 268

Isotopic tritium, 421
Ivanovskiy, Dimitri, 301

J

Jacob, François, 438
James, A. T., 352, 354
James, R. W., 654
Janssen, Pierre, 26
Jeener, Jean, 762
Jeger, Oskar, 263
Joklik, W. K., 619
Joliot, Frédéric, 75, 81, 86, 217, 278
Joliot-Curie, Irène, 75, 81, 86, 223, 267, 278
Jorgensen, S. M., 94
*Journal of Molecular Biology*, 432
Just, G., 120

K

Kaiser, Reinhold, 761
Kaiser Wilhelm Society for the Advancement of Science, 11, 119–120, 122
Kamerlingh Onnes, H., 322
Karle, Jerome, 674, 678
Karrer, Paul, 250, 362
Kasha, M., 489
Kasper, J. V. V., 681, 702
Kealy, T. J., 558, 559, 560
Keilin, David, 436, 598, 599
Kekulé, August, 2, 8, 9, 31, 70, 94
Kendrew, John, 428
Kermack, W. O., 312
Ketenes, 361
Kharasch, Morris, 604, 606
Khorana, H. G., 622
Kiliani, Heinrich, 169
Kilmartin, J. V., 439
Kjeldahl method, 148
Klaproth, Martin Heinrich, 345
Klare, Hermann, 334
Klemperer, William, 688, 691
Klopman, G., 645
Klug, Aaron, 654
Kneser, H. O., 497
Knoop, Franz, 170
Knorr, Ludwig, 115
Kohler, E. P., 463
Kohlrausch, Friedrich Wilhelm Geary, 126
Kolbe, Hermann, 2
Kornberg, Arthur, 619

Kornberg, Roger, 658
Kowarski, Lev, 219
Kramers, H. A., 211, 394
Kraus, Charles A., 502
Krebs, Hans, 183
Krypton, discovery, 27
Kuhn, Alfred, 257
Kuhn, Richard, 248
Kuhn, Werner, 567
Kundt, August Adolph Eduard Eberhardt, 153
Kunitz, Moses, 295
Kwei, George, 688

L

Laboratory manuals, 64
Laboratory safety, 11–12
*Lac* repressor, 628
Ladenburg, Rudolf, 122
Landolt, Hans, 129
Landsteiner, Karl, 372
Lanesterol, 262
Langevin, Paul, 80, 155, 218, 223
Langmuir, Irving, 91, 205, 312, 369
Langmuir adsorption isotherm, 207
Langmuir–Blodgett films, 208
Lanthanide elements, 347
Lappe, Franz, 200
Lapworth, Arthur, 307, 312
Lasers, development, 702–703
Lauterbur, Paul, 763
Lavoisier, Antoine-Laurent, 423
Lawrence, Ernest O., 267, 342, 423
Le Bel, A., 10
Least dissipation of energy, principle, 502
LeBreton, Pierre, 690, 695
Lee, Yuan Tseh, 689, 690, 693, 704
Leghemoglobin, 283–284
Lehman, I. R., 622
Lehn, Jean-Marie, 715, 722
Leloir, Luis, 520
Leuchs, Hans, 311
Leuna Werke, 201
Lewis, G. N., 5, 102, 103, 118, 207, 312, 345, 369, 422, 465, 472, 489, 640
Lewis, Howard B., 380
Lewis–Langmuir theory, 312
LHASA (Logic and Heuristics Applied to Synthetic Analysis), 754
Libby, Willard F., 419, 768
Liebermann, Karl, 115

Liebig, Justus, 32, 33, 71, 128
Ligand-substitution concepts, 663
Lindemann, F. A., 142
Linear combinations of atomic orbitals (LAOCs), 474, 476
Link, Karl Paul, 538
Lipscomb, William, 584, 648
Liquid crystals, 160
Liquid–liquid partition chromatography, 352, 356–357
Liquid-phase reactions, 395
Living polymers, 451
Localization method, 642–643
Lockyer, Joseph Norman, 26
Lodge, Oliver Joseph, 17
Loeb, Jacques, 295
Longuet-Higgins, H. C., 585, 645
Lowry, T. S., 458
Lucas, H. J., 640
Ludwig, Ernst, 152
Luggin, Hans, 116
Lukes, Rudolf, 578
Lunge, Georg, 115
Luther, Robert, 64

**M**

Macrocycles, 725
Macrocyclic compounds, 261
Macromolecular chemistry, 359, 363, 364–365
Macromolecules, 564, 568
Magnetic cooling, 232
Magnetic field, and changes in entropy, 324–325
Magnetic susceptibility, 325
Mahan, Bruce, 694
Mailhe, Alphonse, 90
Maiman, Theodore, 702
Manhattan Project, 345, 346, 420, 422, 473, 475, 502, 679
Mansfield, Peter, 763
Many-membered rings, 261, 450
Marcus, Rudolph A., 766
Marcus theory, 768
Marsden, Ernst, 56
Marshall, John, 101
Martin, A. J. P., 352, 424, 542
Marvel, Carl, 380, 616, 752
Mass action, theory, 3–4, 6, 16, 62
Mass spectrograph, 140, 142–144
Mathews, Joseph H., 103, 160
Maxam, Allan, 627, 628

McClung, R. K., 53
McDonald, J. Douglas, 690, 695
McMillan, Edwin, 338, 345
Meitner, Lise, 12, 275, 276, 278, 279
Meldrum, William Buell, 103
Membrane proteins, crystallization, 731–734
Mendelevium, 348
Mercury isotope separation, 472–473
Merrifield, R. B., 536, 667
Meselson, Matthew, 740
Mesothorium, 136, 274
Messenger RNA, 620
Metal contaminant deactivator, 722–723
Metal hydrides, 39–40
Metal sandwich compounds, 551, 552, 553, 557
Metallic bond, 373
Metallocenes, 557, 560–561
study of, 551–553
Methyl ethers, 238
Mevalonic acid, 574–575
Meyer, Kurt H., 365
Meyer, Lothar, 17
Meyer, Victor, 33
Meyerhoffer, Wilhelm, 6
Michel, Hartmut, 729
Microanalytical methods, 146, 148–149
Microcrystallinity, 443
Microdetermination methods, 148
Microhydrogenation, 250
Microscopic reversibility, principle, 502
Miller, E. G., 547
Miller, John R., 768
Miller, Walter, 690
Miller, Wilhelm von, 152
Millikan, Robert, 472
Miolati, Arturo, 95
Mirror-image isomers, 97–98
Mirsky, Alfred, 371
Mirzabekov, Andrey, 629
Mitchell, Peter, 597
Mittasch, Alwin, 200
Mohr, Ernst, 508
Moissan, Henri, 35, 84, 86, 89
Moissan furnace, 36
Molecular architecture, 514
*Molecular Basis of Evolution, The* (Anfinsen), 533
Molecular-beam chemistry, 691
Molecular beam kinetics, universal machine for, 695
Molecular biology, origins, 156, 158, 304

Molecular medicine, 375, 622
Molecular nitrogen complexes, 664
Molecular orbital (MO) theory, 450, 474, 477, 642
Molecular spectra
  applications, 529
  study, 525–526
*Molecular Spectra and Molecular Structure* (Herzberg), 527
Molecular spectroscopy, 261
Molecular stereodynamics, 691
Molecular structure, 228, 584
Molecular weights, 24
Molecules, electronic structure, 471, 473
Monod, Jacques, 438
Monosaccharides, 238–239, 539
Moore, Stanford, 533, 538, 546
Morgan, Thomas Hunt, 371
Moseley, Harold, 267
Moseley, Henry G. J., 138
Moureu, Charles, 89
Moving boundary electrophoresis, 158, 161
Muetterties, Earl, 650
Muirhead, Hilary, 438
Mulheim atmospheric polyethylene process, 453
Müller, Friedrich von, 187
Muller-Hill, Benno, 628
Mulliken, Robert, 471
Multinuclear compounds, 96
Multistage processes, 494
Murlin, John R., 381
Murphy, George, 212
Myoglobin, 428, 429–430
Myrback, Karl, 177

**N**

Nagai, Kiyoshi, 440
Nageli, Karl von, 43
National Colloid Symposium, 161
Natta, Giulio, 442
Natural products
  structure, 579, 581
  synthesis, 306, 307, 311, 464, 465–467
Neon
  discovery, 27, 142
  separation, 142
Nephelometer, invention, 104
Neptunium, 340, 345
Nernst, Walther, 17, 18, 64, 102, 118–119, 125, 127, 175, 205, 388, 496

Nernst equation, 127
Network theory of interdependent enzyme balance, 390
Neurochemistry, 716
Neutron bombardment, 278
Neutrons, discovery, 219, 224
Newer alchemy, 52
Nichols, J. B., 158, 161, 364
Nilsson, Ragnar, 177
Nitric acid, production, 201
Nitrogen, microdetermination, 148
Nitrogen conversion, 199–200
Nitrogen fixation, 207, 283–284
Nitrogen inversion, 716
Nitrogen problem, solution, 199, 200, 204
NMR spectrometry, 715
Nobelium, 348
Noble gas family, discovery, 23, 26–27
Noise decoupling, 761
Nonclassical ion problem, 608
Nonequilibrium thermodynamics, 590, 592–594
Nonnegativity criterion, 675, 680
Norris, James F., 463, 472, 688, 722
Norrish, Ronald, 479
Northrop, John, 112, 292, 294, 302
Noyes, Arthur A., 369, 472
Nuclear isomerism, 277
Nuclear magnetic resonance (NMR) spectroscopy, 759, 760–763
Nuclear stability, 144
Nucleic acids, sequencing, 636–637
Nucleotide coenzymes, 399
Nucleotides, 9
  synthesis, 399, 403–404
Nuttall, George, 44

**O**

Octet theory, 208
Odorous ketones, 260
Ogston, Alexander, 572, 573
Oils, 70–71
Olszevski, Karol, 26
Onsager, Lars, 129, 500, 591
Onsager limiting law for dilute solutions of electrolytes, 501
Onsager reciprocal relations, 500
Oppenheimer, J. Robert, 221, 340, 372
Opposing reactions, 4
Optical centrifuge, 158
Orbital-control term, 645

Orbital symmetry conservation, 640–650
Orbital theory of chemical reactivity, 639,
    644–646
*Order Out of Chaos* (Prigogine), 595
Organic Age, 691
Organic reaction mechanism, theory,
    308–309
Organic substances, microanalytical
    methods, 146, 148–149
Organic synthesis
    logical approach, 751, 754–755
    organomagnesium halides, 83, 84–85
Organoalkali metal compounds, 450,
    451–453
Organomagnesium halides, 83, 84–85
Organometallic chemistry, 466, 551
Orgel, Leslie, 651
Orthomolecular psychiatry, 376
Osazones, 9
Oscillating chemical reactions, 593
Oser, Johann, 152
Osmium, 200
Osmotic pressure, 1, 4
Ostwald, Wilhelm, 5, 17, 61, 91, 102, 125,
    126, 128, 146, 153, 155, 199
Ostwald, Wolfgang, 365
Ott, Emil, 360
Ourisson, Guy, 715
*Outline of Technical Chemistry on a
    Theoretical Basis* (Haber), 117
*Outlines of Physical Chemistry*
    (Ostwald), 128
Overlap-orientation principle, 643
Owen, Benton B., 5
Owens, R. B., 52
Oxalyl chloride, 361
Oxonium compounds, 33
Oxytocin, synthesis, 380, 383–384

**P**

Packing fraction, 144
Paper chromatography, 354, 357, 408, 424
Pardue, Mary Lou, 746
Partition chromatography, 424, 542
Patterson, T. S., 400
Pauling, Linus, 338, 368, 429, 430, 475,
    584, 642
Pauling's rules, 370
Pauson, P. L., 558, 559, 560
Pearson, R. G., 645
Pechmann, Hans von, 43
Pedersen, Charles, 711, 715, 722

Penicillin, structure, 456, 459
Penicillin chemistry, 309
Peptide bond, 373
Peptides, 11
Periodic table, 16, 26, 89, 101
Perkin, William Henry, Jr., 307
Perrin, Francis, 220
Perrin, Jean, 22, 155, 160, 226
Perutz, Max, 428, 429
Petrie, A. W., 302
Pettersson, Sven Otto, 16
Pfeffer, Wilhelm F. P., 4
Phase stability, principle, 341–342
Phenobarbital, 10
Phenomium ions, 708, 710
Phenylhydrazine, 9, 11
Phenyllithium, 613
Phosphate esters, in metabolic processes,
    183–184
Phosphorous metabolism,
    270–271
Photochemistry, 132
Photosynthesis, 110, 730–731
    reactions, 422, 423–426
Physical biochemistry, 162
Physical chemistry, 3
    journal, 5, 64
    origins, 15, 125
Piezoelectricity, 76
Pimentel, G. C., 702
Pines, Alex, 764
Pirie, Norman, 296, 303
Plant oils, 70–71
Plasma, 208
Platt, John, 475
Plattner, Placidus, 263
Plicque, Jules, 36
Plutonium, 340, 346
Poison gas, 121, 123, 472, 539, 547
Polanyi, John, 689, 700
Polanyi, Michael, 122, 388, 689, 690, 700
Polarography
    applications, 415
    importance, in chemical
        analysis, 416
    invention and development, 412,
        414–415
Polonium, discovery, 75, 77
Polymers
    study, 726
    synthesis, 565, 567
Polymethylene, 260
Polynucleotides, sequencing, 409–410

Polypeptide chains, folding,
534–535, 536
Polypropylene, discovery, 442, 445,
449, 453
Polysaccharides, 239
Pope, William, 480
Popjak, George, 263, 573
Porter, George, 487
Positive electrons, discovery, 218
Positron, 278
Powell, H. M., 457
Pregl, Fritz, 146, 539
Prelog, Vladimir, 261, 263, 571, 578
Price, W. C., 585
Prigogine, Ilya, 590
Primas, Hans, 761
*Principien der Chemie* (Ostwald), 63
*Principles of Polymer Chemistry*
(Flory), 567
Product epimerization, 329
Progesterone, isolation, 253,
255–256, 257
Prostaglandins, synthesis, 755
Protactinium, 277, 278
discovery, 138
Protein chromatography, 320
Protein crystallography, 431,
432, 437
Protein engineering, 440
Protein molecules, structural properties,
538, 540, 546
Protein structure, 158
Proticity, 601
Proton-transfer reactions, 494
Prout, William, 101–102, 143
Puck, Theodore, 739
Pulse and probe method, 488, 489
Pulse labeling, 628
Pummerer, Rudolf, 362
Purcell, Edward, 759
Purine, 9
Pyrroles, synthesis, 189

**Q**

*Quantitative Laws in Biological Chemistry*
(Arrhenius), 20
Quantum mechanical reasoning, 371
Quantum mechanical theory, 57
Quantum mechanics, 212, 473
Quantum theory, 122, 131
Quasi-unimolecular reactions, 387

**R**

Radiation chemistry, 162
Radioactive substances, 49
Radioactive tracers, 267–268, 269
*Radioactive Transformations*
(Rutherford), 54
Radioactivity, 27
artificial, 81, 217, 218, 223, 225, 278
atomic explanation, 78
discovery, 50, 75, 77
disintegration theory, 135
official standard of measurement,
79–80
transformation theory, 54
*Radioactivity* (Rutherford), 54
*Radioactivity* (Soddy), 136
Radiocarbon dating, 420
Radiochemical dating, 277
Radiochemical isotopes, 266
Radiochemistry, 12, 78
Radiology, 80
Radiothorium, 273–274
Radium
discovery, 75, 77, 218
separation process, 79
Radium-D, 267
Radium Institute, 80–81
Radon, discovery, 27
Raman, C. V., 527
Ramsay, William, 23, 134, 135–136,
141, 273
Randall, Merle, 5, 118
Random walk, 681
Ransil, Bernard J., 476
Rare-earth elements, 340, 347
Rau, Hans, 525
Rayleigh, Lord, 25
Reaction center, 730, 734
Reaction chemistry, 661
Reaction dynamics, 689–690, 693,
695–696
Reactions, relaxation methods,
494, 495–496
Reactive recoil, 276
Reactive scattering, 695–696, 697
Reactivity rules, 393
Recombinant DNA, 48, 618, 621–622
Reichstein, Tadeus, 263, 360
Relaxation process, 494, 495–496
Relaxation spectrometry, 498–499
Resonance, 371
Resonance theory, 641, 642

Retrosynthetic analysis, 751, 753–755
Riboflavin, 251
Ribonuclease, 532, 542, 544–545, 548
Rice, Francis, 388
Rice, Oscar K., 767
Rich, Alexander, 657
Richards, Theodore, 100
Richardson, C., 622
Rideal, E. K., 485
Rinde, Herman, 161
RNA, catalysis, 740–741, 743
RNA splicing, 747–748
Roberts, John D., 710
Robinson, Robert, 5, 306, 402, 458, 572
Roentgen rays, 76
Rohdewald, Margarete, 112
Root effect, 439
Roothaan, C. J., 475
Rosaniline, 9
Rose, William C., 380
Rotating-sector technique, 515
Rothenblach, Martin, 276
Roughton, Francis J. W., 436
Rubner, Max, 46
Rudenberg, Klaus, 476
Russell, Alexander, 137
Rutherford, Ernest, 49, 78, 123, 135, 266, 267, 273
Ruzicka, Leopold, 253, 259, 360, 450, 580

S

Sabatier, Paul, 83, 86, 88
Saches, Julius, 423
Sachse, Hermann, 508
Sack, Henri Samuel, 231–232
Safron, Sanford, 690
Saint-Gilles, Leon Pean de, 3
Sakharov, Andrey, 221
Salem, L., 645
Sanger, Frederick, 406, 460, 540, 541, 618, 627, 633, 740
Saussure, Nicholas de, 423
Saytzeff method, 84
Schaefer, Vincent, 208
Scheibe, G., 526
Schenectady Greek, 208
Scherrer, Paul, 231
Schlenk, Fritz, 178
Schlesinger, H. I., 605
Schramm, Gerhard, 256
Schroeder, H. E., 724
Seaborg, Glenn T., 338, 344

Segrè, Emilio, 340
Self-atom polarizability, 642
Semenov, Nikolay, 387, 388, 392, 483
Senderens, Jean Baptiste, 88 89
Sex hormones
  isolation, 253–254, 257
  study, 261, 262
Sheehan, John C., 752
Sickle-cell anemia, 372–373
  hemoglobin, 439
Siedentopf, Henry Friedrich Wilhelm, 154, 159
Slit ultramicroscope, 155
Smith, John, 740
Smoke generator, development, 208
Smoluchowski, Maryan, 155
Smyth, Charles Phelps, 103
Snapback DNA, 746
Soddy, Frederick, 27, 51, 53–54, 58, 78, 134, 142, 275
Sodium borohydride, 606, 607
Solid-phase technique, 670–672
Solubility-product principle, 127
Some Problems of Chemical Kinetics and Reactivity (Semenov), 392
Sommerfeld, Arnold Johannes Wilhelm, 229
Specht, Hugo, 194
Special pair, 734
Spectroscope, 25
Spectroscopy, 33, 465
Stanley, Wendell, 256, 292, 296, 300
Staudinger, Hermann, 259, 359, 443, 564, 616
Steacie, E. W. R., 700
Stein, G., 329
Stein, William, 533, 538, 539, 541, 546
Stereochemistry, 93, 571, 574–575, 579
Stereoelectronic hypothesis, 752
Stereoisomerism, 250
Stereoregularity, 443
Stereospecific polymerization, discovery, 442, 444–445
Stereospecificity, 581
Stern, Otto, 688
Stern–Gerlach experiment, 689
Steroids, 255, 261, 262
  biogenesis, 466
  synthesis, 313, 571
Sterols, 169, 170–171
Stevenson, John, 50
Stewart, John, 671
Stieglitz, Julius, 605

Stock, Alfred, 12, 585
Stokes, R. H., 5
Stone, Irwin, 375
Strassmann, Fritz, 278, 339, 345
Stripping reaction, 689
Strontium, 276, 277
Strutt, John William, *See* Rayleigh, Lord
Subba Rao, B. C., 609
Sugden, T. M., 482
Sulfur, 384
Sumner, James, 112, 288
Supramolecular chemistry, 718–719
Surface-aligned photochemistry, 705
Surface chemistry, 205
Surface free valences, 397–398
Sutin, Norman, 768
Svedberg, The, 155, 158, 315, 364
Synge, R. L. M., 352, 356, 424, 542
Synthetic sugars, 169
Szent-Györgyi, Albert, 239


T


Taube, Henry, 660
Taylor, Ellison, 688
Taylor, W. H., 429
Teller, Edward, 374, 527
Temperatures, accurate measurement, 321
Terpenes, 71–72, 261, 262
Terpinene, identification, 72
Terpinolene, discovery, 72
*Textbook on Stereochemistry* (Wittig), 612
Thalén, Tobias, 16
Theorell, Hugo, 532
*Theoretische Chemie* (Nernst), 128
*Theories of Chemistry* (Arrhenius), 19
*Theories of Solutions* (Arrhenius), 19
Thermal fluctuations, theory, 160
Thermochemistry, 89, 125
Thermodynamic properties, accurate measurement, 321
Thermodynamics, 65
  first law, 130
  mathematical structure, 64
  second law, 66, 130, 592, 595
  third law, 131, 323
  *See also* Heat theorem
*Thermodynamics and the Free Energy of Chemical Substances* (Lewis and Randall), 5

*Thermodynamics of Technical Gas Reactions,* (Haber), 117
Theta point, 568
Thiele, Johannes, 312, 360
Thomsen, Julius, 17, 62, 130
Thomson, H. W., 387
Thomson, J. J., 50, 57, 78, 141, 145, 129, 640
Thomson, William, 56
Thorium, extraction, 138
Thorium compounds, 52
Thorium X, 53–54, 137, 160
Tight-binding theory, 653
Timmermans, Jean, 591
Tiselius, Arne, 315, 357
Tishler, Max, 710
Tobacco mosaic virus (TMV), isolation, 301–302, 303, 655, 656
Todd, A. R., 310, 399
Todd, Margaret, 137
Toeplitz, Otto, 680
Tollens, Bernhard, 579
Tolman, Richard, 369
*Trail of Research in Sulfur Chemistry and Metabolism and Related Fields* (Du Vigneaud), 384
Transcription, 620
Transfer RNA, 620, 628
Transformation, 54
Transition metal–carbene and –carbyne chemistry, 553–554
Transition species, 704
Transition–state theory, 700
Transmethylation, 383
Transmutation, 53, 78, 79, 135, 225
Transulfuration, 382
Transuranium elements, discovery, 338, 340, 344, 345–346, 347–348
*Transuranium Elements,* (Seaborg), 346
Travers, Morris, 26, 142
*Treatise on Inorganic Chemistry* (Moissan), 40
Triangle plots, 703
Triphenylmethane, 30
Triterpenes, 262
Tropine, 109
Tropinone, synthesis, 310–311
Tsoucarius, Georges, 682
Tswett, Michael, 110

## U

Ultracentrifuge, 158, 161, 316, 364
Ultramicroscope, 155–156, 159
Unimolecular reaction kinetics, 387–389
Unsaturated organic compounds, hydrogenation, 88, 90–91
Uranium X, 53
Urbain, Georges, 269
Urease, 289–290, 291
Urey, Harold, 211, 267, 420, 527
Urey–Bradley force field, 212
Uridine diphosphate α–D–glucose (UDPG), discovery, 521–522

## V

Valence, 26, 94
Valence–bond method, 371, 474–475, 641
van Niel, Cornelius, 423
van't Hoff, Jacobus, 1, 10, 17, 27, 64, 125, 128, 130, 176, 393
Vinson, Carl, 302
Virtanen, Artturi, 282
Virus
  isolation, 296, 297
  nucleic acid, 303–304
  research, 256
Viscosity, formula, 364
Vitamin B₁, synthesis, 402
Vitamin B₆, discovery, 251
Vitamin B₁₂
  structure, 456–459–460
  synthesis, 466–467
Vitamin C
  studies, 375–376, 377
  synthesis, 236, 239–240
Vitamin H, 383
Vitamins, 169
Voltammetric method, 412
Votocek, Emil, 579

## W

Waage, Peter, 3, 62, 63
Walden, Paul, 33
Walden inversion, 11
Wallach, Otto, 69
Walton, E. T. S., 58
Warburg, Otto, 251, 423, 600
Warhurst, Ernest, 700
Washburn, Edward, 213

Water–gas equilibrium, 119
Watson, James, 374, 428, 619, 628
Weather modification, 208–209
Weaver, Warren, 431
Weber, William Edward, 126
Weizmann, Chaim, 123
Werder, Fritz von, 168
Werner, Alfred, 93, 661
Werner, Louis B., 346
Werner theory, 95–96
Wetzel, Johannes, 12
Whole-number rule, 143
Whytlaw-Gray, Robert, 27
Wichelhaus, Hans, 70
Wieland, Heinrich, 164, 170, 300, 600
Wien, Wilhelm, 141
Wilkins, Maurice, 428
Wilkinson, Geoffrey, 551, 567
Wilkinson's catalyst, 561
Willard, Hobart H., 103
Williams, John Warren, 233
Williamson, Alexander, 24
Willstätter, Richard, 33, 108, 189, 248, 290–291, 295
Windaus, Adolf, 165, 166, 167, 169, 256
Winkler, C. A., 766
Winstein, Saul, 710, 713
Wipke, Todd, 754
Wittig, Georg, 605, 611
Wittig reaction, 612, 613–616
Wöhler, Friedrich, 69
Woodward, R. B., 263, 311, 462, 552, 559, 572, 644, 649, 710
Woodward–Hoffmann orbital symmetry rules, 467, 649–650
Woolley, D. W., 668
World of neglected dimensions, 161
Wurtz, Charles-Adolphe, 2

## X

Xenon, discovery, 27
X-ray analysis, 261
X-ray crystallography, 433, 435, 437, 457, 515, 552, 588, 654, 694
X-ray diffraction, 158, 228, 231, 303, 443, 584, 656
  nonnegativity criterion, 675

## Y

Yellow enzyme, 251
Ylides, 614, 615

Young, Sydney, 24
Young, William John, 177

## Z

Zaug, Arthur, 747
Zelinsky, N. D., 401

Ziegler, Karl, 443, 445, 446, 449
Ziegler–Natta catalysts, 453–454
Zimmerman, H. E., 645
Zinc fingers proteins, 658
Zincke, Theodor, 335
Zsigmondy, Richard, 151, 159
Zuckerkandl, Emile, 375
Zymase, 46, 47, 177